A Short Course in
Mathematical Methods
with **Maple**

A Short Course in
Mathematical Methods
with Maple

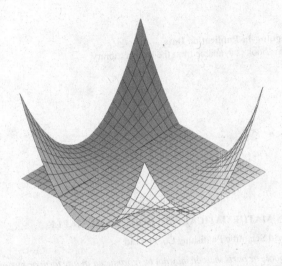

Henrik Aratyn
University of Illinois at Chicago, USA

Constantin Rasinariu
Columbia College Chicago, USA

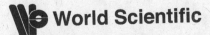

World Scientific

NEW JERSEY · LONDON · SINGAPORE · BEIJING · SHANGHAI · HONG KONG · TAIPEI · CHENNAI

Published by

World Scientific Publishing Co. Pte. Ltd.

5 Toh Tuck Link, Singapore 596224

USA office: 27 Warren Street, Suite 401-402, Hackensack, NJ 07601

UK office: 57 Shelton Street, Covent Garden, London WC2H 9HE

British Library Cataloguing-in-Publication Data
A catalogue record for this book is available from the British Library.

ISBN-13 978-981-256-461-0
ISBN-10 981-256-461-6
ISBN-13 978-981-256-595-2 (pbk)
ISBN-10 981-256-595-7 (pbk)

Printed in Singapore

Preface

This book is based on lecture notes used in a one-semester mathematical methods course for undergraduates at the University of Illinois at Chicago. The course is offered for students who have already taken an introductory physics course and serves as a prerequisite for more advanced courses in electricity and magnetism, classical mechanics, thermal and statistical physics, and quantum mechanics. For many students this course is their first encounter with advanced geometrical and algebraic concepts as well as with more challenging technical problems that require some degree of computational skills. A student's performance in the course is usually a very good indicator of how well he/she will do in advanced undergraduate physics and engineering courses.

The course is accompanied by a computer-based workshop, which guides students through the material using symbolic computation with Maple in the setting of a computer lab. Maple software was found to aid students to conceptualize the mathematical content of the course, and to encourage them to experiment with Mathematics.

A central feature of our book is that it teaches mathematical methods by integrating computer software with a traditional presentation of the material reflecting the pedagogical philosophy of the course. The advent of computers has changed the way people perform research, but, up until now, has had a relatively low impact on classroom instruction. Our book tries to change this status quo by using Maple as an integral component of the teaching process. However, we believe that Maple should not replace traditional mathematical theory (as calculators should not replace knowledge of the multiplication table). Therefore, as authors, we made a conscious decision to split our presentation of the book into two parallel parts: Part I contains an exposition of main topics in mathematical methods and Part II contains a computer-based approach to the same material.

The two parts of our book were designed to complement each other by introducing subject material via two different paths. The use of Maple allows us to omit several technical derivations and facts, thus, providing additional space for broadening the range of problems and applications of mathematical methods. The book contains a variety of problems chosen for their instructional value and physical relevance. Maple can effectively be used to assist students with tedious calculations that are required to solve the problems.

Part I essentially follows the old fashioned "paper and pencil" presentation of mathematical methods. It provides a streamlined and self-contained text of a mathematical methods course that emphasizes concepts that are important from the application perspective. Part I teaches a variety of technical tricks that are needed to successfully apply mathematical methods to solve important physical problems. The emphasis of Part I is on examples; some of the worked out examples illustrate how theorems work, others present their applications. By focusing on conceptual understanding of new ideas and geometrical intuition we attempted to write our book so as to reflect the way that most of physicists and engineers think about mathematics. The mathematical prerequisites are minimal (two semesters of calculus, including knowledge of the chain rule) and no pretense of mathematical rigor is made.

Part II of this book follows a computer-oriented approach to instruction. Topics in Part II are presented in the same order as in Part I. By closely mirroring part I, the second part reinforces it and improves its pedagogy. Maple is used in Part II both as a vehicle to teach mathematics and as a tool to solve mathematical problems. By studying Part II, students gain familiarity with a powerful computer algebra system that they will likely employ in other mathematics, engineering, or science courses. In Part II, Maple provides an advanced visualization environment that is sure to enhance students' grasp of traditional mathematical techniques and heighten their interest in the subject material.

There are a few possible ways in which the book may be used for instruction. One conventional option is a one-semester, four credit hours, undergraduate course covering the first five chapters (and their Maple counterparts, chapters 8-13), with possible omission of material students already know or parts that are too advanced. A much preferable option is a two-semester, three credit hours, undergraduate course which should cover the entire book from the beginning till the end.

Part I, contains seven chapters that cover most relevant topics, such as vector calculus, matrices and the eigenvalue problem, differential equations, power series solutions, Frobenius theorem, orthogonal eigenfunction expansion, Fourier series and the stability of the differential linear systems. The seventh, and final, chapter introduces students to fundamental concepts of nonlinear differential equations. It introduces contemporary and fundamental topics like nonlinear differential equations, chaos and solitons.

Throughout the book we teach students to think about functions as elements of a vector space. An abstract formulation of a vector space is essential for a complete understanding of the Fourier series and, more generally, the Sturm-Liouville problems with orthogonal eigenfunction expansions. This universal way of thinking about various forms of eigenfunctions expansions is especially useful for those students who will go on to take a quantum mechanics course. We introduce an abstract notion of a vector space early in the text and illustrate it by examples of function and polynomial vector spaces. Similarly, the Gram-Schmidt procedure, initially introduced in the setting of three-dimensional space, is extended to the n-dimensional vector space and to the infinite- dimensional vector space of orthogonal eigenfunctions.

The concept of orthogonal curvilinear coordinates is core to the mathematical methods curriculum. A discussion of orthogonal curvilinear coordinates is given in Chapter 1, in considerable detail, and precedes the subject of surface integrals since it can be used in calculation of many surface integrals, which can conveniently be formulated in terms of spherical or cylindrical coordinates.

Complex numbers and functions appear frequently in the text. They are explained in the Appendix A. We have omitted in this project the subject of residue calculus, a traditional goal of chapters on complex variables. Our decision was based on the fact that Maple easily calculates the relevant integrals.

This book was written with LATEX . With a few exceptions, figures were produced either by Maple or PSTricks, a set of powerful macros for drawing high-quality graphs that can be included in TEX or LATEX documents. We are indebted to a very active PSTricks community for making many wonderful macro packages and examples available for public use.

We should say at the outset that there is nothing in this book which is truly original. Although several of the logical connections and pedagogical arguments were invented in the process of writing this book, the very conventional character of the material makes it neither possible to claim originality nor to fully acknowledge our predecessors as generously we would like to. It is impossible, when writing a book of this nature, to give credit where it belongs because of the vast amount of pre-existing material. We profoundly apologize for not quoting countless references and sources, which are available. The only exception is the note after Chapter 7, which lists a few selected sources that influenced us most in the process of writing.

We are very grateful to Yvonne Aratyn for her patient help with editing large parts of Part I of the book and correcting numerous instances of confusing passages and bad grammar. We are indebted to students in Physics 215 classes at the University of Illinois at Chicago for reporting many typos and errors.

Chicago, August 2005

How this book is organized

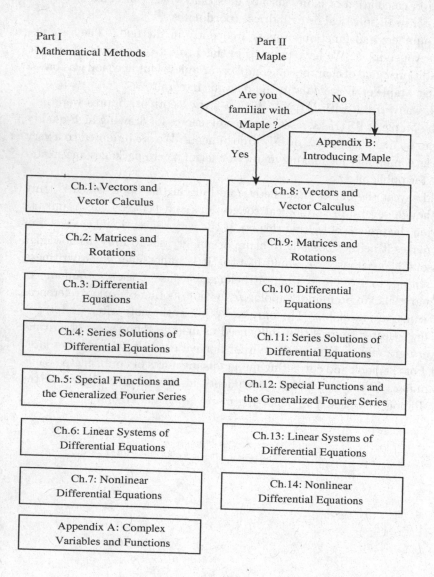

Part I
Mathematical Methods

Part II
Maple

Are you familiar with Maple ?

No

Yes

Appendix B:
Introducing Maple

Ch.1: Vectors and
Vector Calculus

Ch.8: Vectors and
Vector Calculus

Ch.2: Matrices and
Rotations

Ch.9: Matrices and
Rotations

Ch.3: Differential
Equations

Ch.10: Differential
Equations

Ch.4: Series Solutions of
Differential Equations

Ch.11: Series Solutions of
Differential Equations

Ch.5: Special Functions and
the Generalized Fourier Series

Ch.12: Special Functions and
the Generalized Fourier Series

Ch.6: Linear Systems of
Differential Equations

Ch.13: Linear Systems of
Differential Equations

Ch.7: Nonlinear
Differential Equations

Ch.14: Nonlinear
Differential Equations

Appendix A: Complex
Variables and Functions

Contents

Part I

Mathematical Methods

Chapter 1

Vectors and Vector Calculus

VECTORS on a plane or in three dimensional space can be thought of as arrows with magnitude and direction. One can view a collection of vectors satisfying a set of simple algebraic rules as a linear structure called vector space. This gives rise to a more abstract notion of a vector as an element of a vector space making it possible to include more abstract objects like functions and polynomials in the theory of vectors. For vector spaces we discuss topics like: linear dependence and independence, bases and dimension. In case of three dimensional vectors we learn how to project a vector on a given direction and how to construct orthogonal bases.

In order to measure variations of forces and velocities and to evaluate magnetic fluxes we must learn how to differentiate and integrate three dimensional vectors. Vector calculus extends the notion of one-dimensional calculus and provides techniques for taking derivatives of multi-dimensional objects and dealing with line, surface and volume integrals. In this Chapter, we will learn the fundamental theorems of vector calculus and apply them to various coordinate systems.

1.1 Vectors and Vector Spaces

1.1.1 Vectors

Scalar quantities like mass, temperature, charge or pressure are uniquely specified by their magnitude. Physics also requires vector quantities like velocity, force, angular momentum, which have magnitude as well as direction.

The simplest vector quantity is displacement associated with change of position. Every displacement has an initial or starting point and a final point. Let us consider displacements that have a common starting point: the origin. Any point in space can be understood as the final point of a displacement from the origin. We will depict such displacements by an arrow starting at the origin and ending at the final

Vectors and Vector Calculus

point A or B. We will denote such displacements by vectors $\vec{A}$ or $\vec{B}$. Vectors can be multiplied by real numbers (the scalars). If $\lambda > 0$ is a positive real number and $\vec{A}$ is a vector, then $\lambda \vec{A}$ is a vector pointing in the same direction as $\vec{A}$ but λ times as long as $\vec{A}$.

Another important operation is vector addition. Graphically, two vectors are added according to the so-called tip-to-tail method: two vectors $\vec{A}$ and $\vec{B}$ are arranged in such a way that tail of $\vec{B}$ is at the tip of the first, the sum $\vec{A} + \vec{B}$ extends from the tail of $\vec{A}$ to the tip of $\vec{B}$. It is important to notice that the resultant vector $\vec{A} + \vec{B}$ is the diagonal emerging from the origin to the opposite vertex in the parallelogram formed by $\vec{A}$ and $\vec{B}$, as in Figure 1.1. Hence $\vec{A} + \vec{B}$ does not depend on the order of addition and one can put either tip to either tail when adding two vectors. One says that vector addition is commutative, hence:

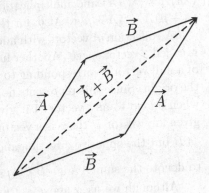

Figure 1.1: Vector addition.

$$\vec{A} + \vec{B} = \vec{B} + \vec{A}. \tag{1.1}$$

The rule for addition of two vectors extends easily to any number of vectors. In the Figure 1.2 below, we add three vectors to illustrate the fact that vector addition is associative. As can be seen in the Figure 1.2, when adding three vectors $\vec{A}$, $\vec{B}$ and $\vec{C}$ it does not matter whether we first add $\vec{A}$ to $\vec{B}$ and then add it to $\vec{C}$ with a result: $\left(\vec{A} + \vec{B}\right) + \vec{C}$ or whether we first add $\vec{B}$ to $\vec{C}$ and then add $\vec{A}$ with the identical result: $\vec{A} + \left(\vec{B} + \vec{C}\right)$.

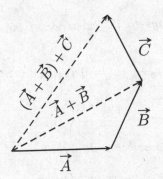
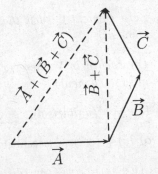

Figure 1.2: Associativity of the vector addition.

Multiplication by a scalar is distributive over vector addition:

$$\lambda(\vec{A} + \vec{B}) = \lambda\vec{A} + \lambda\vec{B}.$$

Furthermore, multiplication of a vector by a product of scalars is associative $(\lambda\mu)\vec{A} = \lambda(\mu\vec{A})$ and multiplication of a vector by a sum of scalar is distributive: $(\lambda + \mu)\vec{A} = \lambda\vec{A} + \mu\vec{A}$. Also, for the special scalar $\lambda = 1$, it holds that $1\vec{A} = \vec{A}$.

A collection of vectors with addition and scalar multiplications as defined above is called a vector space. Another fundamental property of a vector space is existence of a null vector 0 corresponding to the trivial displacement which starts and ends at the origin. Since this represents no displacement at all, the null vector 0 is such that when added to any vector $\vec{A}$ it gives back $\vec{A}$ again; that is, $0 + \vec{A} = \vec{A}$. Similarly, given any vector $\vec{B}$ there is a vector $-\vec{B}$ which is a vector with the same magnitude as $\vec{B}$ but the opposite direction and which satisfies $\vec{B} + \left(-\vec{B}\right) = 0$. It is convenient to denote the sum $\vec{A} + \left(-\vec{B}\right)$ by $\vec{A} - \vec{B}$.

Although we used above a graphical representation of a vector by an arrow in three-dimensional space to write down axioms of a vector space the concept is quite abstract. In general, we will only require the above axioms to be satisfied in order to be able to classify a collection of objects as a vector space. In this way the meaning of a vector generalizes from a three-dimensional arrow to an element of the vector space which satisfies the following definition.

Definition 1.1.1. *A set V is a vector space (over scalars $\alpha, \beta \in K$), if to every pair v, w of elements of V there corresponds a sum $v + w$ which belongs to V and to every pair α, v (with α being in K) there corresponds a product αv which belongs to V, obeying the following properties of multiplication and addition:*

1. *$v + w = w + v$ (commutability of addition).*

2. *$u + (v + w) = (u + v) + w$ (associativity of addition).*

3. *Zero element 0 such that $0 + v = v$ belongs to V (existence of the null element).*

4. *For each v in V there exists $-v$ such that $v + (-v) = 0$ (existence of the inverse element).*

5. *$1 \cdot v = v$ (unitarism).*

6. *$\alpha(\beta v) = (\alpha\beta)v$ (associativity with respect to number multiplication).*

7. *$\alpha(v + w) = \alpha v + \alpha w$ (distributivity with respect to vector addition).*

8. *$(\alpha + \beta)v = \alpha v + \beta v$ (distributivity with respect to number addition).*

If K consists of real scalars then the vector space is called a real vector space.

Furthermore if K consists of complex scalars then the vector space is called a complex vector space.

It is quite customary to apply the notion of a vector space to functions. Here are some examples of vector spaces. Note, that some of these examples deal with functions as elements of an abstract vector space.

☞ **Example 1.1.1.** Denote by $\mathcal{C}[a, b]$ the continuous real-valued functions on the interval $a \leq x \leq b$. We define the addition of two such functions f and g as

$$(f + g)(x) = f(x) + g(x),$$

which yields a continuous function $f + g$. The scalar multiplication by a real scalar a: $(af)(x) = af(x)$ defines a new continuous real-valued function af. With these operations $\mathcal{C}[a, b]$ becomes a vector space. Since we only use real scalars we call $\mathcal{C}[a, b]$ a real vector space.

☞ **Example 1.1.2.** Let $\mathbb{P}^n$ be a set of all polynomials of at most degree n

$$p_n(x) = a_n x^n + a_{n-1} x^{n-1} + \cdots a_1 x + a_0.$$

Under standard addition and multiplication by the scalar $\mathbb{P}^n$ is a vector space.

☞ **Example 1.1.3.** Consider n real numbers $a_1, \ldots, a_n$ aligned in a column

$$\underline{a} = \begin{bmatrix} a_1 \\ a_2 \\ \vdots \\ a_{n-1} \\ a_n \end{bmatrix}$$

and denote by $\mathbb{R}^n$ the set of all columns consisting of n real elements. We define a sum $\underline{a} + \underline{b}$ of two columns $\underline{a}$ and $\underline{b}$ to be a column with i-th element (or component) equal to $a_i + b_i$

$$\underline{a} + \underline{b} = \begin{bmatrix} a_1 + b_1 \\ a_2 + b_2 \\ \vdots \\ a_{n-1} + b_{n-1} \\ a_n + b_n \end{bmatrix}.$$

The product of a column by a real number is a column obtained by multiplying all components of the column by this number

$$\lambda \underline{a} = \begin{bmatrix} \lambda a_1 \\ \lambda a_2 \\ \vdots \\ \lambda a_{n-1} \\ \lambda a_n \end{bmatrix}.$$

It is easy to verify that all axioms of the vector space from Definition (1.1.1) are satisfied if we take as a null element $\underline{0}$ a column consisting of zero components. Moreover, the inverse of $\underline{a}$ is given by

$$-\underline{a} = \begin{bmatrix} -a_1 \\ -a_2 \\ \vdots \\ -a_{n-1} \\ -a_n \end{bmatrix}$$

and it follows that $\underline{a} + (-\underline{a}) = \underline{0}$. Thus, $\mathbb{R}^n$ is a vector space. It is called an Euclidean vector space.

An important generalization of this example is obtained by taking columns with n complex elements. It is straightforward to extend the addition of columns and the product of a column by a complex number to this situation. This gives rise to a vector space $\mathbb{C}^n$ consisting of complex n-column vectors.

1.1.2 Basis Vectors and Components of a Vector

Definition 1.1.2. *A set* $\{\vec{v}_1, \vec{v}_2, ..., \vec{v}_n\}$ *of n nonzero vectors is said to be linearly independent if the relation*

$$\sum_{i=1}^{n} \lambda_i \vec{v}_i = 0 \tag{1.2}$$

has only a trivial solution $\lambda_i = 0$ *for all* $i = 1, 2, ..., n$. *Otherwise, vectors* $\{\vec{v}_1, \vec{v}_2, ..., \vec{v}_n\}$ *are said to be linearly dependent.*

If the system of vectors is linearly dependent, then at least one of the vectors can be expressed as a linear combination of others.

Two non-zero vectors $\vec{u}, \vec{v}$ are linearly dependent if they are "parallel", i.e., if it holds that $\vec{u} = k\vec{v}$ with some scalar k. Indeed, if $\lambda_1 \vec{u} + \lambda_2 \vec{v} = 0$ and $\lambda_1 \neq 0$ then $\vec{u} = -(\lambda_2/\lambda_1)\vec{v}$.

Three non-zero vectors $\vec{u}, \vec{v}, \vec{w}$ are linearly dependent if $\vec{u}$ and $\vec{v}$ are parallel or if $\vec{w}$ is a linear combination of $\vec{u}$ and $\vec{v}$. To check this claim, consider the condition of linear dependence $\lambda_1 \vec{u} + \lambda_2 \vec{v} + \lambda_3 \vec{w} = 0$. Now, if $\lambda_3 = 0$ then $\vec{u}$ and $\vec{v}$ are parallel. If, $\lambda_3 \neq 0$ then $\vec{w} = -(\lambda_1/\lambda_3)\vec{u} - (\lambda_2/\lambda_3)\vec{v}$.

☞ **Example 1.1.4.** Consider 3 vectors

$$\underline{u}_1 = \begin{bmatrix} 0 \\ 2 \\ 1 \end{bmatrix}, \quad \underline{u}_2 = \begin{bmatrix} 1 \\ 3 \\ -4 \end{bmatrix}, \quad \underline{u}_3 = \begin{bmatrix} 2 \\ 0 \\ -1 \end{bmatrix}$$

in $\mathbb{R}^3$. Equation $\sum_{i=1}^{3} \lambda_i \underline{u}_i = 0$ decomposes on

$$\lambda_1 0 + \lambda_2 1 + \lambda_3 2 = 0, \quad \lambda_1 2 + \lambda_2 3 + \lambda_3 0 = 0, \quad \lambda_1 1 + \lambda_2(-4) + \lambda_3(-1) = 0.$$

The only solution is $\lambda_1 = \lambda_2 = \lambda_3 = 0$. Thus, vectors $\underline{u}_1, \underline{u}_2, \underline{u}_3$ are linearly independent and none of the vectors can be written as a linear combination of other vectors.

However, vectors

$$\underline{u}_1 = \begin{bmatrix} 0 \\ 2 \\ 1 \end{bmatrix}, \quad \underline{u}_2 = \begin{bmatrix} 1 \\ 3 \\ -4 \end{bmatrix}, \quad \underline{u}_3 = \begin{bmatrix} 2 \\ 0 \\ -11 \end{bmatrix}$$

are linearly dependent, since $\lambda_1 \underline{u}_1 + \lambda_2 \underline{u}_2 + \lambda_3 \underline{u}_3 = 0$ holds for e.g. $\lambda_1 = 3/2, \lambda_2 = -1, \lambda_3 = 1/2$. Then, we can express the vector $\underline{u}_1$ as a linear combination of $\underline{u}_2$ and $\underline{u}_3$:

$$\underline{u}_1 = \frac{2}{3}\underline{u}_2 - \frac{1}{3}\underline{u}_3.$$

Three linearly dependent vectors in $\mathbb{R}^3$ are called coplanar, see Problems 1.1.8 and 1.1.9 for the definition of a plane. For three co-planar vectors either two of the vectors are parallel or one of them is given by a linear combination of the remaining two vectors.

Definition 1.1.3. *A set $\{\vec{v}_1, \vec{v}_2, ...\}$ of non-zero vectors is said to be a basis for the vector space V if the following two conditions are satisfied:*

1. *The vectors $\{\vec{v}_1, \vec{v}_2, ...\}$ are linearly independent.*

2. *Any vector $\vec{A}$ in V can be written as a linear combination*

$$\vec{A} = \sum_{i=1} \lambda_i \vec{v}_i \qquad (1.3)$$

of the $\{\vec{v}_1, \vec{v}_2, ...\}$ vectors.

☞ **Example 1.1.5.** Consider the vector space $\mathbb{R}^3$. The system of vectors

$$\underline{e}_1 = \begin{bmatrix} 1 \\ 0 \\ 0 \end{bmatrix}, \quad \underline{e}_2 = \begin{bmatrix} 0 \\ 1 \\ 0 \end{bmatrix}, \quad \underline{e}_3 = \begin{bmatrix} 0 \\ 0 \\ 1 \end{bmatrix} \tag{1.4}$$

is a basis. First notice that this system is linearly independent. To see this we consider a linear combination which is equal to zero

$$0 = \sum_{i=1}^{3} \lambda_i \underline{e}_i = 0 = \lambda_1 \begin{bmatrix} 1 \\ 0 \\ 0 \end{bmatrix} + \lambda_2 \begin{bmatrix} 0 \\ 1 \\ 0 \end{bmatrix} + \lambda_3 \begin{bmatrix} 0 \\ 0 \\ 1 \end{bmatrix}.$$

This combination is equivalent to a set of equations $\lambda_1 = 0, \lambda_2 = 0, \lambda_3 = 0$ which has only a trivial zero solution.

Any vector $\underline{v}$ in $\mathbb{R}^3$ can be represented as a linear combination of $\underline{e}_1, \underline{e}_2$ and $\underline{e}_3$:

$$\underline{v} = \begin{bmatrix} v_1 \\ v_2 \\ v_3 \end{bmatrix} = v_1 \begin{bmatrix} 1 \\ 0 \\ 0 \end{bmatrix} + v_2 \begin{bmatrix} 0 \\ 1 \\ 0 \end{bmatrix} + v_3 \begin{bmatrix} 0 \\ 0 \\ 1 \end{bmatrix}.$$

The basis vectors $\underline{e}_1, \underline{e}_2$ and $\underline{e}_3$ from equation (1.4) form the so-called canonical basis of $\mathbb{R}^3$.

Similarly

$$\underline{e}_1 = \begin{bmatrix} 1 \\ 0 \end{bmatrix}, \quad \underline{e}_2 = \begin{bmatrix} 0 \\ 1 \end{bmatrix} \tag{1.5}$$

form the canonical basis of $\mathbb{R}^2$.

Two important properties of vector spaces are:

- If a vector space has a basis of n vectors then any set containing more than n vectors is not linearly independent.

- If a vector space has a basis of n vectors then every basis of V contains n vectors.

These properties ensure that to any vector space one can associate a unique number equal to the number of its basis vectors. This gives rise to the following definition.

Definition 1.1.4. *If a vector space V has a basis consisting of n vectors $\{\vec{v}_1, \vec{v}_2, ..., \vec{v}_n\}$, then V is said to be n-dimensional: $\dim(V) = n$.*

Although every vector space has a basis not every vector space has a finite basis; that is, a basis with a finite number of elements. If a vector space does not possess a finite basis then it is said to be infinite-dimensional.

To determine the dimension of a vector space, it is enough to find a basis for that vector space. The number of basis vectors then equals the dimension of that vector space.

☞ **Example 1.1.6.** In Example (1.1.2) take $n = 3$. $\mathcal{P}_3$ is a vector space consisting of all polynomials on the interval $a \leq x \leq b$ of at most cubic degree. Every such polynomial can be written as a linear combination of elementary monomials:

$$\{1, x, x^2, x^3\}.$$

For instance, the polynomial

$$p(x) = x^3 - 3x^2 + 5$$

has a representation

$$p(x) = 5 \cdot 1 + 0 \cdot x + (-3) \cdot x^2 + 1 \cdot x^3$$

in this basis. Hence $\mathcal{P}_3$ is a finite-dimensional (four-dimensional) vector space. The set $\{1, x, x^2, x^3\}$ is a convenient (but not a unique) choice of a basis in $\mathcal{P}_3$. Other bases one encounters in mathematical physics literature are: $\{1, x, (3x^2 - 1)/2, (5x^3 - 3x)/2\}$, consisting of Legendre polynomials or $\{1, x, (2x^2 - 1), (4x^3 - 3x)\}$ formed by Chebyshev polynomials. These polynomials are defined on the interval $-1 \leq x \leq 1$.

Generally,

$$\{1, x, x^2, x^3, \ldots, x^n\}$$

forms a (canonical) basis in $\mathcal{P}_n$.

Consider a set of polynomials

$$\{1 + x, x + x^3, x^2\}$$

in $\mathcal{P}_3$. Is this a linearly independent set of vectors? To answer this question let us verify whether there exists a non-trivial solution to

$$a_1(1 + x) + a_2(x + x^3) + a_3 x^2 = 0.$$

The polynomial vanishes identically if all its coefficients are zero and hence

$$a_1 = 0, a_1 + a_2 = 0, a_2 = 0, a_3 = 0.$$

This has only a trivial null solution. Hence $\{1 + x, x + x^3, x^2\}$ is linearly independent but it is not a basis. To see this, try to write x^3 as a linear combination of $\{1 + x, x + x^3, x^2\}$.

☞ **Example 1.1.7.** Consider again the vector space $\mathbb{R}^3$. The system of vectors

$$\underline{u}_1 = \begin{bmatrix} 1 \\ 1 \\ 0 \end{bmatrix}, \quad \underline{u}_2 = \begin{bmatrix} 0 \\ 1 \\ 0 \end{bmatrix}$$

is linearly independent but it does not form a basis in $\mathbb{R}^3$. It is not possible to represent the vector $\underline{e}_3$ from equation (1.4) by a linear combination of $\underline{u}_1$ and $\underline{u}_2$. The dimension of $\mathbb{R}^3$ is equal to 3 and two vectors can not form a basis in this vector space.

1.1.3 Three-Dimensional Cartesian Coordinate System and $\mathbb{R}^3$

We will now imagine a space of arrows in a three-dimensional Cartesian coordinate system as depicted in Figure 1.1.3. It is always possible to write a (three-dimensional) vector $\vec{A}$ as:

$$\vec{A} = A_x\hat{i} + A_y\hat{j} + A_z\hat{k} \tag{1.6}$$

in terms of a set of basis vectors $\hat{i}, \hat{j}, \hat{k}$ along the x, y and z-axes and with the unit length. Decomposition (1.6) defines components of the vector $\vec{A}$ as A_x, A_y, A_z. It is convenient to introduce the index notation for components of the vector $\vec{A}$ such that the components of $\vec{A}$ are labeled by numbers instead of directions. In this

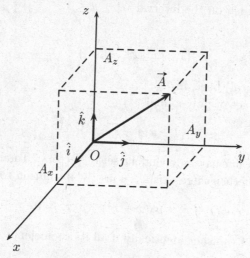

Figure 1.3: The three-dimensional vector and its components.

notation $A_x = A_1, A_y = A_2, A_z = A_3$ and

$$\hat{e}_1 = \hat{i}, \quad \hat{e}_2 = \hat{j}, \quad \hat{e}_3 = \hat{k}.$$

An advantage of index notation is that the decomposition (1.6) now takes a form

$$\vec{A} = A_1\hat{e}_1 + A_2\hat{e}_2 + A_3\hat{e}_3 = \sum_{i=1}^{3} A_i\,\hat{e}_i, \qquad (1.7)$$

which can easily be generalized to any number of dimensions n by $\sum_{i=1}^{3} A_i\,\hat{e}_i \rightarrow \sum_{i=1}^{n} A_i\,\hat{e}_i$.

Alternatively, we may think about a three-dimensional vector as an ordered 3-tuple $\vec{A} = (A_x, A_y, A_z) = (A_1, A_2, A_3)$ of coordinates A_i. For a displacement vector the 3-tuple simply lists the coordinates of the final point.

Rearranging the 3-tuple into a column

$$\underline{A} = \begin{bmatrix} A_1 \\ A_2 \\ A_3 \end{bmatrix}$$

in $\mathbb{R}^3$ replaces decomposition (1.6) or (1.7) by

$$\underline{A} = A_1 \begin{bmatrix} 1 \\ 0 \\ 0 \end{bmatrix} + A_2 \begin{bmatrix} 0 \\ 1 \\ 0 \end{bmatrix} + A_3 \begin{bmatrix} 0 \\ 0 \\ 1 \end{bmatrix}.$$

Thus we may identify the three-dimensional Cartesian coordinate system with the Euclidean space $\mathbb{R}^3$ relating their basis vectors according to

$$\hat{i} = \hat{e}_1 \sim \underline{e}_1 = \begin{bmatrix} 1 \\ 0 \\ 0 \end{bmatrix}, \quad \hat{j} = \hat{e}_2 \sim \underline{e}_2 = \begin{bmatrix} 0 \\ 1 \\ 0 \end{bmatrix}, \quad \hat{k} = \hat{e}_3 \sim \underline{e}_3 = \begin{bmatrix} 0 \\ 0 \\ 1 \end{bmatrix}.$$

Using decomposition along the basis vectors $\hat{e}_i, i = 1, 2, 3$ it is easy to perform algebraically addition and subtraction of vectors, like:

$$\vec{A} + \vec{B} = \sum_{i=1}^{3} A_i\hat{e}_i + \sum_{i=1}^{3} B_i\hat{e}_i = \sum_{i=1}^{3}(A_i + B_i)\hat{e}_i. \qquad (1.8)$$

The components of $\vec{A} + \vec{B}$ are obtained by adding the components of $\vec{A}$ and $\vec{B}$ separately. In the same manner, for example,

$$\vec{A} - 3\vec{B} = \sum_{i=1}^{3} A_i\hat{e}_i - 3\sum_{i=1}^{3} B_i\hat{e}_i = \sum_{i=1}^{3}(A_i - 3B_i)\hat{e}_i.$$

1.1.4   Scalar Product

Let us define the scalar product of two 3-dimensional vectors $\vec{A} = (A_1, A_2, A_3)$ and $\vec{B} = (B_1, B_2, B_3)$ as

$$\vec{A} \cdot \vec{B} = \sum_{i=1}^{3} A_i B_i = A_1 B_1 + A_2 B_2 + A_3 B_3 \,. \tag{1.9}$$

Recall, now the representation of the vector $\vec{A}$ in terms of three basis vectors $\hat{e}_1, \hat{e}_2, \hat{e}_3$: $\vec{A} = \sum_{i=1}^{3} A_i \hat{e}_i$ as in definition (1.6). In this representation the scalar product of two vectors $\vec{A}$ and $\vec{B}$ takes the form

$$\vec{A} \cdot \vec{B} = \sum_{i=1}^{3} \sum_{j=1}^{3} A_i B_j \, \hat{e}_i \cdot \hat{e}_j \,. \tag{1.10}$$

We also need a concept of the length of the vector. The scalar product of the vector with itself defines the norm (or the length) A of the vector $\vec{A}$ through

$$|\vec{A}| = \sqrt{\vec{A} \cdot \vec{A}} = \sqrt{A_1^2 + A_2^2 + A_3^2} \,. \tag{1.11}$$

The square of the norm is often denoted as

$$A^2 = |\vec{A}|^2 = \vec{A} \cdot \vec{A} \,.$$

The following are the fundamental properties of the norm.

- $|\vec{A}| \geq 0$ for all 3-dimensional vectors with $|\vec{A}| = 0$ holding if and only if $\vec{A} = 0$.

- $|\lambda \vec{A}| = |\lambda| \, |\vec{A}|$ for a real number λ and the 3-dimensional vector $\vec{A}$.

- $\left| \vec{A} + \vec{B} \right| \leq \left| \vec{A} \right| + \left| \vec{B} \right|$ for two vectors $\vec{A}$ and $\vec{B}$ (the triangle inequality).

Comparing, expressions (1.9) and (1.10) we see that they agree if the scalar product between two basis vectors is equal to:

$$\hat{e}_i \cdot \hat{e}_j = \begin{cases} 0 & \text{for } i \neq j \\ 1 & \text{for } i = j \end{cases} \tag{1.12}$$

for $i, j = 1, \ldots, 3$.

According to definition (1.11) and (1.12) the basis vectors $\hat{e}_i$, $i = 1, \ldots, 3$ have all unit length in agreement with how they have been initially defined.

Since $\hat{e}_i \cdot \hat{e}_j = 0$ for $i \neq j$ the basis vectors are orthogonal. The orthogonal set of vectors of unit length is being referred to as an orthonormal basis.

It is convenient to define an object δ_{ij}, Kronecker delta, depending on two indices i and j running from 1 to 3. Kronecker delta vanishes for $i \neq j$ and equals one when the two indices are equal:

$$\delta_{ij} = \begin{cases} 0 & \text{for } i \neq j \\ 1 & \text{for } i = j \end{cases}. \tag{1.13}$$

It is convenient to record values of the Kronecker delta in an array

$$\delta_{ij} = \begin{bmatrix} \delta_{11} & \delta_{12} & \delta_{13} \\ \delta_{21} & \delta_{22} & \delta_{23} \\ \delta_{31} & \delta_{32} & \delta_{33} \end{bmatrix} = \begin{bmatrix} 1 & 0 & 0 \\ 0 & 1 & 0 \\ 0 & 0 & 1 \end{bmatrix}.$$

The Kronecker delta compactly describes the rules of the scalar product by one single equation

$$\hat{e}_i \cdot \hat{e}_j = \delta_{ij} \quad i, j = 1, \ldots, 3. \tag{1.14}$$

One of the applications of a scalar product is the concept of projection of a vector on some other directions. Let $\vec{A}$ and $\vec{B}$ be two 3-dimensional vectors. Set

$$\vec{B}_{\parallel} = \frac{\vec{A} \cdot \vec{B}}{\vec{A} \cdot \vec{A}} \vec{A}, \quad \text{and} \quad \vec{B}_{\perp} = \vec{B} - \vec{B}_{\parallel}. \tag{1.15}$$

Then

$$\vec{B}_{\perp} \cdot \vec{A} = \left(\vec{B} - \vec{B}_{\parallel} \right) \cdot \vec{A} = \vec{B} \cdot \vec{A} - \frac{\vec{A} \cdot \vec{B}}{\vec{A} \cdot \vec{A}} \vec{A} \cdot \vec{A} = 0$$

and thus $\vec{B}_{\perp}$ is perpendicular (or orthogonal) to $\vec{A}$ as depicted in Figure 1.1.3.

A vector which can be written as $\lambda \vec{A}$ where λ is a number has the same direction as $\vec{A}$ and is said to be parallel to $\vec{A}$. Equation (1.15) defines a decomposition

$$\vec{B} = \vec{B}_{\perp} + \vec{B}_{\parallel} \tag{1.16}$$

of a vector $\vec{B}$ on parallel and perpendicular vectors to the vector $\vec{A}$. It is an *unique* decomposition. Assume, namely, that the vector $\vec{B}$ can also be written as

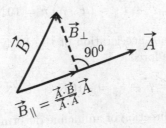

Figure 1.4: Projection of a vector on a given direction.

$\vec{B} = \vec{B}_1 + \vec{B}_2$ with some vectors $\vec{B}_1 = \lambda \vec{A}$ and $\vec{B}_2$ perpendicular to $\vec{A}$, then

$$\vec{B}_1 \cdot \vec{A} = \lambda \vec{A} \cdot \vec{A}$$

and, thus,

$$\vec{B}_1 = \lambda \vec{A} = \frac{\vec{B}_1 \cdot \vec{A}}{\vec{A} \cdot \vec{A}} \vec{A} = \frac{\vec{B} \cdot \vec{A}}{\vec{A} \cdot \vec{A}} \vec{A} = \vec{B}_\parallel$$

since $\vec{B}_1 \cdot \vec{A} = \vec{B} \cdot \vec{A}$. Moreover,

$$\vec{B}_2 = \vec{B} - \vec{B}_1 = \vec{B} - \vec{B}_\parallel = \vec{B}_\perp .$$

The vector $\vec{B}_\parallel$ is called projection of vector $\vec{B}$ on vector $\vec{A}$. It is also denoted by $\text{Proj}_A \vec{B}$. It follows that $\vec{A} \cdot \text{Proj}_A \vec{B} = \vec{A} \cdot \vec{B}$ and $\vec{B} - \text{Proj}_A \vec{B}$ is orthogonal to $\vec{A}$.

☞ **Example 1.1.8.** The concept of vector projection has many applications. Here comes a very basic example from an introductory mechanics class. Think about a body on an inclined plane. Let the plane be inclined by an an-

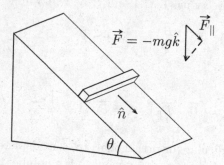

gle $\theta = 30^0$ to the horizontal. Suppose that the direction of an incline is given by a unit vector $\hat{n} = -\frac{1}{2}\hat{k} + \frac{\sqrt{3}}{2}\hat{i}$. Let, furthermore, the mass of the body be $m = 2\text{kg}$. For $g = 10\text{m/s}^2$, the gravitational force is $\vec{F} = -mg\hat{k} = -20\,\text{N}\,\hat{k}$. Projection of the gravitational force $\vec{F}$ in direction of an incline is

$$\text{Proj}_n \vec{F} = \vec{F}_\parallel = \left(\vec{F} \cdot \hat{n} \right) \hat{n} = 10\,\text{N}\,\hat{n}$$

and is used to calculate acceleration of the body along the inclined plane. The component of the force

$$\vec{F}_\perp = \vec{F} - \vec{F}_\parallel = -15\,\text{N}\,\hat{k} - 5\sqrt{3}\,\text{N}\,\hat{i}$$

perpendicular to the direction of an incline determines the normal force $\vec{N} = -\vec{F}_\perp$ applied on the body by the inclined plane.

☞ **Example 1.1.9.** For the 3 vectors $\underline{u}_1, \underline{u}_2, \underline{u}_3$ from Example (1.1.4) the projection of vector $\underline{u}_2$ on vector $\underline{u}_1$ is

$$\text{Proj}_{\underline{u}_1} \underline{u}_2 = \frac{\underline{u}_1 \cdot \underline{u}_2}{\underline{u}_1 \cdot \underline{u}_1} \underline{u}_1 = \frac{2}{5} \begin{bmatrix} 0 \\ 2 \\ 1 \end{bmatrix}$$

as follows from $\underline{u}_1 \cdot \underline{u}_2 = 2$ and $\underline{u}_1 \cdot \underline{u}_1 = 5$. It follows that the vector

$$\underline{u}_2 - \text{Proj}_{\underline{u}_1} \underline{u}_2 = \begin{bmatrix} 1 \\ 3 \\ -4 \end{bmatrix} - \frac{2}{5} \begin{bmatrix} 0 \\ 2 \\ 1 \end{bmatrix} = \frac{1}{5} \begin{bmatrix} 5 \\ 11 \\ -22 \end{bmatrix} \tag{1.17}$$

is orthogonal to $\underline{u}_1$.

Similarly, due to $\underline{u}_1 \cdot \underline{u}_3 = -1$, the projection of the vector $\underline{u}_3$ on the vector $\underline{u}_1$ is found to be

$$\text{Proj}_{\underline{u}_1} \underline{u}_3 = \frac{\underline{u}_1 \cdot \underline{u}_3}{\underline{u}_1 \cdot \underline{u}_1} \underline{u}_1 = -\frac{1}{5} \begin{bmatrix} 0 \\ 2 \\ 1 \end{bmatrix}.$$

One verifies easily that the vector

$$\underline{u}_3 - \text{Proj}_{\underline{u}_1} \underline{u}_3 = \begin{bmatrix} 2 \\ 0 \\ -1 \end{bmatrix} + \frac{1}{5} \begin{bmatrix} 0 \\ 2 \\ 1 \end{bmatrix} = \frac{2}{5} \begin{bmatrix} 5 \\ 1 \\ -2 \end{bmatrix} \tag{1.18}$$

is orthogonal to $\underline{u}_1$.

Vectors $\underline{u}_1, \underline{u}_2 - \text{Proj}_{\underline{u}_1} \underline{u}_2, \underline{u}_3 - \text{Proj}_{\underline{u}_1} \underline{u}_3$ are linearly independent since $\{\underline{u}_1, \underline{u}_2, \underline{u}_3\}$ was a basis and $\text{Proj}_{\underline{u}_1} \underline{u}_2, \text{Proj}_{\underline{u}_1} \underline{u}_3$ are both parallel to $\underline{u}_1$.

Although both vectors $\underline{u}_2 - \text{Proj}_{\underline{u}_1} \underline{u}_2$ and $\underline{u}_3 - \text{Proj}_{\underline{u}_1} \underline{u}_3$ are orthogonal to $\vec{u}_1$ they are not orthogonal to each other.

In Example (1.1.9) we showed how to associate to a basis $\{\vec{u}_1, \vec{u}_2, \vec{u}_3\}$ a new basis $\{\vec{u}_1, \underline{u}_2 - \text{Proj}_{\underline{u}_1} \underline{u}_2, \underline{u}_3 - \text{Proj}_{\underline{u}_1} \underline{u}_3\}$ with only one of the basis vectors, $\vec{u}_1$, being orthogonal to the two other basis vectors. We will now study a procedure which results in a new basis with all three basis vectors being orthogonal to each other. Let $\{\vec{u}_1, \vec{u}_2, \vec{u}_3\}$ be an arbitrary basis for the three-dimensional Cartesian coordinate system ($\mathbb{R}^3$). Let us define three new vectors

$$\vec{w}_1 = \vec{u}_1, \tag{1.19}$$

$$\vec{w}_2 = \vec{u}_2 - \text{Proj}_{w_1} \vec{u}_2 = \vec{u}_2 - \frac{\vec{w}_1 \cdot \vec{u}_2}{\vec{w}_1 \cdot \vec{w}_1} \vec{w}_1, \tag{1.20}$$

$$\vec{w}_3 = \vec{u}_3 - \text{Proj}_{w_1} \vec{u}_3 - \text{Proj}_{w_2} \vec{u}_3 = \vec{u}_3 - \frac{\vec{w}_1 \cdot \vec{u}_3}{\vec{w}_1 \cdot \vec{w}_1} \vec{w}_1 - \frac{\vec{w}_2 \cdot \vec{u}_3}{\vec{w}_2 \cdot \vec{w}_2} \vec{w}_2, \tag{1.21}$$

shown in Figure 1.5. Since $\text{Proj}_{w_1} \vec{u}_2 = c_1 \vec{w}_1$, $\text{Proj}_{w_1} \vec{u}_3 = c_2 \vec{w}_1$ and $\text{Proj}_{w_2} \vec{u}_3 =$

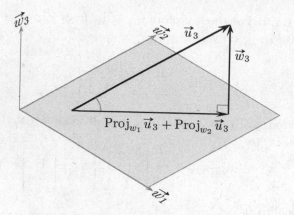

Figure 1.5: Gram-Schmidt procedure for three vectors.

$c_3 \vec{w}_2$ with some numbers c_1, c_2, c_3, the vectors $\vec{w}_1, \vec{w}_2, \vec{w}_3$ are also linearly independent and define a new basis in $\mathbb{R}^3$.

Clearly, $\vec{w}_2$ is orthogonal to $\vec{w}_1$. Therefore $\mathrm{Proj}_{w_2} \vec{u}_3$ being a projection of $\vec{u}_3$ on $\vec{w}_2$ is also orthogonal to $\vec{w}_1$. It follows that $\vec{w}_3$ is orthogonal to $\vec{w}_1$. Moreover,

$$\vec{w}_3 \cdot \vec{w}_2 = \vec{u}_3 \cdot \vec{w}_2 - \frac{\vec{w}_2 \cdot \vec{u}_3}{\vec{w}_2 \cdot \vec{w}_2} \, \vec{w}_2 \cdot \vec{w}_2 = 0 \, .$$

Hence all three basis vectors $\vec{w}_1, \vec{w}_2, \vec{w}_3$ are orthogonal to each other. Such basis is called an orthogonal basis. The construction of the orthogonal basis described by relations (1.19)-(1.21) is called a Gram-Schmidt procedure for a three-dimensional space.

We can simplify the Gram-Schmidt procedure by working with normalized vectors i.e. vectors with an unit norm. Define

$$\vec{e}_1 = \frac{\vec{w}_1}{|\vec{w}_1|} \tag{1.22}$$

such that $\vec{e}_1 \cdot \vec{e}_1 = 1$. Then, replace the definition (1.20) by

$$\vec{\omega}_2 = \vec{u}_2 - (\vec{e}_1 \cdot \vec{u}_2) \, \vec{e}_1 \tag{1.23}$$

and normalize it as in (1.22) to give

$$\vec{e}_2 = \frac{\vec{\omega}_2}{|\vec{\omega}_2|} \, . \tag{1.24}$$

Next, we replace (1.21) by the much simpler expression,

$$\vec{\omega}_3 = \vec{u}_3 - (\vec{e}_1 \cdot \vec{u}_3) \, \vec{e}_1 - (\vec{e}_2 \cdot \vec{u}_3) \, \vec{e}_2 \, . \tag{1.25}$$

As the last step we normalize $\vec{\omega}_3$, yielding

$$\vec{e}_3 = \frac{\vec{\omega}_3}{|\vec{\omega}_3|}. \tag{1.26}$$

The basis vectors $\vec{e}_i$, $i = 1, 2, 3$ obtained in such way are orthonormal, meaning that $\vec{e}_i \cdot \vec{e}_j = \delta_{ij}$ holds. The new basis $\{\vec{e}_1, \vec{e}_2, \vec{e}_3\}$ is called an orthonormal basis. A process of dividing a vector by its norm is called a normalization process. Any non-zero vector can be normalized by dividing it by its norm.

☞ **Example 1.1.10.** Consider 3 linearly independent vectors $\underline{u}_1, \underline{u}_2, \underline{u}_3$ from Example (1.1.4). The relation (1.19) of the Gram-Schmidt procedure gives

$$\underline{w}_1 = \begin{bmatrix} 0 \\ 2 \\ 1 \end{bmatrix}.$$

According to relation (1.20) vector $\underline{w}_2$ coincide with the vector given in equation (1.17) and so

$$\underline{w}_2 = \frac{1}{5} \begin{bmatrix} 5 \\ 11 \\ -22 \end{bmatrix}.$$

For the the last term in equation (1.21), we find

$$\text{Proj}_{w_2} \vec{u}_3 = \frac{1}{63} \begin{bmatrix} 16 \\ 176/5 \\ -352/5 \end{bmatrix}.$$

Subtracting $\text{Proj}_{w_2} \vec{u}_3$ from expression (1.18) yields

$$\underline{w}_3 = \frac{1}{63} \begin{bmatrix} 110 \\ -10 \\ 20 \end{bmatrix}.$$

One verifies readily that $\underline{w}_1, \underline{w}_2, \underline{w}_3$ form an orthogonal basis in $\mathbb{R}^3$.

Let us consider again the vector $\vec{B}_\perp$ defined in relation (1.15). For square of its norm we find an expression

$$|\vec{B}_\perp|^2 = \vec{B}_\perp \cdot \vec{B}_\perp = \left(\vec{B} - \frac{\vec{A} \cdot \vec{B}}{\vec{A} \cdot \vec{A}} \vec{A} \right) \cdot \left(\vec{B} - \frac{\vec{A} \cdot \vec{B}}{\vec{A} \cdot \vec{A}} \vec{A} \right)$$

$$= \vec{B} \cdot \vec{B} - \frac{\vec{A} \cdot \vec{B}}{\vec{A} \cdot \vec{A}} \vec{A} \cdot \vec{B} - \frac{\vec{A} \cdot \vec{B}}{\vec{A} \cdot \vec{A}} \vec{B} \cdot \vec{A} + \left(\frac{\vec{A} \cdot \vec{B}}{\vec{A} \cdot \vec{A}} \right)^2 \vec{A} \cdot \vec{A}$$

$$= \vec{B} \cdot \vec{B} - \frac{\vec{A} \cdot \vec{B}}{\vec{A} \cdot \vec{A}} \vec{B} \cdot \vec{A} \geq 0,$$

where the last inequality follows from the fact that the norm of any vector is always positive. The equality $|\vec{B}_\perp|^2 = 0$ occurs only when the vector $\vec{B}_\perp$ is a null vector. The above formula implies therefore that two vectors $\vec{A}$ and $\vec{B}$ will always satisfy the so-called Cauchy-Schwarz inequality:

$$\left(\vec{A} \cdot \vec{B}\right)^2 \leq \left(\vec{B} \cdot \vec{B}\right)\left(\vec{A} \cdot \vec{A}\right). \tag{1.27}$$

The equality in the Cauchy-Schwarz relation (1.27) holds if and only if $\vec{B}_\perp = 0$ or $\vec{B} = c\vec{A}$ for some non-zero constant c.

1.1.5 Rotation of Vectors

Let us consider a rotation of the Cartesian coordinates by an angle ϕ as shown in Figure 1.6. For simplicity we will only consider a rotation in xy-plane. Let us denote x', y' as new coordinates of the rotated Cartesian coordinate system. Consider displacement vector $\vec{A} = \overrightarrow{OP}$ which starts at the origin O and ends at the point P. The vector $\vec{A}$ as viewed in respectively old and rotated coordinated systems is given by:

$$\vec{A} = A_x\hat{i} + A_y\hat{j} + A_z\hat{k} = A'_x\hat{i}' + A'_y\hat{j}' + A'_z\hat{k}'. \tag{1.28}$$

Let $\vec{A}$ make an angle α with respect to the old x-axis. The x-component of $\vec{A}$

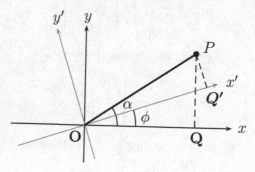

Figure 1.6: Rotation in a plane.

is then equal to $A_x = OQ = OP\cos\alpha = A\cos\alpha$. Similarly, the y-component of $\vec{A}$ is equal to $A_y = A\sin\alpha$ as obtained by projecting $\vec{A}$ on the y-axis. The new coordinates A'_i, $i = 1, 2$ are obtained by projecting $\vec{A}$ on the new x' and y' axis. This yields

$$\begin{aligned} A'_x = OQ' &= A\cos(\alpha - \phi) = A\cos\alpha\cos\phi + A\sin\alpha\sin\phi \\ &= A_x\cos\phi + A_y\sin\phi \end{aligned} \tag{1.29a}$$

and

$$A'_y = A\sin(\alpha - \phi) = A\sin\alpha\cos\phi - A\cos\alpha\sin\phi \qquad (1.29b)$$
$$= -A_x\sin\phi + A_y\cos\phi\,.$$

In the case of the rotation in the xy-plane: $A'_z = A_z$.

We are now ready to check the invariance of the scalar product of vectors under rotation. Given are two vectors $\vec{A}$, $\vec{B}$ which in the rotated coordinate system are given by $\vec{A}' = (A'_x, A'_y, A'_z)$ and $\vec{B}' = (B'_x, B'_y, B'_z)$. Their scalar product is

$$\vec{A}' \cdot \vec{B}' = \sum_{k=1}^{2} A'_k B'_k + A'_3 B'_3\,. \qquad (1.30)$$

Using expressions (1.29a)-(1.29b) for new coordinates of both vectors we obtain

$$\begin{aligned}
\vec{A}' \cdot \vec{B}' &= (A_x\cos\phi + A_y\sin\phi)(B_x\cos\phi + B_y\sin\phi) \\
&+ (-A_x\sin\phi + A_y\cos\phi)(-B_x\sin\phi + B_y\cos\phi) + A_3 B_3 \qquad (1.31) \\
&= \sum_{i=1}^{2} A_i B_i + A_3 B_3 = \vec{A}\cdot\vec{B}\,,
\end{aligned}$$

which shows invariance of the scalar product under rotation of the coordinate systems. Recall expression (1.9) for the scalar product $\vec{A}\cdot\vec{B}$. Since, as shown

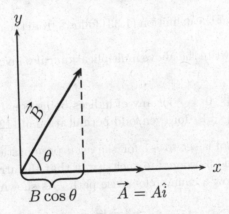

Figure 1.7: The scalar product equals $AB\cos\theta$.

above, this product is invariant under the rotation of the coordinate system we will not change the value of $\vec{A}\cdot\vec{B}$ by rotating the coordinate system to the one in which $\vec{A}$ lies on the x-axis and $\vec{B}$ is in the xy-plane. In this coordinate system, $A_x = A, A_y = 0, A_z = 0, B_x = B\cos\theta$ and we find:

$$\vec{A}\cdot\vec{B} = A_x B_x = AB\cos\theta\,, \qquad (1.32)$$

where θ is the angle between $\vec{A}$ and $\vec{B}$ vector as shown in Figure (1.7). Thus, expression (1.9) for the scalar product agrees with the standard expression for the scalar product of many introductory physics courses.

1.1.6 Vector Product

In the previous section, we have "multiplied" two vectors to obtain a scalar quantity $\vec{A} \cdot \vec{B}$. For two vectors there exists another choice for operation of multiplication and that results in a vector. This operation is called the vector product or cross product and is denoted by $\vec{A} \times \vec{B}$ and equal to:

$$\vec{A} \times \vec{B} = \sum_{i=1}^{3} \sum_{j=1}^{3} A_i B_j \hat{e}_i \times \hat{e}_j = (A_2 B_3 - A_3 B_2)\,\hat{e}_1$$
$$+ (A_3 B_1 - A_1 B_3)\,\hat{e}_2 + (A_1 B_2 - A_2 B_1)\,\hat{e}_3 . \tag{1.33}$$

From definition (1.33) it follows that

$$\hat{e}_1 \times \hat{e}_1 = \hat{e}_2 \times \hat{e}_2 = \hat{e}_3 \times \hat{e}_3 = 0 \tag{1.34a}$$

and

$$\hat{e}_1 \times \hat{e}_2 = \hat{e}_3, \quad \hat{e}_2 \times \hat{e}_3 = \hat{e}_1, \quad \hat{e}_3 \times \hat{e}_1 = \hat{e}_2 . \tag{1.34b}$$

Conversely, one sees that the definition (1.33) follows from the rules of multiplication (1.34a) and (1.34b).

A compact way to write the above multiplication rules involves the permutation symbol ϵ_{ijk}:

$$\epsilon_{ijk} = \left\{ \begin{array}{ll} 0 & \text{for any of indices being equal} \\ \pm 1 & \text{for even/odd permutation of (123)} \end{array} \right. \tag{1.35}$$

The permutation symbol is set to $+1$ for the even permutations of $(1, 2, 3)$ which contain numbers 1, 2 and 3 ordered in such a way that the arrows from 1 to 2 and from 2 to 3 always follow a counterclockwise path. As shown in Figure (1.8) this

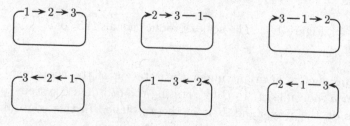

Figure 1.8: Even and odd permutations of $(1, 2, 3)$.

occurs for $(1,2,3), (2,3,1)$ and $(3,1,2)$ and

$$\epsilon_{123} = \epsilon_{231} = \epsilon_{312} = 1.$$

In contrast, for the odd permutations $(2,1,3), (3,2,1), (1,3,2)$ in Figure (1.8) the arrows from 1 to 2 and from 2 to 3 follow a clockwise path. The permutation symbol is assigned a value of -1 for the odd permutations

$$\epsilon_{213} = \epsilon_{321} = \epsilon_{132} = -1.$$

Swapping two neighboring indices changes permutation from even to odd or vice versa and therefore reverses the sign of the permutation symbol:

$$\epsilon_{ijk} = -\epsilon_{jik} = -\epsilon_{ikj} = -\epsilon_{kji}.$$

Thus, a permutation is even if one can obtain given arrangement from the initial arrangement of $(1,2,3)$ by an even number of swappings of the nearest neighbors. In case the given arrangement follows from the initial arrangement of $(1,2,3)$ by performing an odd number of swappings of the nearest neighbors then the permutation is odd.

It is now clear that $\epsilon_{ijk} = 0$ if any two of the indices (i,j,k) are equal to each other since, in this case, swapping of two identical neighbors does not change the arrangement of indices although it produces a minus sign.

We now recover all the multiplication rules (1.34a) and (1.34b) from

$$\hat{e}_i \times \hat{e}_j = \sum_{k=1}^{3} \epsilon_{ijk}\hat{e}_k \tag{1.36}$$

for any $i,j = 1,\ldots,3$. Accordingly,

$$\vec{C} = \vec{A} \times \vec{B} = \sum_{i=1}^{3}\sum_{j=1}^{3}\sum_{k=1}^{3} A_i B_j \epsilon_{ijk}\hat{e}_k \tag{1.37}$$

with components given by $C_k = \sum_{i=1}^{3}\sum_{j=1}^{3} A_i B_j \epsilon_{ijk}$.

Sometimes it is useful rewrite the right hand side of (1.33) in the form of a determinant of a 3×3 matrix. We will take a closer look at determinants in the next chapter. For now it suffices to define an object

$$\begin{vmatrix} a_1 & a_2 & a_3 \\ b_1 & b_2 & b_3 \\ c_1 & c_2 & c_3 \end{vmatrix} = \sum_{i=1}^{3}\sum_{j=1}^{3}\sum_{k=1}^{3} \epsilon_{ijk} a_i b_j c_k$$

$$= a_1(b_2 c_3 - b_3 c_2) + a_2(b_3 c_1 - b_1 c_3) + a_3(b_1 c_2 - b_2 c_1),$$

which due to the antisymmetry of the permutation symbol ϵ_{ijk} changes sign under the exchange of two adjacent rows (or columns):

$$\begin{vmatrix} a_1 & a_2 & a_3 \\ b_1 & b_2 & b_3 \\ c_1 & c_2 & c_3 \end{vmatrix} = - \begin{vmatrix} a_1 & a_2 & a_3 \\ c_1 & c_2 & c_3 \\ b_1 & b_2 & b_3 \end{vmatrix} = - \begin{vmatrix} c_1 & c_2 & c_3 \\ b_1 & b_2 & b_3 \\ a_1 & a_2 & a_3 \end{vmatrix} = - \begin{vmatrix} b_1 & b_2 & b_3 \\ a_1 & a_2 & a_3 \\ c_1 & c_2 & c_3 \end{vmatrix}. \tag{1.38}$$

In terms of a determinant, the vector product (1.33) can be rewritten as:

$$\vec{A} \times \vec{B} = \begin{vmatrix} \hat{e}_1 & \hat{e}_2 & \hat{e}_3 \\ A_1 & A_2 & A_3 \\ B_1 & B_2 & B_3 \end{vmatrix}. \tag{1.39}$$

An advantage of this notation is that it makes the anti-commutativity property $\vec{A} \times \vec{B} = -\vec{B} \times \vec{A}$ of the vector product transparent.

We can use these new formulas to verify the value of the scalar product $\vec{A} \cdot \vec{C} = \vec{A} \cdot (\vec{A} \times \vec{B})$:

$$\vec{A} \cdot \vec{C} = \sum_{l=1}^{3} A_l \hat{e}_l \cdot \sum_{i=1}^{3} \sum_{j=1}^{3} \sum_{k=1}^{3} A_i B_j \epsilon_{ijk} \hat{e}_k = \sum_{i=1}^{3} \sum_{j=1}^{3} \sum_{k=1}^{3} A_i B_j A_k \epsilon_{ijk}$$

$$= \begin{vmatrix} A_1 & A_2 & A_3 \\ A_1 & A_2 & A_3 \\ B_1 & B_2 & B_3 \end{vmatrix}. \tag{1.40}$$

Since the determinant of the matrix with two identical rows as in (1.40) is zero due to antisymmetry relations (1.38) one concludes that the vector $\vec{C} = \vec{A} \times \vec{B}$ is perpendicular to $(\vec{A}, \vec{B})$ plane spanned by the $\vec{A}$ and $\vec{B}$ vectors. $\vec{A}, \vec{B}, \vec{C}$ vectors form a right-handed system defined according the right hand rule shown in Figure (1.9). If you curl the fingers of your right hand through the angle θ from $\vec{A}$ to $\vec{B}$ (where θ is an angle as in (1.32)), then the right thumb points in the direction of $\vec{C} = \vec{A} \times \vec{B}$.

Next, we introduce an identity

$$\sum_{k=1}^{3} \epsilon_{ijk} \epsilon_{nmk} = \delta_{in} \delta_{jm} - \delta_{im} \delta_{jn}. \tag{1.41}$$

To verify this equality it is easiest to just work out the left and right hand sides for all allowed values of indices i, j, n, m (see Problem 1.1.33).

Then it follows that

$$\left(\vec{A} \times \vec{B} \right) \cdot \left(\vec{C} \times \vec{D} \right) = \sum_{k=1} \epsilon_{ijk} \epsilon_{nmk} A_i B_j C_n D_m = \left(\vec{A} \cdot \vec{C} \right) \left(\vec{B} \cdot \vec{D} \right)$$

$$- \left(\vec{A} \cdot \vec{D} \right) \left(\vec{B} \cdot \vec{C} \right).$$

Vectors and Vector Calculus

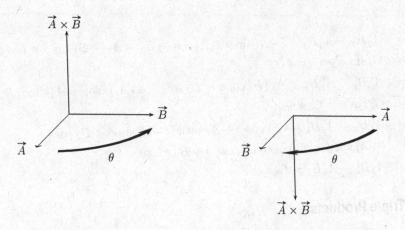

Figure 1.9: The right hand rule.

For $\vec{C} = \vec{A}$ this relation becomes,

$$\left(\vec{A} \times \vec{B}\right) \cdot \left(\vec{A} \times \vec{B}\right) = A^2 B^2 - \left(\vec{A} \cdot \vec{B}\right)^2 = A^2 B^2 \sin^2 \alpha \qquad (1.42)$$

and, therefore, the magnitude of vector $\vec{C} = \vec{A} \times \vec{B}$ is

$$C = AB \sin \theta . \qquad (1.43)$$

Hence, two nonzero vectors $\vec{A}$ and $\vec{B}$ are parallel ($\theta = 0$) if and only if $\vec{A} \times \vec{B} = 0$. Let $\vec{A}$ and $\vec{B}$ be two nonzero and nonparallel vectors. We can consider them to be the sides of a parallelogram. Recall that the area of a parallelogram is equal to a product of its base and its altitude. If A is a base as in Figure (1.1.6) the altitude is $h = B \sin \theta$. Then the magnitude of $\vec{C} = \vec{A} \times \vec{B}$ given in (1.43) is equal to the area of a parallelogram. Similarly, the area of a triangle with sides $\vec{A}$ and $\vec{B}$ is equal to $\frac{1}{2}C$. Let us now show that $\vec{C} = \vec{A} \times \vec{B}$ rotates as a vector when the components

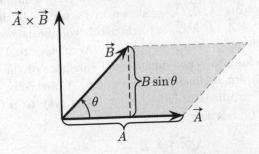

Figure 1.10: The area of a parallelogram is equal to $C = AB \sin \theta$.

of $\vec{A}$ and $\vec{B}$ transform according to (1.29a) and (1.29b). We find, by substitution, that

$$
\begin{aligned}
C_1' &= A_2'B_3' - A_3'B_2' = (-A_1 \sin\phi + A_2 \cos\phi)B_3 - A_3(-B_1 \sin\phi + B_2 \cos\phi) \\
&= C_2 \sin\phi + C_1 \cos\phi,
\end{aligned} \tag{1.44}
$$

$$
\begin{aligned}
C_2' &= A_3'B_1' - A_1'B_3' = A_3(B_1 \cos\phi + B_2 \sin\phi) - (A_1 \cos\phi + A_2 \sin\phi)B_3 \\
&= C_2 \cos\phi - C_1 \sin\phi,
\end{aligned} \tag{1.45}
$$

$$
\begin{aligned}
C_3' &= A_1'B_2' - A_2'B_1' = (A_1 \cos\phi + A_2 \sin\phi)(-B_1 \sin\phi + B_2 \cos\phi) \\
&\quad - (-A_1 \sin\phi + A_2 \cos\phi)(B_1 \cos\phi + B_2 \sin\phi) \\
&= A_1 B_2 - A_2 B_1 = C_3.
\end{aligned} \tag{1.46}
$$

1.1.7 Triple Products

We will now consider triples and mixtures of scalar and vector products such as $\vec{A} \cdot \vec{B} \times \vec{C}$ and $\vec{A} \times (\vec{B} \times \vec{C})$.

The scalar triple product $\vec{A} \cdot \vec{B} \times \vec{C}$, can be derived by noticing that

$$
\hat{e}_i \cdot (\hat{e}_j \times \hat{e}_k) = \epsilon_{ijk} \tag{1.47}
$$

and, therefore,

$$
\vec{A} \cdot \vec{B} \times \vec{C} = \sum_{i=1}^3 \sum_{j=1}^3 \sum_{k=1}^3 \epsilon_{ijk} A_i B_j C_k = \begin{vmatrix} A_1 & A_2 & A_3 \\ B_1 & B_2 & B_3 \\ C_1 & C_2 & C_3 \end{vmatrix}. \tag{1.48}
$$

Thus, the scalar triple product changes sign under the odd permutations of any of two vectors:

$$
\vec{A} \cdot \vec{B} \times \vec{C} = -\vec{A} \cdot \vec{C} \times \vec{B} = -\vec{C} \cdot \vec{B} \times \vec{A} = -\vec{B} \cdot \vec{A} \times \vec{C}. \tag{1.49}
$$

The scalar triple product, $\vec{A} \cdot \vec{B} \times \vec{C}$, is equal to the volume of a parallelepiped spanned by three vectors $\vec{A}, \vec{B}, \vec{C}$, as shown in Figure 1.11. If, the scalar triple product vanishes, i.e. $\vec{A} \cdot \vec{B} \times \vec{C} = 0$, then the volume is zero and the vectors must be co-planar (lying on the same plane). According to Problems 1.1.8 and 1.1.9 such vectors belong to a two-dimensional subspace of the three-dimensional vector space and can not be linearly independent. Conversely, if $\vec{A} \cdot \vec{B} \times \vec{C} \neq 0$ then vectors $\vec{A}, \vec{B}, \vec{C}$ are linearly independent. It is easy to prove this statement explicitly. Consider the relation:

$$
\lambda_1 \vec{A} + \lambda_2 \vec{B} + \lambda_3 \vec{C} = 0 \tag{1.50}
$$

and apply the cross product with $\vec{B}$ on the left hand side of this relation:

$$
\lambda_1 \vec{B} \times \vec{A} + \lambda_3 \vec{B} \times \vec{C} = 0.
$$

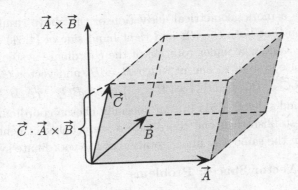

Figure 1.11: The scalar triple product.

Next, take a scalar product of $\vec{A}$ with the left hand side. Since $\vec{A} \cdot \vec{B} \times \vec{A} = 0$ only one term remains:

$$\lambda_3 \, \vec{A} \cdot \vec{B} \times \vec{C} = 0$$

and so $\lambda_3 = 0$ provided that $\vec{A} \cdot \vec{B} \times \vec{C} \neq 0$. Changing the order of multiplication we obtain identities for the remaining two scalars:

$$\lambda_1 \vec{C} \cdot \vec{B} \times \vec{A} = 0 \quad \text{and} \quad \lambda_2 \vec{C} \cdot \vec{A} \times \vec{B} = 0 \, .$$

Due to identities (1.49) all these scalar triple products are different from zero and therefore $\lambda_1 = 0$ and $\lambda_2 = 0$. Thus, vectors $\vec{A}, \vec{B}, \vec{C}$ are linearly independent.

For the vector triple product $\vec{A} \times \left(\vec{B} \times \vec{C} \right)$ we find an identity

$$\vec{A} \times \left(\vec{B} \times \vec{C} \right) = \vec{B} \left(\vec{A} \cdot \vec{C} \right) - \vec{C} \left(\vec{A} \cdot \vec{B} \right) \, . \tag{1.51}$$

This relation follows from the technical identity (1.41) involving the permutation symbol. According to this identity it follows that

$$\vec{A} \times \left(\vec{B} \times \vec{C} \right) = \sum_{i=1}^{3} A_i \hat{e}_i \times \left(\sum_{k,m,n=1}^{3} \epsilon_{kmn} B_m C_n \hat{e}_k \right)$$

$$= \sum_{i,k,m,n=1}^{3} \epsilon_{kmn} A_i B_m C_n \, \hat{e}_i \times \hat{e}_k = \sum_{i,l,m,n=1}^{3} \sum_{k=1}^{3} (\epsilon_{kmn} \epsilon_{kli}) A_i B_m C_n \hat{e}_l$$

$$= \sum_{i,l,m,n=1}^{3} (\delta_{ml} \delta_{ni} - \delta_{mi} \delta_{nl}) A_i B_m C_n \hat{e}_l = \sum_{i=1}^{3} A_i C_i \sum_{m=1}^{3} B_m \hat{e}_m$$

$$- \sum_{i=1}^{3} A_i B_i \sum_{m=1}^{3} C_m \hat{e}_m = \vec{B} \left(\vec{A} \cdot \vec{C} \right) - \vec{C} \left(\vec{A} \cdot \vec{B} \right) \, .$$

There also exists a more geometrical derivation of the above result that is based on an observation that both the left and right hand side of (1.51) are vectors and transform in the same way under rotation of the coordinate system. For example, rotate a coordinate system to one in which $\vec{B} = B\hat{i}$ and vector $\vec{C}$ lies in the xy-plane. Thus, $\vec{B} \times \vec{C} = BC_2\hat{k}$ and $\vec{A} \times \left(\vec{B} \times \vec{C}\right) = \hat{i}A_2BC_2 - \hat{j}A_1BC_2$, which agrees with the right hand side of (1.51) in that specially chosen coordinate system. The vector equality established in one coordinate system holds in general since both sides transform in the same way under rotation of the coordinate system.

◢ Vectors and Vector Spaces: Problems

Problem 1.1.1. A set of vectors $\{\vec{v}_1, \ldots \vec{v}_n\}$ is said to span a vector space V if every vector in V is a linear combination $\sum_{i=1}^{n} c_i \vec{v}_i$ for some scalars $c_1, \ldots, c_n$.
Show that two vectors

$$\vec{v}_1 = \begin{bmatrix} 1 \\ 0 \\ 0 \end{bmatrix}, \quad \vec{v}_2 = \begin{bmatrix} 0 \\ 1 \\ -1 \end{bmatrix}$$

do not span $\mathbb{R}^3$.
Do three vectors

$$\vec{v}_1 = \begin{bmatrix} 1 \\ 0 \\ 0 \end{bmatrix}, \quad \vec{v}_2 = \begin{bmatrix} 0 \\ 1 \\ -1 \end{bmatrix}, \quad \vec{v}_3 = \begin{bmatrix} 1 \\ 1 \\ 1 \end{bmatrix}$$

span $\mathbb{R}^3$?

Problem 1.1.2. Find values of x for which the three vectors

$$\vec{v}_1 = \begin{bmatrix} 1 \\ 0 \\ 0 \end{bmatrix}, \quad \vec{v}_2 = \begin{bmatrix} 0 \\ 1 \\ -1 \end{bmatrix}, \quad \vec{v}_3 = \begin{bmatrix} 1 \\ x \\ 1 \end{bmatrix}$$

are linearly independent.

Problem 1.1.3. A *subspace* of a vector space V is a subset of V that is closed under addition and scalar multiplication.
Show that a set of vectors $\vec{v} = (v_1, v_2, v_3)$, such that $v_1 + v_2 = v_3$, forms a subspace of the vector space $\mathbb{R}^3$ and find its basis and dimension.

Problem 1.1.4. Show that a set of vectors $\vec{v} = (v_1, v_2, v_3)$ such that $v_2 = v_3 - v_1$, forms a subspace of the vector space $\mathbb{R}^3$ and find its basis and dimension. Does a set of vectors $\vec{v} = (v_1, v_2, v_3)$ such that $v_2 = v_3 - v_1 + 3$ form a subspace of the vector space $\mathbb{R}^3$?

Problem 1.1.5. Consider a set of vectors $\vec{v} = (u - 2v, v + u, 3u, v)$ for real scalars u and v. Is it a subspace of $\mathbb{R}^4$? If, yes, find its basis and dimension. Does a set of vectors $\vec{v} = (1, u + v, 3u, v)$ form a subspace of $\mathbb{R}^4$?

Problem 1.1.6. Let $\mathbb{P}^2$ be a vector space of polynomials with degree less than or equal to 2. Find whether the three polynomials

$$p_1 = x^2 + 3, \quad p_2 = x + 1, \quad p_3 = x - 1$$

are linearly independent? Do they form a basis for $\mathbb{P}^2$?

Problem 1.1.7. Can you obtain $p = x^2 - x + 3$ as a linear combination of

$$p_1 = x^2 - 2x - 1, \quad p_2 = x^2 + 1, \quad p_3 = x + 1$$

in $\mathbb{P}^2$? Do p_1, p_2 and p_3 form a basis for $\mathbb{P}^2$?

Problem 1.1.8. A plane containing vector $\vec{r}_0 = x_0\hat{i} + y_0\hat{j} + z_0\hat{k}$ and normal to the vector $\vec{n}$ is swept by all vectors $\vec{r} = x\hat{i} + y\hat{j} + z\hat{k}$ satisfying the relation $(\vec{r} - \vec{r}_0) \cdot \vec{n} = 0$ as shown in Figure 1.12. The vector $\vec{n} = n_1\hat{i} + n_2\hat{j} + n_3\hat{k}$ is

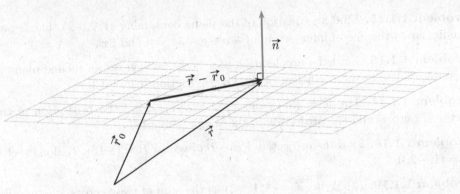

Figure 1.12: The plane and vectors defining the plane.

called a normal vector to a plane. Thus the equation of the plane takes the form $n_1x + n_2y + n_3z = k$, where $k = n_1x_0 + n_2y_0 + n_3z_0$.
  Find the plane through the points $(0, 2, 2)$, $(2, 0, 3)$, $(3, 2, 0)$.

Problem 1.1.9. Show that for every normal vector $\vec{n}$ to the plane the space of vectors $\vec{v} = \vec{r} - \vec{r}_0$ on the plane is a two-dimensional *subspace* of the vector space $\mathbb{R}^3$.

Problem 1.1.10. For two fixed three dimensional vectors $\vec{A}$ and $\vec{B}$ consider a set of vectors $\vec{v}$ which satisfy $\vec{A} \cdot \vec{v} = 0$ and $\vec{B} \cdot \vec{v} = 0$. Does this set of vectors form a vector space? Find its basis and dimension. Show for vectors $\vec{A}$ and $\vec{B}$, which are not parallel, that all vectors in this set must satisfy the condition $\vec{v} \times \vec{n} = 0$ for some vector $\vec{n}$ and find $\vec{n}$ in terms of $\vec{A}$ and $\vec{B}$. Conversely, for a set of vectors $\vec{v}$ which satisfy $\vec{v} \times \vec{n} = 0$ for some vector $\vec{n}$, describe how to find vectors $\vec{A}$ and $\vec{B}$ which are orthogonal to all vectors $\vec{v}$.

Problem 1.1.11. The plane through the points $(0, 2, 1)$, $(-1, 0, 3)$ is described by the equation $n_1 x + n_2 y + n_3 z = 0$. Find the normal vector $\vec{n}$.

Problem 1.1.12. (a) Find the distance from the point $(-2, 3, 1)$ to the plane $4x + 2y - 3z = 8$.

Hint: Use the fact that the distance of point P to the plane is the length of the projection of the vector $P - P'$ on the normal $\vec{n}$, where P' is an arbitrary point on the plane.

(b) Find the point on the plane that is closest to the point $(-2, 3, 1)$.

Problem 1.1.13. Find the angle between the planes $2x + y - z = 8$ and $x - y + z = 14$. (The angle between two intersecting planes is the same as the angle between their normal vectors.)

Problem 1.1.14. Find the area of the triangle if the vertices are $(0, 0, 2)$, $(0, 2, 2)$, $(1, 2, 2)$.

Problem 1.1.15. Find an equation of the plane containing $(1, 2, -1)$ that is perpendicular to the line of intersection of $-x + y - 3z = 4$ and $2x - y + z = 0$.

Problem 1.1.16. Find the acceleration of a body moving on an incline plane in Example (1.1.8), if the coefficient of the kinetic friction is $\mu = 0.3$.

Problem 1.1.17. Let $\vec{u} = (3, 1)$ and $\vec{v} = (2, -1)$. Find the projection of the vector $\vec{u}$ onto $\vec{v}$ and the projection of the vector $\vec{v}$ onto $\vec{u}$.

Problem 1.1.18. Find the projection $\text{Proj}_A \vec{B}$ of vector $\vec{B} = (-1, 1, 2)$ along vector $\vec{A} = (1, -2, 0)$.

Problem 1.1.19. (a) Write $\vec{a} = (3, 1, -2)$ as the sum of two vectors, one parallel $\vec{a}_\parallel$, and one perpendicular $\vec{a}_\perp$ to $\vec{b} = (1, -1, 0)$.

(b) Find a basis vector $\vec{c}$ that can be added to the set $\{\vec{a}_\perp, \vec{b}\}$ to produce an orthogonal basis for $\mathbb{R}^3$.

(c) Let $\vec{v} = (1, 2, 3)$. Find the coordinates of $\vec{v}$ with respect to this orthogonal basis.

Problem 1.1.20. Let $\vec{v}$ be an arbitrary 3-dimensional vector and $\hat{n}$ an arbitrary 3-dimensional unit vector. Show that

$$\vec{v} = \text{Proj}_{\hat{n}} \vec{v} + (\hat{n} \times \vec{v}) \times \hat{n}$$
$$= \hat{n}(\hat{n} \cdot \vec{v}) + (\hat{n} \times \vec{v}) \times \hat{n}.$$

Problem 1.1.21. (a) Check whether the vectors

$$\vec{A} = (-1, 1, -1), \quad \vec{B} = (2, 0, 3), \quad \vec{C} = (1, 2, 1)$$

are linearly independent.

(b) Calculate $\vec{A} \cdot (\vec{B} \times \vec{C})$ for the above three vectors. Is the result consistent with your answer in (a).

(c) Apply the Gram-Schmidt procedure to vectors $\vec{A}, \vec{B}$ and $\vec{C}$.

Problem 1.1.22. Calculate the triple scalar product $\vec{A} \cdot (\vec{B} \times \vec{C})$ for vectors $\vec{A}, \vec{B}$ and $\vec{C}$ such that $\hat{j} \cdot \vec{A} = 0$, $\hat{j} \cdot \vec{B} = 0$ and $\hat{j} \cdot \vec{C} = 0$.

Problem 1.1.23. For $\underline{w}_1, \underline{w}_2, \underline{w}_3$ vectors from Example 1.1.10 calculate $\underline{w}_1 \times \underline{w}_2$. Based on this result propose a simple version of the Gram-Schmidt procedure in three dimensions.

Problem 1.1.24. Consider three vectors

$$\vec{u}_1 = \hat{i} + \hat{j}, \quad \vec{u}_2 = \hat{i} + \hat{k}, \quad \vec{u}_3 = \hat{j} + \hat{k}$$

in three dimensions. Check whether the vectors $\vec{u}_1, \vec{u}_2, \vec{u}_3$ are linearly independent and use the Gram-Schmidt procedure to construct an orthonormal basis $\vec{e}_1, \vec{e}_2, \vec{e}_3$.

Problem 1.1.25. Show that if the vectors $\vec{A}$ and $\vec{B}$ are both perpendicular to the vector $\vec{C}$, then $\left(\vec{A} \times \vec{B}\right) \times \vec{C} = 0$.

Problem 1.1.26. A particle carrying charge q moving along y-axis with velocity $\vec{v} = v_0 \hat{j}$ enters a region with a magnetic field $\vec{B}$. Find all possible values of $\vec{B}$ if a Lorentz force acting on the particle is found to be

$$\vec{F} = q\vec{v} \times \vec{B} = qv_0 \left(2\hat{i} - 3\hat{k}\right)$$

Problem 1.1.27. Show that three mutually orthogonal vectors in three dimensions are linearly independent and therefore form an orthogonal basis in $\mathbb{R}^3$.

Problem 1.1.28. Show that $\vec{A} \times (\vec{B} \times \vec{C}) = \vec{B}(\vec{A} \cdot \vec{C}) - \vec{C}(\vec{A} \cdot \vec{B})$ implies

$$(\vec{A} \times \vec{B}) \cdot (\vec{C} \times \vec{D}) = (\vec{A} \cdot \vec{C})(\vec{B} \cdot \vec{D}) - (\vec{A} \cdot \vec{D})(\vec{B} \cdot \vec{C}).$$

Problem 1.1.29. Prove the parallelogram law, which states that:

$$\left|\vec{A} + \vec{B}\right|^2 + \left|\vec{A} - \vec{B}\right|^2 = 2\left|\vec{A}\right|^2 + 2\left|\vec{B}\right|^2$$

valid for any two three-dimensional vectors $\vec{A}$ and $\vec{B}$.

Problem 1.1.30. For two three-dimensional vectors $\vec{A}$ and $\vec{B}$ prove the Lagrange identity,

$$\left|\vec{A} \times \vec{B}\right|^2 + (\vec{A} \cdot \vec{B})^2 = \left|\vec{A}\right|^2 \left|\vec{B}\right|^2.$$

Use it to prove the Cauchy-Schwarz identity.

Problem 1.1.31. Use the Cauchy-Schwarz inequality (1.27) to prove the triangle inequality:

$$\left|\vec{A} + \vec{B}\right| \leq \left|\vec{A}\right| + \left|\vec{B}\right|$$

for any two three-dimensional vectors $\vec{A}$ and $\vec{B}$.

Problem 1.1.32. Prove the Jacobi identity:

$$\vec{A} \times (\vec{B} \times \vec{C}) + \vec{B} \times (\vec{C} \times \vec{A}) + \vec{C} \times (\vec{A} \times \vec{B}) = 0$$

Problem 1.1.33. For the permutation symbol ϵ_{ijk} show that,

(a) $\sum_{l=1}^{3} \epsilon_{mnl}\epsilon_{ijl} = \delta_{mi}\delta_{nj} - \delta_{mj}\delta_{ni}$, (b) $\sum_{j,k=1}^{3} \epsilon_{mjk}\epsilon_{njk} = 2\delta_{nm}$,

(c) $\sum_{i,j,k=1}^{3} \epsilon_{ijk}\epsilon_{ijk} = 6$.

1.2 Vector Calculus

1.2.1 Vector Calculus: Gradient

Let $\phi(x, y, z)$ be a scalar field. A scalar field is a scalar valued function that associates a number to the position in space represented by three variables x, y, z. The value of a scalar field at any point does note change when the position in space is described by different coordinate systems. Thus, scalar fields are invariant under transformations of coordinate system including rotations of the coordinate systems. As an example consider a function $f(x, y, z) = f(r) = f\left(\sqrt{x^2 + y^2 + z^2}\right)$, which depends on the radial distance r which is clearly invariant under rotations.

The three partial derivatives $\partial\phi/\partial x$, $\partial\phi/\partial y$ and $\partial\phi/\partial z$ transform under rotation of a coordinate system as components of a vector. To see this, recall that under the rotation about z-axis by the angle ϕ the components of $\vec{r}$ transform as

$$\begin{aligned} x' &= x\cos\phi + y\sin\phi, & (1.52) \\ y' &= -x\sin\phi + y\cos\phi. & (1.53) \end{aligned}$$

The old coordinates x, y can be expressed by the new coordinates x', y' via

$$\begin{aligned} x &= x'\cos\phi - y'\sin\phi, & (1.54) \\ y &= x'\sin\phi + y'\cos\phi & (1.55) \end{aligned}$$

obtained from (1.52)-(1.53) by letting $\phi \to -\phi$.

Now,

$$\frac{\partial}{\partial x'} = \frac{\partial x}{\partial x'}\frac{\partial}{\partial x} + \frac{\partial y}{\partial x'}\frac{\partial}{\partial y} + \frac{\partial z}{\partial x'}\frac{\partial}{\partial z}$$

$$= \cos\phi\frac{\partial}{\partial x} + \sin\phi\frac{\partial}{\partial y}$$

and $\partial/\partial x$ transforms under rotations as a coordinate of a vector.

Now we introduce a gradient of the scalar function $\phi(x, y, z)$ defined as:

$$\vec{\nabla}\phi = \hat{i}\frac{\partial\phi}{\partial x} + \hat{j}\frac{\partial\phi}{\partial y} + \hat{k}\frac{\partial\phi}{\partial z} = \sum_{i=1}^{3}\frac{\partial\phi}{\partial x_i}\hat{e}_i, \qquad (1.56)$$

where $(x_1, x_2, x_3) = (x, y, z)$.

We conclude that the components of a gradient transform under rotation like the components of a vector. Thus the gradient of a scalar field defines a vector valued object which transforms under rotation like a vector.

We can think about the gradient $\vec{\nabla}\phi$ as a result of applying a vector differential operator:

$$\vec{\nabla} = \hat{i}\frac{\partial}{\partial x} + \hat{j}\frac{\partial}{\partial y} + \hat{k}\frac{\partial}{\partial z} = \sum_{i=1}^{3}\frac{\partial}{\partial x_i}\hat{e}_i \qquad (1.57)$$

on the function ϕ. $\vec{\nabla}$ is referred to as a gradient operator.

☞ **Example 1.2.1.** For the function $f(x, y, z) = f(r) = f\left(\sqrt{x^2 + y^2 + z^2}\right)$ the gradient of $f(r)$ is:

$$\vec{\nabla}f(r) = \hat{i}\frac{df}{dr}\frac{x}{r} + \hat{j}\frac{df}{dr}\frac{y}{r} + \hat{k}\frac{df}{dr}\frac{z}{r} = \left(\hat{i}\frac{x}{r} + \hat{j}\frac{y}{r} + \hat{k}\frac{z}{r}\right)\frac{df}{dr} \qquad (1.58)$$

where we made use of the identity

$$\frac{\partial f(r)}{\partial x} = \frac{df}{dr}\frac{\partial r}{\partial x} = \frac{df}{dr}\frac{x}{r} \qquad (1.59)$$

following from chain differentiation.

Hence, from equation (1.58):

$$\vec{\nabla}f(r) = \frac{\vec{r}}{r}\frac{df}{dr} = \hat{r}\frac{df}{dr} \qquad (1.60)$$

where $\hat{r} = \vec{r}/r$ is a unit vector in the positive radial direction. Equation (1.60) shows that $\vec{\nabla}f(r)$ is a vector.

1.2.2 Geometrical Interpretation of a Gradient.

Here we will explain the role of a gradient in measuring the rate of change of a function. We will discover that the gradient of the function $\phi(x, y, z)$ is a vector which points in a direction of the maximum increase of ϕ.

Let the argument (x, y, z) of the scalar field $\phi(x, y, z)$ increase in a direction of some unit vector $\hat{u} = (u_1, u_2, u_3)$. This increase is described by:

$$\vec{r} = (x, y, z) = (x_0, y_0, z_0) + (u_1, u_2, u_3)s = \vec{r}_0 + \hat{u}s, \qquad (1.61)$$

where $\vec{r}_0 = (x_0, y_0, z_0)$ is a fixed point and s is a variable parameterizing a straight line through $\vec{r}_0$ in the direction of $\hat{u}$. To describe the variation of ϕ along this line

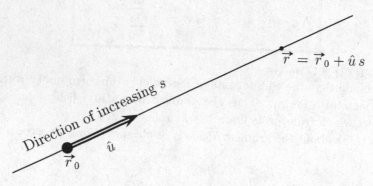

we introduce a function Φ of s:

$$\Phi(s) = \phi(\vec{r}_0 + \hat{u}s).$$

The rate of change of the function Φ indicates how ϕ varies along the line given by equation (1.61). Since Φ is a function of one variable, its rate of change at point $\vec{r}_0 = (x_0, y_0, z_0)$ is given by $\Phi'(0)$. However

$$\Phi'(0) = \lim_{s \to 0} \frac{\phi(\vec{r}_0 + \hat{u}s) - \phi(\vec{r}_0)}{s} = \frac{d\phi}{ds}(\vec{r}_0)$$

and the chain rule yields

$$\Phi'(0) = \frac{d\phi}{ds}(\vec{r}_0) = \frac{\partial\phi}{\partial x}\frac{\partial x}{\partial s} + \frac{\partial\phi}{\partial y}\frac{\partial y}{\partial s} + \frac{\partial\phi}{\partial z}\frac{\partial x}{\partial s} = \frac{\partial\phi}{\partial x}u_1 + \frac{\partial\phi}{\partial y}u_2 + \frac{\partial\phi}{\partial y}u_3 = \vec{\nabla}\phi(\vec{r}_0) \cdot \hat{u}.$$

$$(1.62)$$

This gives rise to the following definition.

Definition 1.2.1. *The rate of change of function $\phi(x, y, z)$ at $\vec{r}_0 = (x_0, y_0, z_0)$ in the direction of a unit vector $\hat{u}$ is given by the directional derivative*

$$D_u\phi(\vec{r}_0) = \frac{d\phi}{ds}(\vec{r}_0).$$

$$(1.63)$$

The directional derivative of a scalar field $\phi(x, y, z)$ in the direction parallel to the x axis is obtained from equation (1.62) by setting $\hat{u} = \hat{i}$. The result agrees with the partial derivative $\partial\phi/\partial x$, which indicates the rate of change of ϕ in the direction parallel to the coordinate axis x. Thus, the directional derivative generalizes a concept of the partial derivative and provides a measure of the rate of change of function ϕ in an arbitrary direction $\hat{u}$.

From (1.62) we see that the projection of the gradient on the given direction $\hat{u}$ describes the rate of change of the scalar field in that direction. The formula for the directional derivative can therefore be written as follows:

$$D_u\phi = \vec{\nabla}\phi \cdot \hat{u}.$$

Suppose α is the angle between the vectors $\vec{\nabla}\phi$ and $\hat{u}$. Then at the point $\vec{r}_0$ it holds that

$$\boxed{D_u\phi = |\vec{\nabla}\phi||\hat{u}|\cos\alpha = |\vec{\nabla}\phi|\cos\alpha.}$$

Choosing the direction $\hat{u}$ to be be parallel with the gradient $\vec{\nabla}\phi$ results in the largest possible value of the directional derivative. Thus, ϕ changes fastest in the direction of $\vec{\nabla}\phi$. In other words, the maximum rate of change of ϕ occurs when $\alpha = 0$ or $\cos\alpha = 1$. The rate of increase in this direction is equal to the magnitude of the gradient vector $|\vec{\nabla}\phi|$.

The directional derivative $D_u\phi$ attains its minimum value when $\alpha = \pi$ or $\cos\alpha = -1$. Then the unit vector $\hat{u}$ is pointing in the direction opposite to the gradient $\vec{\nabla}\phi$. The minimum value of the directional derivative is equal to $-|\vec{\nabla}\phi|$.

☞ **Example 1.2.2.** Let $\phi(x,y,z) = x^2 + 2xy + y^3 + z^2$. The directional derivative of ϕ at $(1,-1,-1)$ in the direction of a unit vector $\hat{u} = a\hat{i} + b\hat{j} + c\hat{k}$, where a, b, c are such that $a^2 + b^2 + c^2 = 1$, is

$$D_u\phi(1,-1,-1) = \left((2x+2y)\hat{i} + (2x+3y^2)\hat{j} + 2z\hat{k}\right) \cdot \hat{u}|_{(1,-1,-1)}$$

$$= 5b - 2c.$$

The maximum value of directional derivative is $\sqrt{25+4} = \sqrt{29}$ and corresponds to the case of vector $\hat{u}$ pointing along the gradient direction.

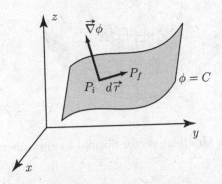

Figure 1.13: The gradient is normal to the contour surface.

The locus of all points (x,y,z) satisfying $\phi(x,y,z) = \phi(x_0,y_0,z_0) = C =$ constant is called a contour surface (or a contour curve in case of two variables

x, y only) of ϕ through the point (x_0, y_0, z_0). Let P_i and P_f be two neighboring points on the contour surface $(\phi(P_f) = \phi(P_i) = C)$ such that $P_f = P_i + (ds)\,\hat{u}$ where $\hat{u}$ is a tangent vector to the contour surface at P_i and where ds is infinitesimally small. Then

$$0 = \phi(P_f) - \phi(P_i) = \phi(P_i + \hat{u}\,ds) - \phi(P_i) = \frac{d\phi(P_i)}{ds}\,ds = \vec{\nabla}\phi(P_i) \cdot \hat{u}\,ds. \quad (1.64)$$

Thus the gradient $\vec{\nabla}\phi$ is perpendicular to the vector $\hat{u}$ tangent to the contour surface $\phi(x, y, z) = C$ at P_i. By varying the point P_i we find that the gradient $\vec{\nabla}\phi$ is normal everywhere to the surface $\phi(x, y, z) = C$. Therefore it follows that

> the gradient is normal to contour surfaces.

☞ **Example 1.2.3.** Let $\phi(x, y) = y - x^2$. The contour curves of ϕ are the parabolas $y = x^2 + C$ The gradient vector $\vec{\nabla}\phi(x, y) = -2x\hat{i} + \hat{j}$ is normal to the parabola $y = x^2 + C$ at any point (x, y).

☞ **Example 1.2.4.** For the function $\phi = r = \sqrt{x^2 + y^2 + z^2}$ the surfaces $\phi = r_i = C_i$ are spherical shells (see Figure (1.14)) and the gradient:

$$\vec{\nabla}\phi = \hat{r}\frac{d\phi}{dr} = \hat{r} \quad (1.65)$$

points in a radial direction, normal everywhere to the spherical shells.

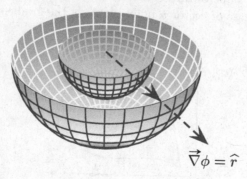

$$\vec{\nabla}\phi = \hat{r}$$

Figure 1.14: Gradient vector normal to the spherical shells.

Let the vector $ds\,\hat{u}$ connect two contour curves $\phi(x, y) = C$ and $\phi(x, y) = C + \Delta C$. Then according to (1.63),

$$\Delta C = \phi\left((x, y) + ds\,\hat{u}\right) - \phi\left((x, y)\right) = D_u\phi\,ds = |\vec{\nabla}\phi||\,ds\hat{u}|\cos\alpha = |\vec{\nabla}\phi|\,ds\cos\alpha.$$

For $\hat{u}$ which points out in the direction of $\vec{\nabla}\phi$, $\cos\alpha = 1$. Accordingly ds is smallest for the fixed change ΔC of ϕ. Thus, the direction of $\vec{\nabla}\phi$ indicates the shortest path to the next contour and the direction of the gradient is where the points on the contour lines are closest together, see Figure 1.15. Thus the greatest rate of change is obtained by moving along the gradient in the direction perpendicular to the contour. Following this direction we arrive at the next contour in the shortest possible distance as shown in Figure 1.15. Figure 1.16 shows the direction of the

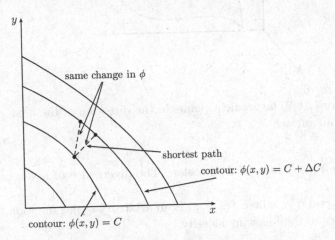

Figure 1.15: The gradient points in the direction of the next contour along the path requiring the shortest possible distance.

gradient for two neighboring contour surfaces in the three-dimensional case.

1.2.3 Vector Calculus: Divergence

A vector field $\vec{V}(x,y,z)$ is a vector valued function that associates a vector to each point (x,y,z) of space. It is convenient to represent it in terms of components V_x, V_y, V_y which are functions of (x,y,z) such that

$$\vec{V}(x,y,z) = \hat{i}V_x(x,y,z) + \hat{j}V_y(x,y,z) + \hat{k}V_z(x,y,z),$$

where $\vec{V}(x,y,z)$ is the value of the vector at the point (x,y,z).

In the last subsection we defined a gradient operator $\vec{\nabla}$. Here we consider its action on a vector field $\vec{V}(x,y,z)$ through the scalar product that defines the divergence of a vector field as follows:

$$\vec{\nabla}\cdot\vec{V} = \frac{\partial V_x}{\partial x} + \frac{\partial V_y}{\partial y} + \frac{\partial V_z}{\partial z} = \sum_{i=1}^{3}\frac{\partial V_i}{\partial x_i}. \tag{1.66}$$

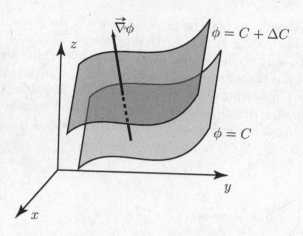

Figure 1.16: The gradient points in the direction of the next contour surface.

Note that the right hand side is a scalar. The divergence of a vector is always a scalar.

For the vector $f\vec{V}$, where $f(x, y, z)$ is an arbitrary scalar function, the product rule generalizes to the following identity

$$
\begin{aligned}
\vec{\nabla} \cdot f\vec{V} &= \sum_{i=1}^{3} \frac{\partial(fV_i)}{\partial x_i} = \sum_{i=1}^{3} f\frac{\partial V_i}{\partial x_i} + \sum_{i=1}^{3} V_i \frac{\partial f}{\partial x_i} \\
&= f\vec{\nabla} \cdot \vec{V} + \left(\vec{\nabla}f\right) \cdot \vec{V}.
\end{aligned}
\tag{1.67}
$$

☞ **Example 1.2.5.** Let $\vec{V} = f(r)\vec{r}$, then according to (1.67) and (1.60) we find:

$$
\vec{\nabla} \cdot f(r)\vec{r} = f\vec{\nabla} \cdot \vec{r} + \frac{df}{dr}\hat{r} \cdot \vec{r} = 3f(r) + r\frac{df}{dr},
\tag{1.68}
$$

where we used that $\vec{\nabla} \cdot \vec{r} = 3$ as follows from definition (1.66). For an interesting case of $f(r) = r^n$ in (1.68) the divergence becomes $\vec{\nabla} \cdot r^n\vec{r} = 3r^n + (n)r^n = (n+3)r^n$. If $n = -3$, then

$$
\vec{\nabla} \cdot \frac{\vec{r}}{r^3} = \vec{\nabla} \cdot \frac{\hat{r}}{r^2} = 0.
\tag{1.69}
$$

This equality is only valid outside the origin since the right hand side of equation (1.69) is only well-defined as long as $r \neq 0$.

The vector field $\vec{V}$ is called divergence free or solenoidal if its divergence vanishes, e.g. $\vec{\nabla} \cdot \vec{V} = 0$.

1.2.4 Vector Calculus: Curl

A vector product of a gradient operator $\vec{\nabla}$ with the vector field $\vec{V}(x,y,z) = \hat{i}V_x(x,y,z) + \hat{j}V_y(x,y,z) + \hat{k}V_z(x,y,z)$ defines the curl of a vector field:

$$
\vec{\nabla} \times \vec{V} = \hat{i}\left(\frac{\partial V_z}{\partial y} - \frac{\partial V_y}{\partial z}\right) + \hat{j}\left(\frac{\partial V_x}{\partial z} - \frac{\partial V_z}{\partial x}\right)
$$
$$
+ \hat{k}\left(\frac{\partial V_y}{\partial x} - \frac{\partial V_x}{\partial y}\right)
$$

(1.70)

or, in terms of the permutation symbol,

$$
\vec{\nabla} \times \vec{V} = \sum_{i=1}^{3}\sum_{j=1}^{3}\sum_{k=1}^{3} \epsilon_{ijk}\frac{\partial V_k}{\partial x_j}\hat{e}_i .
$$

(1.71)

Alternatively, we can express the curl of $\vec{V}$ in terms of the determinant:

$$
\vec{\nabla} \times \vec{V} = \begin{vmatrix} \hat{i} & \hat{j} & \hat{k} \\ \frac{\partial}{\partial x} & \frac{\partial}{\partial y} & \frac{\partial}{\partial z} \\ V_x & V_y & V_z \end{vmatrix},
$$

(1.72)

where the middle row of the determinant consists of the components of the gradient operator.

For the vector $f\vec{V}$ where $f(x,y,z)$ is an arbitrary scalar function one finds that

$$
\vec{\nabla} \times f\vec{V} = \sum_{i=1}^{3}\sum_{j=1}^{3}\sum_{k=1}^{3} \epsilon_{ijk}\frac{\partial(fV_k)}{\partial x_j}\hat{e}_i .
$$

(1.73)

Furthermore it follows from the product rule $\partial(fV_k)/\partial x_j = f\partial(V_k)/\partial x_j + V_k\partial(f)/\partial x_j$ that the expression (1.73) can be rewritten as

$$
\vec{\nabla} \times f\vec{V} = f\vec{\nabla} \times \vec{V} + (\vec{\nabla}f) \times \vec{V} .
$$

(1.74)

☞ **Example 1.2.6.** Let $\vec{V} = f(r)\vec{r}$, then (1.74) yields

$$
\vec{\nabla} \times f(r)\vec{r} = f(r)\vec{\nabla} \times \vec{r} + (\vec{\nabla}f(r)) \times \vec{r} = 0
$$

(1.75)

since both terms on the right hand side of (1.75) vanish. To see it, note that $\vec{\nabla} \times \vec{r} = \sum_{i=1}^{3}\sum_{j=1}^{3}\sum_{k=1}^{3}\epsilon_{ijk}\frac{\partial x_k}{\partial x_j}\hat{e}_i = \sum_{i=1}^{3}\sum_{j=1}^{3}\epsilon_{ijj}\hat{e}_i = 0$ and recall from (1.60) that $\vec{\nabla}f(r)$ is proportional to $\vec{r}$.

The vector field $\vec{V}$ is called *irrotational* if its curl vanishes: $\vec{\nabla} \times \vec{V} = 0$. From (1.75) we see that a vector field $\vec{V} = f(r)\vec{r}$ is irrotational. This class of vector fields contains the gravitational and electrostatic potential proportional to $\vec{r}/r^2$.

☞ **Example 1.2.7.** The curl $\vec{\nabla} \times \vec{V}(\vec{r})$ measures the tendency of the vector field $\vec{V}$ to go around in a circle (or curl) around a point $\vec{r}$. The divergence $\vec{\nabla} \cdot \vec{V}(\vec{r})$ is a measure of how much the vector field $\vec{V}$ spreads out in a radial direction.

For illustration, consider vector fields:

$$\vec{V}_1 = -y\hat{i} + x\hat{j} \quad \text{and} \quad \vec{V}_2 = x\hat{i} + y\hat{j}.$$

We find

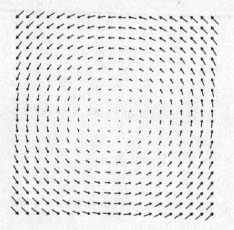

Figure 1.17: Vector field $-y\hat{i} + x\hat{j}$ with vanishing divergence and non-zero curl.

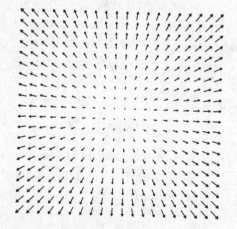

Figure 1.18: Vector field $x\hat{i} + y\hat{j}$ with positive divergence (source) and vanishing curl.

$$\vec{\nabla} \times \vec{V}_1 = 2\hat{k} \quad \text{and} \quad \vec{\nabla} \times \vec{V}_2 = 0$$

and

$$\vec{\nabla} \cdot \vec{V}_1 = 0 \quad \text{and} \quad \vec{\nabla} \cdot \vec{V}_2 = 2.$$

These results are illustrated in Figures 1.17 and 1.18.

If we think about $\vec{V}(\vec{r})$ as velocity of a fluid, then fluid particles close to the point $\vec{r}$ tend to rotate about the axis pointing along the direction of $\vec{\nabla} \times \vec{V}$ and the magnitude of this curl vector is a measure of how fast the particles rotate around the axis. For the vector field $\vec{V}_1$ the fluid analogy is consistent with particles rotating around the origin in the counterclockwise direction (see Figure 1.17). If we define a vector field $\vec{V}_3 = -\vec{V}_1 = y\hat{i} - x\hat{j}$ then $\vec{\nabla} \times \vec{V}_3 = -2\hat{k}$ and the corresponding rotation would be in the clockwise direction.

The irrotational vector field $\vec{V}_2$ with positive value of divergence $\vec{\nabla} \cdot \vec{V}_2$ points outward from the origin along a radial direction. If $\vec{V}_2$ is the velocity of some fluid then the fluid flows out of every closed surface containing the origin and we say that the origin acts as a source of the vector field. In contrast, the vector field $\vec{V}_4 = -x\hat{i} - y\hat{j}$ with negative divergence $\vec{\nabla} \cdot \vec{V}_4 = -2$ would point inward toward the origin (*sink*). Vanishing divergence as in the case of $\vec{V}_1$ means that the vector field has no sources or sinks, and consists only of swirls.

Consider, next the vector field

$$\vec{V}_5 = (x^2 - y^2)\hat{i} - 2xy\hat{j},$$

for which both curl and divergence is zero. $\vec{V}_5$ has no sources or sinks but it does not circle around the origin. Such a configuration is called a *saddle* and it is shown in Figure 1.19. Let us now consider $\vec{V}_6 = (x-y)\hat{i} + (x+y)\hat{j}$. Both the divergence and curl are different from zero. This is a typical example of the whirlpool, shown in Figure 1.20.

1.2.5 Vector Calculus: Successive Applications of the Gradient Operator

The scalar product of two gradient operators acting on the function ϕ is given by

$$
\begin{aligned}
\vec{\nabla} \cdot \vec{\nabla}\phi &= \left(\hat{i}\frac{\partial}{\partial x} + \hat{j}\frac{\partial}{\partial y} + \hat{k}\frac{\partial}{\partial z}\right) \cdot \left(\hat{i}\frac{\partial}{\partial x} + \hat{j}\frac{\partial}{\partial y} + \hat{k}\frac{\partial}{\partial z}\right)\phi \\
&= \frac{\partial^2\phi}{\partial x^2} + \frac{\partial^2\phi}{\partial y^2} + \frac{\partial^2\phi}{\partial z^2}.
\end{aligned}
\tag{1.76}
$$

The square of the gradient operator $\vec{\nabla} \cdot \vec{\nabla} = \vec{\nabla}^2$ is called the Laplace operator and the equation $\vec{\nabla}^2\phi = 0$ is called Laplace's equation.

A Short Course in Mathematical Methods with Maple

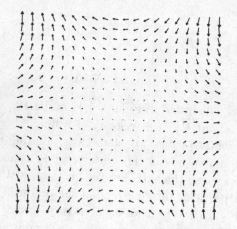

Figure 1.19: A saddle with vanishing divergence and curl.

☞ **Example 1.2.8.** Let $\phi = f(r)$ be a function of radial distance r. Recalling equations (1.60) and (1.68) we obtain

$$
\begin{aligned}
\vec{\nabla}^2 f(r) &= \vec{\nabla} \cdot \frac{1}{r}\frac{df}{dr}\vec{r} = 3\frac{1}{r}\frac{df}{dr} + r\frac{d}{dr}\left(\frac{1}{r}\frac{df}{dr}\right) \\
&= 2\frac{1}{r}\frac{df}{dr} + \frac{d^2 f}{dr^2}.
\end{aligned}
\tag{1.77}
$$

For $f(r) = r^n$, equation (1.77) gives $\vec{\nabla}^2 r^n = n(n+1)r^{n-2}$.

The curl of the gradient of the function ϕ is given by

$$
\vec{\nabla} \times \vec{\nabla}\phi = \sum_{i=1}^{3}\sum_{j=1}^{3}\sum_{k=1}^{3} \epsilon_{ijk} \frac{\partial}{\partial x_j}\frac{\partial}{\partial x_k}\phi \hat{e}_i = 0
\tag{1.78}
$$

which vanishes identically because of antisymmetry of the permutation symbol. Similarly, the divergence of the curl

$$
\vec{\nabla} \cdot \vec{\nabla} \times \vec{V} = \sum_{i=1}^{3}\sum_{j=1}^{3}\sum_{k=1}^{3} \epsilon_{ijk} \frac{\partial}{\partial x_i}\frac{\partial}{\partial x_j}\hat{V}_k = 0
\tag{1.79}
$$

also vanishes for the same reason.

We leave as an exercise to prove that

$$
\vec{\nabla} \times \left(\vec{\nabla} \times \vec{V}\right) = \vec{\nabla}\left(\vec{\nabla} \cdot \vec{V}\right) - \left(\vec{\nabla} \cdot \vec{\nabla}\right)\vec{V}.
\tag{1.80}
$$

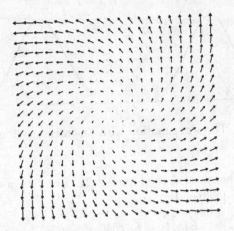

Figure 1.20: A whirlpool with non-zero divergence and non-zero curl.

☞ **Example 1.2.9.** Consider Maxwell equations for magnetic and electric vector fields $\vec{B}$ and $\vec{E}$, respectively:

$$\vec{\nabla} \cdot \vec{B} = 0, \qquad\qquad \vec{\nabla} \cdot \vec{E} = \frac{\rho}{\epsilon_0}, \qquad\qquad (1.81)$$

$$\vec{\nabla} \times \vec{B} = \epsilon_0 \mu_0 \frac{\partial \vec{E}}{\partial t} + \mu_0 \vec{j}, \qquad \vec{\nabla} \times \vec{E} = -\frac{\partial \vec{B}}{\partial t}. \qquad (1.82)$$

In this example, we will study Maxwell equations in empty space with zero charge density, $\rho = 0$, and zero current, $\vec{j} = 0$. The time derivative of the first equation in (1.82) yields:

$$\epsilon_0 \mu_0 \frac{\partial^2 \vec{E}}{\partial t^2} = \vec{\nabla} \times \frac{\partial \vec{B}}{\partial t} = -\vec{\nabla} \times \left(\vec{\nabla} \times \vec{E} \right). \qquad (1.83)$$

Using relation (1.80) and the fact that the electric vector field $\vec{E}$ is solenoidal in free space we obtain the electromagnetic wave equation:

$$\left(\vec{\nabla} \cdot \vec{\nabla} \right) \vec{E} = \epsilon_0 \mu_0 \frac{\partial^2 \vec{E}}{\partial t^2}. \qquad (1.84)$$

For a plane wave moving in the x-direction this reduces to a wave equation

$$\frac{\partial^2 \vec{E}}{\partial x^2} = \frac{1}{c^2} \frac{\partial^2 \vec{E}}{\partial t^2},$$

where the speed of light $c = (\epsilon_0 \mu_0)^{-1/2}$. This equation possesses a general solution $\vec{E}(x,t) = \vec{E}_0 f(x - ct)$ for an arbitrary function f. Such a solution describes the electric field $\vec{E}(x,t)$ propagating as the time t flows with unchanged wave profile given by the shape of function f along the positive

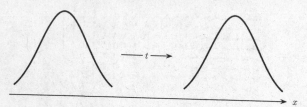

Figure 1.21: The wave motion with an unchanged wave profile.

x-direction at the speed c, see Figure 1.21. Let f be a cosine function, then the electric field is given by

$$\vec{E}(x,t) = \vec{E}_0 \cos(kx - \omega t).$$

This describes the electromagnetic periodic plane wave solution to the wave equation. The wave equation is satisfied when the relation $\omega = ck$ holds for the angular velocity ω and the wave number k. Thus, the electromagnetic periodic plane wave can be rewritten as $\vec{E}(x,t) = \vec{E}_0 \cos\left(k(x - ct)\right)$, from which we can easily see that the electromagnetic periodic plane wave travels with the speed of light c.

♠ Gradient, Divergence, Curl: Problems

Problem 1.2.1. Find the gradient $\vec{\nabla} f$ of $f = e^{x^2 + y^2} \sin(2xy)$. Find the directional derivative of f in the direction of $\hat{u} = (\hat{i} - \hat{j})/\sqrt{2}$ at the point $(1,1)$.

Problem 1.2.2. Find the gradient $\vec{\nabla} f$ of $f = 1/\sqrt{x^2 + y^2 + z^2}$ at the point $(1,-2,1)$. Find the directional derivative of f in the direction of $\hat{u} = (\hat{i} - \hat{j} + \hat{k})/\sqrt{3}$ at that point. Find vectors along which the directional derivative of f at that point is equal to zero.

Problem 1.2.3. Let $f(x,y) = \exp(x^2 y) \sin(x - y)$
(a) In what direction in the xy-plane, is f changing fastest at the point $(x,y) = (0,0)$?
(b) In what directions in the xy-plane, is f changing at 50% of its maximum rate at the point $(x,y) = (\pi/2, 0)$?

Problem 1.2.4. The temperature on the xy-plane is given by the function

$$T(x,y) = 30 + 120 \exp\left(\frac{1}{2}(x-1)^2 + \frac{1}{2}(y-3)^2\right).$$

In what direction one must move to encounter the greatest increase of temperature at the point $(3,2)$.

Problem 1.2.5. Derive an equation of the plane tangent to the sphere $x^2+y^2+z^2 = 6$ at the point $(\sqrt{2}, \sqrt{2}, \sqrt{2})$.

Problem 1.2.6. Find location of points on the ellipsoid $3x^2 + y^2 + 2z^2 = 11$ at which the planes tangent to the ellipsoid are parallel to the plane $x + y + z = 3$. Write down the equations for these planes.

Problem 1.2.7. Show, that

$$\vec{\nabla} \cdot (f\vec{\nabla}g) - \vec{\nabla} \cdot (g\vec{\nabla}f) = f\nabla^2 g - g\nabla^2 f$$

$$\vec{\nabla} \cdot (f\vec{v}) = f\vec{\nabla} \cdot \vec{v} + \vec{v} \cdot \vec{\nabla}f.$$

Problem 1.2.8. Calculate:

$$\vec{\nabla} \frac{1}{|\vec{r} - \vec{r}_0|}$$

for a constant vector $\vec{r}_0 = x_0\hat{i} + y_0\hat{j} + z_0\hat{k}$.

Problem 1.2.9. Prove the identities,

(a) $\vec{\nabla} \cdot (\vec{u} \times \vec{v}) = \vec{v} \cdot \vec{\nabla} \times \vec{u} - \vec{u} \cdot \vec{\nabla} \times \vec{v}$,

(b) $\vec{\nabla} \times (\vec{u} \times \vec{v}) = \left(\vec{\nabla} \cdot \vec{v}\right)\vec{u} + \left(\vec{v} \cdot \vec{\nabla}\right)\vec{u} - \left(\vec{\nabla} \cdot \vec{u}\right)\vec{v} - \left(\vec{u} \cdot \vec{\nabla}\right)\vec{v}$,

(c) $\displaystyle\sum_{i=1}^{3} u_i\vec{\nabla}v_i = \left(\vec{u} \cdot \vec{\nabla}\right)\vec{v} + \vec{u} \times \left(\vec{\nabla} \times \vec{v}\right)$,

(d) $\left(\vec{u} \times \vec{\nabla}\right) \times \vec{v} = \left(\vec{u} \cdot \vec{\nabla}\right)\vec{v} + \vec{u} \times \left(\vec{\nabla} \times \vec{v}\right) - \vec{u}\left(\vec{\nabla} \cdot \vec{v}\right)$,

(e) $\vec{\nabla} \cdot (g\vec{\nabla}f \times f\vec{\nabla}g) = 0$.

Problem 1.2.10. The velocity vector is of the form $\vec{v} = \vec{\omega} \times \vec{r}$ where $\vec{\omega}$ is the constant (rotation) vector. Show that
(a) $\vec{\nabla} \cdot \vec{v} = 0$,
(b) $\left(\vec{\omega} \cdot \vec{\nabla}\right)\vec{r} = \vec{\omega}$,
(c) $\vec{\nabla} \times \vec{v} = 2\vec{\omega}$,
(d) $\vec{\nabla}(\vec{\omega} \cdot \vec{r}) = \vec{\omega}$.

Problem 1.2.11. Use the result from Problem 1.2.10 and the product rules (1.67) and (1.74) to evaluate $\vec{\nabla} \cdot (\vec{\omega} \times \hat{r})$ and $\vec{\nabla} \times (\vec{\omega} \times \hat{r})$, where $\vec{\omega}$ is the constant vector and $\hat{r}$ is the unit vector $\vec{r}/r$.

Problem 1.2.12. The vector potential from the magnetic dipole is given by

$$\vec{A} = \frac{\mu}{4\pi} \frac{\vec{\omega}_0 \times \vec{r}}{r^3}$$

for a constant vector $\vec{\omega}_0$ representing a dipole moment. Use the result from Problem 1.2.10 and the product rule (1.74) to show that the dipole magnetic field is given by

$$\vec{B} = \vec{\nabla} \times \vec{A} = -\frac{\mu}{4\pi}\frac{\vec{\omega}_0}{r^3} + 3\frac{\mu}{4\pi}\frac{\vec{r}(\vec{\omega}_0 \cdot \vec{r})}{r^5}.$$

Problem 1.2.13. Find $\vec{v} \cdot \vec{\nabla} \times \vec{v}$, $\vec{v} \times \vec{\nabla} \times \vec{v}$ for $\vec{v} = xy\hat{i} + yz\hat{j} + zx\hat{k}$.

Problem 1.2.14. Prove the identity

$$\frac{1}{2}\vec{\nabla}\left(\vec{v}^2\right) = \vec{v} \times \left(\vec{\nabla} \times \vec{v}\right) + \left(\vec{v} \cdot \vec{\nabla}\right)\vec{v}.$$

Problem 1.2.15. Prove the identities

$$\left(\vec{a} \cdot \vec{\nabla}\right)\left(\frac{1}{r}\right) = -\frac{\vec{r} \cdot \vec{a}}{r^3}$$

$$\left(\vec{a} \cdot \vec{\nabla}\right)\left(\vec{b} \cdot \vec{\nabla}\right)\left(\frac{1}{r}\right) = \frac{3\left(\vec{r} \cdot \vec{a}\right)\left(\vec{r} \cdot \vec{b}\right) - r^2\left(\vec{a} \cdot \vec{b}\right)}{r^5}$$

for two constant vectors $\vec{a}$ and $\vec{b}$.

Problem 1.2.16. Find $f(x, y, z)$ such that $\vec{\nabla}f(x, y, z) = (x^2 + y^2 + z^2)^{-3/2}(x\hat{i} + y\hat{j} + z\hat{k})$. Find the divergence and curl of $(x^2 + y^2 + z^2)^{-3/2}(x\hat{i} + y\hat{j} + z\hat{k})$ for $(x, y, z) \neq 0$.

Problem 1.2.17. Suppose, that the magnetic moment $\vec{\mu}$ is a constant vector field and the magnetic vector field $\vec{B}$ is both irrotational and solenoidal: $\vec{\nabla} \cdot \vec{B} = 0$, $\vec{\nabla} \times \vec{B} = 0$. Show that the induced force $\vec{F} = \vec{\nabla} \times (\vec{B} \times \vec{\mu})$ can be rewritten as $\vec{F} = \left(\vec{\mu} \cdot \vec{\nabla}\right)\vec{B} = \vec{\nabla}(\vec{B} \cdot \vec{\mu})$. *Hint:* Use identity from Problem 1.2.9.

Problem 1.2.18. (a) Derive continuity equation

$$\frac{\partial \rho}{\partial t} + \vec{\nabla} \cdot \vec{j} = 0$$

from Maxwell equations (1.81)-(1.82).

(b) We associate the field energy, $U = \vec{B}^2/2\mu_0 + \epsilon_0 \vec{E}^2/2$, and the so-called Poynting vector, $\vec{S} = \vec{E} \times \vec{B}/\mu_0$, to the electric and magnetic vector fields. Use Maxwell equations (1.81)-(1.82) to find the quantity

$$\frac{\partial U}{\partial t} + \vec{\nabla} \cdot \vec{S},$$

which describes the rate of doing work as charges are moved in the electric field. *Hint:* Use the first identity from Problem 1.2.9.

1.3 Curvilinear Coordinates

In many physical situations it is useful to introduce curvilinear coordinates that fit the symmetry of a problem.

Let $P = (x_1, x_2, x_3) = (c_1, c_2, c_3)$ be a point. We can describe it as an intersection of the planes $x_1 = c_1$, $x_2 = c_2$ and $x_3 = c_3$ in rectangular Cartesian coordinates.

We can also describe point P as an intersection of three surfaces that form new curvilinear coordinates q_1, q_2, q_3 such that $\vec{r} = \vec{r}(q_1, q_2, q_3)$. The curvilinear coordinate surfaces are defined by $q_1 = $ const, $q_2 = $ const , $q_3 = $ const so that q_i-surface is the surface described by $\vec{r}(q_1, q_2, q_3)$ with one single coordinate q_i held fixed while the two remaining arguments $q_j, q_k, j \neq, k \neq i$ are varied.

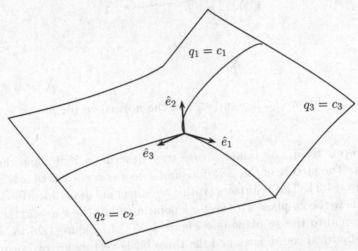

Figure 1.22: Coordinate surfaces and base vectors.

Examples include:

☞ **Example 1.3.1.** Cylindrical coordinates are defined as

$$q_1 = \rho \,;\, q_2 = \varphi \,;\, q_3 = z$$
$$\vec{r}(\rho, \varphi, z) = (\rho \cos \varphi, \rho \sin \varphi, z) \tag{1.85}$$
$$0 \leq \rho < \infty, 0 \leq \varphi \leq 2\pi, -\infty < z < \infty.$$

The subset of the two first cylindrical coordinates $q_1 = \rho$ and $q_2 = \varphi$ defines the so-called polar coordinates on the xy-plane.

We now find the corresponding coordinate surfaces. They are obtained by keeping one of the cylindrical variables:

$$\rho = (x^2 + y^2)^{1/2}, \quad \varphi = \arctan(y/x), \quad z = z$$

constant while varying the two remaining variables.

First, we fix $\rho = (x^2 + y^2)^{1/2} = \text{constant}$ and vary φ and z to find ρ-surfaces as circular cylinders having the z-axis as a common axis. See Figure 1.23 for a ρ-surface and note that vector $\hat{\rho}$ is normal to the ρ-surface. Next,

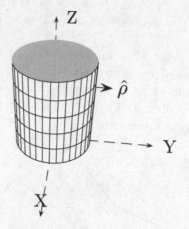

Figure 1.23: ρ-surface and the normal vector $\hat{\rho}$.

we keep φ fixed. As ρ and z vary they describe a half-plane through the z-axis. The picture of the φ-surface and a vector $\hat{\varphi}$ normal to it is illustrated in Figure 1.24. The z-surfaces (with $z = \text{const}$) are easily identified as planes parallel to the xy-plane with vector $\hat{k}$ pointing along the z-axis in the direction orthogonal to the xy-plane (see Figure 1.25). In Figure 1.26 we show that the cylindrical coordinates and the three basis vectors are orthogonal to the ρ-, φ- and z-surfaces.

☞ **Example 1.3.2.** Spherical coordinates are given by

$$
\begin{aligned}
q_1 &= r\,;\, q_2 = \theta\,;\, q_3 = \varphi \\
\vec{r}\,(r, \theta, \varphi) &= (r \sin\theta \cos\varphi,\, r \sin\theta \sin\varphi,\, r \cos\theta) \\
0 &\le r < \infty,\, 0 \le \theta \le \pi,\, 0 \le \varphi \le 2\pi.
\end{aligned}
\tag{1.86}
$$

Here the corresponding coordinate surfaces are obtained by fixing one of spherical variables

$$
r = (x^2 + y^2 + z^2)^{1/2}, \quad \theta = \arccos(z/(x^2 + y^2 + z^2)^{1/2}), \quad \varphi = \arctan(y/x)
$$

to a constant while allowing the two remaining variables to vary.

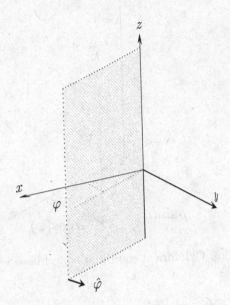

Figure 1.24: φ-surface and the normal vector $\hat{\varphi}$.

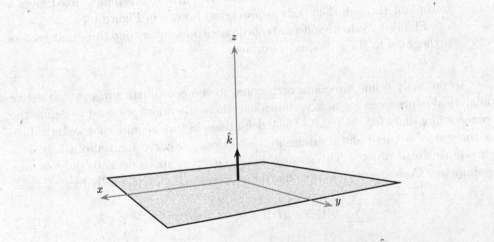

Figure 1.25: z-surface and the normal vector $\hat{k}$.

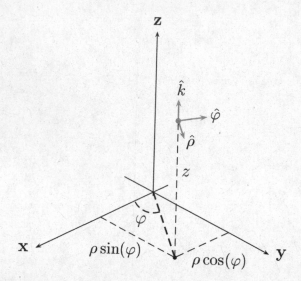

Figure 1.26: Cylindrical coordinates and base vectors.

The $r = (x^2 + y^2 + z^2)^{1/2}$ = constant surface is a surface of a sphere of radius r. It is parametrized by spherical angles θ and φ as shown in Figure 1.28.

For $\theta = \arccos(z/(x^2 + y^2 + z^2)^{1/2})$ = constant we obtain the θ-surface. It is a cone with tip at the origin and axis in the z-direction. Figure 1.27 shows the cone and a normal vector to the surface. Keeping φ fixed gives a half plane through the z-axis as previously shown in Figure 1.24.

Figure 1.28 shows spherical coordinates on a sphere and three unit vectors orthogonal to the r-, θ- and φ-surfaces.

To precisely define directions orthogonal to the coordinate surfaces and derive analytical expressions for unit vectors along these directions we need to define the concept of q_i-lines for $i = 1, 2, 3$. These definitions in turn require that we introduce a concept of a curve and its tangential direction. A curve (and also a line) is a one-dimensional object. This means that a curve in space depends only on one parameter. Consequently, a curve may be represented parametrically by:

$$\vec{r}(t) = \sum_{i=1}^{3} x_i(t)\hat{e}_i = x(t)\hat{i} + y(t)\hat{j} + z(t)\hat{k}, \quad a \leq t \leq b \qquad (1.87)$$

in terms of one parameter t which varies along the curve.

☞ **Example 1.3.3.** Equation $\vec{r}(t) = \hat{i}ct + \hat{j}kt^2$ describes the parabola $kx^2 = c^2 y$ in the xy-plane. For $-1 \leq t \leq 1$ the parabola starts at $(-c, k)$ and ends at (c, k) point. Figure 1.29 shows this parabola for $c = 1, k = 2$.

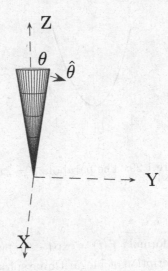

Figure 1.27: A cone with an axis in the z-direction.

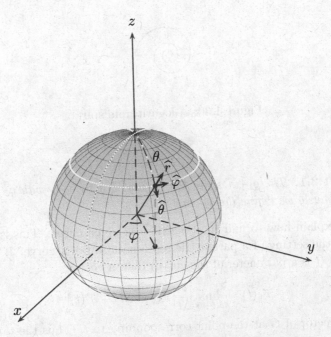

Figure 1.28: Spherical coordinates and base vectors.

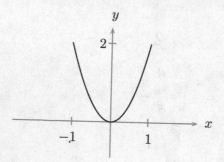

Figure 1.29: The parabola $y = 2x^2$.

☞**Example 1.3.4.** The formula $\vec{r}(t) = \exp(-.01t)\left(\hat{i}\cos(3t) + \hat{j}\sin(3t)\right)$ provides a parametric description of a logarithmic spiral as shown in Figure 1.30. As t ranges from 0 to 2π the curve makes a total of three windings.

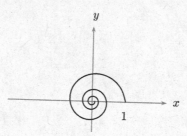

Figure 1.30: A logarithmic spiral.

Definition 1.3.1. *The q_1-lines are the curves $\vec{r}(q_1, q_2, q_3)$ with q_2 and q_3 held fixed. Analogously we define the q_2- and q_3-lines.*

Let us now explain how to find the tangent to a given curve. This is very simple to do within the setting of a parametric representation of a curve. If a curve C is given by $\vec{r}(t)$ (t is a parameter of the curve) then a vector

$$\vec{r}_{\|}(t) = \lim_{\Delta t \to 0} \frac{1}{\Delta t}[\vec{r}(t + \Delta t) - \vec{r}(t)] \tag{1.88}$$

is a tangent vector of C at the point corresponding to t. Thus the tangent to the curve at any point is parallel to the derivative with respect to the parameter t:

$$\frac{\mathrm{d}\vec{r}(t)}{\mathrm{d}t} = \frac{\mathrm{d}x(t)}{\mathrm{d}t}\hat{i} + \frac{\mathrm{d}y(t)}{\mathrm{d}t}\hat{j} + \frac{\mathrm{d}z(t)}{\mathrm{d}t}\hat{k}. \tag{1.89}$$

Accordingly, the unit tangent vector is:

$$\hat{u} = \frac{1}{|\vec{r}_{\parallel}|}\,\vec{r}_{\parallel}\,. \tag{1.90}$$

Note that $\hat{u}$ points in the direction of increasing t.

☞ **Example 1.3.5.** Let C be a semi-circle on the xy-plane starting at $(1,0)$ and ending at $(-1,0)$ as shown in Figure (1.31). A convenient parametrization

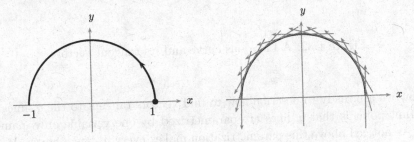

Figure 1.31: A semi-circle and its tangent vectors.

of C is

$$\vec{r}(t) = \hat{i}\cos t + \hat{j}\sin t, \quad 0 \le t \le \pi\,. \tag{1.91}$$

Computing a derivative with respect to the parameter t yields:

$$\frac{d\vec{r}(t)}{dt} = -\hat{i}\sin t + \hat{j}\cos t\,, \tag{1.92}$$

which, in the case of unit circle, is equal to the unit tangent vector $\hat{u}$.

☞ **Example 1.3.6.** Lissajous curves are the family of curves described by the parametric equations

$$x(t) = a\sin(k_x t), \quad y(t) = b\sin(k_y t + \delta), \quad 0 \le t \le 2\pi\,.$$

They describe a combination of two harmonic oscillations along the x and y axes. Figure 1.32 shows the the curve described by

$$\vec{r}(t) = 3\hat{i}\sin 6t + 3\hat{j}\sin(8t + \pi/3)$$

for $0 \le t \le 2\pi$ and the tangent vectors in the direction of

$$\frac{d\vec{r}(t)}{dt} = 18\hat{i}\cos 6t + 18\hat{j}\cos(8t + \pi/3)\,.$$

Figure 1.32: A Lissajous curve and its tangent vectors.

Now we are ready to describe how to find vectors tangent to the q_i-lines. One important point is that q_i-lines are parametrized by one variable only, namely q_i itself. As noticed above this parametrization makes every q_i-line a curve. We have learned in the above paragraph how to determine the tangential direction to a curve in a parametrized representation. Applying this to the q_i-lines we conclude that the tangential directions to the q_i-lines at the point $\vec{r}(q_1, q_2, q_3)$ is given by vectors:

$$\frac{\partial \vec{r}}{\partial q_i}(q_1, q_2, q_3),\tag{1.93}$$

where we compute derivative with respect to the parameter q_i keeping all the remaining q-coordinates fixed.

Accordingly, we define at each point $\vec{r}(q_1, q_2, q_3)$ the unit vector orthogonal to q_i-surface as

$$\hat{e}_i \equiv \left|\frac{\partial \vec{r}}{\partial q_i}\right|^{-1} \frac{\partial \vec{r}}{\partial q_i} = \frac{1}{h_i}\frac{\partial \vec{r}}{\partial q_i} = \frac{1}{h_i}\left(\hat{i}\frac{\partial x}{\partial q_i} + \hat{j}\frac{\partial y}{\partial q_i} + \hat{k}\frac{\partial z}{\partial q_i}\right),\tag{1.94}$$

with coefficients $h_i(q_1, q_2, q_3)$ being equal to the the norm of $\partial \vec{r}/\partial q_i$ vectors:

$$h_i(q_1, q_2, q_3) \equiv \left|\frac{\partial \vec{r}}{\partial q_i}(q_1, q_2, q_3)\right| = \sqrt{\left(\frac{\partial x}{\partial q_i}\right)^2 + \left(\frac{\partial y}{\partial q_i}\right)^2 + \left(\frac{\partial z}{\partial q_i}\right)^2}.\tag{1.95}$$

The coefficients $h_i(q_1, q_2, q_3)$ are called scale factors.

Orthogonal curvilinear coordinates are defined by the condition:

$$\hat{e}_i \cdot \hat{e}_j = \delta_{ij}\tag{1.96}$$

i.e. q_i-lines are pairwise perpendicular to each other. The component V_i of a vector in the $\hat{e}_i$-direction is

$$V_i = \vec{V} \cdot \hat{e}_i \quad \rightarrow \quad \vec{V} = \sum_{i=1}^{3} V_i \cdot \hat{e}_i . \tag{1.97}$$

By the chain rule:

$$dx_i = \sum_{j=1}^{3} \frac{\partial x_i}{\partial q_j} \, dq_j . \tag{1.98}$$

The square of the distance $ds^2 \equiv d\vec{r} \cdot d\vec{r}$ with $d\vec{r} = dx_1 \hat{i} + dx_2 \hat{j} + dx_3 \hat{k}$ becomes

$$ds^2 = \sum_{i=1}^{3} (dx_i)^2 = \sum_{i,j=1}^{3} g_{ij} \, dq_i \, dq_j , \tag{1.99}$$

where g_{ij} is a metric tensor defined by

$$g_{ij} = \frac{\partial x}{\partial q_i} \frac{\partial x}{\partial q_j} + \frac{\partial y}{\partial q_i} \frac{\partial y}{\partial q_j} + \frac{\partial z}{\partial q_i} \frac{\partial z}{\partial q_j} . \tag{1.100}$$

For the q-coordinates it follows from eq.(1.94) and the chain rule that

$$d\vec{r} = \sum_{i=1}^{3} dq_i \frac{\partial \vec{r}}{\partial q_i} = \sum_{i=1}^{3} h_i \, dq_i \, \hat{e}_i = \sum_{i=1}^{3} ds_i \, \hat{e}_i , \tag{1.101}$$

where we introduced products

$$ds_i = h_i \, dq_i , \tag{1.102}$$

having dimension of length (recall that $\hat{e}_i$ are unit vectors).

Hence in the orthogonal curvilinear coordinate system it follows from

$$ds^2 = d\vec{r} \cdot d\vec{r} = (dx)^2 + (dy)^2 + (dz)^2 = h_1^2 (dq_1)^2 + h_2^2 (dq_2)^2 + h_3^2 (dq_3)^2 , \tag{1.103}$$

that

$$g_{ij} = 0 \qquad i \neq j$$

$$g_{ii} = h_i^2 = \sum_{j=1}^{3} \left(\frac{\partial x_j}{\partial q_i} \right)^2 \qquad i = 1, 2, 3 . \tag{1.104}$$

For surface and volume elements we find:

$$dA_{ij} = ds_i \, ds_j = h_i h_j \, dq_i \, dq_j , \tag{1.105}$$

$$d\text{Volume} = ds_1 \, ds_2 \, ds_3 = h_1 h_2 h_3 \, dq_1 \, dq_2 \, dq_3 . \tag{1.106}$$

☞ **Example 1.3.7.** In cylindrical coordinates the q_i-lines are:

ρ-lines: rays starting at the z-axis and parallel to the xy-plane

φ-lines: latitude circles with axis in the direction of the z-axis

z-lines: rays pointing in the direction of the z-axis.

The corresponding basis vectors

$$\hat{\rho} = \hat{e}_1 \,;\, \hat{\varphi} = \hat{e}_2 \,;\, \hat{k} = \hat{e}_3 \qquad\qquad (1.107)$$

are derived as follows. Let $\vec{r}\,(\rho, \varphi, z) = (\rho\cos\varphi, \rho\sin\varphi, z)$, then from the definition (1.94)

$$\frac{\partial \vec{r}}{\partial \rho} = (\cos\varphi, \sin\varphi, 0)\ .$$

The norm of this vector is:

$$h_1 = h_\rho = \sqrt{\cos^2\varphi + \sin^2\varphi} = 1$$

and therefore the unit vector in the ρ-direction is

$$\hat{\rho} = \frac{1}{h_\rho}\frac{\partial \vec{r}}{\partial \rho} = (\cos\varphi, \sin\varphi, 0)\ . \qquad\qquad (1.108)$$

Similarly, we find

$$\frac{\partial \vec{r}}{\partial \varphi} = (-\rho\sin\varphi, \rho\cos\varphi, 0)\ .$$

This time the norm is different from one and given by:

$$h_2 = h_\varphi = \sqrt{\rho^2\sin^2\varphi + \rho^2\cos^2\varphi} = \rho.$$

Consequently

$$\hat{\varphi} = \frac{1}{h_\varphi}\frac{\partial \vec{r}}{\partial \varphi} = (-\sin\varphi, \cos\varphi, 0)\ . \qquad\qquad (1.109)$$

Finally,

$$\frac{\partial \vec{r}}{\partial z} = (0, 0, 1)$$

and

$$h_3 = h_z = 1, \quad\text{and}\quad \hat{k} = (0, 0, 1)\ . \qquad\qquad (1.110)$$

Hence, in this notation:

$$\mathrm{d}\vec{r} = \hat{\rho}\,\mathrm{d}s_\rho + \hat{\varphi}\,\mathrm{d}s_\varphi + \hat{k}\,\mathrm{d}z = \hat{\rho}\,\mathrm{d}\rho + \hat{\varphi}\rho\,\mathrm{d}\varphi + \hat{k}\,\mathrm{d}z\ . \qquad (1.111)$$

☞ **Example 1.3.8.** In spherical coordinates the q_i-lines are:

r-lines: rays starting at the origin
θ-lines: longitude semi circles
φ-lines: latitude circles with their axes in the direction of the z-axis.

Denote:

$$\hat{r} = \hat{e}_1 \; ; \; \hat{\theta} = \hat{e}_2 \; ; \; \hat{\varphi} = \hat{e}_3 \, . \tag{1.112}$$

Then for $\vec{r}(r, \theta, \varphi) = (r \sin\theta \cos\varphi, \, r \sin\theta \sin\varphi, \, r \cos\theta)$ definition (1.94) yields:

$$\frac{\partial \vec{r}}{\partial r} = (\sin\theta \cos\varphi, \, \sin\theta \sin\varphi, \, \cos\theta) \, .$$

The corresponding scale factor is

$$h_1 = h_r = \sqrt{\sin^2\theta \cos^2\varphi + \sin^2\theta \sin^2\varphi + \cos^2\theta} = 1 \, .$$

Thus, $\partial\vec{r}/\partial r$ is a unit vector in the radial direction and defines:

$$\hat{r} = \hat{e}_1 = (\sin\theta \cos\varphi, \, \sin\theta \sin\varphi, \, \cos\theta) \, . \tag{1.113}$$

Next, we compute derivative of $\vec{r}$ with respect to the spherical angle θ:

$$\frac{\partial \vec{r}}{\partial \theta} = (r \cos\theta \cos\varphi, \, r \cos\theta \sin\varphi, \, -r \sin\theta) \, .$$

We evaluate the norm of the above vector to find a scale factor:

$$h_2 = h_\theta = \sqrt{r^2 \cos^2\theta \cos^2\varphi + r^2 \cos^2\theta \sin^2\varphi + r^2 \sin^2\theta} = r$$

leading to the unit vector:

$$\hat{\theta} = \frac{1}{r} \frac{\partial \vec{r}}{\partial \theta} = (\cos\theta \cos\varphi, \, \cos\theta \sin\varphi, \, -\sin\theta) \, . \tag{1.114}$$

Derivative of $\vec{r}$ with respect to the angle φ is equal to a vector

$$\frac{\partial \vec{r}}{\partial \varphi} = (-r \sin\theta \sin\varphi, \, r \sin\theta \cos\varphi, \, 0) \, ,$$

whose norm is

$$h_3 = h_\varphi = \sqrt{r^2 \sin^2\theta \sin^2\varphi + r^2 \sin^2\theta \cos^2\varphi} = r \sin\theta \, .$$

Dividing $\partial\vec{r}/\partial\varphi$ by the scale factor h_3, yields the third unit vector in this basis:

$$\hat{\varphi} = (-\sin\varphi, \, \cos\varphi, \, 0) \, , \tag{1.115}$$

which agrees with the expression obtained in equation (1.109).
Hence, in the notation:

$$\mathrm{d}\vec{r} = \sum_{i=1}^{3} \hat{e}_i \, \mathrm{d}s_i = \hat{r} \, \mathrm{d}r + \hat{\theta} r \, \mathrm{d}\theta + \hat{\varphi} r \sin\theta \, \mathrm{d}\varphi \, , \tag{1.116}$$

the squared distance between two infinitesimally nearby points is:

$$\mathrm{d}\vec{r} \cdot \mathrm{d}\vec{r} = \mathrm{d}x^2 + \mathrm{d}y^2 + \mathrm{d}z^2 = \mathrm{d}r^2 + r^2\,\mathrm{d}\theta^2 + r^2\sin^2\theta\,\mathrm{d}\varphi^2\,.$$

The area element in the $\hat{r}$-direction is:

$$\mathrm{d}A = \mathrm{d}s_2\,\mathrm{d}s_3 = h_\theta h_\varphi\,\mathrm{d}\theta\,\mathrm{d}\varphi = r^2\sin\theta\,\mathrm{d}\theta\,\mathrm{d}\varphi\,. \qquad (1.117)$$

Two area elements in Figure 1.33 span the same solid angle

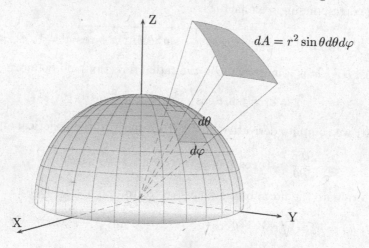

Figure 1.33: The area elements in the $\hat{r}$-direction.

$$\mathrm{d}\Omega = \frac{\mathrm{d}A}{r^2} = \sin\theta\,\mathrm{d}\theta\,\mathrm{d}\varphi\,,$$

which is defined as a surface $\mathrm{d}A$ on a sphere with radius $r = 1$.

The total solid angle is obtained by integrating the infinitesimal solid angle $\mathrm{d}\Omega$ over the complete range of θ and φ:

$$\Omega = \iint \mathrm{d}\Omega = \int_0^\pi \int_0^{2\pi} \sin\theta\,\mathrm{d}\theta\,\mathrm{d}\varphi = 2\pi \int_{-1}^1 \mathrm{d}\cos\theta = 4\pi\,.$$

The volume element is $\mathrm{d}V = \mathrm{d}s_1\,\mathrm{d}s_2\,\mathrm{d}s_3 = r^2\,\mathrm{d}r\,\mathrm{d}\Omega$.

From equations (1.113)-(1.115), we find:

$$\boxed{\begin{aligned}
\hat{r} &= \hat{i}\sin\theta\cos\varphi + \hat{j}\sin\theta\sin\varphi + \hat{k}\cos\theta\,, \\
\hat{\theta} &= \hat{i}\cos\theta\cos\varphi + \hat{j}\cos\theta\sin\varphi - \hat{k}\sin\theta\,, \\
\hat{\varphi} &= -\hat{i}\sin\varphi + \hat{j}\cos\varphi
\end{aligned}} \qquad (1.118)$$

and one can easily recover the expression for a radial vector:

$$\vec{r} = \hat{r} r = \hat{i}\, r \sin\theta \cos\varphi + \hat{j}\, r \sin\theta \sin\varphi + \hat{k}\, r \cos\theta = \hat{i}\, x + \hat{j}\, y + \hat{k}\, z. \quad (1.119)$$

Clearly, the unit vectors $\hat{r}, \hat{\theta}, \hat{\varphi}$ do not have fixed directions in space. Their directions depend on where they are.

It is easy to verify that spherical and cylindrical coordinates are examples of orthogonal curvilinear coordinates.

The $\hat{e}_i$ component of the gradient $\vec{\nabla} f$ at the point $\vec{r}(q_1, q_2, q_3)$ is:

$$\begin{aligned}
\left(\vec{\nabla} f\right)_i &= \hat{e}_i \cdot \vec{\nabla} f = \frac{1}{h_i}\frac{\partial \vec{r}}{\partial q_i} \cdot \vec{\nabla} f \\
&= \frac{1}{h_i}\left(\frac{\partial x}{\partial q_i}\hat{i} + \frac{\partial y}{\partial q_i}\hat{j} + \frac{\partial z}{\partial q_i}\hat{k}\right) \cdot \left(\frac{\partial f}{\partial x}\hat{i} + \frac{\partial f}{\partial y}\hat{j} + \frac{\partial f}{\partial z}\hat{k}\right) \qquad (1.120) \\
&= \frac{1}{h_i}\sum_{j=1}^{3}\frac{\partial x_j}{\partial q_i}\frac{\partial f}{\partial x_j} = \frac{1}{h_i}\frac{\partial f}{\partial q_i}
\end{aligned}$$

and, therefore,

$$\vec{\nabla} f = \sum_{i=1}^{3}\frac{1}{h_i}\frac{\partial f}{\partial q_i}\,\hat{e}_i. \qquad (1.121)$$

For the special case of the function f such that $f = q_i$ for some fixed index i the equation (1.121) yields:

$$\vec{\nabla} q_i = \frac{\hat{e}_i}{h_i}, \qquad (1.122)$$

which is consistent with the fact that the unit vector $\hat{e}_i$ is orthogonal to the coordinate surface $q_i = $ constant.

A consequence of (1.122) is that

$$\vec{\nabla} \times \frac{\hat{e}_i}{h_i} = 0, \qquad (1.123)$$

since a curl of a gradient always vanishes. We will use this identity and the product rule for $\vec{\nabla} \times (f\vec{V})$ to find an expression for a curl of an arbitrary vector $\vec{V} = \sum_i V_i \hat{e}_i$ in an orthogonal curvilinear coordinate system:

$$\vec{\nabla} \times \vec{V} = \sum_i \vec{\nabla} \times \left(V_i h_i \frac{\hat{e}_i}{h_i}\right) = \sum_i \left[\vec{\nabla}(V_i h_i) \times \frac{\hat{e}_i}{h_i} + V_i h_i \vec{\nabla} \times \frac{\hat{e}_i}{h_i}\right]. \qquad (1.124)$$

The second term on the right hand side of the above equation is equal to zero due to equation (1.123). The first term involves expression (1.121) for the gradient of the scalar function $f = V_i h_i$ and the vector product of two vectors.

A curl of a vector field in a general orthogonal curvilinear coordinate system is thus given by:

$$\vec{\nabla} \times \vec{V} = \sum_{i,j,k=1}^{3} \epsilon_{ijk} \frac{1}{h_i h_j} \frac{\partial}{\partial q_i} (V_j h_j) \hat{e}_k \qquad (1.125)$$

or

$$\vec{\nabla} \times \vec{V} = \frac{1}{h_1 h_2 h_3} \begin{vmatrix} \hat{e}_1 h_1 & \hat{e}_2 h_2 & \hat{e}_3 h_3 \\ \frac{\partial}{\partial q_1} & \frac{\partial}{\partial q_2} & \frac{\partial}{\partial q_3} \\ h_1 V_1 & h_2 V_2 & h_3 V_3 \end{vmatrix} = \frac{1}{h_2 h_3} \left[\frac{\partial(h_3 V_3)}{\partial q_2} - \frac{\partial(h_2 V_2)}{\partial q_3} \right] \hat{e}_1$$

$$+ \frac{1}{h_1 h_3} \left[\frac{\partial(h_1 V_1)}{\partial q_3} - \frac{\partial(h_3 V_3)}{\partial q_1} \right] \hat{e}_2 + \frac{1}{h_1 h_2} \left[\frac{\partial(h_2 V_2)}{\partial q_1} - \frac{\partial(h_1 V_1)}{\partial q_2} \right] \hat{e}_3 .$$

$$(1.126)$$

Let's choose three distinct indices $i, j, k = 1, 2, 3$, $i \neq j$, $j \neq k$ and $i \neq k$. Plugging $\vec{V} = \hat{e}_j q_i / h_j$, with fixed i and j, into expression (1.125) we obtain:

$$\vec{\nabla} \times \frac{q_i}{h_j} \hat{e}_j = \sum_{n,m,l=1}^{3} \epsilon_{nml} \frac{1}{h_n h_m} \frac{\partial}{\partial q_n} (V_m h_m) \hat{e}_l$$

$$= \sum_{n,m,l=1}^{3} \epsilon_{nml} \frac{1}{h_n h_m} \frac{\partial}{\partial q_n} \left(\frac{q_i}{h_j} \delta_{mj} h_m \right) \hat{e}_l = \epsilon_{ijk} \frac{\hat{e}_k}{h_i h_j} . \qquad (1.127)$$

Rewriting an arbitrary vector $\vec{V}$ as $\sum_k V_k h_i h_j (\hat{e}_k / h_i h_j)$ we can describe the divergence of $\vec{V}$ as:

$$\vec{\nabla} \cdot \vec{V} = \sum_{k=1}^{3} \vec{\nabla} \cdot V_k h_i h_j \frac{\hat{e}_k}{h_i h_j} = \sum_{k=1}^{3} \left[\left(\vec{\nabla} V_k h_i h_j \right) \cdot \frac{\hat{e}_k}{h_i h_j} + V_k h_i h_j \vec{\nabla} \cdot \frac{\hat{e}_k}{h_i h_j} \right] , \quad (1.128)$$

where we used the product rule (1.67) for divergence.

Since the divergence of the curl is zero it holds from relation (1.127) that $\vec{\nabla} \cdot (\hat{e}_k / h_i h_j) = 0$ for three distinct indices $i, j, k = 1, 2, 3$. Accordingly, the second term on the right hand side of equation (1.128) vanishes. The first term involves the scalar product of the gradient (1.121) for the scalar function $f = V_k h_i h_j$ with the vector $\hat{e}_k / h_i h_j$. This product can be cast in the following form:

$$\vec{\nabla} \cdot \vec{V}(q_1, q_2, q_3) = \left(\vec{\nabla} V_1 h_2 h_3 \right) \cdot \frac{\hat{e}_1}{h_2 h_3} + \left(\vec{\nabla} V_2 h_1 h_3 \right) \cdot \frac{\hat{e}_2}{h_1 h_3} + \left(\vec{\nabla} V_3 h_1 h_2 \right) \cdot \frac{\hat{e}_3}{h_1 h_2}$$

$$= \sum_{i=1}^{3} \frac{1}{h_i} \frac{\partial(V_1 h_2 h_3)}{\partial q_i} \hat{e}_i \cdot \frac{\hat{e}_1}{h_2 h_3} + \sum_{i=1}^{3} \frac{1}{h_i} \frac{\partial(V_2 h_1 h_3)}{\partial q_i} \hat{e}_i \cdot \frac{\hat{e}_2}{h_1 h_3}$$

$$+ \sum_{i=1}^{3} \frac{1}{h_i} \frac{\partial(V_3 h_1 h_2)}{\partial q_i} \hat{e}_i \cdot \frac{\hat{e}_3}{h_1 h_2} . \qquad (1.129)$$

Thus, it follows that for an orthogonal curvilinear coordinate system the divergence of a vector field is:

$$\vec{\nabla} \cdot \vec{V}(q_1, q_2, q_3) = \frac{1}{h_1 h_2 h_3} \left(\frac{\partial(V_1 h_2 h_3)}{\partial q_1} + \frac{\partial(V_2 h_3 h_1)}{\partial q_2} + \frac{\partial(V_3 h_1 h_2)}{\partial q_3} \right).$$

(1.130)

Finally, by substituting $\left(\vec{\nabla} f \right)_i = \frac{1}{h_i} \frac{\partial f}{\partial q_i}$ into eq.(1.130), the Laplace operator $\vec{\nabla} \cdot \vec{\nabla} f$ is given by:

$$\vec{\nabla}^2 f = \frac{1}{h_1 h_2 h_3} \left[\frac{\partial}{\partial q_1} \frac{h_2 h_3}{h_1} \frac{\partial}{\partial q_1} + \frac{\partial}{\partial q_2} \frac{h_1 h_3}{h_2} \frac{\partial}{\partial q_2} + \frac{\partial}{\partial q_3} \frac{h_1 h_2}{h_3} \frac{\partial}{\partial q_3} \right] f.$$

(1.131)

☞ **Example 1.3.9.** Cylindrical coordinates:
The differential operators are given by:

$$\vec{\nabla} f(\rho, \varphi, z) = \hat{\rho} \frac{\partial f}{\partial \rho} + \hat{\varphi} \frac{1}{\rho} \frac{\partial f}{\partial \varphi} + \hat{k} \frac{\partial f}{\partial z},$$

(1.132a)

$$\vec{\nabla} \cdot \vec{V}(\rho, \varphi, z) = \frac{1}{\rho} \frac{\partial(V_\rho \rho)}{\partial \rho} + \frac{1}{\rho} \frac{\partial V_\varphi}{\partial \varphi} + \frac{\partial V_z}{\partial z},$$

(1.132b)

$$\vec{\nabla}^2 f(\rho, \varphi, z) = \frac{1}{\rho} \frac{\partial}{\partial \rho} \left(\rho \frac{\partial f}{\partial \rho} \right) + \frac{1}{\rho^2} \frac{\partial^2 f}{\partial \varphi^2} + \frac{\partial^2 f}{\partial z^2}$$

(1.132c)

and

$$\vec{\nabla} \times \vec{V} = \frac{1}{\rho} \begin{vmatrix} \hat{\rho} & \rho\hat{\varphi} & \hat{k} \\ \frac{\partial}{\partial \rho} & \frac{\partial}{\partial \varphi} & \frac{\partial}{\partial z} \\ V_\rho & \rho V_\varphi & V_z \end{vmatrix}.$$

(1.132d)

☞ **Example 1.3.10.** Spherical coordinates:
The differential operators are given this time by:

$$\vec{\nabla} f(r, \theta, \varphi) = \hat{r} \frac{\partial f}{\partial r} + \hat{\theta} \frac{1}{r} \frac{\partial f}{\partial \theta} + \hat{\varphi} \frac{1}{r \sin \theta} \frac{\partial f}{\partial \varphi},$$

(1.133a)

$$\vec{\nabla} \cdot \vec{V}(r, \theta, \varphi) = \frac{1}{r^2 \sin \theta} \left(\sin \theta \frac{\partial(V_r r^2)}{\partial r} + r \frac{\partial(V_\theta \sin \theta)}{\partial \theta} + r \frac{\partial V_\varphi}{\partial \varphi} \right),$$

(1.133b)

$$\vec{\nabla}^2 f(r, \theta, \varphi) = \frac{1}{r^2 \sin \theta} \left[\sin \theta \frac{\partial}{\partial r} \left(r^2 \frac{\partial f}{\partial r} \right) + \frac{\partial}{\partial \theta} \left(\sin \theta \frac{\partial f}{\partial \theta} \right) + \frac{1}{\sin \theta} \frac{\partial^2 f}{\partial \varphi^2} \right]$$

(1.133c)

and

$$\vec{\nabla} \times \vec{V} = \frac{1}{r^2 \sin\theta} \begin{vmatrix} \hat{r} & r\hat{\theta} & r\sin\theta\,\hat{\varphi} \\ \frac{\partial}{\partial r} & \frac{\partial}{\partial\theta} & \frac{\partial}{\partial\varphi} \\ V_r & rV_\theta & r\sin\theta\,V_\varphi \end{vmatrix}. \tag{1.133d}$$

☞ **Example 1.3.11.** For function $f(r)$ depending on r only, the last two terms in expression (1.133a) for the gradient of f vanish automatically and

$$\vec{\nabla} f(r) = \hat{r}\frac{\partial f}{\partial r}. \tag{1.134}$$

In particular for $f(r) = r^n$ it follows that

$$\vec{\nabla} r^n = \hat{r}\,n r^{n-1}.$$

Inserting $V(r) = \hat{r}f$ into (1.133b) and (1.133d) yields

$$\vec{\nabla} \cdot \hat{r}f(r) = \frac{1}{r^2}\frac{\partial}{\partial r}\left(r^2 f(r)\right) = \frac{2}{r}f(r) + \frac{\partial f}{\partial r} \tag{1.135}$$

and

$$\vec{\nabla} \times \hat{r}f(r) = 0. \tag{1.136}$$

Choosing $f(r) = r^n$ we obtain from (1.135) that

$$\vec{\nabla} \cdot \hat{r}r^n = (n+2)r^{n-1}.$$

⌂ Curvilinear Coordinates: Problems

Problem 1.3.1. Elliptic coordinates ξ, η, are defined in terms of the Cartesian coordinates according to:

$$x = a\xi\eta, \quad y = a\sqrt{(\xi^2 - 1)(1 - \eta^2)}, \quad z = z.$$

Where the new coordinates have ranges $1 \le \xi < \infty$ and $-1 \le \eta \le 1$, and a is a constant (focal length).
(a) Find the unit vectors $\hat{e}_\xi, \hat{e}_\eta$. Do the coordinates ξ, η form an orthogonal system?
(b) What are the scale factors h_ξ, h_η?
(c) Find equations to describe the surfaces of constant ξ and η. What kind of surfaces are they?

Problem 1.3.2. Define toroidal coordinates (η, ξ, ϕ) related to the Cartesian coordinates (x, y, z) by:

$$x = \frac{a}{q} \sinh \eta \cos \phi, \quad y = \frac{a}{q} \sinh \eta \sin \phi, \quad z = \frac{a}{q} \sin \xi,$$

where a is a scale parameter of the dimension of length and q is defined by

$$q = \cosh \eta - \cos \xi$$

The coordinate η ranges from 0 to ∞, while both ξ and ϕ have values between 0 and 2π.

(a) Verify relations:

$$(z - a \cot \xi)^2 + \rho^2 = \frac{a^2}{\sin^2 \xi}$$

$$z^2 + (\rho - a \coth \eta)^2 = \frac{a^2}{\sinh^2 \eta}$$

with $\rho^2 = x^2 + y^2$ and use it to show that the surfaces of constant ξ and η describe, respectively, a sphere and a torus.

(b) Find the unit vectors $\hat{e}_\xi, \hat{e}_\eta$ and $\hat{e}_\phi$. Do the coordinates ξ, η, ϕ form an orthogonal system?

(c) Determine the scale factors h_ξ, h_η and h_ϕ.

(d) Verify the relation:

$$r^2 = x^2 + y^2 + z^2 = a^2 \frac{\cosh \eta + \cos \xi}{\cosh \eta - \cos \xi}$$

and use it to find positions of points $r = 0$ and $r = \infty$ in the toroidal coordinates.

Problem 1.3.3. Consider coordinates, ξ, η and ϕ, related to Cartesian coordinates through the following relations

$$x = \xi \eta \cos \phi \quad ; \quad y = \xi \eta \sin \phi \quad ; \quad z = \frac{(\xi^2 - \eta^2)}{2} .$$

(a) Find the corresponding scale factors h_ξ, h_η and h_ϕ.

(b) Calculate divergence $\vec{\nabla} \cdot \vec{V}$, where $\vec{V} = \sqrt{\xi^2 + \eta^2} \, (\hat{e}_\xi + \hat{e}_\eta)$.

Problem 1.3.4. Consider coordinates, ξ, η and z, related to the Cartesian coordinates through the following relations:

$$x = \frac{1}{2} (\xi^2 - \eta^2) \quad ; \quad y = \xi \eta \quad ; \quad z = z .$$

(a) Find the corresponding scale factors h_ξ, h_η and h_z.

(b) Calculate divergence $\vec{\nabla} \cdot \vec{V}$ and curl $\vec{\nabla} \times \vec{V}$, where $\vec{V} = \sqrt{\xi^2 + \eta^2} \, \hat{e}_\xi$.

Problem 1.3.5. Express the following electric field

$$\vec{E} = q\frac{\hat{r}}{r^2} = -q\vec{\nabla}\frac{1}{r}$$

in the curvilinear coordinate system defined in Problem 1.3.3.

Problem 1.3.6. Expand the unit vector $\hat{i}$ in the spherical unit vectors as shown

$$\hat{i} = \left(\hat{i}\cdot\hat{r}\right)\hat{r} + \left(\hat{i}\cdot\hat{\theta}\right)\hat{\theta} + \left(\hat{i}\cdot\hat{\varphi}\right)\hat{\varphi}$$
$$= \sin\theta\cos\varphi\,\hat{r} + \cos\theta\cos\varphi\,\hat{\theta} - \sin\varphi\,\hat{\varphi}.$$

Obtain similar expressions for $\hat{j}$ and $\hat{k}$.

Problem 1.3.7. Resolve the cylindrical unit vectors, $\hat{\rho}, \hat{\varphi}, \hat{k}$, into their Cartesian components, $\hat{i}, \hat{j}, \hat{k}$, and show that:

$$\frac{\partial\hat{\rho}}{\partial\varphi} = \hat{\varphi} \quad ; \quad \frac{\partial\hat{\varphi}}{\partial\varphi} = -\hat{\rho}.$$

Show that all remaining first derivatives of the cylindrical unit vectors with respect to the cylindrical coordinates vanish.

Problem 1.3.8. Show that $\vec{r} = \rho\hat{\rho} + z\hat{k}$.

Problem 1.3.9. Show that $\vec{\nabla}\cdot\vec{r} = 3$ and $\vec{\nabla}\times\vec{r} = 0$ using cylindrical coordinates.

Problem 1.3.10. Use the results from Problems 1.3.7 and 1.3.8 to show that velocity and acceleration vectors in cylindrical coordinates are given by

$$\vec{v} = \dot{\vec{r}} = \dot{\rho}\,\hat{\rho} + \rho\dot{\varphi}\,\hat{\varphi} + \dot{z}\,\hat{k},$$
$$\vec{a} = \ddot{\vec{r}} = [\ddot{\rho} - \rho\dot{\varphi}^2]\,\hat{\rho} + [2\dot{\rho}\dot{\varphi} + \rho\ddot{\varphi}]\,\hat{\varphi} + \ddot{z}\,\hat{k}.$$

Problem 1.3.11. Express the velocity vector $\vec{v} = -\omega y\hat{i} + \omega x\hat{j}$ with a constant ω in terms of the cylindrical basis vectors and calculate its curl in the cylindrical coordinate system. Compare your results with those from Problem 1.3.10 and show that the velocity vector corresponds to a circular motion in plane with a constant angular velocity.

Problem 1.3.12. Verify that the spherical unit vectors satisfy the following differential relations:

$$\frac{\partial\hat{r}}{\partial\theta} = \hat{\theta}, \qquad\qquad \frac{\partial\hat{r}}{\partial\varphi} = \sin\theta\,\hat{\varphi}, \qquad\qquad \frac{\partial\hat{\theta}}{\partial\theta} = -\hat{r}$$

$$\frac{\partial\hat{\theta}}{\partial\varphi} = \cos\theta\,\hat{\varphi}, \qquad\qquad \frac{\partial\hat{\varphi}}{\partial\theta} = 0, \qquad\qquad \frac{\partial\hat{\varphi}}{\partial\varphi} = -\sin\theta\,\hat{r} - \cos\theta\,\hat{\theta}.$$

Problem 1.3.13. (a) Use the relations from Problem 1.3.12 and the chain rule to show that the derivatives with respect to time of the spherical unit vectors are:

$$\frac{d\hat{r}}{dt} = \dot{\theta}\,\hat{\theta} + \dot{\varphi}\sin\theta\,\hat{\varphi}, \qquad \frac{d\hat{\theta}}{dt} = -\dot{\theta}\,\hat{r} + \dot{\varphi}\cos\theta\,\hat{\varphi}$$

and

$$\frac{d\hat{\varphi}}{dt} = -\dot{\varphi}\sin\theta\,\hat{r} - \dot{\varphi}\cos\theta\,\hat{\theta}.$$

(b) Derive the following expressions:

$$v_r = \dot{r}, \quad v_\theta = r\dot{\theta}, \quad v_\varphi = r\sin\theta\dot{\varphi}, \quad a_r = \ddot{r} - r\dot{\theta}^2 - r\sin^2\theta\dot{\varphi}^2,$$

$$a_\theta = r\ddot{\theta} - 2\dot{r}\dot{\theta} - r\sin\theta\cos\theta\dot{\varphi}^2, \quad a_\varphi = r\sin\theta\ddot{\varphi} + 2\dot{r}\sin\theta\dot{\varphi} + 2r\cos\theta\dot{\theta}\dot{\varphi}$$

for the spherical components of velocity $\vec{v} = \dot{\vec{r}}$ and acceleration $\vec{a} = \ddot{\vec{r}}$ vectors by computing derivatives of vector $\vec{r}(t) = r(t)\hat{r}(t)$ with respect to time.

Problem 1.3.14. Let $\hat{e}_1$ be a unit vector of an arbitrary curvilinear coordinate system. Show that:

(a) $\vec{\nabla} \cdot \hat{e}_1 = \dfrac{1}{h_1 h_2 h_3}\dfrac{\partial(h_2 h_3)}{\partial u_1}$ (b) $\vec{\nabla} \times \hat{e}_1 = \dfrac{1}{h_1}\left[\hat{e}_2\dfrac{\partial h_1}{h_3\partial u_3} - \hat{e}_3\dfrac{\partial h_1}{h_2\partial u_2}\right]$

Problem 1.3.15. Calculate $\vec{\nabla} \cdot \vec{V}(\rho, \varphi, z)$ and $\vec{\nabla} \times \vec{V}$ in cylindrical coordinates for

(a) $\vec{V} = \ln\rho\hat{k}$ (b) $\vec{V} = \ln\rho\hat{\varphi}$

Problem 1.3.16. Using spherical coordinates show that

$$\vec{\nabla}\ln|r| = \frac{\vec{r}}{r^2}.$$

Problem 1.3.17. For a vector field $\vec{V} = \theta\hat{r} + r\hat{\theta}$ evaluate $\vec{\nabla} \times \vec{V}$ and $\vec{\nabla} \cdot (\vec{\nabla} \times \vec{V})$.

Problem 1.3.18. Evaluate

(a) $\vec{r} \cdot \vec{\nabla}r$

(b) $\hat{k} \cdot \vec{\nabla}\theta$

(c) $\vec{\nabla}^2\ln r^2$

Problem 1.3.19. Calculate $\vec{\nabla} \cdot \vec{V}(r, \theta, \varphi)$ and $\vec{\nabla} \times \vec{V}$ in spherical coordinates for

(a) $\vec{V} = f(r)\hat{r}$

(b) $\vec{V} = \hat{\theta}$

(c) $\vec{V} = \hat{\varphi}$

Problem 1.3.20. Show that for a constant B the vector potential

$$\vec{A} = \frac{\hat{\varphi}}{2} Br \sin\theta$$

yields a constant magnetic field $\vec{B} = \vec{\nabla} \times \vec{A} = B\hat{k}$.

Problem 1.3.21. Find a magnetic field $\vec{B} = \vec{\nabla} \times \vec{A}$ for a vector potential $\vec{A} = \hat{\theta} \times \vec{r}$.

Problem 1.3.22. Let the directions of two unit vectors be given by the spherical angles θ_1, φ_1 and θ_2, φ_2, respectively, as shown in Figure (1.34). Show that the cosine of the angle γ between the two unit vectors is given by:

$$\cos\gamma = \cos\theta_1 \cos\theta_2 + \sin\theta_1 \sin\theta_2 \cos(\varphi_1 - \varphi_2).$$

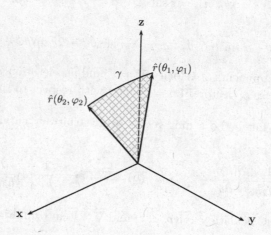

Figure 1.34: Angle γ between two unit vectors $\hat{r}(\theta_i, \varphi_i)$, $i = 1, 2$.

Problem 1.3.23. Using that $\frac{\partial}{\partial x_i} = \hat{e}_i \cdot \vec{\nabla}$ find $\partial/\partial x$, $\partial/\partial y$, $\partial/\partial z$ in terms of spherical and cylindrical coordinates.

Problem 1.3.24. For $\vec{L} = -\mathrm{i}\left(\vec{r} \times \vec{\nabla}\right)$, where $\mathrm{i} = \sqrt{-1}$, show that

(a) $L_z = -\mathrm{i}\left(x\frac{\partial}{\partial y} - y\frac{\partial}{\partial x}\right) = -\mathrm{i}\frac{\partial}{\partial\varphi}$

(b) $\vec{L} = \mathrm{i}\left(\hat{\theta}\frac{1}{\sin\theta}\frac{\partial}{\partial\varphi} - \hat{\varphi}\frac{\partial}{\partial\theta}\right)$

Problem 1.3.25. Determine $L_x = \hat{i} \cdot \vec{L}, L_y = \hat{j} \cdot \vec{L}, L_z = \hat{k} \cdot \vec{L}$ in terms of θ, φ and their derivatives.

Problem 1.3.26. Show that the quantity $L^2 = L_x^2 + L_y^2 + L_z^2$ is given by

$$L^2 = -\frac{1}{\sin\theta}\frac{\partial}{\partial\theta}\left(\sin\theta\frac{\partial}{\partial\theta}\right) - \frac{1}{\sin^2\theta}\frac{\partial^2}{\partial\varphi^2}$$

Problem 1.3.27. Consider the so-called Dirac vector potential

$$\vec{A} = q \frac{\vec{r} \times \vec{n}}{r(r - \vec{r} \cdot \vec{n})},$$

defined for a constant vector $\vec{n}$ and $r \neq \vec{r} \cdot \vec{n}$.

(a) Show that for $\vec{n} = \hat{k}$ the Dirac vector potential can be written in spherical coordinates as

$$\vec{A} = -\frac{q}{r} \frac{\sin \theta}{1 - \cos \theta} \hat{\varphi} \quad \text{for} \quad \theta \neq 0.$$

(b) Show that $\vec{\nabla} \times \vec{A}$ reproduces the magnetic field

$$\vec{B} = q \frac{\vec{r}}{r^3}$$

of a magnetic single pole (monopole) with magnetic charge q.

Problem 1.3.28. (a) Verify, that the vector potential

$$\vec{A}' = \vec{A} + \vec{\nabla} f$$

gives rise to the same magnetic field $\vec{B} = \vec{\nabla} \times \vec{A}'$ as $\vec{A}$. The operation that maps vector potential $\vec{A}$ into $\vec{A}'$ is called a gauge transformation.

(b) Find the scalar field f such that

$$\vec{A}' = \vec{A} + \vec{\nabla} f = -\frac{q}{r} \varphi \sin \theta \, \hat{\theta}$$

for the Dirac vector potential $\vec{A}$ from Problem (1.3.27).

(c) Show by an explicit calculation for the above $\vec{A}'$ that $\vec{B} = \vec{\nabla} \times \vec{A}'$ is that of the magnetic monopole.

Problem 1.3.29. The Newton's second law describing motion of a particle of mass m in a central field can be written as:

$$m \ddot{\vec{r}} = \vec{\nabla} f(r)$$

for some f which is function of r. Show that the particle's angular momentum is conserved, i.e. that $\vec{r} \times \dot{\vec{r}}$ remains constant.

Problem 1.3.30. Use spherical coordinates to evaluate:

$$\iiint_D \exp \left[\left(x^2 + y^2 + z^2 \right)^{3/2} \right] \, dx \, dy \, dz.$$

where D is the region defined by $1 \leq x^2 + y^2 + z^2 \leq 2$ and $z \geq 0$.

Problem 1.3.31. Use cylindrical coordinates to derive the formula $\pi r^2 h/3$ for the volume of a circular cone of base radius r and height h.

Problem 1.3.32. Find the volume of a solid region that lies above the plane $z \geq 1$ and inside the sphere $x^2 + y^2 + z^2 = 4$.

Hint: First compute the volume of the solid "ice cream cone" cut from the sphere by the cone $\cos\theta = 1/2$ and subtract the result of Problem 1.3.31.

1.4   Integral Theorems of Vector Calculus

1.4.1   Vector Calculus: Line Integral

The subject of our study is a line integral $\int_C \vec{V}(\vec{r}) \cdot d\vec{r}$ of a vector function $\vec{V}(\vec{r})$ over a curve C. Here, the path of integration is a one-dimensional object, a curve C for which we have parametrized representation (1.87) in terms of one parameter t, which varies along the curve within the interval $a \leq t \leq b$. Quantities a and b are, respectively, initial and final points of the curve C. If a and b coincide, the curve C is called a closed curve and the corresponding line integral, denoted as $\oint_C \vec{V}(\vec{r}) \cdot d\vec{r}$, is over a closed path.

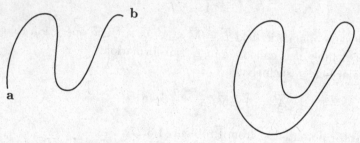

Figure 1.35: Open and closed curves.

Definition 1.4.1. *Using the parametrization (1.87) we define a line integral as an integral over the parameter t:*

$$\int_C \vec{V}(\vec{r}) \cdot d\vec{r} = \int_a^b \vec{V}(\vec{r}(t)) \cdot \frac{d\vec{r}}{dt}\, dt = \int_a^b \sum_{i=1}^{3} V_i(t) \frac{dx_i(t)}{dt}\, dt, \qquad (1.137)$$

where the components of the vector field $\vec{V}(\vec{r})$ depend on parameter t through their dependence on $\vec{r}$.

☞**Example 1.4.1.** A helix is a three-dimensional curve defined by

$$x(t) = \cos(2\pi t), \quad y(t) = \sin(2\pi t), \quad z(t) = t \qquad (1.138)$$

or

$$\vec{r}(t) = \hat{i}\cos(2\pi t) + \hat{j}\sin(2\pi t) + \hat{k}t, \quad t \geq 0.$$

Vectors and Vector Calculus

Note that the z-component is given by the parameter t. A helix has a tangent vector:

$$\frac{\mathrm{d}\vec{r}(t)}{\mathrm{d}t} = -2\pi\hat{i}\sin(2\pi t) + \hat{j}2\pi\cos(2\pi t) + \hat{k}\,.$$

with direction indicated in Figure 1.36.

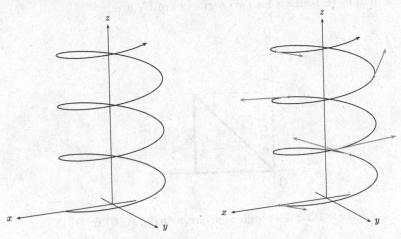

Figure 1.36: Three-dimensional helix and its tangent vector field.

Consider the line integral of the vector field

$$\vec{V} = \hat{i}y - \hat{j}x + \hat{k}$$

along a helix over two revolutions, i.e. $0 \le t \le 2$. As a first step, we substitute parametrization (1.138) into $\vec{V}$ to obtain the value of this vector field along the helix:

$$\vec{V}(\vec{r}(t)) = \hat{i}\sin(2\pi t) - \hat{j}\cos(2\pi t) + \hat{k}\,.$$

Next, we take the scalar product of $\vec{V}(\vec{r}(t))$ with the tangential vector:

$$\vec{V}(\vec{r}(t)) \cdot \frac{\mathrm{d}\vec{r}(t)}{\mathrm{d}t} = -2\pi(\sin^2(2\pi t) + \cos^2(2\pi t)) + 1 = 1 - 2\pi\,.$$

To obtain the line integral we will integrate the above quantity from $t = 0$ to $t = 2$:

$$\int_C \vec{V}(\vec{r}) \cdot \mathrm{d}\vec{r} = \int_0^2 (1 - 2\pi)\,\mathrm{d}t = 2 - 4\pi\,.$$

☞ **Example 1.4.2. The line integral and dependence on the path.** Consider the vector field $\vec{V}(\vec{r}) = 2y\hat{i} + x\hat{j}$ and two different curves C_1 and C_2 with $(x, y) = (0, 0)$ and $(x, y) = (1, 1)$ endpoints . Let C_1 consist of two orthogonal segments of unit length, the horizontal one from $(0, 0)$ to $(0, 1)$ and the vertical one from $(0, 1)$ to $(1, 1)$. The other curve C_2 is taken to be a straight line connecting $(0, 0)$ to $(1, 1)$ and making a 45^0 degree with x-axis. The parameterizations for two segments of C_1 are:

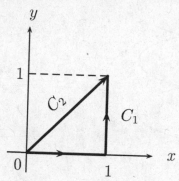

Figure 1.37: Two different curves from $(0, 0)$ to $(1, 1)$.

$$\vec{r}_1(t) = t\hat{i}, \quad 0 \le t \le 1 \tag{1.139}$$
$$\vec{r}_1(s) = 1\hat{i} + s\hat{j}, \quad 0 \le s \le 1. \tag{1.140}$$

The integral of the vector field $\vec{V}(\vec{r})$ over the curve C_1 is then given by:

$$\int_{C_1} \vec{V}(\vec{r}) \cdot d\vec{r} = \int_{t=0}^{1} (0\hat{i} + t\hat{j}) \cdot \hat{i} \, dt + \int_{s=0}^{1} (2s\hat{i} + \hat{j}) \cdot (0\hat{i} + 1\hat{j}) \, ds$$

$$= 0 + \int_{s=0}^{1} ds = 1. \tag{1.141}$$

The parametrization for the curve C_2 is:

$$\vec{r}_2(t) = t\hat{i} + t\hat{j}, \quad 0 \le t \le 1 \tag{1.142}$$

therefore, the integral of the vector field $\vec{V}(\vec{r})$ over the curve C_2 is:

$$\int_{C_2} \vec{V}(\vec{r}) \cdot d\vec{r} = \int_{t=0}^{1} (2t\hat{i} + t\hat{j}) \cdot (\hat{i} + \hat{j}) \, dt = \int_{t=0}^{1} 3t \, dt = \frac{3}{2}. \tag{1.143}$$

The above example shows that, in general, the line integral will dependent on the path connecting two endpoints of the path. For a particular class of vector fields that are gradients of the scalar function path dependence disappears and the line integral depends solely on the integration limits. Consider, namely, the vector field $\vec{V}(\vec{r}) = \vec{\nabla}F(\vec{r})$. Applying definition (1.137) yields:

$$\int_C \vec{\nabla}F(\vec{r}) \cdot \mathrm{d}\vec{r} = \int_a^b \sum_{i=1}^3 \frac{\partial F}{\partial x_i} \frac{\mathrm{d}x_i(t)}{\mathrm{d}t} \,\mathrm{d}t = \int_a^b \frac{\mathrm{d}F}{\mathrm{d}t} \,\mathrm{d}t = F(b) - F(a) , \qquad (1.144)$$

where the chain rule $\frac{\mathrm{d}F}{\mathrm{d}t} = \sum_{i=1}^3 \frac{\partial F}{\partial x_i} \frac{\mathrm{d}x_i(t)}{\mathrm{d}t}$ was used in the derivation.

☞ **Example 1.4.3.** Consider the integral:

$$\int_{(0,1,0)}^{(1,1,1)} \left(3x^2 z\,\mathrm{d}x + 2y\,\mathrm{d}y + x^3\,\mathrm{d}z\right) , \qquad (1.145)$$

which can be rewritten as $\int_C \vec{V}(\vec{r}) \cdot \mathrm{d}\vec{r}$ with $\vec{V}(\vec{r}) = 3x^2 z\hat{i} + 2y\hat{j} + x^3\hat{k}$ and $\mathrm{d}\vec{r} = \mathrm{d}x\hat{i} + \mathrm{d}y\hat{j} + \mathrm{d}z\hat{k}$ for a curve C connecting the endpoints $(0,1,0)$ with $(1,1,1)$. Expression (1.145) does not depend on the path C but only on the endpoints. This property rests on the fact that the vector field $\vec{V}$ can be written as $\vec{V}(\vec{r}) = \vec{\nabla}F(\vec{r})$ with the scalar function $F(\vec{r}) = x^3 z + y^2$. It then follows from equation (1.144) that

$$\int_{(0,1,0)}^{(1,1,1)} \left(3x^2 z\,\mathrm{d}x + 2y\,\mathrm{d}y + x^3\,\mathrm{d}z\right) = F(1,1,1) - F(0,1,0) = 2 - 1 = 1 . \qquad (1.146)$$

1.4.2 Vector Calculus: Surface Integral

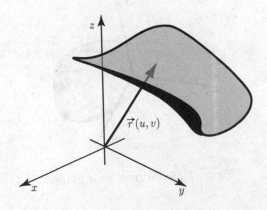

Figure 1.38: Parametrization of a surface.

In order to describe the surface integral we will need to introduce parametrization for surfaces. Since surfaces are the two-dimensional objects their parametrization

$$\vec{r}(u,v) = \sum_{i=1}^{3} x_i(u,v)\hat{e}_i = x(u,v)\hat{i} + y(u,v)\hat{j} + z(u,v)\hat{k} \qquad (1.147)$$

involves two parameters u and v within some range of values: $a_1 \le u \le b_1$, $a_2 \le v \le b_2$. For fixed $u = u_0$, $\vec{r}(u_0, v)$ is a function which maps the line segment $(u_0, v), a_2 \le v \le b_2$ onto a curve in the surface: $v \to \vec{r}(u_0, v)$. The tangent vector $\vec{T}_v(u_0, v_0)$ to this curve at the point $\vec{r}(u_0, v_0)$ is given by:

$$\vec{T}_v(u_0, v_0) = \frac{\partial x}{\partial v}(u_0, v_0)\hat{i} + \frac{\partial y}{\partial v}(u_0, v_0)\hat{j} + \frac{\partial z}{\partial v}(u_0, v_0)\hat{k} = \frac{\partial \vec{r}}{\partial v}(u_0, v_0). \qquad (1.148)$$

Similarly, by keeping $v = v_0$ fixed we obtain a curve $u \to \vec{r}(u, v_0)$. The tangent vector to this curve at the point (u_0, v_0) is

$$\vec{T}_u(u_0, v_0) = \frac{\partial x}{\partial u}(u_0, v_0)\hat{i} + \frac{\partial y}{\partial u}(u_0, v_0)\hat{j} + \frac{\partial z}{\partial u}(u_0, v_0)\hat{k} = \frac{\partial \vec{r}}{\partial u}(u_0, v_0). \qquad (1.149)$$

The partial derivatives $\vec{T}_v$ and $\vec{T}_u$ lie on the tangent plane to the surface S for any point on the surface corresponding to the given value of the parameters u, v. It follows that the vector product:

$$\vec{N} = \vec{T}_u \times \vec{T}_v = \frac{\partial \vec{r}}{\partial u} \times \frac{\partial \vec{r}}{\partial v} \qquad (1.150)$$

is normal to S. Let $\hat{n}\,dA$ be an infinitesimal area element of the surface S with $\hat{n}$ being an unit vector normal to S. Obviously, $|N| = |\partial \vec{r}/\partial u \times \partial \vec{r}/\partial v|$ is the length

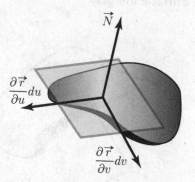

Figure 1.39: Normal to a plane.

of $\vec{N}$ and $\vec{N}/|N|$ is a unit vector. Consider an infinitesimal parallelogram with sides $\vec{r}(u + du, v) - \vec{r}(u, v) \approx (\partial \vec{r}/\partial u)\,du$ and $\vec{r}(u, v + dv) - \vec{r}(u, v) \approx (\partial \vec{r}/\partial v)\,dv$. From construction it follows that $|N|\,du\,dv$ equals the area of the parallelogram.

Thus $|N| \, du \, dv$ is an infinitesimal area element of surface S. Therefore we obtain the relation:

$$\hat{n} \, dA = \frac{\vec{N}}{|N|} \, |N| \, du \, dv = \vec{N} \, du \, dv, \qquad (1.151)$$

which allows us to define the surface integral over S as:

$$\iint\limits_S \vec{V}(\vec{r}) \cdot \hat{n} \, dA = \int_{a_1}^{b_1} \int_{a_2}^{b_2} \vec{V}(\vec{r}(u,v)) \cdot \vec{N}(u,v) \, du \, dv. \qquad (1.152)$$

This integral is often referred to as a flux integral as it describes a flux of a vector field $\vec{V}(\vec{r})$ across S.

Now, we will give examples of standard parametric representations of cylindrical and spherical surfaces

☞ **Example 1.4.4.** Consider the cylinder: $x^2 + y^2 = R^2, -h \le z \le h$ of radius R and height h. The appropriate parametric representation is:

$$\vec{r}(u,v) = R \cos u \, \hat{i} + R \sin u \, \hat{j} + v \hat{k}, \quad -h \le v \le h, \quad 0 \le u \le 2\pi. \qquad (1.153)$$

The parameters u, v are nothing but the cylindrical coordinates φ, z. More-

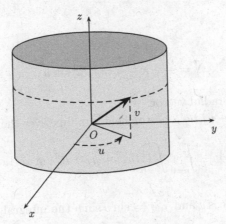

over, the normal to the cylindrical surface is given by the expression

$$\vec{N} = \frac{\partial \vec{r}}{\partial u} \times \frac{\partial \vec{r}}{\partial v} = (-R \sin u \, \hat{i} + R \cos u \, \hat{j}) \times \hat{k} = R \hat{\rho} \qquad (1.154)$$

in which we recognize the radial unit vector $\hat{\rho} = \cos u \, \hat{i} + \sin u \, \hat{j}$ of the cylindrical coordinate system.

☞ **Example 1.4.5.** Consider a sphere S of radius R. The spherical coordinates $u = \theta$ and $v = \varphi$ provide convenient parameters for describing the radial vector $\vec{r}$:

$$\vec{r}(u,v) = R \sin u \cos v \hat{i} + R \sin u \sin v \hat{i} + R \cos u \hat{k}. \qquad (1.155)$$

As u varies from 0 to π and v varies from 0 to 2π the vector $\vec{r}$ sweeps the area of the sphere. The normal to the spherical surface is

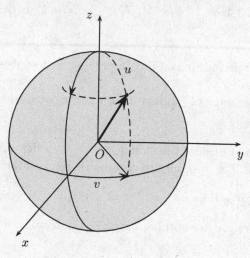

$$\vec{N} = \frac{\partial \vec{r}}{\partial u} \times \frac{\partial \vec{r}}{\partial v} = \hat{r} R^2 \sin u, \qquad (1.156)$$

where $\hat{r}$ is a unit radial vector.

Let $\vec{V} = 5x\hat{i} + y\hat{j}$ be a vector field. Its flux over the sphere S is

$$\iint\limits_{S} \vec{V}(\vec{r}) \cdot \hat{n} \, \mathrm{d}A = \int_{u=0}^{\pi} \int_{v=0}^{2\pi} (5x\hat{i} + y\hat{j}) \cdot R^2 \sin u \hat{r} \, \mathrm{d}u \, \mathrm{d}v. \qquad (1.157)$$

Since $\hat{i} \cdot \hat{r} = \sin u \cos v$ and $\hat{j} \cdot \hat{r} = \sin u \sin v$ the integral takes the form

$$\int_{u=0}^{\pi} \int_{v=0}^{2\pi} R^3 \left(5\sin^2 u \cos^2 v + \sin^2 u \sin^2 v\right) \sin u \, \mathrm{d}u \, \mathrm{d}v$$

$$= R^3 \pi \int_{u=0}^{\pi} 6 \sin^3 u \, \mathrm{d}u = 6R^3 \pi \int_{-1}^{1} (1 - \cos^2 u) \, \mathrm{d}(\cos u) \qquad (1.158)$$

$$= 6R^3 (4\pi/3).$$

One can also use the above parametrization technique to calculate the area of the surface

$$A(S) = \iint\limits_{S} dA = \int_{a_1}^{b_1} \int_{a_2}^{b_2} |\vec{N}|(u,v) \, du \, dv. \qquad (1.159)$$

For the area of the sphere we reproduce the well-known result

$$A(S) = \int_{u=0}^{\pi} \int_{v=0}^{2\pi} R^2 \sin u \, du \, dv = 2\pi R^2 \int_{u=0}^{\pi} \sin u \, du$$
$$= 2\pi R^2 \int_{\cos u=-1}^{\cos u=1} d(\cos u) = 4\pi R^2. \qquad (1.160)$$

☞ **Example 1.4.6.** Let us use formula (1.159) to calculate area of an ellipse. The ellipse with center at the origin and with semi-major axis a and semi-minor axis b (see Figure 1.40) is described by the equation

$$\frac{x^2}{a^2} + \frac{y^2}{b^2} = 1.$$

To parametrize the area enclosed by the ellipse we represent the points inside

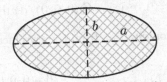

Figure 1.40: An area of an ellipse.

the ellipse by $\vec{r} = a\rho\cos\varphi\,\hat{i} + b\rho\sin\varphi\,\hat{j}$. The polar coordinates ρ, φ take values $0 \le \rho \le 1$, $0 \le \varphi \le 2\pi$. Then $\vec{N}$ takes the form

$$\vec{N} = \frac{\partial\vec{r}}{\partial\rho} \times \frac{\partial\vec{r}}{\partial\varphi} = ab\rho\hat{k}.$$

Plugging $|\vec{N}| = ab\rho$ into formula (1.159) yields

$$A(S) = \int_{\rho=0}^{1} \int_{\varphi=0}^{2\pi} ab\rho \, d\rho \, d\varphi = ab\pi,$$

which reproduces a well-know expression for an area of an ellipse.

1.4.3 Vector Calculus: Divergence Theorem

The divergence of the vector field can be identified with net flux of the vector field out of a closed volume. To illustrate this relation we will consider a small cube of volume $dx\,dy\,dz$ placed at the origin $(0,0,0)$ of the xyz-coordinate system. The

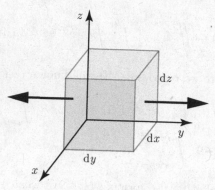

Figure 1.41: A flow out of a box of volume $dx\,dy\,dz$.

net flux $\iint \vec{V} \cdot \hat{n}\,dA$ for the vector field $\vec{V}$ is

$$(V_x(dx,0,0) - V_x(0,0,0))\,dy\,dz + (V_y(0,dy,0) - V_y(0,0,0))\,dx\,dz$$
$$+ (V_z(0,0,dz) - V_z(0,0,0))\,dx\,dy. \tag{1.161}$$

To the first order in dx the contribution from the two faces orthogonal to the x-axis becomes

$$V_x(dx,0,0) - V_x(0,0,0) = \frac{\partial V_x(0,0,0)}{\partial x}\,dx.$$

Analogous results follow for the remaining two terms in (1.161) leading to the net flux:

$$\left(\frac{\partial V_x(0,0,0)}{\partial x} + \frac{\partial V_y(0,0,0)}{\partial y} + \frac{\partial V_z(0,0,0)}{\partial z} \right) dx\,dy\,dz = \vec{\nabla} \cdot \vec{V}(0,0,0)\,dx\,dy\,dz.$$
$$\tag{1.162}$$

Hence, the net flux out of this volume is equal to the divergence of the vector field times the small volume.

Relation (1.162) is an infinitesimal version of the divergence theorem:

$$\oiint_S \vec{V}(\vec{r}) \cdot \hat{n}\,dA = \iiint_V \vec{\nabla} \cdot \vec{V}(\vec{r})\,dV, \tag{1.163}$$

where the integral on the right hand side is a volume integral over a closed volume V bounded by a surface S. Also, $\hat{n}$ is an unit vector normal to the closed surface S. For closed surfaces we follow a convention that the unit normal vector $\hat{n}$ points out of the surface.

The proof of the divergence theorem makes use of relation (1.162) by subdividing the volume V into a large number of tiny sub-volumes such that the union of these sub-volumes approximates the original volume V. For each such tiny sub-volume a

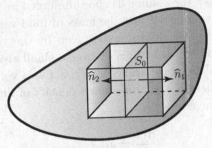

Figure 1.42: A common face between two adjacent sub-volumes with normal vectors $\hat{n}_1 = \hat{n}$ and $\hat{n}_2 = -\hat{n}$.

version of relation (1.162) takes the form:

$$\sum_{\text{faces}} \vec{V} \cdot \hat{n} \, dA = \vec{\nabla} \cdot \vec{V} \, dV, \qquad (1.164)$$

where the sum on the left hand side is over all faces of a tiny sub-volume. We proceed by adding together equations (1.164) and summing over all sub-volumes. Because the normal vector $\hat{n}$ points out of each closed sub-volume, a common face between two adjacent sub-volumes will be associated to two contributions; $\vec{V} \cdot \hat{n} \, dA$ from one sub-volume and $-\vec{V} \cdot \hat{n} \, dA$ from the other one (see Figure 1.42). Hence the contributions to the sum of all interior surfaces cancel and only exterior surfaces give a non-zero contribution to the sum. Thus,

$$\sum_{\text{exterior faces}} \vec{V} \cdot \hat{n} \, dA = \sum_{\text{sub-volumes}} \vec{\nabla} \cdot \vec{V} \, dV. \qquad (1.165)$$

This expression is equivalent to the divergence theorem (1.163).

The following example deals with an application of the divergence theorem for the calculation of the surface integrals.

☞ **Example 1.4.7.** When dealing with closed surfaces, it is often convenient to use the divergence theorem to calculate the surface integral. We will illustrate it by calculating the flux over the sphere S of radius R for the vector field $\vec{V} = 5x\hat{i} + y\hat{j}$. In Example (1.4.5) this was done calculating the surface integrals directly from the definition. Here we use equation (1.163). First, we need to find the divergence of $\vec{V}$: $\vec{\nabla} \cdot \left(5x\hat{i} + y\hat{j}\right) = 5 + 1 = 6$. Hence,

$$\iiint_V \vec{\nabla} \cdot \vec{V} \, dV = 6 \iiint_V dV = 6\frac{4\pi}{3} R^3 \qquad (1.166)$$

in agreement with the result in (1.157).

☞ **Example 1.4.8.** Let the volume dV be submerged in a fluid. The fluid flows in and out of the volume dV and the mass of fluid varies accordingly.

Let the velocity of a fluid flow be given by the vector field $\vec{v}$. If fluid moves a distance $ds = v\,dt$ at time t across a small area $d\vec{A} = dA\hat{n}$, which is perpendicular to $\vec{v}$ (so $\hat{n}$ is parallel to $\vec{v}$), it fills a volume $dA\,ds$ with mass $dm = \rho\,dAv\,dt$, where $\rho(x, y, z)$ is a mass density of fluid. The rate at which the mass crosses the surface is

$$\frac{dm}{dt} = \rho\,dAv.$$

Next, consider a surface, that is not perpendicular to the vector field $\vec{v}$, as shown in Figure 1.43. Then, the mass is transported through this surface at

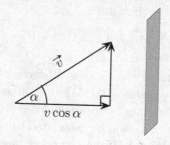

$v\cos\alpha$

Figure 1.43: Only the component of the vector field orthogonal to the surface transports mass across the surface.

a rate:

$$\frac{dm}{dt} = \rho\,d\vec{A}\cdot\vec{v} = \rho\,dA\vec{v}\cdot\hat{n} = \rho\,dAv\cos\alpha,$$

which is proportional to the velocity component $v\cos\alpha$ normal to the surface (α is an angle between $\vec{v}$ and $\hat{n}$). For the volume dV the sum of all mass fluxes through all of its surfaces can be calculated using (1.164):

$$\sum_{\text{faces}} \rho\vec{v}\cdot\hat{n}\,dA = \vec{\nabla}\cdot(\rho\vec{v})\ dV$$

and is equal to a net mass flux through the total surface of dV. When mass accumulates inside dV the corresponding rate of change of mass is positive and related to the net mass flux through

$$\frac{\mathrm{d}m}{\mathrm{d}t} = -\sum_{\text{faces}} \rho\vec{v} \cdot \hat{n}\, \mathrm{d}A = -\vec{\nabla} \cdot (\rho\vec{v})\, \mathrm{d}V\,.$$

Note the minus sign on the right hand side of the above relation. It is due to the fact that when the fluid flows inside the volume $\mathrm{d}V$ the velocity vector $\vec{v}$ is directed opposite to the normal $\hat{n}$, which always points out of $\mathrm{d}V$, as shown in Figure 1.44. Thus, the minus sign compensates a negative value for

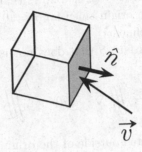

Figure 1.44: The fluid flows into the volume in the direction opposite to the normal pointing out of the volume.

the scalar product $\vec{v} \cdot \hat{n}$ and ensures that the rate of change of mass remains positive for the mass flow into the volume $\mathrm{d}V$. On the other hand, the rate of change of mass inside $\mathrm{d}V$ is dictated by a the rate of change of the fluid density ρ inside $\mathrm{d}V$ through

$$\frac{\mathrm{d}m}{\mathrm{d}t} = \frac{\mathrm{d}}{\mathrm{d}t}(\rho\,\mathrm{d}V) = \frac{\mathrm{d}\rho}{\mathrm{d}t}\,\mathrm{d}V\,.$$

Combining the two above expressions for $\mathrm{d}m/\mathrm{d}t$ leads to the continuity equation

$$\frac{\mathrm{d}\rho}{\mathrm{d}t} + \vec{\nabla} \cdot (\rho\vec{v}) = 0\,,$$

which is valid everywhere in the fluid.

For the constant fluid density ρ the continuity equation reduces to $\vec{\nabla} \cdot \vec{v} = 0$. Thus for the fluid with a constant fluid density the velocity field is divergence free.

1.4.4 Vector Calculus: Gauss' Law

An important application of the divergence theorem (1.163) is Gauss' law that deals with the following vector field

$$\vec{V} = \vec{E} = \frac{q\hat{r}}{4\pi\epsilon_0 r^2} \, .\tag{1.167}$$

The vector field $\vec{V}$ is encountered in Physics as an electric field produced by a point charge q placed at the origin $(0,0,0)$. Recall that $\vec{\nabla} \cdot \vec{r} r^n = (n+3)r^n$ as seen in Example (1.3.11). Therefore for $n = -2$ it follows that $\vec{\nabla} \cdot \hat{r} r^{-2} = 0$ as long $r \neq 0$. Special care is required at the origin since for $r = 0$ we encounter a singularity, which renders divergence ill defined.

Let us consider a closed surface S that does not include the origin. Then, the divergence theorem yields

$$\oiint_S \frac{\hat{r}}{r^2} \cdot \hat{n} dA = \iiint_V \vec{\nabla} \cdot \frac{\hat{r}}{r^2} \, dV = \iiint_V 0 \, dV = 0 \, ,\tag{1.168}$$

since the integrand was calculated outside of the origin.

As a separate exercise, it remains to deal with the case when S includes the origin. For convenience, we choose S to be a sphere with its center at the origin and introduce the auxiliary small spherical cavity S' of radius δ which surrounds the origin. By connecting these two surfaces by an infinitesimally thin tunnel we can

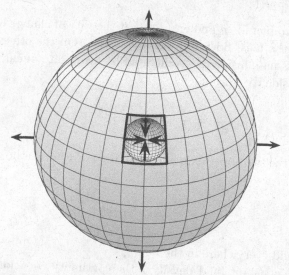

Figure 1.45: Spherical cavity S' inside the sphere S.

think of the sum of S and S', denoted formally by $S + S'$, as a closed surface. The interior of $S + S'$ is a volume between S and S' and does not contain any interior part of S'. Accordingly, the relation (1.168) applies to the $S + S'$ surface:

$$\oiint_{S+S'} \frac{\hat{r}}{r^2} \cdot \hat{n}\, dA = \iint_S \frac{\hat{r}}{r^2} \cdot \hat{n}\, dA + \iint_{S'} \frac{\hat{r}}{\delta^2} \cdot \hat{n}'\, dA' \qquad (1.169)$$

$$= 0\,.$$

The unit normal vector $\hat{n}'$ points in the negative radial direction because it points out of the closed surface, $S + S'$. So $\hat{n}' = -\hat{r}$ and according to (1.156), $\hat{n}'\, dA' = -\hat{r}\delta^2 \sin u\, du\, dv$. Thus, $\delta^{-2}\hat{r} \cdot \hat{n}'\, dA' = -\sin u\, du\, dv$. From equation (1.169) it follows that:

$$\iint_S \frac{\hat{r}}{r^2} \cdot \hat{n}\, dA = -\iint_{S'} \frac{\hat{r}}{\delta^2} \cdot \hat{n}'\, dA' = \int_{u=0}^{\pi} \int_{v=0}^{2\pi} \sin u\, du\, dv = 4\pi \qquad (1.170)$$

for the closed surface S, which includes the origin. We can reformulate expressions (1.168) and (1.170) as Gauss' law:

$$\boxed{\iint_S \vec{E}(\vec{r}) \cdot \hat{n}\, dA = \iint_S \frac{q\hat{r}}{4\pi\epsilon_0 r^2} \cdot \hat{n}\, dA = \begin{cases} \frac{q}{\epsilon_0}, & S \text{ includes origin} \\ 0, & \text{otherwise} \end{cases}} \qquad (1.171)$$

Gauss' law generalizes to

$$\iint_S \vec{E}(\vec{r}) \cdot \hat{n}\, dA = \sum_i q_i/\epsilon_0\,,$$

where the sum $\sum_i$ involves all the charges q_i inside S. For the continuous charge distribution $\rho(\vec{r})$, such that $q = \iiint_V \rho(\vec{r})\, d^3r$, the divergence theorem (1.163) yields

$$\iiint_V \vec{\nabla} \cdot \vec{E}(\vec{r})\, d^3r = \iiint_V \frac{\rho(\vec{r})}{\epsilon_0}\, d^3r$$

or

$$\vec{\nabla} \cdot \vec{E} = \frac{\rho}{\epsilon_0} \qquad (1.172)$$

in which we recognize one of the Maxwell equations (1.81).

Recall from Example (1.3.11) that $\hat{r}/r^2 = -\vec{\nabla}(1/r)$. Using the divergence theorem (1.163), Gauss' law can be rewritten in terms of the volume integral as

$$\iiint_V \vec{\nabla} \cdot \frac{\hat{r}}{r^2}\, d^3r = -\iiint_V \vec{\nabla} \cdot \vec{\nabla}\left(\frac{1}{r}\right) d^3r = \begin{cases} 4\pi, & (0,0,0) \text{ in } V \\ 0, & \text{otherwise} \end{cases} \qquad (1.173)$$

Introduce the so-called three-dimensional Dirac delta function $\delta(\vec{r})$ as:

$$\delta(\vec{r}) = -\frac{1}{4\pi}\vec{\nabla}^2\left(\frac{1}{r}\right) = \frac{1}{4\pi}\vec{\nabla}\cdot\frac{\hat{r}}{r^2}. \tag{1.174}$$

Comparing expression (1.174) with equations (1.173)-(1.174), we see that the delta function satisfies:

$$\delta(\vec{r}) = 0, \text{ for } r \neq 0 \tag{1.175}$$

$$\iiint_V \delta(\vec{r})\,d^3r = 1, \text{ for } (0,0,0) \text{ in } V. \tag{1.176}$$

Since the delta function vanishes outside the origin the only contribution to the volume integral in (1.176) must come from the origin $(x,y,z) = (0,0,0)$. Both properties (1.175) and (1.176) are contained within a single relation:

$$\iiint_V f(\vec{r}')\delta(\vec{r}' - \vec{r})\,d^3r' = \begin{cases} f(\vec{r}) & \text{for } \vec{r} = \vec{r}' \text{ in } V \\ 0 & \text{for } \vec{r} = \vec{r}' \text{ outside } V \end{cases}. \tag{1.177}$$

Indeed, taking the special case of $f(\vec{r}) = 1$ and setting $\vec{r} = 0$ we obtain

$$\iiint_V f(\vec{r}')\delta(\vec{r}')\,d^3r' = \iiint_V \delta(\vec{r}')\,d^3r'$$

$$= \begin{cases} f(0,0,0) = 1 & \text{for } \vec{r}' = (0,0,0) \text{ in } V \\ 0 & \text{for } \vec{r}' = (0,0,0) \text{ outside } V, \end{cases} \tag{1.178}$$

which reproduces properties (1.175) and (1.176).

1.4.5 Vector Calculus: Stokes's Theorem

To understand the geometric interpretation of a curl of the vector field consider the circulation of an infinitesimal rectangle R in Figure (1.46), defined as a closed loop line integral (a line integral of a vector field for which the starting and the ending points coincide). The contribution of all four sides of the rectangle R is:

$$\oint_R \vec{V}\cdot d\vec{r} = \int_{(x_0,y_0)}^{(x_0+dx,y_0)} V_x\,dx + \int_{(x_0+dx,y_0)}^{(x_0+dx,y_0+dy)} V_y\,dy$$

$$+ \int_{(x_0+dx,y_0+dy)}^{(x_0,y_0+dy)} V_x\,dx + \int_{(x_0,y_0+dy)}^{(x_0,y_0)} V_y\,dy. \tag{1.179}$$

By convention, the circulation is always taken in a (positive) counter-clockwise direction.

The first term on the right hand side of (1.179) is equal to $V_x(x_0,y_0)\,dx$. The second term is $V_y(x_0 + dx, y_0)\,dy = (V_y(x_0,y_0) + (\partial V_y/\partial x)(x_0,y_0)\,dx)\,dy$. Similar

Vectors and Vector Calculus

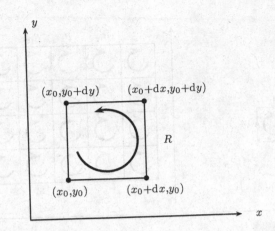

Figure 1.46: Circulation of an infinitesimal rectangle R.

considerations apply to the remaining integrals in equation (1.179). Substituting these derivations into expression (1.179) yields

$$
\begin{aligned}
\oint_R \vec{V} \cdot d\vec{r} &= V_x(x_0, y_0)\, dx + \left(V_y(x_0, y_0) + \frac{\partial V_y}{\partial x}(x_0, y_0)\, dx \right) dy \\
&\quad - \left(V_x(x_0, y_0) + \frac{\partial V_x}{\partial y}(x_0, y_0)\, dy \right) dx - V_y(x_0, y_0)\, dy \\
&= \left(\frac{\partial V_y}{\partial x}(x_0, y_0) - \frac{\partial V_x}{\partial y}(x_0, y_0) \right) dx\, dy \\
&= \left(\vec{\nabla} \times \vec{V} \right)_z (x_0, y_0)\, dx\, dy .
\end{aligned}
\tag{1.180}
$$

This calculation proved equivalence of the line integral of the vector field around a closed infinitesimal loop in a xy-plane with the z-component of the curl of the vector field multiplied by an infinitesimal area encircled by that loop. The result generalizes to any small rectangle R surrounding the infinitesimal plane surface with an area $\hat{n}\, dA$. The argument establishes that the circulation of the given vector field $\vec{V}$ around R is equal to $\vec{\nabla} \times \vec{V} \cdot \hat{n}\, dA$ with the direction of the unit normal vector $\hat{n}$ being positive as determined by the counter-clockwise direction.

Now, consider a finite surface S with boundary C and divide S up into N subregions $S_i, i = 1, \ldots, N$. For large N we may approximate each infinitesimal subregion S_i by a plane area bounded by a rectangle R_i. For each individual rectangle R_i containing area dA_i the counter-clockwise circulation satisfies

$$
\oint_{R_i} \vec{V} \cdot d\vec{r} = \vec{\nabla} \times \vec{V} \cdot \vec{n}\, dA_i .
\tag{1.181}
$$

We sum contributions from all rectangles R_i to obtain a total circulation. Two

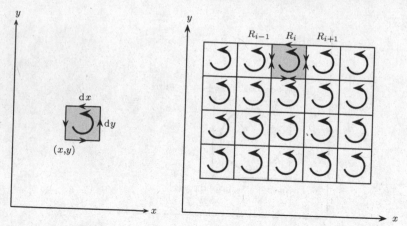

Figure 1.47: Circulations for cells with and without adjacent cells.

adjacent rectangles R_{i-1} and R_i share one interior side. The counter-clockwise circulations on R_{i-1} and R_i lead to integrations in opposite directions on their common interior side. Consequently, the common side will give a zero net contribution to $\sum_{i=1}^{N} \oint_{R_i} \vec{V} \cdot d\vec{r}$, where the sum involves all rectangles R_i covering surface S. This argument applies to every interior side since each such side appears to be a boundary between two adjacent rectangles. Thus only external sides will make a non-zero contribution to the integral. In the limit where N is large we obtain

$$\sum_{i=1}^{N} \oint_{R_i} \vec{V} \cdot d\vec{r} \longrightarrow \oint_C \vec{V} \cdot d\vec{r}, \tag{1.182}$$

which together with

$$\sum_{i=1}^{N} \vec{\nabla} \times \vec{V} \cdot \vec{n} \, dA_i \to \iint_S \vec{\nabla} \times \vec{V} \cdot \vec{n} \, dA \tag{1.183}$$

leads to Stokes's theorem:

$$\oint_C \vec{V} \cdot d\vec{r} = \iint_S \vec{\nabla} \times \vec{V} \cdot \vec{n} \, dA. \tag{1.184}$$

Stokes's theorem says that the flux of $\vec{\nabla} \times \vec{V}$ across any open surface S bounded by the closed curve C is equal to the circulation of $\vec{V}$ around C.

The circulation depends on the direction of integration. A closed curve C that bounds a surface has two possible orientations. We adopt the convention that associates the counterclockwise orientation with a positive orientation and the clockwise orientation with the negative orientation. The choice of orientation for the circulation of the boundary of surface S defines direction of a unit normal vector $\vec{n}$ to S according to the right hand rule as shown in the Figure 1.48. If the surface S and

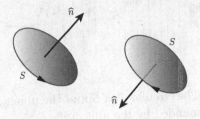

Figure 1.48: Two surfaces with positive counter-clockwise orientations.

its boundary C lie in the xy-plane then Stokes's theorem simplifies to:

$$\oint_C \vec{V} \cdot d\vec{r} = \iint_S \vec{\nabla} \times \vec{V} \cdot \hat{k}\, dx\, dy = \iint_S \left(\vec{\nabla} \times \vec{V}\right)_3 dx\, dy. \tag{1.185}$$

For a two-dimensional vector field $\vec{V} = V_1\hat{i} + V_2\hat{j}$ the third component of $\vec{\nabla} \times \vec{V}$ equals $\partial V_2/\partial x - \partial V_1/\partial y$ and equation (1.185) takes the form:

$$\oint_C (V_1\, dx + V_2\, dy) = \iint_S \left(\frac{\partial V_2}{\partial x} - \frac{\partial V_1}{\partial y}\right) dx\, dy, \tag{1.186}$$

which is referred to as Green's theorem in the plane.

☞ **Example 1.4.9.** If there exist two (or more) different surfaces S and S' that are bounded by the same closed curve C then, according to Stokes's theorem, the fluxes over S and S' are identical:

$$\iint_S \vec{\nabla} \times \vec{V} \cdot \vec{n}\, dA = \iint_{S'} \vec{\nabla} \times \vec{V} \cdot \vec{n}\, dA.$$

We will illustrate this observation for C a unit circle that lies on the xy plane and is centered at the origin and two surfaces S (the circular disk) and S' (the upper half-sphere). Both surfaces are bounded by C as shown in Figure 1.49. The vector field is chosen to be $\vec{V} = (x - y)\hat{i} + (2y - z)\hat{j} + 3z\hat{k}$. Below, we will show that the line integral of $\vec{V}$ along the curve C (see ①) agrees with the fluxes of the curl of $\vec{V}$ through the circular disk S (calculated in ②) and through the upper half-sphere S' (calculated in ③).

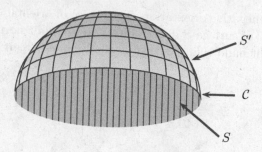

Figure 1.49: The circular disk S and the upper half-sphere S' are both bounded by the unit circle $\mathcal{C}$.

① The line integral. The curve $\mathcal{C}$: $x^2 + y^2 = 1$ is parametrized as in equation (1.91) with the parameter t running from 0 to 2π. Accordingly,

$$\oint_{\mathcal{C}} \vec{V} \cdot d\vec{r} = \int_0^{2\pi} \left((\cos t - \sin t)\, d(\cos t) + 2\sin t\, d(\sin t) \right) = \int_0^{2\pi} \sin^2 t\, dt = \pi.$$

(1.187)

② We choose the area S to be the circular disk $x^2 + y^2 \leq 1$. Then the unit normal vector $\hat{n}$ equals $\hat{k}$. Since $\vec{\nabla} \times \vec{V} = \hat{i} + \hat{k}$, the area integral is given by

$$\iint_S \vec{\nabla} \times \vec{V} \cdot \hat{n}\, dA = \iint_S \left(\hat{i} + \hat{k} \right) \cdot \hat{k}\, dA = \iint_S dA = \pi \qquad (1.188)$$

and is equal to the area of the disk $x^2 + y^2 \leq 1$.

③ We choose the area S' to be the upper half-sphere: $x^2 + y^2 + z^2 = 1$, $z \geq 0$. From (1.156) the area integral is given by:

$$\int_{u=0}^{\pi/2} \int_{v=0}^{2\pi} \left(\hat{i} + \hat{k} \right) \cdot \hat{r} \sin u\, du\, dv = \int_{u=0}^{\pi/2} \int_{v=0}^{2\pi} (\sin u \cos v + \cos u) \sin u\, du\, dv$$

$$= 2\pi \int_0^1 \cos u\, d(\cos u) = \pi.$$

(1.189)

Above, use was made of relations $\hat{i} \cdot \hat{r} = \sin u \cos v$, $\hat{k} \cdot \hat{r} = \cos u$ and $\int_{v=0}^{2\pi} \cos v\, dv = 0$.

☞ **Example 1.4.10.** Calculate the circulation of $\vec{V} = x^2\hat{k}$ over the perimeter of the triangle defined by vertices $(1,0,0)$, $(0,1,0)$ and $(0,0,1)$ as shown in

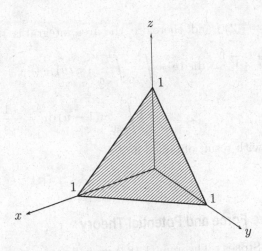

Figure 1.50: Triangle with vertices $(1,0,0)$, $(0,1,0)$ and $(0,0,1)$.

Figure 1.50. The direction of integration is $(1,0,0) \rightarrow (0,1,0) \rightarrow (0,0,1)$. The line integral of $\vec{V}$ along the triangular path $\mathcal{C}$ will be calculated below in ① and compared with the flux, derived in ②, of the curl of $\vec{V}$ through the triangular area bounded by the triangle $\mathcal{C}$.

① The line integral is over $\mathcal{C}$ which consists of three segments:
$\mathcal{C}_1$: $\vec{r}_1(t) = (1-t)\hat{i} + t\hat{j} + 0\hat{k}$ with $0 \leq t \leq 1$ on the xy-plane
$\mathcal{C}_2$: $\vec{r}_2(t) = 0\hat{i} + (1-t)\hat{j} + t\hat{k}$ with $0 \leq t \leq 1$ on the yz-plane
$\mathcal{C}_3$: $\vec{r}_3(t) = t\hat{i} + 0\hat{j} + (1-t)\hat{k}$ with $0 \leq t \leq 1$ on the xz-plane
There is no contribution from $\mathcal{C}_1$ since $\vec{V} \cdot d\vec{r}_1 = 0$. Also, the contribution from $\mathcal{C}_2$ vanishes since $x = 0$ on the yz-plane. The only non-zero contribution is:

$$\int_{\mathcal{C}_3} \vec{V} \cdot d\vec{r}_3 = \int_0^1 t^2 \, d(1-t) = -\int_0^1 t^2 \, dt = -\frac{1}{3}. \tag{1.190}$$

② Let S be the portion of the plane $x + y + z = 1$ in the first octant $x, y, z \geq 0$. Then for $x = u$, $y = v$ we have $z = 1 - u - v$ and the surface is parametrized by $\vec{r}(u,v) = u\hat{i} + v\hat{j} + (1 - u - v)\hat{k}$ with

$$\frac{\partial \vec{r}}{\partial u} = \hat{i} - \hat{k}, \quad \frac{\partial \vec{r}}{\partial v} = \hat{j} - \hat{k} \; \rightarrow \; \vec{N} = \hat{i} + \hat{j} + \hat{k}. \tag{1.191}$$

Also, $\vec{\nabla} \times \vec{V} = -2x\hat{j}$ and, therefore, the area integral is

$$\iint\limits_S \vec{\nabla} \times \vec{V} \cdot \vec{N} \, du \, dv = \int_{u=0}^{1} \int_{v=0}^{1-u} (-2u)\hat{j} \cdot (\hat{i} + \hat{j} + \hat{k}) \, du \, dv$$

$$= -2 \int_{u=0}^{1} u(1-u) \, du = -\frac{1}{3},$$

which agrees with result obtained in ①.

_____

1.4.6   Conservative Force and Potential Theory

As an application of Stokes's Theorem (1.184) we will prove an equivalence between the following three statements concerning the vector field $\vec{F}(\vec{r})$ on some region Γ. We assume that the vector field is smooth (continuous differentiable and with continuous first derivatives) on Γ and that the region Γ is simply connected. We will explain below what it means that the region Γ is simply connected and the reason for requiring this condition here. The three equivalent statements are:

① $\vec{F} = -\vec{\nabla}\phi$ for some single-valued scalar function $\phi(\vec{r})$,

② $\vec{\nabla} \times \vec{F} = 0$,

③ $\oint_C \vec{F} \cdot d\vec{r} = 0$ for every closed curve on Γ.

The vector function $\vec{F}$ which satisfies any one (and hence all) of the above properties is called a *conservative* vector field. The corresponding scalar field ϕ is referred to as a potential function. The minus sign is purely a matter of convention.

Since, the curl of a gradient is identically zero, the property ② follows from the property ① since $\vec{\nabla} \times \vec{F} = -\vec{\nabla} \times \vec{\nabla}\phi = 0$. Also, the property ③ follows since

$$\oint_C \vec{F} \cdot d\vec{r} = -\oint_C \vec{\nabla}\phi \cdot d\vec{r} = -\oint_C d\phi = 0 \tag{1.192}$$

for a single-valued function ϕ and a closed curve C. Thus, we have shown that ① implies both ② and ③. Assume that property ③ holds and let C_1 and C_2 be two paths both starting at A and ending at B. Let $C = C_1 - C_2$ denote a closed path which starts from A and goes to B along C_1 and then back to A along C_2, as shown in Figure 1.51. Since the closed loop integral must vanish we get:

$$\oint_C \vec{F} \cdot d\vec{r} = \left(\int_{C_1} - \int_{C_2} \right) \vec{F} \cdot d\vec{r} = 0. \tag{1.193}$$

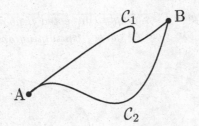

Figure 1.51: Two paths from A to B.

Hence, for any two (simple) curves $\mathcal{C}_1$ and $\mathcal{C}_2$ which share the same endpoints, it holds that

$$\int_{\mathcal{C}_1} \vec{F} \cdot d\vec{r} = \int_{\mathcal{C}_2} \vec{F} \cdot d\vec{r}. \tag{1.194}$$

A curve is called simple if it does not intersect or touch itself.

Thus, for arbitrary two endpoints, A and B, the corresponding line integral of $\vec{F}$ will not depend on the path connecting them. Accordingly, we can associate to that line integral a scalar function ϕ through the formula:

$$\int_A^B \vec{F} \cdot d\vec{r} = \phi(A) - \phi(B), \tag{1.195}$$

where $\int_A^B \vec{F} \cdot d\vec{r} = \int_{\mathcal{C}_i} \vec{F} \cdot d\vec{r}$ for an arbitrary curve $\mathcal{C}_i$ connecting A and B. (Note, that ϕ is defined only up to a constant.)

Let A approach B so that they become separated by an infinitesimal displacement vector $d\vec{r}$. In that limit the relation (1.195) takes the form $\vec{F} \cdot d\vec{r} = -d\phi = -\vec{\nabla}\phi \cdot d\vec{r}$. Accordingly, it holds everywhere that $\left(\vec{F} + \vec{\nabla}\phi\right) \cdot d\vec{r} = 0$. This relation can only hold for all $d\vec{r}$ if $\vec{F} = -\vec{\nabla}\phi$ and the first property is proved. Furthermore, from Stokes's theorem and the property ③, it follows that $0 = \oint_{\mathcal{C}} \vec{F} \cdot d\vec{r} = \iint_S (\vec{\nabla} \times \vec{F}) \cdot \hat{n} \, dA$ for an arbitrary (small) closed loop and therefore $\vec{\nabla} \times \vec{F} = 0$. Hence, we proved that ③ implies ① and ②.

It remains to show that if ② holds then ③ (and automatically ①) holds too. This will be done by invoking Stokes's theorem expressed by equation (1.184). The right hand side of equation (1.184) involves the curl of the vector function integrated over the surface S capping the curve $\mathcal{C}$. If S is entirely inside the region Γ where $\vec{\nabla} \times \vec{F}$ vanishes and $\vec{F}$ is smooth then the circulation of $\vec{F}$ along $\mathcal{C}$ is zero too and we are done with our proof. However, this requires that a closed curve $\mathcal{C}$ in Γ contains points completely within the region Γ. To ensure that this happens we must assume that Γ is simply connected (to understand the significance of this condition consult Example 1.4.13 given below).

Definition 1.4.2. *A region Γ is simply connected if about every point in Γ it is possible to draw a circle (or a closed curve) containing only points of Γ.*

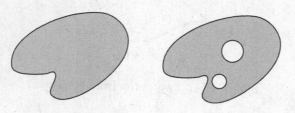

Figure 1.52: Simply and not simply connected regions.

Sometimes it is convenient to characterize the simply connected region Γ by an alternative definition which states that any closed curve in Γ can be shrunk continuously to a single point in Γ without leaving Γ. It is easy to visualize that the surface of a sphere is simply connected, but the surface of a doughnut is not.

☞ **Example 1.4.11.** The gravitational force is given by

$$\vec{F} = -\frac{Gm_1m_2\hat{r}}{r^2} = -\frac{Gm_1m_2\vec{r}}{r^3}. \tag{1.196}$$

Clearly, $\vec{\nabla} \times \vec{F} = 0$, yielding a conservative gravitational force. To find its potential consider the line integral from ∞ to the point a distance r from origin:

$$\phi(r) - \phi(\infty) = -\int_{\infty}^{r} \vec{F} \cdot d\vec{r} = \int_{r}^{\infty} \vec{F} \cdot d\vec{r}. \tag{1.197}$$

As remarked before, formula (1.197) defines a potential up to a constant or relatively to some point in space (in this case ∞). To fix this constant we make a physically plausible assumption that $\phi(\infty) = 0$. Therefore,

$$\phi(r) = -Gm_1m_2 \int_{r}^{\infty} \frac{\hat{r} \cdot d\vec{r}}{r^2} = -Gm_1m_2 \int_{r}^{\infty} \frac{dr}{r^2} = -\frac{Gm_1m_2}{r}. \tag{1.198}$$

☞ **Example 1.4.12.** For the time independent magnetic field $\vec{B}$ the Maxwell equation (1.82) becomes $\vec{\nabla} \times \vec{E} = 0$. Thus, the electric field is conservative and there exists an electrostatic scalar potential ϕ such that

$$\vec{E}(\vec{r}) = -\vec{\nabla}\phi(\vec{r}).$$

Substituting this relation into equation (1.172) yields the Poisson equation

$$\vec{\nabla} \cdot \vec{\nabla}\phi(\vec{r}) = -\frac{\rho(\vec{r})}{\epsilon_0}, \tag{1.199}$$

which plays a central role in electrostatics. Solutions to the Poisson equation determine the electrostatic potentials due to the given charge distribution $\rho(\vec{r})$.

The technique we will use to solve this equation employs Green's function $G(\vec{r}, \vec{r}')$ defined as a function of $\vec{r}$ and $\vec{r}'$, which satisfies

$$\vec{\nabla}^2 G(\vec{r}, \vec{r}') = \delta(\vec{r} - \vec{r}').$$

Generalizing relation (1.174) to

$$-\frac{1}{4\pi}\vec{\nabla}^2 \left(\frac{1}{|\vec{r} - \vec{r}'|} \right) = \delta(\vec{r} - \vec{r}') \tag{1.200}$$

we see that Green's function is given by

$$G(\vec{r}, \vec{r}') = -\frac{1}{4\pi}\frac{1}{|\vec{r} - \vec{r}'|}.$$

Using relation (1.177) that defines the delta function we can rewrite the right hand side of equation (1.199) as

$$\begin{aligned}
-\frac{\rho(\vec{r})}{\epsilon_0} &= -\iiint_V \frac{\rho(\vec{r}')}{\epsilon_0}\delta(\vec{r} - \vec{r}')\,\mathrm{d}^3 r' \\
&= \frac{1}{4\pi}\iiint_V \vec{\nabla}^2 \left(\frac{1}{|\vec{r} - \vec{r}'|} \right) \frac{\rho(\vec{r}')}{\epsilon_0}\,\mathrm{d}^3 r' \\
&= \frac{1}{4\pi}\vec{\nabla}^2 \iiint_V \frac{1}{|\vec{r} - \vec{r}'|}\frac{\rho(\vec{r}')}{\epsilon_0}\,\mathrm{d}^3 r'.
\end{aligned} \tag{1.201}$$

Note that in the above equation we were allowed to pull the $\vec{\nabla}^2$ operator out of the integration since it acts on the position vector $\vec{r}$ and not on the integration variable $\vec{r}'$. Comparing equation (1.201) with the Poisson equation, (1.199), we obtain an expression for the electrostatic potential in terms of an infinite volume integral

$$\phi(\vec{r}) = \frac{1}{4\pi}\iiint_V \frac{1}{|\vec{r} - \vec{r}'|}\frac{\rho(\vec{r}')}{\epsilon_0}\,\mathrm{d}^3 r'. \tag{1.202}$$

Consider the special case where the charge distribution is due to a point charge q located at the origin the charge. Then the distribution $\rho(\vec{r})$ is zero everywhere outside the origin. This is conveniently expressed through

$$\rho(\vec{r}) = q\delta(\vec{r}).$$

Plugging this relation into expression (1.202) we obtain the electrostatic potential due to a point charge:

$$\phi(\vec{r}) = \frac{q}{4\pi} \iiint\limits_{V} \frac{1}{|\vec{r} - \vec{r}'|} \frac{\delta(\vec{r}')}{\epsilon_0} \, d^3 r' = \frac{q}{4\pi\epsilon_0} \frac{1}{r} \tag{1.203}$$

from which we recover Coulomb's law

$$\vec{E}(\vec{r}) = -\vec{\nabla}\phi(\vec{r}) = -\frac{q}{4\pi\epsilon_0} \vec{\nabla}\frac{1}{r} = \frac{q}{4\pi\epsilon_0} \frac{\hat{r}}{r^2} \, . \tag{1.204}$$

Note that in the last calculation we have used the results from example (1.3.11).

☞ **Example 1.4.13.** This example will illustrate the necessity of simply connectedness to conclude that the irrotational vector field is conservative. Let a vector field be described by

$$\vec{F} = -\frac{y}{x^2 + y^2}\hat{i} + \frac{x}{x^2 + y^2}\hat{j} \, .$$

To simplify matters we will use the cylindrical coordinates. Since $x = \rho\cos\varphi$, $y = \rho\sin\varphi$ and $\rho\hat{\varphi} = -\rho\sin\varphi\hat{i} + \rho\cos\varphi\hat{j}$ we can rewrite $\vec{F}$ as:

$$\vec{F} = -\frac{\sin\varphi}{\rho}\hat{i} + \frac{\cos\varphi}{\rho}\hat{j} = \frac{1}{\rho}\hat{\varphi} \, .$$

By plugging the components of $\vec{F}$, $F_\rho = F_z = 0$ and $F_\varphi = 1/\rho$ into expression (1.132d), the curl of $\vec{F}$ in the cylindrical coordinates is given by:

$$\vec{\nabla} \times \vec{F} = \frac{1}{\rho} \begin{vmatrix} \hat{\rho} & \rho\hat{\varphi} & \hat{k} \\ \frac{\partial}{\partial\rho} & \frac{\partial}{\partial\varphi} & \frac{\partial}{\partial z} \\ 0 & 1 & 0 \end{vmatrix} = 0 \, .$$

Therefore, $\vec{F}$ is irrotational. Then is $\vec{F}$ also a conservative vector function? We will see that the answer will depend on whether the region on which we consider $\vec{F}$ is simply connected or not. Two relevant examples are:
1) Γ_1 is the xy-plane with the positive x-axis removed (all points (x, y) without $y = 0, x \geq 0$ half-axis).
2) Γ_2 is the entire xy-plane without the origin $(0,0)$.
It is easy to see that Γ_1 is simply connected but Γ_2 is not because a closed circle around the origin in Γ_2 can not be deformed continuously to a single point in Γ_2. Note, that $\vec{F}$ is smooth (continuous differentiable and with continuous first derivatives) on both Γ_1 and Γ_2.

Recall expression (1.132b):

$$\vec{\nabla}f(\rho, \varphi, z) = \hat{\rho}\frac{\partial f}{\partial\rho} + \hat{\varphi}\frac{1}{\rho}\frac{\partial f}{\partial\varphi} + \hat{k}\frac{\partial f}{\partial z}$$

for the gradient of the scalar function f in cylindrical coordinates. It holds that $\vec{F} = \vec{\nabla} f$ for $f = \varphi + c_0$ where φ is the polar angle and c_0 is a constant.

The polar angle φ is a single-valued scalar function on the simply connected region Γ_1 (for every point in Γ_1 there is only one value of φ). Correspondingly, $\oint_C \vec{F} \cdot d\vec{r} = 0$ for every closed curve on Γ_1. To show that this is no longer the case for Γ_2 we calculate line integral of $\vec{F}$ around the unit circle C in the counterclockwise direction:

$$\oint_C \vec{F} \cdot d\vec{r} = \int_0^{2\pi} \left(-\sin t \hat{i} + \cos t \hat{j} \right) \cdot \left(-\sin t \hat{i} + \cos t \hat{j} \right) dt$$

$$= \int_0^{2\pi} dt = 2\pi \neq 0.$$

Since this closed line integral does not vanish the vector field $\vec{F}$ is not conservative on Γ_2. The polar angle φ is no longer a single valued function on Γ_2 and can no longer serve as a potential for the conservative vector field. Indeed, considering the values of φ along the unit circle C we notice a jump by 2π angle as soon as we cross the positive x-axis, indicating that the polar angle φ is multi-valued on Γ_2.

Helmholtz theorem

Consider, the non-conservative vector field $\vec{G}(\vec{r})$ such that $\vec{\nabla} \times \vec{G} \neq 0$. It is no longer possible to represent the vector field solely as a gradient of a scalar potential. Instead, the following statement known as Helmholtz theorem applies.

> If a vector field $\vec{G}(\vec{r})$ goes to zero faster than $1/r$ as $r \to \infty$ then it can be expressed in an unique way as the sum of the gradient of a scalar potential and the curl of a vector potential:
> $$\vec{G}(\vec{r}) = -\vec{\nabla}\phi(\vec{r}) + \vec{\nabla} \times \vec{A}(\vec{r}).$$
> (1.205)

The behavior of $\vec{G}(\vec{r})$ at infinity makes it possible to use relation (1.200) to represent it in terms of an integral:

$$\vec{G}(\vec{r}) = -\frac{1}{4\pi}\vec{\nabla}^2 \iiint_V \frac{\vec{G}(\vec{r}')}{|\vec{r} - \vec{r}'|} d^3r'. \qquad (1.206)$$

Using the vector identity $\vec{\nabla}^2 \vec{V} = \vec{\nabla}\left(\vec{\nabla}\cdot\vec{V}\right) - \vec{\nabla}\times\left(\vec{\nabla}\times\vec{V}\right)$ (see equation (1.80)) we see that we can cast the right hand side of equation (1.206) in the form of Helmholtz theorem (1.205):

$$\vec{G}(\vec{r}) = \vec{\nabla}\left(-\frac{1}{4\pi}\vec{\nabla}\cdot\iiint_V \frac{\vec{G}(\vec{r}')}{|\vec{r}-\vec{r}'|}\,d^3r'\right) + \vec{\nabla}\times\left(\frac{1}{4\pi}\vec{\nabla}\times\iiint_V \frac{\vec{G}(\vec{r}')}{|\vec{r}-\vec{r}'|}\,d^3r'\right)$$

$$= -\vec{\nabla}\phi(\vec{r}) + \vec{\nabla}\times\vec{A}(\vec{r}).$$

$$(1.207)$$

We can easily obtain alternative integral expressions for the potentials ϕ and $\vec{A}$ defined in (1.207). Applying the gradient operator on both sides of Helmholtz theorem (1.205) gives a scalar potential that satisfies the following relation

$$\vec{\nabla}\cdot\vec{G}(\vec{r}) = -\vec{\nabla}^2\phi(\vec{r}).$$

Using relation (1.200) we arrive at an integral expression:

$$\phi(\vec{r}) = \frac{1}{4\pi}\iiint_V \frac{\vec{\nabla}\cdot\vec{G}(\vec{r}')}{|\vec{r}-\vec{r}'|}\,d^3r'.$$

$$(1.208)$$

Similarly, applying the curl operator on both sides of Helmholtz theorem (1.205) we obtain

$$\vec{\nabla}\times\vec{G}(\vec{r}) = -\vec{\nabla}^2\vec{A}(\vec{r}) + \vec{\nabla}\left(\vec{\nabla}\cdot\vec{A}\right).$$

Note, that we may, without loss of generality, choose $\vec{A}$ to satisfy $\vec{\nabla}\cdot\vec{A} = 0$. This follows from the fact that we can always redefine $\vec{A} \to \vec{A} + \vec{\nabla}f$ without affecting Helmholtz theorem (1.205). For such $\vec{A}$ it holds that $\vec{\nabla}\times\vec{G}(\vec{r}) = -\vec{\nabla}^2\vec{A}(\vec{r})$ and it follows from the same consideration as in relation (1.201) that:

$$\vec{A}(\vec{r}) = \frac{1}{4\pi}\iiint_V \frac{\vec{\nabla}\times\vec{G}(\vec{r}')}{|\vec{r}-\vec{r}'|}\,d^3r'.$$

$$(1.209)$$

Formally, expressions (1.208)-(1.209) remain unaffected by adding to the vector field $\vec{G}(\vec{r}) \to \vec{G}(\vec{r}) + \vec{G}_0(\vec{r})$ an additional vector field $\vec{G}_0(\vec{r})$ whose curl and divergence vanish. This raises a question about the uniqueness of Helmholtz theorem since the above comment implies that $\vec{G}(\vec{r}) + \vec{G}_0(\vec{r})$ would also be represented by the same Helmholtz expression (1.205) as the original vector field $\vec{G}(\vec{r})$. Recall that for Helmholtz theorem to hold $\vec{G}(\vec{r})$ must go to zero at infinity faster than $1/r$. However, there is no vector field that has zero curl and divergence and which satisfies this requirement. To see this, recall that the condition $\vec{\nabla}\times\vec{G}_0(\vec{r}) = 0$ implies that $\vec{G}_0(\vec{r}) = -\vec{\nabla}F(\vec{r})$ with some scalar potential $F(\vec{r})$ which also goes to zero at infinity because of the boundary condition on $\vec{G}_0(\vec{r})$. In addition, $F(\vec{r})$ has to satisfy the Laplace equation

$$\vec{\nabla}^2F(\vec{r}) = 0$$

due to $\vec{G}_0(\vec{r})$ being solenoidal ($\vec{\nabla}\cdot\vec{G}_0 = 0$). Now we will show that there is no scalar function $F(\vec{r})$ that simultaneously satisfies the Laplace equation and goes to zero at infinity. For that purpose we will use the *Green's first identity*. This identity can be obtained from the divergence theorem (1.163) by inserting $\vec{V} = f\vec{\nabla}g$ into equation (1.163), where f, g are two scalar functions. This yields Green's first formula:

$$\oiint_S f\vec{\nabla}g \cdot \hat{n}\,\mathrm{d}A = \iiint_V \vec{\nabla}\cdot\left(f\vec{\nabla}g(\vec{r})\right)\mathrm{d}^3\vec{r}$$

$$\qquad (1.210)$$

$$= \iiint_V \left[f\vec{\nabla}^2 g + \vec{\nabla}f\cdot\vec{\nabla}g\right]\mathrm{d}^3\vec{r}\,.$$

Setting $F = f = g$ in the above equation simplifies it to

$$\oiint_S F\vec{\nabla}F \cdot \hat{n}\,\mathrm{d}A = \iiint_V \left[F\vec{\nabla}^2 F + \vec{\nabla}F\cdot\vec{\nabla}F\right]\mathrm{d}^3\vec{r}\,.$$

For the surface boundary S large enough that $F = 0$ on S, the left hand side becomes zero. Also, the first term on the right hand side vanishes, since $\vec{\nabla}^2 F = 0$. This leaves

$$\iiint_V \vec{\nabla}F \cdot \vec{\nabla}F\,\mathrm{d}^3\vec{r} = 0\,.$$

We recognize, in the integrand $\vec{\nabla}F \cdot \vec{\nabla}F$, the norm $|\vec{\nabla}F|^2$ of the gradient of F. The norm of a vector has to be positive (or zero) everywhere. Therefore, for the above integral to be zero it must hold that $\vec{\nabla}F = 0$ (or F constant) everywhere. Consequently, $\vec{G}_0(\vec{r})$ must be zero everywhere. Hence, the assumption about the behavior of $\vec{G}(\vec{r})$ at infinity ensures the uniqueness of its Helmholtz representation in terms of scalar and vector potentials.

For completeness, we will also introduce *Green's second identity*. It can easily be obtained from the Green's first identity. First, we exchange f and g in relation (1.210). This gives us a new identity. Subtracting the identities from each other yields:

$$\oiint_S \left(f\vec{\nabla}g - g\vec{\nabla}f\right)\cdot\hat{n}\,\mathrm{d}A = \iiint_V \left(f\vec{\nabla}^2 g - g\vec{\nabla}^2 f\right)\mathrm{d}^3\vec{r}\,. \qquad (1.211)$$

Both formulas are useful in proving uniqueness as we have seen above.

◿ Integral Theorems of Vector Calculus: Problems

Problem 1.4.1. Compute $\int_C \vec{F}(\vec{r})\cdot\mathrm{d}\vec{r} = \int_a^b \vec{F}(\vec{r}(t))\cdot(\mathrm{d}\vec{r}/\mathrm{d}t)\,\mathrm{d}t$ for $\vec{F} = 3yx\hat{i} + x^2\hat{j}$ and

(a) curve C being a straight-line segment from $(0,0)$ to $(1,3)$.

(b) curve C being a broken line segment from $(0,0)$ to $(0,3)$ to $(1,3)$.

(c) curve C being a parabola $C : y = 3x^2$, $0 \le x \le 1$.

Repeat calculations in (a),(b) and (c) for $\vec{F} = 2yx\hat{i} + x^2\hat{j}$.

Problem 1.4.2. Evaluate the integral $\int_{(2,1,0)}^{(0,0,0)} \left(x\,dx - ze^{-2yz}\,dy - ye^{-2yz}\,dz \right)$. *Hint:* Write the integral as $\int df$.

Problem 1.4.3. Let $\vec{F} = (z\cos(xz) - x)\hat{i} + y\hat{j} + (x\cos(xz))\hat{k}$ be a vector field in space. Calculate the work done by a particle which follows the trajectory $z = -(x+y)\pi$ under the influence of force $\vec{F}$ from $(0,0,0)$ to $(-1,1/2,\pi/2)$.

Problem 1.4.4. Evaluate the line integral $\int_C \left(3x^2e^{2y}\,dx + 2x^3e^{2y}\,dy + 2z\,dz \right)$, where C is the broken line curve from $(0,0,0)$ to $(1,0,0)$ to $(1,1,1)$.

Problem 1.4.5. As in Example 1.4.13, we consider the vector field:

$$\vec{F} = -\frac{y}{x^2+y^2}\hat{i} + \frac{x}{x^2+y^2}\hat{j}\,.$$

Evaluate a line integral of $\vec{F}$ from $(1,0)$ to $(0,-1)$ along the unit circle.

Problem 1.4.6. Evaluate a line integral $\int_C \vec{F}(\vec{r}) \cdot d\vec{r}$ for the vector field $\vec{F} = yz\hat{i} + xz\hat{j} + xy\hat{k}$ and the curve parametrized by $\vec{r} = t\hat{i} + t^2\hat{j} + t^3\hat{k}$ for $0 \le t \le 1$.

Problem 1.4.7. The three-dimensional spiral is described by

$$\vec{r}(t) = \left(\hat{i}t\cos(2\pi t) + \hat{j}t\sin(2\pi t) + \hat{k}t \right), \quad t \ge 0\,.$$

Here the z-component is given by the parameter t. As t increases the curve encircles larger and larger circles in the xy-plane as seen in Figure 1.53. Evaluate the line

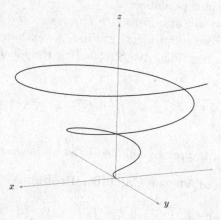

Figure 1.53: Three-dimensional spiral.

integral $\int_C \vec{F}(\vec{r}) \cdot d\vec{r}$ along the spiral over three revolutions ($0 \leq t \leq 3$) for the vector field

$$\vec{F} = \hat{i}zy - \hat{j}zx + \hat{k}z.$$

Problem 1.4.8. Find a parametric representation of the elliptic cylinder: $x^2 + 16z^2 = 16$. Find its normal vector.

Problem 1.4.9. Use polar coordinates (cylindrical coordinates on the plane) to evaluate $\iint_R (x - y) \, dx \, dy$, where the region R is defined by $x^2 + y^2 \leq 4, x \leq 0$.

Problem 1.4.10. Evaluate $\iint_S \vec{F} \cdot \hat{n} \, dA$, where $\vec{F} = e^{xz}\hat{i} + y^2\hat{k}$ and $S : y^2 + z^2 = 1, -3 \leq x \leq 3, z \leq 0$.

Problem 1.4.11. The velocity of a fluid flow is given by the vector field $\vec{v} = y\hat{i} - x\hat{j} + 2(1 - x^2 - y^2)\hat{k}$.
(a) Find the volume of the fluid that flows through an arbitrary closed surface per unit time.
(b) Find the volume of the fluid that flows through a disk $x^2 + y^2 \leq 1$ in the negative z-axis direction per unit time.
(c) Find the volume of the fluid that flows through the upper hemisphere $z \geq 0$, $x^2 + y^2 + z^2 = 1$ per unit time.

Problem 1.4.12. Let S be a surface described by the parametrization

$$\vec{r} = (u^2 + v^2)\hat{i} + (u^2 - v^2)\hat{j} + 2uv\hat{k},$$

where $0 \leq u \leq 1, 0 \leq v \leq 1$. Evaluate $\iint_S \vec{F} \cdot \hat{n} \, dA$ for the vector field $\vec{F} = y\hat{i} + x\hat{j} + z^2\hat{k}$.

Problem 1.4.13. Evaluate $\iint_S \vec{F} \cdot \hat{n} \, dA$, using the divergence theorem for

(a) $\vec{F} = -e^{x^2}\hat{i} + e^{y^2}\hat{j} - e^{z^2}\hat{k}$, S a surface of a box: $0 \leq x \leq 2, 0 \leq y \leq 1, 0 \leq z \leq 1$.
(b) $\vec{F} = x^2\hat{i} + y^3z\hat{j} + \hat{k}$, S the surface of $x^2 + y^2 \leq 4, -3 \leq z \leq 3$
(c) $\vec{F} = x^2\hat{i} + y^3z\hat{j} + \hat{k}$, S the surface of $z^2 + y^2 \leq x^2, 0 \leq x \leq 2$
(d) $\vec{F} = \hat{i}(xz) - \hat{j}(z^2 \exp(z^4 - 3x^6) + y) + \hat{k}(\exp(2x^2) \sin(y^3) + z)$, for S being the unit sphere $x^2 + y^2 + z^2 = 1$.
(e) $\vec{F} = xy^2\hat{i} + x^2y\hat{j} + z^3/3\hat{k}$, S the surface of $x^2 + y^2 + z^2 \leq 4$.
(f) $\vec{F} = 3x\hat{i} + y^3\hat{j} + 9\hat{k}$, S the surface of the cube $-2 \leq x, y, z \leq 2$.
(g) $\vec{F} = x^3\hat{i} + y^3\hat{j} + 9\hat{k}$, S the surface of the vertical cylinder along the z-axis with $-1 \leq z \leq 1$ and radius 2.

Problem 1.4.14. Find flux $\iint_S \vec{F} \cdot \hat{n} \, dA$ through the upper hemisphere $z \geq 0$, $x^2 + y^2 + z^2 = 1$ for $\vec{F} = x^3\hat{i} + y^3\hat{j} + z^y\hat{k}$.

Problem 1.4.15. Verify Stokes's theorem for
(a) $\vec{F} = x^2 y \hat{i} - y^2 x \hat{j}$, and surface S of the rectangle $0 \le x \le 2$, $0 \le y \le 3$.
(b) $\vec{F} = y \hat{i} + x^2 y \hat{j}$ and surface S of the circular disk $x^2 + y^2 \le 1$, $z = 0$.

Problem 1.4.16. Use Stokes's theorem to evaluate the line integral $\int_C \vec{F}(\vec{r}) \cdot d\vec{r}$, counterclockwise around the boundary C of the region R, where $\vec{F} = y^3 \hat{i} + (x^3 + 3xy^2)\hat{j}$, R: $x \le y \le x^2$, $0 \le x \le 1$

Problem 1.4.17. Use Stokes's theorem to evaluate the line integral $\oint_C \vec{V}(\vec{r}) \cdot d\vec{r}$, where C is the triangle in Figure 1.54 and $\vec{V}(\vec{r}) = 2y \hat{i} + x \hat{j}$. Compare with the result obtained in Example 1.4.2.

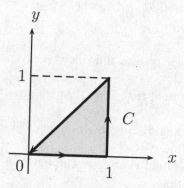

Figure 1.54: A counter-clockwise triangle.

Problem 1.4.18. Use Stokes's theorem to evaluate the line integral $\oint_C \vec{F}(\vec{r}) \cdot d\vec{r}$ for $\vec{F} = x^2 \hat{i} + y^2 \hat{j} + \hat{k}$, along the triangular path from $(1,0,0)$ to $(0,1,0)$ to $(0,0,1)$ and back to $(1,0,0)$.

Problem 1.4.19. Use Stokes's theorem to evaluate the line integral $\oint_C \vec{F}(\vec{r}) \cdot d\vec{r}$, for $\vec{F} = y^2 z \hat{i} + xz^2 \hat{j} - zx^2 \hat{k}$ clockwise around C
(a) being a circle $x^2 + y^2 = R^2$, $z = h > 0$,
(b) being a circle $z^2 + y^2 = R^2$, $x = h > 0$.

Problem 1.4.20. Use Stokes's theorem to evaluate the line integral $\oint_C \vec{F}(\vec{r}) \cdot d\vec{r}$, counterclockwise around C being an intersection of the unit sphere $x^2 + y^2 + z^2 = 1$ and the plane $x + y + z = 0$ for $\vec{F}$, where $\vec{F} = (y - z)\hat{i} + (z - x)\hat{j} + (x - y)\hat{k}$.

Problem 1.4.21. Use Stokes's theorem to evaluate the line integral $\oint_C \vec{F}(\vec{r}) \cdot d\vec{r}$, around C being an intersection of the cylinder $x^2 + y^2 = 1$ and the parabolic sheet $z = y^2$ for $\vec{F} = 2yz \hat{i} + zx \hat{j} + xy \hat{k}$.

Problem 1.4.22. (a) Use Stokes's theorem to evaluate the line integral $\oint_C (\vec{\omega} \times \vec{r}) \cdot d\vec{r}$, around C being a unit circle $x^2 + y^2 = 1$ for an arbitrary constant vector $\vec{\omega}$.

(b) Evaluate the surface integral $\iint_S (\vec{\omega} \times \vec{r}) \cdot \hat{n} \, dA$, for an arbitrary constant vector $\vec{\omega}$ and surface S of the unit disk $x^2 + y^2 \leq 1$. Apply the divergence theorem to the closed surface consisting of the upper half-sphere S' together with the unit disk S.

Problem 1.4.23. Use Green's theorem (1.186) in the plane to evaluate the line integral

$$\oint_C x^2 \, dx + (x^3 + 4y) \, dy,$$

where C is the circle $x^2 + y^2 = 4$.

Problem 1.4.24. Use Green's theorem (1.186) in the plane to evaluate the line integral $\oint_C \vec{F}(\vec{r}) \cdot d\vec{r}$, around the closed path C shown in Figure 1.55 for $\vec{F} = (x^2 - y^2)\hat{i} + 2xy\hat{j}$.

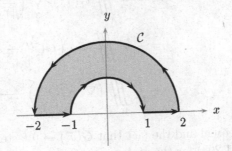

Figure 1.55: Path C used in Problem 1.4.24.

Problem 1.4.25. Evaluate

$$\oint_C (e^x + 2y^2) \, dx + (2x - 2xy) \, dy,$$

where C is the boundary of the half-disk $x^2 + y^2 \leq 4$, $y \geq 0$ oriented counterclockwise.

Problem 1.4.26. Find potentials $f(\rho, \varphi, z)$ corresponding to the vector fields $\vec{V}(\rho, \varphi, z)$:

(a) $\vec{V} = (z + 2\rho)\hat{\rho} + \rho\hat{k}$

(b) $\vec{V} = \hat{\rho}\tan\varphi + \dfrac{\hat{\varphi}}{\cos^2\varphi}$

Problem 1.4.27. Apply the infinitesimal versions of Stokes' and divergence theorems for surface S, its boundary C and volume V:

$$\vec{n} \cdot \left(\vec{\nabla} \times \vec{V} \right) = \lim_{|S| \to 0} \frac{1}{|S|} \oint_C \vec{V} \cdot d\vec{r}, \tag{1.212}$$

$$\vec{\nabla} \cdot \vec{V} = \lim_{|V| \to 0} \frac{1}{|V|} \oiint_S \vec{V} \cdot \hat{n} \, dA \tag{1.213}$$

to obtain in an alternative way expressions (1.126) and (1.130) for the curl and divergence of the vector field $\vec{V}$ in an orthogonal curvilinear coordinate system.

Problem 1.4.28. Show that one can rewrite ϕ from eq. (1.207) as follows:

$$\frac{1}{4\pi} \vec{\nabla} \cdot \iiint_V \frac{\vec{G}(\vec{r}\,')}{|\vec{r} - \vec{r}\,'|} \, d^3r' = \frac{1}{4\pi} \iiint_V \vec{G}(\vec{r}\,') \cdot \vec{\nabla} \left(\frac{1}{|\vec{r} - \vec{r}\,'|} \right) d^3r'$$

$$= -\frac{1}{4\pi} \iiint_V \vec{G}(\vec{r}\,') \cdot \vec{\nabla}' \left(\frac{1}{|\vec{r} - \vec{r}\,'|} \right) d^3r'$$

$$= -\frac{1}{4\pi} \iiint_V \vec{\nabla}' \cdot \left(\vec{G}(\vec{r}\,') \frac{1}{|\vec{r} - \vec{r}\,'|} \right) d^3r'$$

$$+ \frac{1}{4\pi} \iiint_V \left(\vec{\nabla}' \cdot \vec{G}(\vec{r}\,') \right) \frac{1}{|\vec{r} - \vec{r}\,'|} \, d^3r'.$$

Use the divergence theorem and the fact that $\vec{G}(\vec{r}) \to 0$ faster than $1/r$ as $r \to \infty$ to prove the formula (1.208) for the scalar potential ϕ from eq. (1.207).

Problem 1.4.29. Show that $\vec{\nabla} \cdot \vec{A}(\vec{r}) = 0$ if $\vec{A}(\vec{r})$ is given by

$$\vec{A}(\vec{r}) = \frac{1}{4\pi} \iiint_V \frac{\vec{\omega} \times \vec{r}\,'}{|\vec{r} - \vec{r}\,'|} \, d^3r',$$

where $\vec{\omega}$ is a constant vector and the integral is over all space.

Chapter 2

Matrices and Rotations

MATRICES are blocks of numbers used to represent transformations between vector spaces that preserve their linear structure. In this chapter, we will learn how matrices multiply vectors and other matrices and we will study their properties. Solving the eigenvalue problem for a matrix is very important and motivates much of the exposition given here. The eigenvalue problem is related to the fundamental problem of matrix diagonalization. Eigenvectors facilitate the choice of the basis in which the matrix is diagonal while the eigenvalues carry a basis independent information about the matrix. The key role is played in many applications by special types of matrices like orthogonal and unitary matrices and symmetric and hermitian matrices and will study their properties in detail.

Important examples of linear transformations involve rotations in three dimensional space. The study of rotations and rotation matrices will lead us naturally to a systematic study of symmetries within the group theoretical framework.

2.1 Matrices: Basic Facts

A matrix is an ordered set of numbers listed in a rectangular form:

$$\underline{\underline{A}} = \begin{bmatrix} a_{11} & a_{12} & \cdots & a_{1n} \\ a_{21} & a_{22} & \cdots & a_{2n} \\ \vdots & \vdots & & \vdots \\ a_{m1} & a_{m2} & \cdots & a_{mn} \end{bmatrix} = [a_{ij}]_{1 \le i \le m, 1 \le j \le n} . \tag{2.1}$$

The matrix $\underline{\underline{A}}$ has m rows and n columns. We say it is a $m \times n$ matrix. We denote the element on the i-th row and j-th column with a_{ij}.

If a matrix $\underline{\underline{A}}$ has equal number of rows and columns $(n = m)$ we say it is a square matrix or $n \times n$ matrix. In this case, the number of rows or columns n is called an order or degree of $\underline{A}$. In a square matrix the elements a_{ij}, with $i = j$, are called diagonal elements. A square matrix containing only non-zero diagonal elements is called a diagonal matrix.

A $n \times 1$ matrix is a matrix with one column consisting of n entries:

$$\underline{v} = \begin{bmatrix} v_1 \\ v_2 \\ \vdots \\ v_n \end{bmatrix}. \tag{2.2}$$

A $n \times 1$ matrix is called a column vector or n-column vector. Similarly, a $1 \times n$ matrix is a matrix with only one row consisting of n entries:

$$\begin{bmatrix} v_1 & v_2 & \cdots & v_{1n} \end{bmatrix}. \tag{2.3}$$

A $1 \times n$ matrix is called a row vector or n-row vector.

A square matrix $\underline{A}$ is upper (lower) triangular provided that $a_{ij} = 0$ for all $i > j$ (all $j > i$). $\underline{A}$ is strictly upper (lower) triangular provided that $a_{ij} = 0$ for all $i \geq j$ (all $j \geq i$).

Zero matrix $\underline{0}$ is such that its matrix elements are $a_{ij} = 0$ for each i and j.

The following is a list of fundamental properties of matrices and operations which can be performed on matrices:

① Matrix equality $\underline{A} = \underline{B}$ if $a_{ij} = b_{ij}$ for each i and j.

② Matrix addition and subtraction $\underline{A} + \underline{B} = [a_{ij} + b_{ij}]$ is:

- Commutative, meaning $\underline{A} + \underline{B} = \underline{B} + \underline{A}$.
- Associative, meaning $\underline{A} + (\underline{B} + \underline{C}) = (\underline{A} + \underline{B}) + \underline{C}$.

③ Multiplication by a number follows from $\lambda \underline{A} = [\lambda a_{ij}]$.

④ Matrix multiplication $\underline{A}\,\underline{B} = \underline{C}$, is such that $c_{jk} = \sum_{i=1}^{m} a_{ji} b_{ik}$ are matrix elements of $\underline{C}$. If $\underline{A}$ is an $n \times m$ matrix and $\underline{B}$ is an $m \times p$ matrix then $\underline{C}$ is an $n \times p$ matrix obtained as follows:

$$\mathbf{j}\begin{bmatrix} a_{11} & a_{12} & \cdots & a_{1m} \\ a_{21} & a_{22} & \cdots & a_{2m} \\ \cdot & \cdot & \cdots & \cdot \\ \boxed{a_{j1}\ \ a_{j2}\ \ \cdots\ \ a_{jm}} \\ \cdot & \cdot & \cdots & \cdot \\ \cdot & \cdot & \cdots & \cdot \\ \cdot & \cdot & \cdots & \cdot \\ a_{n1} & a_{n2} & \cdots & a_{nm} \end{bmatrix}$$

$$\mathbf{n \times m}$$

$$\begin{bmatrix} b_{11} & b_{12} & \cdots & \overset{\mathbf{k}}{b_{1k}} & \cdots & b_{1p} \\ b_{21} & b_{22} & \cdots & b_{2k} & \cdots & b_{2p} \\ \cdot & \cdot & \cdots & \cdot & \cdots & \cdot \\ \cdot & \cdot & \cdots & \cdot & \cdots & \cdot \\ \cdot & \cdot & \cdots & \cdot & \cdots & \cdot \\ b_{m1} & b_{m2} & \cdots & b_{mk} & \cdots & b_{mp} \end{bmatrix}$$

$$\mathbf{m \times p}$$

$$
= \; \mathbf{j} \begin{bmatrix} c_{11} & c_{12} & \cdots & c_{1k} & \cdots & c_{1p} \\ c_{21} & c_{22} & \cdots & c_{2k} & \cdots & c_{2p} \\ \cdot & \cdot & \cdots & \cdot & \cdots & \cdot \\ c_{j1} & c_{j2} & \cdots & \boxed{c_{jk}} & \cdots & c_{jp} \\ \cdot & \cdot & \cdots & \cdot & \cdots & \cdot \\ \cdot & \cdot & \cdots & \cdot & \cdots & \cdot \\ \cdot & \cdot & \cdots & \cdot & \cdots & \cdot \\ c_{n1} & c_{n2} & \cdots & c_{nk} & \cdots & c_{np} \end{bmatrix}
$$

$$\mathbf{k}$$ (over the c_{1k} column)

$$\mathbf{n \times p}$$

Matrix multiplication is

- Distributive: $\underline{A}\,(\underline{B}+\underline{C}) = \underline{A}\,\underline{B}+\underline{A}\,\underline{C}$.

- Associative: $\underline{A}\,(\underline{B}\,\underline{C}) = (\underline{A}\,\underline{B})\,\underline{C}$.

- not commutative, meaning that, in general, $\underline{A}\,\underline{B} \neq \underline{B}\,\underline{A}$.

To a matrix $\underline{A}$ one can associate several other related matrices as described below.

① **Transpose matrix:** If $\underline{A}$ is an $n \times m$ matrix, its transpose, denoted $\underline{A}^T$, is the $m \times n$ matrix:

$$\underline{A}^T = [a_{ij}]^T = [a_{ji}] \;\rightarrow\; \text{if } \underline{A} = \begin{bmatrix} 1 & 2 \\ 3 & 7 \end{bmatrix} \text{ then } \underline{A}^T = \begin{bmatrix} 1 & 3 \\ 2 & 7 \end{bmatrix}. \tag{2.4}$$

In words, transposition interchanges the rows and columns of $\underline{A}$ to columns and rows of the transpose matrix $\underline{A}^T$. Obviously, $\left(\underline{A}^T\right)^T = \underline{A}$. Under transposition, the product of matrices behaves as follows

$$\left(\underline{A}\,\underline{B}\right)^T = \underline{B}^T\,\underline{A}^T.$$

The transpose of a n-column vector

$$\underline{v} = \begin{bmatrix} v_1 \\ v_2 \\ \vdots \\ v_{1n} \end{bmatrix}$$

is a n-row vector

$$\underline{v}^T = \begin{bmatrix} v_1 & v_2 & \cdots & v_{1n} \end{bmatrix}.$$

② **Complex conjugate matrix:** If $\underline{\underline{A}}$ is an $n \times m$ matrix, its complex conjugate matrix $\underline{\underline{A}}^*$ is given by:

$$\underline{\underline{A}}^* = [a_{ij}^*] \;\rightarrow\; \text{if } \underline{\underline{A}} = \begin{bmatrix} 2i & 1+3i \\ 4 & 8 \end{bmatrix} \text{ then } \underline{\underline{A}}^* = \begin{bmatrix} -2i & 1-3i \\ 4 & 8 \end{bmatrix}. \qquad (2.5)$$

Matrix elements of $\underline{\underline{A}}^*$ are obtained by the complex conjugation of matrix elements of $\underline{\underline{A}}$.

③ **Adjoint matrix:** The adjoint (or Hermitian conjugate) matrix $\underline{\underline{A}}^\dagger$ is obtained from $\underline{\underline{A}}$ by the combined operations of complex conjugation and transposition:

$$\underline{\underline{A}}^\dagger = [a_{ij}]^\dagger = [a_{ji}^*] \;\rightarrow\; \text{if } \underline{\underline{A}} = \begin{bmatrix} 2i & 1+3i \\ 4 & 8 \end{bmatrix} \text{ then } \underline{\underline{A}}^\dagger = \begin{bmatrix} -2i & 4 \\ 1-3i & 8 \end{bmatrix}. \qquad (2.6)$$

④ **Inverse matrix:** If $\underline{\underline{A}}$ is square matrix, and there is a matrix $\underline{\underline{B}}$ such that $\underline{\underline{A}}\,\underline{\underline{B}} = \underline{\underline{I}}$, where $\underline{\underline{I}}$ is the identity matrix

$$\underline{\underline{I}} = \begin{bmatrix} 1 & & & & \\ & 1 & & \mathbf{0} & \\ & & \ddots & & \\ & \mathbf{0} & & 1 & \\ & & & & 1 \end{bmatrix}, \qquad (2.7)$$

having matrix elements $a_{ij} = \delta_{ij}$, then we say that $\underline{\underline{A}}$ is invertible or nonsingular. We call $\underline{\underline{B}}$ the inverse of $\underline{\underline{A}}$, and denote it by $\underline{\underline{A}}^{-1}$. For 2×2 matrices it holds that

$$\text{if } \underline{\underline{A}} = \begin{bmatrix} a & b \\ c & d \end{bmatrix} \text{ then } \underline{\underline{A}}^{-1} = \frac{1}{ad-bc} \begin{bmatrix} d & -b \\ -c & a \end{bmatrix}, \text{ provided } ad-bc \neq 0. \quad (2.8)$$

For example:

$$\text{if } \underline{\underline{A}} = \begin{bmatrix} 1 & 2 \\ 2 & 3 \end{bmatrix} \text{ then } \underline{\underline{A}}^{-1} = \begin{bmatrix} -3 & 2 \\ 2 & -1 \end{bmatrix}. \qquad (2.9)$$

If $\underline{\underline{A}}\,\underline{\underline{B}} = I$ and $\underline{\underline{C}}\,\underline{\underline{A}} = I$ then $\underline{\underline{B}} = \underline{\underline{C}} = \underline{\underline{A}}^{-1}$. This relation follows from the distributive property of matrix multiplication:

$$\underline{\underline{B}} = \underline{\underline{I}}\,\underline{\underline{B}} = \left(\underline{\underline{C}}\,\underline{\underline{A}}\right)\underline{\underline{B}} = \underline{\underline{C}}\left(\underline{\underline{A}}\,\underline{\underline{B}}\right) = \underline{\underline{C}}\,\underline{\underline{I}} = \underline{\underline{C}}. \qquad (2.10)$$

If a matrix is not invertible, we say it is singular or non-invertible. Not all square matrices are invertible, i.e., have inverses. For instance the matrix $\underline{\underline{A}}$ from (2.8) is singular for $ad-bc = 0$.

For the $n \times n$ matrix $\underline{\underline{A}}$ the sum of its diagonal elements is called a trace of matrix $\underline{\underline{A}}$ and is denoted by

$$\operatorname{Tr}\left(\underline{\underline{A}}\right) = \sum_{i=1}^{n} a_{ii}. \tag{2.11}$$

Some of fundamental properties of trace are:

$$\operatorname{Tr}\left(\underline{\underline{A}}\,\underline{\underline{B}}\right) = \operatorname{Tr}\left(\underline{\underline{B}}\,\underline{\underline{A}}\right), \quad \operatorname{Tr}\left(\underline{\underline{A}} + \underline{\underline{B}}\right) = \operatorname{Tr}\left(\underline{\underline{A}}\right) + \operatorname{Tr}\left(\underline{\underline{B}}\right),$$
$$\operatorname{Tr}\left(\underline{\underline{A}}^{T}\right) = \operatorname{Tr}\left(\underline{\underline{A}}\right), \qquad \operatorname{Tr}\left(c\underline{\underline{A}}\right) = c\operatorname{Tr}\left(\underline{\underline{A}}\right).$$

2.1.1 The Vector Space of the Column Vectors. The Inner Product.

A special case of matrix multiplication is $\underline{\underline{A}}\,\underline{v} = \underline{w}$, where $\underline{\underline{A}}$ is an $n \times m$ matrix, $\underline{v}$ is an m-column vector, and $\underline{w}$ is an n-column vector:

$$
\begin{bmatrix}
a_{11} & a_{12} & \cdots & a_{1m} \\
a_{21} & a_{22} & \cdots & a_{2m} \\
\cdot & \cdot & \cdots & \cdot \\
\boxed{a_{j1}} & a_{j2} & \cdots & a_{jm} \\
\cdot & \cdot & \cdots & \cdot \\
\cdot & \cdot & \cdots & \cdot \\
a_{n1} & a_{n2} & \cdots & a_{nm}
\end{bmatrix}
\begin{bmatrix}
v_1 \\
v_2 \\
\vdots \\
\vdots \\
v_m
\end{bmatrix}
=
\begin{bmatrix}
w_1 \\
w_2 \\
\cdot \\
\boxed{w_j} \\
\cdot \\
\cdot \\
w_n
\end{bmatrix}.
$$

The formula for the components of the n-vector is $w_j = a_{j1}v_1 + \cdots + a_{jm}v_m$, $j = 1, \ldots . n$.

We can think of matrix vector multiplication (with an $n \times m$ matrix) as a linear operation, i.e. an operation that satisfies

$$\underline{\underline{A}}c\underline{v} = c\underline{\underline{A}}\,\underline{v}, \quad \underline{\underline{A}}\left(\underline{v}_1 + \underline{v}_2\right) = \underline{\underline{A}}\,\underline{v}_1 + \underline{\underline{A}}\,\underline{v}_2 \tag{2.12}$$

while transforming m-vectors into n-vectors.

For row and column vectors the matrix multiplication

$$
\begin{bmatrix} v_1 & v_2 & \cdots & v_{1n} \end{bmatrix}
\begin{bmatrix}
w_1 \\
w_2 \\
\vdots \\
w_n
\end{bmatrix}
= \sum_{i=1}^{n} v_i w_i \tag{2.13}
$$

generalizes the scalar product of three dimensional vectors defined for vectors on the $\mathbb{R}^3$ vector space to n-dimensional vectors on the $\mathbb{R}^n$ vector space. Accordingly, we can associate two n-column vectors $\underline{v}$ and $\underline{w}$ to a generalized scalar product, referred to as an inner product $\langle \underline{v} | \underline{w} \rangle$, by transposing the first column vector $\underline{v}$ to a row $\underline{v}^T = [v_1, v_2, \ldots, v_n]$ and multiplying it with $\underline{w}$ according to the matrix multiplication as in (2.13):

$$\langle \underline{v} | \underline{w} \rangle = \underline{v}^T \underline{w} = \sum_{i=1}^{n} v_i w_i . \tag{2.14}$$

The corresponding norm of the n-dimensional vector is given by

$$\|\underline{v}\| = \sqrt{\langle \underline{v} | \underline{v} \rangle}. \tag{2.15}$$

Thus, the matrix element $c_{jk} = \sum_{i=1}^{m} a_{ji} b_{ik}$ of $\underline{C} = \underline{A}\,\underline{B}$ appears to be a scalar product

$$c_{jk} = \langle \underline{a}_j | \underline{b}_k \rangle = \begin{bmatrix} a_{j1} & \cdots & a_{jm} \end{bmatrix} \begin{bmatrix} b_{1k} \\ \vdots \\ b_{mk} \end{bmatrix} = a_{j1} b_{1k} + a_{j2} b_{2k} + \cdots + a_{jm} b_{mk}$$

of vectors

$$\underline{a}_j = \begin{bmatrix} a_{j1} \\ \vdots \\ a_{jm} \end{bmatrix} \quad \text{and} \quad \underline{b}_k = \begin{bmatrix} b_{1k} \\ \vdots \\ b_{mk} \end{bmatrix}.$$

The canonical basis on $\mathbb{R}^n$ is given by the column vectors

$$\underline{e}_1 = \begin{bmatrix} 1 \\ 0 \\ \vdots \\ 0 \\ 0 \end{bmatrix}, \quad \underline{e}_2 = \begin{bmatrix} 0 \\ 1 \\ 0 \\ \vdots \\ 0 \end{bmatrix}, \quad \ldots, \quad \underline{e}_n = \begin{bmatrix} 0 \\ 0 \\ \vdots \\ 0 \\ 1 \end{bmatrix}, \tag{2.16}$$

and, therefore, the basis element $\underline{e}_i$ is a column vector whose components are 1 on the i-th row and zero everywhere else. Clearly

$$\langle \underline{e}_i | \underline{e}_j \rangle = \delta_{ij}$$

and, hence, the canonical basis of column vectors $\underline{e}_i$, $i = 1, \ldots, n$ form an orthonormal system, i.e. system of orthogonal vectors of unit norm.

Suppose that we have a basis $\{\underline{v}_1, \ldots, \underline{v}_n\}$ for $\mathbb{R}^n$. We can always find an orthonormal basis for $\mathbb{R}^n$ in terms of $\{\underline{v}_1, \ldots, \underline{v}_n\}$ using a generalization of the Gram-Schmidt process introduced in Chapter 1 for $\mathbb{R}^3$. Since our goal is an orthonormal basis we start by setting $\underline{\omega}_1 = \underline{v}_1$ and defining:

$$\hat{\underline{e}}_1 = \frac{\underline{\omega}_1}{\|\underline{\omega}_1\|} \tag{2.17}$$

to ensure that $\langle \hat{\underline{e}}_1 | \hat{\underline{e}}_1 \rangle = 1$. Then, we replace $\underline{v}_2$ by a vector

$$\underline{\omega}_2 = \underline{v}_2 - \langle \hat{\underline{e}}_1 | \underline{v}_2 \rangle \hat{\underline{e}}_1 , \tag{2.18}$$

orthogonal to $\hat{\underline{e}}_1$. Now, normalize it as in (2.17):

$$\hat{\underline{e}}_2 = \frac{\omega_2}{\|\underline{\omega}_2\|}. \tag{2.19}$$

We are ready to generalize this process for higher i in a successive way:

$$
\begin{aligned}
\underline{\omega}_3 &= \underline{v}_3 - \left[\langle \hat{\underline{e}}_1 | \underline{v}_3 \rangle \, \hat{\underline{e}}_1 + \langle \hat{\underline{e}}_2 | \underline{v}_3 \rangle \, \hat{\underline{e}}_2 \right], \quad \rightarrow \quad \hat{\underline{e}}_3 = \frac{\omega_3}{\|\underline{\omega}_3\|} \\
\underline{\omega}_4 &= \underline{v}_4 - \left[\langle \hat{\underline{e}}_1 | \underline{v}_4 \rangle \, \hat{\underline{e}}_1 + \langle \hat{\underline{e}}_2 | \underline{v}_4 \rangle \, \hat{\underline{e}}_2 + \langle \hat{\underline{e}}_3 | \underline{v}_4 \rangle \, \hat{\underline{e}}_3 \right], \quad \rightarrow \quad \hat{\underline{e}}_4 = \frac{\omega_4}{\|\underline{\omega}_4\|} \\
\dots &= \dots \\
\underline{\omega}_i &= \underline{v}_i - \left[\langle \hat{\underline{e}}_1 | \underline{v}_i \rangle \, \hat{\underline{e}}_1 + \langle \hat{\underline{e}}_2 | \underline{v}_i \rangle \, \hat{\underline{e}}_2 + \dots + \langle \hat{\underline{e}}_{i-1} | \underline{v}_i \rangle \, \hat{\underline{e}}_{i-1} \right] \\
&= \underline{v}_i - \sum_{k=1}^{i-1} \langle \hat{\underline{e}}_k | \underline{v}_i \rangle \, \hat{\underline{e}}_k, \quad \rightarrow \quad \hat{\underline{e}}_i = \frac{\omega_i}{\|\underline{\omega}_i\|}
\end{aligned}
\tag{2.20}
$$

and so on. The sum $\sum_{k=1}^{i-1} \langle \hat{\underline{e}}_k | \underline{v}_i \rangle \, \hat{\underline{e}}_k$ defines a projection of $\underline{v}_i$ on a vector space spanned by $\hat{\underline{e}}_1, \dots \hat{\underline{e}}_{i-1}$ and $\underline{\omega}_i$ is obtained from $\underline{v}_i$ by subtracting this projection. From the relation, $\langle \underline{e}_k | \underline{e}_l \rangle = \delta_{kl}$ for $k, l = 1, \dots, i-1$, it follows that for any $1 \leq l \leq i-1$

$$
\begin{aligned}
\langle \hat{\underline{e}}_l | \underline{\omega}_i \rangle &= \langle \hat{\underline{e}}_l | \underline{v}_i \rangle - \sum_{k=1}^{i-1} \langle \hat{\underline{e}}_k | \underline{v}_i \rangle \langle \hat{\underline{e}}_l | \hat{\underline{e}}_k \rangle \\
&= \langle \hat{\underline{e}}_l | \underline{v}_i \rangle - \sum_{k=1}^{i-1} \langle \hat{\underline{e}}_k | \underline{v}_i \rangle \, \delta_{kl} = \langle \hat{\underline{e}}_l | \underline{v}_i \rangle - \langle \hat{\underline{e}}_l | \underline{v}_i \rangle = 0.
\end{aligned}
$$

Thus, the Gram-Schmidt algorithm yields a vector $\underline{\omega}_i$ orthogonal to vectors $\hat{\underline{e}}_1, \dots, \hat{\underline{e}}_{i-1}$. Normalizing $\underline{\omega}_i$ gives a set of i orthonormal vectors $\hat{\underline{e}}_1, \dots, \hat{\underline{e}}_i$. Continuing this process leads to an orthonormal basis $\{\hat{\underline{e}}_1, \dots, \hat{\underline{e}}_n\}$ for $\mathbb{R}^n$.

Since we allow the matrix elements to be complex as in (2.5) we need to introduce an inner product on the complex vector space. A complex vector space satisfies the same postulates as the real vector space, except that the scalars are composed of complex numbers as opposed to real numbers. Hence, linear combinations are complex linear combinations involving complex scalars. The vector space consists of vectors of the form $[z_1, \dots, z_n]^T$ where each entry z_i is a complex number. This vector space is denoted by $\mathbb{C}^n$ and is still spanned by the canonical basis column vectors $\underline{e}_i$, $i = 1, \dots, n$ provided we take linear combinations involving complex scalars. One of the main differences between real and complex vector spaces is in the way one defines an inner product. Recall, that the inner (scalar) product should give rise to the concept of a length, or the norm of a vector. Therefore, the norm must be positive (or at least non-negative) real number. To generate a non-negative real number we define the following inner product on $\mathbb{C}^n$:

$$\langle \underline{z} | \underline{w} \rangle = \underline{z}^\dagger \underline{w} = \sum_{i=1}^{n} z_i^* w_i , \qquad (2.21)$$

where $\underline{z} = [z_1, z_2, ..., z_n]^T$ and $\underline{w} = [w_1, w_2, ..., w_N]^T$ are two complex n-th column vectors. It follows that the corresponding norm

$$||\underline{z}||^2 = \langle \underline{z} | \underline{z} \rangle = \sum_{i=1}^{n} z_i^* z_i = \sum_{i=1}^{n} |z_i|^2 \qquad (2.22)$$

is indeed a non-negative real number.

2.1.2 Writing Vectors in terms of an Orthonormal Basis

Here, we show that an advantage can be gained by using an orthonormal basis in some vector space when expanding an arbitrary vector in terms of the basis vectors. Our consideration are quite general, but for simplification, we will think about the vector space as $\mathbb{R}^n$. Suppose, $\{\underline{\omega}_1, ..., \underline{\omega}_n\}$ is an orthogonal basis, similar to the one obtained above via the Gram-Schmidt procedure. Then for an arbitrary column vector $\underline{v}$ in $\mathbb{R}^n$ we have the representation:

$$\underline{v} = \frac{\langle v | \omega_1 \rangle}{\langle \omega_1 | \omega_1 \rangle} \, \underline{\omega}_1 + \ldots + \frac{\langle v | \omega_n \rangle}{\langle \omega_n | \omega_n \rangle} \, \underline{\omega}_n . \qquad (2.23)$$

To prove this identity we recall that a vector $\underline{v}$ always has an unique decomposition

$$\underline{v} = \sum_{i=1}^{n} c_i \, \underline{\omega}_i$$

with respect to the basis $\{\underline{\omega}_1, ..., \underline{\omega}_n\}$. We multiply this relation with vector $\underline{\omega}_k$ for a fixed k, $1 \le k \le n$ and use that $\langle \omega_i | \omega_k \rangle = 0$ for $i \ne k$. We obtain

$$\langle v | \omega_k \rangle = \sum_{i=1}^{n} c_i \langle \omega_i | \omega_k \rangle = c_k \langle \omega_k | \omega_k \rangle \longrightarrow c_k = \frac{\langle v | \omega_k \rangle}{\langle \omega_k | \omega_k \rangle} ,$$

which yields the desired result for an arbitrary index k.

In particular, if $\{\underline{\hat{e}}_1, ..., \underline{\hat{e}}_n\}$ is an orthonormal basis in $\mathbb{R}^n$, then things simplify even more. We represent a vector $\underline{v}$ in $\mathbb{R}^n$

$$\underline{v} = c_1 \underline{\hat{e}}_1 + \ldots + c_n \underline{\hat{e}}_n = \sum_{i=1}^{n} c_i \underline{\hat{e}}_i \qquad (2.24)$$

by a linear combination of the orthonormal basis elements. We find the coefficients c_i of this linear combination by multiplying relation (2.24) by one of the basis vectors

$\hat{\underline{e}}_k$

$$\langle \hat{\underline{e}}_k | \underline{v} \rangle = \sum_{i=1}^{n} c_i \langle \hat{\underline{e}}_k | \hat{\underline{e}}_i \rangle = \sum_{i=1}^{n} c_i \delta_{ki} = c_k .$$

Thus, for every vector $\underline{v}$ in $\mathbb{R}^n$:

$$\underline{v} = \langle \hat{\underline{e}}_1 | \underline{v} \rangle \, \hat{\underline{e}}_1 + \langle \hat{\underline{e}}_1 | \underline{v} \rangle \, \hat{\underline{e}}_2 + \ldots + \langle \hat{\underline{e}}_n | \underline{v} \rangle \, \hat{\underline{e}}_n \qquad (2.25)$$

for an orthonormal basis $\{\hat{\underline{e}}_1, \ldots, \hat{\underline{e}}_n\}$ in $\mathbb{R}^n$.

☞ Matrices: warm up problems

Problem 2.1.1. Let

$$\underline{\underline{A}} = \begin{bmatrix} 1 & 1 \\ 1 & 1 \end{bmatrix}$$

Show that for a positive integer n:

$$\underline{\underline{A}}^n = 2^{n-1} \underline{\underline{A}} .$$

Problem 2.1.2. Verify that the inverse of

$$\underline{\underline{A}} = \begin{bmatrix} a_{11} & a_{12} \\ a_{21} & a_{22} \end{bmatrix} \qquad (2.26)$$

is given by

$$\underline{\underline{A}}^{-1} = \frac{1}{|\underline{\underline{A}}|} \begin{bmatrix} a_{22} & -a_{12} \\ -a_{21} & a_{11} \end{bmatrix} . \qquad (2.27)$$

Problem 2.1.3. Show that $(\underline{\underline{A}}^{-1})^T = (\underline{\underline{A}}^T)^{-1}$.

Problem 2.1.4. Show that for any matrix $\underline{\underline{A}}$ the matrices defined as

$$\underline{\underline{A}}_s = \underline{\underline{A}} + \underline{\underline{A}}^T ; \quad \underline{\underline{A}}_a = \underline{\underline{A}} - \underline{\underline{A}}^T$$

behave under transposition as

$$\underline{\underline{A}}_s{}^T = \underline{\underline{A}}_s ; \quad \underline{\underline{A}}_a{}^T = -\underline{\underline{A}}_a .$$

Problem 2.1.5. For the square matrix $\underline{\underline{A}}$ verify that

$$\mathrm{Tr}\left(\underline{\underline{A}}\right) = \mathrm{Tr}\left(\underline{\underline{A}}^T\right) , \quad \mathrm{Tr}\left(\underline{\underline{A}}^n\right) = \mathrm{Tr}\left(\left(\underline{\underline{A}}^T\right)^n\right)$$

for any integer $n > 0$.

Problem 2.1.6. (a) Show that

$$\langle \underline{A}\,\underline{x}\,|\,\underline{y}\rangle = \langle \underline{x}\,|\,\underline{A}^T\,\underline{y}\rangle$$

for the $m \times n$ real matrix $\underline{A}$ and $\underline{x}$ and $\underline{y}$ being two column vectors in $\mathbb{R}^n$ and $\mathbb{R}^m$, respectively.

(b) Show that

$$\langle \underline{A}\,\underline{x}\,|\,\underline{y}\rangle = \langle \underline{x}\,|\,\underline{A}^\dagger\,\underline{y}\rangle$$

for the $m \times n$ complex matrix $\underline{A}$ where $\underline{x}$ and $\underline{y}$ are two column vectors in $\mathbb{C}^n$ and $\mathbb{C}^m$, respectively.

Problem 2.1.7. Represent a vector

$$\underline{v} = \begin{bmatrix} 1 \\ 4 \\ 2 \\ 3 \end{bmatrix}$$

by a linear combination of vectors

$$\underline{e}_1 = \frac{1}{\sqrt{2}}\begin{bmatrix} 1 \\ 0 \\ 1 \\ 0 \end{bmatrix},\quad \underline{e}_2 = \frac{1}{\sqrt{2}}\begin{bmatrix} 1 \\ 0 \\ -1 \\ 0 \end{bmatrix},\quad \underline{e}_3 = \frac{1}{\sqrt{2}}\begin{bmatrix} 0 \\ 1 \\ 0 \\ 1 \end{bmatrix},\quad \underline{e}_4 = \frac{1}{\sqrt{2}}\begin{bmatrix} 0 \\ -1 \\ 0 \\ 1 \end{bmatrix}$$

in $\mathbb{R}^4$.

2.1.3 Change of the Coordinate System. The Rotation Matrix

Let V_n and V_m be n and m-dimensional vector spaces, respectively. Also, let n unit vectors $\{\widehat{e}_1, \ldots, \widehat{e}_n\}$ be a basis in V_n and m unit vectors $\{\widehat{e}_1', \ldots, \widehat{e}_m'\}$ be a basis in V_m. Thus, for every vector $\vec{v}$ in V_n there are unique scalars $x_i, i = 1, \ldots, n$ such that

$$\vec{v} = \sum_{i=1}^{n} x_i\widehat{e}_i \tag{2.28}$$

and for every vector $\vec{V}$ in V_m there are unique scalars $x_i', i = 1, \ldots, m$ such that

$$\vec{V} = \sum_{j=1}^{m} x_j'\widehat{e}_j'. \tag{2.29}$$

A function A, which maps V_n to V_m, $A : V_n \to V_m$, is a linear transformation if

$$\begin{aligned} A(\vec{v} + \vec{w}) &= A(\vec{v}) + A(\vec{w}) & (2.30) \\ A(a\vec{v}) &= a\,A(\vec{v}) & (2.31) \end{aligned}$$

for any two vectors $\vec{v}, \vec{w}$ in V_n and a scalar a.

It follows from definitions (2.30)-(2.31) that

$$A\left(\sum_{i=1}^{k} a_i \vec{v}_i\right) = \sum_{i=1}^{k} a_i A\left(\vec{v}_i\right) \qquad (2.32)$$

for any linear combination $\sum_{i=1}^{k} a_i \vec{v}_i$ of vectors in V_n.

Above, we saw that the matrix multiplication of vectors by a matrix automatically satisfied the linearity condition (2.12) equivalent to relations (2.30)-(2.31) and, therefore, provided an example of a linear transformation. Conversely, an action of a general linear transformation can be expressed by matrix multiplication, as we will show below.

For vector $\vec{v}$ from (2.28) consider the corresponding vector

$$A\left(\vec{v}\right) = \sum_{j=1}^{m} x'_j \hat{e}'_j \qquad (2.33)$$

in V_m.

According to relation (2.32) and (2.28) it holds that

$$A\left(\vec{v}\right) = \sum_{i=1}^{n} x_i A\left(\hat{e}_i\right) \qquad (2.34)$$

due to linearity of map A.

Each vector $A\left(\hat{e}_i\right)$ is in V_m and has the unique representation,

$$A\left(\hat{e}_i\right) = \sum_{j=1}^{m} a_{ji} \hat{e}'_j, \qquad (2.35)$$

according to the decomposition in (2.29). Scalars a_{ji} are therefore defined by action of A on the basis vectors $\vec{e}_i$. Plugging relation (2.35) into the linear combination in (2.34) we obtain

$$A\left(\vec{v}\right) = \sum_{i=1}^{n}\sum_{j=1}^{m} a_{ji} x_i \hat{e}'_j = \sum_{j=1}^{m} \left(\sum_{i=1}^{n} a_{ji} x_i\right) \hat{e}'_j. \qquad (2.36)$$

Comparing (2.33) with (2.36), yields

$$x'_j = \sum_{i=1}^{n} a_{ji} x_i = a_{j1} x_1 + a_{j2} x_2 + \cdots + a_{jn} x_n, \quad j = 1, \ldots, m \qquad (2.37)$$

or

$$
\begin{bmatrix} x_1' \\ x_2' \\ \vdots \\ x_m' \end{bmatrix} = \begin{bmatrix} a_{11} & a_{12} & \cdots & a_{1n} \\ a_{21} & a_{22} & \cdots & a_{2n} \\ \vdots & \vdots & \cdots & \vdots \\ a_{m1} & a_{m2} & \cdots & a_{mn} \end{bmatrix} \begin{bmatrix} x_1 \\ x_2 \\ \vdots \\ x_n \end{bmatrix} \triangleq \underline{\underline{A}} \begin{bmatrix} x_1 \\ x_2 \\ \vdots \\ x_n \end{bmatrix} \tag{2.38}
$$

in matrix notation. Thus, (2.38) associates a matrix representation with a linear map. For linear map $A : V_n \to V_m$ we obtain a matrix $\underline{\underline{A}}$ with m rows and n columns. While a linear map transforms vectors from one vector space to another the corresponding matrix transforms their vector components. The advantage of matrix notation is that it allows us to handle successive transformations through matrix multiplication. For two linear transformations $A : V_n \to V_m$ and $B : V_m \to V_p$ with respective vectors,

$$
A\left(\widehat{e}_i\right) = \sum_{j=1}^{m} a_{ji} \widehat{e}_j', \qquad B\left(\widehat{e}_j'\right) = \sum_{k=1}^{p} b_{kj} \widehat{e}_k'', \quad i = 1, \ldots, n, \quad j = 1, \ldots, m,
$$

we find that

$$
B\left(A\left(\widehat{e}_i\right)\right) = \sum_{j=1}^{m} a_{ji} B\left(\widehat{e}_j'\right)
$$

$$
= \sum_{j=1}^{m} a_{ji} \sum_{k=1}^{p} b_{kj} \widehat{e}_k''
$$

$$
= \sum_{k=1}^{p} \left(\sum_{j=1}^{m} b_{kj} a_{ji} \right) \widehat{e}_k''.
$$

Thus, for the combined linear map BA we obtain

$$
BA\left(\widehat{e}_i\right) = \sum_{k=1}^{p} c_{ki} \widehat{e}_k'', \quad c_{ki} = \sum_{j=1}^{m} b_{kj} a_{ji}.
$$

Hence, the combined linear transformation BA has a matrix representation given by the matrix product $\underline{\underline{B}}\,\underline{\underline{A}}$. A convenience of employing matrix notation is now transparent.

As an example of the linear map we will now discuss the three-dimensional vector rotation in the matrix notation. Let $\widehat{i}, \widehat{j}, \widehat{k}$ be a set of unit orthogonal vectors associated with orthogonal coordinate systems with axes x, y, z. Let x', y', z' denote axes of a coordinate system obtained from x, y, z by rotation and let $\widehat{i}', \widehat{j}', \widehat{k}'$ be its set of unit orthogonal vectors of the rotated coordinate system.

The arbitrary vector $\vec{V}$ can be expressed in rotated and un-rotated coordinate systems as:

$$\vec{V} = V_x\hat{i} + V_y\hat{j} + V_z\hat{k}, \tag{2.39}$$
$$\vec{V} = V_{x'}\hat{i'} + V_{y'}\hat{j'} + V_{z'}\hat{k'}. \tag{2.40}$$

The coefficient $V_{x'}$ of $\vec{V}$ in (2.40) can be obtained by projecting $\vec{V}$ on the unit vector $\hat{i'}: V_{x'} = \vec{V} \cdot \hat{i'}$. This projection is obtained by taking a scalar product of the right hand side of (2.40) with $\hat{i'}$ and using $\hat{i'} \cdot \hat{i'} = 1, \hat{i'} \cdot \hat{j'} = 0, \hat{i'} \cdot \hat{k'} = 0$. Since similar relations hold for all directions, we have

$$V_{x'} = \vec{V} \cdot \hat{i'} = (\hat{i} \cdot \hat{i'})V_x + (\hat{j} \cdot \hat{i'})V_y + (\hat{k} \cdot \hat{i'})V_z, \tag{2.41}$$
$$V_{y'} = \vec{V} \cdot \hat{j'} = (\hat{i} \cdot \hat{j'})V_x + (\hat{j} \cdot \hat{j'})V_y + (\hat{k} \cdot \hat{j'})V_z, \tag{2.42}$$
$$V_{z'} = \vec{V} \cdot \hat{k'} = (\hat{i} \cdot \hat{k'})V_x + (\hat{j} \cdot \hat{k'})V_y + (\hat{k} \cdot \hat{k'})V_z. \tag{2.43}$$

These relations can be rewritten compactly in matrix notation as:

$$\begin{bmatrix} V_{x'} \\ V_{y'} \\ V_{z'} \end{bmatrix} = \begin{bmatrix} \hat{i} \cdot \hat{i'} & \hat{j} \cdot \hat{i'} & \hat{k} \cdot \hat{i'} \\ \hat{i} \cdot \hat{j'} & \hat{j} \cdot \hat{j'} & \hat{k} \cdot \hat{j'} \\ \hat{i} \cdot \hat{k'} & \hat{j} \cdot \hat{k'} & \hat{k} \cdot \hat{k'} \end{bmatrix} \begin{bmatrix} V_x \\ V_y \\ V_z \end{bmatrix}, \tag{2.44}$$

which defines a rotation matrix with a generic matrix element of the type $\hat{j} \cdot \hat{i'} = \cos\alpha_{yx'}$ that is equal to the cosine of the angle $\alpha_{yx'}$ between y and x'.

Since the rotated basis vector $\hat{i'}$ is equal to

$$\hat{i'} = (\hat{i} \cdot \hat{i'})\hat{i} + (\hat{j} \cdot \hat{i'})\hat{j} + (\hat{k} \cdot \hat{i'})\hat{k}, \tag{2.45}$$

with similar basis vectors for the remaining two directions, the relation (2.44) extends to the basis vectors:

$$\begin{bmatrix} \hat{i'} \\ \hat{j'} \\ \hat{k'} \end{bmatrix} = \begin{bmatrix} \hat{i} \cdot \hat{i'} & \hat{j} \cdot \hat{i'} & \hat{k} \cdot \hat{i'} \\ \hat{i} \cdot \hat{j'} & \hat{j} \cdot \hat{j'} & \hat{k} \cdot \hat{j'} \\ \hat{i} \cdot \hat{k'} & \hat{j} \cdot \hat{k'} & \hat{k} \cdot \hat{k'} \end{bmatrix} \begin{bmatrix} \hat{i} \\ \hat{j} \\ \hat{k} \end{bmatrix}. \tag{2.46}$$

Equation (2.46) defines a change of the canonical basis vectors under rotation of coordinate axes in the matrix formulation.

☞ **Example 2.1.1.** Let us rotate the coordinate system by an angle ϕ about the z-axis. The new orthogonal unit vectors $\hat{i'}$, $\hat{j'}$ make an angle ϕ with the old unit vectors $\hat{i}$, $\hat{j}$ as seen in the Figure 2.1. In this case, the direction cosines are easily obtained as:

$$\hat{i} \cdot \hat{i'} = \hat{j} \cdot \hat{j'} = \cos\phi, \tag{2.47}$$
$$\hat{j} \cdot \hat{i'} = -\hat{i} \cdot \hat{j'} = \sin\phi, \tag{2.48}$$
$$\hat{k} \cdot \hat{k'} = 1, \tag{2.49}$$

A Short Course in Mathematical Methods with Maple

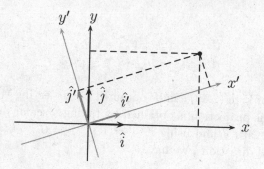

Figure 2.1: Vector rotation.

which reproduces the rotation matrix:

$$\underline{\underline{R_z}}(\phi) = \begin{bmatrix} a_{11} & a_{12} & a_{13} \\ a_{21} & a_{22} & a_{23} \\ a_{31} & a_{32} & a_{33} \end{bmatrix} = \begin{bmatrix} \cos\phi & \sin\phi & 0 \\ -\sin\phi & \cos\phi & 0 \\ 0 & 0 & 1 \end{bmatrix}. \tag{2.50}$$

From relation (2.46) the primed basis vectors are related to the unprimed basis vectors by

$$\begin{aligned} \widehat{i'} &= \widehat{i}\cos\phi + \widehat{j}\sin\phi \\ \widehat{j'} &= \widehat{j}\cos\phi - \widehat{i}\sin\phi \end{aligned} \tag{2.51} \tag{2.52}$$

and $\widehat{k'} = \widehat{k}$.

✎ **Change of coordinate system. The rotation matrix: Problems**

Problem 2.1.8. Find a 2×2 matrix corresponding to the change of the basis:

$$\begin{bmatrix} 1 \\ 2 \end{bmatrix}, \begin{bmatrix} 5 \\ 1 \end{bmatrix} \longrightarrow \begin{bmatrix} 1 \\ 3 \end{bmatrix}, \begin{bmatrix} 1 \\ 1 \end{bmatrix}$$

Do the same for

$$\begin{bmatrix} 1 \\ 2 \end{bmatrix}, \begin{bmatrix} 5 \\ 1 \end{bmatrix} \longrightarrow \begin{bmatrix} 1 \\ 4 \end{bmatrix}, \begin{bmatrix} 3 \\ 1 \end{bmatrix}.$$

Problem 2.1.9. Let $\underline{v}_1, \underline{v}_2$ be two vectors that form a basis in $\mathbb{R}^2$. Furthermore, let A be a linear transformation of $\mathbb{R}^2$ such that

$$A\left(a_1\underline{v}_1 + a_2\underline{v}_2\right) = \left(2a_1 + 3a_2\right)\underline{v}_1 + \left(4a_1 - a_2\right)\underline{v}_2.$$

(a) Find the matrix representation $\underline{\underline{A}}$ of A with respect to the basis $\{\underline{v}_1, \underline{v}_2\}$.

(b) For

$$\underline{v}_1 = \begin{bmatrix} 1 \\ 2 \end{bmatrix}, \quad \underline{v}_2 = \begin{bmatrix} 5 \\ 3 \end{bmatrix}$$

find the matrix representation $\underline{A}'$ of A with respect to the canonical basis

$$\underline{e}_1 = \begin{bmatrix} 1 \\ 0 \end{bmatrix}, \quad \underline{e}_2 = \begin{bmatrix} 0 \\ 1 \end{bmatrix}.$$

Problem 2.1.10. Find a matrix which corresponds to the following change in the basis

$$\{1, x, x^2, x^3\} \longrightarrow \{2, 3x - 1, 4x^2 - x, x^3 - 3x + 5\}$$

in the space of polynomials of degree at most 3. Use this matrix to rewrite the polynomial, $6 + 4x - 7x^2 + 8x^3$, in the new basis.

Problem 2.1.11. Show that

$$\underline{\underline{R}}_x(\phi) = \begin{bmatrix} 1 & 0 & 0 \\ 0 & \cos\phi & \sin\phi \\ 0 & -\sin\phi & \cos\phi \end{bmatrix}, \quad \underline{\underline{R}}_y(\phi) = \begin{bmatrix} \cos\phi & 0 & -\sin\phi \\ 0 & 1 & 0 \\ \sin\phi & 0 & \cos\phi \end{bmatrix} \quad (2.53)$$

and describe the coordinate changes under rotations by angle ϕ around the x and y-axes.

Problem 2.1.12. Show that the coefficients of $\underline{\underline{R}}_z(\phi)$ matrix from (2.50) and $\underline{\underline{R}}_x(\phi), \underline{\underline{R}}_y(\phi)$ matrices from (2.53) satisfy the relation

$$\sum_{k=1}^{3} a_{ki} a_{kj} = \delta_{ij}, \quad (2.54)$$

where δ_{ij} is the Kronecker delta.

Problem 2.1.13. Show that for the matrix:

$$\underline{\underline{R}} = \begin{bmatrix} \cos\theta & -\sin\theta \\ \sin\theta & \cos\theta \end{bmatrix}, \quad (2.55)$$

that describes a rotation of the two-dimensional coordinate system by angle θ, it holds that

$$\underline{\underline{R}}^n = \begin{bmatrix} \cos n\theta & -\sin n\theta \\ \sin n\theta & \cos n\theta \end{bmatrix}. \quad (2.56)$$

2.1.4 Determinant and Matrix Inversion

Suppose that $\underline{\underline{A}}$ is a $n \times n$ matrix. For $n = 2$, the determinant of

$$\underline{\underline{A}} = \begin{bmatrix} a_{11} & a_{12} \\ a_{21} & a_{22} \end{bmatrix} \tag{2.57}$$

is given by

$$\left| \underline{\underline{A}} \right| = a_{11}a_{22} - a_{12}a_{21}. \tag{2.58}$$

If $n = 3$, the determinant of

$$\underline{\underline{A}} = \begin{bmatrix} a_{11} & a_{12} & a_{13} \\ a_{21} & a_{22} & a_{23} \\ a_{31} & a_{32} & a_{33} \end{bmatrix} \tag{2.59}$$

is given by

$$\left| \underline{\underline{A}} \right| = a_{11}a_{22}a_{33} + a_{12}a_{23}a_{31} + a_{13}a_{21}a_{32} - a_{11}a_{23}a_{32} - a_{12}a_{21}a_{33}$$

$$- a_{13}a_{22}a_{31} = \sum_{i,j,k=1}^{3} \epsilon_{ijk} a_{1i} a_{2j} a_{3k} = \sum_{i,j,k=1}^{3} \epsilon_{ijk} a_{i1} a_{j2} a_{k3} \tag{2.60}$$

For a general $n \times n$ matrix $\underline{\underline{A}}$, the determinant of $\underline{\underline{A}}$ is

$$\left| \underline{\underline{A}} \right| = \sum_{i_1,i_2,\ldots,i_n=1}^{n} \epsilon_{i_1,i_2,\ldots,i_n} a_{1i_1} a_{2i_2} \cdots a_{ni_n}, \tag{2.61}$$

where $\epsilon_{i_1,i_2,\ldots,i_n}$ is the n-dimensional permutation symbol which we will define after we first introduce a concept of permutation. Let Π be an ordered set of n elements. A one to one transformation σ of the set Π onto itself is called a permutation of Π. For example, let $\Pi = (1, 2, 3, 4)$. Define a permutation σ as $\sigma(1, 2, 3, 4) = (2, 4, 1, 3)$. The permutation, which transforms $(2, 4, 1, 3)$ back to $(1, 2, 3, 4)$ is the inverse permutation of σ. A permutation σ such that it interchanges two elements while keeping all others fixed, is called a transposition. In the above example of $\Pi = (1, 2, 3, 4)$, the permutation $\sigma(1, 2, 3, 4) = (1, 4, 3, 2)$ is a transposition. Every permutation can be expressed as a sequence of transpositions. A permutation which can be expressed as an even number of transpositions is called an even permutation. A permutation that can be expressed as an odd number of transpositions is called an odd permutation. The inverse permutation of σ can be expressed by the same number of transpositions as of σ. The sign of an even permutation σ is $\text{sgn}(\sigma) = +1$. The sign of an odd permutation σ is $\text{sgn}(\sigma) = -1$.

The permutation symbol is defined as

$$\epsilon_{i_1,i_2,\ldots,i_n} = \begin{cases} +1, & \text{if } (i_1, i_2, \ldots, i_n) \text{ is an even permutation of } (1, 2, \ldots, n) \\ -1, & \text{if } (i_1, i_2, \ldots, i_n) \text{ is an odd permutation of } (1, 2, \ldots, n) \\ 0 & \text{if any index is repeated} \end{cases}.$$

For the permutation σ such that $\sigma(1,2,...,n) = (\sigma(1),\sigma(2),...,\sigma(n)) = (i_1,i_2,...,i_n)$ the permutation symbol is equal to the sign of permutation. Hence, we can rewrite the definition of the determinant $|\underline{A}|$ in definition (2.61) as a sum over all $n!$ permutations of $\Pi = (1,2,3,...,n)$ leading to:

$$|\underline{A}| = \sum_\sigma \text{sgn}(\sigma)\, a_{1\sigma(1)} a_{2\sigma(2)} \cdots a_{n\sigma(n)} \, . \tag{2.62}$$

Note that each term of the sum in (2.62) involves each row and each column only once.

An important consequence of definition (2.62) is that determinant of a product of two matrices is equal to the product of determinants of each matrix:

$$|\underline{A}\,\underline{B}| = |\underline{A}|\,|\underline{B}| = |\underline{B}\,\underline{A}| \, . \tag{2.63}$$

We provide this result without a proof.

For the transpose matrix $\underline{A}^T$, expression (2.62) becomes

$$|\underline{A}^T| = \sum_\sigma \text{sgn}(\sigma)a_{1\sigma(1)}^T a_{2\sigma(2)}^T \cdots a_{n\sigma(n)}^T = \sum_\sigma \text{sgn}(\sigma)a_{\sigma(1)1} a_{\sigma(2)2} \cdots a_{\sigma(n)n} \, .$$

Let $(\sigma(1),\sigma(2),...,\sigma(n))$ be a permutation of $(1,2,...,n)$. The inverse permutation σ' of σ transforms $(\sigma(1),\sigma(2),...,\sigma(n))$ back to $(1,2,...,n)$ and, therefore, $\sigma'(\sigma(i)) = i$ for $i = 1,...,n$. The inverse permutation must have exactly the same number of transpositions as σ. Thus, $\text{sgn}(\sigma) = \text{sgn}(\sigma')$ and

$$|\underline{A}^T| = \sum_\sigma \text{sgn}(\sigma)a_{\sigma(1)\sigma'(\sigma(1))} a_{\sigma(2)\sigma'(\sigma(2))} \cdots a_{\sigma(n)\sigma'(\sigma(n))}$$

$$= \sum_\sigma \text{sgn}(\sigma')a_{1\sigma'(1)} a_{2\sigma'(2)} \cdots a_{n\sigma'(n)} \, . \tag{2.64}$$

Because the set of all permutations is identical to a set of all inverse permutations we conclude that:

$$|\underline{A}^T| = |\underline{A}| \, . \tag{2.65}$$

☞ **Example 2.1.2.** Consider matrix $\underline{A}$ and its transpose $\underline{A}^T$ given in (2.4). Then

$$|\underline{A}| = \begin{vmatrix} 1 & 2 \\ 3 & 7 \end{vmatrix} = 1\cdot7 - 2\cdot3 = 1$$

and

$$|\underline{A}^T| = \begin{vmatrix} 1 & 3 \\ 2 & 7 \end{vmatrix} = 1\cdot7 - 3\cdot3 = 1 \, .$$

Let, σ' be a permutation transposing only i and j:

$$\sigma'(1, ..., i, ..., j, ..., n) = (1, ..., j, ..., i, ..., n).$$

Let $\underline{\underline{A}}'$ be obtained by interchanging the i-th and j-th column of $\underline{\underline{A}}$. Thus, it holds $a'_{ki} = a_{kj}$ and, more generally, $a'_{k\sigma(p)} = a_{k\sigma'(\sigma(p))}$ for arbitrary k, p and any permutation σ. For $\left| \underline{\underline{A}}' \right|$ it holds that

$$
\begin{aligned}
\left| \underline{\underline{A}}' \right| &= \sum_\sigma \operatorname{sgn}(\sigma)\, a'_{1\sigma(1)} a'_{2\sigma(2)} \cdots a'_{n\sigma(n)} \\
&= \sum_\sigma \operatorname{sgn}(\sigma)\, a_{1\sigma'(\sigma(1))} a_{2\sigma'(\sigma(2))} \cdots a_{n\sigma'(\sigma(n))} \\
&= -\sum_{\sigma'\sigma} \operatorname{sgn}(\sigma'\sigma)\, a_{1\sigma'(\sigma(1))} a_{2\sigma'(\sigma(2))} \cdots a_{n\sigma'(\sigma(n))} \\
&= -\left| \underline{\underline{A}} \right|.
\end{aligned}
\tag{2.66}
$$

The minus sign is due to $\operatorname{sgn}(\sigma') = -1$ and accordingly $\operatorname{sgn}(\sigma'\sigma) = -\operatorname{sgn}(\sigma)$. The conclusion in (2.66) follows from the fact that the set of all permutations σ is identical to the set of all permutations $\sigma'\sigma$. Hence, when we transpose two columns in $\underline{\underline{A}}$, the determinant $\left| \underline{\underline{A}} \right|$ changes its sign. If we write the matrix $\underline{\underline{A}}$ in terms of its column vectors $\underline{a}_i, i = 1, \ldots, n$ as

$$\underline{\underline{A}} = \begin{bmatrix} \underline{a}_1 & \underline{a}_2 & \cdots & \underline{a}_n \end{bmatrix} \tag{2.67}$$

then the statement that was just proved reads as

$$\left| \underline{a}_1 \cdots \underline{a}_i \cdots \underline{a}_j \cdots \underline{a}_n \right| = -\left| \underline{a}_1 \cdots \underline{a}_j \cdots \underline{a}_i \cdots \underline{a}_n \right|.$$

Similarly, when we transpose two rows in $\underline{\underline{A}}$, the determinant $\left| \underline{\underline{A}} \right|$ changes its sign. In particular, if $\underline{\underline{A}}$ contains two identical columns or rows, then the determinant vanishes:

$$\left| \underline{a}_1 \cdots \underline{a}_i \cdots \underline{a}_i \cdots \underline{a}_n \right| = 0. \tag{2.68}$$

☞ **Example 2.1.3.** Consider matrix $\underline{\underline{A}}$ from equation (2.4). Interchanging two rows yields

$$\begin{vmatrix} 1 & 2 \\ 3 & 7 \end{vmatrix} \rightsquigarrow \begin{vmatrix} 3 & 7 \\ 1 & 2 \end{vmatrix} = 3 \cdot 2 - 7 \cdot 1 = -1 = -\left| \underline{\underline{A}} \right|$$

while interchanging two columns gives

$$\begin{vmatrix} 1 & 2 \\ 3 & 7 \end{vmatrix} \rightsquigarrow \begin{vmatrix} 2 & 1 \\ 7 & 3 \end{vmatrix} = 2 \cdot 3 - 1 \cdot 7 = -1 = -\left| \underline{\underline{A}} \right|.$$

Now consider a fixed i-th row in a matrix $\underline{A}$. Each term of the sum over permutations in (2.62) contains exactly one of the matrix elements $a_{i1}, a_{i2}, a_{i3}, \ldots, a_{in}$. Thus, $|\underline{A}|$ can be written as a linear polynomial in $a_{i1}, a_{i2}, a_{i3}, \ldots, a_{in}$. Denote the coefficients of a_{ij}, $j = 1, \ldots, n$ by C_{ij}. The coefficient C_{ij} is called a cofactor of a_{ij} and the expansion of the determinant of $\underline{A}$ follows as

$$|\underline{A}| = a_{i1}C_{i1} + a_{i2}C_{i2} + \cdots + a_{in}C_{in} = \sum_{j=1}^{n} a_{ij}C_{ij}, \qquad (2.69)$$

and is valid for any i such that $1 \leq i \leq n$.

The cofactor C_{ij} cannot depend on elements of the i-th row and elements of the j-th column due to the fact that each term of the sum (2.62) involves each row and each column only once. Hence, C_{ij} must only contain elements from a $n-1 \times n-1$ submatrix, which is obtained from $\underline{A}$ by deleting the i-th row and the j-th column.

☞ **Example 2.1.4.** In terms of cofactors we can rewrite the determinant (2.60) as

$$\begin{aligned} |\underline{A}| &= a_{11}C_{11} + a_{12}C_{12} + a_{13}C_{13} = a_{11}(a_{22}a_{33} - a_{32}a_{23}) \\ &+ a_{12}(a_{23}a_{31} - a_{21}a_{33}) + a_{13}(a_{21}a_{32} - a_{22}a_{31}). \end{aligned} \qquad (2.70)$$

Comparing equations (2.69) and (2.61) we find that

$$C_{ij} = \sum_{\substack{\sigma \\ \sigma(i)=j}} \epsilon_{i_1, i_2, \ldots, i_n}\, a_{1i_1} a_{2i_2} \cdots \widehat{a_{ij}} \cdots a_{ni_n}, \qquad (2.71)$$

where the summation is over all permutations $\sigma : (1, 2, \ldots, i, \ldots, n) \to (i_1, i_2, \ldots, i_n)$ such that $\sigma(i) = i_i = j$. Our notation is such that the caret above a symbol means that it is omitted and, therefore, the factor a_{ij} is omitted from the product $a_{1i_1} a_{2i_2} \cdots \widehat{a_{ik}} \cdots a_{ni_n}$.

Next, introduce a so-called minor M_{ij}, which is a real number equal to the determinant of a submatrix of dimension $n-1 \times n-1$ obtained from $\underline{A}$ by crossing out the i-th row and the j-th column:

$$M_{ij} = \sum_{\substack{i_1, i_2, \ldots, i_{n-1} \\ i_k \neq j, k=1, \ldots n-1}}^{n-1} \epsilon_{i_1, i_2, \ldots, i_{n-1}}\, a_{1i_1} a_{2i_2} \cdots a_{n-1\, i_{n-1}}. \qquad (2.72)$$

Here, the summation is over permutations of the ordered sets of $n-1$ numbers which do not include i with values in the ordered sets $(i_1, i_2, \ldots, i_{n-1})$ for $i_k \neq j$ and $k = 1, \ldots, n-1$.

☞ **Example 2.1.5.** To calculate M_{22} for $\underline{\underline{A}}$ as in (2.59), we omit 2-nd row and 2-nd column as shown below

$$M_{22} = \begin{vmatrix} a_{11} & \vdots & a_{13} \\ \cdots & \vdots & \cdots \\ a_{31} & \vdots & a_{33} \end{vmatrix} = \begin{vmatrix} a_{11} & a_{13} \\ a_{31} & a_{33} \end{vmatrix} = a_{11}a_{33} - a_{13}a_{31}. \tag{2.73}$$

In terms of minors we can rewrite the determinant in (2.60) (and in (2.70)) as

$$|\underline{\underline{A}}| = a_{11}M_{11} - a_{12}M_{12} + a_{13}M_{13} = \sum_{i=1}^{3} a_{1i}(-1)^{1+i}M_{1i}. \tag{2.74}$$

This represents a minor expansion along the first row.

In order to relate the expression for the cofactor C_{ij} in (2.71) to the minor M_{ij} it is enough to permute the initial order set $(1, 2, \ldots, i, \ldots, n)$ into $(i, 1, 2, \ldots, \widehat{i}, \ldots, n)$ simultaneously transforming the permutation symbol $\epsilon_{i_1,i_2,\ldots,i_n}$ in (2.71) into $\epsilon_{j,i_1,i_2,\ldots,\widehat{j},\ldots,i_n}$. These two operations effectively cast the sum $\sum_\sigma \mathrm{sgn}(\sigma)$ over all allowed permutations in (2.71) to the one in (2.72). The first operation results in factor $(-1)^{i-1}$ since it involves $i-1$ successive transpositions. The second operation is achieved after $j-1$ successive transpositions and results in a factor $(-1)^{j-1}$. Altogether we find

$$C_{ij} = (-1)^{i+j}M_{ij}. \tag{2.75}$$

To summarize, for an $n \times n$ matrix

$$\underline{\underline{A}} = \begin{bmatrix} a_{11} & a_{12} & \cdots & a_{1\,j-1} & a_{1j} & a_{1j+1} & \cdots & a_{1n} \\ a_{21} & a_{22} & \cdots & a_{2\,j-1} & a_{2j} & a_{2j+1} & \cdots & a_{2n} \\ \cdot & \cdot & \cdots & \cdot & \vdots & \cdot & \cdots & \cdot \\ a_{i-1\,1} & a_{i-1\,2} & \cdots & a_{i-1\,j-1} & a_{i-1\,j} & a_{i-1\,j+1} & \cdots & a_{i-1\,n} \\ a_{i1} & a_{i2} & \cdots & a_{i\,j-1} & a_{ij} & a_{i\,j+1} & \cdots & a_{in} \\ a_{i+1\,1} & a_{i+1\,2} & \cdots & a_{i+1\,j-1} & a_{i+1\,j} & a_{i+1\,j+1} & \cdots & a_{i+1\,n} \\ \cdot & \cdot & \cdots & \cdot & \vdots & \cdot & \cdots & \cdot \\ a_{n1} & a_{n2} & \cdots & a_{n\,j-1} & a_{nj} & a_{n\,j+1} & \cdots & a_{nn} \end{bmatrix}$$

we associate the cofactor C_{ij} of the element a_{ij} as the product of the sign factor $(-1)^{i+j}$ with the determinant of the $(n-1) \times (n-1)$ matrix that is obtained by removing the i-th row and j-th column of $\underline{\underline{A}}$:

$$C_{ij} = (-1)^{i+j} \begin{vmatrix} a_{11} & \cdots & a_{1\,j-1} & \vdots & a_{1j+1} & \cdots & a_{1n} \\ a_{21} & \cdots & a_{2\,j-1} & \vdots & a_{2j+1} & \cdots & a_{2n} \\ \cdot & \cdots & \cdot & \vdots & \cdot & \cdots & \cdot \\ a_{i-1\,1} & \cdots & a_{i-1\,j-1} & \vdots & a_{i-1\,j+1} & \cdots & a_{i-1\,n} \\ \cdots\cdots & & \cdots\cdots & \vdots & \cdots\cdots & & \cdots\cdots \\ a_{i+1\,1} & \cdots & a_{i+1\,j-1} & \vdots & a_{i+1\,j+1} & \cdots & a_{i+1\,n} \\ \cdot & \cdots & \cdot & \vdots & \cdot & \cdots & \cdot \\ a_{n1} & \cdots & a_{n\,j-1} & \vdots & a_{n\,j+1} & \cdots & a_{nn} \end{vmatrix} \quad \leftarrow i\text{-th row}$$

$$\uparrow$$
$$j\text{-th}$$
$$\text{column}$$

The signs produced by the factor $(-1)^{i+j}$ follow an alternating pattern of plus and minus signs, which can be inscribed on $n \times n$ matrix as follows:

$$\begin{bmatrix} + & - & + & - & + & \cdots \\ - & + & - & + & - & \cdots \\ + & - & + & - & + & \cdots \\ - & + & - & + & - & \cdots \\ \vdots & \vdots & \vdots & \vdots & \vdots & \ddots \end{bmatrix}.$$

☞ **Example 2.1.6.** Here, we illustrate how to associate the cofactor, C_{33}, to the matrix element, a_{33}, of the 4×4 matrix:

$$\begin{bmatrix} a_{11} & a_{12} & a_{13} & a_{14} \\ a_{21} & a_{22} & a_{23} & a_{24} \\ a_{31} & a_{32} & a_{33} & a_{34} \\ a_{41} & a_{42} & a_{43} & a_{44} \end{bmatrix} \rightsquigarrow C_{33} = (-1)^{3+3} \begin{vmatrix} a_{11} & a_{12} & a_{14} \\ a_{21} & a_{22} & a_{24} \\ a_{41} & a_{42} & a_{44} \end{vmatrix}.$$

In summary, the expansion in (2.69) provides a practical method to calculate the determinant of an arbitrary $n \times n$ square matrix $\underline{\underline{A}}$ by expanding it by minors along the i-th row:

$$| \underline{\underline{A}} | = \sum_{j=1}^{n} a_{ij} C_{ij} = \sum_{j=1}^{n} a_{ij} (-1)^{i+j} M_{ij} \qquad (2.76)$$

for i between 1 and n, $1 \le i \le n$.

From our earlier discussion (leading to (2.65)) it follows that each property valid for the rows of a matrix will also hold for the columns and vice versa. Hence, the minor expansion is valid for an expansion along an arbitrary column as well as along an arbitrary row and, the determinant of the $n \times n$ square matrix $\underline{\underline{A}}$ is also given by the following expansions by minors along the i-th column:

$$|\underline{\underline{A}}| = \sum_{j=1}^{n} a_{ji} C_{ji} = \sum_{j=1}^{n} a_{ji}(-1)^{i+j} M_{ji}. \tag{2.77}$$

Thus, if matrix $\underline{\underline{A}}$ has a complete row (or column) of zeroes then $|\underline{\underline{A}}| = 0$.

☞ **Example 2.1.7.** Consider the 3×3 matrix

$$\underline{\underline{A}} = \begin{bmatrix} 1 & 1 & 0 \\ 1 & 2 & 1 \\ 0 & 3 & 1 \end{bmatrix}. \tag{2.78}$$

The minor expansions along the three rows are (starting from the top row):

$$1 \cdot \begin{vmatrix} 2 & 1 \\ 3 & 1 \end{vmatrix} - 1 \cdot \begin{vmatrix} 1 & 1 \\ 0 & 1 \end{vmatrix} + 0 \cdot \begin{vmatrix} 1 & 2 \\ 0 & 3 \end{vmatrix} = -2$$

$$-1 \cdot \begin{vmatrix} 1 & 0 \\ 3 & 1 \end{vmatrix} + 2 \cdot \begin{vmatrix} 1 & 0 \\ 0 & 1 \end{vmatrix} - 1 \cdot \begin{vmatrix} 1 & 1 \\ 0 & 3 \end{vmatrix} = -2$$

$$0 \cdot \begin{vmatrix} 1 & 0 \\ 2 & 1 \end{vmatrix} - 3 \cdot \begin{vmatrix} 1 & 0 \\ 1 & 1 \end{vmatrix} + 1 \cdot \begin{vmatrix} 1 & 1 \\ 1 & 2 \end{vmatrix} = -2.$$

The same results from minor expansions along the three columns (starting from the left column):

$$1 \cdot \begin{vmatrix} 2 & 1 \\ 3 & 1 \end{vmatrix} - 1 \cdot \begin{vmatrix} 1 & 0 \\ 3 & 1 \end{vmatrix} + 0 \cdot \begin{vmatrix} 1 & 0 \\ 2 & 1 \end{vmatrix} = -2$$

$$-1 \cdot \begin{vmatrix} 1 & 1 \\ 0 & 1 \end{vmatrix} + 2 \cdot \begin{vmatrix} 1 & 0 \\ 0 & 1 \end{vmatrix} - 3 \cdot \begin{vmatrix} 1 & 0 \\ 1 & 1 \end{vmatrix} = -2$$

$$0 \cdot \begin{vmatrix} 1 & 2 \\ 0 & 3 \end{vmatrix} - 1 \cdot \begin{vmatrix} 1 & 1 \\ 0 & 3 \end{vmatrix} + 1 \cdot \begin{vmatrix} 1 & 1 \\ 1 & 2 \end{vmatrix} = -2.$$

One can use expansion by minors to show that the identity matrix $\underline{\underline{I}}$, shown in expression (2.7), has unit determinant:

$$|\underline{\underline{I}}| = 1.$$

Similarly, the determinant of a diagonal matrix with n matrix elements $a_{ij} = a_i \delta_{ij}$ is the product of the diagonal elements, $a_1 a_2 \cdots a_n$.

More generally, it follows from successive applications of minor expansions to the first columns that the determinant of an upper triangular matrix

$$\underline{A} = \begin{bmatrix} a_{11} & a_{12} & \cdots & \cdots & a_{1n} \\ 0 & a_{22} & \cdots & \cdots & a_{2n} \\ 0 & 0 & a_{33} & \cdots & a_{3n} \\ \vdots & \vdots & \ddots & \ddots & \vdots \\ 0 & 0 & \cdots & 0 & a_{nn} \end{bmatrix}$$

is equal to the product of the diagonal elements $a_{11} a_{22} \cdots a_{nn}$.

Compose a matrix $\underline{\tilde{A}}$ by multiplying the i-th row of $\underline{A}$ by a real number K. Then $\tilde{a}_{i,j} = K a_{i,j}$ for each j and fixed i. For i', different from i, it holds that $\tilde{a}_{i',j} = a_{i',j}$. Using the minor expansion along the i-th row we conclude that under multiplication of a row in $\underline{A}$ with a real number K the determinant $|\underline{A}|$ changes to $K|\underline{A}|$. Similarly, under multiplication of a column in

$$\underline{A} = \begin{bmatrix} \underline{a}_1 & \cdots & \underline{a}_i & \cdots & \underline{a}_n \end{bmatrix}$$

with a real number K, the determinant of

$$\begin{bmatrix} \underline{a}_1 & \cdots & K\underline{a}_i & \cdots & \underline{a}_n \end{bmatrix}$$

is

$$\begin{vmatrix} \underline{a}_1 & \cdots & K\underline{a}_i & \cdots & \underline{a}_n \end{vmatrix} = K \begin{vmatrix} \underline{a}_1 & \cdots & \underline{a}_i & \cdots & \underline{a}_n \end{vmatrix}.$$

Now, consider the matrix

$$\underline{\tilde{A}} = \begin{bmatrix} \underline{a}_1 & \cdots & \underline{a}_i + \underline{b}_i & \cdots & \underline{a}_n \end{bmatrix},$$

which has the sum of two column vectors $\underline{a}_i$ and $\underline{b}_i$ in the position of its i-th column. Using a minor expansion along the i-th column we arrive at the identity

$$\begin{vmatrix} \underline{a}_1 & \cdots & \underline{a}_i + \underline{b}_i & \cdots & \underline{a}_n \end{vmatrix} = \begin{vmatrix} \underline{a}_1 & \cdots & \underline{a}_i & \cdots & \underline{a}_n \end{vmatrix} \\ + \begin{vmatrix} \underline{a}_1 & \cdots & \underline{b}_i & \cdots & \underline{a}_n \end{vmatrix}. \tag{2.79}$$

Thus, the determinant of a sum of two matrices, which differ only by one column (or row), is equal to the sum of determinants of each matrix.

Recalling relation (2.68) it is now easy to prove the following statement:

> Adding a multiple of one column (or row) to another column (or row) does not change the determinant.

☞ **Example 2.1.8.** Consider matrix $\underline{\underline{A}}$ from (2.4). Multiplying the first row by 3 and subtracting from the second row yields

$$\begin{bmatrix} 1 & 2 \\ 3 & 7 \end{bmatrix} \overset{-3}{\underset{+}{\longleftarrow}} \rightsquigarrow \underline{\underline{A}}' = \begin{bmatrix} 1 & 2 \\ 0 & 1 \end{bmatrix},$$

with $|\underline{\underline{A}}'| = 1 \cdot 1 - 2 \cdot 0 = 1$. Similarly, multiplying the first column by 2 and subtracting from the second column results in a matrix

$$\overset{-2\ +}{\begin{bmatrix} 1 & 2 \\ 3 & 7 \end{bmatrix}} \rightsquigarrow \underline{\underline{A}}'' = \begin{bmatrix} 1 & 0 \\ 3 & 1 \end{bmatrix},$$

with $|\underline{\underline{A}}''| = 1 \cdot 1 - 2 \cdot 0 = 1$.

We conclude that the determinant is invariant under the addition of a linear combination of other rows to a row and the addition of a linear combination of other columns to a column:

$$\begin{vmatrix} \underline{a}_1 & \cdots & \underline{a}_{i-1} & \underline{a}_i + \sum_{j \neq i} c_j \underline{a}_j & \underline{a}_{i+1} & \cdots & \underline{a}_n \end{vmatrix}$$
$$= \begin{vmatrix} \underline{a}_1 & \cdots & \underline{a}_{i-1} & \underline{a}_i & \underline{a}_{i+1} & \cdots & \underline{a}_n \end{vmatrix}.$$

Consider a sum $\sum_{j=1}^{n} a_{ij} C_{kj}$, which for $k = i$ is identical to the cofactor expansion in (2.76). Comparing with $|\underline{\underline{A}}| = \sum_{j=1}^{n} a_{kj} C_{kj}$ we find that $\sum_{j=1}^{n} a_{ij} C_{kj}$ is equal to the determinant of a matrix obtained from $\underline{\underline{A}}$ by replacing the k-th row by its i-th row. For $k \neq i$, that matrix will have two identical rows and its determinant must vanish. Accordingly, we can generalize (2.76) to

$$\delta_{ik} |\underline{\underline{A}}| = \sum_{j=1}^{n} a_{ij} C_{kj}. \tag{2.80}$$

Let $\underline{\underline{C}}$ be a matrix with matrix element c_{ij} equal to the cofactor C_{ij}. Expression (2.80) can be rewritten, using the product of matrices, as

$$\delta_{ik} |\underline{\underline{A}}| = \sum_{j=1}^{n} a_{ij} C_{jk}^T \quad \longrightarrow \quad \underline{\underline{I}} |\underline{\underline{A}}| = \underline{\underline{A}} \, \underline{\underline{C}}^T, \tag{2.81}$$

where $\underline{\underline{C}}^T$ is a transpose of the cofactor matrix $\underline{\underline{C}}$. For $|\underline{\underline{A}}| \neq 0$ dividing both sides of equation (2.81) by the determinant $|\underline{\underline{A}}|$ yields:

$$\underline{\underline{I}} = \frac{\underline{\underline{A}} \, \underline{\underline{C}}^T}{|\underline{\underline{A}}|}. \tag{2.82}$$

Thus, the following theorem holds:

The matrix $\underline{\underline{A}}$ is nonsingular if and only if its determinant is non-zero, $|\underline{\underline{A}}| \neq 0$. For the nonsingular matrix $\underline{\underline{A}}$ its inverse is given by the expression

$$\underline{\underline{A}}^{-1} = \frac{\underline{\underline{C}}^T}{|\underline{\underline{A}}|} . \tag{2.83}$$

For an invertible square matrix $\underline{\underline{A}}$ it follows from (2.63) by taking $\underline{\underline{B}} = \underline{\underline{A}}^{-1}$ that

$$|\underline{\underline{A}}^{-1}| = |\underline{\underline{A}}|^{-1} . \tag{2.84}$$

The concept of determinants can be used to develop a simple method to solve $n \times n$ system of algebraic equations

$$a_{11}x_1 + a_{12}x_2 + \cdots + a_{1n}x_n = b_1$$
$$a_{21}x_1 + a_{22}x_2 + \cdots + a_{2n}x_n = b_2$$
$$\vdots \tag{2.85}$$
$$a_{n1}x_1 + a_{n2}x_2 + \cdots + a_{nn}x_n = b_n .$$

In matrix notation, these algebraic equations can be cast into

$$\underline{\underline{A}}\underline{x} = \underline{b}, \quad \underline{x} = \begin{bmatrix} x_1 \\ x_2 \\ \vdots \\ x_n \end{bmatrix}, \quad \underline{b} = \begin{bmatrix} b_1 \\ b_2 \\ \vdots \\ b_n \end{bmatrix} . \tag{2.86}$$

If the n-column vector $\underline{b} = 0$ the equation is called *homogeneous*, if $\underline{b} \neq 0$ then equation (2.86) is called *non-homogeneous*.

Let us consider the case of $|\underline{\underline{A}}| \neq 0$ with an inverse $\underline{\underline{A}}^{-1}$ of $\underline{\underline{A}}$. We multiply both sides of (2.86) by $\underline{\underline{A}}^{-1}$, the solution is

$$\underline{x} = \underline{\underline{A}}^{-1}\underline{b} = \frac{\underline{\underline{C}}^T \underline{b}}{|\underline{\underline{A}}|} \tag{2.87}$$

or in components

$$x_i = \frac{\sum_{j=1}^{n} b_j C_{ji}}{|\underline{\underline{A}}|} . \tag{2.88}$$

Thus, we can find solution $\underline{x}$ once an inverse $\underline{\underline{A}}^{-1}$ of $\underline{\underline{A}}$ is known. In this case, if $\underline{b} = 0$ then a trivial solution $\underline{x} = 0$ is the only solution to the homogeneous problem.

The numerator of (2.88) bears resemblance to the cofactor expansion $|\underline{\underline{A}}| = \sum_{j=1}^{n} a_{ji} C_{ji}$ along the i-th column. This comparison reveals that the numerator can be viewed as the cofactor expansion of the matrix obtained from $\underline{\underline{A}}$ by replacing its i-th column by the column $\underline{b}$. Equation (2.88) is referred to as Cramer's rule. A unique solution to algebraic equation (2.85) for $\Delta \equiv |\underline{\underline{A}}| \neq 0$ is given by

$$x_1 = \frac{\Delta_1}{\Delta}, \; x_2 = \frac{\Delta_2}{\Delta}, \; \ldots, \; x_i = \frac{\Delta_i}{\Delta}, \ldots, x_n = \frac{\Delta_n}{\Delta}, \tag{2.89}$$

where

$$\Delta_i = |\underline{a}_1 \; \cdots \; \underline{a}_{i-1} \; \underline{b} \; \underline{a}_{i+1} \; \cdots \; \underline{a}_n|$$

is the determinant of the matrix formed by replacing the i-th column $\underline{a}_i$ of the matrix $\underline{\underline{A}}$ by the column vector $\underline{b}$ from (2.86).

☞ **Example 2.1.9.** Cramer's rule for 2 equations with 2 unknowns. If $\Delta = ad - bc \neq 0$ the system

$$\begin{array}{c} ax + by = e \\ cx + dy = f \end{array} \quad \rightarrow \quad \begin{bmatrix} a & b \\ c & d \end{bmatrix} \begin{bmatrix} x \\ y \end{bmatrix} = \begin{bmatrix} e \\ f \end{bmatrix}$$

has the unique solution

$$x = \frac{\Delta_1}{\Delta}, \quad y = \frac{\Delta_2}{\Delta},$$

where

$$\Delta_1 = \begin{vmatrix} e & b \\ f & d \end{vmatrix} \quad \text{and} \quad \Delta_2 = \begin{vmatrix} a & e \\ c & f \end{vmatrix}.$$

⚖ Determinant and Matrix Inversion: Problems

Problem 2.1.14. Let $\underline{\underline{A}} = \begin{bmatrix} 1 & 2 & 3 \\ 4 & 5 & 6 \\ 1 & 1 & 2 \end{bmatrix}$. Compute $|\underline{\underline{A}}|$ using a minor expansion.

Problem 2.1.15. Apply expression (2.83) to a 2×2 matrix $\underline{\underline{A}}$ to verify (2.8) and (2.84).

Problem 2.1.16. Show that

$$|k\underline{\underline{A}}| = k^n |\underline{\underline{A}}|,$$

where $\underline{\underline{A}}$ is any $n \times n$ matrix and k is a constant.

Problem 2.1.17. For the permutation symbol ϵ_{ijk} introduced in expression (1.35), show that,

$$\epsilon_{ijk}\epsilon_{mnl} = \begin{vmatrix} \delta_{im} & \delta_{in} & \delta_{il} \\ \delta_{jm} & \delta_{jn} & \delta_{jl} \\ \delta_{km} & \delta_{kn} & \delta_{kl} \end{vmatrix}, \quad i, j, k, m, n, l = 1, \ldots, 3.$$

2.1.5 Rank of a Matrix

We will need the following useful condition for the existence of an inverse matrix $\underline{\underline{A}}^{-1}$ of the $n \times n$ matrix $\underline{\underline{A}}$.

The inverse $\underline{\underline{A}}^{-1}$ exists if and only if all n columns (or all n rows) of $\underline{\underline{A}}$ are linearly independent.

Let us write the matrix $\underline{\underline{A}}$ as a row consisting of n column vectors:

$$\underline{\underline{A}} = \begin{bmatrix} \underline{a}_1 & \cdots & \underline{a}_n \end{bmatrix}, \quad \text{with} \quad \underline{a}_i = \begin{bmatrix} a_{1i} \\ a_{2i} \\ \vdots \\ a_{ni} \end{bmatrix} . \tag{2.90}$$

If all n column vectors $\underline{a}_i$, with $i = 1, .., n$ are linearly independent then

$$c_1 \underline{a}_1 + c_2 \underline{a}_2 + \ldots + c_n \underline{a}_n = 0 \tag{2.91}$$

only permits a trivial solution

$$c_1 = c_2 = \ldots = c_n = 0 .$$

Note that equation (2.91) can be given a matrix form of $\underline{\underline{A}} \underline{c} = 0$ with the column vector $\underline{c}$:

$$\underline{c} = \begin{bmatrix} c_1 \\ c_2 \\ \vdots \\ c_n \end{bmatrix} . \tag{2.92}$$

If $\underline{\underline{A}}$ has an inverse then $\underline{c} = \underline{\underline{A}}^{-1} \underline{\underline{A}} \underline{c} 0$ and the column vectors $\underline{a}_i$ are linearly independent.

Conversely, if the n column vectors $\underline{a}_i$ with $i = 1, .., n$ are linearly independent then it follows that $|\underline{\underline{A}}| \neq 0$. We will show that using operations on the columns that do not change the value of the determinant of $\underline{\underline{A}}$. Choose (if necessary by transposition) the last column $\underline{a}_n$ to be such that the last element $a_{nn} \neq 0$. At least one column vector must have a non-zero last element or the n-column vectors are not linearly independent. By subtracting a multiple of $\underline{a}_n$ from $\underline{a}_i$, $i < n$ columns one can arrange the last row to be $\begin{bmatrix} 0 & 0 & \cdots & a_{nn} \end{bmatrix}$ without changing the value of the determinant. Let $\underline{a}_{n-1}$ be the next-to-last column with matrix element $a_{n-1n-1} \neq 0$. It is always possible to arrange for a non-zero matrix element in that position because of the assumption that the column vectors are linearly independent. Consider the $n-1$-th row. By subtracting a multiple of $\underline{a}_{n-1}$ from all $\underline{a}_i$, $i < n-1$ columns one can arrange for the next-to-last row to have $[0, 0, \ldots, a_{n-1n-1}]$ among its first $n-1$ entries. Continuing this process one arrives at the *upper-triangular* matrix of the form:

$$\underline{\underline{A}}' = \begin{bmatrix} a'_{11} & \cdots & \cdots & \cdots & a_{1n} \\ 0 & a'_{22} & \cdots & \cdots & a_{2n} \\ 0 & 0 & a'_{33} & \cdots & a_{3n} \\ \vdots & 0 & \ddots & \ddots & \vdots \\ 0 & \cdots & 0 & 0 & a_{nn} \end{bmatrix} \tag{2.93}$$

in which all entries $a_{i<j}$ below the diagonal are zero. By construction, it holds that $|\underline{\underline{A}}'| = |\underline{\underline{A}}|$. Now, we will reverse the process and subtract the multiple of the first column in the matrix in (2.93) from the remaining $n-1$ columns located to the right of the first column to ensure that the first row takes the form $\begin{bmatrix} a'_{11} & 0 & \cdots & 0 \end{bmatrix}$. In the resulting matrix, the second column becomes

$$\underline{a}'_2 = \begin{bmatrix} 0 \\ a'_{22} \\ 0 \\ \vdots \\ 0 \end{bmatrix}$$

and can be used to eliminate all elements $a'_{2j}, j > 2$ from the second row. Eventually, we end up with the diagonal matrix $\underline{\underline{A}}''$ with determinant $|\underline{\underline{A}}''| = |\underline{\underline{A}}| = a'_{11} a'_{22} \cdots a'_{nn}$. This expression can only be zero if one of the diagonal elements a'_{ii} vanishes. However, that amounts to the obtaining a null i-th column as a result of subtracting linear combinations of other columns. Clearly, this can not happen for the n linearly independent column vectors and so the determinant $|\underline{\underline{A}}| \neq 0$ and $\underline{\underline{A}}^{-1}$ exists.

☞ **Example 2.1.10.** Let us illustrate the proof by considering the 3×3 matrix $\underline{\underline{A}}$ from (2.78). $\underline{\underline{A}}$ contains three linearly independent column vectors $\underline{a}_i$, $i = 1, 2, 3$. Multiplying the third column by 3 and subtracting it from the second column gives

$$\begin{bmatrix} 1 & 1 & 0 \\ 1 & 2 & 1 \\ 0 & 3 & 1 \end{bmatrix} \rightsquigarrow \begin{bmatrix} 1 & 1 & 0 \\ 1 & -1 & 1 \\ 0 & 0 & 1 \end{bmatrix}. \tag{2.94}$$

Adding the second column to the first yields an upper triangular matrix,

$$\begin{bmatrix} 1 & 1 & 0 \\ 1 & -1 & 1 \\ 0 & 0 & 1 \end{bmatrix} \rightsquigarrow \begin{bmatrix} 2 & 1 & 0 \\ 0 & -1 & 1 \\ 0 & 0 & 1 \end{bmatrix}. \tag{2.95}$$

Now, dividing the first column by 2 and subtracting it from the second results in

$$
\begin{bmatrix} 2 & 1 & 0 \\ 0 & -1 & 1 \\ 0 & 0 & 1 \end{bmatrix} \rightsquigarrow \begin{bmatrix} 2 & 0 & 0 \\ 0 & -1 & 1 \\ 0 & 0 & 1 \end{bmatrix} . \tag{2.96}
$$

Finally, the second column is added to the last, producing the diagonal matrix:

$$
\begin{bmatrix} 2 & 0 & 0 \\ 0 & -1 & 1 \\ 0 & 0 & 1 \end{bmatrix} \rightsquigarrow \begin{bmatrix} 2 & 0 & 0 \\ 0 & -1 & 0 \\ 0 & 0 & 1 \end{bmatrix} . \tag{2.97}
$$

The determinant of the matrix in (2.97) is equal to $2 \cdot (-1) \cdot 1 = -2$ which agrees with the determinant of $\underline{\underline{A}}$ in (2.78). All the steps of the proof can be repeated for rows instead of columns.

The maximum number of linearly independent row vectors of a matrix $\underline{\underline{A}}$ is called the row rank of $\underline{\underline{A}}$. The column rank of a matrix $\underline{\underline{A}}$ is the maximum number of linearly independent column vectors in $\underline{\underline{A}}$.

These two numbers, the row rank and column rank of $\underline{\underline{A}}$ are equal. This common number is called the rank of $\underline{\underline{A}}$ and denoted by rank $\underline{\underline{A}}$.

Let us consider a general $m \times n$ matrix $\underline{\underline{A}}$. The matrix $\underline{\underline{A}}$ consists of m rows and can be represented as an array of m row vectors:

$$
\underline{\underline{A}} = \begin{bmatrix} \underline{a}_1^T \\ \underline{a}_2^T \\ \vdots \\ \underline{a}_m^T \end{bmatrix}, \quad \underline{a}_j^T = \begin{bmatrix} a_{j1} & a_{j2} & \cdots & a_{jn} \end{bmatrix}, \quad j = 1, \ldots, m .
$$

Imagine that the row rank of $\underline{\underline{A}}$ is equal to some number r such that $r \le m$. Then, there are r row vectors $\underline{v}_1^T, \underline{v}_2^T, \ldots, \underline{v}_r^T$ among all row vectors $\underline{a}_1^T, \underline{a}_2^T, \ldots, \underline{a}_m^T$ which span all the row vectors:

$$
\underline{a}_j^T = \sum_{i=1}^{r} c_{ji} \, \underline{v}_i^T = \sum_{i=1}^{r} c_{ji} \begin{bmatrix} v_{i1} & v_{i2} & \cdots & v_{in} \end{bmatrix}, \quad j = 1, \ldots, m \tag{2.98}
$$

or in matrix notation

$$
\underline{\underline{A}} = \begin{bmatrix} c_{11} & c_{12} & \cdots & c_{1r} \\ c_{21} & c_{22} & \cdots & c_{2r} \\ \vdots & \vdots & \cdots & \vdots \\ c_{m1} & c_{m2} & \cdots & c_{mr} \end{bmatrix} \begin{bmatrix} v_{11} & v_{12} & \cdots & v_{1n} \\ v_{21} & v_{22} & \cdots & v_{2n} \\ \vdots & \vdots & \cdots & \vdots \\ v_{r1} & v_{r2} & \cdots & v_{rn} \end{bmatrix} = \underline{\underline{C}}\,\underline{\underline{V}} \qquad (2.99)
$$

with $m \times r$ matrix $\underline{\underline{C}}$ containing the coefficients of the row vector expansion in (2.98) and $r \times n$ matrix $\underline{\underline{V}}$ with r rows consisting of row vectors $\underline{v}_1, \ldots, \underline{v}_r$. Transposing both sides of equation (2.99) yields

$$
\underline{\underline{A}}^T = \left(\underline{\underline{C}}\,\underline{\underline{V}} \right)^T = \underline{\underline{V}}^T \underline{\underline{C}}^T .
$$

with $r \times m$ matrix $\underline{\underline{C}}^T$. Thus, the row vectors of the matrix $\underline{\underline{A}}^T$ will be spanned by at most r row vectors of $\underline{\underline{C}}^T$ and the row rank of $\underline{\underline{A}}^T$ is at most equal to r. However, by construction, the row vectors of $\underline{\underline{A}}^T$ are the column vectors of $\underline{\underline{A}}$ and so the column rank of $\underline{\underline{A}}$ is at most equal to r. Thus, for any matrix $\underline{\underline{A}}$ we have proved that

$$
\text{column rank of } \underline{\underline{A}} \leq \text{row rank of } \underline{\underline{A}}. \qquad (2.100)
$$

Repeating the same steps of the above argument for transpose matrix $\underline{\underline{A}}^T$ we get

$$
\text{column rank of } \underline{\underline{A}}^T \leq \text{row rank of } \underline{\underline{A}}^T
$$

or

$$
\text{row rank of } \underline{\underline{A}} \leq \text{column rank of } \underline{\underline{A}}. \qquad (2.101)
$$

Inequalities (2.100) and (2.101) imply that

$$
\text{column rank of } \underline{\underline{A}} = \text{column rank of } \underline{\underline{A}} = \text{rank}\,\underline{\underline{A}}.
$$

☞ **Example 2.1.11.** The matrix

$$
\underline{\underline{A}} = \begin{bmatrix} 1 & 1 & 0 & 0 \\ 0 & 0 & 2 & 3 \\ 1 & 1 & 2 & 3 \end{bmatrix}
$$

has row rank 2. The first two rows of $\underline{\underline{A}}$ are linearly independent while the third row is a sum of the first two top rows. The column rank is also 2. The first and the third columns are linearly independent. The second column is identical to the first column and the last column can be obtained from the third one via multiplication by a factor 3/2.

It is convenient to use the concept of rank to rephrase the condition for the existence of an inverse to the $n \times n$ matrix $\underline{\underline{A}}$:

The $n \times n$ matrix $\underline{\underline{A}}$ is invertible if and only if rank $\underline{\underline{A}} = n$.

We can use the above arguments to settle the question of linear independence of n-column vectors $\underline{v}_1, \underline{v}_2, \ldots, \underline{v}_n$ in terms of the determinant calculation.

It follows that $\underline{v}_1, \underline{v}_2, \ldots, \underline{v}_n$ are linearly independent if and only if the $n \times n$ matrix $\underline{\underline{A}} = \begin{bmatrix} \underline{v}_1 & \underline{v}_2 & \cdots & \underline{v}_n \end{bmatrix}$ is non-singular, i.e. the determinant $\begin{vmatrix} \underline{v}_1 & \underline{v}_2 & \cdots & \underline{v}_n \end{vmatrix}$ is different from zero.

⚓ Rank of a matrix: Problems

Problem 2.1.18. Find the rank of the matrix:

$$\begin{bmatrix} -1 & 4 & 13 \\ 3 & -12 & -39 \\ -2 & 8 & 26 \end{bmatrix}. \tag{2.102}$$

Problem 2.1.19.
(a) Show that n column vectors $\underline{v}_1, \ldots, \underline{v}_n$ are linearly dependent if and only if the determinant

$$|\underline{\underline{W}}| = \begin{vmatrix} \langle v_1 | v_1 \rangle & \langle v_1 | v_2 \rangle & \cdots & \langle v_1 | v_n \rangle \\ \langle v_2 | v_1 \rangle & \langle v_2 | v_2 \rangle & \cdots & \langle v_2 | v_n \rangle \\ \vdots & \vdots & \cdots & \vdots \\ \langle v_n | v_1 \rangle & \langle v_n | v_2 \rangle & \cdots & \langle v_n | v_n \rangle \end{vmatrix}$$

vanishes.
Hint: Consider a product of $\underline{v}_i$ with $\sum_{j=1}^{n} c_j v_j$ for $i = 1, \ldots, n$ and rewrite the result in matrix notation as $\underline{\underline{W}} \, \underline{c}$ with the column vector $\underline{c}$ as in equation (2.92).
(b) Use the above result to show that n orthogonal vectors must be linearly independent.
(c) Show that you can write the matrix $\underline{\underline{W}}$ from question (a) as

$$\underline{\underline{W}} = \underline{\underline{V}}^T \underline{\underline{V}}$$

for matrix $\underline{\underline{V}} = \begin{bmatrix} \underline{v}_1 & \cdots & \underline{v}_n \end{bmatrix}$ and that, therefore, vectors $\underline{v}_1, \ldots, \underline{v}_n$ are linearly dependent if and only if the determinant of $\underline{\underline{V}}$ vanishes in accordance with the main result of subsection 2.1.5.

2.2 The Eigenvalue Problem

In general, a vector $\underline{v}$ changes its direction when acted on by a matrix $\underline{\underline{A}}$. In some exceptional cases, the new vector $\underline{\underline{A}} \underline{v}$ lies along the same direction as $\underline{v}$. In such cases, the vector $\underline{\underline{A}}\,\underline{v}$ can be obtained by multiplying $\underline{v}$ by a number. These special cases are described by the following definition:

Definition 2.2.1. *A number λ is an eigenvalue of the square matrix $\underline{\underline{A}}$ if there exists a non-zero vector $\underline{v}$ such that*

$$\underline{\underline{A}}\,\underline{v} = \lambda\underline{v}. \tag{2.103}$$

Any such $\underline{v}$ is called an eigenvector of $\underline{\underline{A}}$ corresponding to the eigenvalue λ.

☞ **Example 2.2.1.** Recall the rotation matrix $\underline{\underline{R}}_z(\phi)$ from equation (2.50), which describes rotations around the z-axis by angle ϕ. Let us find a vector $\underline{v}$ that remains unaffected by such rotations. Clearly, this is equivalent to finding a solution to the equation $\underline{\underline{R}}_z(\phi)\underline{v} = \underline{v}$ or $\left(\underline{\underline{R}}_z(\phi) - \underline{\underline{I}}\right)\underline{v} = 0$. Therefore, we have to solve

$$\begin{bmatrix} \cos\phi - 1 & \sin\phi & 0 \\ -\sin\phi & \cos\phi - 1 & 0 \\ 0 & 0 & 0 \end{bmatrix} \begin{bmatrix} v_1 \\ v_2 \\ v_3 \end{bmatrix} = \begin{bmatrix} v_1(\cos\phi - 1) + v_2\sin\phi \\ -v_1\sin\phi + v_2(\cos\phi - 1) \\ 0 \end{bmatrix} = 0$$

for any angle ϕ. It is easy to see that the solution is provided by $v_1 = v_2 = 0$ and arbitrary v_3. Indeed, it makes sense that the rotation around the z-axis maps everything on the z-axis onto itself.

In other words, only a vector

$$\underline{v} = v_0 \begin{bmatrix} 0 \\ 0 \\ 1 \end{bmatrix},$$

which lies along the z-axis, is invariant under rotations generated by the matrix $\underline{\underline{R}}_z(\phi)$. Any such vector $\underline{v}$ is the eigenvector of the rotation matrix $\underline{\underline{R}}_z(\phi)$ with eigenvalue $+1$.

If λ is an eigenvalue of a square $n \times n$ matrix $\underline{\underline{A}}$ then the corresponding eigenvector $\underline{v}$ must be a nontrivial solution to the homogeneous linear system of equations given by

$$\left(\underline{\underline{A}} - \lambda\underline{\underline{I}}\right)\underline{v} = 0.$$

The nontrivial solution $\underline{v}$ exists if and only if the matrix $\underline{\underline{A}} - \lambda\underline{\underline{I}}$ is singular. Thus, the nontrivial solution to the eigenvalue problem exists if and only if

$$\boxed{\,\left|\,\underline{\underline{A}} - \lambda\underline{\underline{I}}\,\right| = 0.\,} \tag{2.104}$$

This equation is called the characteristic equation of $\underline{A}$ while the polynomial $|\underline{A} - \lambda \underline{I}|$ is called the characteristic polynomial of $\underline{A}$. Hence, the eigenvalue λ is a root of the characteristic polynomial $|\underline{A} - \lambda \underline{I}|$.

Using a cofactor expansion along the first row we obtain an expression

$$|\underline{A} - \lambda \underline{I}| = (a_{11} - \lambda)C_{11} + a_{12}C_{12} + \cdots + a_{1n}C_{1n}, \tag{2.105}$$

where the first cofactor C_{11} is a polynomial of order $n - 1$ in λ while each of the cofactors $C_{12}, C_{13}, \ldots, C_{1n}$ contains at most $n - 2$-nd power of λ. Repeated use of the minor expansions yields:

$$|\underline{A} - \lambda \underline{I}| = (a_{11} - \lambda)(a_{22} - \lambda) \cdots (a_{nn} - \lambda) + \text{terms of order } n - 2 \text{ and lower}. \tag{2.106}$$

The first term on the right hand side of (2.106) equals

$$(a_{11} - \lambda)(a_{22} - \lambda) \cdots (a_{nn} - \lambda) = (-1)^n \lambda^n + (-1)^{n-1}(a_{11} + a_{22} + \cdots + a_{nn})\lambda^{n-1}$$
$$+ \text{ terms of order } n - 2 \text{ and lower}$$
$$\tag{2.107}$$

The characteristic polynomial is a polynomial of order n in λ and, therefore, has n roots $\lambda_1, \lambda_2, \ldots, \lambda_n$. Based on this observation we can factorize the characteristic polynomial as

$$|\underline{A} - \lambda \underline{I}| = (\lambda_1 - \lambda)(\lambda_2 - \lambda) \cdots (\lambda_n - \lambda). \tag{2.108}$$

Expanding the right hand side of (2.108) in λ we obtain

$$|\underline{A} - \lambda \underline{I}| = (-1)^n \lambda^n + (-1)^{n-1}(\lambda_1 + \lambda_2 + \cdots + \lambda_n)\lambda^{n-1} + \cdots + \lambda_1 \lambda_2 \cdots \lambda_n, \tag{2.109}$$

where we only showed explicitly terms of order n, $n - 1$ and zero in λ. Setting $\lambda = 0$ on both sides of (2.109) we obtain

$$|\underline{A} - \lambda \underline{I}|\Big|_{\lambda=0} = |\underline{A}| = \lambda_1 \lambda_2 \cdots \lambda_n. \tag{2.110}$$

Thus, the product of all eigenvalues of $\underline{A}$ is equal to the determinant of $\underline{A}$.

We now turn our attention to the term containing λ^{n-1} on the right hand side of (2.109). Comparing expansions in (2.109) and (2.107) we make the following relation

$$\lambda_1 + \lambda_2 + \cdots + \lambda_n = a_{11} + a_{22} + \cdots + a_{nn}. \tag{2.111}$$

We recognize on the right hand side a trace of matrix $\underline{A}$ as defined in (2.11). Thus, we have shown that:

$$\text{Tr}(\underline{A}) = \sum_{i=1}^{n} \lambda_i \tag{2.112}$$

☞ **Example 2.2.2.** Consider

$$\underline{\underline{A}} = \begin{bmatrix} 0 & 1 \\ 1 & 0 \end{bmatrix}. \tag{2.113}$$

Then $\mathrm{Tr}\left(\underline{\underline{A}}\right) = 0$ and $|\underline{\underline{A}}| = -1$. Therefore, $\lambda_1 + \lambda_2 = 0, \lambda_1\lambda_2 = -1$. Thus, the eigenvalues of $\underline{\underline{A}}$ are $1, -1$.

Note that in the above example both eigenvalues of real matrix $\underline{\underline{A}}$ are real. However, the real matrices do not necessarily have real eigenvalues, as discussed in the next example.

☞ **Example 2.2.3.** Consider a real 3×3 matrix

$$\underline{\underline{A}} = \begin{bmatrix} 1 & 0 & 0 \\ 1 & 0 & -1 \\ 0 & 1 & 0 \end{bmatrix}. \tag{2.114}$$

The eigenvalues are solutions of the characteristic equation

$$|\underline{\underline{A}} - \lambda\underline{\underline{I}}| = \begin{vmatrix} 1-\lambda & 0 & 0 \\ 1 & -\lambda & -1 \\ 0 & 1 & -\lambda \end{vmatrix} = (1-\lambda)(\lambda^2 + 1) = 0. \tag{2.115}$$

The matrix $\underline{\underline{A}}$ has three eigenvalues $\lambda_1 = 1, \lambda_2 = i, \lambda_3 = -i$. Note that two complex eigenvalues $\pm i$ are complex conjugate of each other.

The eigenvector

$$\underline{v}_1 = \begin{bmatrix} v_1 \\ v_2 \\ v_3 \end{bmatrix}$$

corresponding to the eigenvalue λ_1 satisfies the linear equation $\left(\underline{\underline{A}} - \lambda_1\underline{\underline{I}}\right)\underline{v}_1 = 0$ or,

$$\begin{bmatrix} 0 & 0 & 0 \\ 1 & -1 & -1 \\ 0 & 1 & -1 \end{bmatrix} \begin{bmatrix} v_1 \\ v_2 \\ v_3 \end{bmatrix} = 0.$$

Clearly, the solution is

$$\underline{v}_1 = \begin{bmatrix} 2 \\ 1 \\ 1 \end{bmatrix}. \tag{2.116}$$

Similarly, the eigenvectors for the remaining eigenvalues are:

$$\text{for } \lambda_2 = i, \ \underline{v}_2 = \begin{bmatrix} 0 \\ i \\ 1 \end{bmatrix} \quad \text{and for } \lambda_3 = -i, \ \underline{v}_3 = \begin{bmatrix} 0 \\ -i \\ 1 \end{bmatrix}. \tag{2.117}$$

Notice that all eigenvalues $1, i, -i$ are distinct and that all three eigenvectors are linearly independent. This outcome is in agreement with a general theorem which we will prove below.

In general, complex eigenvalues of a real matrix always appear in pairs and are relatively complex conjugate to each other. This is a general feature that follows from the fact that the eigenvalues of the real matrices are roots of polynomials having real coefficients. We recall from basic algebra that complex roots of polynomials which have real coefficients can be grouped in mutually conjugate pairs. Below, we will learn that real eigenvalues are guaranteed for real matrices, that are invariant under matrix transposition.

☞ **Example 2.2.4.** Consider a system of two masses m_1 and m_2 connected by two springs with spring constants k_1 and k_2. For the displacements y_1, y_2 of

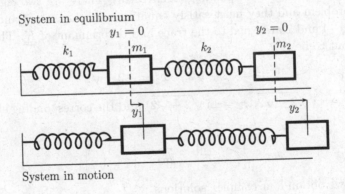

Figure 2.2: Two masses on two springs.

m_1, m_2 from their positions of equilibrium, shown in Figure 2.2, Newton's equation of motion yields

$$m_1 \frac{d^2 y_1}{dt^2} = -k_1 y_1 - k_2(y_1 - y_2)$$

$$m_2 \frac{d^2 y_2}{dt^2} = -k_2(y_2 - y_1).$$

In matrix notation the equations of motion become

$$\frac{d^2}{dt^2}\begin{bmatrix} y_1 \\ y_2 \end{bmatrix} = \begin{bmatrix} -\frac{k_1+k_2}{m_1} & \frac{k_2}{m_1} \\ \frac{k_2}{m_2} & -\frac{k_2}{m_2} \end{bmatrix}\begin{bmatrix} y_1 \\ y_2 \end{bmatrix} = \underline{\underline{K}}\begin{bmatrix} y_1 \\ y_2 \end{bmatrix}.$$

Note that the matrix $\underline{\underline{K}}$ is invariant under transposition ($\underline{\underline{K}} = \underline{\underline{K}}^T$) for $m_1 = m_2$. For values $m_1 = m_2 = 1$ and $k_1 = 2, k_2 = 1$ we obtain for the equations of motion

$$\frac{d^2}{dt^2}\begin{bmatrix} y_1 \\ y_2 \end{bmatrix} = \begin{bmatrix} -3 & 1 \\ 1 & -1 \end{bmatrix}\begin{bmatrix} y_1 \\ y_2 \end{bmatrix} = \underline{\underline{K}}\begin{bmatrix} y_1 \\ y_2 \end{bmatrix}.$$

We try an oscillating solution of the form

$$\underline{y} = \underline{y}_0 e^{\omega t},$$

with some constant vector $\underline{y}_0$. Plugging this expression into the equations of motion allows us to cast them in the form of the eigenvalue equation for the matrix $\underline{\underline{K}}$

$$\underline{\underline{K}}\,\underline{y}_0 = \begin{bmatrix} -3 & 1 \\ 1 & -1 \end{bmatrix}\underline{y}_0 = \omega^2 \underline{y}_0.$$

In the above equation ω^2 plays the role of the eigenvalue λ while vector $\underline{y}_0$ appears to be the corresponding eigenvector. There are two eigenvalues in this problem and they must satisfy equations $\omega_1^2 + \omega_2^2 = -4$ and $\omega_1^2\omega_2^2 = 2$ where -4 and 2 are equal to the trace and determinant of $\underline{\underline{K}}$. The solutions turn out to be

$$\omega_1^2 = -2 + \sqrt{2}, \quad \omega_2^2 = -2 - \sqrt{2}$$

(or $\omega_1 = \pm i\sqrt{2 - \sqrt{2}}, \omega_2 = \pm i\sqrt{2 + \sqrt{2}}$) and the corresponding eigenvectors are

$$\underline{y}_{01} = \begin{bmatrix} 1 \\ 1 + \sqrt{2} \end{bmatrix}, \quad \underline{y}_{02} = \begin{bmatrix} 1 \\ 1 - \sqrt{2} \end{bmatrix}.$$

Thus, we obtain four complex solutions

$$\underline{y}(t) = \underline{y}_{01}\exp\left(\pm it\sqrt{2 - \sqrt{2}}\right), \quad \underline{y}(t) = \underline{y}_{02}\exp\left(\pm it\sqrt{2 + \sqrt{2}}\right),$$

describing harmonic oscillations of two masses on two springs. The general oscillation is given by a superposition of these four complex solutions and can be written as

$$\begin{aligned} \underline{y}(t) &= a_1\underline{y}_{01}\cos\left(t\sqrt{2 - \sqrt{2}}\right) + b_1\underline{y}_{01}\sin\left(t\sqrt{2 - \sqrt{2}}\right) \\ &+ a_2\underline{y}_{02}\cos\left(t\sqrt{2 + \sqrt{2}}\right) + b_2\underline{y}_{02}\sin\left(t\sqrt{2 + \sqrt{2}}\right), \end{aligned}$$

with four complex coefficients a_1, b_1, a_2, b_2 which can be determined from information about initial locations and speeds of two oscillators.

Note that, in general, there are at most n distinct eigenvalues of $\underline{\underline{A}}$. The number of distinct eigenvalues provides information about linear independence of corresponding eigenvectors.

> Eigenvectors corresponding to distinct eigenvalues are linearly independent.

Suppose that $\lambda_1, \ldots, \lambda_k$, $k \leq n$ are distinct eigenvalues of a $n \times n$ matrix $\underline{\underline{A}}$, and let $\underline{v}_i$ be an eigenvector of $\underline{\underline{A}}$ with the eigenvalue λ_i, $\underline{\underline{A}}\underline{v}_i = \lambda_i\underline{v}_i$, $i = 1, \ldots, k$. We will show that the k-eigenvectors $\underline{v}_1, \ldots, \underline{v}_k$ are linearly independent by method of induction. The induction method starts with the case $k = 1$ but for only one eigenvector the statement follows trivially. Next, assume that the statement holds for $k - 1$ distinct eigenvalues $\lambda_1, \ldots, \lambda_{k-1}$ and suppose that the relation

$$c_1\underline{v}_1 + \cdots + c_k\underline{v}_k = 0$$

holds. Now, apply the matrix $\left(\underline{\underline{A}} - \lambda_k\underline{\underline{I}}\right)$ on both sides of the above relation to get:

$$\left(\underline{\underline{A}} - \lambda_k\underline{\underline{I}}\right)\left(c_1\underline{v}_1 + \cdots + c_k\underline{v}_k\right) = c_1(\lambda_1 - \lambda_k)\underline{v}_1 + \cdots + c_{k-1}(\lambda_{k-1} - \lambda_k)\underline{v}_{r-1} + 0 = 0.$$

By induction hypothesis, $\underline{v}_1, \ldots, \underline{v}_{k-1}$ are linearly independent. Hence,

$$c_1(\lambda_1 - \lambda_k) = c_2(\lambda_2 - \lambda_k) = \ldots = c_{k-1}(\lambda_{k-1} - \lambda_k) = 0.$$

the assumption that all $\lambda_i - \lambda_k \neq 0$ for $i = 1, \ldots, k - 1$ implies that all coefficients $c_i, i = 1, \ldots k - 1$ must be zero. Then from the relation, $c_1\underline{v}_1 + \cdots + c_k\underline{v}_k = 0$, we conclude that $c_k\underline{v}_k = 0$ and c_k must vanish too. Thus, all the vectors $\underline{v}_1, \ldots, \underline{v}_k$ must be linearly independent.

2.2.1 Some Applications of the Eigenvalue Problem, Diagonalization

Here, we describe the use of eigenvectors to diagonalize a square $n \times n$ matrix $\underline{\underline{A}}$, which possesses n *linearly independent* eigenvectors (that happens, for instance when $\underline{\underline{A}}$ has n distinct eigenvalues).

Let $\underline{v}_i, i = 1, \ldots, n$ be linearly independent solutions of the eigenvalue problem:

$$\underline{\underline{A}}\,\underline{v}_i = \lambda_i\,\underline{v}_i, \quad i = 1, \ldots, n. \tag{2.118}$$

Let matrix $\underline{\underline{V}}$ be constructed out of the eigenvectors $\underline{v}_i$ in such a way that the columns of $\underline{\underline{V}}$ are taken to be eigenvectors of $\underline{\underline{A}}$:

$$\underline{\underline{V}} = \begin{bmatrix} \underline{v}_1 & \underline{v}_2 & \cdots & \underline{v}_n \end{bmatrix}. \tag{2.119}$$

The square matrix $\underline{\underline{V}}$ has rank n because all its column vectors are assumed to be linearly independent. Thus, there exists the inverse matrix $\underline{\underline{V}}^{-1}$.

Now, we will apply a similarity transformation on $\underline{\underline{A}}$:

$$\underline{\underline{A}} \; \rightarrow \; \underline{\underline{V}}^{-1}\underline{\underline{A}}\underline{\underline{V}} \equiv \underline{\underline{D}}. \tag{2.120}$$

The resulting matrix $\underline{\underline{D}}$ is diagonal. T show that this this is true, consider

$$\underline{\underline{A}}\underline{\underline{V}} = \begin{bmatrix} \underline{\underline{A}}\underline{v}_1 & \underline{\underline{A}}\underline{v}_2 & \cdots & \underline{\underline{A}}\underline{v}_n \end{bmatrix} = \begin{bmatrix} \lambda_1\underline{v}_1 & \lambda_2\underline{v}_2 & \cdots & \lambda_n\underline{v}_n \end{bmatrix}$$

$$= \underline{\underline{V}} \begin{bmatrix} \lambda_1 & 0 & 0 & 0 \\ 0 & \lambda_2 & 0 & 0 \\ 0 & 0 & \ddots & 0 \\ 0 & 0 & 0 & \lambda_n \end{bmatrix}, \tag{2.121}$$

where we used the fact that multiplying $\underline{\underline{V}}$ by $\underline{\underline{A}}$ amounts to replacing each column vector $\underline{v}_i$ by $\underline{\underline{A}}\underline{v}_i$. Since $\underline{v}_i$ is a solution of the eigenvalue problem (2.118) this column vector is equal to $\lambda_i\underline{v}_i$. It follows that

$$\underline{\underline{D}} = \underline{\underline{V}}^{-1}\underline{\underline{A}}\underline{\underline{V}} = \underline{\underline{V}}^{-1}\underline{\underline{V}} \begin{bmatrix} \lambda_1 & 0 & 0 & 0 \\ 0 & \lambda_2 & 0 & 0 \\ 0 & 0 & \ddots & 0 \\ 0 & 0 & 0 & \lambda_n \end{bmatrix} = \begin{bmatrix} \lambda_1 & 0 & 0 & 0 \\ 0 & \lambda_2 & 0 & 0 \\ 0 & 0 & \ddots & 0 \\ 0 & 0 & 0 & \lambda_n \end{bmatrix}.$$

If the above operation of to transform a matrix by a similarity transformation to a diagonal matrix is possible, then we say that the matrix is *diagonalizable*. Therefore, an $n \times n$ matrix that has n linearly independent eigenvectors is diagonalizable.

☞ **Example 2.2.5.** Consider the 2×2 matrix

$$\underline{\underline{A}} = \begin{bmatrix} 4 & -6 \\ 1 & -1 \end{bmatrix}.$$

The characteristic equation

$$\begin{vmatrix} 4-\lambda & -6 \\ 1 & -1-\lambda \end{vmatrix} = \lambda^2 - 3\lambda + 2 = (\lambda - 1)(\lambda - 2) = 0$$

yields two distinct eigenvalues, $\lambda_1 = 1$ and $\lambda_2 = 2$. Two independent eigenvectors

$$\underline{v}_1 = \begin{bmatrix} 2 \\ 1 \end{bmatrix}, \quad \underline{v}_2 = \begin{bmatrix} 3 \\ 1 \end{bmatrix}$$

satisfy the eigenvalue equations

$$\underline{\underline{A}}\underline{v}_1 = \underline{v}_1, \quad \underline{\underline{A}}\underline{v}_2 = 2\underline{v}_2.$$

Define the 2×2 matrix $\underline{V}$ as

$$\underline{V} = [\underline{v}_1 \quad \underline{v}_2] = \begin{bmatrix} 2 & 3 \\ 1 & 1 \end{bmatrix}.$$

The matrix $\underline{V}$ has rank 2. Subsequently, there exists an inverse matrix

$$\underline{V}^{-1} = \begin{bmatrix} -1 & 3 \\ 1 & -2 \end{bmatrix}.$$

Applying the similarity transformation defined in (2.120) one obtains the diagonal matrix

$$\begin{bmatrix} -1 & 3 \\ 1 & -2 \end{bmatrix} \begin{bmatrix} 4 & -6 \\ 1 & -1 \end{bmatrix} \begin{bmatrix} 2 & 3 \\ 1 & 1 \end{bmatrix} = \begin{bmatrix} 1 & 0 \\ 0 & 2 \end{bmatrix},$$

with eigenvalues 1 and 2 as diagonal elements.

✎ The eigenvalue problem: Problems

Problem 2.2.1. Calculate the eigenvalues and eigenvectors of the following matrices:

(a) $\sigma_1 = \begin{bmatrix} 0 & 1 \\ 1 & 0 \end{bmatrix}$

(b) $\sigma_2 = \begin{bmatrix} 0 & -i \\ i & 0 \end{bmatrix}$

(c) $\begin{bmatrix} 0 & 0 & 1 \\ 0 & 0 & 1 \\ 1 & 1 & 0 \end{bmatrix}$

(d) $\begin{bmatrix} i & 0 & 0 \\ 0 & 0 & i \\ 0 & i & 0 \end{bmatrix}$

Problem 2.2.2. Verify that the characteristic polynomial of a 3×3 matrix $\underline{\underline{A}}$ can be written as:

$$|\underline{\underline{A}} - \lambda \underline{\underline{I}}| = -\lambda^3 + \mathrm{Tr}\,(\underline{\underline{A}})\, \lambda^2 - \mathrm{Tr}\,(\underline{\underline{C}})\, \lambda + |\underline{\underline{A}}|,$$

where $\underline{\underline{C}}$ is the cofactor matrix of a 3×3 matrix $\underline{\underline{A}}$ (a matrix with matrix element c_{ij} equal to the the determinant of the 2×2 matrix formed from eliminating the i-th row and j-th column from $\underline{\underline{A}}$).

Problem 2.2.3. Let $\lambda_1, \ldots, \lambda_n$ be eigenvalues of an $n \times n$ matrix, $\underline{\underline{A}}$. Find eigenvalues of the matrix $\underline{\underline{A}}'$ where

$$\underline{\underline{A}}' = \underline{\underline{S}}^{-1} \underline{\underline{A}} \underline{\underline{S}}$$

and $\underline{S}$ is a nonsingular $n \times n$ matrix.

Problem 2.2.4. Let $\underline{\underline{A}}' = \underline{\underline{S}}^{-1} \underline{\underline{A}} \underline{\underline{S}}$. Show that $\mathrm{Tr}\,\underline{\underline{A}}' = \mathrm{Tr}\,\underline{\underline{A}}$ and $|\underline{\underline{A}}'| = |\underline{\underline{A}}|$.

Problem 2.2.5. We have shown in the text that a matrix which has n linearly independent eigenvectors is diagonalizable. Show the inverse, namely, that the $n \times n$ diagonalizable matrix $\underline{\underline{A}}$ has n linearly independent eigenvectors. *Hint:* Suppose that $\underline{\underline{D}} = \underline{V}^{-1} \underline{\underline{A}} \underline{V}$ and show that the column vectors of $\underline{V}$ coincide with the eigenvectors of $\underline{\underline{A}}$.

Problem 2.2.6. Show that the two matrices found in Problem 2.1.9 are related by the similarity transformation, $\underline{\underline{A}}' = \underline{\underline{S}}^{-1} \underline{\underline{A}} \underline{\underline{S}}$, where $\underline{\underline{S}}$ is a matrix that maps the canonical basis $\{\underline{e}_1, \underline{e}_2\}$ into the basis $\{\underline{v}_1, \underline{v}_2\}$.

Problem 2.2.7. Find a basis of eigenvectors and diagonalize:

(a) $\begin{bmatrix} 5 & 4 \\ 1 & 2 \end{bmatrix}$
(b) $\begin{bmatrix} 5 & 8 \\ 4 & 1 \end{bmatrix}$

Problem 2.2.8. (a) Find a basis of eigenvectors for

$$\underline{\underline{A}} = \frac{1}{2} \begin{bmatrix} 3 & -3 \\ -3 & 3 \end{bmatrix}$$

and diagonalize it by a similarity transformation.
(b) Use a similarity transformation from (a) to find $\underline{\underline{A}}^n$, where n is a positive integer.

Problem 2.2.9. (a) For the matrices:

$$\underline{\underline{B}} = \begin{bmatrix} 13 & -16 \\ 9 & -11 \end{bmatrix}, \quad \underline{\underline{V}} = \begin{bmatrix} -2 & 3 \\ 3 & -4 \end{bmatrix}$$

find matrix $\underline{\underline{C}} = \underline{\underline{V}} \underline{\underline{B}} \underline{\underline{V}}^{-1}$.
(b) Solve the eigenvalue problem for the matrix $\underline{\underline{B}}$. Can this matrix be diagonalized?
(c) Use the similarity transformation from (a) to show that

$$\underline{\underline{B}}^n = \begin{bmatrix} 1 + 12n & -16n \\ 9n & 1 - 12n \end{bmatrix}.$$

Problem 2.2.10. Find a basis of eigenvectors and diagonalize:

(a) $\begin{bmatrix} 1 & 0 & -3/2 \\ 0 & 1 & 3 \\ 0 & 2 & 2 \end{bmatrix}$
(b) $\begin{bmatrix} 1 & 2 & 2 \\ 1 & 2 & -1 \\ -1 & 1 & 4 \end{bmatrix}$

Problem 2.2.11. (a) Find eigenvalues of the upper-triangular matrix

$$\begin{bmatrix} 2 & 5 & 4 \\ 0 & 3 & 13 \\ 0 & 0 & 1 \end{bmatrix}$$

(b) Show that the triangular matrices have their eigenvalues appearing along the diagonal.

Problem 2.2.12. Let $\underline{A}$ be a 3×3 matrix with eigenvalues $\lambda_1 = -3$, $\lambda_2 = 2$, $\lambda_3 = 0$ and corresponding eigenvectors $\underline{v}_1$, $\underline{v}_2$, $\underline{v}_3$.
(a) Find $\underline{v}$ such that $\underline{A}\underline{v} = \underline{v}_1 - \underline{v}_2$ in terms of $\underline{v}_1$, $\underline{v}_2$, $\underline{v}_3$.
(b) Is matrix $\underline{\underline{X}} = \underline{I} + \underline{A}$ invertible? If yes find the determinant and trace of $\underline{X}^{-1}$.

Problem 2.2.13. Show that a square singular (non-invertible) matrix $\underline{A}$ will have at least one (non-zero) eigenvector with zero eigenvalue ($\underline{A}\underline{v} = 0$). Is $\underline{A}$ invertible if it has an eigenvector with zero eigenvalue?

2.2.2 Orthogonal and Unitary Matrices

Definition 2.2.2. *A real square square matrix $\underline{A}$ is called orthogonal if its transpose $\underline{A}^T$ is equal to its inverse:*

$$\underline{A}^T = \underline{A}^{-1} \quad \rightarrow \quad \underline{A}^T \underline{A} = \underline{I}. \tag{2.122}$$

The transformation of a column vector, $\underline{x}$, in the following relation,

$$\underline{x}' = \underline{A}\underline{x}, \tag{2.123}$$

by an orthogonal matrix $\underline{A}$ is called an orthogonal transformation. Its significance stems from the fact that the scalar product of vectors defined in (2.14) is kept invariant under the orthogonal transformation. Let $\underline{x}'$, $\underline{y}'$ be two transformed vectors, such that

$$\underline{x}' = \underline{A}\underline{x}, \quad \underline{y}' = \underline{A}\underline{y}. \tag{2.124}$$

Then, the inner product of vectors, $\underline{x}$ and $\underline{y}$ is preserved under orthogonal transformation as seen from:

$$\langle \underline{x}'|\underline{y}' \rangle = (\underline{A}\underline{x})^T \underline{A}\underline{y} = \underline{x}^T \underline{A}^T \underline{A}\underline{y} = \underline{x}^T \underline{I}\underline{y} = \langle \underline{x}|\underline{y} \rangle. \tag{2.125}$$

The above relation implies that the norm of the vector $\|\underline{x}\| = \sqrt{\langle \underline{x}|\underline{x} \rangle}$ is also preserved under orthogonal transformation.

Recall the orthonormal canonical basis vectors on $\mathbb{R}^n$ as defined in (2.16). The transformed column vector

$$\underline{e}_i' = \underline{A}\underline{e}_i = \underline{a}_i \tag{2.126}$$

is equal to the i-th column of the transforming matrix, where

$$\underline{A} = [\underline{a}_1 \quad \underline{a}_2 \quad \cdots \quad \underline{a}_n].$$

If matrix $\underline{A}$ is orthogonal then we obtain, from (2.125),

$$\delta_{ij} = \langle \underline{e}_i|\underline{e}_j \rangle = \langle \underline{e}_i'|\underline{e}_j' \rangle = \underline{a}_i^T \underline{a}_j = \langle \underline{a}_i|\underline{a}_j \rangle. \tag{2.127}$$

We conclude by the following observation.

The column vectors of the orthogonal matrix must form an orthonormal system:

$$\langle \underline{a}_i | \underline{a}_j \rangle = \delta_{ij}\,.$$

The above result can also be obtained directly from the fact that $\underline{\underline{A}}^T \underline{\underline{A}} = \underline{\underline{I}}$. Representing the matrix $\underline{\underline{A}}^T$ in terms of the row vectors $\underline{a}_i^T$, $i = 1, \ldots, n$ as

$$\underline{\underline{A}}^T = \begin{bmatrix} \underline{a}_1^T \\ \underline{a}_2^T \\ \vdots \\ \underline{a}_n^T \end{bmatrix}$$

we get

$$\underline{\underline{A}}^T \underline{\underline{A}} = \begin{bmatrix} \underline{a}_1^T \underline{a}_1 & \underline{a}_1^T \underline{a}_2 & \cdots & \underline{a}_1^T \underline{a}_n \\ \underline{a}_2^T \underline{a}_1 & \underline{a}_2^T \underline{a}_2 & \cdots & \underline{a}_2^T \underline{a}_n \\ \vdots & \vdots & \cdots & \vdots \\ \underline{a}_n^T \underline{a}_1 & \underline{a}_n^T \underline{a}_2 & \cdots & \underline{a}_n^T \underline{a}_n \end{bmatrix} = \begin{bmatrix} 1 & 0 & \cdots & 0 \\ 0 & 1 & \ddots & \vdots \\ \vdots & \ddots & \ddots & 0 \\ 0 & \cdots & 0 & 1 \end{bmatrix},$$

which, for the matrix elements, amounts to

$$\left(\underline{\underline{A}}^T \underline{\underline{A}} \right)_{ij} = \sum_{k=1}^n a_{ki} a_{kj} = \underline{a}_i^T \underline{a}_j = \delta_{ij}, \quad i, j = 1, \ldots, n\,.$$

Thus, the matrix $\underline{\underline{A}}$ is orthogonal if and only if

$$\underline{a}_i^T \underline{a}_j = \delta_{ij}, \quad i, j = 1, \ldots, n\,,$$

meaning that the column vectors of $\underline{\underline{A}}$ form an orthonormal system.

☞ **Example 2.2.6.** Consider the 3×3 matrix

$$\underline{\underline{A}} = \frac{1}{2} \begin{bmatrix} 1 & \sqrt{2} & 1 \\ -\sqrt{2} & 0 & \sqrt{2} \\ 1 & -\sqrt{2} & 1 \end{bmatrix}\,.$$

The three column vectors:

$$\frac{1}{2} \begin{bmatrix} 1 \\ -\sqrt{2} \\ 1 \end{bmatrix}, \quad \frac{1}{2} \begin{bmatrix} \sqrt{2} \\ 0 \\ -\sqrt{2} \end{bmatrix}, \quad \frac{1}{2} \begin{bmatrix} 1 \\ \sqrt{2} \\ 1 \end{bmatrix}$$

are orthogonal to each other and all have length one. The same holds for the row vectors of $\underline{\underline{A}}$. Thus, $\underline{\underline{A}}$ is an orthogonal matrix.

Now, let us assign complex values to the matrix elements of $\underline{A}$. For complex square matrices, a concept of the orthogonal matrix generalizes to:

Definition 2.2.3. *A square matrix $\underline{U}$ is called unitary if*

$$\underline{U}^\dagger = \underline{U}^{-1} \quad \rightarrow \quad \underline{U}^\dagger \underline{U} = \underline{I}. \tag{2.128}$$

For the real matrix $\underline{U}$ its adjoint is equal to its transpose $\underline{U}^\dagger = \underline{U}^T$ and definition (2.128) reproduces the definition of an orthogonal matrix.

Hence, the concept of unitarity seems to generalize the concept of orthogonality to complex matrices. This generalization involves complex vector spaces instead of real vector spaces. In an analogy with a discussion of orthogonal matrices and real vector spaces we will see that the inner product defined on the complex vector space $\mathbb{C}^n$ is kept invariant by transformations that are generated by unitary transformations. Let two complex column vectors $\underline{z}', \underline{w}'$ be derived from vectors $\underline{z}, \underline{w}$ via transformation by $\underline{U}$:

$$\underline{z}' = \underline{U}\,\underline{z}, \qquad \underline{w}' = \underline{U}\,\underline{w}. \tag{2.129}$$

The inner product (2.21) of transformed complex column vectors becomes

$$\langle \underline{z}' | \underline{w}' \rangle = (\underline{U}\,\underline{z})^\dagger \underline{U}\,\underline{w} = \underline{z}^\dagger \underline{U}^\dagger \underline{U}\,\underline{w} = \underline{z}^\dagger \underline{I}\,\underline{w} = \langle \underline{z} | \underline{w} \rangle, \tag{2.130}$$

and is preserved under unitary transformation.

The concept of an orthonormal system of real vectors generalizes to the unitary system:

$$\langle \underline{a}_i | \underline{a}_j \rangle = \underline{a}_i^\dagger \underline{a}_j = \delta_{ij}, \quad i,j = 1,\ldots,n \tag{2.131}$$

that is defined with respect to the inner product (2.21) on the complex vector space $\mathbb{C}^n$. The column vectors of the unitary matrices form an unitary system analogous to the column vectors of the orthogonal matrices forming an orthonormal system.

☞ **Example 2.2.7.** The 3×3 matrix

$$\underline{U} = \begin{bmatrix} i & 0 & 0 \\ 0 & -i & 0 \\ 0 & 0 & i \end{bmatrix},$$

is unitary. One way to see it is to notice that the column vectors of $\underline{U}$,

$$\begin{bmatrix} i \\ 0 \\ 0 \end{bmatrix}, \begin{bmatrix} 0 \\ -i \\ 0 \end{bmatrix}, \begin{bmatrix} 0 \\ 0 \\ i \end{bmatrix},$$

form an unitary system.

Now, consider the eigenvalue problem for an unitary matrix

$$\underline{U}\,\underline{z} = \lambda \underline{z}. \tag{2.132}$$

Applying the adjoint operation on both sides of relation (2.132) yields

$$(\underline{U}\,\underline{z})^{\dagger} = (\lambda \underline{z})^{\dagger} = \lambda^{*}\underline{z}^{\dagger}. \tag{2.133}$$

The inner product of $\underline{U}\,\underline{z}$ with itself gives

$$\langle \underline{U}\,\underline{z} | \underline{U}\,\underline{z} \rangle = (\underline{U}\,\underline{z})^{\dagger}\underline{U}\,\underline{z} = \lambda^{*}\lambda\,\underline{z}^{\dagger}\underline{z} = |\lambda|^{2}\,\|\underline{z}\|^{2}. \tag{2.134}$$

On the other hand, we know from (2.130) that the right hand side of (2.134) is equal to $\langle \underline{z}|\underline{z} \rangle = \|\underline{z}\|^{2}$ and, therefore,

$$\|\underline{z}\|^{2} = |\lambda|^{2}\|\underline{z}\|^{2}. \tag{2.135}$$

Dividing both sides of (2.135) by $\|\underline{z}\|^{2} \neq 0$ reveals that all eigenvalues of the unitary matrix have an absolute value given by:

$$\lambda^{*}\lambda = 1. \tag{2.136}$$

This property is shared by eigenvalues of an orthogonal matrix. In addition, since an orthogonal matrix has real entries, the characteristic polynomial will have real coefficients and, thus, its roots (which are eigenvalues) must be either real or complex conjugate in pairs. The possible real eigenvalues of an orthogonal matrix are then $+1$ or -1. The complex eigenvalues appear in pairs which are of the type $e^{i\theta}, e^{-i\theta}$ with the real parameter θ.

We know from the definition of an unitary (and orthogonal) matrix that its determinant can not vanish since the inverse matrix exists. The possible values of such determinants turn out to be very restricted as shown by the following theorem.

The determinant of an unitary matrix has absolute value 1.

The proof is based on determinant properties (2.63) and (2.84) and the relation $|\underline{A}| = |\underline{A}^{T}|$. A simple calculation yields:

$$1 = |\underline{U}\,\underline{U}^{-1}| = |\underline{U}\,\underline{U}^{\dagger}| = |\underline{U}|\,(|\underline{U}^{T}|)^{*} = |\underline{U}|\,(|\underline{U}|)^{*}, \tag{2.137}$$

which proves the theorem.

For an orthogonal matrix $\underline{A}$ this theorem shows that its real determinant $|\underline{A}|$ can only take two values $+1$ or -1. The orthogonal matrix $\underline{A}$ whose determinant $|\underline{A}|$ is equal $+1$ is called a special orthogonal matrix.

◿ **Orthogonal and unitary matrices: Problems**

Problem 2.2.14. (a) Find the matrix $\underline{\underline{O}}$ which transforms the Cartesian unit vectors into the spherical unit vectors according to

$$\begin{bmatrix} \hat{r} \\ \hat{\theta} \\ \hat{\varphi} \end{bmatrix} = \underline{\underline{O}} \begin{bmatrix} \hat{i} \\ \hat{j} \\ \hat{k} \end{bmatrix}$$

These unit vectors are shown in Figure 2.3. Their mutual relation is given by equation (1.118).

(b) Show that $\underline{\underline{O}}$ is an orthogonal matrix and find its inverse, which maps $\hat{r}, \hat{\theta}, \hat{\varphi}$ into $\hat{i}, \hat{j}, \hat{k}$.

(c) Show that

$$\underline{v} = \begin{bmatrix} \sin(\theta) + \cos(\varphi) \\ \cos(\theta) + \sin(\varphi) \\ \cos(\theta + \varphi) \end{bmatrix}$$

is an eigenvector of $\underline{\underline{O}}$ with eigenvalue, $\lambda = 1$.

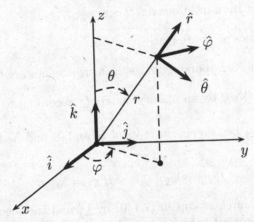

Figure 2.3: Cartesian and spherical unit vectors.

Problem 2.2.15. Show that the upper triangular matrix that is also orthogonal must be diagonal.

2.2.3 Hermitian and Symmetric Matrices

Definition 2.2.4. *A square matrix $\underline{\underline{H}} = [a_{ij}]_{1 \le i,j \le n}$ is called a Hermitian (or self adjoint) matrix if it is equal to its adjoint matrix $\underline{\underline{H}}^\dagger$:*

$$\underline{\underline{H}} = \underline{\underline{H}}^\dagger = \left(\underline{\underline{H}}^T\right)^* \quad \text{or} \quad a_{ij} = a_{ji}^*. \tag{2.138}$$

☞ **Example 2.2.8.** Define three Pauli matrices as:

$$\sigma_1 = \begin{bmatrix} 0 & 1 \\ 1 & 0 \end{bmatrix}, \qquad \sigma_2 = \begin{bmatrix} 0 & -i \\ i & 0 \end{bmatrix}, \qquad \sigma_3 = \begin{bmatrix} 1 & 0 \\ 0 & -1 \end{bmatrix}. \qquad (2.139)$$

All three Pauli matrices are Hermitian, satisfying:

$$\sigma_i = \sigma_i^\dagger, \quad i = 1, 2, 3$$

Let us compute the eigenvalues of σ_2. From

$$\text{Tr}(\sigma_2) = \lambda_1 + \lambda_2, \quad \mid \sigma_2 \mid = -1 = \lambda_1 \lambda_2$$

we find that $\lambda_1 = 1, \lambda_2 = -1$. The fact that the complex matrix σ_2 has real eigenvalues is not a coincidence but a consequence of σ_2 being Hermitian.

The hermiticity property has a number of important consequences for the related system of eigenvalues and eigenvectors. The following theorem holds:

Let $\underline{\underline{H}}$ be an $n \times n$ Hermitian matrix. Then,

① All eigenvalues of $\underline{\underline{H}}$ are real.

② Eigenvectors corresponding to distinct eigenvalues are orthogonal.

③ $\underline{\underline{H}}$ is diagonalized by an unitary matrix $\underline{\underline{U}}$ according to $\underline{\underline{U}}^\dagger \underline{\underline{H}} \underline{\underline{U}} = \underline{\underline{D}}$.

Proof. Let $\underline{v}_i$ and $\underline{v}_j$ be two eigenvectors of $\underline{\underline{H}}$ and λ_i and λ_j their corresponding eigenvalues:

$$\underline{\underline{H}}\underline{v}_i = \lambda_i \underline{v}_i, \qquad \underline{\underline{H}}\underline{v}_j = \lambda_j \underline{v}_j. \qquad (2.140)$$

Multiplying the first of equations in (2.140) by $\underline{v}_j$ and the second of equations in (2.140) by $\underline{v}_i$ with respect to the inner product (2.21) yields:

$$\langle \underline{v}_j \mid \underline{\underline{H}}\underline{v}_i \rangle = \lambda_i \langle \underline{v}_j \mid \underline{v}_i \rangle, \qquad \langle \underline{v}_i \mid \underline{\underline{H}}\underline{v}_j \rangle = \lambda_j \langle \underline{v}_i \mid \underline{v}_j \rangle. \qquad (2.141)$$

Note that the complex conjugation of the inner product $\langle \underline{x} | \underline{\underline{H}} \underline{y} \rangle$ gives

$$\langle \underline{x} \mid \underline{\underline{H}}\underline{y} \rangle^* = \left(\sum_{k,l=1}^{n} \underline{x}_k^* a_{kl} \underline{y}_l \right)^* = \sum_{k,l=1}^{n} \underline{y}_l^* a_{kl}^* \underline{x}_k = \sum_{k,l=1}^{n} \underline{y}_l^* a_{lk}^{T*} \underline{x}_k = \langle \underline{y} \mid \underline{\underline{H}}\underline{x} \rangle,$$

$$(2.142)$$

where a_{kl} are matrix elements of $\underline{\underline{H}}$. In the last equation we used the fact that the matrix $\underline{\underline{H}}$ is Hermitian. Applying the complex conjugation on the first equation in (2.141) we obtain:

$$\langle \underline{v}_j \mid \underline{\underline{H}}\underline{v}_i \rangle^* = \langle \underline{v}_i \mid \underline{\underline{H}}\underline{v}_j \rangle = \lambda_i^* \langle \underline{v}_i \mid \underline{v}_j \rangle. \qquad (2.143)$$

Subtracting the above result from the second equation in (2.141) yields:

$$(\lambda_j - \lambda_i^*)\langle \underline{v}_i \mid \underline{v}_j \rangle = 0 \quad i, j = 1, \ldots, n \tag{2.144}$$

Now, consider the case $i = j$. Equation (2.144) becomes

$$(\lambda_i - \lambda_i^*)\langle \underline{v}_i \mid \underline{v}_i \rangle = 0 \quad \rightarrow \quad \lambda_i - \lambda_i^* = 0 \quad i = 1, \ldots, n, \tag{2.145}$$

resulting from the fact that the norm is not zero, $\langle \underline{v}_i | \underline{v}_i \rangle \neq 0$, for the non-trivial eigenvector $\underline{v}_i$. Accordingly, all eigenvalues of the Hermitian matrix $\underline{H}$ are real.

Let $i \neq j$ and assume that $\lambda_i \neq \lambda_j$. Then, relation (2.144) gives

$$(\lambda_j - \lambda_i)\langle \underline{v}_i \mid \underline{v}_j \rangle = 0 \quad \rightarrow \quad \langle \underline{v}_i \mid \underline{v}_j \rangle = 0. \tag{2.146}$$

Hence, the eigenvectors corresponding to distinct eigenvalues of the Hermitian matrix $\underline{H}$ are orthogonal to each other.

We still need to consider the case of $\lambda_i = \lambda_j$ for two different eigenvectors $\underline{v}_i$ and $\underline{v}_j$. This case is called the degenerate case and it is quite common in physical systems exhibiting some form of symmetry. The corresponding eigenvectors $\underline{v}_i$ and $\underline{v}_j$ are called degenerate eigenvectors. In the degenerate case, equation (2.146) does not automatically guarantee orthogonality of eigenvectors $\underline{v}_i$ and $\underline{v}_j$.

For the degenerate case with two eigenvectors $\underline{v}_i$ and $\underline{v}_j$ that belong to the same eigenvalue $\lambda_j = \lambda_i$, the vector $\underline{v}_j' = \underline{v}_j + c\underline{v}_i$ with arbitrary constant, c, is still an eigenvector of $\underline{H}$ with the same eigenvalue $\lambda_j = \lambda_i$. Moreover, assigning $c = -\langle \underline{v}_j|\underline{v}_i \rangle / \langle \underline{v}_i|\underline{v}_i \rangle$ we obtain for that value of c that

$$\langle \underline{v}_i \mid \underline{v}_j' \rangle = \langle \underline{v}_i \mid \underline{v}_j \rangle + c\langle \underline{v}_i \mid \underline{v}_i \rangle = 0, \tag{2.147}$$

meaning that redefined degenerate eigenvectors are orthogonal to each other.

It is a fact, which we state without proof, that an $n \times n$ Hermitian matrix possesses n linearly independent eigenvectors. Using this fact and the Gram-Schmidt procedure it can be shown that in the general degenerate case, it is possible to arrange degenerate eigenvectors to be orthogonal to each other by redefining them as linear combinations of the linearly independent degenerate eigenvectors.

Hence, it holds that all eigenvectors of the $n \times n$ Hermitian matrix can be defined in such a way that they form an n-dimensional orthogonal basis $\{\underline{v}_1, \ldots, \underline{v}_n\}$ on the complex vector space $\mathbb{C}^n$. This fact allows us to write down a representation for a vector $\underline{u}$ from $\mathbb{C}^n$ in terms of eigenvectors of Hermitian matrix of the form:

$$\underline{u} = \frac{\langle \underline{u}|\underline{v}_1 \rangle}{\langle \underline{v}_1|\underline{v}_1 \rangle} \underline{v}_1 + \ldots + \frac{\langle \underline{u}|\underline{v}_n \rangle}{\langle \underline{v}_n|\underline{v}_n \rangle} \underline{v}_n.$$

We note that this expansion is identical to an orthogonal expansion in equation (2.23). By normalizing each eigenvector according to,

$$\underline{v}_k \quad \rightarrow \quad \hat{\underline{e}}_k = \frac{1}{||\underline{v}_k||} \underline{v}_k,$$

we arrive at an unitary system, as in equation (2.131), on the complex vector space $\mathbb{C}^n$. Accordingly, the matrix constructed out of the eigenvectors $\underline{\hat{e}}_i$ of the Hermitian matrix as follows:

$$\underline{U} = \begin{bmatrix} \underline{\hat{e}}_1 & \underline{\hat{e}}_2 & \cdots & \underline{\hat{e}}_n \end{bmatrix} \tag{2.148}$$

is an unitary matrix. Performing the similarity transformation (2.120) diagonalizes $\underline{\underline{H}}$ according to $\underline{U}^{-1}\,\underline{\underline{H}}\,\underline{U} = \underline{U}^\dagger\,\underline{\underline{H}}\,\underline{U} = \underline{\underline{D}}$, which finishes the proof.

For a real square matrix, $\underline{\underline{A}} = [a_{ij}]_{1\le i,j\le n}$, the condition (2.138) simplifies to

$$\underline{\underline{A}}^T = \underline{\underline{A}} \quad \text{or} \quad a_{ij} = a_{ji}. \tag{2.149}$$

The real square matrix $\underline{\underline{A}}$ which satisfies (2.149) is called a symmetric matrix.

It follows that all eigenvalues of the symmetric matrix are real and eigenvectors of distinct eigenvalues are orthogonal. The whole set of eigenvectors of the symmetric matrix can be chosen to form an orthogonal system. Accordingly, the symmetric matrix is diagonalizable by a similarity transformation with an orthogonal matrix $\underline{\underline{Q}}$, such that

$$\underline{\underline{Q}}^{-1}\,\underline{\underline{A}}\,\underline{\underline{Q}} = \underline{\underline{Q}}^T\,\underline{\underline{A}}\,\underline{\underline{Q}} = \underline{\underline{D}}. \tag{2.150}$$

☞ **Example 2.2.9.** In this example we will find the eigenvalues and eigenvectors of the matrix

$$\underline{\underline{A}} = \begin{bmatrix} 1 & 1 & 1 \\ 1 & 1 & 1 \\ 1 & 1 & 1 \end{bmatrix}. \tag{2.151}$$

The rows and columns of $\underline{\underline{A}}$ are identical. Thus, the matrix is symmetric and its eigenvalues must be real. Clearly, $\underline{\underline{A}}$ has rank equal to 1 and, therefore, is singular. For that reason the characteristic polynomial $|\,\underline{\underline{A}} - \lambda\underline{\underline{I}}\,|$ must have $\lambda = 0$ as a root. In fact it is a double root as follows from solving the characteristic equation

$$\begin{vmatrix} 1-\lambda & 1 & 1 \\ 1 & 1-\lambda & 1 \\ 1 & 1 & 1-\lambda \end{vmatrix} = (1-\lambda)^3 - 3(1-\lambda) + 2 = 0. \tag{2.152}$$

The cubic equation in (2.152) has roots

$$\lambda_1 = \lambda_2 = 0, \quad \lambda_3 = 3.$$

We say that the repeated root $\lambda_1 = \lambda_2 = 0$ is a degenerate eigenvalue since it has more than one eigenvector. To find its eigenvectors we have to solve the corresponding eigenvalue equation:

$$\begin{bmatrix} 1 & 1 & 1 \\ 1 & 1 & 1 \\ 1 & 1 & 1 \end{bmatrix} \underline{a} = \begin{bmatrix} 1 & 1 & 1 \\ 1 & 1 & 1 \\ 1 & 1 & 1 \end{bmatrix} \begin{bmatrix} a_1 \\ a_2 \\ a_3 \end{bmatrix} = 0.$$

The vector components a_1, a_2, a_3 of the eigenvector $\underline{a}$ must satisfy the equation,

$$a_1 + a_2 + a_3 = 0.$$

A general solution can be written as

$$\underline{a} = \begin{bmatrix} a_1 \\ a_2 \\ a_3 \end{bmatrix} = \begin{bmatrix} u \\ v \\ -u - v \end{bmatrix} = u \begin{bmatrix} 1 \\ 0 \\ -1 \end{bmatrix} + v \begin{bmatrix} 0 \\ 1 \\ -1 \end{bmatrix},$$

where u, v are two real parameters. Thus, solutions to the eigenvalue equation for $\lambda = 0$ span a two-dimensional space (a plane) and vectors

$$\begin{bmatrix} 1 \\ 0 \\ -1 \end{bmatrix}, \quad \begin{bmatrix} 0 \\ 1 \\ -1 \end{bmatrix}$$

can be chosen for its basis. Because we found the space of eigenvectors to be two-dimensional we say that $\lambda = 0$ is a double degenerate eigenvalue. A different choice for the basis of solutions to the eigenvalue equation for $\lambda = 0$ could be

$$\underline{v}_1 = \frac{1}{\sqrt{2}} \begin{bmatrix} 1 \\ 0 \\ -1 \end{bmatrix}, \quad \underline{v}_2 = \frac{1}{\sqrt{6}} \left(2 \begin{bmatrix} 0 \\ 1 \\ -1 \end{bmatrix} - \begin{bmatrix} 1 \\ 0 \\ -1 \end{bmatrix} \right) = \frac{1}{\sqrt{6}} \begin{bmatrix} -1 \\ 2 \\ -1 \end{bmatrix}.$$

An advantage of this new basis is that it is orthonormal:

$$\langle \underline{v}_1 | \underline{v}_2 \rangle = 0, \quad \langle \underline{v}_1 | \underline{v}_1 \rangle = 1, \quad \langle \underline{v}_2 | \underline{v}_2 \rangle = 1.$$

To find the eigenvector $\underline{v}_3$ corresponding to $\lambda_3 = 3$ we have to solve the eigenvalue equation:

$$\begin{bmatrix} 1 - \lambda_3 & 1 & 1 \\ 1 & 1 - \lambda_3 & 1 \\ 1 & 1 & 1 - \lambda_3 \end{bmatrix} \begin{bmatrix} a_1 \\ a_2 \\ a_3 \end{bmatrix} = \begin{bmatrix} -2 & 1 & 1 \\ 1 & -2 & 1 \\ 1 & 1 & -2 \end{bmatrix} \begin{bmatrix} a_1 \\ a_2 \\ a_3 \end{bmatrix} = 0,$$

which is equivalent to three linear equations:

$$\begin{aligned} -2a_1 + a_2 + a_3 &= 0 \\ a_1 - 2a_2 + a_3 &= 0 \\ a_1 + a_2 - 2a_3 &= 0 \end{aligned}$$

with three unknowns a_1, a_2, a_3. This system has a one parameter solution $a_1 = a_2 = a_3$. The space of solutions to the eigenvalue equation $\underline{\underline{A}} \underline{v}_3 = \lambda_3 \underline{v}_3$

corresponding to the single root λ_3 is one dimensional. We choose $\underline{v}_3$ to be a unit vector:

$$\underline{v}_3 = \frac{1}{\sqrt{3}} \begin{bmatrix} 1 \\ 1 \\ 1 \end{bmatrix}.$$

Because $\lambda_3 \neq \lambda_1 = \lambda_2$ the eigenvector $\underline{v}_3$ is automatically orthogonal to $\underline{v}_1$ and $\underline{v}_2$. Hence, we could have solved the eigenvalue equation $\underline{\underline{A}}\,\underline{v}_3 = \lambda_3 \underline{v}_3$ by choosing vector $\underline{v}_3$ as the cross product $\underline{v}_1 \times \underline{v}_2$, which is automatically orthogonal to the plane spanned by the degenerated eigenvectors associated to $\lambda = 0$.

Now, construct the orthogonal matrix:

$$\underline{\underline{Q}} = \begin{bmatrix} \underline{v}_1 & \underline{v}_2 & \underline{v}_3 \end{bmatrix} = \begin{bmatrix} \frac{1}{\sqrt{2}} & -\frac{1}{\sqrt{6}} & \frac{1}{\sqrt{3}} \\ 0 & \frac{2}{\sqrt{6}} & \frac{1}{\sqrt{3}} \\ -\frac{1}{\sqrt{2}} & -\frac{1}{\sqrt{6}} & \frac{1}{\sqrt{3}} \end{bmatrix}.$$

It follows from construction that the similarity transformation (2.150) diagonalizes the symmetric matrix $\underline{\underline{A}}$:

$$\underline{\underline{Q}}^T \underline{\underline{A}}\,\underline{\underline{Q}} = \begin{bmatrix} \lambda_1 & 0 & 0 \\ 0 & \lambda_2 & 0 \\ 0 & 0 & \lambda_3 \end{bmatrix} = \begin{bmatrix} 0 & 0 & 0 \\ 0 & 0 & 0 \\ 0 & 0 & 3 \end{bmatrix}.$$

☞ **Example 2.2.10.** This example involves the transformation of quadratic forms to principal axes.

Define a quadratic form as

$$Q = \langle \underline{x} | \underline{\underline{A}}\,\underline{x} \rangle = \underline{x}^T \underline{\underline{A}}\,\underline{x} = \sum_{i=1}^{n} \sum_{j=1}^{n} x_i a_{ij} x_j \qquad (2.153)$$

for a symmetric matrix $\underline{\underline{A}}$. Let $\underline{\underline{Q}}$ be an orthogonal matrix, which diagonalizes $\underline{\underline{A}}$. Thus, $\underline{\underline{A}} = \underline{\underline{Q}}\,\underline{\underline{D}}\,\underline{\underline{Q}}^T$. Substituting this relation into the expression for the quadratic form Q yields:

$$Q = \underline{x}^T \underline{\underline{Q}}\,\underline{\underline{D}}\,\underline{\underline{Q}}^T \underline{x} = \underline{y}^T \underline{\underline{D}}\,\underline{y} = \lambda_1 y_1^2 + \lambda_2 y_2^2 + \cdots + \lambda_n y_n^2, \qquad (2.154)$$

in terms of the new rotated column vector $\underline{y} = \underline{\underline{Q}}^T \underline{x}$. The transformation $\underline{x} \to \underline{y}$ which rotates the quadratic form to the diagonal expression (2.154) is called a transformation to the principal axes. Consider equation

$$5x_1^2 - 3x_1 x_2 + x_2^2 = 1, \qquad (2.155)$$

which can be rewritten as

$$\underline{x}^T \underline{\underline{A}} \, \underline{x} = \begin{bmatrix} x_1 & x_2 \end{bmatrix} \begin{bmatrix} 5 & -\frac{3}{2} \\ -\frac{3}{2} & 1 \end{bmatrix} \begin{bmatrix} x_1 \\ x_2 \end{bmatrix} = 1 . \tag{2.156}$$

The matrix $\underline{\underline{A}}$ defined above has eigenvalues $\lambda_1 = 1/2, \lambda_2 = 11/2$ with the

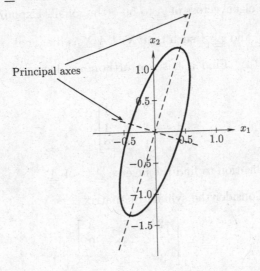

Figure 2.4: Ellipse with major axis $\sqrt{2}$ and minor axis $\sqrt{2/11}$.

corresponding orthonormal eigenvectors

$$\underline{v}_1 = \frac{1}{\sqrt{10}} \begin{bmatrix} 1 \\ 3 \end{bmatrix} \quad \text{and} \quad \underline{v}_2 = \frac{1}{\sqrt{10}} \begin{bmatrix} -3 \\ 1 \end{bmatrix} .$$

In terms of the principal axes, equation (2.155), is rewritten as

$$\frac{1}{2} y_1^2 + \frac{11}{2} y_2^2 = 1 . \tag{2.157}$$

The principal axes are obtained by rotating $[x_1, x_2]$ to $[y_1, y_2]$ through an action of the matrix $\underline{\underline{Q}}^T$

$$\begin{bmatrix} y_1 \\ y_2 \end{bmatrix} = \underline{\underline{Q}}^T \underline{x} = \frac{1}{\sqrt{10}} \begin{bmatrix} 1 & -3 \\ 3 & 1 \end{bmatrix}^T \begin{bmatrix} x_1 \\ x_2 \end{bmatrix} = \frac{1}{\sqrt{10}} \begin{bmatrix} 1 & 3 \\ -3 & 1 \end{bmatrix} \begin{bmatrix} x_1 \\ x_2 \end{bmatrix} .$$

Thus, $y_1 = (x_1 + 3x_2)/\sqrt{10}$ and $y_2 = (-3x_1 + x_2)/\sqrt{10}$ define two principal axes of an ellipse (2.155) as shown in Figure 2.4.

⬧ **Hermitian and symmetric matrices: Problems**

Problem 2.2.16. (a) Find eigenvalues and the corresponding eigenvectors of

$$\underline{\underline{A}} = \begin{bmatrix} 2 & 0 & 1 \\ 0 & 1 & 0 \\ 1 & 0 & 2 \end{bmatrix}.$$

(b) Can you choose eigenvectors of $\underline{\underline{A}}$ to be orthogonal? Explain your answer.

(c) Find matrices $\underline{\underline{X}}$ and $\underline{\underline{X}}^{-1}$ such that $\underline{\underline{X}}^{-1}\underline{\underline{A}}\,\underline{\underline{X}}$ is diagonal.

Problem 2.2.17. (a) Find a basis of orthonormal eigenvectors and use it to diagonalize:

$$\underline{\underline{A}} = \begin{bmatrix} 2 & 1 & 1 \\ 1 & 2 & 1 \\ 1 & 1 & 2 \end{bmatrix}.$$

(b) Use the diagonalization to find the inverse $\underline{\underline{A}}^{-1}$ of $\underline{\underline{A}}$.

Problem 2.2.18. Consider the symmetric matrix

$$\underline{\underline{A}} = \begin{bmatrix} -24 & 7 & 3 \\ 7 & -18 & 9 \\ 3 & 9 & -8 \end{bmatrix}.$$

(a) Verify that

$$\underline{v_1} = \begin{bmatrix} 1 \\ 2 \\ 3 \end{bmatrix} \quad \text{and} \quad \underline{v_2} = \begin{bmatrix} 1 \\ 1 \\ -1 \end{bmatrix}$$

are eigenvectors for $\underline{\underline{A}}$ and find their corresponding eigenvalues.
(b) Find a third eigenvector $\underline{v_3}$ for $\underline{\underline{A}}$ and its eigenvalue.
(c) Find constants c_1, c_2, c_3 such that

$$\begin{bmatrix} 2 \\ 1 \\ 3 \end{bmatrix} = c_1\underline{v_1} + c_2\underline{v_2} + c_3\underline{v_3}.$$

Problem 2.2.19. Let $\underline{\underline{H}}$ be a 3×3 Hermitian matrix with characteristic polynomial $(\lambda+1)(\lambda-2)(\lambda-\lambda_3)$ and trace $\operatorname{Tr}\left(\underline{\underline{H}}\right) = -2$. The (normalized) eigenvectors $\underline{v_1}, \underline{v_2}$ of eigenvalues $\lambda_1 = -1$ and $\lambda_2 = 2$ are

$$\underline{v_1} = \frac{1}{\sqrt{3}} \begin{bmatrix} 1 \\ 1 \\ 1 \end{bmatrix}, \quad \underline{v_2} = \frac{1}{\sqrt{2}} \begin{bmatrix} 1 \\ -1 \\ 0 \end{bmatrix}.$$

(a) Find the eigenvalue λ_3.
(b) Find the corresponding eigenvector $\underline{v_3}$.
(c) Find the matrix $\underline{\underline{H}}$.

Problem 2.2.20. Let $\underline{\underline{F}}$ be a 3×3 Hermitian matrix with the characteristic polynomial $(\lambda - x)(\lambda - x)(\lambda - 3)$ where x is some unknown number. The determinant of $\underline{\underline{F}}$ is 12. The normalized eigenvector $\underline{v}$ of the eigenvalue 3 is

$$\underline{v} = \frac{1}{\sqrt{6}} \begin{bmatrix} 1 \\ 1 \\ 2 \end{bmatrix}.$$

(a) Find possible values of x and the corresponding eigenvectors of $\underline{\underline{F}}$.
(b) Find $\underline{\underline{F}}$. How many different symmetric matrices that satisfy the above conditions can you find?

Problem 2.2.21. Show that the eigenvalues of a Hermitian and unitary matrix will only take two values: $+1$ and -1. Check that the statement for matrices in Example 2.2.8 is true.

Problem 2.2.22. (a) Show, that for any vector $\underline{u}$ in $\mathbb{C}^n$ it holds that

$$\underline{\underline{H}}\,\underline{u} = \lambda_1 \frac{\langle u|v_1 \rangle}{\langle v_1|v_1 \rangle}\, \underline{v}_1 + \ldots + \lambda_n \frac{\langle u|v_n \rangle}{\langle v_n|v_n \rangle}\, \underline{v}_n, \qquad (2.158)$$

for a $n \times n$ Hermitian matrix $\underline{\underline{H}}$ with orthogonal eigenvectors $\{\underline{v}_1, \ldots, \underline{v}_n\}$.
(b) Let $\underline{\underline{H}}$ be a 3×3 Hermitian matrix with eigenvalues $1, 2$ and -1 and corresponding eigenvectors

$$\underline{v}_1 = \begin{bmatrix} -1 \\ 0 \\ 1 \end{bmatrix}, \quad \underline{v}_2 = \begin{bmatrix} 1 \\ 1 \\ 1 \end{bmatrix} \quad \text{and} \quad \underline{v}_3 = \begin{bmatrix} 1 \\ -2 \\ 1 \end{bmatrix}.$$

Apply the right hand side of formula (2.158) on the three canonical basis vectors:

$$\begin{bmatrix} 1 \\ 0 \\ 0 \end{bmatrix}, \quad \begin{bmatrix} 0 \\ 1 \\ 0 \end{bmatrix}, \quad \begin{bmatrix} 0 \\ 0 \\ 1 \end{bmatrix}$$

and find matrix $\underline{\underline{H}}$.

Problem 2.2.23. (a) Show that two commuting Hermitian matrices can be diagonalized by the same unitary transformation. Matrices $\underline{\underline{A}}$ and $\underline{\underline{B}}$ are said to commute when

$$[\underline{\underline{A}}, \underline{\underline{B}}] = \underline{\underline{A}}\,\underline{\underline{B}} - \underline{\underline{B}}\,\underline{\underline{A}} = 0.$$

(b) Find one orthogonal matrix that diagonalizes both matrices:

$$\underline{\underline{A}} = \begin{bmatrix} 1 & 2 & 0 \\ 2 & 1 & 0 \\ 0 & 0 & -3 \end{bmatrix} \quad \text{and} \quad \underline{\underline{B}} = \begin{bmatrix} -1 & 3 & 0 \\ 3 & -1 & 0 \\ 0 & 0 & 4 \end{bmatrix}.$$

Problem 2.2.24. Show that for a Hermitian matrix $\underline{\underline{H}}$ it holds that

$$\langle \underline{x} | \underline{\underline{H}} \, \underline{y} \rangle = \langle \underline{\underline{H}} \, \underline{x} | \underline{y} \rangle,$$

with respect to the inner product $\langle \underline{x} | \underline{y} \rangle$ on $\mathbb{C}^n$ defined in equation (2.21). Next, show that $\langle \underline{x} | \underline{\underline{H}} \, \underline{x} \rangle$ is real for any vector $\underline{x}$ in $\mathbb{C}^n$.

Problem 2.2.25. The exponential, $\exp\left(\underline{\underline{A}}\right)$, of a matrix $\underline{\underline{A}}$ is defined by a power series expansion:

$$e^{\underline{\underline{A}}} = \underline{\underline{I}} + \underline{\underline{A}} + \frac{1}{2!}\underline{\underline{A}}^2 + \cdots + \frac{1}{n!}\underline{\underline{A}}^n + \cdots = \sum_{k=0}^{\infty} \frac{1}{k!}\underline{\underline{A}}^k. \qquad (2.159)$$

(a) Show that $\exp\left(-\underline{\underline{A}}\right)$ is an inverse of exponential matrix $\exp\left(\underline{\underline{A}}\right)$ for any square matrix $\underline{\underline{A}}$.

(b) Show that $\underline{\underline{O}} = \exp\left(\underline{\underline{A}}\right)$ is orthogonal if $\underline{\underline{A}}$ is skew-symmetric, meaning that $\underline{\underline{A}}^T = -\underline{\underline{A}}$.

(c) Show that $\underline{\underline{U}} = \exp\left(i\,\underline{\underline{H}}\right)$ is unitary if $\underline{\underline{H}}$ is Hermitian.

Problem 2.2.26. For the Hermitian matrix $\underline{\underline{H}}$ show that $\left| e^{\underline{\underline{H}}} \right| = e^{\operatorname{Tr} \underline{\underline{H}}}$.

Problem 2.2.27. For two square matrices A and B verify the Baker-Hausdorff formula:

$$e^{\underline{\underline{A}}} \, \underline{\underline{B}} \, e^{-\underline{\underline{A}}} = \underline{\underline{B}} + [\underline{\underline{A}}, \underline{\underline{B}}] + \frac{1}{2}[\underline{\underline{A}}, [\underline{\underline{A}}, \underline{\underline{B}}]] + \cdots$$

Problem 2.2.28. (a) Show that two Hermitian matrices that have the same eigenvalues are related by a unitary similarity transformation.

(b) Illustrate your proof for the above statement by finding an orthogonal matrix $\underline{\underline{P}}$ that maps two symmetric matrices

$$\underline{\underline{A}} = \begin{bmatrix} 2 & -2 \\ -2 & -1 \end{bmatrix} \quad \text{and} \quad \underline{\underline{B}} = \frac{1}{2}\begin{bmatrix} 1 & -5 \\ -5 & 1 \end{bmatrix}$$

into each other via the similarity transformation $\underline{\underline{P}}^T \underline{\underline{A}} \, \underline{\underline{P}} = \underline{\underline{B}}$.

Problem 2.2.29. Consider a quadratic form $x_1^2 - 4x_1 x_2 + x_2^2 = 2$. Find a transformation $\underline{x} \to \underline{y}$ that rotates the quadratic form to the diagonal expression $-y_1^2 + 3y_2^2 = 2$.

Problem 2.2.30. Derive a quadratic form in x and y that describes an intersection of the unit sphere $x^2 + y^2 + z^2 = 1$ and the plane $x + y + z = 0$ and find a rotation to the principal axes.

Problem 2.2.31. (a) Show that the $n \times n$ symmetric matrix

$$\underline{\underline{K_n}} = \begin{bmatrix} 2 & -1 & 0 & 0 & 0 & \cdots & 0 \\ -1 & 2 & -1 & 0 & 0 & \cdots & 0 \\ 0 & -1 & 2 & -1 & 0 & \cdots & 0 \\ 0 & 0 & -1 & 2 & -1 & \cdots & 0 \\ & & & \cdot & & \cdot & \\ 0 & 0 & \cdots & 0 & -1 & 2 & -1 \\ 0 & 0 & 0 & \cdots & 0 & -1 & 2 \end{bmatrix}$$

has n eigenvectors:

$$\underline{v_k} = \begin{bmatrix} \sin\left[k\pi/(n+1)\right] \\ \sin\left[2k\pi/(n+1)\right] \\ \vdots \\ \sin\left[nk\pi/(n+1)\right] \end{bmatrix}, \quad k = 1, \ldots, n \qquad (2.160)$$

with eigenvalues $\lambda_k = 2\left(1 - \cos\dfrac{k\pi}{n+1}\right)$.

(b) Use (a) to evaluate:

$$\sum_{k=1}^{n} \cos\frac{k\pi}{n+1}.$$

Problem 2.2.32. (a) Write down equations of motion for three identical masses connected in series to four identical springs as shown in Fig. 2.5. *Hint:* Compare with Example 2.2.4.

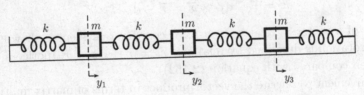

Figure 2.5: Three masses connected in series to four springs.

(b) Show that the frequencies of oscillations can be obtained by solving the eigenvalue problem of the 3×3 matrix $\underline{\underline{K_3}}$ from Problem 2.2.31.

(c) Verify explicitly that the eigenvectors $\underline{v_1}, \underline{v_2}$ and $\underline{v_3}$ of $\underline{\underline{K_3}}$ given by relation (2.160) are orthogonal to each other.

Problem 2.2.33. (a) Show that if $\underline{\underline{A}}$ is a real $m \times n$ matrix then the $n \times n$ matrix $\underline{\underline{A}}^T \underline{\underline{A}}$ has real, no negative eigenvalues.

(b) Find eigenvalues of $\underline{\underline{A}}^T\underline{\underline{A}}$ for

$$\underline{\underline{A}} = \begin{bmatrix} 1 & 1 & 0 \\ 0 & 0 & 1 \\ 0 & 0 & -1 \end{bmatrix}.$$

Problem 2.2.34. Let $\underline{\underline{A}}$ be a $n \times n$ non-Hermitian matrix and $\underline{\underline{A}}^\dagger$ its Hermitian conjugate. Show that if $\underline{\underline{A}}\underline{v}_i = \lambda_i\underline{v}_i$ and $\underline{\underline{A}}^\dagger\underline{u}_i = \rho_i\underline{u}_i$, $i = 1, \ldots, n$ then $\lambda_i = \rho_j^*$ if $\langle \underline{v}_i | \underline{u}_j \rangle \neq 0$ and $\langle \underline{v}_i | \underline{u}_j \rangle = 0$ if $\lambda_i \neq \rho_j^*$.

2.3   Rotation Transformations and Special Orthogonal Matrices

We start this section by studying a class of infinitesimal rotations of the coordinate system and showing how to conveniently realize them in terms of matrix multiplication. Introduce a vector

$$\mathrm{d}\vec{\phi} = \mathrm{d}\phi\,\vec{\omega}, \tag{2.161}$$

which describes a rotation by an infinitesimal angle $\mathrm{d}\phi$ about an axis of rotation that points in the direction of the vector $\vec{\omega}$. The magnitude of the vector $\mathrm{d}\vec{\phi}$ is equal to the infinitesimal angle $\mathrm{d}\phi$. The vector $\vec{\omega}$ is chosen be a unit vector satisfying

$$\vec{\omega} \cdot \vec{\omega} = \omega_1^2 + \omega_2^2 + \omega_3^2 = 1.$$

Its role is solely to indicate the direction of the axis of rotation.

An infinitesimal rotation of the coordinate system by an angle $\mathrm{d}\phi$ about an axis of rotation $\vec{\omega}$ results in a new coordinate system with the new position vector $\vec{r}'$ given in terms of the old position vector $\vec{r}$ through

$$\vec{r}' = \vec{r} - \mathrm{d}\vec{\phi} \times \vec{r} = \vec{r} - \mathrm{d}\phi\,\vec{\omega} \times \vec{r}. \tag{2.162}$$

A rotation of a vector $\vec{r}$ relative to a fixed coordinate system is shown in Figure 2.6. That rotation is described by relations $\vec{r}' = \vec{r} + \mathrm{d}\vec{\phi} \times \vec{r}$ (notice a difference in sign when compared with equation (2.162)).

It is convenient to rewrite the vector product in terms of matrix multiplication. This operation involves a notion of a skew-symmetric matrix. A square matrix $\underline{\underline{\Omega}}$ which satisfies condition $\underline{\underline{\Omega}}^T = -\underline{\underline{\Omega}}$ is called a skew-symmetric matrix. The nine matrix elements ω_{ij} of a 3×3 skew-symmetric matrix $\underline{\underline{\Omega}}$ satisfy relations $\omega_{ij} = -\omega_{ji}$ for all $i,j = 1,2,3$. Consequently, $\omega_{ii} = 0$, i.e. all three diagonal elements of $\underline{\underline{\Omega}}$ vanish. The remaining six off-diagonal elements obey three conditions; $\omega_{13} = -\omega_{31}$, $\omega_{23} = -\omega_{32}$ and $\omega_{12} = -\omega_{21}$. Thus, a 3×3 skew-symmetric matrix $\underline{\underline{\Omega}}$ possesses only three independent matrix elements, which can be chosen as follows

$$\underline{\underline{\Omega}} = \begin{bmatrix} 0 & -\omega_3 & \omega_2 \\ \omega_3 & 0 & -\omega_1 \\ -\omega_2 & \omega_1 & 0 \end{bmatrix}. \tag{2.163}$$

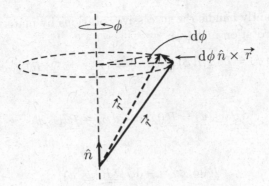

Figure 2.6: Rotation by an angle $\mathrm{d}\phi$ about axis of rotation $\hat{n}$.

Associating these three matrix elements with components of a three-dimensional vector $\vec{\omega} = (\omega_1, \omega_2, \omega_3)$ via the relation

$$\vec{\omega} = (\omega_1, \omega_2, \omega_3) \iff \underline{\underline{\Omega}} = \begin{bmatrix} 0 & -\omega_3 & \omega_2 \\ \omega_3 & 0 & -\omega_1 \\ -\omega_2 & \omega_1 & 0 \end{bmatrix} \tag{2.164}$$

introduces a general, one to one, correspondence between the three-dimensional vectors and the 3×3 skew-symmetric matrices. One recovers the components of a vector $\vec{\omega}$ from a skew-symmetric matrix $\underline{\underline{\Omega}}$ via the relation, $\omega_i = -\frac{1}{2} \sum_{j,k=1}^{3} \epsilon_{ijk} \omega_{jk}$. Conversely, the matrix elements ω_{ij} of $\underline{\underline{\Omega}}$ are reproduced from the relation, $\omega_{ij} = -\sum_{k=1}^{3} \epsilon_{ijk} \omega_k$.

In this setting a vector product $\vec{\omega} \times \vec{v}$ can be rewritten as a matrix multiplication of the column vector $\underline{v}$ by a skew-symmetric matrix $\underline{\underline{\Omega}}$:

$$\underline{\underline{\Omega}}\,\underline{v} = \begin{bmatrix} 0 & -\omega_3 & \omega_2 \\ \omega_3 & 0 & -\omega_1 \\ -\omega_2 & \omega_1 & 0 \end{bmatrix} \begin{bmatrix} v_1 \\ v_2 \\ v_3 \end{bmatrix} = \begin{bmatrix} \omega_2 v_3 - \omega_3 v_2 \\ \omega_3 v_1 - \omega_1 v_3 \\ \omega_1 v_2 - \omega_2 v_1 \end{bmatrix} = \begin{bmatrix} (\vec{\omega} \times \vec{v})_1 \\ (\vec{\omega} \times \vec{v})_2 \\ (\vec{\omega} \times \vec{v})_3 \end{bmatrix} = \underline{\omega} \times \underline{v}.$$
$$\tag{2.165}$$

The infinitesimal rotation (2.162) by a small angle $\mathrm{d}\phi$ about $\vec{\omega}$ can be written in matrix notation as

$$\underline{r}' = \underline{r} - \mathrm{d}\underline{\phi} \times \underline{r} = \underline{r} - \mathrm{d}\phi\,\underline{\underline{\Omega}}\,\underline{r}. \tag{2.166}$$

Now, define $\underline{\underline{R}}(\underline{\omega}, \mathrm{d}\phi)$ as an infinitesimal rotation matrix

$$\underline{\underline{R}}(\underline{\omega}, \mathrm{d}\phi) = \underline{\underline{I}} - \mathrm{d}\phi\,\underline{\underline{\Omega}} \tag{2.167}$$

that is associated to the infinitesimal angle (2.161) and to the skew-symmetric matrix $\underline{\underline{\Omega}}$, which is related to $\underline{\omega}$ as in (2.164). The advantage of the above notation is that we can elegantly handle the successive rotations by matrix multiplication of the corresponding rotation matrices:

$$\underline{\underline{R}}(\underline{\omega}, \phi + d\phi) = \underline{\underline{R}}(\underline{\omega}, d\phi)\, \underline{\underline{R}}(\underline{\omega}, \phi) = (I - d\phi\,\underline{\underline{\Omega}})\, \underline{\underline{R}}(\underline{\omega}, \phi). \tag{2.168}$$

It follows that

$$d\underline{\underline{R}}(\underline{\omega}, \phi) = \underline{\underline{R}}(\underline{\omega}, \phi + d\phi) - \underline{\underline{R}}(\underline{\omega}, \phi)$$

can be rewritten as

$$d\underline{\underline{R}}(\underline{\omega}, \phi) = -d\phi\, \underline{\underline{\Omega}}\, \underline{\underline{R}}(\underline{\omega}, \phi). \tag{2.169}$$

A solution to the above differential equation with an initial condition $\underline{\underline{R}}(0) = I$ is given by:

$$\underline{\underline{R}}(\underline{\omega}, \phi) = e^{-\phi\underline{\underline{\Omega}}}, \tag{2.170}$$

where we have formally introduced an exponential of a matrix defined by a power series expansion as in (2.159). Accordingly,

$$\underline{\underline{R}}(\underline{\omega}, \phi) = e^{-\phi\underline{\underline{\Omega}}} = \sum_{n=0}^{\infty} \frac{(-1)^n}{n!}\left(\phi\underline{\underline{\Omega}}\right)^n.$$

Equation (2.170) provides a formula for the rotation matrix that describes rotation by an angle ϕ about an axis of rotation $\vec{\omega}$.

Note that it follows from equation (2.170) that

$$\underline{\underline{R}}^{\dagger}(\underline{\omega}, \phi) = e^{\phi\underline{\underline{\Omega}}} = \underline{\underline{R}}^{-1}(\underline{\omega}, \phi).$$

Furthermore, since $\exp(\underline{\underline{A}})\exp(-\underline{\underline{A}}) = \underline{\underline{I}}$ the rotation matrix $\underline{\underline{R}}(\underline{\omega}, \phi)$ is unitary. Since, $\underline{\underline{R}}(\underline{\omega}, \phi) = \exp\left(-\phi\underline{\underline{\Omega}}\right)$ is real it must also be orthogonal (real, unitary matrices are by definition orthogonal).

The finite version of the coordinate transformation (2.166) takes the form:

$$\begin{aligned}
\underline{r}' &= \underline{\underline{R}}(\underline{\omega}, \phi)\,\underline{r} = e^{-\phi\underline{\underline{\Omega}}}\underline{r} = \sum_{n=0}^{\infty} \frac{(-\phi)^n}{n!}\, \underline{\underline{\Omega}}^n\, \underline{r} \\
&= \underline{r} - \phi\,\underline{\underline{\Omega}}\,\underline{r} + \frac{\phi^2}{2}\left(-\underline{\underline{\Omega}}\right)^2\underline{r} + \frac{\phi^3}{3!}\left(-\underline{\underline{\Omega}}\right)^3\underline{r} + \cdots.
\end{aligned} \tag{2.171}$$

Recall, that

$$\underline{\underline{\Omega}}\,\underline{r} = \underline{\omega} \times \underline{r},$$

as shown in (2.165). Applying this identity successively yields

$$\begin{aligned}
\underline{\underline{\Omega}}^2\,\underline{r} &= \underline{\underline{\Omega}}\,\underline{\omega} \times \underline{r} = \underline{\omega} \times (\underline{\omega} \times \underline{r}) \\
\left(-\underline{\underline{\Omega}}\right)^3\underline{r} &= -\underline{\underline{\Omega}}\,\underline{\omega} \times (\underline{\omega} \times \underline{r}) = -\underline{\omega} \times (\underline{\omega} \times (\underline{\omega} \times \underline{r})) \\
&= -\underline{\omega}\,(\underline{\omega} \cdot (\underline{\omega} \times \underline{r})) - (\underline{\omega} \times \underline{r})(\underline{\omega} \cdot \underline{\omega}) = \underline{\omega} \times \underline{r}.
\end{aligned} \tag{2.172}$$

The above relations generalizes to higher powers as follows

$$(-\underline{\underline{\Omega}})^{2k+1}\underline{r} = (-1)^{k+1}\underline{\omega} \times \underline{r} = (-1)^{k+1}\underline{\underline{\Omega}}\,\underline{r} \tag{2.173}$$

$$\underline{\underline{\Omega}}^{2k}\underline{r} = (-1)^{k+1}\underline{\omega} \times (\underline{\omega} \times \underline{r}) = (-1)^{k+1}\underline{\underline{\Omega}}^2\,\underline{r}$$

allowing us to calculate the expansion in (2.171) as:

$$\begin{aligned}
\underline{r}' &= \underline{r} - \sum_{k=0}^{\infty} \frac{(-1)^k \phi^{2k+1}}{(2k+1)!}\underline{\omega} \times \underline{r} + \sum_{k=1}^{\infty} \frac{(-1)^{k+1}\phi^{2k}}{(2k)!}(\underline{\omega} \times (\underline{\omega} \times \underline{r})) \\
&= \underline{r} - \underline{\omega} \times \underline{r}\sin\phi + (\underline{\omega} \times (\underline{\omega} \times \underline{r}))(1 - \cos\phi)\,, \tag{2.174}
\end{aligned}$$

from which we can derive the so-called angle and axis parameterization of a rotation matrix.

The angle and axis parameterization

$$\underline{\underline{R}}(\underline{\omega},\phi) = e^{-\phi\underline{\underline{\Omega}}} = \underline{\underline{I}} - \underline{\underline{\Omega}}\sin\phi + \underline{\underline{\Omega}}^2(1 - \cos\phi)\,, \tag{2.175}$$

specifies a rotation matrix in terms of the axis of rotation given by a unit vector $\underline{\omega}$ and an angle of rotation ϕ about that axis:

☞ **Example 2.3.1.** Let us consider a rotation about the z-axis with $\vec{\omega} = \hat{k}$. In this case, (2.164) becomes

$$(0,0,1) \iff \underline{\underline{\Omega}} = \begin{bmatrix} 0 & -1 & 0 \\ 1 & 0 & 0 \\ 0 & 0 & 0 \end{bmatrix},$$

which, when inserted into (2.175), gives

$$\begin{aligned}
\underline{\underline{R}}(\hat{k},\phi) &= \underline{\underline{I}} - \begin{bmatrix} 0 & -1 & 0 \\ 1 & 0 & 0 \\ 0 & 0 & 0 \end{bmatrix}\sin\phi + \begin{bmatrix} -1 & 0 & 0 \\ 0 & -1 & 0 \\ 0 & 0 & 0 \end{bmatrix}(1 - \cos\phi) \\
&= \begin{bmatrix} \cos\phi & \sin\phi & 0 \\ -\sin\phi & \cos\phi & 0 \\ 0 & 0 & 1 \end{bmatrix}.
\end{aligned} \tag{2.176}$$

The above solution reproduces the rotation matrix (2.50).

The rotation matrix $\underline{\underline{R}}(\underline{\omega},\phi)$ has an inverse obtained by changing the sign of $\vec{\phi} = \phi\vec{\omega}$:

$$\underline{\underline{R}}^{-1}(\underline{\omega},\phi) = \underline{\underline{R}}(\underline{\omega},-\phi) = e^{\phi\underline{\underline{\Omega}}} = \underline{\underline{I}} + \underline{\underline{\Omega}}\sin\phi + \underline{\underline{\Omega}}^2(1 - \cos\phi)\,. \tag{2.177}$$

It describes a rotation by an angle ϕ but in the opposite direction.

For a rotation of a vector $\underline{v}$ relative to a fixed coordinate system we obtain:

$$
\begin{aligned}
\underline{v}' &= \underline{\underline{R}}^{-1}(\underline{\omega}, \phi)\,\underline{v} = \underline{v} + \underline{\omega} \times \underline{v}\sin\phi + \underline{\omega} \times (\underline{\omega} \times \underline{v})\,(1 - \cos\phi) \\
&= \underline{v}\cos\phi + \underline{\omega} \times \underline{v}\sin\phi + \underline{\omega}(\underline{\omega}\cdot\underline{v})\,(1 - \cos\phi)\,,
\end{aligned}
\tag{2.178}
$$

where we used the vector identity $\underline{\omega} \times (\underline{\omega} \times \underline{v}) = \underline{\omega}(\underline{\omega}\cdot\underline{v}) - \underline{v}$, valid for the unit vector $\underline{\omega}$.

Let us study the eigenvalue problem for the rotation matrix $\underline{\underline{R}}^{-1}(\underline{\omega}, \phi)$ that rotates a vector relative to a fixed coordinate system. According to equation (2.178) the eigenvalue problem $\underline{\underline{R}}^{-1}(\underline{\omega}, \phi)\underline{v} = \lambda\underline{v}$ takes the form

$$
\underline{v}\cos\phi + \underline{\omega} \times \underline{v}\sin\phi + \underline{\omega}(\underline{\omega}\cdot\underline{v})\,(1 - \cos\phi) = \lambda\underline{v}\,.
\tag{2.179}
$$

Assume that $\underline{\omega}\cdot\underline{v} \neq 0$ and multiply both sides of (2.179) by vector $\underline{\omega}$. Only the first and the third of the terms on the left hand side of (2.179) give a non-zero contribution. Upon cancellation of terms with $\cos\phi$ we obtain the relation $\underline{\omega}\cdot\underline{v} = \lambda\underline{\omega}\cdot\underline{v}$. Hence, in the case of $\underline{\omega}\cdot\underline{v} \neq 0$ the eigenvalue is $\lambda = 1$. The corresponding eigenvector is $\underline{v} = \underline{\omega}$ as seen from the fact that $\underline{\underline{R}}^{-1}(\underline{\omega}, \phi)\underline{\omega} = \underline{\omega}$.

It remains to consider vector $\underline{a}$ such that $\underline{a}\cdot\underline{\omega} = 0$. Write $\underline{b} = \underline{a} \times \underline{\omega}$. It follows that $\underline{\omega} \times \underline{b} = -\underline{\omega}(\underline{a}\cdot\underline{\omega}) + \underline{a}(\underline{\omega}\cdot\underline{\omega}) = \underline{a}$. Successively plugging $\underline{v} = \underline{a}$ and $\underline{v} = \underline{b}$ into (2.178) yields:

$$
\underline{\underline{R}}^{-1}(\underline{\omega}, \phi)\,\underline{a} = \underline{a}\cos\phi - \underline{b}\sin\phi, \quad \underline{\underline{R}}^{-1}(\underline{\omega}, \phi)\,\underline{b} = \underline{b}\cos\phi + \underline{a}\sin\phi\,.
\tag{2.180}
$$

Thus, in a basis given by vectors $\underline{b}, \underline{a}, \underline{v}$ the rotation matrix $\underline{\underline{R}}^{-1}(\underline{\omega}, \phi)$ becomes equivalent to the rotation (2.176) (see Problem (2.3.13) for an explicit construction of vectors $\underline{a}, \underline{b}$ in terms $\underline{\omega}$).

Define two vectors $\underline{v}_\pm = \underline{a} \pm i\,\underline{b}$ that are orthogonal to $\underline{v}$ and each other. From (2.180) it follows that

$$
\underline{\underline{R}}^{-1}(\underline{\omega}, \phi)\underline{v}_\pm = \lambda_\pm\underline{v}_\pm, \quad \lambda_\pm = \exp(\pm i\,\phi)
\tag{2.181}
$$

and $\underline{v}_\pm$ are two remaining eigenvectors of $\underline{\underline{R}}^{-1}(\underline{\omega}, \phi)$. The orthogonal vectors $\underline{\omega}, \underline{v}_+, \underline{v}_-$ form the basis of $\mathbb{R}^3$ in which the $\underline{\underline{R}}^{-1}(\underline{\omega}, \phi)$ matrix takes the simple diagonal form:

$$
\underline{\underline{R}}^{-1}(\underline{\omega}, \phi) = \begin{bmatrix} 1 & 0 & 0 \\ 0 & e^{i\phi} & 0 \\ 0 & 0 & e^{-i\phi} \end{bmatrix}.
\tag{2.182}
$$

Note that $\lambda_+\lambda_- = \lambda_+\lambda_+^* = 1$. Thus the product of all three eigenvalues of $\underline{\underline{R}}^{-1}(\underline{\omega}, \phi)$ is equal to one and, hence, the determinant is $\mid \underline{\underline{R}}^{-1}(\underline{\omega}, \phi)\mid = \lambda_1\lambda_+\lambda_- = 1$. Accordingly,

the rotation matrix $\underline{\underline{R}}(\underline{\omega}, \phi)$ is a special orthogonal matrix, i.e., a matrix whose inverse is equal to its transpose and with a determinant equal to unity.

Furthermore, $\underline{v}_+^\dagger \underline{v}_- = 0$ and all three eigenvectors are orthogonal with respect to the inner product $\langle \underline{v} | \underline{w} \rangle = \underline{v}^\dagger \underline{w}$.

Conversely, every special orthogonal matrix $\underline{\underline{A}}$ can be identified with some rotation matrix $\underline{\underline{R}}^{-1}(\underline{\omega}, \phi)$, which transforms vectors according to (2.178). Let $\lambda_1, \lambda_2, \lambda_3$ be eigenvalues of a special orthogonal matrix $\underline{\underline{A}}$. Their product is

$$\lambda_1 \lambda_2 \lambda_3 = | \underline{\underline{A}} | = 1 .$$

According to equation (2.136) $|\lambda_i| = 1$, $i = 1, 2, 3$ and the complex eigenvalues appear in pairs consisting of complex conjugate numbers. Therefore, at least one eigenvalue must be real and equal to one. Hence, we can assume that the eigenvalues are given by $\lambda_1 = 1$, $\lambda_2 = e^{i\alpha}$, $\lambda_3 = e^{-i\alpha}$ for some real parameter α. Let $\underline{v}_i$, $i = 1, 2, 3$ be eigenvectors corresponding to λ_i, $i = 1, 2, 3$. They are orthogonal to each other due to the relation

$$\langle \underline{\underline{A}}\,\underline{v}_j | \underline{\underline{A}}\,\underline{v}_i \rangle = \langle \lambda_j \underline{v}_j | \lambda_i \underline{v}_i \rangle = \lambda_i \lambda_j^* \langle \underline{v}_j | \underline{v}_i \rangle \qquad (2.183)$$
$$= \langle \underline{v}_j | \underline{\underline{A}}^T \underline{\underline{A}}\,\underline{v}_i \rangle = \langle \underline{v}_j | \underline{v}_i \rangle .$$

Since $\lambda_i \lambda_j^* \neq 1$ for $i \neq j$, it follows that $\langle \underline{v}_j | \underline{v}_i \rangle = 0$ for $i \neq j$. We can normalize eigenvectors $\underline{v}_i$ so that they form an orthonormal set of vectors. By definition, $\underline{\underline{A}}\,\underline{v}_1 = \underline{v}_1$. Therefore, the eigenvector $\underline{v}_1$ remains invariant under transformation by $\underline{\underline{A}}$ and defines an axis of rotation $\underline{\omega}$ under action of $\underline{\underline{R}}^{-1}(\underline{\omega}, \phi)$. Furthermore, the eigenvalue equations $\underline{\underline{A}}\,\underline{v}_2 = e^{i\alpha} \underline{v}_2$ and $\underline{\underline{A}}\,\underline{v}_3 = e^{-i\alpha} \underline{v}_3$ become identical to those in (2.181) using $\alpha = \phi$ and $\underline{v}_+ = \underline{v}_2$ and $\underline{v}_- = \underline{v}_3$. Let's define an orthogonal 3×3 matrix $\underline{\underline{V}} = [\underline{v}_1 \quad \underline{v}_2 \quad \underline{v}_3]$. The diagonalization procedure yields:

$$\underline{\underline{V}}^T \underline{\underline{A}} \, \underline{\underline{V}} = \begin{bmatrix} 1 & 0 & 0 \\ 0 & e^{i\phi} & 0 \\ 0 & 0 & e^{-i\phi} \end{bmatrix}$$

or

$$\underline{\underline{A}} = \underline{\underline{V}} \begin{bmatrix} 1 & 0 & 0 \\ 0 & e^{i\phi} & 0 \\ 0 & 0 & e^{-i\phi} \end{bmatrix} \underline{\underline{V}}^T .$$

Compare the above matrix with expression (2.182). We conclude that a special orthogonal matrix $\underline{\underline{A}}$ coincides with a rotation matrix $\underline{\underline{R}}^{-1}(\underline{\omega}, \phi)$, when both matrices are derived in a basis of its eigenvectors. This identification relates the axis of rotation $\underline{\omega}$, in $\underline{\underline{R}}^{-1}(\underline{\omega}, \phi)$, to the eigenvector of $\underline{\underline{A}}$ corresponding to $\lambda_1 = 1$ and the angle ϕ to the eigenvalues λ_2 and λ_3 via relation $\cos\phi = (\lambda_2 + \lambda_3)/2$. In a basis given by vectors $(\underline{v}_2 - \underline{v}_3)/2i$, $(\underline{v}_2 + \underline{v}_3)/2$ and $\underline{v}_1$ the special orthogonal matrix $\underline{\underline{A}}$ becomes equivalent to a rotation (2.176) with $\alpha = \phi$.

☞ **Example 2.3.2.** Consider the 3×3 orthogonal matrix

$$\underline{\underline{A}} = \frac{1}{2} \begin{bmatrix} 1 & \sqrt{2} & 1 \\ -\sqrt{2} & 0 & \sqrt{2} \\ 1 & -\sqrt{2} & 1 \end{bmatrix}$$

from Example (2.2.6). Since $|\underline{\underline{A}}| = 1$ the matrix $\underline{\underline{A}}$ is a special orthogonal matrix and, therefore, it can be identified with some rotation matrix $\underline{\underline{R}}^{-1}(\underline{\omega}, \phi)$. This can be accomplished by finding the axis of rotation $\vec{\omega}$ and the angle of rotation ϕ about $\vec{\omega}$. First, we solve the eigenvalue problem of $\underline{\underline{A}}$. The characteristic polynomial:

$$|\underline{\underline{A}} - \lambda \underline{\underline{I}}| = -\lambda + \lambda^2 + 1 - \lambda^3 = (\lambda - 1)(\lambda - i)(\lambda + i)$$

has roots $1, +i, -i$. The eigenvector corresponding to $\lambda = 1$ is

$$\underline{v} = \begin{bmatrix} 1 \\ 0 \\ 1 \end{bmatrix}$$

and defines the unit vector

$$\underline{\omega} = \frac{1}{\sqrt{2}} \begin{bmatrix} 1 \\ 0 \\ 1 \end{bmatrix}$$

along the axis of rotation of the corresponding rotation matrix. The two remaining eigenvectors

$$\underline{v}_+ = \begin{bmatrix} -1 \\ -i\sqrt{2} \\ 1 \end{bmatrix}, \quad \underline{v}_- = \begin{bmatrix} -1 \\ i\sqrt{2} \\ 1 \end{bmatrix},$$

satisfy the eigenvalue problems

$$\underline{\underline{A}}\underline{v}_+ = i\underline{v}_+ = e^{i\pi/2}\underline{v}_+$$

and

$$\underline{\underline{A}}\underline{v}_- = -i\underline{v}_- = e^{-i\pi/2}\underline{v}_-.$$

Therefore, the angle of rotation generated by the matrix $\underline{\underline{A}}$ is $\phi = \pi/2$. As in the discussion surrounding equations (2.180) and (2.181) we can define a vector

$$\underline{a} = \frac{1}{2}(\underline{v}_+ + \underline{v}_-) = \begin{bmatrix} -1 \\ 0 \\ 1 \end{bmatrix},$$

which is orthogonal to $\underline{n}$. Another vector in the plane orthogonal to $\underline{\omega}$ is

$$\underline{b} = \underline{a} \times \underline{\omega} = -\frac{i}{2}\left(\underline{v}_+ - \underline{v}_-\right) = \begin{bmatrix} 0 \\ -\sqrt{2} \\ 0 \end{bmatrix}.$$

These two orthogonal vectors rotates into each other via multiplication by the matrix $\underline{\underline{A}}$:

$$\underline{\underline{A}}\,\underline{a} = -\underline{b}, \quad \underline{\underline{A}}\,\underline{b} = \underline{a},$$

consistent with relations (2.181) for $\phi = \pi/2$, which appears to be the angle of rotation in the plane orthogonal to the axis of rotation $\underline{\omega}$.

⚓ Rotation transformations and special orthogonal matrices: Problems

Problem 2.3.1. Show that equation (2.178) leads to a familiar result for $\vec{\omega} = \hat{k}$.

Problem 2.3.2. Show that the rotation matrix $\underline{\underline{R}}^{-1}\,(\underline{\omega},\,\phi)$ from equation (2.177) transforms an arbitrary vector $\vec{a}$ according to:

$$\vec{a} \longrightarrow \vec{a}' = \underline{\underline{R}}\,(\underline{\omega},\,\phi)\,(\vec{a}) = \vec{a}_\parallel + \vec{a}_\perp \cos\phi + (\vec{\omega} \times \vec{a}_\perp)\sin\phi,$$

where $\vec{a} = \vec{a}_\parallel + \vec{a}_\perp$, with $\vec{a}_\parallel$ parallel and $\vec{a}_\perp$ perpendicular to a given unit vector $\vec{\omega}$.

Problem 2.3.3. (a) Suppose a unit vector $\hat{n}$ defines the rotation axis. As explained in Chapter 1 (see, for instance, Problem 1.1.20), an arbitrary vector $\vec{v}$ can be decomposed into components parallel and perpendicular to the rotation axis as follows:

$$\vec{v} = \vec{v}_\parallel + \vec{v}_\perp$$
$$= \hat{n}\,(\hat{n}\cdot\vec{v}) + (\hat{n}\times\vec{v})\times\hat{n}.$$

Only $\vec{v}_\perp$ is affected by a rotation about $\hat{n}$ by an angle ϕ, which maps vector $\vec{v}$ to vector $\vec{v}'$ as shown in Figure 2.7. Decompose vector $\vec{v}'_\perp$ on two orthogonal vectors $\vec{v}_\perp$ and $\hat{n}\times\vec{v}$ and find an expression for $\vec{v}' = \vec{v}_\parallel + \vec{v}'_\perp$. Compare your result with Problem 2.3.2.

(b) The vector $\vec{v} = \hat{j}$ is rotated by $\phi = \pi/3$ about an axis along the direction

$$\begin{bmatrix} 1 \\ -1 \\ 1 \end{bmatrix}.$$

Use (a) to find the coordinates of the new vector.

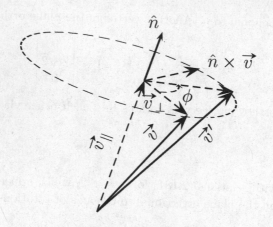

Figure 2.7: Rotation by an angle ϕ about $\hat{n}$.

Problem 2.3.4. Find a rotation matrix $\underline{\underline{R}}$ corresponding to a rotation by $\phi = \pi/3$ about an axis of a rotation along the direction

$$\begin{bmatrix} 1 \\ -1 \\ 1 \end{bmatrix}.$$

Problem 2.3.5. Prove that if

$$\underline{u} = \begin{bmatrix} u_1 \\ u_2 \\ u_3 \end{bmatrix} \iff \underline{\underline{U}} = \begin{bmatrix} 0 & -u_3 & u_2 \\ u_3 & 0 & -u_1 \\ -u_2 & u_1 & 0 \end{bmatrix}$$

and

$$\underline{v} = \begin{bmatrix} v_1 \\ v_2 \\ v_3 \end{bmatrix} \iff \underline{\underline{V}} = \begin{bmatrix} 0 & -v_3 & v_2 \\ v_3 & 0 & -v_1 \\ -v_2 & v_1 & 0 \end{bmatrix}$$

then

$$\underline{u} \times \underline{v} = \underline{\underline{U}}\,\underline{v} = -\underline{\underline{V}}\,\underline{u}$$

and

$$\underline{\underline{U}}\,\underline{\underline{V}} = \underline{v}\,\underline{u}^T - (\underline{u} \cdot \underline{v})\,\underline{\underline{I}},$$

where $\underline{v}\,\underline{u}^T$ is the matrix $[v_i u_j]$ obtained by multiplying a column $\underline{v}$ with a row $\underline{u}^T$.

Problem 2.3.6. Show that for the two skew-symmetric matrices $\underline{\underline{U}}$ and $\underline{\underline{V}}$ from Problem 2.3.5 holds the following relation:

$$\mathrm{Tr}\left(\underline{\underline{U}}\,\underline{\underline{V}}\right) = -2\,\underline{u}^T\underline{v}.$$

Problem 2.3.7. Consider a special orthogonal matrix $\underline{\underline{A}}$ which is only infinitesimally different from the identity matrix:

$$\underline{\underline{A}} = \underline{\underline{I}} + \epsilon \underline{\underline{M}}, \tag{2.184}$$

with parameter ϵ so small that the quadratic term ϵ^2 may be neglected when compared with ϵ. Show that the matrix $\underline{\underline{M}}$ must be skew-symmetric for $\underline{\underline{A}}$ to be orthogonal ($\underline{\underline{A}}^T \underline{\underline{A}} = \underline{\underline{I}}$). Show that for any vector $\underline{v}$

$$\underline{\underline{A}} \underline{v} = \underline{v} + \epsilon \underline{m} \times \underline{v},$$

where vector $\vec{m}$ has components given by $m_i = -\frac{1}{2\epsilon} \sum_{j,k=1}^{3} \epsilon_{ijk} a_{jk}$, where a_{jk} are matrix entries of $\underline{\underline{A}}$. Find an angle and an axis of infinitesimal rotation corresponding to the above matrix $\underline{\underline{A}}$.

Problem 2.3.8. Using equations (2.175) and (2.177) derive the following relation

$$2 \sin \phi \, \underline{\Omega} = \underline{\underline{R}}^T (\underline{\omega}, \phi) - \underline{\underline{R}} (\underline{\omega}, \phi)$$

and use it to prove that the axis of rotation $\underline{\omega}$ associated to the special orthogonal matrix $\underline{\underline{A}} = [a_{ij}]$ is given by the relation

$$\underline{\omega} = \frac{1}{2 \sin \phi} \begin{bmatrix} a_{23} - a_{32} \\ a_{31} - a_{13} \\ a_{12} - a_{21} \end{bmatrix},$$

valid for the angle of rotation $\phi (\neq \pi)$ given by

$$\cos \phi = \frac{1}{2} \left(\text{Tr} \left(\underline{\underline{A}} \right) - 1 \right).$$

Problem 2.3.9. (a) Use the result from Problems 2.2.2 and 2.3.8 to show that the characteristic polynomial of a rotation matrix $\underline{\underline{A}}$ can be written as:

$$\left| \underline{\underline{A}} - \lambda \underline{\underline{I}} \right| = -\lambda^3 + \text{Tr} \left(\underline{\underline{A}} \right) \lambda^2 - \text{Tr} \left(\underline{\underline{A}} \right) \lambda + 1 = - (\lambda - 1) \left[\lambda^2 - 2\lambda \cos \phi + 1 \right].$$

(b) Verify that the eigenvalues of a rotation matrix $\underline{\underline{A}}$ are given by:

$$\lambda_1 = 1, \quad \lambda_{2,3} = \cos \phi \pm i \sin \phi = e^{\pm i \phi}.$$

Problem 2.3.10. Find the angle that principal axes y_1, y_2, shown in Figure 2.4 in Example (2.2.10), make with coordinates x_1, x_2.

Problem 2.3.11. Show that the angle of rotation ϕ for the matrix $\underline{\underline{Q}}$, from Problem 2.2.14, which transforms the Cartesian unit vectors into spherical unit vectors is given by:

$$\cos \phi = \frac{1}{2} \left(\sin(\theta + \varphi) - 1 \right).$$

Problem 2.3.12. Show that an eigenvector of a special orthogonal matrix $\underline{\underline{A}}$ with eigenvalue $\lambda = 1$ is also an eigenvector of $\underline{\underline{A}}^T$ with the same eigenvalue $\lambda = 1$. Based on the above, show that the axis of rotation of $\underline{\underline{A}}$ is parallel to the vector $\vec{w}$ that is associated to $\underline{\underline{\Omega}} = \underline{\underline{A}} - \underline{\underline{A}}^T$ via correspondence (2.164). This observation agrees with the result obtained in the problem 2.3.8 but is based on a different argument.

Problem 2.3.13. Let $\underline{\underline{\Omega}}$ be a 3×3 skew-symmetric matrix and $\underline{w}$ the 3-dimensional vector related to each other via (2.164). Consider the eigenvalue problem $\underline{\underline{\Omega}}\, v_i = \underline{w} \times \underline{v_i} = \lambda_i\, \underline{v_i}$, $i = 1, 2, 3$ and denote by $w = \sqrt{w_1^2 + w_2^2 + w_1^2} = \sqrt{-\frac{1}{2}\operatorname{Tr}\left(\underline{\underline{\Omega}}^2\right)}$ the length of $\underline{w}$.

(a) Show that $0, \pm i\,w$ are the eigenvalues of $\underline{\underline{\Omega}}$.

(b) Verify that vectors:

$$\underline{a} = \frac{w}{w_1^2 + w_2^2}\begin{bmatrix} w_2 \\ -w_1 \\ 0 \end{bmatrix}, \qquad \underline{b} = \frac{1}{w_1^2 + w_2^2}\begin{bmatrix} -w_1 w_3 \\ -w_2 w_3 \\ w_1^2 + w_2^2 \end{bmatrix}, \qquad \text{for } w_1^2 + w_2^2 \neq 0,$$

satisfy conditions:

$$\underline{\underline{\Omega}}\,\underline{a} = -w\,\underline{b}, \qquad \underline{\underline{\Omega}}\,\underline{b} = w\,\underline{a}$$

and

$$\vec{w} \cdot \vec{a} = 0, \quad \vec{w} \cdot \vec{b} = 0, \quad \vec{a} \cdot \vec{b} = 0.$$

(c) Find the eigenvectors $\underline{v}, \underline{v}_\pm$ such that:

$$\underline{\underline{\Omega}}\,\underline{v} = 0, \qquad \underline{\underline{\Omega}}\,\underline{v}_\pm = \pm i\,w\,\underline{v}_\pm$$

in terms of $\underline{w}, \underline{a}$ and $\underline{b}$.

(d) For $w = 1$ solve the eigenvalue problem for the matrix exponential, $\exp\left(\phi\,\underline{\underline{\Omega}}\right)$.

Problem 2.3.14. Let vector $\vec{w}$ be associated to the skew-symmetric matrix $\underline{\underline{\Omega}}$ via the correspondence (2.164). Show that if $\underline{w}$ satisfies equations

$$\frac{\partial \underline{w}}{\partial x_i} = \underline{\underline{V_i}}\,\underline{w}, \qquad i = 1, 2, 3$$

with some skew-symmetric matrices $\underline{\underline{V_1}}, \underline{\underline{V_2}}, \underline{\underline{V_3}}$ then the corresponding matrix $\underline{\underline{\Omega}}$ satisfies:

$$\frac{\partial}{\partial x_i}\underline{\underline{\Omega}} = \underline{\underline{V_i}}\,\underline{\underline{\Omega}} - \underline{\underline{\Omega}}\,\underline{\underline{V_i}}, \qquad i = 1, 2, 3.$$

Also, show that $\operatorname{Tr}\left(\Omega^2\right)$ does not depend on x, y, z.

Problem 2.3.15. Given is the 3×3 matrix

$$\underline{\underline{R}} = \begin{bmatrix} 1/4 & 3/4 & -\sqrt{3/8} \\ 3/4 & 1/4 & \sqrt{3/8} \\ \sqrt{3/8} & -\sqrt{3/8} & -1/2 \end{bmatrix}.$$

(a) Show that $\underline{\underline{R}}$ is a rotation matrix.

(b) Find the eigenvector of $\underline{\underline{R}}$ that corresponds to the eigenvalue $\lambda = 1$.

(c) Find the axis of rotation $\vec{\omega}$ and the angle of rotation ϕ about $\vec{\omega}$ corresponding to matrix $\underline{\underline{R}}$.

Problem 2.3.16. Given is the 3×3 matrix

$$\underline{\underline{B}} = \frac{1}{9} \begin{bmatrix} 1 & -4 & 8 \\ 8 & 4 & 1 \\ -4 & 7 & 4 \end{bmatrix}.$$

(a) Show that $\underline{\underline{B}}$ is a rotation matrix.

(b) Find the eigenvector of $\underline{\underline{B}}$ that corresponds to the eigenvalue $\lambda = 1$.

(c) Find the axis of rotation $\vec{\omega}$ and the angle of rotation ϕ about $\vec{\omega}$ corresponding to matrix $\underline{\underline{B}}$.

Problem 2.3.17. Find a rotation matrix which rotates vectors on a plane $x+y+z = 4$ by π.

Problem 2.3.18. Find a rotation matrix which rotates a vector

$$\begin{bmatrix} 1 \\ 1 \\ 1 \end{bmatrix} \quad \text{into the vector} \quad \begin{bmatrix} 1 \\ -1 \\ -1 \end{bmatrix}.$$

Problem 2.3.19. Find the angle and axis of rotation for the rotation that rotates the x axis to the y axis, the y axis to the z axis, and the z axis to the x axis.

Problem 2.3.20. Show that, according to the angle and axis parametrization (2.175), the matrix elements of the rotation matrix are given by

$$\underline{\underline{R}}_{ij}(\underline{\omega}, \phi) = \sum_{k=1}^{3} \epsilon_{ijk}\omega_k \sin \phi + [\delta_{ij} - \omega_i\omega_j] \cos \phi + \omega_i\omega_j.$$

Problem 2.3.21. For $\underline{\omega}' = \underline{\underline{Q}}\underline{\omega}$, with a rotation matrix $\underline{\underline{Q}}$, show that

$$\underline{\underline{R}}(\underline{\omega}', \phi) = \underline{\underline{Q}}\,\underline{\underline{R}}(\underline{\omega}, \phi)\,\underline{\underline{Q}}^T.$$

2.4 Group Theory and Rotations

In the previous section we identified rotations with special orthogonal matrices. Such matrices preserve the length. If the dynamics of a physical system is determined by a potential that depends only on the distance between masses then the equations of motion remain invariant under rotation and the physical system is unchanged. Rotations are one of many types of symmetry transformations that simplify study of the physical problems. In general, a symmetry transformation belongs to some group and a group theory allows a systematic study and classification of symmetries being encountered in nature.

Definition 2.4.1. *A set G of finite or infinitely many elements g_i, $i = 1, 2, 3, \ldots$ is called a group with respect to some operation, $\circ$, if the following conditions are fulfilled:*

① *Closure: $g_i \circ g_j$ is in G for any g_i, g_j in G .*

② *The operation, $\circ$, is associative: $(g_i \circ g_j) \circ g_k = g_i \circ (g_j \circ g_k)$.*

③ *There exists an identity element I in G such that $I \circ g_i = g_i \circ I = g_i$ for any g_i in G.*

④ *For every g_i in G there exists an inverse g_i^{-1} in G such that $g_i \circ g_i^{-1} = g_i^{-1} \circ g_i = I$.*

The "multiplication" operation, $\circ$, is a binary composition that denotes an abstract prescription for associating a third element to an ordered pair of two group elements. The closure axiom ensures that this third element is inside the group. Contrary to ordinary multiplication, the composition, $\circ$, does not need to be commutative (in general $g_i \circ g_j \neq g_j \circ g_i$). The group with the commutative operation, $\circ$, is called an Abelian group.

☞ **Example 2.4.1.** The integers

$$\ldots, -4, -3, -2, -1, 0, 1, 2, 3, 4, \ldots$$

form a group under addition, with the composition operation being ordinary addition. Indeed, closure is ensured by the fact that the sum of any two integers is an integer. Also, addition is clearly an associative operation. 0 is the identity element and $-n$ is the inverse of n.

☞ **Example 2.4.2.** A very simple, but physically relevant, example of a group is the group of translations of functions of the position vector.

Consider a space translation operator $T(\vec{a})$ defined by the equation

$$T(\vec{a})f(\vec{r}) = f(\vec{r} + \vec{a}).$$

For infinitesimal translations,

$$T(\delta \vec{a}) f(\vec{r}) = f(\vec{r} + \delta \vec{a}) = f(\vec{r}) + \delta \vec{a} \cdot \vec{\nabla} f(\vec{r}) \rightsquigarrow T(\delta \vec{a}) = 1 + \delta \vec{a} \cdot \vec{\nabla}.$$

Translation by $\vec{a} + \delta \vec{a}$ is equal to two successive translations by $\vec{a}$ and $\delta \vec{a}$:

$$T(\vec{a} + \delta \vec{a}) = T(\delta \vec{a}) T(\vec{a}) = (1 + \delta \vec{a} \cdot \vec{\nabla}) T(\vec{a})$$

and, therefore,

$$\mathrm{d} T(\vec{a}) = T(\vec{a} + \delta \vec{a}) - T(\vec{a}) = \delta \vec{a} \cdot \vec{\nabla} T(\vec{a}).$$

Choosing the solution as

$$T(\vec{a}) = e^{\vec{a} \cdot \vec{\nabla}}$$

ensures that a condition for zero displacement satisfies the identity, $T(0) = 1$.

A set of translation operators forms a group under the multiplication:

$$T(\vec{a}) T(\vec{b}) = e^{\vec{a} \cdot \vec{\nabla}} e^{\vec{b} \cdot \vec{\nabla}} = e^{(\vec{a} + \vec{b}) \cdot \vec{\nabla}} = T(\vec{a} + \vec{b}).$$

Obviously, $T(-\vec{a})$ is an inverse of $T(\vec{a})$ since $T(\vec{a}) T(-\vec{a}) = T(\vec{a} - \vec{a}) = T(0) = 1$.

The multiplication is commutative due to:

$$T(\vec{b}) T(\vec{a}) = T(\vec{b} + \vec{a}) = T(\vec{a} + \vec{b})$$

and, therefore, the group of translations is Abelian.

☞ **Example 2.4.3.** Let $O(n)$ denote a set of all $n \times n$ orthogonal matrices. Such matrices satisfy all requirements of a group. In this case, the composition is matrix multiplication, which is associative. Multiplication of two orthogonal matrices leads to a third, which also is orthogonal and, thus, the closure requirement is satisfied. The $n \times n$ identity matrix $\underline{\underline{I}}$ belongs to the $O(n)$ group, and each orthogonal matrix $\underline{\underline{O}}$ has an inverse $\underline{\underline{O}}^T$ inside the group. For $n = 3$, the group of three-dimensional orthogonal transformations $O(3)$, consists of rotations for which $|\underline{\underline{O}}| = 1$ and reflections for which $|\underline{\underline{O}}| = -1$.

2.4.1 SO(3) Group

The rotation matrices form a subgroup SO (3) of $O(3)$ consisting of all matrices in $O(3)$ such that $| \underline{\underline{A}} | = 1$. SO(3) is called a special orthogonal group. It is a group of rotations in three dimensions.

The orthogonal 3×3 matrices that represent rotations $\underline{r}' = \underline{\underline{R}}(\hat{n}, \phi) \underline{r}$ have a simple form when the rotation axis $\hat{n}$ coincides with x, y or z directions of the coordinate axes. In the spirit of correspondence (2.164), we associate three Hermitian matrices:

$$\underline{\underline{J_1}} = \mathrm{i} \begin{bmatrix} 0 & 0 & 0 \\ 0 & 0 & -1 \\ 0 & 1 & 0 \end{bmatrix}, \quad \underline{\underline{J_2}} = \mathrm{i} \begin{bmatrix} 0 & 0 & 1 \\ 0 & 0 & 0 \\ -1 & 0 & 0 \end{bmatrix}, \quad \underline{\underline{J_3}} = \mathrm{i} \begin{bmatrix} 0 & -1 & 0 \\ 1 & 0 & 0 \\ 0 & 0 & 0 \end{bmatrix} \tag{2.185}$$

to vectors $\hat{i}, \hat{j}$ and $\hat{k}$ The imaginary number i in front of matrices was inserted to ensure that the matrices are Hermitian.

Comparing the above Hermitian matrices with (2.164), we can write:

$$\underline{\underline{\Omega}} = -\mathrm{i} \sum_{i=1}^{3} \omega_i \, \underline{\underline{J_i}} = -\mathrm{i} \left(\vec{\omega} \cdot \underline{\underline{\vec{J}}} \right) \tag{2.186}$$

and use it to rewrite the angle and axis parameterization (2.175) for the rotation matrix in terms of the Hermitian matrices $\underline{\underline{J_i}}$, $i = 1, 2, 3$:

$$\underline{\underline{R}}(\underline{\omega}, \phi) = e^{\mathrm{i}\vec{\phi} \cdot \underline{\underline{\vec{J}}}} = \underline{\underline{I}} + \mathrm{i}\vec{\omega} \cdot \underline{\underline{\vec{J}}} \sin\phi + \left(\mathrm{i}\vec{\omega} \cdot \underline{\underline{\vec{J}}} \right)^2 (1 - \cos\phi), \quad \vec{\phi} = \phi\vec{\omega}. \tag{2.187}$$

For a rotation of angle α about the z-axis we find that the rotation matrix becomes

$$\underline{\underline{R}}(\hat{k}, \alpha) = e^{\mathrm{i}\alpha \underline{\underline{J_3}}} = \underline{\underline{I}} + \mathrm{i}\underline{\underline{J_3}} \sin\alpha + \left(\mathrm{i}\underline{\underline{J_3}} \right)^2 (1 - \cos\alpha) = \begin{bmatrix} \cos\alpha & \sin\alpha & 0 \\ -\sin\alpha & \cos\alpha & 0 \\ 0 & 0 & 1 \end{bmatrix}. \tag{2.188}$$

Similarly, for rotations about the x and y axes we get

$$\underline{\underline{R}}(\hat{i}, \phi) = e^{\mathrm{i}\phi \underline{\underline{J_1}}} = \underline{\underline{I}} + \mathrm{i}\underline{\underline{J_1}} \sin\phi + \left(\mathrm{i}\underline{\underline{J_1}} \right)^2 (1 - \cos\phi) = \begin{bmatrix} 1 & 0 & 0 \\ 0 & \cos\phi & \sin\phi \\ 0 & -\sin\phi & \cos\phi \end{bmatrix}.$$

$$\underline{\underline{R}}(\hat{j}, \psi) = e^{\mathrm{i}\psi \underline{\underline{J_2}}} = \underline{\underline{I}} + \mathrm{i}\underline{\underline{J_2}} \sin\psi + \left(\mathrm{i}\underline{\underline{J_2}} \right)^2 (1 - \cos\psi) = \begin{bmatrix} \cos\psi & 0 & -\sin\psi \\ 0 & 1 & 0 \\ \sin\psi & 0 & \cos\psi \end{bmatrix}. \tag{2.189}$$

Rotations about the same axis commute, for instance:

$$\underline{\underline{R}}(\hat{k}, \frac{\pi}{2}) \underline{\underline{R}}(\hat{k}, \frac{\pi}{3}) = \underline{\underline{R}}(\hat{k}, \frac{\pi}{3}) \underline{\underline{R}}(\hat{k}, \frac{\pi}{2}).$$

However, is important to note that, in general, rotation matrices corresponding to different axes do not commute. For instance:

$$\underline{\underline{R}}(\hat{k}, \frac{\pi}{2}) \underline{\underline{R}}(\hat{i}, \frac{\pi}{2}) = \begin{bmatrix} 0 & 1 & 0 \\ -1 & 0 & 0 \\ 0 & 0 & 1 \end{bmatrix} \begin{bmatrix} 1 & 0 & 0 \\ 0 & 0 & 1 \\ 0 & -1 & 0 \end{bmatrix} = \begin{bmatrix} 0 & 0 & 1 \\ -1 & 0 & 0 \\ 0 & -1 & 0 \end{bmatrix}$$

Matrices and Rotations

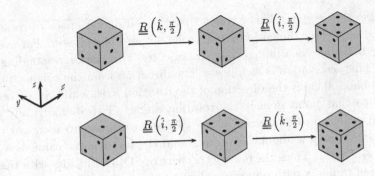

Figure 2.8: Two non-commutative rotations.

differs from

$$\underline{\underline{R}}(\hat{i},\,\frac{\pi}{2})\,\underline{\underline{R}}(\hat{k},\,\frac{\pi}{2}) = \begin{bmatrix} 1 & 0 & 0 \\ 0 & 0 & 1 \\ 0 & -1 & 0 \end{bmatrix} \begin{bmatrix} 0 & 1 & 0 \\ -1 & 0 & 0 \\ 0 & 0 & 1 \end{bmatrix} = \begin{bmatrix} 0 & 1 & 0 \\ 0 & 0 & 1 \\ 1 & 0 & 0 \end{bmatrix}.$$

The fact that rotations do not commute is illustrated in Figure (2.8), which shows two successive $\pi/2$ rotations of a six-sided die performed in different order. In one case, the rotation about z-axis was applied before the rotation about x-axis, in the other case the order was reversed. Due to non-commutativity of rotations, the SO(3) group is a non-Abelian group.

The orthogonality condition $\underline{\underline{A}}^T \underline{\underline{A}} = \underline{I}$ gives rise to nine equations $\underline{a}_i^T \underline{a}_j = \delta_{ij}$ with $i, j = 1, 2, 3$ for the column vectors of matrix $\underline{\underline{A}}$. Due to the fact that $\underline{a}_i^T \underline{a}_j = \underline{a}_j^T \underline{a}_i$ and $\delta_{ij} = \delta_{ji}$ only six among those nine equations are independent. Hence, the nine real matrix elements of $\underline{\underline{A}}$ are subjected to six independent conditions. Left are three independent real parameters that fully describe every element of the SO(3) group. The same conclusion can be reached by identifying SO(3) with a group of three dimensional rotations. We have shown that every rotation matrix is characterized by an angle of rotation and an axis of rotation. The angle and axis parametrization (2.175) relates an SO(3) rotation matrix to an angle of rotation, ϕ, and the unit vector indicating the direction of the rotation axis $\vec{\omega}$. The endpoint of the unit vectors $\vec{\omega}$ sweeps the unit sphere described by only two real parameters, e.g. an azimuthal angle φ and a spherical angle θ. In this spherical parametrization, the components of $\vec{\omega}$ are given by $(\omega_1, \omega_2, \omega_3) = (\sin\theta\cos\varphi, \sin\theta\sin\varphi, \cos\theta)$ and the parameter space of SO(3) can be represented as

$$(\phi\sin\theta\cos\varphi, \phi\sin\theta\sin\varphi, \phi\cos\theta) \,.$$

Thus, the above three real parameters suffice to describe the SO(3) group. Since a rotation of $\phi > \pi$ is equivalent to a rotation of $\phi - \pi$ in the opposite direction, the rotation angle ϕ is in the range $0 \le \phi \le \pi$. The spherical angles are in the range:

$0 \leq \varphi \leq 2\pi$, $0 \leq \theta \leq \pi$. This ensures that the parameter space of rotations can be pictured as a three dimensional ball (solid sphere) of radius π. For every point within the ball there is a unique way to associate it to the corresponding element of SO(3). This association is as follows: The direction from the origin to the point within the ball indicates the direction of the rotation axis, while the distance from the origin to that point defines the rotation angle. Two diametrically opposite (antipodal) points on the surface of the sphere correspond to identical rotations (rotations by π and $-\pi$ about $\vec{\omega}$ are equal and the latter is the same as a rotation of π about $-\vec{\omega}$ axis). Thus, the parameter space of SO(3) coincides with the interior of a sphere of radius π with opposite points on the surface identified.

Recall from definition 1.4.2 that a region is called simply connected region if any closed curve (loop) can be shrunk continuously to a single point within that region. Consider a curve within three dimensional ball associated to SO(3) made out of rotations about some fixed direction $\vec{\omega}$ by angles ranging from $-\pi$ to π. Antipodal points $-\pi\vec{\omega}$ and $\pi\vec{\omega}$ on the surface of the sphere that are associated to identical rotations are equivalent. Thus, the curve that touches the surface at one point will emerge on the opposite side of the sphere. Two diametrically opposite points on the surface will remain on the opposite sides and the curve cannot be deformed continuously to the point. Therefore, SO(3) is not simply connected.

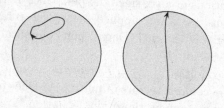

Figure 2.9: The loop inside the sphere can be deformed continuously to a point. However, the loop touching the boundary cannot be deformed continuously to a point.

Two different SO(3) rotations that lie the same distance ϕ from origin correspond to two different directions let say $\vec{\omega}$ and $\vec{\omega'}$. According to Problem (2.3.21), if $\underline{Q}$ is a special orthogonal matrix that rotates $\vec{\omega}$ into $\vec{\omega'}$ then the corresponding rotation matrices are connected by the similarity transformation,

$$\underline{\underline{R}}(\underline{\omega'}, \phi) = \underline{Q}\,\underline{\underline{R}}(\underline{\omega}, \phi)\,\underline{Q}^T.$$

In physics, it is often convenient to use *the Euler angle parametrization* of a SO(3) group. The three Euler angles (α, β, γ) have ranges $0 \leq \alpha \leq 2\pi$, $0 \leq \beta \leq \pi$ and $0 \leq \gamma \leq 2\pi$. The Euler angles are associated with three successive counterclockwise rotations:

ⓐ Rotation through an angle α about z-axis

$$\underline{\underline{R}}\left(\widehat{k}, \alpha\right) = \exp\left(i\alpha \underline{\underline{J_z}}\right) = \begin{bmatrix} \cos\alpha & \sin\alpha & 0 \\ -\sin\alpha & \cos\alpha & 0 \\ 0 & 0 & 1 \end{bmatrix},$$

which transforms the x, y, z coordinates to $x', y', z' = z$ coordinates.

ⓑ Rotation through an angle β about y'-axis

$$\underline{\underline{R}}\left(\widehat{j}', \beta\right) = \exp\left(i\beta \underline{\underline{J_{y'}}}\right),$$

which transforms the x', y', z' coordinates to $x'', y'' = y', z''$ coordinates.

ⓒ Finally, a rotation through an angle γ about z''-axis

$$\underline{\underline{R}}\left(\widehat{k}'', \gamma\right) = \exp\left(i\gamma \underline{\underline{J_{z''}}}\right),$$

which transforms the x'', y'', z'' coordinates to $x''', y''', z''' = z''$ coordinates.

The Euler angles and the corresponding rotations are shown in Figure 2.10

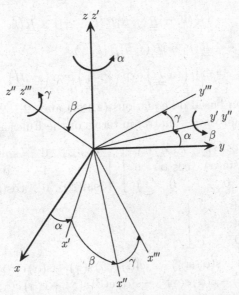

Figure 2.10: Euler angles α, β, γ.

The combined rotation matrix representing the three Euler rotations is then given by

$$\underline{\underline{R}}\left(\alpha, \beta, \gamma\right) = \underline{\underline{R}}\left(\widehat{k}', \gamma\right)\underline{\underline{R}}\left(\widehat{j}', \beta\right)\underline{\underline{R}}\left(\widehat{k}, \alpha\right) = \exp\left(i\gamma \underline{\underline{J_{z'}}}\right)\exp\left(i\beta \underline{\underline{J_{y'}}}\right)\exp\left(i\alpha \underline{\underline{J_z}}\right). \tag{2.190}$$

Since the first rotation $\underline{R}(\widehat{k}, \alpha)$ rotates the unit vector $\widehat{j}$ onto $\widehat{j}'$ we have, according to Problem (2.3.21), a relation:

$$\underline{R}(\widehat{j}', \beta) = \underline{R}(\widehat{k}, \alpha)\underline{R}(\widehat{j}, \beta)\underline{R}^{-1}(\widehat{k}, \alpha) = \underline{R}(\widehat{k}, \alpha)\underline{R}(\widehat{j}, \beta)\underline{R}(\widehat{k}, -\alpha).$$

Similarly, since $\underline{R}(\widehat{j}', \beta)\underline{R}(\widehat{k}, \alpha)$ rotates x, y, z coordinates into the x'', y'', z'' coordinates we obtain a relation

$$\begin{aligned}
\underline{R}(\widehat{k}'', \gamma) &= \underline{R}(\widehat{j}', \beta)\underline{R}(\widehat{k}, \alpha)\underline{R}(\widehat{k}, \gamma)\underline{R}(\widehat{k}, -\alpha)\underline{R}(\widehat{j}', -\beta) \\
&= \left[\underline{R}(\widehat{k}, \alpha)\underline{R}(\widehat{j}, \beta)\underline{R}(\widehat{k}, -\alpha)\right]\underline{R}(\widehat{k}, \alpha)\underline{R}(\widehat{k}, \gamma)\underline{R}(\widehat{k}, -\alpha) \\
&\quad \times \left[\underline{R}(\widehat{k}, \alpha)\underline{R}(\widehat{j}, -\beta)\underline{R}(\widehat{k}, -\alpha)\right] \\
&= \underline{R}(\widehat{k}, \alpha)\underline{R}(\widehat{j}, \beta)\underline{R}(\widehat{k}, \gamma)\underline{R}(\widehat{j}, -\beta)\underline{R}(\widehat{k}, -\alpha).
\end{aligned}$$

Accordingly, the relation for the combined Euler rotation simplifies to

$$\begin{aligned}
\underline{R}(\alpha, \beta, \gamma) &= \underline{R}(\widehat{k}', \gamma)\underline{R}(\widehat{j}', \beta)\underline{R}(\widehat{k}, \alpha) \\
&= \left[\underline{R}(\widehat{k}, \alpha)\underline{R}(\widehat{j}, \beta)\underline{R}(\widehat{k}, \gamma)\underline{R}(\widehat{j}, -\beta)\underline{R}(\widehat{k}, -\alpha)\right] \\
&\quad \times \left[\underline{R}(\widehat{k}, \alpha)\underline{R}(\widehat{j}, \beta)\underline{R}(\widehat{k}, -\alpha)\right] \times \underline{R}(\widehat{k}, \alpha) \\
&= \underline{R}(\widehat{k}, \alpha)\underline{R}(\widehat{j}, \beta)\underline{R}(\widehat{k}, \gamma) \\
&= \exp\left(i\alpha\, \underline{J_z}\right)\exp\left(i\beta\, \underline{J_y}\right)\exp\left(i\gamma\, \underline{J_z}\right).
\end{aligned}$$

An explicit calculation based on equations (2.188) and (2.189) gives the following parametrization of the rotation matrix in terms of the Euler angles:

$$\begin{aligned}
\underline{R}(\alpha, \beta, \gamma) &= \begin{bmatrix} \cos(\alpha) & \sin(\alpha) & 0 \\ -\sin(\alpha) & \cos(\alpha) & 0 \\ 0 & 0 & 1 \end{bmatrix} \begin{bmatrix} \cos(\beta) & 0 & -\sin(\beta) \\ 0 & 1 & 0 \\ \sin(\beta) & 0 & \cos(\beta) \end{bmatrix} \\
&\quad \times \begin{bmatrix} \cos(\gamma) & \sin(\gamma) & 0 \\ -\sin(\gamma) & \cos(\gamma) & 0 \\ 0 & 0 & 1 \end{bmatrix} \\
&= \begin{bmatrix} c(\alpha)\,c(\beta)\,c(\gamma) - s(\alpha)\,s(\gamma) & c(\alpha)\,c(\beta)\,s(\gamma) + s(\alpha)\,c(\gamma) & -c(\alpha)\,s(\beta) \\ -s(\alpha)\,c(\beta)\,c(\gamma) - c(\alpha)\,s(\gamma) & -s(\alpha)\,c(\beta)\,s(\gamma) + c(\alpha)\,c(\gamma) & s(\alpha)\,s(\beta) \\ s(\beta)\,c(\gamma) & s(\beta)\,s(\gamma) & c(\beta) \end{bmatrix},
\end{aligned}$$

$$(2.191)$$

where we used the abbreviations $c(\alpha) = \cos(\alpha)$ and $s(\alpha) = \sin(\alpha)$.

2.4.2 The Infinitesimal Rotations

According to (2.167) the three Hermitian matrices $\underline{J}_i, i = 1, 2, 3$ given in (2.185) define an infinitesimal rotation matrix $\underline{R}(\underline{\omega}, \mathrm{d}\,\phi)$, which describes a rotation by an infinitesimal angle $d\phi$ about an axis of rotation that points in the direction of the vector $\vec{\omega}$. Identifying the vector $\vec{\omega}$ with three orthogonal unit vectors $\hat{i}, \hat{j}, \hat{k}$ we obtain, respectively:

$$\underline{R}(\hat{i}, d\phi) = \begin{bmatrix} 1 & 0 & 0 \\ 0 & 1 & d\phi \\ 0 & -d\phi & 1 \end{bmatrix} = \underline{I} + i\, d\phi \underline{J_1}, \tag{2.192}$$

$$\underline{R}(\hat{j}, d\psi) = \begin{bmatrix} 1 & 0 & -d\psi \\ 0 & 1 & 0 \\ d\psi & 0 & 1 \end{bmatrix} = \underline{I} + i\, d\psi \underline{J_2}, \tag{2.193}$$

$$\underline{R}(\hat{k}, d\alpha) = \begin{bmatrix} 1 & d\alpha & 0 \\ -d\alpha & 1 & 0 \\ 0 & 0 & 1 \end{bmatrix} = \underline{I} + i\, d\alpha \underline{J_3}. \tag{2.194}$$

In this sense, the matrices $\underline{J_i}$ generate infinitesimal rotations. Now, we can investigate the question of whether the infinitesimal rotations commute with each other. For illustration, we perform infinitesimal rotation about the z-axis followed by an infinitesimal rotation about the y-axis:

$$\underline{R}(\hat{j}, d\psi)\underline{R}(\hat{k}, d\alpha) = \left(\underline{I} + i\, d\psi \underline{J_2} \right)\left(\underline{I} + i\, d\alpha \underline{J_3} \right)$$

$$= \underline{I} + i\, d\psi \underline{J_2} + i\, d\alpha \underline{J_3} - d\psi\, d\alpha \underline{J_2}\,\underline{J_3}.$$

For the same rotations performed in the opposite order we obtain,

$$\underline{R}(\hat{k}, d\alpha)\underline{R}(\hat{j}, d\psi) = \underline{I} + i\, d\psi \underline{J_2} + i\, d\alpha \underline{J_3} - d\psi\, d\alpha \underline{J_3}\,\underline{J_2}.$$

The difference is equal to

$$\underline{R}(\hat{j}, d\psi)\underline{R}(\hat{k}, d\alpha) - \underline{R}(\hat{k}, d\alpha)\underline{R}(\hat{j}, d\psi) = d\psi\, d\alpha \left(\underline{J_3}\,\underline{J_2} - \underline{J_2}\,\underline{J_3} \right). \tag{2.195}$$

Definition 2.4.2. *For two matrices $\underline{A}$ and $\underline{B}$ an expression*

$$[\underline{A}, \underline{B}] = \underline{A}\,\underline{B} - \underline{B}\,\underline{A} \tag{2.196}$$

is called a commutator.

In terms of a commutator we can rewrite relation (2.195) as

$$\underline{R}(\hat{j}, d\psi)\underline{R}(\hat{k}, d\alpha) - \underline{R}(\hat{k}, d\alpha)\underline{R}(\hat{j}, d\psi) = d\psi\, d\alpha \left[\underline{J_3}, \underline{J_2} \right]. \tag{2.197}$$

Similarly,

$$\underline{R}(\hat{i}, d\phi)\underline{R}(\hat{j}, d\psi) - \underline{R}(\hat{i}, d\phi)\underline{R}(\hat{j}, d\psi) = d\psi\, d\phi \left[\underline{J_2}, \underline{J_1} \right] \tag{2.198}$$

and

$$\underline{R}(\hat{i}, d\phi)\underline{R}(\hat{k}, d\alpha) - \underline{R}(\hat{i}, d\phi)\underline{R}(\hat{k}, d\alpha) = d\alpha\, d\phi \left[\underline{J_3}, \underline{J_1} \right]. \tag{2.199}$$

A non-zero value of a commutator $\left[\underline{\underline{J_i}}, \underline{\underline{J_j}}\right]$ confirms that the rotation group is non-Abelian, e.g. a rotation about x_i-axis plus a rotation about the x_j-axis does not equal a rotation about the x_j-axis plus a rotation about the x_i-axis.

From definition (2.185) we get

$$\underline{\underline{J_1}}\,\underline{\underline{J_2}} = -\begin{bmatrix} 0 & 0 & 0 \\ 0 & 0 & -1 \\ 0 & 1 & 0 \end{bmatrix}\begin{bmatrix} 0 & 0 & 1 \\ 0 & 0 & 0 \\ -1 & 0 & 0 \end{bmatrix} = -\begin{bmatrix} 0 & 0 & 0 \\ 1 & 0 & 0 \\ 0 & 0 & 0 \end{bmatrix}$$

and

$$\underline{\underline{J_2}}\,\underline{\underline{J_1}} = -\begin{bmatrix} 0 & 0 & 1 \\ 0 & 0 & 0 \\ -1 & 0 & 0 \end{bmatrix}\begin{bmatrix} 0 & 0 & 0 \\ 0 & 0 & -1 \\ 0 & 1 & 0 \end{bmatrix} = -\begin{bmatrix} 0 & 1 & 0 \\ 0 & 0 & 0 \\ 0 & 0 & 0 \end{bmatrix}.$$

Thus, the commutator, $\underline{\underline{J_1}}\,\underline{\underline{J_2}} - \underline{\underline{J_2}}\,\underline{\underline{J_1}}$, is equal to

$$\left[\underline{\underline{J_1}}, \underline{\underline{J_2}}\right] = \begin{bmatrix} 0 & 1 & 0 \\ -1 & 0 & 0 \\ 0 & 0 & 0 \end{bmatrix} = i\,\underline{\underline{J_3}}.$$

In a similar way, we find two other commutators

$$\left[\underline{\underline{J_2}}, \underline{\underline{J_3}}\right] = i\,\underline{\underline{J_1}}, \qquad \left[\underline{\underline{J_3}}, \underline{\underline{J_1}}\right] = i\,\underline{\underline{J_2}}.$$

The permutation symbol ϵ_{ijk} can be used to cast all the three commutators in a compact form

$$\boxed{\left[\underline{\underline{J_i}}, \underline{\underline{J_j}}\right] = i\sum_{k=0}^{3} \epsilon_{ijk}\,\underline{\underline{J_k}}.}$$

Thus, a non-vanishing commutator of two generators of infinitesimal rotation is equal to another generator. This result indicates that the commutator operation exhibits a closure property similar to the one possessed by the product operation, $\circ$, in the definition of a group.

This closure property extends to any two matrices that are linear combinations of $\underline{\underline{J_1}}$, $\underline{\underline{J_2}}$ and $\underline{\underline{J_3}}$ meaning that a commutator of these two matrices will be a linear combination of $\underline{\underline{J_1}}$, $\underline{\underline{J_2}}$ and $\underline{\underline{J_3}}$. This follows by the linearity property of a commutator:

$$\left[a\underline{\underline{A}} + b\underline{\underline{B}}, \underline{\underline{C}}\right] = a\left[\underline{\underline{A}}, \underline{\underline{C}}\right] + b\left[\underline{\underline{B}}, \underline{\underline{C}}\right].$$

Thus, the commutator of two elements of a vector space generated by $\underline{\underline{J_1}}$, $\underline{\underline{J_2}}$ and $\underline{\underline{J_3}}$ is itself an element of a vector space generated by $\underline{\underline{J_1}}$, $\underline{\underline{J_2}}$ and $\underline{\underline{J_3}}$. In other words, commutation closes on such vector space. which makes it space an example of a Lie algebra:

Definition 2.4.3. *A Lie algebra is a vector space V equipped with a skew-symmetric bilinear map:*

$$(v_1, v_2) \mapsto [v_1, v_2] = -[v_2, v_1],$$

which satisfies the so-called Jacobi identity

$$[v_1, [v_2, v_3]] + [v_3, [v_1, v_2]] + [v_2, [v_3, v_1]] = 0 \qquad (2.200)$$

for any three vectors v_1, v_2 and v_3 in V.

The commutator of two matrices, i.e., $[\underline{A}, \underline{B}]$, where $\underline{A}$ and $\underline{B}$ are two square matrices, satisfies the above requirements for a skew-symmetric bilinear map since it automatically obeys the Jacobi identity (see Problem (2.4.8)). A Lie algebra spanned by generators J_1, J_2 and J_3 of rotations is called a $\mathfrak{so}(3)$ algebra and consists of all skew-symmetric 3×3 matrices.

Now, consider a function f of x, y and z. Under rotation by $\underline{R}(\hat{i}, \mathrm{d}\,\phi)$, from equation (2.192), the argument of f transforms as

$$\begin{bmatrix} x \\ y \\ z \end{bmatrix} \rightsquigarrow \begin{bmatrix} x' \\ y' \\ z' \end{bmatrix} = \underline{R}(\hat{i}, \mathrm{d}\,\phi) \begin{bmatrix} x \\ y \\ z \end{bmatrix} = \begin{bmatrix} x \\ y + z\,\mathrm{d}\phi \\ z - y\,\mathrm{d}\phi \end{bmatrix}.$$

We are interested in the quantity

$$\mathrm{d}f = f(x', y', z') - f(x, y, z), \qquad (2.201)$$

which measures a change in f under infinitesimal rotation (2.192). Expanding $\mathrm{d}f$ in a Taylor series we obtain

$$\mathrm{d}f = f(x, y + z\,\mathrm{d}\phi, z - y\,\mathrm{d}\phi) - f(x, y, z) = f(x, y, z)$$
$$+ \mathrm{d}\phi \left(z\frac{\partial}{\partial y} - y\frac{\partial}{\partial z} \right) f(x, y, z) - f(x, y, z)$$
$$= -\mathrm{i}\,\mathrm{d}\phi\, L_1\, f(x, y, z),$$

where

$$L_1 = -\mathrm{i} \left(y\frac{\partial}{\partial z} - z\frac{\partial}{\partial y} \right) \qquad (2.202)$$

is the first component of the angular momentum operator:

$$\vec{L} = -\mathrm{i} \left(\vec{r} \times \vec{\nabla} \right). \qquad (2.203)$$

Similarly, under the infinitesimal rotation $\underline{R}(\hat{j}, d\,\psi)$ from equation (2.193) we find

$$\begin{bmatrix} x \\ y \\ z \end{bmatrix} \rightsquigarrow \begin{bmatrix} x' \\ y' \\ z' \end{bmatrix} = \underline{R}(\hat{j}, d\,\psi) \begin{bmatrix} x \\ y \\ z \end{bmatrix} = \begin{bmatrix} x - z d\psi \\ y \\ z + x d\psi \end{bmatrix}$$

and

$$df = f(x - z\,d\psi, y, z + x\,d\psi) - f(x,y,z) = f(x,y,z)$$
$$+ d\psi \left(x\frac{\partial}{\partial z} - z\frac{\partial}{\partial x} \right) f(x,y,z) - f(x,y,z) = -i\,d\psi L_2\,f(x,y,z),$$

with the second component of the angular momentum operator $\vec{L}$:

$$L_2 = -i\left(z\frac{\partial}{\partial x} - x\frac{\partial}{\partial z} \right). \tag{2.204}$$

Finally, the infinitesimal transformation

$$\begin{bmatrix} x \\ y \\ z \end{bmatrix} \rightsquigarrow \begin{bmatrix} x' \\ y' \\ z' \end{bmatrix} = \underline{\underline{R}}(\hat{k}, d\alpha) \begin{bmatrix} x \\ y \\ z \end{bmatrix} = \begin{bmatrix} x + y\,d\alpha \\ y - x\,d\alpha \\ z \end{bmatrix}$$

induces

$$df = f(x + y\,d\alpha, y - x\,d\alpha, z) - f(x,y,z) = -i\,d\alpha L_3\,f(x,y,z),$$

where

$$L_3 = -i\left(x\frac{\partial}{\partial y} - y\frac{\partial}{\partial x} \right). \tag{2.205}$$

All three components can be combined in one compact formula:

$$L_i = -i\sum_{j,k=1}^{3} \epsilon_{ijk} x_j \frac{\partial}{\partial x_k}, \tag{2.206}$$

which uses the permutation symbol ϵ_{ijk}.

We see that the generators $\underline{J_i}$, $i = 1, 2, 3$ of rotations in three-dimensional space give rise to angular momentum operators that generate corresponding transformations of functions of the three-dimensional space coordinates. If the function is a scalar function (e.g. $f(x,y,z) = x^2 + y^2 + z^2$) then it remains invariant under rotation. Thus, $df = 0$ or

$$L_i f(x,y,z) = 0.$$

Now, we examine the commutation relations of the angular momentum operators. Extending the definition of the commutator to differential operators we find that:

$$[L_1, L_2] = L_1 L_2 - L_2 L_1 = \left(z\frac{\partial}{\partial x} - x\frac{\partial}{\partial z} \right)\left(y\frac{\partial}{\partial z} - z\frac{\partial}{\partial y} \right)$$
$$- \left(y\frac{\partial}{\partial z} - z\frac{\partial}{\partial y} \right)\left(z\frac{\partial}{\partial x} - x\frac{\partial}{\partial z} \right) = x\frac{\partial}{\partial z}z\frac{\partial}{\partial y} - y\frac{\partial}{\partial z}z\frac{\partial}{\partial x}$$
$$= x\frac{\partial}{\partial y} - y\frac{\partial}{\partial x} = i L_3.$$

In a similar manner,

$$[L_3 , L_1] = i L_2$$
$$[L_2 , L_3] = i L_1 ,$$

which can easily be combined in the single commutation relation

$$\left[L_i , L_j\right] = i \sum_{k=1}^{3} \epsilon_{ijk} L_k .$$

Thus, angular momentum operators satisfy the $\mathfrak{so}(3)$ algebra.

2.4.3 SU(2) Group

SU(2) is a group of special unitary 2×2 matrices. A special unitary matrix is a matrix whose inverse equals its hermitian conjugate and has a determinant equal to unity. Recall that determinant of an unitary matrix satisfies $|\underline{\underline{U}}||\underline{\underline{U}}|^* = 1$ and, therefore, is equal to a phase, $\exp(i\alpha)$. For special unitary matrices this phase must be equal to 1.

Let a 2×2 matrix be

$$\underline{\underline{U}} = \begin{bmatrix} \alpha & \beta \\ \gamma & \delta \end{bmatrix} , \tag{2.207}$$

with its adjoint given by

$$\underline{\underline{U}}^\dagger = \begin{bmatrix} \alpha^* & \gamma^* \\ \beta^* & \delta^* \end{bmatrix} = \underline{\underline{U}}^{-1} = \begin{bmatrix} \delta & -\beta \\ -\gamma & \alpha \end{bmatrix} . \tag{2.208}$$

Imposing the conditions $\underline{\underline{U}}\,\underline{\underline{U}}^\dagger = \underline{\underline{I}}$ and $|\underline{\underline{U}}| = 1$ we find that a general SU(2) matrix must have the form

$$\underline{\underline{U}}(\alpha, \beta) = \begin{bmatrix} \alpha & \beta \\ -\beta^* & \alpha^* \end{bmatrix} , \tag{2.209}$$

with the complex parameters, $\alpha = a_1 + i a_2$ and $\beta_1 + i b_2$, that satisfy the three-dimensional sphere condition $|\alpha|^2 + |\beta|^2 = a_1^2 + a_2^2 + b_1^2 + b_2^2 = 1$. Thus the parameter space of SU(2) group is a three-sphere (a surface of the four-dimensional solid sphere (ball)). Using this parametrization, we directly check that matrices $\underline{\underline{U}}(\alpha, \beta)$ satisfy all properties of a group. A closure of the product follows from the matrix relation:

$$\underline{\underline{U}}(\alpha_1, \beta_1) \, \underline{\underline{U}}(\alpha_2, \beta_2) = \underline{\underline{U}}\left(\alpha_1\alpha_2 - \beta_1\beta_2^*, \alpha_1\beta_2 + \alpha_2^*\beta_1\right) , \tag{2.210}$$

with condition $|\alpha_1\alpha_2 - \beta_1\beta_2^*|^2 + |\alpha_1\beta_2 + \alpha_2^*\beta_1|^2 = 1$, that follow directly from the determinant property $|\underline{\underline{U}}(\alpha_1, \beta_1) \, \underline{\underline{U}}(\alpha_2, \beta_2)| = |\underline{\underline{U}}(\alpha_1, \beta_1)| \, |\underline{\underline{U}}(\alpha_2, \beta_2)| = 1.$

The identity matrix, corresponds in this parametrization to the choice $\alpha = 1, \beta = 0$,

$$\underline{\underline{U}}(1,0) = \begin{bmatrix} 1 & 0 \\ 0 & 1 \end{bmatrix}.$$

The existence of an inverse is ensured by the unitarity property $\underline{\underline{U}}(\alpha,\beta)\,\underline{\underline{U}}^\dagger(\alpha,\beta) = \underline{\underline{I}}$. Finally, associativity follows from the fact that multiplications of matrices are associative.

A useful parametrization for the three-sphere condition $|\alpha|^2 + |\beta|^2 = a_1^2 + a_2^2 + b_1^2 + b_2^2 = 1$ is obtained by setting

$$a_1 = \cos\left(\frac{1}{2}\phi\right), \quad a_2 = n_z \sin\left(\frac{1}{2}\phi\right), \quad b_1 = n_y \sin\left(\frac{1}{2}\phi\right), \quad b_2 = n_x \sin\left(\frac{1}{2}\phi\right)$$

in terms of three parameters; an angle ϕ and a unit vector $\hat{n}$ (recall that a unit vector has only two independent components). A general element of SU(2) can be written in this parametrization as

$$\underline{\underline{U}}(\hat{n},\phi) = e^{i\frac{\phi}{2}\hat{n}\cdot\vec{\sigma}}, \tag{2.211}$$

in terms of Pauli matrices σ_i, $i = 1, 2, 3$ defined in equation (2.139) in Example (2.2.8). A useful property of Pauli matrices is,

$$\sigma_i \sigma_j = \delta_{ij}\underline{\underline{I}} + i\sum_{k=1}^{3} \epsilon_{ijk}\,\sigma_k. \tag{2.212}$$

It follows that

$$\left(\vec{\sigma}\cdot\vec{a}\right)\left(\vec{\sigma}\cdot\vec{b}\right) = \left(\vec{a}\cdot\vec{b}\right)\underline{\underline{I}} + i\vec{\sigma}\left(\vec{a}\times\vec{b}\right). \tag{2.213}$$

Expanding the exponential on the right hand side of equation (2.211) in power series yields

$$e^{i\frac{\phi}{2}\hat{n}\cdot\vec{\sigma}} = \underline{\underline{I}} + i\frac{\phi}{2}\hat{n}\cdot\vec{\sigma} + \frac{1}{2!}(i\frac{\phi}{2})^2(\hat{n}\cdot\vec{\sigma})^2 + \frac{1}{3!}(i\frac{\phi}{2})^3(\hat{n}\cdot\vec{\sigma})^3 + \frac{1}{4!}(i\frac{\phi}{2})^4(\hat{n}\cdot\vec{\sigma})^4 + \ldots$$

We notice that

$$(\hat{n}\cdot\vec{\sigma})^2 = (\hat{n}\cdot\vec{\sigma})(\hat{n}\cdot\vec{\sigma}) = \underline{\underline{I}}\,\hat{n}\cdot\hat{n} + +i\vec{\sigma}\cdot(\hat{n}\times\hat{n}) = \underline{\underline{I}}$$

is a simple application of formula (2.213). The power series becomes:

$$e^{i\frac{\phi}{2}\hat{n}\cdot\vec{\sigma}} = \underline{\underline{I}} + i\frac{\phi}{2}\hat{n}\cdot\vec{\sigma} + \frac{1}{2!}(i\frac{\phi}{2})^2\underline{\underline{I}} + \frac{1}{3!}(i\frac{\phi}{2})^3\hat{n}\cdot\vec{\sigma} + \frac{1}{4!}(i\frac{\phi}{2})^4\underline{\underline{I}} + \ldots$$

$$= \underline{\underline{I}}\left(1 + \frac{1}{2!}(i\frac{\phi}{2})^2 + \frac{1}{4!}(i\frac{\phi}{2})^4 + \ldots\right) + i\hat{n}\cdot\vec{\sigma}\left(\frac{\phi}{2} - \frac{1}{3!}(\frac{\phi}{2})^3 + \ldots\right)$$

$$= \underline{\underline{I}}\cos\frac{\phi}{2} + i\hat{n}\cdot\vec{\sigma}\sin\frac{\phi}{2}.$$

Thus,

$$\underline{\underline{U}}(\hat{n}, \phi) = \underline{\underline{I}} \cos\frac{\phi}{2} + i\,\hat{n} \cdot \vec{\sigma} \sin\frac{\phi}{2}$$

$$= \begin{bmatrix} \cos\frac{\phi}{2} + i\,n_z \sin\frac{\phi}{2} & (n_y + i\,n_x)\sin\frac{\phi}{2} \\ (-n_y + i\,n_x)\sin\frac{\phi}{2} & \cos\frac{\phi}{2} - i\,n_z \sin\frac{\phi}{2} \end{bmatrix}, \tag{2.214}$$

where

$$\hat{n} \cdot \vec{\sigma} = n_x \sigma_1 + n_y \sigma_2 + n_z \sigma_3 = \begin{bmatrix} 0 & n_x \\ n_x & 0 \end{bmatrix} + \begin{bmatrix} 0 & -i\,n_y \\ i\,n_y & 0 \end{bmatrix} + \begin{bmatrix} n_z & 0 \\ 0 & -n_z \end{bmatrix}$$

$$= \begin{bmatrix} n_z & n_x - i\,n_y \\ n_x + i\,n_y & -n_z \end{bmatrix}.$$

Since

$$\left|\cos\frac{\phi}{2} + i\,n_z \sin\frac{\phi}{2}\right|^2 + \left|(-n_y + i\,n_x)\sin\frac{\phi}{2}\right|^2 = \cos^2\frac{\phi}{2} + \sin^2\frac{\phi}{2}\left(n_x^2 + n_y^2 + n_z^2\right) = 1$$

we have verified the parameterization given in (2.211). This parametrization is reminiscent of the angle and axis parameterization of the rotation group we saw earlier.

Inserting an infinitesimal angle $d\phi$ into parametrization (2.214) we find

$$\underline{\underline{U}}(\hat{n}, d\phi) = \underline{\underline{I}} + \frac{1}{2}i\,\hat{n} \cdot \vec{\sigma}\,d\phi = \underline{\underline{I}} + i\left(n_x \frac{\sigma_1}{2} + n_y \frac{\sigma_2}{2} + n_z \frac{\sigma_3}{2}\right) d\phi.$$

The generators of SU(2) are, therefore, matrices $J_i = \sigma_i/2$. As seen from the multiplicative relation (2.212) they satisfy the $\mathfrak{so}(3)$ algebra:

$$\left[\frac{\sigma_i}{2}, \frac{\sigma_j}{2}\right] = \frac{1}{4}\left(\sigma_i \sigma_j - \sigma_j \sigma_i\right) = \frac{1}{4}\left(i\sum_{k=1}^{3}\epsilon_{ijk}\sigma_k - i\sum_{k=1}^{3}\epsilon_{jik}\sigma_k\right)$$

$$= i\sum_{k=1}^{3}\epsilon_{jik}\frac{\sigma_k}{2}.$$

Thus, the generators of the SU(2) group satisfy the same commutation relations as those of SO(3). Therefore, we say that the groups, SU(2) and SO(3), have the same Lie algebra. How do they compare as groups? Although elements of the SU(2) group are 2×2 matrices, each SU(2) matrix can be shown to generate a rotation in 3-dimensional space. To show this point we look for simplicity on a rotation about the z-axis by the angle α:

$$\begin{bmatrix} x' \\ y' \\ z' \end{bmatrix} = \underline{\underline{R}}(\hat{k}, \alpha)\begin{bmatrix} x \\ y \\ z \end{bmatrix} = \begin{bmatrix} \cos\alpha & \sin\alpha & 0 \\ -\sin\alpha & \cos\alpha & 0 \\ 0 & 0 & 1 \end{bmatrix}\begin{bmatrix} x \\ y \\ z \end{bmatrix}. \tag{2.215}$$

There exists another way to perform the same rotation using matrix algebra. Let us arrange the coordinates (x, y, z) into a 2×2 complex matrix:

$$\vec{r} \cdot \vec{\sigma} = x\sigma_1 + y\sigma_2 + z\sigma_3 = \begin{bmatrix} z & x - iy \\ x + iy & -z \end{bmatrix}$$

and perform a similarity transformation:

$$\vec{r}' \cdot \vec{\sigma} = \underline{U}\left(e^{i\alpha/2}, 0\right) \vec{r} \cdot \vec{\sigma} \, \underline{U}^{\dagger}\left(e^{i\alpha/2}, 0\right)$$

using a parametrization of $\underline{U}$ from equation (2.209). Explicitly, the similarity transformation yields

$$\begin{bmatrix} z' & x' - iy' \\ x' + iy' & -z' \end{bmatrix} = \begin{bmatrix} e^{i\alpha/2} & 0 \\ 0 & e^{-i\alpha/2} \end{bmatrix} \begin{bmatrix} z & x - iy \\ x + iy & -z \end{bmatrix} \begin{bmatrix} e^{-i\alpha/2} & 0 \\ 0 & e^{i\alpha/2} \end{bmatrix}$$

and leads to an identical expression for the transformed coordinates (x', y', z') as obtained above in equation (2.215) by the SO(3) rotation involving the $\underline{R}(\widehat{k}, \alpha)$ matrix. One can show that for every matrix $\underline{U}(\alpha, \beta)$ there exists an SO(3) rotation $\underline{R}$ matrix such that the rotation $\vec{r}' = \underline{R}\vec{r}$ agrees with the similarity transformation $\vec{r}' \cdot \vec{\sigma} = \underline{U}(\alpha, \beta)\vec{r} \cdot \vec{\sigma}\underline{U}^{\dagger}(\alpha, \beta)$. Thus, the SU(2) group describes rotations. However, this construction represents a two to one mapping of the elements of SU(2) onto those of SO(3). One sees that $\underline{U}(\alpha, \beta)$ and $\underline{U}(-\alpha, -\beta)$ correspond, via transformation $\vec{r} \cdot \vec{\sigma} \to \underline{U}\vec{r} \cdot \vec{\sigma}\,\underline{U}^{\dagger}$, to the same 3×3 matrix of SO(3). Thus, SU(2) and SO(3) groups have the same local structure but different global properties. The SU(2) group, which is called the covering group of SO(3), is twice as large as SO(3). This property is also seen from the fact that $\underline{U}$ does not return to itself after 2π rotation. Rather, it becomes $-\underline{U}$ through:

$$\underline{U}(\hat{n}, \phi + 2\pi) = -\underline{U}(\hat{n}, \phi) \, .$$

In contrast, in SO(3) there is no distinction between ϕ and $\phi + 2\pi$ rotations. This provides one more way to see that there are two elements of SU(2) for each element of SO(3).

✑ Group Theory and Rotations

Problem 2.4.1. Show that the upper triangular 3×3 matrices

$$\begin{bmatrix} 1 & a & b \\ 0 & 1 & c \\ 0 & 0 & 1 \end{bmatrix}$$

with real parameters a, b, c form a group.

Problem 2.4.2. Show that a set of all complex-valued 2×2 matrices having determinant $+1$ forms a group with group composition law identified with matrix multiplication. This group is called the special linear group $SL(2)$.

Problem 2.4.3. Show that a set of all real-valued 2×2 matrices $\underline{M}$ that satisfy the relation:

$$\underline{M}^T \sigma_3 \underline{M} = \sigma_3, \qquad \sigma_3 = \begin{bmatrix} 1 & 0 \\ 0 & -1 \end{bmatrix}$$

forms a group with matrix multiplication as group composition law. This group is called the two-dimensional Lorentz group. Show that matrices

$$\begin{bmatrix} \cosh\alpha & \sinh\alpha \\ \sinh\alpha & \cosh\alpha \end{bmatrix}, \quad \begin{bmatrix} \cosh\alpha & -\sinh\alpha \\ \sinh\alpha & -\cosh\alpha \end{bmatrix}$$

and

$$\begin{bmatrix} -\cosh\alpha & \sinh\alpha \\ \sinh\alpha & -\cosh\alpha \end{bmatrix}, \quad \begin{bmatrix} -\cosh\alpha & -\sinh\alpha \\ \sinh\alpha & \cosh\alpha, \end{bmatrix}$$

with real parameter α, all belong to the two-dimensional Lorentz group. Note that replacing σ_3 with identity $\underline{I}$ leads to the defining relation for the group O(2) of the 2×2 orthogonal matrices.

Problem 2.4.4. Consider the equilateral triangle with center at the origin, as shown below. The 2×2 orthogonal matrices in O(2) that leave the equilateral triangle unchanged form the dihedral group D_3.

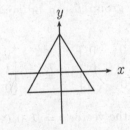

(a) Show that

$$\begin{bmatrix} \cos\phi & \sin\phi \\ -\sin\phi & \cos\phi \end{bmatrix}, \quad \text{with} \quad \phi = \frac{2\pi}{3}$$

belongs to D_3.
(b) Show that

$$\begin{bmatrix} -1 & 0 \\ 0 & 1 \end{bmatrix}$$

belongs to D_3.
(c) Find all 6 elements of the D_3 group by forming the products of the above two matrices. Don't forget the identity matrix.

Problem 2.4.5. (a) Find all elements of the tetrahedral group T that consists of 12 rotation matrices that rotate the four vectors

$$\begin{bmatrix} 1 \\ 1 \\ 1 \end{bmatrix}, \quad \begin{bmatrix} 1 \\ -1 \\ -1 \end{bmatrix}, \quad \begin{bmatrix} -1 \\ -1 \\ 1 \end{bmatrix}, \quad \begin{bmatrix} -1 \\ 1 \\ -1 \end{bmatrix}$$

onto each other (for each rotation matrix R, its inverse R^T will also belong to the group). The endpoints of these four vectors are located at the four vertices of a

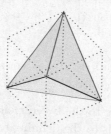

Figure 2.11: A regular tetrahedron.

regular tetrahedron (see Figure 2.11), which remain invariant under rotations from the tetrahedral group T.

(b) Show that the tetrahedral group T can be generated by taking powers and products of

$$\underline{\underline{R}}_1 = \begin{bmatrix} 1 & 0 & 0 \\ 0 & -1 & 0 \\ 0 & 0 & -1 \end{bmatrix}, \quad \underline{\underline{R}}_2 = \begin{bmatrix} 0 & 0 & 1 \\ 1 & 0 & 0 \\ 0 & 1 & 0 \end{bmatrix}.$$

Problem 2.4.6. Suppose that the vector $\underline{v} = [1, 0, 0]^T$ is rotated about the x-axis by $45°$ followed by a $45°$ rotation about the z-axis.

(a) Find the resulting vector.

(b) Next, reverse the order of rotations and compare the new result to the one obtained in (a).

(c) Find vector $\underline{w}$ for which the result of applying the above two rotations is independent of the order in which they are performed.

Problem 2.4.7. Verify the cyclicity property of trace which says that

$$\mathrm{Tr}\left(\underline{\underline{A}}\left[\underline{\underline{B}}, \underline{\underline{C}}\right]\right) = \mathrm{Tr}\left(\left[\underline{\underline{C}}, \underline{\underline{A}}\right]\underline{\underline{B}}\right)$$

Problem 2.4.8. Verify the Jacobi identity:

$$\left[\underline{\underline{A}}, \left[\underline{\underline{B}}, \underline{\underline{C}}\right]\right] + \left[\underline{\underline{B}}, \left[\underline{\underline{C}}, \underline{\underline{A}}\right]\right] + \left[\underline{\underline{C}}, \left[\underline{\underline{A}}, \underline{\underline{B}}\right]\right] = 0$$

for an arbitrary set of three square matrices $\underline{\underline{A}}$, $\underline{\underline{B}}$, $\underline{\underline{C}}$.

Problem 2.4.9. Use the results from Problem 2.3.8 to prove that the angle ϕ of the rotation of matrix $\underline{\underline{R}}\,(\alpha, \beta, \gamma)$ from equation (2.191) is given in terms of Euler angles by

$$\cos(\tfrac{1}{2}\phi) = \cos(\tfrac{1}{2}\beta)\cos\left(\tfrac{1}{2}(\alpha + \gamma)\right).$$

Problem 2.4.10. Show that the matrix elements of the generators $\underline{\underline{J_i}}$, $i = 1, 2, 3$ of $\mathfrak{so}(3)$ can be obtained from the single formula:

$$\left(\underline{\underline{J_i}}\right)_{jk} = -i\,\epsilon_{ijk}.$$

Problem 2.4.11. Show that the Jacobi identity for the $\mathfrak{so}(3)$ algebra is equivalent to the following relation:

$$\sum_{n=1}^{3} (\epsilon_{jkn}\epsilon_{inm} + \epsilon_{kin}\epsilon_{jnm} + \epsilon_{ijn}\epsilon_{knm}) = 0,$$

for the permutation symbol ϵ_{ijk}. Verify, explicitly, that the permutation symbol satisfies the above relation.

Problem 2.4.12. (a) Show that

$$\underline{\underline{J}}^2 = \underline{\underline{J_1}}^2 + \underline{\underline{J_2}}^2 + \underline{\underline{J_3}}^2$$

is a Casimir operator for $\mathfrak{so}(3)$ meaning that it commutes with all three generators $\underline{\underline{J_i}}$, $i = 1, 2, 3$.
(b) Show that $\vec{L}^2 = L_1^2 + L_2^2 + L_3^2$ commutes with all three angular momentum operators L_i, $i = 1, 2, 3$.

Problem 2.4.13. Show that for two skew-symmetric matrices $\underline{\underline{U}}$ and $\underline{\underline{V}}$ from Problem (2.3.5) and $\vec{\underline{\underline{J}}}$ from equation (2.185) it holds that

$$[\underline{\underline{U}}, \underline{\underline{V}}] = -i\,(\underline{u} \times \underline{v}) \cdot \vec{\underline{\underline{J}}} = \begin{bmatrix} 0 & -(\underline{u} \times \underline{v})_3 & (\underline{u} \times \underline{v})_2 \\ (\underline{u} \times \underline{v})_3 & 0 & -(\underline{u} \times \underline{v})_1 \\ -(\underline{u} \times \underline{v})_2 & (\underline{u} \times \underline{v})_1 & 0 \end{bmatrix}.$$

Problem 2.4.14. Verify that $\underline{\underline{U}}^\dagger\,(\alpha, \beta) = \underline{\underline{U}}\,(\alpha^*, -\beta)$ and use this to verify unitarity on the basis of the multiplication law (2.210).

Problem 2.4.15. For the Pauli matrices:

$$\sigma_1 = \begin{bmatrix} 0 & 1 \\ 1 & 0 \end{bmatrix}, \qquad \sigma_2 = \begin{bmatrix} 0 & -i \\ i & 0 \end{bmatrix}, \qquad \sigma_3 = \begin{bmatrix} 1 & 0 \\ 0 & -1 \end{bmatrix}$$

show that $\sigma_i\sigma_j = \delta_{ij}\underline{\underline{I}} + i\sum_{k=1}^{3}\epsilon_{ijk}\,\sigma_k$.

Problem 2.4.16. Show that

$$\left[\vec{\sigma} \cdot \vec{a}, \vec{\sigma} \cdot \vec{b}\right] = 2\mathrm{i} \left(\vec{a} \times \vec{b}\right) \cdot \vec{\sigma}.$$

Problem 2.4.17. (a) Show that the group SO(n) of special orthogonal matrices has $n(n-1)/2$ independent real parameters.

(b) Show that the group SU(n) of special unitary matrices has n^2-1 independent real parameters.

Problem 2.4.18. Show that the similarity transformation

$$\vec{r}' \cdot \vec{\sigma} = \underline{\underline{U}}\left(\hat{e}_i, \phi\right) \vec{r} \cdot \vec{\sigma} \, \underline{\underline{U}}^\dagger\left(\hat{e}_i, \phi\right), \quad \underline{\underline{U}}\left(\hat{e}_i, \phi\right) = \underline{\underline{I}} \cos\frac{\phi}{2} + \mathrm{i}\,\sigma_i \sin\frac{\phi}{2}$$

corresponds to rotation

$$\vec{r}' = \underline{\underline{R}}\left(\hat{e}_i, \phi\right) \vec{r}, \qquad \underline{\underline{R}}\left(\hat{e}_i, \phi\right) = e^{\mathrm{i}\,\phi\,J_i}.$$

Problem 2.4.19. Show that the determinant of $\vec{r} \cdot \vec{\sigma}$ is

$$\left|\, \vec{r} \cdot \vec{\sigma} \,\right| = -(x^2 + y^2 + z^2)$$

and, therefore, the length

$$x'^2 + y'^2 + z'^2 = x^2 + y^2 + z^2$$

is preserved under the similarity transformation,

$$\vec{r} \cdot \vec{\sigma} \longrightarrow \vec{r}' \cdot \vec{\sigma} = \underline{\underline{U}}\,\vec{r} \cdot \vec{\sigma}\,\underline{\underline{U}}^\dagger,$$

by a unitary matrix $\underline{\underline{U}}$.

Problem 2.4.20. (a) Show that property (2.212) of Pauli matrices implies

$$\mathrm{Tr}\left(\sigma_i \sigma_j\right) = 2\delta_{ij} \quad \text{and} \quad \mathrm{Tr}\left(\vec{\sigma} \cdot \vec{a}\,\vec{\sigma} \cdot \vec{b}\right) = 2\,\vec{a} \cdot \vec{b}$$

(b) For $\vec{r}' = \underline{\underline{R}}\vec{r}$, show that relation

$$\vec{r}' \cdot \vec{\sigma} = \underline{\underline{U}}\,\vec{r} \cdot \vec{\sigma}\,\underline{\underline{U}}^\dagger$$

implies

$$\underline{\underline{R}}_{ij} = \frac{1}{2}\,\mathrm{Tr}\left(\underline{\underline{U}}\,\sigma_j\,\underline{\underline{U}}^\dagger\,\sigma_i\right)$$

for the matrix coefficients of the rotation matrix $\underline{\underline{R}}$.

Chapter 3

Differential Equations

D IFFERENTIAL equations are ubiquitous in science. Many fundamental problems in physics and engineering are formulated as differential equations. Therefore, it is of great interest to develop methods for finding solutions to the differential equations. This chapter presents differential equations of first- and second-order and standard methods of obtaining their solutions.

In general, a differential equation is an equation that contains one or more unknown functions and one or more of their derivatives. The order of the differential equation is the order of the highest derivative that occurs in the equation. Thus, a differential equation of order-n is a differential equation which relates a function to its derivatives of up to the n-th order. The differential equation is said to be linear if it does not contain higher powers of the unknown functions and their derivatives. Here, we will only study differential equations with a single unknown function and its derivatives. The second-order differential equations encountered in this chapter will all be linear.

3.1 First Order Differential Equations

In this section, we will study first-order differential equations. A first-order differential equation involves one unknown function y of a variable x and its derivative $y' = \frac{\mathrm{d}y}{\mathrm{d}x}$ and variable x. It can be be written as

$$y' = \frac{\mathrm{d}y}{\mathrm{d}x} = F(x, y) = -\frac{M(x, y)}{N(x, y)}, \qquad (3.1)$$

where $F(x, y)$, $M(x, y)$, and $N(x, y)$ are functions of x and y. Alternatively, we can write a first-order differential equation as

$$N(x, y)\, dy + M(x, y)\, dx = 0.\tag{3.2}$$

There are many forms of function $F(x, y)$ for which the solution of equation (3.1) can be found in a straightforward way.

3.1.1 Separable Equations

If

$$\frac{dy}{dx} = F(x, y) = -\frac{M(x)}{N(y)}\tag{3.3}$$

then the differential equation is a *separable* equation. Multiplying by dx and $N(y)$, we cast the equation into the following form:

$$N(y)\, dy = -M(x)\, dx,\tag{3.4}$$

where each side depends only on either x or y. This separated form readily allows for integration of each side:

$$\int N(y)\, dy = -\int M(x)\, dx.\tag{3.5}$$

☞**Example 3.1.1.** A basic separable equation is given by $y' = ky$ with constant k. Manipulating both sides as discussed above we find $\int dy/y = \ln y = \int k\, dx = kx + C$ or $y(x) = \exp(kx + C)$. By defining the constant $c = \exp(C)$ we obtain a classic exponential solution $y(x) = c \exp(kx)$.

☞**Example 3.1.2.** Consider $y' = \frac{x}{y(1+x^2)}$. This is a separable equation and the operations that lead to (3.5) produce the equation:

$$\int y\, dy = \frac{1}{2}y^2 = \int \frac{x\, dx}{(1 + x^2)} = \frac{1}{2}\ln(1 + x^2) + C,\tag{3.6}$$

where C is an integration constant that is fixed by providing an initial condition. The initial condition specifies a value of $y(x)$ at some point $x = x_0$. Let us choose the initial condition to be $y(0) = 1$, which yields $C = 1/2$. The solution to (3.6) becomes $y = \sqrt{\ln(1 + x^2) + 1}$.

☞**Example 3.1.3.** An equation is homogeneous if it can be written as $y' = F(y/x)$. A homogeneous equation can be cast in the separable form upon the change of variables, $z = y/x$. The derivative of $y = zx$ with respect to x gives

$$y' = z'x + z = F(z) \quad \rightarrow \quad x\frac{dz}{dx} = F(z) - z , \tag{3.7}$$

which can be treated by a separation of variables as,

$$\int \frac{dz}{F(z) - z} = \int \frac{dx}{x} \quad \rightarrow \quad \int \frac{dz}{F(z) - z} = \ln x + C . \tag{3.8}$$

Thus, for $y' = (y + x)/x$, $F(z) = z + 1$. Plugging $F(z)$ into the integration in equation (3.8) yields $z = y/x = \ln x + C$ or $y = x \ln x + Cx$.

3.1.2 Exact First Order Differential Equations

In general, the first-order differential equation (3.1) is of the non-separable form and requires more sophisticated techniques in order to solve it.

One special case occurs when the expression on the left hand side of equation (3.2) can be rewritten as a total derivative of some two-dimensional function. Recall that for $\phi(x, y)$, which is function of two variables, the difference $\phi(x + \delta x, y + \delta y) - \phi(x, y)$ in the limit $\delta x, \delta y \to 0$ equals the total derivative:

$$d\phi(x, y) = \frac{\partial \phi}{\partial x}(x, y)\, dx + \frac{\partial \phi}{\partial y}(x, y)\, dy . \tag{3.9}$$

Equation (3.2) is called *exact* when there exists a two-dimensional function $\phi(x, y)$ such that

$$d\phi(x, y) = N(x, y)\, dy + M(x, y)\, dx \tag{3.10}$$

or

$$M(x, y) = \frac{\partial \phi}{\partial x}(x, y) \tag{3.11}$$

$$N(x, y) = \frac{\partial \phi}{\partial y}(x, y) . \tag{3.12}$$

In this case, the contour curve defined by solution to the algebraic equation $\phi(x, y) = c$ with an arbitrary constant c, is the solution to the differential equation. There is a simple test to decide when the first-order differential equation is exact. Since for a continuous and differentiable function $\phi(x, y)$ the mixed derivatives

$$\frac{\partial^2 \phi}{\partial x \partial y}(x, y) = \frac{\partial^2 \phi}{\partial y \partial x}(x, y) \tag{3.13}$$

must be equal the following relations

$$\frac{\partial}{\partial x}\left(\frac{\partial \phi}{\partial y}\right) = \frac{\partial N}{\partial x}, \quad \frac{\partial}{\partial y}\left(\frac{\partial \phi}{\partial x}\right) = \frac{\partial M}{\partial y} \tag{3.14}$$

have to be equal as well. Thus, the condition

$$\frac{\partial N}{\partial x} = \frac{\partial M}{\partial y} \qquad (3.15)$$

must be satisfied in order for the first-order differential equation to be exact and possess a solution of the type $\phi(x, y) = c$. If condition (3.15) is satisfied, then it is possible to find a function $\phi(x, y)$ such that relations (3.11)-(3.12) are satisfied. In this case, solving the equation is equivalent to integrating equation (3.11) or (3.12) in order to get an explicit expression for the function $\phi(x, y)$. Let us demonstrate the technique by integrating, for example, equation (3.11) (in a specific situation one chooses one of the two equations, (3.11) or (3.12), which appear to have a simpler form). Integrating equation (3.11) over x yields:

$$\phi(x, y) = \int^x M(x', y)\, \mathrm{d}x' + c_1(y), \qquad (3.16)$$

where the integration constant c_1 does not depend on the integration variable x but may depend, in principle, on the remaining variable y. To find the dependence of $c_1(y)$ on y we plug expression (3.16) into equation (3.11):

$$\frac{\mathrm{d}c_1(y)}{\mathrm{d}y} = N(x, y) - \int^x \frac{\partial M(x', y)}{\partial y}\, \mathrm{d}x'. \qquad (3.17)$$

Note that the right hand side of the above equation is a function of y only. It is easy to verify that by taking the derivative with respect to x. The result is zero due to condition (3.15). The explicit form of $c_1(y)$ can now be obtained by a simple integration in y.

☞**Example 3.1.4.** Consider:

$$\frac{\mathrm{d}y}{\mathrm{d}x} = \frac{(x - y)^2}{(x - y)^2 + 2}, \qquad (3.18)$$

which is of the form (3.2) with $N(x, y) = (x-y)^2 + 2$ and $M(x, y) = -(x-y)^2$. Since

$$\frac{\partial N}{\partial x} = \frac{\partial M}{\partial y} = 2(x - y) \qquad (3.19)$$

equation (3.18) is exact and it suffices to find ϕ that satisfies equations (3.11)-(3.12), which, in this case, are

$$\frac{\partial \phi}{\partial x} = -(x - y)^2, \quad \frac{\partial \phi}{\partial y} = (x - y)^2 + 2, \qquad (3.20)$$

to obtain a solution to the differential equation (3.18).

Integrating the left equation in (3.20) with respect to x we get

$$\phi(x, y) = -\int^x (x - y)^2\, \mathrm{d}x = -\frac{(x - y)^3}{3} + c_1(y), \qquad (3.21)$$

where the integration constant c_1 obtained from integration in variable x may depend on a variable y. Expression (3.21) for ϕ must agree with the right equation in (3.20). Taking a derivative with respect to y on both sides of (3.21) yields

$$\frac{\partial \phi}{\partial y} = (x - y)^2 + \frac{dc_1(y)}{dy}, \tag{3.22}$$

which agrees with the right equation in (3.20) provided that $dc_1(y)/dy = 2$ or $c_1 = 2y$. Hence, the solution is

$$\phi(x, y) = -\frac{1}{3}(x - y)^3 + 2y = c, \tag{3.23}$$

for an arbitrary constant c.

If condition (3.15) is not satisfied then the equation is not exact. In many cases, one is able to turn the given differential equation into an exact equation by multiplication with an appropriate function $\alpha(x, y)$, which is referred to as an integrating factor.

Assume that equation (3.2) is not exact. We will determine a function $\alpha(x, y)$, such that the equation

$$\alpha(x, y)N(x, y)\,dy + \alpha(x, y)M(x, y)\,dx = 0 \tag{3.24}$$

is exact, meaning that condition (3.15) holds for the modification of equation (3.24), i.e.:

$$\frac{\partial(\alpha N)}{\partial x} = \frac{\partial(\alpha M)}{\partial y}. \tag{3.25}$$

For convenience, we will choose the integrating factor to be a function of x only. Then, equation (3.25) becomes:

$$\frac{\alpha'(x)}{\alpha(x)} = \frac{1}{N}\left(\frac{\partial M}{\partial y} - \frac{\partial N}{\partial x}\right) \tag{3.26}$$

and is only consistent if the right hand side of the above equation is a function of x only. In such case, we can rewrite the right hand side of the above equation as

$$\left(\frac{\partial M}{\partial y} - \frac{\partial N}{\partial x}\right)/N = g(x)$$

for some function g of x. The integrating factor α can be obtained by integrating both sides of

$$\frac{\alpha'(x)}{\alpha(x)} = g(x)$$

resulting in

$$\alpha(x) = c\,e^{\int^x g\,\mathrm{d}x}. \tag{3.27}$$

If $(\partial M/\partial y - \partial N/\partial x)/M$ is a function of y only then the integrating factor α depends solely on y and we find it after integrating $(\partial M/\partial y - \partial N/\partial x)/M$ over y.

☞ **Example 3.1.5.** Consider:

$$N\,\mathrm{d}y + M\,\mathrm{d}x = (e^y + \frac{1}{x})\,\mathrm{d}y + \frac{e^y}{x}\,\mathrm{d}x = 0. \tag{3.28}$$

Since

$$\frac{\partial N}{\partial x} = -\frac{1}{x^2} \neq \frac{\partial M}{\partial y} = \frac{e^y}{x} \tag{3.29}$$

equation (3.28) is not exact. However, the combination $(\partial M/\partial y - \partial N/\partial x)/N$ is a function of x only and defines the integrating factor according to

$$\frac{\alpha'(x)}{\alpha(x)} = \frac{1}{N}\left(\frac{\partial M}{\partial y} - \frac{\partial N}{\partial x}\right) = \frac{\frac{e^y}{x} + \frac{1}{x^2}}{e^y + \frac{1}{x}} = \frac{1}{x}. \tag{3.30}$$

Integrating the above equation yields $\alpha(x) = cx$ and equation (3.28) becomes exact after multiplication by x:

$$xN\,\mathrm{d}y + xM\,\mathrm{d}x = (xe^y + 1)\,\mathrm{d}y + e^y\,\mathrm{d}x = 0. \tag{3.31}$$

We can easily verify that the above equation is exact. We can proceed by using the method developed for the exact equations to solve equation (3.31).

3.1.3   Linear First Order Differential Equations

The differential equation (3.1) is called a linear equation if the function $F(x, y)$ is linear in y, meaning that

$$F(x, y) = -p(x)y(x) + q(x).$$

The corresponding differential equation has the form:

$$\boxed{y'(x) + p(x)y(x) = q(x)} \tag{3.32}$$

for some functions p and q which depend on x only. When the "source term" q is non-zero $q \neq 0$, equation (3.32) is called a non-homogeneous linear differential equation. With a zero "source term" $q = 0$ equation (3.32) becomes a homogeneous linear differential equation:

$$y'(x) + p(x)y(x) = 0 \quad \rightarrow \quad \frac{y'}{y} = -p(x), \tag{3.33}$$

which is separable. We will denote the solution to the homogeneous equation (3.33)
obtained by integration by $y_0(x)$:

$$y_0(x) = Ce^{-\int^x p(x')\,dx'} = Ce^{-P(x)}, \tag{3.34}$$

where we have introduced the function P, which is an integral of $p(x)$:

$$\frac{dP(x)}{dx} = p(x). \tag{3.35}$$

Function P is defined up to an integration constant, which can be absorbed into
the constant C in (3.34).

We will search for a solution of the non-homogeneous equation (3.32) that is of
the form

$$y(x) = y_0(x)f(x) \tag{3.36}$$

where $f = y_0^{-1}y$ is yet an unknown function to be determined by requiring that y
satisfies equation (3.32). The derivative of f is given by

$$\frac{df}{dx} = \frac{d(y_0^{-1}y)}{dx} = \frac{dy_0^{-1}}{dx}y + y_0^{-1}\frac{dy}{dx}. \tag{3.37}$$

The function $y_0^{-1}(x) = C^{-1}\exp\left(P(x)\right)$ satisfies equation $(y_0^{-1})'(x) - p(x)y_0^{-1}(x) = 0$
and, therefore,

$$\frac{df}{dx} = y_0^{-1}p(x)y(x) + y_0^{-1}\frac{dy}{dx} = y_0^{-1}q(x), \tag{3.38}$$

where the last equality follows from the fact that $y(x)$ is required to satisfy equation
(3.32).

The above equation shows that y_0^{-1} is an integrating factor for equation (3.32)
since multiplication by y_0^{-1} transforms the left hand side of (3.32) into the derivative
of the function f.

Integrating both sides of (3.38) yields the following expression for f:

$$f(x) = \int^x y_0^{-1}q(x)\,dx + k \quad \rightarrow \quad y(x) = y_0(x)f(x) = y_0(x)\left(\int^x y_0^{-1}q(x)\,dx + k\right), \tag{3.39}$$

where k is an integration constant.

Plugging y_0 from equation (3.34) into the last expression we obtain:

$$y(x) = e^{-P(x)}\left(\int^x e^{P(x)}q(x)\,dx + k\right). \tag{3.40}$$

This equation defines the solution, $y(x)$, up to an integration constant k, which
can be fixed by in terms of the value y takes at some initial point x_0. Relation
$y(x_0) = k_0$ with some constant k_0 is called the initial condition.

☞**Example 3.1.6.** Consider a linear differential equation (3.32) with $p(x) = 1 + 1/x$ and the source $1/x$:

$$y' + (1 + \frac{1}{x})y = \frac{1}{x},$$
(3.41)

with the initial condition: $y(x_0 = 1) = 0$. First, we determine the function $P = \int p \, dx$ equal to $P(x) = x + \ln x$, which leads to the following relations

$$e^P = xe^x \quad \text{and} \quad q(x)e^P = e^x.$$
(3.42)

Plugging the above equations into formula (3.40) and solving for y yields:

$$y(x) = \frac{1}{x}e^{-x}(\int^x e^x \, dx + k) = \frac{1}{x} + k\frac{1}{x}e^{-x}.$$
(3.43)

The constant k is found by imposing the initial condition $y(1) = 1 + k\exp(-1) = 0$, which yields the final expression for the solution y:

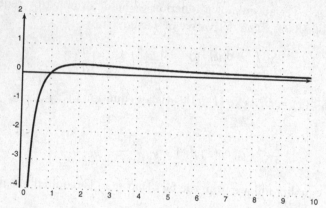

Figure 3.1: The curve representing $y(x) = \frac{1}{x}(1 - e^{-x+1})$.

$$y(x) = \frac{1}{x}(1 - e^{-x+1}).$$
(3.44)

The behavior of the solution is shown in Figure 3.1.

Some non-linear first-order differential equations can be transformed into the linear form. One class of such equations is Bernoulli's differential equation. It is of the form

$$y'(x) + p(x)y(x) = q(x)y^n(x),$$
(3.45)

where $n \neq 0, 1$. It is possible to reduce Bernoulli's equation to a linear one by setting $u = 1/y^{n-1}$ and using the fact that $\frac{du}{dx} = (1-n)y'/y^n$. Dividing Bernoulli's equation by y^n and substituting function $u(x)$ yields a linear equation:

$$\frac{y'}{y^n} + \frac{p(x)}{y^{n-1}} = q(x) \qquad \rightarrow \qquad \frac{du}{dx} + (1-n)p(x)u = (1-n)q(x),$$

which can be solved by a standard method.

☞ **Example 3.1.7.** The differential equation $y' + y = y^2$ is Bernoulli's equation. Dividing both sides of this equation by y^2 and inserting $u = 1/y$, we obtain

$$\frac{du}{dx} = u - 1,$$

which is a linear equation with the solution $u = 1 + ke^x$ or

$$y = \frac{1}{u} = \frac{1}{1 + ke^x}.$$

An integration constant k can be fixed by setting the value of y at some initial point x_0 to be $y(x_0) = k_0$. Indeed, for a given x_0 and $k_0 \neq 0$, the equation

$$\frac{1}{1 + ke^{x_0}} = k_0$$

has the unique solution $k = (1 - k_0)e^{-x_0}/k_0$. If $k_0 = 0$ then $y = 0$ is the only solution to the differential equation. Thus, the problem

$$y' + y = y^2, \quad y(x_0) = k_0$$

has a unique solution for any x_0 and k_0.

✍ First Order Differential Equations: Problems

Problem 3.1.1. Solve the following differential equation:

$$y' = \frac{x}{1 + x^2} \tan y.$$

Problem 3.1.2. Solve the following differential equation:

$$e^x y y' = e^{-y} - e^{x-y}.$$

Problem 3.1.3. Solve the following differential equation:

$$y' = \frac{x^2}{1 + y^2}, \quad y(0) = -1.$$

Problem 3.1.4. Solve the following differential equation:

$$y' = \frac{dy}{dx} = \frac{x^2 - xy + y^2}{x^2}, \quad y(1) = 0.$$

Hint: Rewrite both sides of the equation in terms of $u = y/x$.

Problem 3.1.5. Solve the differential equation:

$$xy' + (2 + x)y = 0$$

by

(a) separation of variables
(b) method of integrating factors.

Problem 3.1.6. Solve the following differential equation:

$$xy' + y + 4 = 0.$$

Problem 3.1.7. Solve the differential equation

$$y'(x) - xy(x) = -x^3,$$

with the initial condition $y(0) = 1$.

Problem 3.1.8. Solve the differential equation

$$y'(x) + y(x)\tan(x) = \frac{1}{\cos(x)}, \quad -\frac{\pi}{2} < x < \frac{\pi}{2},$$

with the initial condition $y(\pi/6) = 0$.

Problem 3.1.9. Solve the differential equation

$$xy'(x) - y(x) = x^3 e^x,$$

with initial condition $y(1) = 0$.

Problem 3.1.10. Solve the differential equation

$$y'(x) + 2y(x) = xe^{-2x},$$

with initial condition $y(0) = 1$.

Problem 3.1.11. Find a solution to:

$$y' + \frac{2}{x}y = 4x.$$

Problem 3.1.12. Solve:

$$x(x - 1)y'(x) + y(x) = x$$

for $x \neq 0$ and $x \neq 1$.

Problem 3.1.13. The equation for an LR circuit connected to a voltage source is

$$L\frac{dI}{dt} + RI = V(t).$$

Solve the above differential equation for

(a) $V(t) = 1, \; I(0) = 0.$ (b) $V(t) = e^t, \; I(0) = 0.$

Problem 3.1.14. A skydiver with mass $m = 50$ kg jumps out of an airplane. His parachute opens immediately and the air resistance gives rise to a retarding force $-k\vec{v}(t)$, which is proportional to the instantaneous velocity $\vec{v}(t)$, with $k = 10$ N $\cdot$ sec/meter. The acceleration due to gravity is constant with gravitational acceleration, $g = 10$ meter/sec^2.

(a) Calculate the speed $v(t)$ of the skydiver with the initial condition $v(0) = 0$.

(b) Find the maximum speed v_∞ at which the skydiver will fall.

(c) How long time it will take for the skydiver to reach $0.9\,v_\infty$ and what distance will he travel during that time?

Problem 3.1.15. A stone of mass $m = 2$ kg is thrown vertically upward from a height of 1 meter above the ground. The magnitude of stone's initial upward velocity is $v = 5$ meter/sec. The air resistance is $\vec{F} = -k\vec{v}(t)$ with $k = 8$ N $\cdot$ sec/meter. The gravitational acceleration is $g = 10$ meter/sec^2.

(a) Find the speed of the stone as a function of time t when it rises.

(b) Find the time when the stone starts falling back to the ground.

(c) Find the maximum height of the stone.

Problem 3.1.16. Suppose that the resistance of water on a boat produces a drag force proportional to the square of its speed $-kv^2(t)$ with constant k. The motor is turned off at time $t = 0$ when the boat is at the origin ($x = 0$) traveling at speed v_0 in the $+x$ direction.

(a) Write down the differential equation of motion for the boat and find $v(t)$ as function of time, mass m, v_0 and k.

(b) Find the distance $x(t_1)$ traveled by the boat at time t_1 when $v(t_1) = v_0/2$.

Problem 3.1.17. Consider a falling stone of mass m that is released from rest at time $t = 0$ and is subject to the equation of motion

$$m\frac{dv}{dt} = mg - kv^2,$$

where k is a measure of air resistance and g is gravitational acceleration.

(a) Use separation of variables to find speed $v(t)$.

(b) Find the terminal speed v_∞, which v approaches as $t \to \infty$.

(c) Find the distance x as a function of time t and show that for $t \to \infty$ the distance approaches $v_\infty t - (m/k)\ln 2$.

Problem 3.1.18. The radioactive nuclei decay follows the relation:

$$\frac{dN(t)}{dt} = R - \lambda N(t),$$

where N is the number of radioactive atoms, λ is the decay probability per second and R is the constant number of radioactive atoms produced per second in a sample. Find N as a function of time for the initial condition $N(0) = 0$.

Problem 3.1.19. The radioactive nuclei decay of two different nuclides is described by the following system of equations:

$$\frac{dN_1}{dt} = -\lambda_1 N_1, \qquad \frac{dN_2}{dt} = \lambda_1 N_1 - \lambda_2 N_2.$$

Find N_2 as a function of time for the initial conditions $N_1(0) = N_0$ and $N_2(0) = 0$.

Problem 3.1.20. Solve the differential equation

$$y'(x) + 2y(x) = \begin{cases} x & \text{when} \quad x \le 0 \\ 0 & \text{when} \quad x > 0 \end{cases}$$

with the initial condition $y(0) = 1$.

Problem 3.1.21. Solve $y' + y = y^2$ by separation of variables and compare your solution to that obtained in Example 3.1.7, where this equation was treated as a Bernoulli equation.

3.2 Second-Order Differential Equations

The differential equation

$$y'' = \frac{d^2 y}{dx^2} = F(x, y, y') \tag{3.46}$$

is called a second-order differential equation. In classical mechanics, Newton's second law of motion, $m\ddot{x} = F(t, x, \dot{x})$, is an example of a second-order differential equation in one dimension with force F.

3.2.1 Second Order Homogeneous Linear Differential Equation with Constant Coefficients

For $F(x, y, y') = -\frac{b}{a}y'(x) - \frac{c}{a}y(x)$ with constants a, b, c the equation (3.46) takes the form of a second-order homogeneous linear differential equation with constant coefficients:

$$\boxed{ay'' + by' + cy = 0.} \tag{3.47}$$

The term homogeneous refers to the absence of any function (source) on the right hand side. As a result, a constant zero function $y = 0$ is always a trivial solution to (3.47).

Let us consider a special case with constant coefficients $b = 0$ and $c = -1$. Thus, we can rewrite equation (3.47) as

$$y'' - y = 0, \tag{3.48}$$

or $y'' = y$. This equation possesses two solutions y_1 and y_2 given by the exponential functions:

$$y_1(x) = c_1 e^x, \qquad y_2(x) = c_2 e^{-x}. \tag{3.49}$$

The presence of two solutions (instead of one as was the case for the first-order differential equation) is an important feature of the second-order equations. In contrast with the first-order equation the second order differential equation requires two initial conditions for the values of y and y'. Accordingly, the solution has to contain two constants, like c_1 and c_2 in equation (3.49), in order to match the two initial conditions.

Let us return to the general form of a second-order homogeneous linear differential equation (3.47) with constant coefficients. Motivated by the previous example, we will try a solution of the exponential type

$$y(x) = c_0 \, e^{\lambda x}. \tag{3.50}$$

The parameter λ is a constant that is to be determined by the quadratic equation:

$$a\lambda^2 + b\lambda + c = 0, \tag{3.51}$$

which is the characteristic equation for a second-order homogeneous linear differential equation with constant coefficients. The characteristic equation (3.51) is obtained by plugging $y(x)$ from equation (3.50) into equation (3.47). The roots for the quadratic equation (3.51) are

$$\lambda_1 = \frac{-b + \sqrt{b^2 - 4ac}}{2a} \tag{3.52}$$

and

$$\lambda_2 = \frac{-b - \sqrt{b^2 - 4ac}}{2a}. \tag{3.53}$$

From the quadratic formulas (3.52)-(3.53), we deduce that:

① If the discriminant $\Delta \equiv b^2 - 4ac > 0$ then both roots λ_1 and λ_2 are real and distinct.

② If $\Delta < 0$ then there are no real roots and both λ_1 and λ_2 are complex.

③ If $\Delta = 0$ then there is only one root, namely $\lambda = -b/2a$.

Let us consider case ①. In that case, we obtain two solutions from equation (3.50) and (3.52)-(3.53):

$$y_1(x) = c_1 e^{\lambda_1 x}, \qquad y_2(x) = c_2 e^{\lambda_2 x}. \tag{3.54}$$

Note, that, since $\lambda_1 \neq \lambda_2$, we cannot obtain the function $y_2(x)$ from $y_1(x)$ by multiplication with a constant.

As seen in Chapter 1, the concept of the linear independence of functions can be formulated as the concept of linear independence of vectors.

Two functions $f_1(x)$ and $f_2(x)$ are called *linearly independent if neither one is a constant multiple of the other. As a result of this definition, the equation*

$$k_1 f_1(x) + k_2 f_2(x) = 0 \tag{3.55}$$

possesses only the trivial (zero) solution for constant parameters k_1, k_2.

☞ **Example 3.2.1.** Let us consider two continuous functions $f_1(x) = \cos x$ and $f_2(x) = \cos 2x$ that are both defined in the interval $-\pi \leq x \leq \pi$. The set of continuous functions on $-\pi \leq x \leq \pi$ defines a vector space according to the general definition 1.1.1 given in Chapter 1. Furthermore, functions f_1 and f_2 are linearly independent as follows from the following argument: If equation $k_1 \cos x + k_2 \cos 2x = 0$ holds for all x between $-\pi$ and π then the values $x = 0, \pi/2$ yield

$$k_1 + k_2 = 0, \quad -k_2 = 0.$$

These equations only have one solution, $k_1 = k_2 = 0$, which is the trivial solution.

The two solutions from equation (3.54) are linearly independent. Note that any linear combination $y(x) = c_1 y_1(x) + c_2 y_2(x)$ of the two solutions from (3.54) will also be a solution of the differential equation (3.47). The differential equation (3.47) is a linear equation and the above linear combination $y(x)$ is called a general solution. To fix constants c_1, c_2 in the general solution $y(x)$ one needs to specify two initial conditions. Those conditions are usually obtained by assigning some values to $y(x_0)$ and $y'(x_0)$ at some initial point x_0.

☞ **Example 3.2.2.** Solve $y'' + 2y' - 3y = 0$ with $y(0) = 0$ and $y'(0) = 2$.

As a first step, one always tries the exponential solution $y(x) = c \exp(\lambda x)$, which leads to the characteristic equation:

$$\lambda^2 + 2\lambda - 3 = (\lambda - 1)(\lambda + 3) = 0. \tag{3.56}$$

The factorization reveals $\lambda_1 = 1, \lambda_2 = -3$ as roots of the characteristic equation. Thus, the general solution is:

$$y(x) = c_1 e^x + c_2 e^{-3x}. \tag{3.57}$$

By applying the initial conditions we find that

$$y(0) = c_1 + c_2 = 0 \quad \text{and} \quad y'(0) = c_1 - 3c_2 = 2. \tag{3.58}$$

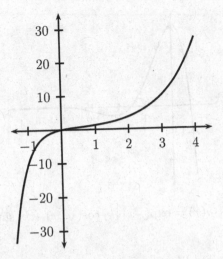

Figure 3.2: $y(x) = \frac{1}{2}e^x - \frac{1}{2}e^{-3x}$.

We find that $c_1 = 1/2$ and $c_2 = -1/2$ and, therefore,

$$y(x) = \frac{1}{2}e^x - \frac{1}{2}e^{-3x} \tag{3.59}$$

is the unique solution to the problem and it's behavior is shown in Figure 3.2. Equation (3.58) clearly shows that for two initial conditions are necessary to determine the constants c_1, c_2.

Now, we move to case ② for which the discriminant $\Delta = b^2 - 4ac$ is negative. In that case, the two solutions of the characteristic equation become complex and can be written as:

$$\lambda_{\pm} = \rho \pm i\beta = -\frac{b}{2a} \pm i\frac{\sqrt{4ac - b^2}}{2a}. \tag{3.60}$$

The general solution of equation (3.47) takes the form:

$$y(x) = c_1 e^{(\rho + i\beta)x} + c_2 e^{(\rho + i\beta)x} = e^{\rho x}\left[(c_1 + c_2)\cos(\beta x) + i(c_1 - c_2)\sin(\beta x)\right] \tag{3.61}$$

and can be rewritten as $y(x) = \exp(\rho x)\left[k_1 \cos(\beta x) + k_2 \sin(\beta x)\right]$ with constants k_1, k_2 that can be determined by two initial conditions.

☞ **Example 3.2.3.** Solve $y'' + 2y' + 3y = 0$ with $y(0) = 2\sqrt{2}$ and $y'(0) = \sqrt{2}$.
Solving the characteristic equation yields $\lambda_{\pm} = -1 \pm i\sqrt{2}$. Accordingly, the general solution is

$$y(x) = e^{-x}\left[k_1 \cos(\sqrt{2}x) + k_2 \sin(\sqrt{2}x)\right]. \tag{3.62}$$

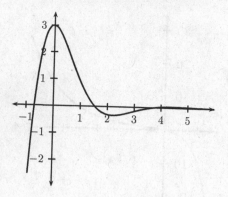

Figure 3.3: $y(x) = \exp(-x)\left[k_1 \cos(\sqrt{2}x) + k_2 \sin(\sqrt{2}x)\right]$.

Imposing initial conditions amounts to solving a pair of linear equations $k_1 = 2\sqrt{2}$ and $-k_1 + \sqrt{2}k_2 = \sqrt{2}$. The solution $y(x) = \exp(-x)\left[2\sqrt{2}\cos(\sqrt{2}x) + 3\sin(\sqrt{2}x)\right]$ is shown in Figure 3.3.

What remains is case ③ with $\Delta = 0$ or $b^2 = 4ac$ and only one double root $\lambda = -b/2a$ of the characteristic equation. Thus, $y_1 = \exp(-bx/2a)$ is a solution although not a general one since the second linearly independent solution is missing. The procedure to find the missing solution is as follows. Try $y_2 = x\exp(-bx/2a)$, then

$$y_2' = e^{-bx/2a} - \frac{bx}{2a}e^{-bx/2a},$$

$$y_2'' = -\frac{b}{a}e^{-bx/2a} + \frac{b^2x}{4a^2}e^{-bx/2a}$$

and inserting these values into equation (3.47) yields

$$ay_2'' + by_2' + cy_2 = -be^{-bx/2a} + \frac{b^2x}{4a}e^{-bx/2a} + be^{-bx/2a} - \frac{b^2x}{2a}e^{-bx/2a}$$

$$+ cxe^{-bx/2a} = e^{-bx/2a}\left(x\left(\frac{b^2}{4a} - \frac{b^2}{2a} + c\right) - b + b\right) = 0,$$

where we used that $-b^2/4a + c = 0$. Hence, y_2 is also a solution and it is clearly linearly independent of y_1. Thus, a general solution is given in this case by

$$y(x) = c_1 e^{-bx/2a} + c_2 xe^{-bx/2a}. \tag{3.63}$$

☞ **Example 3.2.4.** Solve $y'' + 2y' + y = 0$ with $y(0) = 2$ and $y'(0) = 2$. The double root is $\lambda = -b/2a = -1$ and according to equation (3.63), the general solution is given by:

$$y(x) = c_1 e^{-x} + c_2 xe^{-x}. \tag{3.64}$$

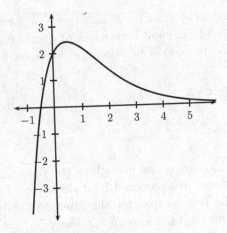

Figure 3.4: $y(x) = 2e^{-x} + 4xe^{-x}$.

Imposing the initial conditions yields equations $c_1 = 2, -c_1 + c_2 = 2$ with the corresponding solution:

$$y(x) = 2e^{-x} + 4xe^{-x}, \tag{3.65}$$

whose graph is shown in Figure 3.4.

It is useful to introduce the operator notation:

$$L[f] = f''(x) + p(x)f'(x) + q(x)f(x) \tag{3.66}$$

with $p(x), q(x)$ as continuous functions on some interval (a, b) for x such that $a < x < b$.

Based on examples of solutions to differential equations with constant coefficients we suspect that two initial conditions are required to specify a unique solution of equation $L[f] = 0$ for a linear operator L from (3.66). A general argument goes as follows. Rewrite $L[f] = 0$ as

$$\frac{d^2 f(x)}{dx^2} = -p(x)\frac{df(x)}{dx} - q(x)f(x).$$

Differentiating both sides of this equation with respect to x gives $d^3 f(x)/dx^3$ in terms of $d^2 f(x)/dx^2$ and lower derivatives of $f(x)$. Since $d^2 f(x)/dx^2$ is known in terms of $f'(x)$ and $f(x)$ we can eliminate it from the expression for $d^3 f(x)/dx^3$. Thus, $d^3 f(x)/dx^3$ can be expressed in terms of $f'(x)$ and $f(x)$. Differentiating once more on both sides of this equation gives an expression for $d^4 f(x)/dx^4$ in terms of $d^3 f(x)/dx^3$ and lower derivatives of $f(x)$. Eliminating $d^3 f(x)/dx^3$ and $f''(x)$ in terms of $f'(x)$ and $f(x)$ leads to an expression for $d^4 f(x)/dx^4$ in terms of $f'(x)$ and

$f(x)$. Continuing this process we arrive at an expression for an arbitrary derivative of $f(x)$ given by $f'(x)$ and $f(x)$ and derivatives of known functions $p(x)$ and $q(x)$. Now, recall the Taylor expansion of function f about some fixed point x_0:

$$f(x) = f(x_0) + (x - x_0)\frac{\mathrm{d}f(x)}{\mathrm{d}x}\Big|_{x=x_0} + \frac{1}{2}(x - x_0)^2\frac{\mathrm{d}^2 f(x)}{\mathrm{d}x^2}\Big|_{x=x_0} + \cdots$$
$$+ \frac{1}{n!}(x - x_0)^n\frac{\mathrm{d}^n f(x)}{\mathrm{d}x^n}\Big|_{x=x_0} + \cdots$$

valid for variables x, which are closed enough to the point x_0 for the series on the right hand side to converge (see section 4.1 for details on Taylor expansion). We conclude that due to the Taylor expansion the function $f(x)$ which solves $L[f] = 0$ is fully determined by initial conditions $f'(x_0)$ and $f(x_0)$ (values of f' and f at a fixed point x_0) and functions $p(x)$ and $q(x)$.

> The results we have gathered above support the following *existence and unique-ness* theorem which states that a system composed of a homogeneous second-order differential equation
>
> $$L[y] = y''(x) + p(x)y'(x) + q(x)y(x) = 0 \qquad (3.67)$$
>
> and the initial conditions
>
> $$y(x_0) = y_0 \quad \text{and} \quad y'(x_0) = y_0', \qquad (3.68)$$
>
> with constants y_0 and y_0', possesses a unique solution.

The crucial property of the operator L is its *linearity*:

$$L[c_1 f_1 + c_2 f_2] = c_1 L[f_1] + c_2 L[f_2]. \qquad (3.69)$$

Hence, if both $y_1(x)$ and $y_2(x)$ are solutions of $L[y] = 0$ then any linear combination $c_1 y_1 + c_2 y_2$ is also a solution for arbitrary constants c_1, c_2. This ambiguity no longer exists if we impose the initial conditions

$$c_1 y_1(x_0) + c_2 y_2(x_0) = y_0 \qquad (3.70)$$
$$c_1 y_1'(x_0) + c_2 y_2'(x_0) = y_0' \qquad (3.71)$$

on the linear combination $c_1 y_1 + c_2 y_2$. Solving these two equations for the constants c_1, c_2 yields:

$$c_1 = \frac{y_0 y_2'(x_0) - y_0' y_2(x_0)}{y_1(x_0)y_2'(x_0) - y_1'(x_0)y_2(x_0)} \qquad (3.72)$$

$$c_2 = \frac{y_0' y_1(x_0) - y_0 y_1'(x_0)}{y_1(x_0)y_2'(x_0) - y_1'(x_0)y_2(x_0)}. \qquad (3.73)$$

These relations fix the values of the constants c_1, c_2 provided the denominator of both fractions in (3.72) and (3.73):

$$W[y_1, y_2](x_0) \equiv \begin{vmatrix} y_1(x_0) & y_2(x_0) \\ y_1'(x_0) & y_2'(x_0) \end{vmatrix} = y_1(x_0)y_2'(x_0) - y_1'(x_0)y_2(x_0) \qquad (3.74)$$

is different from zero: $W[y_1, y_2](x_0) \neq 0$. The expression in (3.74) is called the Wronskian of y_1 and y_2 at x_0. Since neither c_1 or c_2 in (3.72) and (3.73) are well-defined for the vanishing Wronskian we must require that $W[y_1, y_2] \neq 0$ for the point x_0 in the interval $a < x < b$. It turns out that this condition implies that two solutions y_1 and y_2 of $L[y] = 0$, are linearly independent, as we will show below.

Solutions y_1 and y_2 of $L[y] = 0$ are linearly dependent if and only if $W[y_1, y_2](x_0) = 0$ at some x_0 in the interval $a < x < b$.

Proof. If two linearly dependent functions are proportional to each other then $y_1(x) = ky_2(x)$ for $a < x < b$. It follows that the Wronskian is identically zero since $W[y_1, y_2] = y_1y_2' - y_2y_1' = ky_2y_2' - y_2ky_2' = 0$. Conversely, let $W[y_1, y_2] = 0$ at some x_0 in the interval $a < x < b$. Consider the system of equations:

$$0 = c_1y_1(x_0) + c_2y_2(x_0), \qquad (3.75)$$
$$0 = c_1y_1'(x_0) + c_2y_2'(x_0), \qquad (3.76)$$

which in the matrix form reads

$$\begin{bmatrix} y_1(x_0) & y_2(x_0) \\ y_1'(x_0) & y_2'(x_0) \end{bmatrix} \begin{bmatrix} c_1 \\ c_2 \end{bmatrix} = 0. \qquad (3.77)$$

We recognize the Wronskian from equation (3.74) in the determinant of the matrix on the left hand side in equation (3.77). Since the Wronskian was assumed to be zero, the matrix

$$\begin{bmatrix} y_1(x_0) & y_2(x_0) \\ y_1'(x_0) & y_2'(x_0) \end{bmatrix}$$

must be singular and there exists a non-zero solution (c_1, c_2) to the algebraic equations (3.75) or (3.77). Now, define $y(x) = c_1y_1(x) + c_2y_2(x)$ for all x in the interval $a < x < b$. This function satisfies $L[y] = 0$ due to linearity of L. Due to equations (3.75) and (3.76) it also satisfies the initial conditions $y(x_0) = 0, y'(x_0) = 0$. However, the constant function $Y(x) = 0$ will also satisfy both the differential equation and the same initial conditions. Due to the uniqueness theorem both solutions must be identical and, therefore, $y(x) = c_1y_1(x) + c_2y_2(x) = 0$ for all x in the interval $a < x < b$. Hence, the functions y_1, y_2 are linearly dependent, which concludes the proof.

The important question is what happens if $W[y_1, y_2](x_0) \neq 0$ at some point x_0 for two solutions y_1 and y_2. Is it a sufficient condition for the linear independence of y_1 and y_2? The following observation will help in answering this question.

The Wronskian $W[y_1, y_2](x)$ satisfies

$$W[y_1, y_2](x) = W[y_1, y_2](x_0)e^{-\int_{x_0}^{x} p(t)\, dt}, \tag{3.78}$$

where x_0 is some arbitrary chosen point on the interval $a < x < b$. Thus, if the Wronskian $W[y_1, y_2](x_0) = 0$ is zero at some point x_0 of the interval $a < x < b$ then it is identically zero, $W[y_1, y_2](x) = 0$, on the whole interval $a < x < b$.

Proof. For two solutions of $L[y_1] = L[y_2] = 0$, the second-order differential equations for y_1 and y_2 are:

$$y_1''(x) + p(x)y_1'(x) + q(x)y_1(x) = 0, \quad \text{and} \quad y_2''(x) + p(x)y_2'(x) + q(x)y_2(x) = 0. \tag{3.79}$$

Multiplying the first of these equations by $y_2(x)$ and the second by $y_1(x)$ and subtracting one the other we obtain

$$(y_1 y_2'' - y_2 y_1'') + p(x)(y_1 y_2' - y_2 y_1') = 0. \tag{3.80}$$

Since, the derivative of the Wronskian satisfies the following relation:

$$W'[y_1, y_2](x) = \frac{d}{dx}(y_1(x)y_2'(x) - y_1'(x)y_2(x))$$

$$= y_1(x)y_2''(x) - y_1''(x)y_2(x) = \begin{vmatrix} y_1(x) & y_2(x) \\ y_1''(x) & y_2''(x) \end{vmatrix} \tag{3.81}$$

equation (3.80) can be rewritten as:

$$W'[y_1, y_2](x) + p(x)W[y_1, y_2](x) = 0. \tag{3.82}$$

This is a separable first-order differential equation, which can easily be integrated to yield

$$W[y_1, y_2](x) = Ce^{-\int_{x_0}^{x} p(t)\, dt}. \tag{3.83}$$

Setting x in (3.83) equal to the lower limit of integration, x_0, yields an integration constant C equal to the Wronskian $W[y_1, y_2](x_0)$. Thus, the solution to the Wronskian is

$$\boxed{W[y_1, y_2](x) = W[y_1, y_2](x_0)\, e^{-\int_{x_0}^{x} p(t)\, dt},} \tag{3.84}$$

which yields the desired result.

Thus, if the Wronskian is different from zero at some point x_0, i.e. $W[y_1, y_2](x_0) \neq 0$, then the Wronskian is different from zero everywhere on the whole interval. This outcome is not possible if the two solutions y_1, y_2 are linearly dependent.

> Hence, if $W[y_1, y_2] \neq 0$ at some point x_0 in the interval between a and b then the solutions y_1 and y_2 are linearly independent between a and b.

Let us consider two solutions y_1 and y_2 of the second-order differential equation $L[y_1] = L[y_2] = 0$, such that their Wronskian is different from zero at some point x_0 in the interval $a < x < b$ i.e. $W[y_1, y_2](x_0) \neq 0$. It follows that the solution $c_1 y_1(x) + c_2 y_2(x)$ is a general solution of $L[y] = 0$ (3.67), meaning that any solution $Y(x)$ of (3.67) can be written as a linear combination $c_1 y_1(x) + c_2 y_2(x)$ for some constants c_1, c_2.

In other words, two linearly independent solutions y_1 and y_2 form a *basis* for the equation $L[y] = 0$, which amounts to saying that every solution Y of $L[y] = 0$ can be rewritten as $Y(x) = c_1 y_1(x) + c_2 y_2(x)$.

Now, we will prove the above statement. Let x_0 be some point in the interval $a < x < b$. Let us try to find two non-zero constants c_1, c_2, such that two following initial conditions

$$Y(x_0) = c_1 y_1(x_0) + c_2 y_2(x_0), \tag{3.85}$$
$$Y'(x_0) = c_1 y_1'(x_0) + c_2 y_2'(x_0), \tag{3.86}$$

hold. Let's rewrite equations (3.85)-(3.86) in matrix form as

$$\begin{bmatrix} y_1(x_0) & y_2(x_0) \\ y_1'(x_0) & y_2'(x_0) \end{bmatrix} \begin{bmatrix} c_1 \\ c_2 \end{bmatrix} = \begin{bmatrix} Y(x_0) \\ Y'(x_0) \end{bmatrix}. \tag{3.87}$$

Since y_1 and y_2 are linearly independent, the determinant $W[y_1, y_2](x_0)$ of the above matrix is different from zero. Thus, there exists a non-trivial solution to equations (3.85)-(3.86) for the constants c_1 and c_2 that is given by:

$$\begin{bmatrix} c_1 \\ c_2 \end{bmatrix} = \begin{bmatrix} y_1(x_0) & y_2(x_0) \\ y_1'(x_0) & y_2'(x_0) \end{bmatrix}^{-1} \begin{bmatrix} Y(x_0) \\ Y'(x_0) \end{bmatrix},$$

which agrees with expressions (3.72)-(3.73) for $Y(x_0) = y_0$ and $Y'(x_0) = y_0'$. Thus, both solutions $c_1 y_1(x) + c_2 y_2(x)$ and $Y(x)$ satisfy the same initial conditions at x_0 and, according to the uniqueness theorem, they must be identical. Therefore, $Y(x) = c_1 y_1(x) + c_2 y_2(x)$ for all x in the interval $a < x < b$.

3.2.2 Wronskian Representation of the Second-order Differential Equation

Here, we will show that the linear second-order differential operator $L[y]$ from equation (3.66) can be fully described by two linearly independent solutions, y_1 and y_2, of equation $L[y] = 0$ through expressions involving determinants of 2×2 and 3×3 matrices. We begin by establishing determinant representations for the coefficients functions $p(x)$ and $q(x)$ that appear in the differential equation $L[y] = y'' + p(x)y' + q(x)y = 0$.

According to equation (3.82) the function $p(x)$ is given by

$$p(x) = -\left(\ln W[y_1, y_2]\right)'(x) = -\frac{1}{W[y_1, y_2](x)}\frac{d}{dx}W[y_1, y_2](x)$$

$$= -\frac{1}{W[y_1, y_2](x)}\begin{vmatrix} y_1(x) & y_2(x) \\ y_1''(x) & y_2''(x) \end{vmatrix}, \tag{3.88}$$

where we used relation (3.81) to simplify the above formula.

To derive a similar expression for the coefficient $q(x)$ we turn our attention to equations in (3.79). This time, we multiply the left equation by $y_2'(x)$ and the right by $y_1'(x)$. Subtracting the resulting equations from each other we obtain

$$(y_2'(x)y_1''(x) - y_1'(x)y_2''(x)) + q(x)(y_1(x)y_2'(x) - y_2(x)y_1'(x)) = 0. \tag{3.89}$$

Thus, the function $q(x)$ is given by the following relation

$$q(x) = \frac{1}{W[y_1, y_2](x)}\begin{vmatrix} y_1'(x) & y_2'(x) \\ y_1''(x) & y_2''(x) \end{vmatrix}. \tag{3.90}$$

Plugging expressions (3.88) and (3.90) into the formula for $L[y]$ we obtain

$$L[y] = y''(x) + p(x)y'(x) + q(x)y(x)$$

$$= \frac{1}{W[y_1, y_2](x)}\left(y''(x)W[y_1, y_2](x)\right.$$

$$\left. - y'(x)\begin{vmatrix} y_1(x) & y_2(x) \\ y_1''(x) & y_2''(x) \end{vmatrix} + y(x)\begin{vmatrix} y_1'(x) & y_2'(x) \\ y_1''(x) & y_2''(x) \end{vmatrix}\right) \tag{3.91}$$

$$= \frac{1}{W[y_1, y_2]}W_3[y_1, y_2, y],$$

where $W_3[y_1, y_2, y]$ generalizes the regular Wronskian to the determinant of a 3×3 matrix given by:

$$W_3[y_1, y_2, y] = \begin{vmatrix} y_1(x) & y_2(x) & y(x) \\ y_1'(x) & y_2'(x) & y'(x) \\ y_1''(x) & y_2''(x) & y''(x) \end{vmatrix}. \tag{3.92}$$

The last equality in expression (3.91) follows from the minor expansion of $W_3[y_1, y_2, y]$ along the last column. We notice that the determinant $W_3[y_1, y_2, c_1 y_1 + c_2 y_2]$ vanishes identically because the last column is a linear combination of the first two columns. Consequently, expression (3.91) implies that $L[y] = 0$ for any function $y(x)$, which is a linear combination of solutions $y_1(x)$ and $y_2(x)$.

3.2.3 Reduction of Order or How to Obtain a Missing Second Solution

Suppose that $y_1(x)$ is a solution of equation $y''(x) + p(x)y'(x) + q(x)y(x) = 0$. In this section we will present a procedure for finding the remaining missing solution $y_2(x)$.

We will begin by setting $y_2(x) = u(x)y_1(x)$, where $u(x)$ is an unknown function. Computing the first two derivatives of $u(x)y_1(x)$ with respect to x yields:

$$(uy_1)' = u'y_1 + uy_1', \quad (uy_1)'' = u''y_1 + 2u'y_1' + uy_1''.$$

We will plug $y_2 = uy_1$ into expression (3.88) for the coefficient function $p(x)$. We investigate the Wronskian of y_1 and $y_2 = uy_1$ given by:

$$W[y_1, uy_1] = \begin{vmatrix} y_1 & uy_1 \\ y_1' & u'y_1 + uy_1' \end{vmatrix} = \begin{vmatrix} y_1 & uy_1 \\ y_1' & uy_1' \end{vmatrix} + \begin{vmatrix} y_1 & 0 \\ y_1' & u'y_1 \end{vmatrix}$$

$$= \begin{vmatrix} y_1 & 0 \\ y_1' & u'y_1 \end{vmatrix} = u'y_1^2,$$

where we calculated the determinant according to rules we learned in subsection 2.1.4. In particular, we omitted the determinant having two columns proportional to each other.

Next, we examine the determinant in the nominator in expression (3.88) for $p(x)$. For $y_2 = uy_1$ we find:

$$\begin{vmatrix} y_1 & uy_1 \\ y_1'' & (uy_1)'' \end{vmatrix} = \begin{vmatrix} y_1 & uy_1 \\ y_1'' & u''y_1 + 2u'y_1' + uy_1'' \end{vmatrix} = \begin{vmatrix} y_1 & uy_1 \\ y_1'' & uy_1'' \end{vmatrix}$$

$$+ \begin{vmatrix} y_1 & 0 \\ y_1'' & u''y_1 + 2u'y_1' \end{vmatrix} = \begin{vmatrix} y_1 & 0 \\ y_1'' & u''y_1 + 2u'y_1' \end{vmatrix}$$

$$= y_1 \left(u''y_1 + 2u'y_1' \right),$$

where again we omitted the determinant having two columns proportional to each other.

Thus, for $y_2 = uy_1$ relation (3.88) simplifies to

$$p(x) = -\frac{1}{u'(x)y_1^2(x)} y_1(x) \left(u''(x)y_1(x) + 2u'(x)y_1'(x) \right)$$

$$= -\frac{1}{u'(x)y_1^2(x)} \frac{d}{dx} \left(u'(x)y_1^2(x) \right) = -\frac{d}{dx} \ln\left(u'(x)y_1^2(x) \right).$$

(3.93)

Integrating both sides of equation (3.93) and applying the exponential function yields

$$u'(x) = \frac{1}{y_1^2(x)} e^{-\int^x p(x')\,dx'},$$

(3.94)

Therefore, the second solution given by the product of y_1 and u follows after additional integration of equation (3.94):

$$y_2(x) = u(x)y_1(x) = y_1(x) \int^x \frac{1}{y_1^2(x')} e^{-\int^{x'} p(x'') \, dx''} \, dx' . \qquad (3.95)$$

Since $u(x)$ is a non-trivial function of x, $y_2(x)$ is linearly independent of $y_1(x)$.

☞ **Example 3.2.5.** Consider the differential equation:

$$xy'' - y' - \frac{3}{x}y = 0, \qquad (3.96)$$

with one known solution $y_1(x) = 1/x$. First, we will put equation (3.96) in the standard form of the equation $y''(x) + p(x)y'(x) + q(x)y(x) = 0$ by dividing it by x. This yields $p(x) = -1/x$ and $\exp(-\int p(x) \, dx) = x$. Inserting these results into equation (3.95) gives:

$$\begin{aligned} y_2(x) = u(x)y_1(x) &= y_1(x) \int^x \frac{1}{y_1^2(x')} e^{-\int^{x'} p(x'') \, dx''} \, dx' \\ &= \frac{1}{x} \int^x (x')^3 \, dx' = \frac{x^3}{4} . \end{aligned}$$

Therefore, the second solution is given up to a constant by $y_2(x) = x^3$. Verify, as an exercise, that both $y_1(x) = 1/x$ and $y_2(x) = x^3$ are solutions of equation (3.96).

_____

3.2.4   The Non-Homogeneous Equations

The differential equation

$$y''(x) + p(x)y'(x) + q(x)y(x) = r(x), \qquad (3.97)$$

with a non-zero source term on the right hand side, represented by some function $r(x)$, is called the non-homogeneous second-order differential equation. The presence of a non-zero "source-term" is motivated by many physical applications in which a source is often played by a driving force behind a mechanical system whose motion is described by the second-order differential equation.

A general solution $y(x)$ to the non-homogeneous equation is defined as a linear combination of two terms

$$y(x) = y_h(x) + y_p(x), \qquad (3.98)$$

where the first term $y_h(x) = c_1 y_1(x) + c_2 y_2(x)$ is a homogeneous solution to the second-order differential equation (3.67) containing two arbitrary constants c_1, c_2,

Differential Equations

while the second term $y_p(x)$ is a particular solution to the non-homogeneous equation. An unique solution is obtained from the general solution (3.98) by fixing the values of the constants c_1, c_2 according to the initial conditions.

☞ **Example 3.2.6.** Solve $y''(x) + y'(x) - 2y(x) = 5\exp(3x)$ with $y(0) = 2$ and $y'(0) = -3$.

The exponential solution $y(x) = c\exp(\lambda x)$ leads to the characteristic equation:

$$\lambda^2 + \lambda - 2 = (\lambda - 1)(\lambda + 2) = 0, \tag{3.99}$$

with roots of the characteristic equation $\lambda_1 = 1, \lambda_2 = -2$. The homogeneous solution is:

$$y_h(x) = c_1 e^x + c_2 e^{-2x}. \tag{3.100}$$

For the particular solution $y_p(x)$ we will assign the function $y_p(x) = C\exp(3x)$. Inserting it into the differential equation, we get

$$y_p''(x) + y_p'(x) - 2y_p(x) = Ce^{3x}\left(3^2 + 3 - 2\right) = 10Ce^{3x} = 5\exp(3x).$$

Hence, $C = 1/2$. Therefore, the general solution is:

$$y(x) = c_1 e^x + c_2 e^{-2x} + \frac{1}{2}e^{3x}.$$

Applying the initial conditions we find that

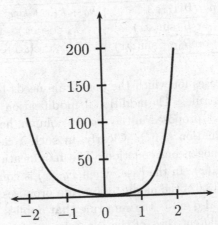

Figure 3.5: $y(x) = -\frac{1}{2}e^x + 2e^{-2x} + \frac{1}{2}e^{3x}$.

$$y(0) = c_1 + c_2 + \frac{1}{2} = 2 \quad \text{and} \quad y'(0) = c_1 - 2c_2 + \frac{3}{2} = -3, \tag{3.101}$$

which implies that $c_1 = -1/2, c_2 = 2$. This choice gives the unique solution

$$y(x) = -\frac{1}{2}e^x + 2e^{-2x} + \frac{1}{2}e^{3x} \tag{3.102}$$

shown in Figure 3.5.

3.2.5 Method of Undetermined Coefficients

In the last section we learned how to solve the non-homogeneous equation by choosing the particular solution to be of the same exponential form as the source (e.g. $r(x) = 5\exp(3x)$ in Example 3.2.6). The method of undetermined coefficients extends this approach to a number of scenarios. The basic rule can be written in the form of table 3.1. The rule determines the functional form of the particular solution $y_p(x)$ given in the second column. The constants $k, k_1, \ldots$ have to be found from the value of the constant R_0 given in the first column and a specific form of the differential equation. In the last two rows, the functions $y_p(x)$ will, in general, contain both sine and cosine functions even when the source function $r(x)$ contains only a sine or cosine function.

Table 3.1: Method of undetermined coefficients.

$r(x)$	$y_p(x)$
$R_0 e^{\gamma x}$	$k e^{\gamma x}$
$R_0 x^n, \ (n = 0, 1, \ldots)$	$k_0 + k_1 x + k_2 x^2 + \cdots + k_n x^n$
$R_0 \cos(\omega x)$ or $R_0 \sin(\omega x)$	$k_1 \cos(\omega x) + k_2 \sin(\omega x)$
$R_0 e^{\lambda x} \cos(\omega x)$ or $R_0 e^{\lambda x} \sin(\omega x)$	$e^{\lambda x}\left(k_1 \cos(\omega x) + k_2 \sin(\omega x)\right)$

There are special cases for which the basic rule needs to be augmented by an additional modification rule. The additional modification rule applies when the particular solution $y_p(x)$, proposed in the second column, happens to be a solution of the homogeneous equation (3.67). Clearly, in such a case, $y_p(x)$ fails to be a solution of the non-homogeneous equation. The modification rule states that the proposed solution is $xy_p(x)$. In the case, when $xy_p(x)$ is equal to a solution of the homogeneous equation (3.67) the modification rule proposes $x^2 y_p(x)$ as a solution.

Furthermore, there also exists the sum rule that applies when $r(x)$ is a sum of expressions from the different rows of the left column of the Table 3.1. In such a case, the rule advises to formulate a solution $y_p(x)$ as the sum of solutions from the corresponding rows of the second column of the Table 3.1.

☞ **Example 3.2.7.** To illustrate the modification rule, consider the equation

$$y''(x) + y'(x) - 2y(x) = e^x .$$

In equation (3.100), we found the solution $y_h(x) = c_1 e^x + c_2 e^{-2x}$. According to the modification rule, since $r(x) = \exp(x)$ coincides with one of the solutions to the homogeneous equation, we need to consider $y_p(x) = kx \exp(x)$. In this case, $y_p'(x) = k \exp(x) + xk \exp(x)$ and $y_p''(x) = 2k \exp(x) + xk \exp(x)$. Thus,

$$y_p''(x) + y_p'(x) - 2y_p(x) = (1 + 1 - 2)kxe^x + (2+1)ke^x = 3ke^x = e^x$$

and k must be equal to $1/3$. Because the method required us to find the value of the coefficient, k, it is called the method of undetermined coefficients.

Thus, the general solution is:

$$y(x) = y_h(x) + y_p(x) = c_1 e^x + c_2 e^{-2x} + \frac{1}{3}xe^x.$$

Next, we turn our attention to the following equation

$$y'' + 2y' + y = \exp(-x). \tag{3.103}$$

We have seen the homogeneous version of this equation in Example 3.2.4 with the solution $y_h(x) = c_1 e^{-x} + c_2 x e^{-x}$. Therefore, neither e^{-x} or xe^{-x} is a candidate for the solution $y_p(x)$ to the non-homogeneous equation. According to the modification rule, we propose $y_p(x) = Cx^2 e^{-x}$ as the particular solution to the equation (3.103) and take its first and second derivative:

$$y_p'(x) = C\left(2xe^{-x} - x^2 e^{-x}\right), \quad y_p''(x) = C\left(2e^{-x} - 4xe^{-x} + x^2 e^{-x}\right).$$

Consequently,

$$y_p'' + 2y_p' + y_p = C\left(2e^{-x} - 4xe^{-x} + x^2 e^{-x}\right) + 2C\left(2xe^{-x} - x^2 e^{-x}\right)$$
$$+ Cx^2 e^{-x} = 2Ce^{-x} = e^{-x} \quad \longrightarrow \quad C = \frac{1}{2}.$$

The above application of the modification rule yields

$$y(x) = y_h(x) + y_p(x) = c_1 e^{-x} + c_2 x e^{-x} + \frac{1}{2}x^2 e^{-x}$$

as the general solution to equation (3.103).

3.2.6 General Method for Particular Solutions.

Now, we present a formula for a particular solution $y_p(x)$ that applies to all types of source functions, including those $r(x)$ that do not explicitly appear in table 3.1. The formula is giving by:

$$y_p(x) = -y_1(x) \int^x \frac{y_2(t)r(t)}{W(t)}\, dt + y_2(x) \int^x \frac{y_1(t)r(t)}{W(t)}\, dt, \tag{3.104}$$

where y_1, y_2 form a basis for solutions of the homogeneous equation and $W(t) = W[y_1, y_2](t)$ is their Wronskian. For two independent solutions to the homogeneous equation it holds that $W[y_1, y_2] \neq 0$ and, therefore, the denominator in (3.104) is never zero, ensuring that the expression is well-defined.

The drawback of the proposed solution is that it requires an integration in order to obtain $y_p(x)$. However, the clear advantage of formula (3.104) is that, in principle, it holds for an arbitrary non-homogeneous differential equation, (3.97), with an arbitrary source function $r(x)$.

To verify equation (3.104) we compute the first two derivatives of the proposed function $y_p(x)$. A straightforward calculation and use of definition (3.74) of the Wronskian establishes that

$$y_p'(x) = -y_1'(x) \int^x \frac{y_2(t)r(t)}{W(t)}\, dt + y_2'(x) \int^x \frac{y_2(t)r(t)}{W(t)}\, dt,$$

$$y_p''(x) = -y_1''(x) \int^x \frac{y_2(t)r(t)}{W(t)}\, dt + y_2''(x) \int^x \frac{y_2(t)r(t)}{W(t)}\, dt + r(x).$$

Now, we are ready to plug the function $y_p(x)$, defined in equation (3.104), and its derivatives $y_p'(x)$ and $y_p''(x)$ into the determinant $W_3[y_1, y_2, y_p]$, which was introduced in (3.92). Using basic properties of determinants given in subsection 2.1.4, we obtain

$$W_3[y_1, y_2, y_p] = \begin{vmatrix} y_1(x) & y_2(x) & y_p(x) \\ y_1'(x) & y_2'(x) & y_p'(x) \\ y_1''(x) & y_2''(x) & y_p''(x) \end{vmatrix}$$

$$= -\int^x \frac{y_2(t)r(t)}{W(t)}\, dt \begin{vmatrix} y_1(x) & y_2(x) & y_1(x) \\ y_1'(x) & y_2'(x) & y_1'(x) \\ y_1''(x) & y_2''(x) & y_1''(x) \end{vmatrix}$$

$$+ \int^x \frac{y_1(t)r(t)}{W(t)}\, dt \begin{vmatrix} y_1(x) & y_2(x) & y_2(x) \\ y_1'(x) & y_2'(x) & y_2'(x) \\ y_1''(x) & y_2''(x) & y_2''(x) \end{vmatrix}.$$

$$+ \begin{vmatrix} y_1(x) & y_2(x) & 0 \\ y_1'(x) & y_2'(x) & 0 \\ y_1''(x) & y_2''(x) & r(x) \end{vmatrix}.$$

The first two terms contain determinants with two repeated columns and, therefore, vanish identically. The third term, when computed using minor expansion along the last column, yields the expression $r(x)\, W[y_1, y_2](x)$.

It is now easy to show that $y_p(x)$ given by formula (3.104) is the particular solution of the non-homogeneous second-order differential equation (3.97) due to

$$y_p''(x) + p(x)y_p'(x) + q(x)y_p(x) = \frac{1}{W[y_1, y_2]} W_3[y_1, y_2, y]$$

$$= \frac{1}{W[y_1, y_2]} r(x)\, W[y_1, y_2] = r(x)\,. \tag{3.105}$$

☞ **Example 3.2.8.** Consider the non-homogeneous differential equation

$$y'' + y = \sin^2(x)\,. \tag{3.106}$$

We can take $y_1(x) = \cos(x)$ and $y_2(x) = \sin(x)$ as a basis for solutions to the homogeneous equation $y'' + y = 0$. The corresponding Wronskian is $W[y_1, y_2] = y_1 y_2' - y_2 y_1' = \cos^2 x + \sin^2 x = 1$. Accordingly, equation (3.104) simplifies to

$$y_p(x) = -\cos(x) \int^x \sin^3(t)\, dt + \sin(x) \int^x \cos(t)\sin^2(t)\, dt\,. \tag{3.107}$$

The integration in equation (3.107) can be performed using convenient changes of integration variables:

$$\int^x \cos(t)\sin^2(t)\, dt = \int^x \sin^2(t)\, d\sin(t) = \frac{1}{3}\sin^3(x)\,,$$

$$\int^x \sin^3(t)\, dt = -\int^x (1 - \cos^2(t))\, d\cos(t) = -\cos(x) + \frac{1}{3}\cos^3(x)\,.$$

Plugging these results back into the expression for $y_p(x)$ in (3.107) yields a solution to equation (3.106):

$$y_p(x) = \cos^2(x) - \frac{1}{3}\cos^4(x) + \frac{1}{3}\sin^4(x) = \frac{1}{3} + \frac{1}{3}\cos^2(x)$$

$$= \frac{1}{2} + \frac{1}{6}\cos(2x)\,. \tag{3.108}$$

The method we developed in this subsection is designed to help solve the second-order differential equation when the source function is not among the functions in the left column of table (3.1). At first sight, $r(x) = \sin^2(x)$ appears to be missing among the functions in the left column of the table (3.1). However, it is not difficult to see that via a trigonometric identity we can rewrite $r(x) = \sin^2(x)$ as $r(x) = 1/2 - \cos(2x)/2$. Writing $r(x) = c_1(x) + c_2(x)$ with $c_1(x) = 1/2$ and $c_2(x) = -\cos(2x)/2$ we see that we can apply the sum rule. Since $Y_1(x) = 1/2$ solves $y'' + y = c_1$ and $Y_1(x) = \cos(2x)/6$ solves $y'' + y = c_2$ the solution $y_p = Y_1 + Y_2$ obtained using the modified basic rule of the method of undetermined coefficients agrees with the result in equation (3.108).

3.2.7   Complex Method for the Exponential Source Term

Quite often physical applications involve the non-homogeneous differential equation with constant coefficients and a periodic trigonometric source term. Quite generally, such equations can be written as

$$Ay''(x) + By'(x) + Cy(x) = R_0 e^{i\omega x}, \tag{3.109}$$

where A, B, C and R_0 are real constants.

For $r(x) = R_0 \exp(i\omega x)$, Table (3.1) yields as a particular solution:

$$y_p(x) = K e^{i\omega x} \tag{3.110}$$

with, in general, a complex constant K to be determined by plugging $y_p(x)$ back into equation (3.109). This gives

$$\left(-A\omega^2 + i B\omega + C\right) K e^{i\omega x} = R_0 e^{i\omega x}. \tag{3.111}$$

After dividing by the exponential function $\exp(i\omega x)$ and the constant factor $(-A\omega^2 + i B\omega + C)$ one obtains:

$$
\begin{aligned}
K &= \frac{R_0}{-A\omega^2 + i B\omega + C} = \frac{R_0}{-A\omega^2 + i B\omega + C} \frac{-A\omega^2 - i B\omega + C}{-A\omega^2 - i B\omega + C} \\
&= \frac{R_0(C - A\omega^2)}{B^2\omega^2 + (C - A\omega^2)^2} + i \frac{R_0(-B\omega)}{B^2\omega^2 + (C - A\omega^2)^2}.
\end{aligned} \tag{3.112}
$$

Hence, the real and imaginary parts of the constant $K = K_1 + i K_2$ are

$$K_1 = \frac{R_0(C - A\omega^2)}{B^2\omega^2 + (C - A\omega^2)^2}, \qquad K_2 = -\frac{R_0 B\omega}{B^2\omega^2 + (C - A\omega^2)^2}. \tag{3.113}$$

The complex method still applies when either cosine or sine periodic source term replaces the complex exponential function $\exp(i\omega x) = \cos(\omega x) + i \sin(\omega x)$ on the right hand side of equation (3.109). For instance, taking the real part of equation (3.109) yields

$$A\,\mathrm{Re}\,(y''(x)) + B\,\mathrm{Re}\,(y'(x)) + C\,\mathrm{Re}\,(y(x)) = R_0 \cos\omega x.$$

Thus, it follows that the real part of the particular solution to equation (3.109), $\mathrm{Re}\,(y_P(x))$, solves the same differential equation as equation (3.109) but with the cosine source term, $\cos\omega x$, instead of the exponential function $\exp(i\omega x)$. Similar argument shows that $\mathrm{Im}\,(y_P(x))$ solves the differential equation of the same type as equation (3.109) but with the sine source, $\sin\omega x$.

Thus, by taking the real part of the particular solution $y_p(x)$ to equation (3.110) we obtain a solution to equation $Ay''(x) + By'(x) + Cy(x) = R_0 \cos(\omega x)$, which is given by:

$$y_p^{\text{re}}(x) \;=\; \text{Re}\,(Ke^{\mathrm{i}\omega x}) = K_1\cos(\omega x) - K_2\sin(\omega x) \tag{3.114}$$

$$=\; R_0\frac{(C-A\omega^2)}{B^2\omega^2+(C-A\omega^2)^2}\cos(\omega x) + R_0\frac{B\omega}{B^2\omega^2+(C-A\omega^2)^2}\sin(\omega x)\,.$$

Similarly, the solution to $Ay''(x)+By'(x)+Cy(x) = R_0\sin(\omega x)$ will be obtained by projecting the particular solution $y_p(x)$ to equation (3.110) on its imaginary part:

$$y_p^{\text{im}}(x) \;=\; \text{Im}\,(Ke^{\mathrm{i}\omega x}) = K_1\sin(\omega x) + K_2\cos(\omega x) \tag{3.115}$$

$$=\; R_0\frac{(C-A\omega^2)}{B^2\omega^2+(C-A\omega^2)^2}\sin(\omega x) - R_0\frac{B\omega}{B^2\omega^2+(C-A\omega^2)^2}\cos(\omega x)\,.$$

☞ **Example 3.2.9.** Consider $y''+y'-y = 10\cos x$. The corresponding complex equation is $y''+y'-y = 10\exp(\mathrm{i}x)$, with the particular solution $y_p(x) = k\exp(\mathrm{i}x)$. Inserting this solution back into the complex equation $y''+y'-y = 10\exp(\mathrm{i}x)$ gives $(-1+\mathrm{i}-1)k\exp(\mathrm{i}x) = 10\exp(\mathrm{i}x)$. Thus, the constant k is

$$k = \frac{10}{-2+\mathrm{i}} = \frac{10}{-2+\mathrm{i}}\,\frac{-2-\mathrm{i}}{-2-\mathrm{i}} = -4 - 2\mathrm{i}\,.$$

Correspondingly, the particular solution to the original equation $y''+y'-y = 10\cos x$ is given by

$$y_p^{\text{re}}(x) = \text{Re}\,\big((-4-2\mathrm{i}\,)e^{\mathrm{i}x}\big) = -4\cos(x) + 2\sin(x)\,.$$

The complex function $y_p(x)$ also contains a solution to equation $y''+y'-y = 10\sin x$. This time the complex method instructs us to take the imaginary part of $y_p(x)$:

$$y_p^{\text{im}}(x) = \text{Im}\,\big((-4-2\mathrm{i}\,)e^{\mathrm{i}x}\big) = -2\cos(x) - 4\sin(x)\,.$$

3.2.8 Complex Method for the Forced Oscillations

Consider mass m on an elastic string with an elastic constant k, as the one shown in Figure 3.6. The friction proportional to the speed is represented by the damping

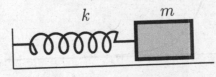

Figure 3.6: Mass m on an elastic string with the spring constant k.

force. In addition, there is a periodic driving force $F = F_0 \cos(\omega t)$ acting on mass m, where t is time. The second-order Newtonian equation of motion gives:

$$m\ddot{y}(t) = F_0 \cos(\omega t) - c\dot{y}(t) - ky(t),$$ (3.116)

where term $-ky(t)$ is the spring force and $-c\dot{y}(t)$ is the damping force. The constants k and c are called the spring constant and the damping coefficient, respectively. By moving the damping force and the spring force terms to the left hand side of equation (3.116) we obtain a conventional form of equation $Ay''(x) + By'(x) + Cy(x) = R_0 \cos(\omega x)$, namely

$$m\ddot{y}(t) + c\dot{y}(t) + ky(t) = F_0 \cos(\omega t).$$ (3.117)

It describes the motion of a mechanical system subject to a periodic elastic spring force and a frictional damping force. Let us first look for the homogeneous solutions to eq. (3.117). These solutions have the exponential form $\exp(\lambda t)$ with constant λ, determined by the characteristic equation:

$$m\lambda^2 + c\lambda + k = 0, \quad \lambda_{1,2} = \frac{-c}{2m} \pm \sqrt{\frac{c^2}{4m^2} - \frac{k}{m}}.$$ (3.118)

Solutions to the above characteristic equation fall into three categories.

① the *over-damped* case:

$$\frac{c^2}{4m^2} - \frac{k}{m} > 0.$$

The roots λ_1, λ_2 from (3.118) are both real and negative and the homogeneous solution:

$$y_h(t) = A_1 e^{\lambda_1 t} + A_2 e^{\lambda_2 t}$$

goes to zero for $t \to \infty$ without oscillations as shown in Figure 3.7.

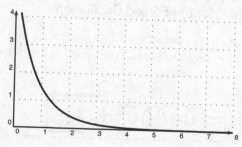

Figure 3.7: The *over-damped* case.

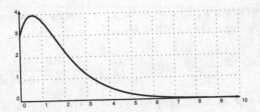

Figure 3.8: The *critical* case.

② the *critically damped* case:

$$\frac{c^2}{4m} = k$$

with double root $\lambda = \lambda_1 = \lambda_2 = -c/2m$. The homogeneous solution

$$y_h(t) = A_1 e^{-ct/2m} + A_2 t e^{-ct/2m}$$

will converge to zero for large t as shown in Figure 3.8.

③ the *under-damped* case:

$$\frac{c^2}{4m^2} - \frac{k}{m} < 0.$$

Then $\lambda_{1,2} = -c/2m \pm \mathrm{i}\Omega$ with $\Omega = \sqrt{k/m - c^2/4m^2}$. The homogeneous solution shown in Figure 3.9 is given by:

$$y_h(t) = A_1 e^{-ct/2m} \cos(\Omega t) + A_2 e^{-ct/2m} \sin(\Omega t)$$

and describes damped oscillations.

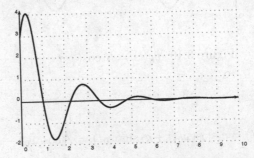

Figure 3.9: The *under-damped* case.

The particular solution y_p follows from y_p^{re} given in equation (3.114) upon substituting $R_0 \to F_0, A \to m, B \to c, C \to k$ and $x \to t$:

$$y_p(t) = F_0 \frac{(k - m\omega^2)}{c^2\omega^2 + (k - m\omega^2)^2} \cos(\omega t) + F_0 \frac{c\omega}{c^2\omega^2 + (k - m\omega^2)^2} \sin(\omega t). \quad (3.119)$$

In the limit $c \to 0$ equation (3.117) becomes $m\ddot{y}(t) + ky(t) = F_0 \cos(\omega t)$. After dividing by m and introducing the natural frequency of the spring:

$$\omega_0 = \sqrt{\frac{k}{m}}. \quad (3.120)$$

we arrive at equation

$$\ddot{y}(t) + \omega_0^2 y(t) = \frac{F_0}{m} \cos(\omega t). \quad (3.121)$$

Assigning $c = 0$ in formula (3.119), simplifies the particular solution to

$$y_p(t) = \frac{F_0}{m(\omega_0^2 - \omega^2)} \cos(\omega t), \quad (3.122)$$

which solves equation (3.121) that describes friction free undamped oscillations. The general solution is obtained by adding the homogeneous solution

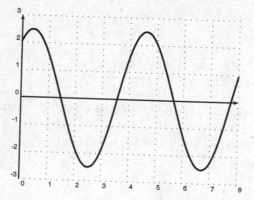

Figure 3.10: $y_h(t) = K \cos(\omega_0 t - \delta)$.

$$y_h(t) = c_1 \cos(\omega_0 t) + c_2 \sin(\omega_0 t) = A \cos(\omega_0 t - \delta)$$

of $\ddot{y}(t) + \omega_0^2 y(t) = 0$ to the particular solution $y_p(t)$. This yields

$$y(t) = A \cos(\omega_0 t - \delta) + \frac{F_0}{m(\omega_0^2 - \omega^2)} \cos(\omega t). \quad (3.123)$$

Here A and phase δ are two constants that can be found by solving the initial conditions. They are related to constants c_1, c_2 in $y_h(t) = c_1 \cos(\omega_0 t) + c_2 \sin(\omega_0 t)$ by $c_1 = A \cos \delta$ and $c_2 = A \sin \delta$.

For the initial conditions $y(0) = y'(0) = 0$, the solution $y(t)$ from (3.123) becomes

$$y(t) = \frac{F_0}{m(\omega_0^2 - \omega^2)} (\cos(\omega t) - \cos(\omega_0 t)) . \tag{3.124}$$

Using the trigonometric identity, $2 \sin \alpha \sin \beta = \cos(\alpha - \beta) - \cos(\alpha + \beta)$, we can

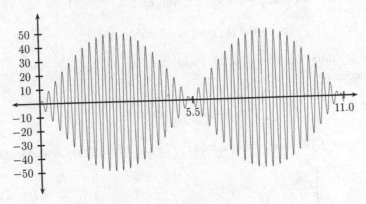

Figure 3.11: Beats resulting from superposition of two waves with slightly different frequencies.

rewrite equation (3.124) as

$$y(t) = \frac{2F_0}{m(\omega_0^2 - \omega^2)} \sin\left(\frac{\omega_0 + \omega}{2} t\right) \sin\left(\frac{\omega_0 - \omega_0}{2} t\right) . \tag{3.125}$$

As ω approaches ω_0 the amplitude, $F_0/m(\omega_0^2 - \omega^2)$, goes to infinity. This phenomenon is called a resonance. For a system approaching resonance, $\omega \sim \omega_0$, the expression, $(\omega_0 - \omega)/2$, becomes small as compared to the much larger expression, $(\omega_0 + \omega)/2$. Thus, the term, $\sin\left(\frac{\omega_0 + \omega}{2} t\right)$, undergoes rapid oscillations (small period) while the term, $\sin\left(\frac{\omega_0 - \omega}{2} t\right)$, involves much slower oscillations (large period). Effectively, the oscillations have a slowly changing amplitude given by

$$\frac{2F_0}{m(\omega_0^2 - \omega^2)} \sin\left(\frac{\omega_0 - \omega_0}{2} t\right) , \tag{3.126}$$

with rapid oscillations due to the term, $\sin\left(\frac{\omega_0 + \omega}{2} t\right)$. This behavior gives rise to the beats shown in Figure 3.11. Due to formula (3.123), the oscillation $y(t)$ in equation (3.125) can be viewed as an superposition of two vibrations with slightly different frequencies, ω and ω_0.

Note that expressions (3.122) and (3.123) are only valid for $\omega \neq \omega_0$. To have a closer look at the resonance situation with $\omega = \omega_0$ we consider equation (3.117) without damping, $c = 0$, and with the driving force oscillating with frequency ω equal to system's natural frequency ω_0:

$$\ddot{y}(t) + \omega_0^2 y(t) = \frac{F_0}{m} \cos(\omega_0 t) . \tag{3.127}$$

In the case of resonance, the particular solution coincides with the homogeneous solution and we need to apply the modification rule. According to the modification rule, the proposed particular solution is given by

$$y_p(t) = t\left(k_1 \cos(\omega_0 t) + k_2 \sin(\omega_0 t)\right) . \tag{3.128}$$

Plugging $y_p(t)$ into equation (3.127), yields $k_1 = 0$ and $k_2 = F_0/2m\omega_0$. Thus, the solution becomes:

$$y_p(t) = t\frac{F_0}{2m\omega_0} \sin(\omega_0 t) , \tag{3.129}$$

shown in Figure 3.12. The amplitude, $tF_0/2m\omega_0$, of these oscillations grows with

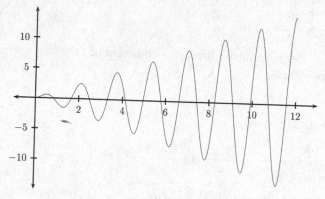

Figure 3.12: $y_p(t) = t\frac{F_0}{2m\omega_0} \sin(\omega_0 t)$.

time, which illustrates the concept of resonance.

For the case with non-zero damping coefficient $c \neq 0$ in equation (3.117), the method of undetermined coefficients leads to a solution of the form,

$$y_p(t) = A_1 \cos(\omega t) + A_2 \sin(\omega t) . \tag{3.130}$$

After inserting (3.130) back into the equation of motion (3.117) one finds that

$$y_p(t) = \frac{F_0}{m^2(\omega_0^2 - \omega^2)^2 + c^2\omega^2} \left(m(\omega_0^2 - \omega^2) \cos(\omega t) + c\omega \sin(\omega t)\right) . \tag{3.131}$$

It is interesting to note that even a small but non-zero damping coefficient c prevents solution $y_p(t)$ from going to infinity when the resonance condition $\omega_0 = \omega$ is met.

3.2.9 Complex Method for Electric Circuits

Consider the LRC circuit from Figure 3.13, which has a periodic sinusoidal EMF $E(t) = E_0 \sin(\omega t)$ applied to it. Ohm's law reads

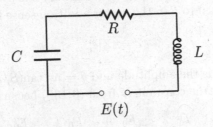

Figure 3.13: *LRC*-circuit

$$LI' + RI + \frac{1}{C} \int I \, dt = E_0 \sin(\omega t).$$ (3.132)

Differentiating the above equation with respect to time yields:

$$LI'' + RI' + \frac{1}{C} I = E_0 \omega \cos(\omega t).$$ (3.133)

Let's consider the complex version of Ohm's law (3.133):

$$LI'' + RI' + \frac{1}{C} I = E_0 \omega e^{i\omega t}.$$ (3.134)

Taking the real part of equation (3.134) we obtain

$$L \operatorname{Re}(I'') + R \operatorname{Re}(I') + \frac{1}{C} \operatorname{Re}(I) = E_0 \omega \cos(\omega t),$$

which shows that the real part $\operatorname{Re}(I)$ of the solution to equation (3.134) solves the original equation (3.133).

Next, we search for the particular solution $I_p(t) = K \exp(i\omega t)$ to the complex version (3.134) of Ohm's law. Inserting $I_p(t)$ into equation (3.134) gives

$$\left(-\omega^2 + i\omega R + \frac{1}{C} \right) K e^{i\omega t} = E_0 \omega e^{i\omega t},$$ (3.135)

which yields an amplitude, K:

$$K = \frac{E_0}{-\left(\omega L - \frac{1}{\omega C} \right) + i R}.$$ (3.136)

Let us introduce the reactance $S = \omega L - 1/\omega C$ and the complex impedance

$$Z = R + i\left(\omega L - \frac{1}{\omega C}\right) = R + iS. \tag{3.137}$$

It is convenient to represent Z graphically on the complex plane. In Figure 3.14, we consider the case with $\omega L > 1/\omega C$. Subtracting $1/\omega C$ from ωL along the imaginary axis yields reactance S pointing along the imaginary axis. Adding the resistance R (that lies on the real axis) to S in the vector addition sense leads to an expression:

$$Z = |Z|e^{i\theta}, \tag{3.138}$$

where $|Z| = \sqrt{R^2 + S^2}$ is the amplitude and $\theta = \arctan(S/R)$ the phase. In terms of these new quantities, the constant K from (3.136) becomes

$$K = \frac{E_0}{-S + iR} = \frac{E_0}{iZ} = -i\frac{E_0}{Z} = -i\frac{E_0}{|Z|}e^{-i\theta}. \tag{3.139}$$

The projection on the real part of $I_p(t) = K\exp(i\omega t)$ is the physical current we

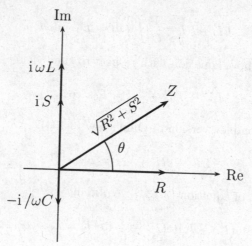

Figure 3.14: Graphical representation of the impedance Z.

observe. It is given by

$$I_p^{\mathrm{re}} = \mathrm{Re}\left(Ke^{i\omega t}\right) = \mathrm{Re}\left(-i\frac{E_0}{|Z|}e^{-i\theta}e^{i\omega t}\right) = \mathrm{Re}\left(-i\frac{E_0}{|Z|}e^{i(\omega t-\theta)}\right)$$

$$= \mathrm{Re}\left(-i\frac{E_0}{|Z|}(\cos(\omega t-\theta) + i\sin(\omega t-\theta))\right) = \frac{E_0}{\sqrt{R^2+S^2}}\sin(\omega t-\theta). \tag{3.140}$$

☞ **Example 3.2.10.** In the case of a pure R-circuit, the reactance S is zero and, therefore, $|Z| = R$ and $\theta = 0$. The current

$$I_p^{re} = \frac{E_0}{R} \sin(\omega t)$$

is in phase with the voltage, $E_0 \sin(\omega t)$.

In the case of a circuit that only contains an inductance L, the reactance is $S = \omega L$ and the complex impedance is $Z = i\omega L$. Thus, the current

$$I_p^{re} = \mathrm{Re}\left(-i\frac{E_0}{Z}e^{i\omega t}\right) = -\frac{E_0}{\omega L}\cos(\omega t) = \frac{E_0}{\omega L}\sin\left(\omega t - \frac{\pi}{2}\right)$$

lags behind the voltage by a phase $\pi/2$.

Finally, for the C-circuit, the complex impedance is $Z = -i/\omega C$ and the current

$$I_p^{re} = \mathrm{Re}\left(-i\frac{E_0}{Z}e^{i\omega t}\right) = E_0\omega C \cos(\omega t) = E_0\omega C \sin\left(\omega t + \frac{\pi}{2}\right)$$

leads the voltage by a phase $\pi/2$.

⚂ **Second Order Differential Equations: Problems**

Problem 3.2.1. Show that the Wronskian associated to the equation, $y'' + q(x)y = 0$, is a constant.

Problem 3.2.2. Show that two solutions of $y'' + p(x)y' + q(x)y = 0$ that are zero at the same point cannot form a basis of solutions.

Problem 3.2.3. Use relation (3.91) to find the differential equation that has e^{2x}, e^{-3x} as solutions.

Problem 3.2.4. Show that the linear first-order differential equation $y'(x) + p(x)y(x) = 0$ can be expressed in terms of the Wronskian as:

$$\frac{W[y_0, y](x)}{y_0(x)} = 0,$$

where $y_0(x)$ is a solution. Using the basic properties of a determinant, show that $y(x) = y_0(x)\int^x q(t)/y_0(t)\,dt$ is a solution to:

$$\frac{W[y_0, y](x)}{y_0(x)} = q(x).$$

Problem 3.2.5. Three functions $f_1(x)$, $f_2(x)$ and $f_3(x)$ are called linearly independent if the equation

$$k_1 f_1(x) + k_2 f_2(x) + k_3 f_3(x) = 0$$

only has the trivial (zero) solution for constant parameters k_1, k_2, k_3.

By means of the result obtained in equation (3.91), show that $y'' + p(x)y' + q(x)y = 0$ cannot have three independent solutions.

Problem 3.2.6. Solve the following system of equations:

$$9y'' + 42y' + 49y = 0, \quad y(0) = 1, \quad y'(0) = -2.$$

Problem 3.2.7. Solve the following system of equations:

$$4y'' + 24y' + 37y = 0, \quad y(0) = 0, \quad y'(0) = 1.$$

Problem 3.2.8. Consider a spring/mass system damped by a frictional force with an equation of motion described by the following initial value problem:

$$\frac{1}{3}y'' + 2y' + 3y = 0, \quad y(0) = 1, \quad y'(0) = 3.$$

Find the maximum displacement of the mass.

Problem 3.2.9. Equation

$$x^2 y'' + xy' + (x^2 - \frac{1}{4})y = 0$$

has the solution $y_1 = x^{-1/2} \sin x$. Find the second independent solution y_2 and show that y_1 and y_2 are linearly independent.

Problem 3.2.10. Equation

$$(x^2 - 1)y'' - 2xy' + 2y = 0$$

has a solution $y_1 = x$. Find the second independent solution y_2.

Problem 3.2.11. Equation

$$(x + 1)y'' - (x - 1)y' - y = 0$$

has a solution $y_1 = (x + 1)^{-1}$. Find the second independent solution y_2.

Problem 3.2.12. The second-order homogeneous differential equation has a solution $y_1 = (1 + x)^{-1}$. Find the second solution y_2, such that the Wronskian $W[y_1, y_2]$ is a constant. Find the corresponding differential equation.

Problem 3.2.13. Two functions $-2 + \exp(-2x)$ and $-2 + 3\exp(-x)$ are found to be solutions to $y'' + by' + cy = r(x)$, where b and c are constants. Find b, c and the function $r(x)$.

Problem 3.2.14. Verify, in each case, that y_p is a solution to the given differential equations and find the general solution:

$$y'' - y' - 2y = 3e^{2x}, \quad y_p = x\,e^{2x}$$

$$y'' + 6y' + 9y = -5\sin x, \quad y_p = -\frac{2}{5}\sin x + \frac{3}{10}\cos x.$$

Problem 3.2.15. Show that if y_1 is a solution to $y'' + p(x)y' + q(x)y = r_1(x)$ and y_2 is a solution to $y'' + p(x)y' + q(x)y = r_2(x)$, then $y_1 + y_2$ is a solution to $y'' + p(x)y' + q(x)y = r_1(x) + r_2$.

Problem 3.2.16. Given that one solution to

$$y'' + \frac{1}{x}y' - \frac{n^2}{x^2}y = 0$$

is $y(x) = x^n$, find the second solution. Using this result, find a particular solution to the non-homogeneous equation,

$$x^2 y'' + xy' - y = \frac{1}{1-x}.$$

Problem 3.2.17. Let $y_1(x) = \exp(x^2)$ and $y_2(x) = x^2 + 1$ be solutions to $y'' + p(x)y' + q(x)y = 0$ for $x > 0$.
(a) Find the Wronskian of y_1, y_2 and use it to verify that that y_1, y_2 are linearly independent.
(b) Find coefficients $p(x)$ and $q(x)$.
(c) For the source function, $r(x) = x^4(1 + x^2)^{-1}$, find the particular solution to $y'' + p(x)y' + q(x)y = r(x)$.

Problem 3.2.18. Find a general solution to

$$y''(x) + 3y'(x) - 4y(x) = x^2 - 2e^{3x} + 10\cos(2x).$$

Problem 3.2.19. Find a general solution to

$$y''(x) - 4y'(x) + 3y(x) = xe^{2x} + e^{3x}.$$

Problem 3.2.20. Find unique solutions to the differential equations,
(a) $y''(x) - 2y'(x) + 5y(x) = e^x \sin x$
(b) $y''(x) - 2y'(x) + y(x) = 14xe^x$
with the initial conditions $y(0) = 1$, $y'(0) = 0$.

Problem 3.2.21. Use formula (3.104) to find a general solution to the differential equation

$$y'' + 8y' - 9y = xe^x.$$

Problem 3.2.22. Use formula (3.104) to find the general solution to the differential equation $2y'' - 3y' + y = r(x)$, with
(a) $r(x) = xe^{-x}$.
(b) $r(x) = \cos(x)e^x$.

Problem 3.2.23. Use the formula (3.104) to find a general solution to the differential equation

$$y'' + 3y' + 2y = \cos(e^x).$$

Problem 3.2.24. Solve the differential equation $y'' + 4y' + 4y = r(x)$, for
(a) $r(x) = e^{-2x}$ with initial conditions $y(0) = 1$ and $y'(0) = 0$.
(b) $r(x) = 2xe^{-2x}$.

Problem 3.2.25. Solve:

$$y''(x) + 4y(x) = \begin{cases} x^2 & \text{if} \quad x \le 0 \\ 0 & \text{if} \quad x > 0, \end{cases}$$

with initial conditions $y(0) = 1, y'(0) = 0$.
   Hint: At $x = 0$, both y and y' must be continuous.

Problem 3.2.26. Find a particular solution to:

$$y'' + 4y' + \frac{17}{4}y = 15\sin(t/2)$$

and

$$y'' + 4y' + \frac{17}{4}y = 15\cos(t/2).$$

Hint: First, solve the complex equation $y'' + 4y' + (17/4)y = 15e^{it/2}$. Then project it onto the real and imaginary parts.

Problem 3.2.27. (a) Use the complex method to find a particular solution of $y'' + 4y' + 13y = \cos x$.
(b) Solve the system given by the differential equation $y'' + 4y' + 13y = \cos x$ with initial conditions $y(0) = 0, y'(0) = 1$.

Problem 3.2.28. Find solution to each of the differential equations

(a) $y'' - 6y' + 9y = 4\sin(3x)$
(b) $y'' + 9y = 4\sin(3x)$

with initial conditions $y(0) = 0, y'(0) = 0$.

Chapter 4

Series Solutions of Differential Equations

FOR differential equations with non-constant coefficients, that often arise in physical and engineering applications, it is often necessary to obtain solutions in the form of series expansions.

In solving the partial differential equations, like Laplace's and Helmholtz's equations by method of separation of variables certain ordinary differential equations emerge with solutions that are special functions in mathematical physics. Here we shall study Legendre polynomials which are solutions of Legendre's equation obtained by studying power series expansions around the regular point and Bessel functions derived from series expansions around the singular point of Bessel's equation.

4.1   Introduction to the Power Series Method

We begin with some definitions.

Definition 4.1.1. *A power series in x, centered at the point x_0, is an infinite series of the form:*

$$\sum_{n=0}^{\infty} a_n(x - x_0)^n = a_0 + a_1(x - x_0) + a_2(x - x_0)^2 + \cdots \qquad (4.1)$$

with constant coefficients a_n.

Definition 4.1.2. *A finite sum $S_N = \sum_{n=0}^{N} a_n(x - x_0)^n$ is said to be a partial sum of the power series (4.1).*

The concept of a partial sum of a power series is used to define the convergence of a power series. The power series (4.1) is said to converge at $x = x_0$ if $\lim_{N \to \infty} S_N$ exists at $x = x_0$ for the sequence $S_n, n = 0, 1, 2. \ldots$. At points where the series converges the series sum is defined as the limit of the partial sums, $S(x) = \lim_{N \to \infty} S_N$.

The effectiveness of the power series method rests on the expansion a function $f(x)$ with derivatives defined , at all orders, in an interval around x_0 in Taylor power series expansion $f(x) = \sum_{n=0}^{\infty} a_n (x - x_0)^n$ centered around x_0. The first order Taylor expansion is a linear approximation of a function in the neighborhood of a given point by an equation of a tangent line at that point. The first order Taylor expansion is illustrated in Figure 4.1. The Figure shows the result of a graphical

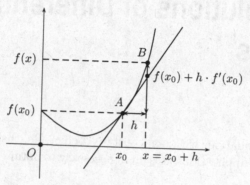

Figure 4.1: Linear approximation of a function.

approximation of $f(x_0 + h)$ by $f(x_0) + h f'(x_0)$, for small values of $h = x - x_0$. Since the derivative of f at $x = x_0$ is given by

$$f'(x_0) = \lim_{h \to 0} \frac{f(x_0 + h) - f(x_0)}{h}$$

then for small h,

$$f'(x_0) = \frac{f(x_0 + h) - f(x_0)}{h} + O(h) ,$$

where the term $O(h)$ encompasses all the terms that go to zero as $h \to 0$. Therefore, it is possible to factor h out and write $O(h) = h \times X$ with some unknown factor X. Multiplying the derivative by h we obtain

$$h f'(x_0) = f(x_0 + h) - f(x_0) + h O(h)$$

or

$$f(x_0 + h) = f(x_0) + h f'(x_0) + h^2 \times X = f(x_0) + h f'(x_0) + O(h^2) ,$$

where the term $O(h^2)$ stands for all the terms of the form $h^2 \times X$, which go to zero as fast as h^2.

Continuing this process leads to the Taylor approximation of the n-th order:

$$f(x) = f(x_0 + h) = f(x_0) + hf'(x_0) + \frac{1}{2}h^2 f''(x_0) + \ldots + \frac{1}{n!}f^{(n)}(x_0) + O(h^{n+1}),$$

where

$$f^{(n)}(x_0) = \frac{d^n f}{dx^n}\Big|_{x=x_0}$$

and the remainder $O(h^{n+1})$ is of the order $n+1$ in h. As n increases the remainder gets smaller for small h and the Taylor approximation becomes more accurate.

Eventually, by taking limit $n \to \infty$ we arrive at a Taylor series, which expresses the function $f(x)$ by the power series in $x - x_0$:

$$\begin{aligned}
f(x) &= f(x_0) + \frac{f'(x_0)}{1!}(x - x_0) + \frac{f''(x_0)}{2!}(x - x_0)^2 + \ldots \\
&+ \frac{f^{(n)}(x_0)}{n!}(x - x_0)^n + \ldots = \sum_{n=0}^{\infty} \frac{f^{(n)}(x_0)}{n!}(x - x_0)^n .
\end{aligned} \tag{4.2}$$

This power series is valid for the values of x for which the series converges. Typically, this is the case in the vicinity of x_0, meaning that equation $|x - x_0| < R$ holds for the x for which the power series is convergent for some positive constant R. That constant R is called a radius of convergence of the power series. We have the following definition.

Definition 4.1.3. *The positive constant R is said to be a radius of convergence of the power series (4.1) if the power series converges for all x such that $|x - x_0| < R$. All x such that $|x - x_0| < R$ define a convergence interval.*

In order to define values of the variable x for which the series converges we must find its radius of convergence, R. There exists a number of methods for finding the radius of convergence R for the given power series (4.1). We will be using one of two main formulas, namely:

$$R = \frac{1}{\lim_{n \to \infty} \left|\frac{a_{n+1}}{a_n}\right|} \quad \text{or} \quad R = \frac{1}{\lim_{n \to \infty} \sqrt[n]{|a_n|}} \tag{4.3}$$

in order to determine R.

☞ **Example 4.1.1.** For the power series, $\sum_{n=0}^{\infty} \frac{(-1)^n}{n+3}(x - 3)^n$, the radius of convergence is:

$$\frac{1}{R} = \lim_{n \to \infty} \left|\frac{a_{n+1}}{a_n}\right| = \lim_{n \to \infty} \frac{n+3}{n+4} = 1 \quad \to \quad R = 1 \tag{4.4}$$

and the power series is convergent for x such that $|x - 3| < 1$ or $2 < x < 4$.

☞ **Example 4.1.2.** Some well-known Taylor series representations along with their convergence intervals are:

$$e^x = \sum_{n=0}^{\infty} \frac{x^n}{n!} = 1 + x + \frac{x^2}{2!} + \frac{x^3}{3!} + \cdots, \text{ for } -\infty < x < \infty \quad (4.5a)$$

$$\sin(x) = \sum_{n=0}^{\infty} (-1)^n \frac{x^{2n+1}}{(2n+1)!} = x - \frac{x^3}{3!} + \frac{x^5}{5!} - \cdots, \quad (4.5b)$$
$$\text{for } -\infty < x < \infty$$

$$\cos(x) = \sum_{n=0}^{\infty} (-1)^n \frac{x^{2n}}{(2n)!} = 1 - \frac{x^2}{2!} + \frac{x^4}{4!} - \cdots, \quad (4.5c)$$
$$\text{for } -\infty < x < \infty$$

$$\sinh(x) = \sum_{n=0}^{\infty} \frac{x^{2n+1}}{(2n+1)!} \text{ for } -\infty < x < \infty \quad (4.5d)$$

$$\cosh(x) = \sum_{n=0}^{\infty} \frac{x^{2n}}{(2n)!} \text{ for } -\infty < x < \infty \quad (4.5e)$$

$$\ln(1+x) = \sum_{n=0}^{\infty} (-1)^n \frac{1}{n+1} x^{n+1} = x - \frac{x^2}{2} + \frac{x^3}{3} - \cdots, \quad (4.5f)$$
$$\text{for } -1 < x \leq 1.$$

All the Taylor series in (4.5a)-(4.5e) have infinite convergence intervals. They define the functions on the right hand sides of (4.5a)-(4.5e) for all values of x.

☞ **Example 4.1.3.** In this example we will apply a Taylor expansion to the function $f(x) = (1+x)^r$ with some constant r. The first few derivatives of f are

$$f'(x) = r(1+x)^{r-1},$$
$$f''(x) = r(r-1)(1+x)^{r-2},$$
$$f'''(x) = r(r-1)(r-2)(1+x)^{r-3}\ldots.$$

A derivative of f of arbitrary order is

$$f^{(k)}(x) = r(r-1)\cdots(r-k+1)(1+x)^{r-k}.$$

If the power r is a positive integer n then

$$f^{(k)}(x) = \frac{n!}{(n-k)!}(1+x)^{n-k}, \ k \leq n,$$

and $f^{(k)}(x) = 0$ for $k = n+1, n+2, \ldots$. Thus, the Taylor expansion $f(x)$

around $x_0 = 0$ becomes

$$(1+x)^n = \sum_{k=0}^{n} \frac{1}{k!} \frac{n!}{(n-k)!} x^k$$

$$= 1 + nx + \binom{n}{2} x^2 + \cdots + \binom{n}{k} x^k + \cdots + x^n,$$

which is a binomial expansion.

For r, which is a non-positive integer, the Taylor series expansion for $f(x)$ is

$$f(x) = (1+x)^r = 1 + rx + \frac{r(r-1)}{2!} x^2 + \frac{r(r-1)(r-2)}{3!} x^3 + \cdots$$

$$= \sum_{k=0}^{\infty} \frac{1}{k!} r(r-1) \cdots (r-k+1) x^k$$

and becomes a power series expansion with convergence radius, $R = 1$.

Accordingly, for $r = -1/2$, the Taylor expansion around zero is

$$\frac{1}{\sqrt{1+x}} = 1 - \frac{1}{2} x + \frac{1}{2!} \left(-\frac{1}{2}\right) \left(-\frac{3}{2}\right) x^2$$

$$+ \frac{1}{3!} \left(-\frac{1}{2}\right) \left(-\frac{3}{2}\right) \left(-\frac{5}{2}\right) x^3 + \cdots = \sum_{k=0}^{\infty} \left(-\frac{1}{2}\right)^k \frac{(2k-1)!!}{k!} x^k,$$

where the double factorial (noted by two exclamation marks) of the odd number $2k - 1$ is defined as the product of all odd integers ranging from 1 to a number $2k - 1$:

$$(2k-1)!! = 1 \cdot 3 \cdot 5 \cdots (2k-1).$$

Similarly, for $k \geq 1$,

$$(2k)!! = 2 \cdot 4 \cdots 6 \cdots 2k.$$

For $r = 1/2$, the Taylor expansion around zero is

$$\sqrt{1+x} = 1 + \frac{1}{2} x + \frac{1}{2!} \left(\frac{1}{2}\right) \left(-\frac{1}{2}\right) x^2 + \frac{1}{3!} \left(\frac{1}{2}\right) \left(-\frac{1}{2}\right) \left(-\frac{3}{2}\right) x^3 + \cdots$$

$$= 1 - \sum_{k=1}^{\infty} \left(-\frac{1}{2}\right)^k \frac{(2k-3)!!}{k!} x^k,$$

For $r = -1$, the Taylor expansion around zero is

$$\frac{1}{1+x} = \sum_{n=0}^{\infty} (-x)^n = 1 - x + x^2 - x^3 + \cdots.$$

Reversing the sign in front of x gives the geometric series

$$\frac{1}{1-x} = \sum_{n=0}^{\infty} x^n = 1 + x + x^2 + x^3 + \dots.$$

Definition 4.1.4. *A function $f(x)$ is said to be analytic at a point $x = x_0$ if it admits an expansion of a power series in $x - x_0$ with a positive radius of convergence.*

The functions $\exp(x)$, $\sin(x)$, $\cos(x)$, $\sinh(x)$, $\cosh(x)$ and $(1+x)^r$ seen in the above examples, are analytic at $x = 0$.

Some fundamental properties of power series are:

1. $\sum_{n=k}^{\infty} a_n(x-x_0)^n + \sum_{n=k}^{\infty} b_n(x-x_0)^n = \sum_{n=k}^{\infty} (a_n+b_n)(x-x_0)^n$, for $k \geq 0$.

2. Suppose that $\sum_{n=0}^{\infty} a_n(x - x_0)^n = \sum_{n=0}^{\infty} b_n(x - x_0)^n$ for all $|x - x_0| < R$, where $R > 0$. Then $a_n = b_n$. This implies that if $\sum_{n=0}^{\infty} a_n(x-x_0)^n = 0$ then $a_n = 0$, which easily follows from the fact that $(x - x_0)^n$ for $n = 0, 1, 2, \dots$, define linearly independent polynomials.

3. We can differentiate a power series term by term within the radius of convergence. If $y(x) = \sum_{n=0}^{\infty} a_n(x - x_0)^n$ then $y'(x) = a_1 + 2a_2(x - x_0) + \dots = \sum_{n=1}^{\infty} na_n(x - x_0)^{n-1}$ and $y''(x) = 2a_2 + 6a_3(x - x_0) + \dots = \sum_{n=2}^{\infty} n(n - 1)a_n(x - x_0)^{n-2}$.

4. We can shift the dummy (summation) variable without changing the power series. For example,

$$y''(x) = \sum_{n=2}^{\infty} n(n - 1)a_n(x - x_0)^{n-2} = \sum_{n=0}^{\infty} (n + 2)(n + 1)a_{n+2}(x - x_0)^n.$$

4.1.1   Power Series Method, a Warm up Example

Now, we will illustrate the power series method by the following basic example. Let us search for solutions to the equation

$$y''(x) - y(x) = 0 \tag{4.6}$$

in the form of the following power series:

$$y(x) = \sum_{i=0}^{\infty} a_i x^i. \tag{4.7}$$

By inserting the right hand side of expression (4.7) into equation (4.6) we obtain:

$$\sum_{i=2}^{\infty} i(i-1)a_i x^{i-2} - \sum_{i=0}^{\infty} a_i x^i = \sum_{i=0}^{\infty} [(i+2)(i+1)a_{i+2} - a_i]x^i = 0. \qquad (4.8)$$

For this identity to hold, the coefficient of each term x^i for all $i \geq 0$ must be identically equal to zero. In this way, we obtain the *recurrence relation*, valid for coefficients a_i:

$$(i+2)(i+1)a_{i+2} = a_i, \quad i = 0, 1, 2, \ldots. \qquad (4.9)$$

The above relation recursively determines all the higher coefficients of the power series (4.7) in terms of the first two coefficients, a_0 and a_1. Since these coefficients are determined by initial conditions, $y(0) = a_0$ and $y'(0) = a_1$, the solution is entirely determined by the initial conditions through the recurrence relation (4.9). For $a_0 \neq 0$, we easily deduce from the recurrence relations, that

$$\frac{a_{2n}}{a_0} = \frac{a_{2n}}{a_{2n-2}} \frac{a_{2n-2}}{a_{2n-4}} \cdots \frac{a_2}{a_0} = \frac{1}{2n(2n-1)} \frac{1}{(2n-2)(2n-3)} \cdots \frac{1}{2 \cdot 1} = \frac{1}{(2n)!} \qquad (4.10)$$

and, therefore, all coefficients with even indices are:

$$a_{2n} = \frac{a_0}{(2n)!}. \qquad (4.11)$$

The above result remains valid for all values of a_0. Similarly, for $a_1 \neq 0$ we find that

$$\frac{a_{2n+1}}{a_1} = \frac{a_{2n+1}}{a_{2n-1}} \frac{a_{2n-1}}{a_{2n-3}} \cdots \frac{a_3}{a_1} = \frac{1}{(2n+1)(2n)} \frac{1}{(2n-1)(2n-2)} \cdots \frac{1}{3 \cdot 2} = \frac{1}{(2n+1)!}$$

and, consequently, all coefficients with odd indices are:

$$a_{2n+1} = \frac{a_1}{(2n+1)!}. \qquad (4.12)$$

Again, the above result remains generally valid for all values of a_1.

Therefore, the general power series solution to equation (4.6) is:

$$y(x) = a_0 \sum_{n=0}^{\infty} \frac{x^{2n}}{(2n)!} + a_1 \sum_{n=0}^{\infty} \frac{x^{2n+1}}{(2n+1)!} = a_0 \cosh x + a_1 \sinh x$$

$$= \frac{a_0 + a_1}{2} e^x + \frac{a_0 - a_1}{2} e^{-x}, \qquad (4.13)$$

in which we recognize the well-known exponential solutions to equation (4.6).

4.2   Power Series Method, Expansion around a Regular Point

In this section, we consider homogeneous differential equations of the second order:

$$R(x)y''(x) + P(x)y'(x) + Q(x)y(x) = 0, \tag{4.14}$$

with coefficients $R(x), P(x)$, and $Q(x)$ as functions of x. For $R(x) \neq 0$, equation (4.14) may be rewritten in a more conventional form as,

$$y''(x) + p(x)y'(x) + q(x)y(x) = 0, \tag{4.15}$$

with coefficients $p(x) = P(x)/R(x)$ and $q(x) = Q(x)/R(x)$.

Here, we consider coefficients $p(x)$ and $q(x)$, which can be written in terms of a power series. In general, a point x_0 is called a regular point of equation (4.15) if the coefficient functions $p(x)$ and $q(x)$ are analytic at x_0. Thus,

$$p(x) = \sum_{i=0}^{\infty} p_i(x - x_0)^i, \quad q(x) = \sum_{i=0}^{\infty} q_i(x - x_0)^i, \tag{4.16}$$

with positive radii of convergence, R_p and R_q. If $R_0 = \min(R_p, R_q) > 0$ then both power series in (4.16) are simultaneously convergent for $|x - x_0| < R_0$. A point that is not an ordinary point is said to be a singular point of the equation.

The following theorem states the existence of a power series solution.

> If $p(x)$ and $q(x)$ are analytic at x_0 then there exists a general power series solution to equation (4.15) of the form $y(x) = C_1y_1(x) + C_2y_2(x)$ with linearly independent solutions y_1, y_2, analytic at x_0. Thus, both are of the form $\sum_{i=1}^{\infty} a_i(x - x_0)^i$ with a radius of convergence R equal to the distance from x_0 to the closest singular point.

For equations written in the form of equation (4.14) x_0 is a regular point if $R(x_0) \neq 0$ and $P(x)/R(x), Q(x)/R(x)$ are analytic at $x = x_0$. In this context, the following information is very useful: For a function

$$p(x) = \frac{P(x)}{R(x)},$$

which is a ratio of two polynomial functions $P(x)$ and $R(x)$, it holds that if $R(x_0) \neq 0$, then

- $p(x)$ has a power series expansion around $x = x_0$.

- The radius of convergence of this power series around x_0 is equal to the distance from x_0 to the nearest zero of $R(x)$.

☞ **Example 4.2.1.** Consider the differential equation:

$$(x^2 - 4)\, y'' + (x^2 + 3)\, y' + (x + 2)y = 0. \tag{4.17}$$

Thus $R(x) = (x^2 - 4)$, $P(x) = x^2 + 3$, $Q(x) = x + 2$, $p(x) = \frac{x^2+3}{x^2-4}$ and $q(x) = \frac{x+2}{x^2-4} = \frac{1}{x-2}$. Since $R(x) = 0$ at $x = \pm 2$ the singular points are $x = \pm 2$. The coefficient functions $p(x)$ and $q(x)$ can be expanded as a geometric series around $x = 0$ for $|x| < 2$. Thus, $x = 0$ is a regular point.

4.2.1 Legendre's Differential Equation

Legendre's differential equation has the following form:

$$\boxed{(1 - x^2)y'' - 2xy' + n(n+1)y = 0,} \tag{4.18}$$

where n is a real number. In the present situation, $R(x) = (1 - x^2)$, $P(x) = -2x$ and $Q(x) = n(n+1)$ are all analytic at $x = x_0 = 0$. The singular points of Legendre's differential equation are $x = \pm 1$, where $R(x) = 0$. Hence, there exists a general power series solution to (4.18) of the form $y = \sum_{m=0}^{\infty} a_m x^m$. Then

$$y' = \sum_{m=1}^{\infty} m a_m x^{m-1}, \quad \text{and} \quad y'' = \sum_{m=2}^{\infty} m(m-1) a_m x^{m-2}. \tag{4.19}$$

Substituting y' and y'' into equation (4.18) yields

$$\sum_{m=2}^{\infty} m(m-1) a_m x^{m-2} - \sum_{m=2}^{\infty} m(m-1) a_m x^m - 2 \sum_{m=1}^{\infty} m a_m x^m + n(n+1) \sum_{m=0}^{\infty} a_m x^m = 0. \tag{4.20}$$

Next, we rewrite the linear combination of the power series in (4.20) as one power series $\sum_{m=0}^{\infty} A_m x^m$. We recall that such powers series vanishes if and only if $A_m = 0$ for all powers m of x. Therefore, the method requires that we put equal to zero a sum of all coefficients of x^m, that we collect from equation (4.20). We start with the lowest power, namely the term x^0 in equation (4.20). By splitting all the constant terms from all the other terms in expression (4.20) we obtain an expression for A_0, which reads:

$$(2a_2 + n(n+1)a_0) + (\ldots)\, x^1 + \ldots = 0 \;\; \rightarrow \;\; A_0 = 2a_2 + n(n+1)a_0 = 0. \tag{4.21}$$

Next, we find a coefficient A_1 of x. It is given by:

$$A_1 = 3 \cdot 2a_3 - 2a_1 + n(n+1)a_1 = 0. \tag{4.22}$$

Generally,

$$\begin{aligned} A_k &= (k+2)(k+1)a_{k+2} - k(k-1)a_k - 2ka_k + n(n+1)a_k \\ &= (k+2)(k+1)a_{k+2} + (-k(k-1) - 2k + n(n+1))\,a_k = 0 \quad (4.23) \end{aligned}$$

or

$$(k+2)(k+1)a_{k+2} + (n-k)\,(n+k+1)\,a_k = 0. \qquad (4.24)$$

This gives rise to a recursion formula:

$$a_{k+2} = -\frac{(n-k)(n+k+1)}{(k+2)(k+1)}\,a_k \qquad k = 0,1,2,3,\ldots \qquad (4.25)$$

Note that the above recursion relation links coefficient a_k with a_{k+2}. Thus, the index increases by 2 every time we use relation (4.25). Successive use of the recursion formula (4.25) leads to a string of relations for higher coefficients in terms of the coefficient a_0:

$$\begin{aligned} a_2 &= -\frac{(n+1)n}{2}a_0, \\ a_4 &= -\frac{(n-2)(n+3)}{4\cdot 3}a_2 = \frac{(n-2)(n+3)(n+1)n}{4!}a_0,\ldots, \qquad (4.26) \end{aligned}$$

which determine coefficients with even indices $a_2, a_4, \ldots$ in terms of a_0.

Another string of relations follows from formula (4.25) for coefficients with odd indices $a_3, a_5, \ldots$ that are obtained in terms of a_1:

$$\begin{aligned} a_3 &= -\frac{(n-1)(n+2)}{3!}a_1, \\ a_5 &= -\frac{(n-3)(n+4)}{5\cdot 4}a_3 = \frac{(n-3)(n-1)(n+2)(n+4)}{5!}a_1,\ldots. \qquad (4.27) \end{aligned}$$

Hence, the power series

$$\begin{aligned} y(x) = a_0 &\left[1 - \frac{n(n+1)}{2}x^2 + \frac{(n-2)(n+3)n(n+1)}{4!}x^4 + \ldots\right] \\ &+ a_1\left[x - \frac{(n-1)\,(n+2)}{6}x^3 + \frac{(n-3)(n+4)(n-1)(n+2)}{5!}x^5 + \ldots\right] \qquad (4.28) \end{aligned}$$

splits into two terms:

$$y(x) = a_0 y_1(x) + a_1 y_2(x), \qquad (4.29)$$

where y_1 and y_2 are even and odd functions, respectively, with arbitrary constants a_0 and a_1. Consequently, y_1 and y_2 are two linearly independent solutions of Legendre's equation, written in equation (4.18). The solution $y(x)$ of Legendre's equation is called a Legendre function. It is defined for x within the interval of convergence of $y(x)$. Points $x = 1$ and $x = -1$ are singular and, accordingly, the radius of convergence R of the power series $y(x)$ is equal to the distance from $x_0 = 0$ to $x = 1$

or $x = -1$, which gives $R = 1$. Hence, y_1 and y_2 converge for $-1 < x < 1$ and $y(x)$ is a general solution within this convergence interval.

In physics, Legendre's equation naturally appears in a treatment of Helmholtz's equation when we employ method of separation of variables in spherical coordinates as explained in subsection 4.4.1. Legendre's equation emerges in subsection 4.4.1 in equation (4.132) with an argument $x = \cos\theta$, where the spherical angle θ is between 0 and π. Correspondingly, $x = \cos\theta$ ranges between -1 and 1 and it is of interest, to obtain a solution to Legendre's equation for $|x| = 1$. This requires that the power series y_1 or y_2 reduce to a polynomial of finite degree. This happens when the parameter n in (4.18) is a non-negative integer. In such case, the right hand side of the recurrence relation (4.25) is

$$-a_k \frac{(n-k)(n+k+1)}{(k+2)(k+1)} \tag{4.30}$$

and becomes identically zero for $k = n$ resulting in $a_{n+2} = 0$. The recurrence relation implies that all the higher coefficients must also be zero. Thus, $a_{n+4} = a_{n+6} = \ldots = 0$. Hence, if n is an even integer, the power series y_1 truncates and y_1 becomes a polynomial $P_n(x)$ of degree n. Similarly, if n is an odd integer, the power series y_2 truncates and y_2 becomes a polynomial $P_n(x)$ of degree n. The finite degree polynomials $P_n(x)$ are referred to as Legendre polynomials.

☞ **Example 4.2.2.** For $n = 0$, Legendre's differential equation is

$$(1 - x^2)y'' - 2xy' = 0, \tag{4.31}$$

where the coefficient $a_2 = -\frac{(n+1)n}{2}a_0$ vanishes and subsequently $a_4 = 0$, $a_6 = 0$, ... The solution $y_1(x)$ from equation (4.28) for $n = 0$ is $y_1(x) = P_0(x) = 1$. The remaining power series in (4.28) is:

$$y_2(x) = x + \frac{2}{6}x^3 + \frac{4!}{5!}x^5 + \ldots \tag{4.32}$$

Alternatively, the second solution $y_2(x)$ can be obtained by the method of reduction of order through the formula

$$y_2(x) = y_1(x) \int^x \frac{1}{y_1^2(x')} e^{-\int^{x'} p(x'')\, dx''}\, dx'. \tag{4.33}$$

Plugging $y_1(x) = 1$ and $p(x) = -2x/(1-x^2) = -1/(1+x) + 1/(1-x)$ into the above expression we obtain

$$y_2(x) = \int^x \exp\left(-\int^{x'}\left(-\frac{1}{(1+z)} + \frac{1}{(1-z)}\right) dz\right) dx'$$

$$= \frac{1}{2}\ln\frac{(1+x)}{(1-x)}.$$

The first three terms of the Taylor expansion of this function agree with the terms in (4.32).

For $n = 1$ in equation (4.18), $a_3 = 0$ and subsequently $a_5 = 0$, $a_7 = 0$, ... Thus, $y_2 = a_1 x$, is a Legendre polynomial of degree 1. For $n = 2$, $a_4 = 0$, $a_6 = 0$, $a_8 = 0$, ... and $y_1 = a_0 + a_2 x^2$, is Legendre polynomial of degree 2.

We will explicitly construct a Legendre polynomial $P_n(x)$ of degree n using an inverted version of the recursion relation (4.25), where

$$a_k = -\frac{(k+2)(k+1)}{(n-k)(n+k+1)} a_{k+2} \qquad k = n-2, n-4, \ldots \qquad (4.34)$$

It follows that we can express all lower coefficients a_k with $k < n$ in terms of the highest coefficient a_n. The highest coefficient a_n is a constant that is chosen according to a convention. The value of a_n, which we will use is

$$a_n = \begin{cases} 1 & \text{for } n = 0 \\ \frac{(2n)!}{2^n(n!)^2} = \frac{1 \cdot 3 \cdot 5 \cdots (2n-1)}{n!} & n = 1, 2, 3, \ldots \end{cases} \qquad (4.35)$$

Later, we will see that this normalization ensures that all Legendre polynomials will be equal to 1 at $x = 1$. Thus, $P_n(x = 1) = 1$ for all $n = 0, 1, 2, \ldots$.

From relation (4.34) and definition (4.35), we obtain

$$\begin{aligned} a_{n-2} &= -\frac{n(n-1)}{2(2n-1)} a_n = -\frac{n(n-1)(2n)!}{2(2n-1)2^n(n!)^2} \\ &= -\frac{n(n-1)(2n)(2n-1)(2n-2)!}{2(2n-1)2^n n(n-1)! n(n-1)(n-2)!} \\ &= -\frac{(2n-2)!}{2^n(n-1)!(n-2)!} \end{aligned} \qquad (4.36)$$

for $k = n - 2$. Applying relation (4.34) once more yields

$$\begin{aligned} a_{n-4} &= -\frac{(n-2)(n-3)}{4(2n-3)} a_{n-2} = \frac{(n-2)(n-3)(2n-2)!}{4(2n-3)2^n(n-1)!(n-2)!} \\ &= \frac{(n-2)(n-3)(2n-2)(2n-3)(2n-4)!}{4(2n-3)2^n(n-1)!(n-2)(n-3)(n-4)!} \\ &= \frac{(2n-4)!}{2^n 2!(n-2)!(n-4)!}. \end{aligned} \qquad (4.37)$$

Expressions (4.36) and (4.37) give way to a general formula for the coefficients of Legendre polynomial. One easily verifies that:

$$a_{n-2m} = (-1)^m \frac{(2n-2m)!}{2^n m!(n-m)!(n-2m)!} \qquad (4.38)$$

agrees with (4.35), (4.36), and (4.37) for $m = 0, m = 1$ and $m = 2$, respectively. In the Legendre polynomial $P_n(x) = a_n x^n + a_{n-2} x^{n-2} + a_{n-4} x^{n-4} + \ldots$, the generic term is $a_{n-2m}x^{n-2m}$ with coefficient a_{n-2m} from (4.38). Thus, we can write

$$P_n(x) = \sum_{m=0}^{M} (-1)^m \frac{(2n-2m)!}{2^n m! (n-m)! (n-2m)!} x^{n-2m} . \qquad (4.39)$$

The upper limit M determines the term with the lowest power of x. We have seen that the even Legendre polynomial starts with a constant and the odd Legendre polynomial starts with a term proportional to x. Correspondingly, for even n, the upper limit is $M = n/2$ for which it holds that $x^{n-2M} = 1$, while, for odd n, the upper limit must be $M = (n-1)/2$ which is such that $x^{n-2M} = x$.

☞ **Example 4.2.3.** For $n = 0$, we clearly have $P_0(x) = 1$. For $n = 1$, the summation in equation (4.39) contains only one term equal to x and so the Legendre polynomial is $P_1(x) = x$. For $n = 2$,

$$P_2(x) = \frac{1 \cdot 3}{2!} x^2 - \frac{2!}{2^2 1! 0!} = \frac{1}{2} \left(3x^2 - 1 \right) . \qquad (4.40)$$

Similarly,

$$P_3(x) = \frac{1}{2} \left(5x^3 - 3x \right) . \qquad (4.41)$$

One readily confirms that in all of the above examples $P_n(x = 1) = 1$ and $P_n(x = -1) = (-1)^n$. The first three Legendre polynomials are shown in Fig. 4.2.

✐ Power Series solutions of Differential Equations: Problems

Problem 4.2.1. Determine the radius of convergence of

$$\sum_{n=1}^{\infty} \frac{(-1)^{n+1}}{3n} x^{n+2} .$$

Problem 4.2.2. Recall the special relativity formula for the total energy of an object with rest-mass m_0 moving at the speed v:

$$E = \frac{m_0 c^2}{\sqrt{1 - \frac{v^2}{c^2}}} .$$

Use binomial expansion from Example 4.1.3 to find the kinetic energy $K = E - m_0 c^2$ up to the sixth order in v.

Problem 4.2.3. Apply the power series method to solve:

$$y'' - 9y = 0 .$$

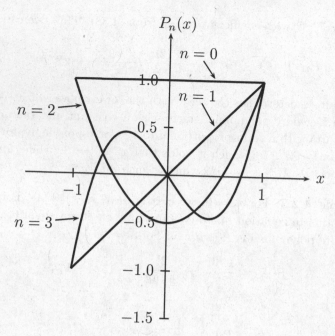

Figure 4.2: Legendre polynomials $P_0(x)$, $P_1(x)$, $P_2(x)$ and $P_3(x)$.

Problem 4.2.4. (a) Apply the power series method to solve:

$$(1 - x^2)y'' - 6xy' - 6y = 0.$$

(b) Identify the power series solutions with expansions of known functions.

Problem 4.2.5. Apply the power series method to find two solutions of the Airy's differential equation:

$$y'' + xy = 0.$$

Problem 4.2.6. (a) Insert $y(x) = u(x)(x^2 - 1)^{-2}$ into the differential equation in Problem 4.2.4 and reduce it to a differential equation for $u(x)$.
(b) Find solutions for $u(x)$ by the power series method.

Problem 4.2.7. Using

$$P_n(x) = \sum_{m=0}^{M} (-1)^m \frac{(2n - 2m)!}{2^n m!(n - m)!(n - 2m)!} x^{n-2m} \quad M = n/2 \text{ or } (n-1)/2,$$

show that $P_n(-x) = (-1)^n P_n(x)$ and $P_n'(-x) = (-1)^{n+1} P_n'(x)$.

Problem 4.2.8. Verify the Rodrigues formula:

$$P_n(x) = \frac{1}{2^n n!} \frac{d^n}{dx^n} \left[(x^2 - 1)^n \right].$$

Hint: Apply the binomial theorem to $(x^2 - 1)^n$ and apply differentiation $\mathrm{d}^n / \mathrm{d}x^n$ on the result.

Problem 4.2.9. Show that

$$G(x, t) = \frac{1}{\sqrt{1 - 2tx + t^2}} = \sum_{n=0}^{\infty} P_n(x) t^n.$$

Hint: Use the binomial expansion:

$$(1 - v)^{-\frac{1}{2}} = \sum_{n=0}^{\infty} \frac{(2n)!}{2^{2n}(n!)^2} v^n,$$

with $v = 2tx - t^2$, and study the term involving t^n.

Problem 4.2.10. Find the generating function $F(x, t)$ such that

$$F(x, t) = \sum_{n=0}^{\infty} \frac{t^{n+1}}{n+1} P_n(x).$$

Problem 4.2.11. Evaluate:

$$I = \int_{-1}^{1} (x^2 - 1)^n \, \mathrm{d}x = \int_{-1}^{1} (x - 1)^n (x + 1)^n \, \mathrm{d}x.$$

Hint: Change the integration variable from x to $\xi = x + 1$ and use successive integration by parts to obtain:

$$I = (-1)^p \frac{n \cdots (n - p + 1)}{(n + 1) \cdots (n + p)} \int_0^2 (\xi - 2)^{n-p} \xi^{n+p} \, \mathrm{d}\xi, \quad 0 \leq p \leq n.$$

The result follows by taking $p = n$.

Problem 4.2.12. Show that Legendre polynomials satisfy the orthogonality relation

$$\int_{-1}^{1} P_n(x) P_m(x) \, \mathrm{d}x = \frac{2}{2n+1} \delta_{n,m}, \quad n = 0, 1, \ldots$$

by two different methods:
(a) Use the Rodrigues formula from Problem (4.2.8) and successive integration by parts to arrive at the integration formula derived in Problem (4.2.11).
(b) Use the generating function $G(x, t)$ derived in Problem 4.2.9.

Problem 4.2.13. Show that:

$$P_n(1) = 1.$$

Problem 4.2.14. Show that:

$$P_n(-1) = (-1)^n.$$

Problem 4.2.15. Show that:

$$P_{2n+1}(0) = 0.$$

Problem 4.2.16. Obtain the recursion formula

$$(n+1)P_{n+1}(x) = (2n+1)xP_n(x) - nP_{n-1}(x).$$

Hint: Differentiate $G(x,t)$ with respect to t and compare coefficients of t^n.

Problem 4.2.17. Obtain the recursion formula

$$nP_n(x) = xP_n'(x) - P_{n-1}'(x).$$

Hint: Differentiate $G(x,t)$ with respect to x, compare coefficients of x^n and use the recurrence relation from Problem 4.2.16.

Problem 4.2.18. Let $\vec{r}_1$ and $\vec{r}_2$ be coordinates of two points in space. Use Problem 4.2.9 to show that:

$$\frac{1}{r} = \frac{1}{\sqrt{r_1^2 + r_2^2 - 2r_1 r_2 \cos\theta}} = \frac{1}{r_2}\sum_{m=0}^{\infty} P_m(\cos\theta)\left(\frac{r_1}{r_2}\right)^m,$$

where θ is the angle between $\vec{r}_1$ and $\vec{r}_2$ and $\vec{r} = \vec{r}_2 - \vec{r}_1$.

Problem 4.2.19. (a) Prove the identity:

$$\sum_{m=0}^{\infty} P_m(\cos\theta) = \frac{1}{2\sin\theta/2}.$$

(b) Show that

$$\int_0^{\pi} d\theta\, \frac{\sin\theta\cos^2\theta}{\sin\theta/2} = \frac{28}{15}.$$

Problem 4.2.20. This problem deals with the charge density $\sigma(\theta)$ induced on a metallic grounded sphere due to an external point charge, q. Consider a sphere of radius R with its center at the origin. Let the external point charge be on the z-axis, a distance z $(z > R)$ from the origin. The formula for a surface charge density induced by q on the spherical surface is given by

$$\sigma(\theta) = -\frac{qt}{4\pi R^2}\frac{1 - t^2}{(1 - 2tx + t^2)^{3/2}},$$

where $t = R/z$, $x = \cos\theta$ and θ is the angle between the z-axis and a position vector of a point lying on a sphere.

(a) Show that

$$\frac{1 - t^2}{(1 - 2tx + t^2)^{3/2}} = G(x,t) + 2t\frac{\partial G(x,t)}{\partial t}.$$

for the generating function $G(x,t)$ defined in Problem 4.2.9.

(b) Using the result from (a), expand the charge density $\sigma(\theta)$ in terms of Legendre polynomials.

Problem 4.2.21. Represent the polynomial $10x^3 - 3x^2 - 5x - 1$ in terms of Legendre polynomials.

Problem 4.2.22. Obtain the first three terms of

$$f(x) = \begin{cases} 0 & \text{if} \quad -1 < x < 0 \\ 1 & \text{if} \quad 0 < x < 1 \end{cases}$$

in terms of Legendre polynomials.

Problem 4.2.23. Find values of the following integrals:

(a) $\int_{-1}^{1} x P_n(x) P_{n-1} \, dx$ (b) $\int_{-1}^{1} x^2 P_n^2(x) \, dx$

Hint: Use the recurrence relations.

Problem 4.2.24. Given are two functions:

$$f(x) = x\left(5x^3 - 3x\right), \qquad g(x) = \frac{1}{2}P_2(x) + \frac{1}{4}P_6(x).$$

(a) Represent $f(x)$ in terms of Legendre polynomials.
Evaluate the integrals:

(b) $\int_{-1}^{1} f(x)g(x) \, dx$ (c) $\int_{-1}^{1} g(x)g(x) \, dx$
(d) $\int_{-1}^{1} x f(x) \, dx$

(e) Represent $F(x) = x g(x)$ in terms of Legendre polynomials.

Problem 4.2.25. (a) Given is the differential equation

$$-\frac{d^2\psi(x)}{dx^2} + x^2\psi(x) = \frac{2E}{\hbar\omega}\psi(x)$$

which describes the harmonic oscillator in quantum mechanics. The coefficient $2E/\hbar\omega$ is a constant. Show that the function $y(x)$, defined as: $\psi(x) = y(x)\exp(-x^2/2)$, will satisfy

$$y'' - 2xy' + \lambda y = 0$$

where $\lambda = (2E/\hbar\omega) - 1$.

(b) Use the power series expansion $y = \sum_{n=0}^{\infty} a_n x^n$ around the regular point $x = 0$ to obtain the recurrence relation

$$a_{n+2} = \frac{2n - \lambda}{(n+2)(n+1)} a_n, \quad n \geq 0$$

and find the power series expansion for $y(x) = a_0 y_{even}(x) + a_1 y_{odd}(x)$ up to the seventh power of x.

(c) Show that for $\lambda = 2n$, where n is a non-negative even/odd integer, that the even/odd power series expansion $y_{even}(x)/y_{odd}(x)$ terminates and becomes a polynomial, which we denote by y_n. Find the polynomial $y_n(x)$ for $n = 0, 1, 2, 3, 4$.

Problem 4.2.26. For the Hermite polynomials defined as:

$$H_0 = 1, \quad H_n(x) = (-1)^n e^{x^2} \frac{d^n}{dx^n}(e^{-x^2}), \quad n = 1, 2, \ldots$$

show that

$$H_1(x) = 2x, \quad H_2(x) = 4x^2 - 2, \quad H_3(x) = 8x^3 - 12x, \quad H_4(x) = 16x^4 - 48x^2 + 12$$

and compare with $y_n(x)$, obtained in the previous problem.

Problem 4.2.27. Show that

$$G(x, t) = e^{2xt - t^2} = \sum_{n=0}^{\infty} H_n(x) \frac{t^n}{n!}.$$

Problem 4.2.28. Show that the Hermite polynomials satisfy the relation

$$H_{n+1}(x) = 2x H_n(x) - H_n'(x).$$

Problem 4.2.29. Differentiate the generating function $G(x, t) = e^{2xt - t^2}$ with respect to x to show that

$$H_n'(x) = 2n H_{n-1}(x).$$

and use it to prove that the Hermite polynomials satisfy the differential equation

$$H_n''(x) - 2x H_n'(x) + 2n H_n(x) = 0.$$

Problem 4.2.30. Show that $\varphi_n = H_n(x) e^{-x^2/2}$ is a solution to

$$\varphi_n'' + (2n + 1 - x^2)\varphi_n = 0,$$

which is the time-independent Schrödinger equation for the harmonic oscillator problem.

Problem 4.2.31. Show that the Hermite polynomials satisfy the orthogonality relation

$$\int_{\infty}^{-\infty} e^{-x^2} H_n(x) H_m(x) \, dx = 2^n n! \sqrt{\pi} \delta_{nm}.$$

4.3   Frobenius' Method, Expansion around a Singular Point

In this section, we consider a series solution to equation (4.15) expanded around point x_0, which is a singular point of that equation. More specifically, we are interested in an expansion around the singular point in equation (4.15), which represents a special form of singularity with a solution of the form $\sum_{m=0}^{\infty} a_m x^{m+r}$. To define the concept of such singular point we rewrite the differential equation (4.15) in the following form

$$y'' + \frac{p_0(x)}{x} y' + \frac{q_0(x)}{x^2} y = 0, \quad p_0(x) = xp(x), \quad q_0(x) = x^2 q(x). \tag{4.42}$$

For functions $p_0(x)$ and $q_0(x)$ analytic at $x = 0$, the point $x = 0$ is said to be a *regular singular point* of the differential equation (4.42) and *Frobenius' theorem* states that there exists **at least one** solution of the form

$$y = x^r \sum_{m=0}^{\infty} a_m x^m = \sum_{m=0}^{\infty} a_m x^{m+r}, \tag{4.43}$$

where r is a real (or complex) number and $a_0 \neq 0$. If $p_0(x)$ and $q_0(x)$ have power series which converge for $0 < x < R$ with a radius of convergence R then the power series (4.43) is also convergent on the interval $0 < x < R$.

Note that the expansion in (4.43) is a power series only if r is a positive integer, otherwise, it is a series solution. The Frobenius' theorem guarantees the existence of one series solution to the differential equation (4.42) in the form given in (4.43). The second independent solution may also have the form $x^{r'} \sum_{m=0}^{\infty} a_m x^m$ (with possible different constant $r' \neq r$) or, alternatively, it may contain an additional logarithmic term $\ln x$. Since $\ln x$ is not defined at $x = 0$, it does not have a series centered at $x = 0$.

☞ **Example 4.3.1.** Consider the Euler equation

$$x^2 y'' + p_0 xy' + q_0 y = 0, \quad p_0 = \text{const}, q_0 = \text{const}, \tag{4.44}$$

which has a singular point at $x = 0$. It is a regular singular point as seen by a comparison with equation (4.42) for constant functions p_0, q_0, which clearly are analytic according to Definition 4.1.4. We will try a solution, $y(x) = x^r \sum_{m=0}^{\infty} a_m x^m$, of the form given in (4.43). Inserting expressions

$$y'(x) = \sum_{m=0}^{\infty} a_m (r+m) x^{m+r-1} \tag{4.45}$$

$$y''(x) = \sum_{m=0}^{\infty} a_m (r+m)(r+m-1) x^{m+r-2} \tag{4.46}$$

back into the Euler equation (4.44), we obtain

$$\sum_{m=0}^{\infty} a_m x^{m+r} \left[(r+m)(r+m-1) + (r+m)p_0 + q_0\right]$$

$$= x^r a_0 \left[r(r-1) + p_0 r + q_0\right] + x^{r+1} a_1 \left[r(r+1) + p_0(r+1) + q_0\right] + \dots$$

$$= 0.$$

Since a_0 is assumed to be different from zero, the constant r must satisfy equation:

$$r(r-1) + p_0 r + q_0 = 0 \tag{4.47}$$

so that the coefficient of x^r is zero. Then, the next term $+x^{r+1} a_1 \left(r(r+1) + p_0(r+1) + q_0\right)$ vanishes only for $a_1 = 0$. Similarly, we are forced to set $a_m = 0$, $m > 0$ to eliminate all remaining coefficients of x^{r+m} with $m > 0$. Then, the solution has the form $y(x) = x^r$ with r satisfying equation (4.47). Equation (4.47) is reminiscent of the characteristic equation (3.51), which determined a solution to the homogeneous differential second order with constant coefficients in Chapter 3. Roots of equation (4.47) are

$$r_i = \frac{1}{2}\left(-(p_0 - 1) \pm \sqrt{(p_0 - 1)^2 - 4q_0}\right), \quad i = 1, 2 \tag{4.48}$$

and there are three special cases to be considered. For $(p_0 - 1)^2 - 4q_0 > 0$, two distinct real roots follow from equation (4.48) and the general solution is given by $y(x) = c_1 x^{r_1} + c_2 x^{r_2}$. For $(p_0 - 1)^2 - 4q_0 < 0$, the roots $r_i = \alpha \pm i\beta$ are complex and the corresponding solutions becomes $x^{r_i} = x^\alpha x^{\pm i\beta} = x^\alpha \exp(\pm i\beta \ln x)$. Finally, in the case where $(p_0 - 1)^2 - 4q_0 = 0$, equation (4.47) possesses the double root $r = -(p_0 - 1)/2$. In addition to solution x^r, there also exists a second solution $\ln(x)x^r$, which can be obtained by the method of reduction. All three cases can be summarized as follows

$$y(x) = \begin{cases} c_1 x^{r_1} + c_2 x^{r_2} & r_1 \neq r_2, \ r_1, r_2 \ \text{real} \\ (c_1 + c_2 \ln(x))x^r & r_1 = r_2 = r, \ r \ \text{real} \\ (c_1 \sin(\beta \ln x) + c_2 \cos(\beta \ln x)) x^\alpha & r_{1,2} = \alpha \pm i\beta \ r_1, r_2 \ \text{complex} \end{cases} \tag{4.49}$$

Let us go back to the general equation (4.42) rewritten in a form similar to the Euler equation (4.44)

$$x^2 y'' + x p_0(x) y' + q_0(x) y = 0. \tag{4.50}$$

Since functions $p_0(x)$ and $q_0(x)$ are assumed to be analytic at $x = 0$, we can expand them around zero as

$$p_0(x) = \sum_{m=0}^{\infty} p_0^{(m)} x^m = p_0^{(0)} + p_0^{(1)} x^1 + \dots,$$

$$q_0(x) = \sum_{m=0}^{\infty} q_0^{(m)} x^m = q_0^{(0)} + q_0^{(1)} x^1 + \dots. \tag{4.51}$$

Plugging expansions (4.51), (4.45), and (4.46) into the differential equation (4.50), we get

$$x^r \left(r(r-1)a_0 + \dots\right) + \left(p_0^{(0)} + p_0^{(1)} x + \dots\right) x^r (r a_0 + \dots)$$

$$+ \left(q_0^{(0)} + q_0^{(1)} x + \dots\right) x^r (a_0 + \dots) = 0 \tag{4.52}$$

for the few lowest order terms. Setting the coefficient of $a_0 x^r$ in (4.52) equal to zero yields the indicial equation:

$$r(r-1) + p_0^{(0)} r + q_0^{(0)} = 0, \tag{4.53}$$

which is used to determine the constant r. It is a generalization of equation (4.47), which we encountered above in our study of the Euler equation. Solutions r_1 and r_2 of the indicial equation (4.53) are called the indicial roots.

As in the case of equation (4.47), we encounter three different kinds of solutions of the indicial equation:

① If $r_1 \neq r_2$ and $r_1 - r_2$ not an integer then equation (4.42) has a basis of solutions:

$$y_1(x) = x^{r_1} \sum_{m=0}^{\infty} a_m x^m \tag{4.54}$$

$$y_2(x) = x^{r_2} \sum_{m=0}^{\infty} A_m x^m \tag{4.55}$$

② If $r_1 = r_2 = r$ then the solutions are

$$y_1(x) = x^r \sum_{m=0}^{\infty} a_m x^m \tag{4.56}$$

$$y_2(x) = y_1(x) \ln x + x^r \sum_{m=0}^{\infty} A_m x^m \tag{4.57}$$

③ If $r_1 \neq r_2$ and $r_1 - r_2$ is a positive integer then solutions are:

$$y_1(x) = x^{r_1} \sum_{m=0}^{\infty} a_m x^m \qquad\qquad (4.58)$$

$$y_2(x) = k y_1(x) \ln x + x^{r_2} \sum_{m=0}^{\infty} A_m x^m, \qquad\qquad (4.59)$$

where k is a constant (not necessarily different from zero) that is determined by the differential equation.

In all the above cases, the general solution is $y(x) = c_1 y_1(x) + c_2 y_2(x)$.

☞ **Example 4.3.2.** Consider the cases of indicial roots:
- Distinct roots, which do not differ by an integer:

$$x^2 y'' - \frac{1}{2} x y' + \frac{1}{2} y = 0. \qquad\qquad (4.60)$$

The indicial equation $r^2 - 3r/2 + 1/2 = 0$ has two distinct roots $r_1 = 1$ and $r_2 = 1/2$ and two solutions:

$$y_1(x) = x^1, \quad y_2(x) = x^{1/2}.$$

- Double root:

$$x^2 y'' + 3 x y' + y = 0. \qquad\qquad (4.61)$$

The indicial equation $r^2 + 2r + 1 = 0$ has the double root, $r_1 = r_2 = r = -1$. Hence, it has a solution:

$$y_1(x) = x^{-1}.$$

The second solution can be obtained by the method of reduction of order. We set $y_2 = u(x) y_1(x)$ and plug it into equation (4.61). The auxiliary function $u(x)$ must satisfy $x u'' + u' = 0$, which yields $u'(x) = c x^{-1}$ or $u(x) = c \ln x + c_0$. Thus, the second solution is

$$y_2(x) = x^{-1} \ln x = y_1(x) \ln x.$$

- Distinct roots, which differ by an integer (case of no logarithmic term. i.e. $k = 0$ in equation (4.59)):

$$x^2 y'' - 4 x y' + 4 y = 0. \qquad\qquad (4.62)$$

The indicial equation $r^2 - 5r + 4 = 0$ has two distinct roots $r_1 = 1$ and $r_2 = 4$ and two solutions:

$$y_1(x) = x^1, \quad y_2(x) = x^4.$$

• Distinct roots, which differ by an integer (with the logarithmic term, i.e. $k \neq 0$ in equation (4.59)):

$$xy'' + y = 0. \tag{4.63}$$

When $p_0^{(0)} = q_0^{(0)} = 0$ the indicial equation is $r(r-1) = 0$ with two roots $r = 1, 0$. The series solution $y_1(x) = x \sum_m a_m x^m$ corresponds to the highest root, $r_1 = 1$. Plugging it into equation (4.63) yields

$$\sum_{m=1}^{\infty} a_m (m+1)m \, x^m + \sum_{m=0}^{\infty} a_m x^{m+1} = \sum_{m=1}^{\infty} \left[a_m(m+1)m + a_{m-1} \right] x^m = 0$$

along with a recurrence relation:

$$a_m = \frac{-1}{m(m+1)} a_{m-1} = \frac{(-1)^m}{m!(m+1)!} a_0, \qquad m = 1, 2, \ldots.$$

Then the first solution is:

$$y_1(x) = x - \frac{1}{2}x^2 + \frac{1}{12}x^3 - \frac{1}{144}x^4 + \ldots \tag{4.64}$$

At first, it may seem natural to associate the second solution to the series $\sum_m A_m x^m$, corresponding to the lower root, $r = 0$. The problem with this approach becomes evident when we plug $y_2(x) = \sum_{m=0}^{\infty} A_m x^m$ into equation (4.63) obtaining:

$$\sum_{m=2}^{\infty} A_m m(m-1)x^{m-1} + \sum_{m=0}^{\infty} A_m x^m$$

$$= A_0 + \sum_{m=1}^{\infty} \left[A_{m+1}m(m+1) + A_m \right] x^m = 0,$$

which requires $A_0 = 0$, which contradicts the basic assumption of the method. It is easy to see what goes wrong in such a case: If $A_0 = 0$, then the series $y_2(x) = \sum_{m=1}^{\infty} A_m x^m$ can be rewritten as $y_2(x) = x \sum_{m=0}^{\infty} A_{m+1} x^m$, which is the same series solution as in $y_1(x)$. Therefore, for $A_0 = 0$ the recurrence relations leads us back to the first solution.

The remedy is to propose as a second solution:

$$y_2(x) = k y_1(x) \ln x + \sum_{m=0}^{\infty} A_m x^m, \tag{4.65}$$

in agreement with (4.59). Substituting this solution into the differential equation (4.63) yields:

$$-\frac{k y_1}{x} + 2k y_1' + \sum_{m=2}^{\infty} A_m m(m-1)x^{m-1} + \sum_{m=0}^{\infty} A_m x^m = 0. \tag{4.66}$$

Plugging the first solution (4.64) into (4.66) yields up to quadratic terms:

$$(-k + \frac{1}{2}kx - k\frac{1}{12}x^2) + (2k - 2kx + \frac{1}{2}kx^2) + (2A_2x + 3!A_3x^2)$$

$$+ (A_0 + A_1x + A_2x^2) + O(x^3) = 0.$$

Thus, $k = -A_0$ and $A_2 = -A_1 + 3k/4$. Via the recursion relations we can determine all the higher coefficients A_n in terms of A_0 and A_1. In this case, k must remain different from zero since taking $k = -A_0 = 0$ reduces the solution $y_2(x)$ in (4.65) to $y_1(x)$.

4.3.1   Bessel's Differential Equation

In this section, we apply Frobenius' method to Bessel's differential equation,

$$\boxed{x^2 y'' + xy' + \left(x^2 - \nu^2\right) y = 0,} \tag{4.67}$$

where ν is a real number. Since this equation can be cast in the form of equation (4.42) with the analytic functions $p_0(x) = 1$ and $q_0(x) = x^2 - \nu^2$, $x = 0$ is a regular singular point. By Frobenius' theorem, there exists at least one solution of the form: $y = x^r \sum_{m=0}^{\infty} a_m x^m$.

Since $p_0^{(0)} = 1$ and $q_0^{(0)} = -\nu^2$ the indicial equation is:

$$r(r-1) + p_0^{(0)}r + q_0^{(0)} = r(r-1) + r - \nu^2 = r^2 - \nu^2 = 0, \tag{4.68}$$

with two indicial roots given by the real numbers

$$r_1 = \nu, \qquad r_2 = -\nu.$$

Their difference is:

$$r_1 - r_2 = 2\nu. \tag{4.69}$$

Let us assume that $\nu > 0$ and take $r = r_1 = \nu$ to be the larger root. In order to solve the Bessel's equation using Frobenius' method, we consider a solution of the form $y = \sum_{m=0}^{\infty} a_m x^{m+\nu}$ with

$$y' = \sum_{m=0}^{\infty} a_m(m+\nu)x^{m+\nu-1} \text{ and } y'' = \sum_{m=0}^{\infty} a_m(m+\nu)(m+\nu-1)x^{m+\nu-2}.$$

Plugging these functions into Bessel's equation (4.67) yields:

$$\sum_{m=0}^{\infty} a_m(m+\nu)(m+\nu-1)x^{m+\nu} + \sum_{m=0}^{\infty} a_m(m+\nu)x^{m+\nu} + \sum_{m=0}^{\infty} a_m x^{m+\nu+2} \quad (4.70)$$

$$ - \sum_{m=0}^{\infty} \nu^2 a_m x^{m+\nu} = 0.$$

Replacing m by $n = m+2$ in the third summation then yields $\sum_{m=0}^{\infty} a_m x^{m+\nu+2} = \sum_{n=2}^{\infty} a_{n-2} x^{n+\nu}$. Replacing m by $n = m$ in the remaining first, second, and last summation in equation (4.70) we get

$$\sum_{n=0}^{\infty} a_n \left[(n+\nu)(n+\nu-1) + (n+\nu) - \nu^2 \right] x^{n+\nu} + \sum_{n=2}^{\infty} a_{n-2} x^{n+\nu} \quad (4.71)$$

$$= (1+2\nu) a_1 x^{\nu+1} + \sum_{n=2}^{\infty} \left[n(n+2\nu)a_n + a_{n-2} \right] x^{n+\nu} = 0.$$

Since coefficients of all powers of x in (4.71) have to be identically zero we obtain

$$a_1 = 0 \quad (4.72)$$

along with the following recurrence relation

$$a_n = \frac{-1}{n(n+2\nu)} a_{n-2}, \quad \text{for n} = 2, 3, \ldots. \quad (4.73)$$

Since $a_1 = 0$ then $a_3 = 0$, ..., $a_{2k+1} = 0$, for $k = 1, 2, \ldots.$ Next, consider even indices $n = 2k$ with $k = 1, 2, \ldots$:

$$a_{2k} = \frac{-1}{2k(2k+2\nu)} a_{2k-2} = \frac{-1}{2^2 k(k+\nu)} a_{2k-2}, \quad k = 1, 2, \ldots. \quad (4.74)$$

Therefore, $a_2 = \frac{-1}{2^2(1)(1+\nu)} a_0$, $a_4 = \frac{-1}{2^2(2)(2+\nu)} a_2 = \frac{(-1)^2}{2^2(2)\, 2!(2+\nu)(1+\nu)} a_0$, and so forth. The general term is

$$a_{2k} = \frac{(-1)^k}{2^{2k} k!(k+\nu)(k-1+\nu)\ldots(1+\nu)} a_0, \quad k = 1, 2, \ldots. \quad (4.75)$$

Thus, we obtain the following solution to Bessel's equation (4.67):

$$y_1(x) = \sum_{k=0}^{\infty} a_{2k} x^{2k+\nu} = a_0 \sum_{k=0}^{\infty} \frac{(-1)^k}{2^{2k} k!(k+\nu)(k-1+\nu)\cdots(1+\nu)} x^{2k+\nu}. \quad (4.76)$$

Note that $x = 0$ is the only singular point of Bessel's equation and the polynomials p_0 and q_0 are analytic everywhere. According to Frobenius' theorem, the radius of convergence $R = \infty$ and the series in (4.76) is convergent for all real x.

For the integer indicial root $\nu = n$ the solution becomes a power series expansion. In this case we propose the coefficient a_0 to be of the following form

$$a_0 = \frac{1}{2^n n!}.$$ (4.77)

We obtain a particular solution to Bessel's equation denoted by the symbol $J_n(x)$:

$$J_n(x) = \sum_{k=0}^{\infty} \frac{(-1)^k}{k!(n+k)!} \left(\frac{x}{2}\right)^{2k+n}.$$ (4.78)

The function $J_n(x)$ is called an n-th order Bessel function of the first kind. For

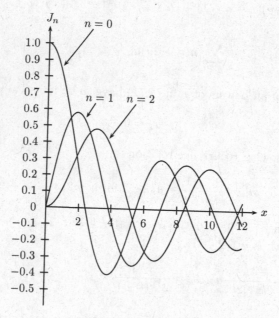

Figure 4.3: Bessel functions $J_0(x), J_1(x), J_2(x)$.

example, a Bessel function of order zero is equal to:

$$J_0(x) = \sum_{k=0}^{\infty} \frac{(-1)^k}{(k!)^2} \left(\frac{x}{2}\right)^{2k}.$$ (4.79)

One sees that $J_0(0) = 1$ while all Bessel functions of the first kind of order $n \geq 1$ will vanish at the origin and, therefore, $J_n(0) = 0$ for $n = 1, 2, \ldots$. Their graphs for $n = 0, 1, 2$ is shown in Figure 4.3.

To simplify expressions for the coefficients of Bessel functions for the general order ν we introduce the *Gamma function* defined as:

$$\Gamma(\nu) = \int_0^\infty t^{\nu-1} e^{-t} \, dt. \tag{4.80}$$

The basic properties of the Gamma function are:

$$\Gamma(\nu+1) = \nu \Gamma(\nu) \tag{4.81}$$

and

$$\Gamma(1) = \int_0^\infty e^{-t} \, dt = 1. \tag{4.82}$$

Those properties easily follow from the integral expression in equation (4.80). For the integer argument of the Gamma we get from (4.81) that

$$\Gamma(n+1) = n!. \tag{4.83}$$

The above result illustrates why the Gamma function is sometimes called the generalized factorial.

☞ **Example 4.3.3.** Here, we evaluate $\Gamma\left(\frac{1}{2}\right)$ and $\Gamma\left(\frac{5}{2}\right)$.

$$\Gamma\left(\frac{1}{2}\right) = \int_0^\infty e^{-t} t^{-1/2} \, dt = 2 \int_0^\infty e^{-t} \, dt^{1/2} = 2 \int_0^\infty e^{-x^2} \, dx = \sqrt{\pi}. \tag{4.84}$$

One can derive the last integral in the above equation using a trick involving polar coordinates:

$$\left(\Gamma\left(\frac{1}{2}\right)\right)^2 = 4 \int_0^\infty e^{-x^2} \, dx \int_0^\infty e^{-y^2} \, dy = 4 \int_0^\infty \int_0^\infty e^{-(x^2+y^2)} \, dx \, dy$$

$$= 4 \int_0^\infty \int_0^{\pi/2} e^{-\rho^2} \rho \, d\rho \, d\phi = 4 \frac{\pi}{2} \int_0^\infty e^{-\rho^2} \, d\left(\frac{\rho^2}{2}\right) = \pi.$$

Now, using the recursion relation (4.81) one obtains: $\Gamma\left(\frac{5}{2}\right) = \Gamma\left(\frac{3}{2}+1\right) = \frac{3}{2}\Gamma\left(\frac{3}{2}\right) = \frac{3}{2}\Gamma\left(1+\frac{1}{2}\right) = \frac{3}{2}\left(\frac{1}{2}\right)\Gamma\left(\frac{1}{2}\right) = \frac{3}{4}\sqrt{\pi}.$

The concept of a Gamma function as a generalized factorial suggests a natural way to generalize the choice in (4.77) to

$$a_0 = \frac{1}{2^\nu \Gamma(\nu+1)} \tag{4.85}$$

for the indicial root ν, which is not an integer. Given the above expression for a_0 the solution $y_1(x)$ in (4.76) becomes

$$
\begin{aligned}
y_1(x) = J_\nu(x) &= \sum_{k=0}^{\infty} \frac{(-1)^k}{2^{2k+\nu} k! (k+\nu)(k-1+\nu)\cdots(1+\nu)\Gamma(\nu+1)} x^{2k+\nu} \\
&= \sum_{k=0}^{\infty} \frac{(-1)^k}{k! \Gamma(\nu+k+1)} \left(\frac{x}{2}\right)^{2k+\nu}.
\end{aligned}
\tag{4.86}
$$

$J_\nu(x)$ is called the Bessel function of the first kind of order ν.

According to Frobenius' method (see relations (4.54)-(4.59)), a general solution to Bessel's equation will fall into three different classes depending on whether the difference, $r_1 - r_2 = 2\nu$, is non-integer, zero, or integer. Now, we will consider all three cases.

Case 1. Let $2\nu \neq 1, 2, 3, \ldots$ then the difference, $r_1 - r_2$, is not an integer. According to (4.54)-(4.55) and (4.69), we get two independent series solutions to Bessel's equation and the general homogeneous solution can be written as:

$$
y = c_1 J_\nu(x) + c_2 J_{-\nu}(x),
\tag{4.87}
$$

which is valid for all $x \neq 0$ and arbitrary constants c_1, c_2. The second solution $J_{-\nu}$ (x) to Bessel's equation is obtained from equation (4.86) by letting $\nu \to -\nu$ (in agreement with (4.55)):

$$
J_{-\nu}(x) = \sum_{k=0}^{\infty} \frac{(-1)^k}{k! \, \Gamma(k-\nu+1)} \left(\frac{x}{2}\right)^{2k-\nu}.
\tag{4.88}
$$

Case 2. The second case concerns $\nu = 0$ with a double root $r_1 = r_2 = 0$. Thus, the first solution is $J_0(x)$ as given in equation (4.79). According to the Frobenius's theorem and equation (4.57), the second solution is given by an expression with logarithmic singularity at $x = 0$:

$$
y_2(x) = J_0(x) \ln x + u_2(x),
\tag{4.89}
$$

with function $u_2(x)$ given by the series, $u_2(x) = \sum_{m=0}^{\infty} A_m x^m$. Plugging $y_2(x)$ into Bessel's equation of order $\nu = 0$:

$$
x^2 y'' + x y' + x^2 y = 0
$$

one obtains $(x J_0'' + J_0' + x J_0) \ln x + 2 J_0' + x u_2'' + u_2' + x u_2 = 0$. Since the Bessel function J_0 satisfies Bessel's equation of the zero order, the coefficient in front of $\ln x$ vanishes and $u_2(x)$ must be a solution of the equation:

$$
x u_2'' + u_2' + x u_2 + 2 J_0' = 0.
\tag{4.90}
$$

Accordingly, the coefficients A_m of the series expansion u_2 must satisfy

$$A_1 x + \sum_{m=2}^{\infty} \left(m^2 A_m + A_{m-2} \right) x^m + 4 \sum_{k=0}^{\infty} \frac{(-1)^k k}{(k!)^2} \left(\frac{x}{2} \right)^{2k} = 0.$$

One can show that the coefficients A_{2k-1} with odd indices must vanish, $A_{2k-1} = 0$, $k = 1, 2, \ldots$, and the remaining coefficients with even indices satisfy the recurrence relation:

$$-k^2 A_{2k} = \frac{1}{4} A_{2k-2} + \frac{(-1)^k k}{4^k (k!)^2}, \quad k = 1, 2, \ldots. \tag{4.91}$$

Setting $A_0 = 0$ implies that $A_2 = 1/4$ and leads to the following solution to equation (4.91)

$$A_{2k} = \frac{(-1)^{k+1}}{4^k (k!)^2} \left(1 + \frac{1}{2} + \cdots + \frac{1}{k} \right), \quad k = 1, 2, \ldots. \tag{4.92}$$

Thus, the second solution, $y_2(x)$, to Bessel's equation of order zero is found. One conventionally denotes $y_2(x)$ by $N_0(x)$ given by:

$$N_0(x) = J_0(x) \ln x + \sum_{k=1}^{\infty} \frac{(-1)^{k+1}}{(k!)^2} \left(1 + \frac{1}{2} + \cdots + \frac{1}{k} \right) \left(\frac{x}{2} \right)^{2k}, \tag{4.93}$$

which is referred to as Neumann's function of order zero. It is quite common to introduce the Bessel function of the second kind of order zero as a linear combination of $N_0(x)$ and $J_0(x)$:

$$\begin{aligned} Y_0(x) &= \frac{2}{\pi} \left[N_0(x) - (\ln 2 - \gamma) J_0(x) \right] \\ &= \frac{2}{\pi} J_0(x) \left[\gamma + \ln \left(\frac{x}{2} \right) \right] - \frac{2}{\pi} \sum_{k=1}^{\infty} \frac{(-1)^k}{2^{2k} (k!)^2} \left(1 + \frac{1}{2} + \ldots + \frac{1}{k} \right) x^{2k}, \end{aligned} \tag{4.94}$$

where

$$\gamma = \lim_{k \to \infty} \left(1 + \frac{1}{2} + \cdots + \frac{1}{k} - \ln k \right) = 0.57721566 \tag{4.95}$$

is the Euler's constant. In this way, one can re-write the general solution as:

$$y = c_1 J_0(x) + c_2 Y_0(x).$$

Case 3. The last case involves a positive integer $2\nu = N$. This case splits in two separate sub-cases depending on whether N is odd or even. First, consider N even and ν an integer, for example, $\nu = n$,. Then J_n and J_{-n} are linearly dependent due to the identity

$$J_{-n}(x) = (-1)^n J_n(x). \tag{4.96}$$

Identity (4.96) can be proved as follows:

$$J_{-n}(x) = \sum_{m=0}^{\infty} \frac{(-1)^m}{2^{2m-n}m!(m-n)!} x^{2m-n} = \sum_{m=n}^{\infty} \frac{(-1)^m}{2^{2m-n}m!(m-n)!} x^{2m-n}.$$

Let $m = s + n$ $(s = m - n)$. Then,

$$J_{-n}(x) = \sum_{s=0}^{\infty} \frac{(-1)^{s+n}}{2^{2(s+n)-n}(s+n)!(s+n-n)!} x^{2(s+n)-n} = \sum_{s=0}^{\infty} \frac{(-1)^{s+n}}{2^{2s+n}(s+n)!s!} x^{2s+n}$$

$$= (-1)^n \sum_{m=0}^{\infty} \frac{(-1)^m}{2^{2m+n}(m+n)!m!} x^{2m+n} = (-1)^n J_n(x).$$

Thus, we need to find a second linearly independent solution for $\nu = n$. It is given by Neumann's function of order n:

$$N_n(x) = J_n(x) \ln x - \frac{1}{2} \sum_{k=0}^{n-1} \frac{(n-k-1)!}{k!} \left(\frac{x}{2}\right)^{2k-n}$$

$$+ \frac{1}{2} \sum_{k=0}^{\infty} \frac{(-1)^{k+1}}{k!(k+n)!} \left(1 + \frac{1}{2} + \cdots + \frac{1}{k} + 1 + \frac{1}{2} + \cdots + \frac{1}{k+n}\right) \left(\frac{x}{2}\right)^{2k+n}.$$

$$(4.97)$$

Again, as in the case of $n = 0$, we will replace N_n by a Bessel function of the second kind. Let

$$Y_n(x) = \frac{2}{\pi} [N_n(x) - (\ln 2 - \gamma) J_n(x)] \qquad (4.98)$$

define Bessel functions of the second kind of order n. These functions are shown in Figure 4.4 for $n = 0, 1, 2$. We notice the logarithmic singularity at $x = 0$ due to the term, $J_n(x) \ln x$, in equation (4.97). Then, the general solution to Bessel's equation is $y = c_1 J_n(x) + c_2 Y_n(x)$ for the integer n.

Next, define the Bessel function of the second kind of arbitrary order ν as

$$Y_\nu(x) = \frac{1}{\sin(\nu\pi)} [J_\nu(x) \cos(\nu\pi) - J_{-\nu}(x)], \qquad (4.99)$$

where ν may or may not be an integer. Since $\sin(\nu) = 0$ if ν is an integer $Y_\nu(x)$ is not defined for integer values of ν. However, if we apply L'Hospital's rule then we can show that the limit does exist as ν approaches an integer value. Thus, when ν is an integer, we define $Y_n(x)$ as the limit of $Y_\nu(x)$ as ν approaches n:

$$Y_n(x) = \lim_{\nu \to n} Y_\nu(x), \quad n = 0, 1, \ldots \qquad (4.100)$$

We can show that the function Y_n, obtained as a limit of functions $Y_\nu(x)$ defined

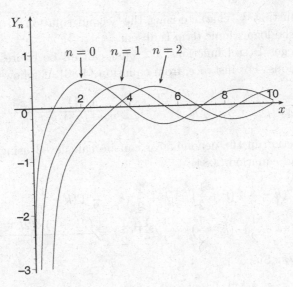

Figure 4.4: The first three Bessel functions of the second kind $Y_0(x)$, $Y_1(x)$ and $Y_2(x)$.

in (4.99), is equal to

$$Y_n(x) = \frac{2}{\pi} J_n(x) \left[\gamma + \ln\left(\frac{x}{2}\right) \right] - \frac{x^{-n}}{\pi} \sum_{k=0}^{n-1} \frac{(n-k-1)!}{2^{2k-n}k!} x^{2k}$$

$$- \frac{x^n}{\pi} \sum_{k=0}^{\infty} \frac{(-1)^k}{2^{2k+n}k!(k+n)!} \left(\left(1 + \frac{1}{2} + ... + \frac{1}{k}\right) + \left(1 + \frac{1}{2} + ... + \frac{1}{k+n}\right) \right) x^{2k},$$

and agrees with Y_n in equation (4.98).

Thus, $J_\nu(x)$ and $Y_\nu(x)$ are linearly independent solutions to Bessel's equation and it follows that the general solution is given by the formula:

$$y = c_1 J_\nu(x) + c_2 Y_\nu(x), \qquad (4.101)$$

where $Y_\nu(x)$ defined in (4.99) is a Bessel function of the second kind of order ν .

It remains to consider the sub-case where $2\nu = N$ with odd integer N. Here we are dealing with half-integer orders $\nu = \pm 1/2, \pm 3/2, \pm 5/2, \ldots$. In this case, equation (4.99) can be written as

$$Y_\nu(x) = \frac{1}{\sin(\nu\pi)} \left[J_\nu(x)\cos(\nu\pi) - J_{-\nu}(x) \right] = (-1)^{2\nu} J_{-\nu}(x), \qquad (4.102)$$

since $\cos[(n+1/2)\pi] = 0$ and $\sin[(n+1/2)\pi] = (-1)^n$. Accordingly, the general solution $y(x) = c_1 J_{n+1/2}(x) + c_2 J_{-n-1/2}(x)$ is given by Bessel's functions of the first order only. It is a pure coincidence, that for the half integer order of Bessel's equation, the constant k, characterizing the second solution in equation (4.59), vanishes. Thus, the logarithmic term is absent.

The half integer Bessel functions $J_{\pm n/2}(x)$ can all be expressed in terms of elementary functions. For instance, from equation (4.86), it follows that:

$$J_{\frac{1}{2}}(x) = \sum_{k=0}^{\infty} \frac{(-1)^k}{k!\,\Gamma(k+3/2)} \left(\frac{x}{2}\right)^{2k+1/2} = \sum_{k=0}^{\infty} \frac{(-1)^k x^{2k+1} x^{-1/2}}{2^{2k+1} 2^{-1/2} k!\,\Gamma(k+3/2)}. \qquad (4.103)$$

The Gamma function in the denominator can be calculated using the recurrence relation (4.81) and equation (4.84):

$$\begin{aligned}
\Gamma(k+3/2) &= (k+\tfrac{1}{2})\Gamma(k+\tfrac{1}{2}) = (k+\tfrac{1}{2})(k-\tfrac{1}{2})\Gamma(k-\tfrac{1}{2}) \\
&= (k+\tfrac{1}{2})\cdot(k-\tfrac{1}{2})\cdots\tfrac{3}{2}\cdot\tfrac{1}{2}\Gamma(\tfrac{1}{2}) = \frac{(2k+1)\cdot(2k-1)\cdots 3\cdot 1}{2^{k+1}}\sqrt{\pi},
\end{aligned}$$

where $\Gamma(1/2) = \sqrt{\pi}$. Since

$$2^{2k+1}k! = 2^{k+1} 2k\cdot(2k-2)\cdots 4\cdot 2,$$

the term $\Gamma(k+3/2)2^{2k+1}k!$ in the denominator of equation (4.103) is equal to

$$\Gamma(k+3/2)2^{2k+1}k! = (2k+1)!\sqrt{\pi}.$$

Accordingly, expression (4.103) becomes

$$J_{\frac{1}{2}}(x) = \sqrt{\frac{2}{\pi x}} \sum_{k=0}^{\infty}\sum_{k=0}^{\infty} \frac{(-1)^k x^{2k+1}}{(2k+1)!},$$

and recognizing in the above power series the Taylor expansion for the $\sin x$ function we obtain

$$J_{\frac{1}{2}}(x) = \sqrt{\frac{2}{\pi x}} \sin x. \qquad (4.104)$$

4.3.2 Properties of Bessel Functions

Bessel functions have their own recursion relations:

$$\begin{aligned}
(x^{\nu} J_{\nu}(x))' &= x^{\nu} J_{\nu-1}(x), & (4.105) \\
(x^{-\nu} J_{\nu}(x))' &= -x^{-\nu} J_{\nu+1}(x). & (4.106)
\end{aligned}$$

Let's verify the identity (4.105):

$$\begin{aligned}
(x^{\nu} J_{\nu}(x))' &= \frac{d}{dx} \sum_{k=0}^{\infty} \frac{(-1)^k 2^{\nu}}{k!\,\Gamma(\nu+k+1)} \left(\frac{x}{2}\right)^{2(k+\nu)} = \sum_{k=0}^{\infty} \frac{(-1)^k 2^{\nu}(k+\nu)}{k!\,\Gamma(\nu+k+1)} \left(\frac{x}{2}\right)^{2k+2\nu-1} \\
&= \sum_{k=0}^{\infty} \frac{(-1)^k 2^{\nu}}{k!\,\Gamma(\nu+k)} \left(\frac{x}{2}\right)^{2k+2\nu-1} = x^{\nu} J_{\nu-1}. \qquad (4.107)
\end{aligned}$$

By adding and subtracting relations (4.105)-(4.106) after multiplying these relations by $x^{-\nu}$ and x^{ν} we obtain:

$$J_{\nu-1}(x) + J_{\nu+1}(x) = \frac{2\nu}{x} J_{\nu}(x), \qquad (4.108)$$

$$J_{\nu-1}(x) - J_{\nu+1}(x) = 2J_{\nu}'(x). \qquad (4.109)$$

Identity (4.108) allows us to recursively obtain Bessel functions of order $\nu + p$ ($p > 0$ and integer) from known expressions for $J_{\nu}(x)$ and $J_{\nu-1}(x)$. For example, one can obtain all Bessel functions $J_n(x)$ of order n from expressions for Bessel functions $J_0(x), J_1(x)$.

☞ **Example 4.3.4.** Recall equation (4.104) and apply (4.105) to $\nu = 1/2$. Then,

$$\left(\sqrt{x} J_{\frac{1}{2}}(x)\right)' = \left(\sqrt{x}\sqrt{\frac{2}{\pi x}}\sin x\right)' = \sqrt{\frac{2}{\pi}}\cos x = \sqrt{x} J_{-\frac{1}{2}}(x).$$

Accordingly,

$$J_{-\frac{1}{2}}(x) = \sqrt{\frac{2}{x\pi}}\cos x.$$

Successive use of recurrence relations leads to expressions for all Bessel functions, $J_{n+1/2}$ of the half-integer order, in terms of sines, cosines, and powers of x.

4.3.3 Reduction to Bessel's Equation

Quite often, a solution to some complicated differential equation can be found by a reduction to a corresponding Bessel's equation. Below follows a general statement regarding such a reduction.

Consider equation:

$$x^2\frac{d^2 y}{dx^2} + (2\beta+1)x\frac{dy}{dx} + (k\alpha)^2 x^{2\alpha}y + \left[\beta^2 - (\nu\alpha)^2\right]y = 0 \qquad (4.110)$$

with non-zero constants α, β, k and ν. Equation (4.110) has the general solution

$$y(x) = c_1\frac{J_{\nu}(kx^{\alpha})}{x^{\beta}} + c_2\frac{Y_{\nu}(kx^{\alpha})}{x^{\beta}}, \qquad (4.111)$$

which is written in terms of Bessel functions $J_{\nu}(z), Y_{\nu}(z)$.

Recall that $J_{\nu}(z)$ and $Y_{\nu}(z)$ are solutions to Bessel's equation of order ν:

$$z^2\frac{d^2 u(z)}{dz^2} + z\frac{du(z)}{dz} + \left(z^2 - \nu^2\right)u(z) = 0.$$

To establish a link between equation (4.110) and Bessel's equation, set $z = kx^\alpha$. The identities below follow from the chain-rule:

$$\frac{du}{dx} = \frac{dz}{dx}\frac{du}{dz} = k\alpha x^{\alpha-1}\frac{du}{dz}$$

$$\frac{d^2u}{dx^2} = \frac{dz}{dx}\frac{d}{dz}\frac{du}{dx} = \frac{\alpha - 1}{x}\frac{du}{dx} + (k\alpha)^2 x^{2(\alpha-1)}\frac{d^2u}{dz^2}.$$

Therefore, the function u must satisfy

$$x^2\frac{d^2u}{dx^2} + x\frac{du}{dx} + \left((k\alpha)^2 x^{2\alpha} - (\nu\alpha)^2\right)u = 0.$$

Next, we introduce a new function,

$$y(x) = \frac{u(z)}{x^\beta} = \frac{u(kx^\alpha)}{x^\beta}.$$
(4.112)

It follows from technical relations,

$$\frac{dy}{dx} = -\beta x^{-\beta-1}u + x^{-\beta}\frac{du}{dx}$$

$$\frac{d^2y}{dx^2} = \beta(\beta+1)x^{-\beta-2}u - 2\beta x^{-\beta-1}\frac{du}{dx} + x^{-\beta}\frac{d^2u}{dx^2},$$

that $y(x)$ satisfies equation (4.110). Due to the definition (4.112), in terms of Bessel function u, we conclude that $y(x)$ must be given by a linear combination as in equation (4.111).

☞ **Example 4.3.5.** Equation

$$x^2\frac{d^2y}{dx^2} + 2\frac{dy}{dx} + \frac{1}{16}\sqrt{x}\,y = 0$$

agrees with equation (4.110) for $\beta = 1/2, \alpha = 1/4, k = 1, \nu = 2$ and, therefore, its solution is $y = c_1 J_2(x^{1/4})/\sqrt{x} + c_2 Y_2(x^{1/4})/\sqrt{x}$.

Next, consider the equation

$$x^2\frac{d^2y}{dx^2} + \frac{1}{2}x\frac{dy}{dx} + xy = 0.$$

It is obtained from equation (4.110) with $\beta = -1/4, \alpha = 1/2, k = 2$ and $\nu = 1/2$. Then, according to relation (4.111), its solution is given by

$$y = c_1 x^{1/4} J_{1/2}(2\sqrt{x}) + c_2 x^{1/4} J_{-1/2}(2\sqrt{x})$$

$$= c_1\sqrt{\frac{2}{\pi}}\sin\left(2\sqrt{x}\right) + c_2\sqrt{\frac{2}{\pi}}\cos\left(2\sqrt{x}\right).$$

The simplicity of this result indicates that the initial differential equation assumes the form of a harmonic equation under the change of variables, $x \to z = 2\sqrt{x}$. Indeed, the chain rule gives

$$\frac{dy}{dx} = \frac{dz}{dx}\frac{dy}{dz} = \frac{2}{z}\frac{dy}{dz}$$

and

$$\frac{d^2y}{dx^2} = \frac{4}{z^2}\frac{d^2y}{dz^2} - \frac{4}{z^3}\frac{dy}{dz}.$$

Plugging these two results into equation, $x^2 d^2y/dx^2 + (x/2)dy/dx + xy = 0$, we arrive at

$$\frac{d^2y}{dz^2} + y = 0$$

with familiar harmonic solutions, $y(z) = \sin(z)$ and $y(z) = \cos(z)$.

The differential equations in problems (4.3.19)- (4.3.27) all belong to a very large class of second order differential equations which can be reduced to Bessel's equation by the method presented in this subsection. Therefore, a solution to these problems can be obtained in each case by fixing the values of constants α, β, ν and k to ensure an agreement with equation (4.110) as it was done in the above example.

◢ Method of Frobenius: Problems

Problem 4.3.1. Identify the series obtained by the Frobenius' method applied to the following equations:

(a) $(x-1)^2 y'' + (x-1)y' - 4y = 0$ (b) $x^2 y'' + x^3 y' + (x^2 - 2)y = 0$

as expansions of known functions.

Problem 4.3.2. Use equation (3.104) to find a general solution to Euler equation:

$$x^2 y'' + xy' + y = x.$$

Problem 4.3.3. Find the basis of solutions to:

$$y'' - \frac{1}{x}y' + \frac{1}{x^2}y = 0$$

by making a comparison to a solution to:

$$y' - \frac{1}{x}y = 1.$$

How does your result agree with the method of Frobenius?

Problem 4.3.4. Identify the series obtained by the method of Frobenius applied to the following equations:

(a) $x^2 y'' + xy' + (x^2 - \frac{1}{4})y = 0$ (b) $(x^2-1)x^2 y'' - (x^2+1)xy' + (x^2+1)y = 0$

as expansions of known functions.

Problem 4.3.5. Consider the differential equation

$$y''(x) + \left(\frac{2}{x} + x\right) y'(x) + y(x) = 0.$$

(a) Find the indicial equation and check whether roots differ by an integer.
(b) Find out whether the logarithmic term is present by looking for series solution corresponding to the lower of two roots of the indicial equation.
(c) Find the expression for a second solution in the form of an integral.

Problem 4.3.6. Consider the differential equation

$$x^2 y''(x) + 4xy'(x) + (x^2 + 2) y(x) = 0.$$

(a) Find the indicial equation. How many series solutions are guaranteed by the method of Frobenius?
(b) Find series solutions corresponding to both roots.
(c) Identify the series solutions as expansions of known functions.

Problem 4.3.7. Consider the differential equation

$$y''(x) - \left(\frac{3}{x} + 2x\right) y'(x) + 4y(x) = 0.$$

(a) Find the indicial equation. How many series solutions are guaranteed by the method of Frobenius?
(b) Find series solutions corresponding to both roots.
(c) Identify the series solutions as expansions of known functions.

Problem 4.3.8. For the differential equation

$$y''(x) - \frac{2}{x}y'(x) + \left(1 + \frac{2}{x^2}\right) y(x) = 0,$$

(a) Find solutions to the indicial equation.
(b) Find series solutions corresponding to both roots and identify them with known functions.
(c) Find a particular solution to

$$y''(x) - \frac{2}{x}y'(x) + \left(1 + \frac{2}{x^2}\right) y(x) = \frac{x}{\sin^3(x)}.$$

Problem 4.3.9. For the differential equation

$$y''(x) + \left(\frac{1}{x} - \frac{x}{2}\right) y'(x) + y(x) = 0,$$

(a) Find solutions to the indicial equation.
(b) How many series solutions are there according to the Frobenius theorem? Explain your answer.
(c) Find a recurrence relation and use it to find a closed expression for a series solution.

Problem 4.3.10. Show that between any two consecutive positive zeros of $J_n(x)$ there is precisely one zero of $J_{n+1}(x)$.
Hint: Use recurrence relations for Bessel functions.

Problem 4.3.11. Using the recurrence relations for Bessel functions verify that

$$J_{3/2}(x) = \sqrt{\frac{2}{x\pi}} \left(\frac{\sin x}{x} - \cos x\right)$$

and

$$J_{-3/2}(x) = -\sqrt{\frac{2}{x\pi}} \left(\frac{\cos x}{x} + \sin x\right).$$

Problem 4.3.12. A generating function $G(x,t)$ for Bessel functions is defined by

$$G(x,t) = \sum_{n=-\infty}^{\infty} J_n(x)\, t^n = \sum_{n=-\infty}^{\infty} \sum_{k=0}^{\infty} \frac{(-1)^k}{k!(n+k)!} \left(\frac{x}{2}\right)^{2k+n} t^n,$$

where for the $n < 0$ terms in the above relation, we assume that $1/(k+n)! = 1/\Gamma(k+n+1)$, which is zero for $n < k$. Show that by making use of the change of variables from n to $m = n+k$ we can eliminate the parameter n and rewrite $G(x,t)$ in terms of m as:

$$G(x,t) = \sum_{k=0}^{\infty} \sum_{m=0}^{\infty} \frac{(-1)^k}{k!m!} \left(\frac{x}{2}\right)^{k+m} t^{m-k}.$$

Next, show that

$$G(x,t) = \exp\left[\frac{x}{2}\left(t - \frac{1}{t}\right)\right].$$

Problem 4.3.13. Show that

$$J_n(x+y) = \sum_{k=-\infty}^{\infty} J_k(x) J_{n-k}(y),$$

by considering the product of generating functions $G(x,t)$ and $G(y,t)$ defined in Problem 4.3.12.

Problem 4.3.14. Use the expression for the generating function $G(x,t)$ from Problem 4.3.12 to deduce the recursion relations:

$$J_n'(x) = \frac{1}{2}\left[J_{n-1}(x) - J_{n+1}(x)\right]$$

and

$$J_{n+1}(x) = \frac{x}{2n+2}\left[J_n(x) + J_{n+2}(x)\right].$$

This method provides an alternative way of proving relations (4.108) and (4.109) for $\nu = n$.

Problem 4.3.15. (a) Find a closed expression for

$$F(x) = \sum_{n=-\infty}^{\infty} J_n(x)\, e^{inx}$$

and find the value of $F(x)$ at $x = \pi/2$.
(b) Verify that

$$\sum_{n=0}^{\infty} J_{2n+1}(x)\sin(2n+1)x = \frac{1}{2}\sin(x\sin x)$$

and

$$\sum_{n=0}^{\infty} J_{2n}(x)\cos(2nx) = \frac{1}{2}\left(\cos(x\sin x) + J_0(x)\right).$$

(c) Show that

$$\int_0^{2\pi} e^{i\,(x\sin\alpha - n\alpha)}\, d\alpha = \int_0^{2\pi} \cos(n\alpha - x\sin\alpha)\, d\alpha = 2\pi J_n(x).$$

Problem 4.3.16. Show that

$$J_0'(x) = -J_1(x) \quad \text{and} \quad (xJ_1(x))' = xJ_0(x).$$

Problem 4.3.17. Evaluate

$$\int_0^a \frac{J_1^2(x)}{x}\, dx$$

using the recurrence relations for Bessel functions.

Problem 4.3.18. Let $y_1(x)$ and $y_2(x)$ be two linearly independent solutions to Bessel's equation of order ν and let $W[y_1, y_2](x)$ be their Wronskian. Using Bessel's equation show that

$$W[y_1, y_2](x) = \frac{c}{x},$$

with some constant, c. Suppose, that $y_1(x) = a_0 + a_1 x + \ldots$ is a power series. Then, show that for $y_2(x) = y_1(x) \ln x + \sum_{m=0}^{\infty} A_m x^m$ it is possible to choose coefficients $A_m, m = 1, 2, \ldots$ to reproduce the above form of the Wronskian with $c \neq 0$.

Problem 4.3.19. (a) Use the method of Frobenius to find a series solution to:

$$x^2 y'' + 4xy' + xy = 0.$$

(b) Reduce the above equation to Bessel's differential equation by setting $y = u(z)/z^3$ with $z = 2\sqrt{x}$.
(c) Find a general solution in terms of Bessel functions.

Problem 4.3.20. Consider the differential equation

$$2x^2 y'' + 3xy' + (2x^2 - 1) y = 0.$$

(a) Find solutions to the indicial equation. How many series solutions are there according to Frobenius theorem?
(b) Find the solution up to the third order.
(c) Reduce the above equation to Bessel's differential equation and find its general solution in terms of Bessel functions by using the results in subsection (4.3.3).

Problem 4.3.21. Consider the differential equation $4x^2 y'' + 4xy' + (x - \frac{1}{9}) y = 0$.
(a) Set $z = \sqrt{x}$ and reduce the differential equation to Bessel's equation in terms of z, $\frac{dy}{dz}$, and $\frac{d^2 y}{dz^2}$.
(b) Find a general solution in terms of Bessel functions and check your result by comparing with equation (4.110).

Problem 4.3.22. Find a general solution to the differential equation from Problem (4.3.8) in terms of Bessel functions by using the results in subsection (4.3.3).

Problem 4.3.23. Find a general solution to the equation, $9x^2 y'' + 9xy' + (9x^2 - \frac{1}{4}) y = 0$ in terms of Bessel functions.

Problem 4.3.24. Find a solution to the equation, $xy'' + (1 - 2n) y' + xy = 0$. Find a solution to the equation: $xy'' - y' + xy = 0$.

Problem 4.3.25. (a) Reduce the equation, $xy'' + y = 0$ to Bessel's differential equation for $u(z)$, where $z = 2\sqrt{x}$ and $y = zu(z)$.
(b) Find a general solution to the differential equation given in (a) in terms of Bessel functions and confirm your result by comparing it with equation (4.110).

Problem 4.3.26. Find a general solution to

$$x^2 y'' + \frac{1}{2} xy' + \frac{1}{16}(x^{1/2} + \frac{15}{16}) y = 0$$

in terms of Bessel functions.

Problem 4.3.27. Find a general solution to

$$xy'' + (1 + 2n)y' + xy = 0$$

in terms of Bessel functions.

4.4 Method of Separation of Variables; Helmholtz's and Laplace's Equations

In this section, we will study a special class of solutions to Helmholtz's equation,

$$\vec{\nabla}^2 u(\vec{r}) = -k^2 u(\vec{r}).$$ (4.113)

By setting $k = 0$ on the right hand side of the Helmholtz's equation we obtain Laplace's equation,

$$\vec{\nabla}^2 u(\vec{r}) = \frac{\partial^2 u(\vec{r})}{\partial x^2} + \frac{\partial^2 u(\vec{r})}{\partial y^2} + \frac{\partial^2 u(\vec{r})}{\partial z^2} = 0.$$ (4.114)

Notice the appearance of partial derivatives in differential equations (4.113)-(4.114). Both Helmholtz's and Laplace's equations are partial differential equations. The method of separation of variables reduces the problem of solving these partial differential equations to that of solving a set of ordinary differential equations that we are already familiar with.

The special class of solutions studied in this section consists of functions, $u(\vec{r}) = u(x, y, z)$, that can be factorized as a product

$$u(x, y, z) = X(x)Y(y)Z(z),$$ (4.115)

of three functions X, Y, Z each a function of x, y, and z, respectively. Not all solutions can be written in this form. Below we will generalize these solutions by forming linear combinations of functions from equation (4.115).

For functions $u(x, y, z)$ whose dependence on coordinates separates as in (4.115) the right hand side of equation (4.114) becomes:

$$\frac{d^2 X(x)}{dx^2} Y(y)Z(z) + \frac{d^2 Y(y)}{dy^2} X(x)Z(z) + \frac{d^2 Z(z)}{dz^2} X(x)Y(y).$$

Substituting the above expression into the right hand side of equation (4.113) and dividing by $u = XYZ$ yields

$$\frac{1}{X(x)} \frac{d^2 X(x)}{dx^2} + \frac{1}{Y(y)} \frac{d^2 Y(y)}{dy^2} + \frac{1}{Z(z)} \frac{d^2 Z(z)}{dz^2} = -k^2.$$

The term, $(1/X(x)) \, \mathrm{d}^2 X(x)/\mathrm{d}x^2$, depends only on x. On the other hand, the following relation

$$\frac{1}{X(x)} \frac{\mathrm{d}^2 X(x)}{\mathrm{d}x^2} = -k^2 - \frac{1}{Y(y)} \frac{\mathrm{d}^2 Y(y)}{\mathrm{d}y^2} - \frac{1}{Z(z)} \frac{\mathrm{d}^2 Z(z)}{\mathrm{d}z^2}$$

shows that $(1/X(x)) \, \mathrm{d}^2 X(x)/\mathrm{d}x^2$ can only be a function of y and z since the right hand side exhibits dependence only on y and z. The only way to reconcile this contradiction is to set

$$\frac{1}{X(x)} \frac{\mathrm{d}^2 X(x)}{\mathrm{d}x^2} = -k_1^2, \qquad (4.116)$$

for some constant k_1. In future discussions, we will refer to constants appearing in such a setting as separation constants.

Thus,

$$\frac{1}{Y(y)} \frac{\mathrm{d}^2 Y(y)}{\mathrm{d}y^2} + \frac{1}{Z(z)} \frac{\mathrm{d}^2 Z(z)}{\mathrm{d}z^2} = -k^2 + k_1^2 .$$

Similar arguments as above yield

$$\frac{1}{Y(y)} \frac{\mathrm{d}^2 Y(y)}{\mathrm{d}y^2} = -k_2^2, \qquad \frac{1}{Z(z)} \frac{\mathrm{d}^2 Z(z)}{\mathrm{d}z^2} = -k_3^2, \qquad (4.117)$$

with separation constants, k_2 and k_3. The separation constants are related to each other through

$$k_1^2 + k_2^2 + k_3^2 = k^2 . \qquad (4.118)$$

Thus, for functions that separate as in (4.115) the problem of solving Helmholtz' equation simplifies to finding solutions to the ordinary differential equations (4.116)-(4.117). Solutions to these equations are easily found to be:

$$X(x) = b_1 e^{ik_1 x} + c_1 e^{-ik_1 x} \qquad (4.119)$$
$$Y(y) = b_2 e^{ik_2 y} + c_2 e^{-ik_2 y} \qquad (4.120)$$
$$Z(z) = b_3 e^{ik_3 z} + c_3 e^{-ik_3 z} . \qquad (4.121)$$

The product of these solutions gives a general separable solution to Helmholtz' equation in the form,

$$u(\vec{r}) = B e^{i \vec{k} \cdot \vec{r}} . \qquad (4.122)$$

Vector $\vec{k}$ has a magnitude equal to $k = \sqrt{\vec{k}^2}$. As long as the magnitude k of $\vec{k}$ is kept fixed we can reproduce all exponential solutions contained in equations (4.119)-(4.121) by changing the signs of components k_1, k_2, k_3 in the exponent $i\,(k_1 x + k_2 y + k_3 z)$ of $u(\vec{r})$.

4.4.1 Separation in Spherical Coordinates

In spherical coordinates, the Helmholtz' equation for the function $u(r, \theta, \varphi)$ takes the form

$$\frac{1}{r^2}\frac{\partial}{\partial r}\left(r^2\frac{\partial u}{\partial r}\right) + \frac{1}{r^2\sin\theta}\left[\frac{\partial}{\partial\theta}\left(\sin\theta\frac{\partial u}{\partial\theta}\right) + \frac{1}{\sin\theta}\frac{\partial^2 u}{\partial\varphi^2}\right] + k^2 u = 0. \qquad (4.123)$$

We seek solutions which separate as follows:

$$u(r, \theta, \varphi) = R(r)Y(\theta, \varphi). \qquad (4.124)$$

Substituting $u = RY$ into equation (4.123) and dividing by RY yields

$$\frac{1}{Rr^2}\frac{d}{dr}\left(r^2\frac{dR}{dr}\right) + \frac{1}{r^2 Y\sin\theta}\left[\frac{\partial}{\partial\theta}\left(\sin\theta\frac{\partial Y}{\partial\theta}\right) + \frac{1}{\sin\theta}\frac{\partial^2 Y}{\partial\varphi^2}\right] = -k^2.$$

After multiplication by r^2, we can further separate dependence on r and angular variables θ, φ:

$$\frac{1}{Y\sin\theta}\left[\frac{\partial}{\partial\theta}\left(\sin\theta\frac{\partial Y}{\partial\theta}\right) + \frac{1}{\sin\theta}\frac{\partial^2 Y}{\partial\varphi^2}\right] = -\frac{1}{R}\frac{d}{dr}\left(r^2\frac{dR}{dr}\right) - k^2 r^2.$$

The left hand side depends on θ, φ while the right hand side depends only on r. This dependence implies that

$$\frac{1}{Y\sin\theta}\left[\frac{\partial}{\partial\theta}\left(\sin\theta\frac{\partial Y}{\partial\theta}\right) + \frac{1}{\sin\theta}\frac{\partial^2 Y}{\partial\varphi^2}\right] = -\gamma \qquad (4.125)$$

and

$$\frac{1}{Rr^2}\frac{d}{dr}\left(r^2\frac{dR}{dr}\right) + k^2 - \frac{\gamma}{r^2} = 0,$$

for some separation constant γ.

First, we deal with the case of the non-zero constant, k. Introducing the new variable, $\sigma = rk$, we can rewrite the radial equation for R as

$$\frac{d^2 R}{d\sigma^2} + \frac{2}{\sigma}\frac{dR}{d\sigma} + \left(1 - \frac{\gamma}{\sigma^2}\right)R = 0. \qquad (4.126)$$

We replace the function R by a function B related to R through

$$R = \frac{1}{\sqrt{\sigma}}B.$$

In terms of function B, equation (4.126) assumes the form

$$\frac{d^2 R}{d\sigma^2} + \frac{1}{\sigma}\frac{dB}{d\sigma} + \left(1 - \frac{\gamma + 1/4}{\sigma^2}\right)B = 0,$$

in which we recognize Bessel's equation of order $\nu = \gamma + 1/4$ with the following general solution

$$B(\sigma) = c_1 J_{\gamma+1/4}(\sigma) + c_2 Y_{\gamma+1/4}(\sigma),$$

which is written in terms of Bessel functions of the first and second kind. Accordingly, the function B can be written as a function of k and r,

$$R(kr) = \frac{c_1}{\sqrt{kr}} J_{\gamma+1/4}(kr) + \frac{c_2}{\sqrt{kr}} Y_{\gamma+1/4}(kr).$$

For Laplace's equation with $k = 0$, the radial equation for R takes the simple form,

$$\frac{d}{dr}\left(r^2 \frac{dR}{dr}\right) - \gamma R = 0. \tag{4.127}$$

The above expression is Euler's equation with a general solution

$$R(r) = c_1 r^{m+} + c_2 r^{m-},$$

where

$$m_\pm = \frac{1}{2}\left(-1 \pm \sqrt{1+4\gamma}\right),$$

are roots of a quadratic equation $m(m-1) + 2m - \gamma = 0$.

Now, let us go back to the angular part in Helmholtz' equation, (4.125), and further separate function $Y(\theta, \varphi)$ in θ and φ dependence as follows,

$$Y(\theta, \varphi) = \Theta(\theta)\Phi(\varphi).$$

Plugging the above relation into equation (4.125) yields

$$\frac{1}{\Theta \sin\theta} \frac{\partial}{\partial\theta}\left(\sin\theta \frac{\partial\Theta}{\partial\theta}\right) + \frac{1}{\Phi \sin^2\theta} \frac{\partial^2\Phi}{\partial\varphi^2} = -\gamma.$$

By introducing one more separation constant, m, we can rewrite the above equation as two differential equations. One for variable φ:

$$\frac{1}{\Phi} \frac{\partial^2\Phi}{\partial\varphi^2} = -m^2 \tag{4.128}$$

and the other for θ:

$$\frac{1}{\sin\theta} \frac{\partial}{\partial\theta}\left(\sin\theta \frac{\partial\Theta}{\partial\theta}\right) + \left(\gamma - \frac{m^2}{\sin^2\theta}\right)\Theta = 0. \tag{4.129}$$

Equation (4.128) is easy to integrate. The solution has the simple form:

$$\Phi(\varphi) = C_1 e^{im\varphi} + C_2 e^{-im\varphi}. \tag{4.130}$$

Equation (4.129) can be related to Legendre's equation after a change of coordinates, $\theta \to x = \cos \theta$, and noticing that $P(x) = \Theta(\theta)$ satisfies:

$$\frac{d\Theta}{d\theta} = \frac{dP}{dx}\frac{dx}{d\theta} = -\sin\theta\frac{dP}{dx} = -(1 - x^2)^{1/2}\frac{dP}{dx}.$$

Using the above identity, we can rewrite equation (4.129) in terms of new coordinates as

$$\frac{d}{dx}\left[(1 - x^2)\frac{dP}{dx}\right] + \left[\gamma - \frac{m^2}{1 - x^2}\right]P = 0. \tag{4.131}$$

Let's consider the solution $Y(\theta, \varphi)$, which is independent of φ. For this to be true, m in equation (4.130) has to be zero. In this case, equation (4.131) reduces to

$$\frac{d}{dx}\left[(1 - x^2)\frac{dP}{dx}\right] + \gamma P = 0, \tag{4.132}$$

which is Legendre's equation. The range of $x = \cos \theta$ is $-1 \le x \le 1$ since θ is between 0 and π. Recall from subsection 4.2.1 that in order for the series solution to equation (4.132) to be finite at the endpoints $x = \pm 1$, the series must terminate. This requires that $\gamma = n(n + 1)$ for some integer n. The resulting polynomial solutions to (4.132) are Legendre polynomials $P_n(x)$. Correspondingly, for $\gamma = n(n + 1)$, the radial equation (4.127) has two independent solutions, $R(r) = r^n$ and $R(r) = r^{-(n+1)}$. Then, the general φ-independent solution to the Laplace's equation on the (r, θ)-plane is given by a linear combination:

$$R(r)Y(\theta) = \sum_{n=0}^{\infty}\left[a_n r^n + b_n r^{-(n+1)}\right]P_n(\cos\theta). \tag{4.133}$$

☞ **Example 4.4.1.** We illustrate the last result with an example from electrostatics. Imagine a point charge q located at the point $z = z_0$ on the z-axis. The electrostatic potential at the arbitrary position r due to charge q is

$$V(r, \theta) = \frac{q}{4\pi\epsilon_0}\frac{1}{|\vec{r} - z_0\hat{k}|} = \frac{q}{4\pi\epsilon_0}\frac{1}{(r^2 - 2rz_0\cos\theta + z^2)^{1/2}}, \tag{4.134}$$

where θ is the angle between the position vector $\vec{r}$ and the z-axis.

The electric field is a conservative field that satisfies $\vec{\nabla} \times \vec{E} = 0$. Thus, it possesses a potential related to the electric field via $\vec{E} = -\vec{\nabla}V$. In a region without charges, the electric field $\vec{E}$ satisfies $\vec{\nabla} \cdot \vec{E} = 0$. For the electrostatic potential V, this amounts to Laplace's equation $\vec{\nabla}^2 V = 0$. Therefore, we expect that the Coulomb expression (4.134) for the electrostatic potential as a function of two variables r, θ should agree with that of solutions to the two-dimensional Laplace equation (4.133) on the (r, θ)-plane.

Next, we expand expression (4.134) in Legendre polynomials. Let, $x = \cos\theta$ and $t = z_0/r$ for $r > z_0$. Then, the potential

$$V(r,\theta) = \frac{q}{4\pi\epsilon_0 r} \frac{1}{(1 - 2tx + t^2)^{1/2}}$$

can be rewritten as

$$V(r,\theta) = \frac{q}{4\pi\epsilon_0 r} G(x,t),$$

in terms of the Legendre generating function $G(x,t) = \sum_{n=0}^{\infty} P_n(x)t^n$, which was defined in Problem 4.2.9. By expanding the generating function for $r > z_0$ (or $t < 1$), we obtain the following expression for the electrostatic potential:

$$V(r,\theta) = \frac{q}{4\pi\epsilon_0} \sum_{n=0}^{\infty} z_0^n r^{-(n+1)} P_n(\cos\theta).$$

For $r < z_0$, we define t as $t = r/z_0 < 1$ and derive an expression for $V(r,\theta)$ by a similar technique to the developed above,

$$V(r,\theta) = \frac{q}{4\pi\epsilon_0} \sum_{n=0}^{\infty} r^n z_0^{-(n+1)} P_n(\cos\theta).$$

These formulas agree with solutions (4.133) to the two-dimensional Laplace equation.

☞ **Example 4.4.2.** Now, consider a conducting uncharged sphere of radius R placed inside a uniform external electric field $\vec{E} = E_0\hat{k}$ in such a way that the origin of the coordinate system coincides with the center of the sphere. We also assume that the conducting sphere is grounded ($V(r,\theta) = 0$, $r \le R$). Because of the symmetry of the problem, we ignore variable φ. Now, we can expand the potential $V(r,\theta)$ like we did in equation (4.133) using the fact that $V(r,\theta)$ is a solution of Laplace's equation. In the region outside the conductor, $r > R$, we can write

$$V(r,\theta) = \sum_{n=0}^{\infty} \left[b_n r^n + c_n r^{-(n+1)} \right] P_n(\cos\theta), \quad r > R.$$

Note that

$$V(r,\theta) \longrightarrow -E_0 z + C = -E_0 r \cos\theta + C,$$

as $r \to \infty$. Next, we use the fact that expression $V = -E_0 z + C$, with an arbitrary constant C, gives rise to an electric field $\vec{E} = -\vec{\nabla}V = E_0\hat{k}$. By recalling that $P_1(\cos\theta) = \cos\theta$ it becomes clear that the only non-vanishing

b_n's are $b_1 = -E_0$ and $b_0 = C$. Moreover, the potential, $V(r, \theta)$, has to vanish on the surface of the sphere:

$$0 = V(R, \theta) = C - E_0 R P_1(\cos\theta) + \sum_{n=0}^{\infty} c_n R^{-(n+1)} P_n(\cos\theta).$$

The Legendre polynomials are linearly independent and, therefore, the coefficient in front of every Legendre polynomial has to be zero. Thus, $c_0 = -RC$, $c_1 = R^3 E_0$, and $c_n = 0$ for $n \geq 2$.

Therefore, we obtain the following expression

$$V(r, \theta) = C\left(1 - \frac{R}{r}\right) - E_0\left(r - \frac{R^3}{r^2}\right)\cos\theta.$$

The term, $C(1 - R/r)$, makes the following contribution to $\vec{\nabla}^2 V(r, \theta)$,

$$\vec{\nabla}^2 C\left(1 - \frac{R}{r}\right) = \frac{1}{r^2}\frac{\partial}{\partial r}r^2\frac{\partial}{\partial r}C\left(1 - \frac{R}{r}\right) = C\frac{R}{r^2},$$

where we performed calculation in spherical coordinates. Thus, constant C contributes to the charge distribution on the sphere and, therefore, we need to set $C = 0$ for the uncharged sphere. Therefore, the final solution is

$$V(r, \theta) = -E_0\left(r - \frac{R^3}{r^2}\right)\cos\theta.$$

In the general case for equation (4.131) with non-zero m such that $-n \leq m \leq n$, the solutions are the associated Legendre polynomials:

$$P_n^m(x) = \frac{(-1)^n}{2^n n!}(1 - x^2)^{m/2}\frac{d^{n+m}}{dx^{n+m}}(1 - x^2)^n. \qquad (4.135)$$

For $m > 0$, we can write this definition in a simpler form:

$$P_n^m(x) = (1 - x^2)^{m/2}\frac{d^m}{dx^m}P_n(x),$$

which gives the associated Legendre polynomials in terms of Legendre polynomials.

By plugging $\gamma = n(n+1)$ into equation (4.126) and multiplying by σ^2 we obtain a spherical Bessel equation

$$\sigma^2\frac{d^2 R}{d\sigma^2} + 2\sigma\frac{dR}{d\sigma} + (\sigma^2 - n(n+1))R = 0, \qquad (4.136)$$

which agrees with equation (4.110) for $\beta = 1/2$, $\alpha = k = 1$ and $\nu^2 = 1/4 + n(n+1) = (n + 1/2)^2$. This relation leads to one of two independent solutions to the spherical Bessel equation in the form $cJ_{n+1/2}(\sigma)/\sqrt{x}$ written in terms of Bessel equation of order $n + 1/2$ and an arbitrary constant c. Setting the constant c equal to $\sqrt{\pi/2}$

reproduces a conventional choice known as the spherical Bessel function of order n:

$$j_n(\sigma) = \left(\frac{\pi}{2\sigma}\right)^{1/2} J_{n+\frac{1}{2}}(\sigma) . \tag{4.137}$$

4.4.2 Separation in Cylindrical Coordinates

In cylindrical coordinates, Helmholtz's equation takes the form:

$$\vec{\nabla}^2 f(\rho,\varphi,z) + k^2 f(\rho,\varphi,z) = \frac{1}{\rho}\frac{\partial}{\partial\rho}\left(\rho\frac{\partial f}{\partial\rho}\right) + \frac{1}{\rho^2}\frac{\partial^2 f}{\partial\varphi^2} + \frac{\partial^2 f}{\partial z^2} + k^2 f(\rho,\varphi,z) = 0 .$$

$$\tag{4.138}$$

This equation has a separable solutions in the form:

$$f(\rho,\varphi,z) = R(\rho)\Phi(\varphi)Z(z) .$$

Upon substitution of the separated solution in Helmholtz's equation and dividing by $f = R\Phi Z$ we get:

$$\frac{1}{R\rho}\frac{\partial}{\partial\rho}\left(\rho\frac{\partial R}{\partial\rho}\right) + \frac{1}{\Phi\rho^2}\frac{\partial^2\Phi}{\partial\varphi^2} + \frac{1}{Z}\frac{\partial^2 Z}{\partial z^2} + k^2 = 0 .$$

Using the method separation, we can split the above equation in three one-dimensional ordinary differential equations:

$$\frac{1}{Z}\frac{d^2 Z}{dz^2} + k^2 = \beta^2 ,$$

$$\frac{1}{\Phi}\frac{d^2\Phi}{d\varphi^2} = -m^2 , \tag{4.139}$$

and

$$\frac{1}{R\rho}\frac{d}{d\rho}\left(\rho\frac{dR}{d\rho}\right) - \frac{m^2}{\rho^2} + \beta^2 = 0 ,$$

in terms of new separation constants, m and β. The first two equations in z and φ have the following simple exponential solutions

$$Z(z) = C_1 e^{\sqrt{\beta^2 - k^2} z} + C_2 e^{-\sqrt{\beta^2 - k^2} z}$$

and

$$\Phi(\varphi) = K_1 e^{i\,m\varphi} + K_2 e^{-i\,m\varphi} . \tag{4.140}$$

Requiring (as one routinely does in physics) that $\Phi(\varphi)$ is a single-valued function on the interval $0 \le \varphi \le 2\pi$ (i.e. $\Phi(0) = \Phi(2\pi)$) forces the index m to be an integer.

For $\beta \ne 0$ the third equation can be rewritten as:

$$\frac{d^2 R}{dx^2} + \frac{1}{x}\frac{dR}{dx} + \left(1 - \frac{m^2}{x^2}\right)R = 0$$

after a suitable change of coordinate $\rho \to x = \beta\rho$. Upon multiplying the above equation with x^2 we obtain the standard Bessel's equation of order m:

$$x^2 \frac{d^2 R}{dx^2} + x \frac{dR}{dx} + (x^2 - m^2) R = 0,$$

which sometimes is also called a cylindrical Bessel equation. Recall that this equation has two linearly independent solutions, $J_m(x)$ and $Y_m(X)$.

For $\beta = 0$, the equation for $R(\rho)$ reduces to equation

$$\frac{1}{R\rho} \frac{d}{d\rho} \left(\rho \frac{dR}{d\rho} \right) - \frac{m^2}{\rho^2} = 0, \tag{4.141}$$

which has solutions $R = c_1 \rho^m + c_2 \rho^{-m}$ for $m \neq 0$ and $R = c_1 + c_2 \ln \rho$ for $m = 0$.

Now, consider a much simpler problem of Laplace's equation ($k = 0$ in equation (4.138)). We choose to seek solutions of this equation, which are independent of z and, therefore, satisfy

$$\frac{1}{\rho} \frac{\partial}{\partial \rho} \left(\rho \frac{\partial f}{\partial \rho} \right) + \frac{1}{\rho^2} \frac{\partial^2 f}{\partial \varphi^2} = 0.$$

In addition, we restrict the problem by considering solutions of the form

$$f(\rho, \varphi) = R(\rho)\Phi(\varphi)$$

for which we find

$$\frac{\rho}{R} \frac{\partial}{\partial \rho} \left(\rho \frac{\partial R}{\partial \rho} \right) = -\frac{1}{\Phi} \frac{\partial^2 \Phi}{\partial \varphi^2} = C,$$

where C is a separation constant. We only allow single-valued solutions (i.e. $\Phi(0) = \Phi(2\pi)$) of the Φ equation

$$\frac{\partial^2 \Phi}{\partial \varphi^2} = -C\Phi. \tag{4.142}$$

As discussed upon evaluating equation (4.140), the solution $\Phi(\varphi) = A_\pm \exp\left(\pm i \sqrt{C} \varphi \right)$ of the φ-equation will only be single-valued when the separation constant C is equal to a square of some integer $m = 0, 1, 2, \ldots$. In such case, it is convenient to label $\Phi(\varphi)$ by index m and express it by trigonometric functions as in:

$$\Phi_m(\varphi) = a_m \cos(m\varphi) + b_m \sin(m\varphi) \quad m = 0, 1, 2, 3, \ldots.$$

The remaining R equation:

$$\rho \frac{\partial}{\partial \rho} \left(\rho \frac{\partial R}{\partial \rho} \right) = m^2 R$$

is identical to equation (4.141) and has solutions, $R = c_1\rho^m + c_2\rho^{-m}$ for $m \neq 0$ and $R = c_1 + c_2 \ln\rho$ for $m = 0$. Thus, we found that solutions to the two-dimensional Laplace equation, $\vec{\nabla}^2 f(\rho, \varphi) = 0$, can be given in terms of cylindrical harmonics:

$$1, \quad \ln\rho, \quad \rho^{\pm m}\cos(m\varphi), \quad \rho^{\pm m}\sin(m\varphi),$$

where $m = 1, 2, 3, \ldots$. The linear combination of these cylindrical harmonics are used as solutions to Laplace's equation.

☞ **Example 4.4.3.** We seek solutions to the two-dimensional Laplace equation $\vec{\nabla}^2 f(\rho, \varphi) = 0$ in the region $\rho \leq a$ with the boundary condition $f(a, \varphi) = F(\varphi)$ for some function F of φ between 0 and 2π. For solutions $f(a, \varphi)$, whose dependence on coordinates separates, the function $F(\varphi)$ must satisfy equation (4.139) or (4.142).

We can represent solution $f(\rho, \varphi)$ as a linear combination of cylindrical harmonics:

$$f(\rho, \varphi) = A_0 + \sum_{m=1}^{\infty} \left(A_m \rho^m \cos(m\varphi) + B_m \rho^m \sin(m\varphi) \right),$$

where we omitted those cylindrical harmonics that have a singularity at the origin $\rho = 0$, which is part of the region.

Setting $\rho = a$ gives the expansion

$$F(\varphi) = A_0 + \sum_{m=1}^{\infty} \left(A_m a^m \cos(m\varphi) + B_m a^m \sin(m\varphi) \right)$$

in terms of sine and cosine functions for the solution $F(\varphi)$ of differential equation (4.139). This is the so-called Fourier series expansion, which we will study in Chapter 5.

⬧ **Method of separation of variables: Problems**

Problem 4.4.1. Use separation of variables to find a solution to equation

$$\frac{\partial^2 f(x, y)}{\partial x^2} = 2\frac{\partial f(x, y)}{\partial y},$$

such that $f(0, y) = 0$.

Problem 4.4.2. Use separation of variables to find a solution to equation:

$$x\frac{\partial f(x, y)}{\partial x} = -y\frac{\partial f(x, y)}{\partial y}.$$

Problem 4.4.3. A conducting sphere of radius R is centered at the origin. The sphere is grounded ($V = 0$) and surrounded concentrically by a spherical shell of radius $2R$. The potential on the spherical shell is independent of φ and satisfies $V(2R, \theta) = V_0 P_2(\cos \theta)$. Find the potential $V(r, \theta)$ in the region between the spherical surfaces, $R < r < 2R$.

Problem 4.4.4. Find a solution $F(x, y)$ to the two-dimensional Laplace equation, $\vec{\nabla}^2 F(x, y) = 0$, in the region $0 \le x \le \pi$ and $0 \le y < \infty$ with initial conditions

$$F(0, y) = 0, \quad F(\pi, y) = 0, \quad \lim_{y \to \infty} F(x, y) = 0.$$

Problem 4.4.5. Find a solution $y(t, x)$ to the heat equation:

$$\frac{\partial y}{\partial t} = \kappa \frac{\partial^2 y}{\partial x^2} \quad\quad 0 < x < \pi, t > 0$$
$$y(t, 0) = 0, \quad y(t, \pi) = 0 \quad\quad t > 0$$
$$y(0, x) = f(x) \quad\quad 0 < x < \pi.$$

The above system consists of a differential equation with initial conditions. That system describes the heat distribution in a finite metal bar of length π. The initial conditions are chosen in such a way as to ensure a fixed null temperature at the endpoints, 0 and π. The function $f(x)$ gives the initial temperature at time $t = 0$.

Chapter 5

Special Functions and the Generalized Fourier Series

FOURIER series provides an expansion of a periodic function (that may have discontinuities) in terms of sine and cosine functions. In this Chapter, we will study the Fourier series method from a general perspective of expansions in terms of orthogonal eigenfunctions. We will treat Fourier series as a special case of Sturm-Liouville theory, which deals with differential equations that can be written in a self-adjoint form. The solutions to the self-adjoint differential equation form an orthogonal basis in some function space defined by appropriate boundary conditions. These solutions are analogous to eigenvectors of an $n \times n$ Hermitian matrix that provide an orthogonal basis of $\mathbb{R}^n$ vector space. In the case of Fourier series, the set of trigonometric functions $\{1, \cos nx, \sin nx\}$ for $n = 1, 2, \ldots$ is an orthogonal basis on the interval $\pi \le x \le \pi$ and the theory of Fourier series is a natural extension of the representation of a vector in terms of an orthogonal basis.

5.1 Sturm-Liouville Theory and the Orthogonal Functions Expansion

5.1.1 Introduction, Vibrations of the String

Here, we consider a system of equations that governs the transverse vibrations of a finite string of length π, with endpoints kept at rest. The system contains the one-dimensional wave equation (5.1a), boundary conditions (5.1b) and initial conditions

(5.1c)-(5.1d):

$$\frac{\partial^2 y}{\partial t^2} = \kappa^2 \frac{\partial^2 y}{\partial x^2} \quad 0 < x < \pi, t > 0, \tag{5.1a}$$

$$y(t,0) = 0 \quad y(t,\pi) = 0 \quad t > 0, \tag{5.1b}$$

$$\frac{\partial y(0,x)}{\partial t} = 0, \quad 0 < x < \pi, \tag{5.1c}$$

$$y(0,x) = f(x) \quad 0 \le x \le \pi. \tag{5.1d}$$

The initial displacement at time $t = 0$ is given by a function $f(x)$. As in Chapter 4, we will look for solutions in the factorized form $y(t,x) = T(t)X(x)$. Plugging such structure into the differential equation (5.1a) yields:

$$T''X = \kappa^2 T X''$$

or, after dividing by $\kappa^2 T X$,

$$\frac{X''(x)}{X(x)} = \frac{T''(t)}{\kappa^2 T(t)}.$$

The left hand side is a function of x alone, while the right hand side is a function of t only. Thus, both sides must be equal to some constant, λ, so that

$$\frac{T''(t)}{T(t)} = \kappa^2 \lambda \tag{5.2}$$

and

$$\frac{X''(x)}{X(x)} = \lambda.$$

The above equations are ordinary differential equations, which can easily be solved.

Thus, the separation technique leads to the ordinary second order differential equation:

$$X''(x) - \lambda X(x) = 0, \tag{5.3}$$

with x on the interval $0 < x < \pi$ and function $X(x)$ subject to the boundary conditions:

$$X(0) = 0 \quad \text{and} \quad X(\pi) = 0. \tag{5.4}$$

Equation (5.3) is known as a *harmonic oscillator equation*. The parameter λ is an unknown constant variable, which we will try to determine by imposing the boundary conditions (5.4). For $\lambda > 0$, the solution is $X(x) = A \exp(\sqrt{\lambda}x) + B \exp(-\sqrt{\lambda}x)$ but the condition (5.4) requires $A = B = 0$. Thus, this solution fails to solve both equations (5.3) and (5.4). If $\lambda = 0$, then the solution becomes $X(x) = A + Bx$. As before, $A = B = 0$ appears to be the only solution to the condition (5.4). Hence, the only remaining possibility is a negative parameter λ. We write $\lambda = -k^2$ and obtain

$$X(x) = A_k \cos(kx) + B_k \sin(kx).$$

Applying the boundary condition (5.4) yields $A_k = 0$ and $B_k \sin(k\pi) = 0$. The constant B_k can not be zero, for then $X(x)$ would vanish everywhere and solution would be trivial. Hence, $\sin(k\pi) = 0$, which implies

$$k = n, \quad n = 1, 2, 3, \cdots$$

and

$$\lambda = -n^2.$$

These values of λ are said to be *eigenvalues*. The allowed solutions to the system of equations (5.3)-(5.4) are found to be

$$X_n(x) = B_n \sin(nx). \tag{5.5}$$

The functions $X_n(x)$ are said to be *eigenfunctions* of the system of equations (5.3)-(5.4). We will refer to the system of equations (5.3)-(5.4) as an *eigenvalue problem*. Thus, the first few eigenvalue-eigenfunction pairs of solutions to the above eigenvalue problem are:

$$\lambda_1 = -1 \quad \longleftrightarrow \quad y_1(x) = B_1 \sin(x)$$
$$\lambda_2 = -4 \quad \longleftrightarrow \quad y_2(x) = B_2 \sin(2x)$$
$$\lambda_3 = -9 \quad \longleftrightarrow \quad y_3(x) = B_3 \sin(3x)$$
$$\vdots$$

Solution to the ordinary differential equation in t (5.2) with $\lambda = -n^2$, that also satisfies the condition (5.1c), $T'(0) = 0$, must be of the form:

$$T(t) = C \cos(n\kappa t),$$

where C is an arbitrary constant.

Combining all the derived results, we find that a family of functions

$$c_n \cos(n\kappa t) \sin(nx), \quad n = 1, 2, 3, \cdots$$

satisfies the system (5.1a)-(5.1d). Likewise, the infinite series of functions

$$y(t, x) = \sum_{n=1}^{\infty} c_n \cos(n\kappa t) \sin(nx)$$

will satisfy the differential equation(5.1a) and all the associated boundary and initial conditions. The terms in the above series have the initial value at the time $t = 0$ equal to $c_n \sin(nx)$. Is their sum equal to the initial displacement of the string given by function $f(x)$? The work by D'Alembert, Euler, Bernoulli and Fourier showed that any reasonable function $f(x)$ on the interval $0 \le x \le \pi$ with the boundary conditions $f(0) = 0$ and $f(\pi) = 0$, could indeed be expanded in terms of the functions $\sin(nx)$ as

$$f(x) = \sum_{n=1}^{\infty} c_n \sin(nx), \qquad (5.6)$$

with some appropriate numbers c_n, which are referred to as the Fourier coefficients of $f(x)$.

The above results represent a fundament of the Fourier series method, that is a most widely known and applied example of the eigenfunction expansion.

5.1.2 The Sturm-Liouville Eigenvalue Problem

In this subsection, we will study a general class of differential equations with solutions subject to boundary conditions. These solutions are the eigenfunctions entering similar series expansions as Fourier series shown in (5.6). Sturm-Liouville theory is a study of the eigenvalue problem involving differential equation:

$$\frac{\mathrm{d}}{\mathrm{d}x}\left(p(x)\frac{\mathrm{d}y(x)}{\mathrm{d}x}\right) + q(x)y(x) + \lambda w(x)y(x)$$
$$= p\frac{\mathrm{d}^2 y}{\mathrm{d}x^2} + \frac{\mathrm{d}p}{\mathrm{d}x}\frac{\mathrm{d}y}{\mathrm{d}x} + (q + \lambda w)y = 0, \qquad (5.7)$$

generalizing equation (5.3). Here p, q, w are real functions of x belonging to the interval $a \le x \le b$, with endpoints, a and b. In addition, the function $w(x)$ called the *weight function* is assumed positive on the whole interval $a \le x \le b$.

It is convenient to introduce a differential operator:

$$\mathcal{L} = p(x)\frac{\mathrm{d}^2}{\mathrm{d}x^2} + \frac{\mathrm{d}p(x)}{\mathrm{d}x}\frac{\mathrm{d}}{\mathrm{d}x} + q(x), \qquad (5.8)$$

which allows us to rewrite (5.7) as an eigenvalue problem:

$$\mathcal{L}[y] = -\lambda w y. \qquad (5.9)$$

with eigenfunction y.

☞ **Example 5.1.1.** The Legendre equation $(1 - x^2)y'' - 2xy' + n(n+1)y = 0$ acquires a Sturm-Liouville form (5.7) $\left((1 - x^2)y'\right)' + n(n+1)y = 0$ with $p(x) = 1 - x^2, q(x) = 0$, the weight function $w(x) = 1$ and the eigenvalue $\lambda = n(n+1)$.

It remains to associate the boundary conditions to the Sturm-Liouville problem. The boundary conditions will be chosen in such a way as to ensure that the operator $\mathcal{L}$ satisfies condition

$$\int_a^b f^* \mathcal{L}[g] \, \mathrm{d}x = \int_a^b g \mathcal{L}[f^*] \, \mathrm{d}x. \tag{5.10}$$

Definition 5.1.1. *The operator $\mathcal{L}$ which satisfies property (5.10) is called a self-adjoint operator.*

We will see that the Sturm-Liouville operator $\mathcal{L}$ becomes self-adjoint for any pair of functions f, g such that they satisfy appropriate boundary conditions. To find these boundary conditions we use integration by parts to rewrite the left hand side of (5.10) as

$$\begin{aligned}
\int_a^b f^* \mathcal{L}[g] \, \mathrm{d}x &= \int_a^b \left(f^*(pg')' + f^* qg \right) \mathrm{d}x \\
&= \int_a^b \left(-f^{*\prime} pg' + f^* qg \right) \mathrm{d}x + [pf^* g']_a^b \\
&= \int_a^b \left((pf^{*\prime})'g + f^* qg \right) \mathrm{d}x + [pf^* g' - pf^{*\prime} g]_a^b. \tag{5.11}
\end{aligned}$$

The last integrand in equation (5.11) agrees with the integrand on the right hand side of relation (5.10). Hence, relation (5.10) will hold for functions f, g such that the surface term $[pf^* g' - pf^{*\prime} g]_a^b$ vanishes or alternatively

$$p(a) \left(f^*(a)g'(a) - f^{*\prime}(a)g(a) \right) = p(b) \left(f^*(b)g'(b) - f^{*\prime}(b)g(b) \right). \tag{5.12}$$

These are the appropriate boundary conditions which ensure that the Sturm-Liouville operator $\mathcal{L}$ is self-adjoint.

One class of solutions to equation (5.12) is given by the boundary conditions of the form $\alpha_1 f(a) + \alpha_2 f'(a) = 0$, $\beta_1 f(b) + \beta_2 f'(b) = 0$ and $\alpha_1 g(a) + \alpha_2 g'(a) = 0$, $\beta_1 g(b) + \beta_2 g'(b) = 0$ for which both sides of (5.12) vanish.

The conditions (5.12) simplify for $p(x)$ such that $p(a) = p(b)$ and become:

$$f^*(a)g'(a) = f^*(b)g'(b), \quad f^{*\prime}(a)g(a) = f^{*\prime}(b)g(b). \tag{5.13}$$

In particular, they are satisfied for f and g, which obey the so-called periodic boundary conditions:

$$f(a) = f(b), \quad f'(a) = f'(b), \quad g(a) = g(b), \quad g'(a) = g'(b). \tag{5.14}$$

☞ **Example 5.1.2.** The operator $\mathcal{L} = \frac{\mathrm{d}^2}{\mathrm{d}x^2}$ from the harmonic oscillator equation (5.3) is a Sturm-Liouville operator on the interval $a \leq x \leq b$, with $p(x) = 1, q(x) = 0$ and the unit weight function $w(x) = 1$.

If $a = 0$ and $b = \pi$ and, moreover, we choose boundary conditions as in equation (5.4) then the conditions (5.13) are trivially satisfied by all allowed functions from eq. (5.5) and the operator $\mathcal{L}$ is self-adjoint.

Even greater simplification takes place for the function $p(x)$ which vanishes at the endpoints a and b. In this case, $p(a)f(a) = p(b)f(b) = 0$ and $p(a)g(a) = p(b)g(b) = 0$ for any function f and g which are finite at the endpoints a and b. Consequently, both sides of relation (5.12) are equal to zero.

It is convenient to introduce the integral

$$\boxed{\langle f \mid g \rangle = \int_a^b f^*(x)g(x)w(x)\, dx,} \tag{5.15}$$

defined for two functions f, g on the interval $a \leq x \leq b$. We call this integral an inner product of f and g. If c is any complex number the following rules hold:

$$\langle f \mid cg \rangle = c\langle f \mid g \rangle, \tag{5.16}$$
$$\langle cf \mid g \rangle = c^*\langle f \mid g \rangle, \tag{5.17}$$
$$\langle f \mid g \rangle^* = \langle g \mid f \rangle. \tag{5.18}$$

The inner product of function f with itself

$$\langle f \mid f \rangle = \int_a^b f^*(x)f(x)w(x)\, dx = \int_a^b |f(x)|^2 w(x)\, dx \geq 0 \tag{5.19}$$

is (strictly) positive as long function f is different from zero. It is therefore natural to refer to $\langle f \mid f \rangle^{1/2}$ as a norm $\|f\|$ of function f. Quantity

$$\|f\| = \sqrt{\langle f \mid f \rangle}$$

shares with a norm of a vector a common property of always being positive. It only becomes zero when the function is zero.

Another common property of vector spaces endowed with the inner product is the Cauchy-Schwartz inequality:

$$|\langle f \mid g \rangle| \leq \|f\|\|g\|. \tag{5.20}$$

The proof of the Cauchy-Schwartz inequality is given in Problem (5.1.1).

The self-adjoint operators have three important properties listed below.

① The eigenvalues of self-adjoint operators are real.

Let y satisfy the eigenvalue equation (5.9) with the eigenvalue λ: $\mathcal{L}[y] = -\lambda w y$. It also holds that $\mathcal{L}[y^*] = -\lambda^* w y^*$. Then

$$\lambda \langle y \mid y \rangle = \int_a^b y^*(x)\lambda y(x)w(x)\, dx = -\int_a^b y^*(x)\mathcal{L}[y(x)]\, dx \tag{5.21}$$

and due to the self-adjointness property (5.10)

$$\lambda \langle y \,|\, y \rangle = - \int_a^b \mathcal{L}[y^*]y(x)\,\mathrm{d}x = \int_a^b \lambda^* y^*(x)y(x)w(x)\,\mathrm{d}x = \lambda^* \langle y \,|\, y \rangle. \quad (5.22)$$

By subtracting the right hand side from the left hand side we obtain a reality condition for the eigenvalue λ:

$$(\lambda - \lambda^*)\,\|y\|^2 = 0 \quad \rightarrow \quad \lambda = \lambda^*, \quad (5.23)$$

since $\|y\|^2 > 0$ for the non-zero eigenfunction y.

② The eigenfunctions of the self-adjoint operator (corresponding to the distinct eigenvalues) are orthogonal with respect to the inner product (5.15).

Let y_1 and y_2 be eigenfunctions of the eigenvalue equation (5.9) with distinct eigenvalues λ_1 and λ_2 such that $\mathcal{L}[y_1] = -\lambda_1 w y_1$ and $\mathcal{L}[y_2] = -\lambda_2 w y_2$. Consider:

$$\lambda_2 \langle y_1 \,|\, y_2 \rangle = - \int_a^b y_1^* \mathcal{L}[y_2(x)]\,\mathrm{d}x = - \int_a^b \mathcal{L}[y_1^*]y_2(x)\,\mathrm{d}x = \lambda_1^* \langle y_1 \,|\, y_2 \rangle. \quad (5.24)$$

However, we have already proved that all eigenvalues of $\mathcal{L}$ are real. Therefore, it follows that:

$$(\lambda_1 - \lambda_2)\,\langle y_1 \,|\, y_2 \rangle = 0. \quad (5.25)$$

For $\lambda_1 \neq \lambda_2$, this equation implies that

$$\langle y_1 \,|\, y_2 \rangle = 0. \quad (5.26)$$

③ Completeness of eigenfunctions of the self-adjoint operator.

The eigenfunctions $y_n(x)$ of self-adjoint operator $\mathcal{L}$ form a complete set of functions, meaning that sufficiently nice (at least piecewise continuous, see Definition 5.2.1) function $f(x)$ can be expanded in terms of these eigenfunctions as

$$f(x) = \sum_{n=0}^{\infty} a_n y_n(x), \quad (5.27)$$

The sum is over all eigenfunctions $y_n(x)$ of self-adjoint operator $\mathcal{L}$ which satisfy the eigenvalue problem $\mathcal{L}[y_n] = -\lambda_n w y_n$. The series is called an eigenfunction expansion or generalized Fourier series. Depending on the eigenfunctions used in the expansion in (5.27) it is common to refer to the series as Fourier-sine, Fourier-cosine, Fourier-Legendre, Fourier-Bessel, etc.

The coefficients a_n of the eigenfunction expansion are obtained by taking the inner product of y_k with function f:

$$\langle y_k \mid f \rangle = \sum_{n=0}^{\infty} a_n \langle y_k \mid y_n \rangle = \sum_{n=0}^{\infty} a_n \delta_{n\,k} \|y_k\|^2 = a_k \|y_k\|^2 \qquad (5.28)$$

or

$$a_k = \frac{\langle y_k \mid f \rangle}{\langle y_k \mid y_k \rangle} = \frac{\langle y_k \mid f \rangle}{\|y_k\|^2}. \qquad (5.29)$$

In deriving expression (5.28) we used orthogonality of the eigenfunctions:

$$\langle y_k \mid y_n \rangle = \begin{cases} 0 & \text{if } k \neq n, \\ \|y_k\|^2 & \text{if } k = n. \end{cases} \qquad (5.30)$$

This relation can be rewritten more compactly using the Kronecker's delta as:

$$\langle y_k \mid y_n \rangle = \delta_{k,n} \|y_k\|^2. \qquad (5.31)$$

It follows from the completeness relation (5.27) and the orthogonality relation that the square of the norm of function f is given by the so-called Parseval's equation

$$\int_a^b |f(x)|^2 w(x)\,\mathrm{d}x = \langle f \mid f \rangle = \sum_{n=0}^{\infty} \sum_{m=0}^{\infty} a_n^* a_m \langle y_n, \, y_m \rangle = \sum_{n=0}^{\infty} |a_n|^2. \qquad (5.32)$$

The presence of the inner product (5.15) allows use of the Gram-Schmidt orthogonalization procedure to turn the linearly independent functions into orthogonal functions. Carrying the discussion in Chapter 2 over to the vector space of functions endowed with the inner product (5.15) we write down the induction process of the Gram-Schmidt orthogonalization algorithm as the following set of recurrence expressions:

$$f_1(x) = g_1(x)$$

$$f_n(n) = g_n(x) - \sum_{k=1}^{n-1} \frac{\langle f_k \mid g_n \rangle}{\langle f_k \mid f_k \rangle} f_k(x), \quad n = 2, 3, \ldots,$$

which map the set $\{g_1(x), \ldots, g_n(x)\}$ of the linearly independent functions into the set $\{f_1(x), \ldots, f_n(x)\}$ of functions orthogonal with respect to the inner product (5.15). The final step of the Gram-Schmidt orthogonalization procedure:

$$f_i(x) \longrightarrow \hat{f}_i(x) = \frac{1}{\sqrt{\langle f_i \mid f_i \rangle}} f_i(x), \quad i = 1, \ldots, n$$

ensures normalization of each member of the orthonormal basis $\{\hat{f}_1(x), \ldots, \hat{f}_n(x)\}$. We will illustrate this procedure for the particular choice of $w(x) = 1$ in definition (5.15) of the inner product.

☞ **Example 5.1.3.** Consider the basis $\{1, x, x^2, ..., x^n\}$ for the infinite-dimensional vector space $\mathbb{P}^n$ of polynomials (of up to the n-th degree). Let

$$\langle f|g \rangle = \int_{-1}^{1} f(x)g(x)\, dx$$

be the inner product over $\mathbb{P}^n$. We will apply the Gram-Schmidt algorithm to the basis $\{1, x, x^2, ..., x^n\}$ in order to obtain the first four elements f_0, f_1, f_2, f_3 of the orthonormal basis for $\mathbb{P}^n$ with respect to the above inner product.

Clearly, $f_1(x) = 1$ with $\langle f_1|f_1 \rangle = 2$. Next, due to $\langle 1|x \rangle = 0$ we obtain $f_2(x) = x$. To obtain $f_3(x)$ we subtract from x^2 its components in $f_1(x)$ and $f_2(x)$ directions

$$f_3(x) = x^2 - \frac{\langle x^2|1 \rangle}{\langle 1|1 \rangle} 1 - \frac{\langle x^2|x \rangle}{\langle x|x \rangle} x$$

$$= x^2 - \frac{1}{3}.$$

Note, that $\int_{-1}^{1} x^{2k+1}\, dx = 0$ for any integer, k, which simplifies calculation.

As the next step, we compute $f_4(x)$ by subtracting components of x^3 in directions of $f_1(x)$, $f_2(x)$ and $f_3(x)$:

$$f_4(x) = x^3 - \frac{\langle x^3|1 \rangle}{\langle 1|1 \rangle} 1 - \frac{\langle x^3|x \rangle}{\langle x|x \rangle} x - \frac{\langle x^3|x^2 - \frac{1}{3} \rangle}{\langle x^2 - \frac{1}{3}|x^2 - \frac{1}{3} \rangle} \left(x^2 - \frac{1}{3} \right)$$

$$= x^3 - \frac{3}{5}x.$$

The functions obtained in this way are proportional to the Legendre polynomials $P_n(x), n = 0, 1, 2, \ldots$. This should not surprise us since we learned in Chapter 4 that Legendre polynomials are orthogonal to each other, meaning that $\int_{-1}^{1} P_n(x)P_m(x)\, dx = 0$ for $n \neq m$.

Explicitly, $f_1(x) = P_0(x)$, $f_2(x) = P_1(x)$, $f_3(x) = (2/3)P_2(x)$ and $f_4(x) = (2/5)P_3(x)$. It is possible to show that similar proportionality relations hold for the higher Legendre polynomials as well. This provides a new characterization of Legendre polynomials up to the n-th order as elements of an orthogonal basis for $\mathbb{P}^n$ with respect to the inner product (5.15) with $w(x) = 1$.

The orthonormal basis elements are obtained by normalizing the functions obtained above. This yields:

$$\hat{f}_1 = \frac{1}{\sqrt{2}}, \quad \hat{f}_2 = \sqrt{\frac{3}{2}}x, \quad \hat{f}_3 = \frac{1}{2}\sqrt{\frac{5}{2}}(3x^2 - 1), \quad \hat{f}_4 = \frac{1}{2}\sqrt{\frac{7}{2}}(5x^3 - 3x)$$

for the first four elements of the orthonormal Gram-Schmidt basis of $\mathbb{P}^n$ with respect to the inner product (5.15) with $w(x) = 1$.

5.1.3 Fourier-Legendre Series

We have seen in Example (5.1.1) that Legendre's equation

$$(1 - x^2)y'' - 2xy' + n(n+1)y = 0,$$

can be cast in a Sturm-Liouville form:

$$\frac{d}{dx}\left((1 - x^2)y'\right) = -\lambda y. \tag{5.33}$$

Here, the eigenvalue is $\lambda = n(n+1)$ and the eigenfunctions on $-1 \leq x \leq 1$ are the Legendre polynomials, $P_n(x)$. Since the function $p(x) = 1 - x^2$ vanishes at the endpoints $x = \pm 1$, the boundary conditions (5.12) are satisfied by the eigenfunctions that are finite at $x = \pm 1$. Recall that the Legendre polynomials $P_n(x)$ are normalized so that $P_n(1) = 1$. One also finds that $P_n(-1) = (-1)^n$. Hence, the boundary conditions (5.12) are satisfied and the orthogonality

$$\int_{-1}^{1} P_n(x)P_m(x)\,dx = 0, \quad n \neq m$$

of Legendre polynomials follows automatically. In Problem 4.2.8 we have introduced Rodrigues' formula:

$$P_n(x) = \frac{1}{2^n n!}\frac{d^n}{dx^n}\left[(x^2 - 1)^n\right]. \tag{5.34}$$

for Legendre polynomials. It is convenient to use the above formula to derive several basic properties of Legendre polynomials. Here, we will use Rodrigues' formula to derive the recurrence relations for Legendre polynomials. Consider the polynomial $(x^2 - 1)^n$, which enters the Rodrigues' formula (5.34). Two successive derivations of $(x^2 - 1)^n$ yield

$$\frac{d}{dx}(x^2 - 1)^n = 2nx(x^2 - 1)^{n-1}, \tag{5.35}$$

$$\frac{d^2}{dx^2}(x^2 - 1)^n = 2n\left[(2n - 1)(x^2 - 1)^{n-1} + 2(n - 1)(x^2 - 1)^{n-2}\right]. \tag{5.36}$$

Now, we insert these identities into Rodrigues' formula (5.34) for $P_n(x)$ to obtain relations

$$P_n(x) = \frac{1}{2^n n!}\frac{d^{n-1}}{dx^{n-1}}\left[2nx(x^2 - 1)^{n-1}\right]$$

$$= \frac{1}{2^n n!}\frac{d^{n-2}}{dx^{n-2}}\left[2n[(2n - 1)(x^2 - 1)^{n-1} + 2(n - 1)(x^2 - 1)^{n-2}]\right]. \tag{5.37}$$

To perform differentiation in relations (5.37) we will use a generalized product rule:

$$\frac{d^k}{dx^k}(fg) = \sum_{i=0}^{k} \frac{k!}{i!(k-i)!} \frac{d^i}{dx^i}(f) \frac{d^{k-i}}{dx^{k-i}}(g).$$
(5.38)

For the special case of $f(x) = x$ and $k = n - 1$ the above product rule yields a simpler expression

$$\frac{d^{n-1}}{dx^{n-1}}(xg) = x\frac{d^{n-1}}{dx^{n-1}}(g) + (n-1)\frac{d^{n-2}}{dx^{n-2}}(g),$$

which allows us to rewrite the first of relations in equation (5.37) as

$$P_n(x) = \frac{1}{2^{n-1}(n-1)!} \left[x\frac{d^{n-1}}{dx^{n-1}}(x^2-1)^{n-1} + (n-1)\frac{d^{n-2}}{dx^{n-2}}(x^2-1)^{n-1} \right]$$
(5.39)
$$= xP_{n-1}(x) + \frac{n-1}{2^{n-1}(n-1)!}\frac{d^{n-2}}{dx^{n-2}}(x^2-1)^{n-1}.$$

The second of relations in equation (5.37) can be rewritten, using Rodrigues' formula, as

$$P_n(x) = \frac{1}{2^{n-1}(n-1)!} \left[(2n-1)\frac{d^{n-2}}{dx^{n-2}}(x^2-1)^{n-1} + 2(n-1)\frac{d^{n-2}}{dx^{n-2}}(x^2-1)^{n-2} \right]$$
$$= \frac{2n-1}{2^{n-1}(n-1)!}\frac{d^{n-2}}{dx^{n-2}}(x^2-1)^{n-1} + P_{n-2}(x).$$
(5.40)

Eliminating the common factor $d^{n-2}(x^2-1)^{n-1}/dx^{n-2}$ from equations (5.39) and (5.40) produces a basic recurrence relation satisfied by Legendre polynomials:

$$nP_n(x) = (2n-1)xP_{n-1}(x) - (n-1)P_{n-2}(x).$$
(5.41)

Differentiating both sides of (5.40) gives another fundamental recurrence relation:

$$P_n'(x) = P_{n-2}'(x) + \frac{2n-1}{2^{n-1}(n-1)!}\frac{d^{n-1}}{dx^{n-1}}(x^2-1)^{n-1}$$
(5.42)
$$= P_{n-2}'(x) + (2n-1)P_{n-1}(x).$$

The recurrence relation (5.41) can be used to obtain a norm of Legendre polynomial:

$$\|P_n\|^2 = \int_{-1}^{1} P_n^2(x)\, \mathrm{d}x = \int_{-1}^{1} P_n(x) \frac{(2n-1)xP_{n-1}(x) - (n-1)P_{n-2}(x)}{n}\, \mathrm{d}x$$

$$= \frac{2n-1}{n} \int_{-1}^{1} xP_n(x)P_{n-1}(x)\, \mathrm{d}x$$

$$= \frac{2n-1}{n} \int_{-1}^{1} \frac{(n+1)P_{n+1}(x) + nP_{n-1}(x)}{2n+1} P_{n-1}(x)\, \mathrm{d}x$$

$$= \frac{2n-1}{2n+1} \int_{-1}^{1} P_{n-1}^2(x)\, \mathrm{d}x = \frac{2n-1}{2n+1}\|P_{n-1}\|^2,$$

where we used twice the recurrence relation (5.41) as well as orthogonality of Legendre polynomials. It follows that

$$\|P_n\|^2 = \frac{2n-1}{2n+1}\|P_{n-1}\|^2 = \frac{2n-1}{2n+1}\frac{2n-3}{2n-1}\|P_{n-2}\|^2 = \ldots = \frac{2}{2n+1},$$

since $\int_{-1}^{1} P_0^2(x)\, \mathrm{d}x = \int_{-1}^{1} \mathrm{d}x = 2$.

The above results can be summarized in the orthogonality relation:

$$\langle P_n \mid P_m \rangle = \int_{-1}^{1} P_n P_m\, \mathrm{d}x = \frac{2}{2n+1}\delta_{mn}, \tag{5.43}$$

where we used that the weight function is $w(x) = 1$ for Legendre's equation.

The eigenfunction expansion (5.27) in terms of the Legendre polynomials takes a form of the so-called Fourier-Legendre series:

$$f(x) = \sum_{n=0}^{\infty} a_n P_n(x), \tag{5.44}$$

with coefficients, which can determined according to (5.29) by expression:

$$a_n = \frac{\langle P_n \mid f \rangle}{\langle P_n \mid P_n \rangle} = \frac{2n+1}{2} \int_{-1}^{1} P_n(x)f(x)\, \mathrm{d}x, \quad n = 0, 1, 2, \ldots \tag{5.45}$$

☞ **Example 5.1.4.** Let $f(x) = x^3 - 2x^2 + 3$ then the formula (5.45) yields the following Fourier-Legendre series for $f(x)$:

$$f(x) = \frac{7}{3}P_0(x) + \frac{3}{5}P_1(x) - \frac{4}{3}P_2(x) + \frac{2}{5}P_3(x). \tag{5.46}$$

☞ **Example 5.1.5.** Let $f(x)$ be a step function

$$f(x) = \begin{cases} 1 & \text{for } 0 \leq x \leq 1 \\ 0 & \text{for } -1 \leq x < 0 \end{cases}. \tag{5.47}$$

The coefficients of the Fourier-Legendre series follow from the formula (5.45):

$$a_n = \frac{2n+1}{2} \int_0^1 P_n(x)\, \mathrm{d}x, \quad n = 0, 1, 2, \ldots$$

This integral can be evaluated by integrating Legendre's differential equation (5.33):

$$n(n+1) \int_0^1 P_n(x)\, \mathrm{d}x = - \int_0^1 \frac{\mathrm{d}}{\mathrm{d}x} \left((1-x^2)P_n'(x)\right)\, \mathrm{d}x \tag{5.48}$$
$$= - \left[(1-x^2)P_n'(x)\right]_0^1 = P_n'(0)\,.$$

The recursion formula $nP_n(x) = xP_n'(x) - P_{n-1}'(x)$ taken at $x = 0$ yields $P_n'(0) = -(n+1)P_{n+1}(0)$. Hence,

$$\int_0^1 P_n(x)\, \mathrm{d}x = -\frac{1}{n}P_{n+1}(0) \quad \rightarrow \quad a_n = -\frac{2n+1}{2n}P_{n+1}(0), \quad n > 0\,.$$

From Rodrigues' formula (5.34) we derive for $k > 0$ that

$$P_{2k-1}(0) = 0, \quad P_{2k}(0) = (-1)^k \frac{(2k)!}{2^{2k}(k!)^2} = (-1)^k \frac{1 \cdot 3 \cdots (2k-1)}{2 \cdot 4 \cdots 2k}\,.$$

This leads to

$$a_1 = -\frac{3}{2}P_2(0) = \frac{3}{4}$$

and

$$a_{2k} = 0, \quad a_{2k+1} = (-1)^k \frac{(4k+3)}{(4k+4)} \frac{1 \cdot 3 \cdots (2k-1)}{2 \cdot 4 \cdots 2k}, \quad k > 0\,.$$

Thus, $a_2 = 0, a_3 = -7/16, a_4 = 0$ and so on. The remaining a_0 coefficient can be directly calculated by substituting $P_0(x) = 1$ into the integral:

$$a_0 = \frac{1}{2} \int_0^1 1\, \mathrm{d}x = \frac{1}{2}\,.$$

5.1.4 Fourier-Bessel Series

Recall Bessel's differential equation of order n (where n is an integer):

$$x^2 J_n''(x) + x J_n'(x) + \left(x^2 - n^2\right) J_n(x) = 0, \tag{5.49}$$

with solutions given by the Bessel functions $J_n(x)$. Upon substitution $x \rightarrow kx$, where k is a constant, this equation becomes

$$(kx)^2 \frac{\mathrm{d}^2 J_n(kx)}{\mathrm{d}(kx)^2} + kx \frac{\mathrm{d}J_n(kx)}{\mathrm{d}(kx)} + \left((kx)^2 - n^2\right) J_n(kx) = 0, \tag{5.50}$$

or

$$x^2 J_n''(kx) + x J_n'(kx) + \left(k^2 x^2 - n^2\right) J_n(kx) = 0. \tag{5.51}$$

Finally, by dividing equation (5.51) by x and introducing $\lambda = k^2$ we are able to rewrite it in a Sturm-Liouville form:

$$[x J_n'(kx)]' + \left(-\frac{n^2}{x} + \lambda x\right) J_n(kx) = 0, \tag{5.52}$$

with $p(x) = x$, $q(x) = -n^2/x$ and $w(x) = x$. To turn the Sturm-Liouville operator from equation (5.52) into the self-adjoint operator we need to ensure that the appropriate boundary conditions are satisfied. Let equation (5.52) define the Sturm-Liouville problem on the interval $0 \le x \le R$ with some positive constant R. Since, $p(0) = 0$ the boundary condition at the lower endpoint of the interval amounts to requiring that $J_n(0)$ is finite, which is the case. At the upper endpoint $x = R$ it holds that $p(R) = R \ne 0$. To satisfy the boundary condition (5.12) it is enough to require that for each n in (5.52) the solutions $J_n(k_{mn}x) = J_n(\sqrt{\lambda_{mn}}x)$ vanish at $x = R$. In other words for each fixed n, the quantities $k_{mn}R$ are equal to the zeros of the Bessel function $J_n(x)$ of order n. For each n, there are infinitely many corresponding eigenvalues λ_{mn} with $k_{mn} = \sqrt{\lambda_{mn}}$ since each function $J_n(x)$ has an infinite number of the non-negative zeros.

For $n = 0, 1, 2, 3$, the first four solutions to the boundary condition $J_n(k_{mn}R) = 0$ at $x = R$, are displayed in the accompanying table:

$J_n(k_{mn}R) = 0$				
$k_{mn}R$	$m = 1$	$m = 2$	$m = 3$	$m = 4$
$n = 0$	2.405	5.520	8.654	11.798
$n = 1$	3.832	7.016	10.173	13.323
$n = 2$	5.135	8.417	11.620	14.796
$n = 3$	6.379	9.760	13.017	16.224

Accordingly, define $\varphi_m(x) = J_n(k_{mn}x)$, where m labels the zero's of a Bessel function of order n. This defines $\varphi_m(x)$ as orthogonal eigenfunctions of the Sturm-Liouville problem

$$[x \varphi_m'(x)]' + \left(-\frac{n^2}{x} + k_{mn}^2 x\right) \varphi_m(x) = 0, \quad \text{for} \quad 0 \le x \le R, \tag{5.53}$$

with $\varphi_m(R) = J_n(k_{mn}R) = 0$. The orthogonality condition with respect to the weight function $w(x) = x$ takes a form

$$\langle \varphi_m \mid \varphi_l \rangle = \int_0^R \varphi_m(x)\varphi_l(x)x\,dx \tag{5.54}$$

$$= \int_0^R J_n(k_{mn}x)J_n(k_{ln}x)x\,dx = 0, \quad \text{for} \quad m \neq l,$$

which is valid for each fixed integer n. For $m = l$, the result is:

$$\|\varphi_m\|^2 = \int_0^R J_n(k_{mn}x)J_n(k_{mn}x)x\,dx = \frac{R^2}{2}J_{n+1}^2(k_{mn}R). \tag{5.55}$$

This result is obtained as follows. First, we multiply the equation (5.52) by $2xJ_n'(kx)$ to obtain:

$$\left[(xJ_n'(kx))^2 \right]' + (-n^2 + k^2x^2) \left[J_n^2(kx) \right]' = 0. \tag{5.56}$$

Next, we integrate both terms of the above equation over x from 0 to R. The first term is a total derivative and the straightforward integration yields:

$$\int_0^R dx \left[(xJ_n'(kx))^2 \right]' = \left[(xJ_n'(kx))^2 \right]_0^R = R^2 \left(J_n'(kR) \right)^2. \tag{5.57}$$

We use the integration by parts to integrate the second term of equation (5.56) as follows:

$$\int_0^R dx \, (-n^2 + k^2x^2) \left[J_n^2(kx) \right]' = \left[(-n^2 + k^2x^2) J_n^2(kx) \right]_0^R \tag{5.58}$$

$$- 2k^2 \int_0^R dx \, J_n^2(kx).$$

Let k be equal to k_{mn}, such that $J_n(k_{mn}R) = 0$. Then:

$$\left[(-n^2 + k_{mn}^2 x^2) J_n^2(k_{mn}x) \right]_0^R = -n^2 J_n^2(0) = 0,$$

since $J_n(0) = 0$ for $n > 0$. The sum of expressions (5.57) and (5.58) is zero and, therefore,

$$\int_0^R dx \, J_n^2(kx) = \frac{R^2}{2k_{mn}^2} \left(J_n'(k_{mn}R) \right)^2. \tag{5.59}$$

The recursion relation $xJ_n'(x) = -xJ_{n+1}(x) + nJ_n(x)$ changes into $xJ_n'(kx) = -kxJ_{n+1}(kx) + nJ_n(kx)$ under a transformation $x \to kx$. Thus, $RJ_n'(k_{mn}R) = -k_{mn}RJ_{n+1}(k_{mn}R)$ and

$$\int_0^R dx \, J_n^2(kx) = \frac{R^2}{2}J_{n+1}^2(k_{mn}R), \tag{5.60}$$

which proves the expression (5.55).

The Fourier-Bessel orthogonal expansion has a form

$$f(x) = \sum_{m=1}^{\infty} a_m \varphi_m(x) = \sum_{m=1}^{\infty} a_m J_n(k_{mn}x). \tag{5.61}$$

Notice, that the Fourier-Bessel series is valid for a fixed order n and the summation is over the index m, which labels zeros of the Bessel function $J_n(k_{nm}R)$. Expression for the Fourier-Bessel coefficients follows from the general relation (5.29) and are given by

$$a_m = \frac{\langle \varphi_m \mid f \rangle}{\|\varphi_m\|^2} = \frac{2}{R^2 J_{n+1}^2(k_{mn}R)} \int_0^R J_n(k_{mn}x) f(x) \, x \, dx. \tag{5.62}$$

5.1.5 Hermite's Equation

Hermite's equation is given by:

$$y''(x) - 2xy'(x) + 2ny(x) = 0. \tag{5.63}$$

By multiplying it with $\exp(-x^2)$ and using that

$$\left[e^{-x^2} y'(x)\right]' = e^{-x^2} y''(x) - 2xe^{-x^2} y'(x). \tag{5.64}$$

Hermite's equation can be expressed in a Sturm-Liouville form:

$$\left[e^{-x^2} y'(x)\right]' + 2ne^{-x^2} y(x) = 0. \tag{5.65}$$

The solutions to this equation are Hermite polynomials

$$H_n(x) = (-1)^n e^{x^2} \frac{d^n}{dx^n} (e^{-x^2}), \quad n = 1, 2, \ldots. \tag{5.66}$$

The first four Hermite polynomials are

$$H_1(x) = 2x, \quad H_2(x) = 4x^2 - 2, \quad H_3(x) = 8x^3 - 12x, \quad H_4(x) = 16x^4 - 48x^2 + 12.$$

It follows from relation (5.66) that the Hermite polynomials satisfy a recurrence relation

$$H_{n+1}(x) = 2xH_n(x) - H_n'(x) = \left(2x - \frac{d}{dx}\right) H_n(x). \tag{5.67}$$

It can also be shown that

$$H_n'(x) = 2nH_{n-1}(x). \tag{5.68}$$

Differentiating the recurrence relation (5.67) with respect to x yields $H_{n+1}'(x) = 2xH_n'(x) + 2H_n(x) - H_n''(x)$. Eliminating $H_{n+1}'(x)$ by substituting relation (5.68) shows that the Hermite polynomials satisfy equation (5.63).

With the function $p(x) = \exp\left(-x^2\right)$ going rapidly to zero for $x \to \pm\infty$ the boundary conditions are satisfied and the Hermite polynomials are orthogonal on $-\infty \leq x \leq \infty$ with respect to the weight function $w(x) = \exp\left(-x^2\right)$. Thus,

$$\langle H_n \mid H_m \rangle = \int_{-\infty}^{\infty} e^{-x^2} H_n(x) H_m(x) \, \mathrm{d}x = 0, \quad \text{for} \quad n \neq m \qquad (5.69)$$

is ensured by self-adjointness of the underlying Sturm-Liouville problem (5.65). For $n = m$, the above integral gives the norm of the Hermite polynomials,

$$\|H_n\|^2 = \int_{-\infty}^{\infty} e^{-x^2} H_n^2(x) \, \mathrm{d}x. \qquad (5.70)$$

To calculate the above integral we use the recurrence relation (5.67) to first rewrite it as:

$$\|H_n\|^2 = \int_{-\infty}^{\infty} e^{-x^2} \left(2x H_{n-1}(x) - H_{n-1}'(x)\right) H_n(x) \, \mathrm{d}x$$

and then use integration by parts and the recurrence relation (5.68) to obtain:

$$\|H_n\|^2 = \int_{-\infty}^{\infty} e^{-x^2} H_{n-1}(x) H_n'(x) \, \mathrm{d}x = 2n \int_{-\infty}^{\infty} e^{-x^2} H_{n-1}(x) H_{n-1}(x) \, \mathrm{d}x.$$

Hence $\|H_n\|^2 = 2n\|H_{n-1}\|^2$. This relation can be successively applied to yield:

$$\int_{-\infty}^{\infty} e^{-x^2} H_n^2(x) \, \mathrm{d}x = 2^n n! \int_{-\infty}^{\infty} e^{-x^2} H_0^2(x) \, \mathrm{d}x$$

$$= 2^n n! \int_{-\infty}^{\infty} e^{-x^2} \, \mathrm{d}x = 2^n n! \sqrt{\pi}, \qquad (5.71)$$

where in the last step we used the formula (4.84). Therefore,

$$\langle H_n \mid H_m \rangle = \int_{-\infty}^{\infty} e^{-x^2} H_n(x) H_m(x) \, \mathrm{d}x = 2^n n! \sqrt{\pi} \delta_{nm}. \qquad (5.72)$$

The coefficients a_n in the Fourier-Hermite orthogonal series

$$f(x) = \sum_{n=0}^{\infty} a_n H_n(x)$$

are given by

$$a_n = \frac{1}{2^n n! \sqrt{\pi}} \int_{-\infty}^{\infty} e^{-x^2} H_n(x) f(x) \, \mathrm{d}x. \qquad (5.73)$$

✍ **Problems: Sturm Liouville theory and the orthogonal functions expansion.**

Problem 5.1.1. Consider the norm of the function $f(x)+\lambda g(x)$ with some constant λ. It holds due to the positive property of the norm that:

$$\langle f + \lambda g \,|\, f + \lambda g \rangle \geq 0$$

for any constant λ. Prove Cauchy-Schwartz inequality (5.20) by choosing in the above inequality

$$\lambda = \frac{\langle g \,|\, f \rangle}{\langle g \,|\, g \rangle}, \quad \lambda^* = \frac{\langle f \,|\, g \rangle}{\langle g \,|\, g \rangle} .$$

Problem 5.1.2. Consider the basis $\{1, x, x^2, ..., x^n\}$ for the infinite-dimensional vector space $\mathbb{P}^n$ of polynomials (of up to n-th degree). Let

$$\langle f|g \rangle = \int_a^b f(x)g(x)w(x)\,\mathrm{d}x$$

be an inner product over $\mathbb{P}^n$ with the weight function $w(x)$. Apply the Gram-Schmidt algorithm to the basis $\{1, x, x^2, ..., x^n\}$ in order to obtain the first four elements f_0, f_1, f_2, f_3 of the orthonormal basis for $\mathbb{P}^n$ with respect to the inner product for the following choices:
(a) $w(x) = \exp(-x^2)$ with $a = -\infty, b = \infty$. Show that the four polynomials f_0, f_1, f_2, f_3 are proportional to the first four Hermite polynomials $H_n(x)$. *Hint:* Use the integral formulas

$$\int_0^\infty e^{-x^2}\,\mathrm{d}x = \frac{1}{2}\sqrt{\pi}, \quad \int_0^\infty e^{-x^2}x^2\,\mathrm{d}x = \frac{1}{4}\sqrt{\pi}, \quad \int_0^\infty e^{-x^2}x^4\,\mathrm{d}x = \frac{3}{8}\sqrt{\pi}, \quad (5.74)$$

where the first identity follows from expression (4.84) for $\Gamma\left(\frac{1}{2}\right)$ and the other two can be obtained by integration by parts.
(b) $w(x) = \exp(-x)$ with $a = 0, b = \infty$. Show that the four polynomials f_0, f_1, f_2, f_3 are proportional to the first four Laguerre polynomials $L_n(x)$, such that

$$L_n(x) = e^x \frac{\mathrm{d}^n}{\mathrm{d}x^n} x^n e^{-x} .$$

Hint: $\int_0^\infty x^m e^{-x}\,\mathrm{d}x = m!$, for $m = 0, 1, 2, ...$

Problem 5.1.3. Consider the vector space consisting of real-valued continuous functions on the interval $-\pi \leq x \leq \pi$ with the inner product

$$\langle f|g \rangle = \int_{-\pi}^{\pi} f(x)g(x)\,\mathrm{d}x .$$

Apply the Gram-Schmidt orthogonalization procedure to the linearly independent functions $1, x, \sin x$.

Problem 5.1.4. Show that the differential equation $p(x)y'' + r(x)y' + (q(x) + \lambda w(x))y = 0$ can be transformed to self-adjoint form by multiplying it by the integrating factor $F(x) = (1/p(x)) \exp \left(\int^x (r/p) \, dx \right)$.

Problem 5.1.5. Rewrite Chebyshev's differential equation

$$(1 - x^2) \frac{d^2y}{dx^2} - x \frac{dy}{dx} + n^2 y = 0, \quad -1 < x < 1$$

as a standard Sturm-Liouville problem (5.7).

Problem 5.1.6. Rewrite Laguerre differential equation

$$xL''(x) + (1 - x)L'_n(x) + nL_n(x) = 0, \quad 0 < x < \infty$$

as a standard Sturm-Liouville problem (5.7). Show that Laguerre's polynomials, $L_n(x)$, satisfy the orthogonality conditions:

$$\int_0^\infty L_n(x)L_m(x)e^{-x} \, dx = 0, \quad n \neq m.$$

Problem 5.1.7. A polynomial $f(x)$ satisfies:

$$\int_{-1}^1 f(x) \frac{1}{\sqrt{1 - 2xt + t^2}} \, dx = \frac{1}{2} - \frac{1}{3}t + \frac{2}{5}t^2 + 2t^3,$$

for $|t| < 1$. Evaluate the integrals:

(a) $\int_{-1}^1 xf(x) \, dx$ \qquad (b) $\int_{-1}^1 P_2(x)f(x) \, dx$

(c) Represent $f(x)$ in terms of Legendre polynomials on the interval $-1 \leq x \leq 1$.
(d) Find $f(x)$ as a function of x.

Problem 5.1.8. Find the eigenvalues and eigenfunctions of

$$y'' + \lambda^2 y = 0,$$

for the boundary conditions:
(a) $y(0) = y(L) = 0$.
(b) $y(0) = y'(L) = 0$.
(c) $y'(0) = y(L) = 0$.
(d) $y'(0) = y'(L) = 0$.
(e) $y(0) = y(L)$.
(f) $y'(0) = y'(L)$.

Problem 5.1.9. Find the eigenvalues and eigenfunctions of the following Sturm-Liouville problem:

$$(e^{-4x}y')' + e^{-4x}(\lambda + 4)y = 0, \quad y(0) = 0, \quad y(\pi) = 0.$$

Hint: Set $y = e^{2x}u$.

Problem 5.1.10. Find the eigenvalues and eigenfunctions of the following Sturm-Liouville problem defined on $1 < x < e$:

$$(xy')' + \frac{\lambda}{x} y = 0, \quad y(1) = 0, \quad y(e) = 0.$$

Hint: Introduce the variable $z = \ln x$.

5.2 Harmonic Oscillator Equation, Periodic Sturm-Liouville Problem and Fourier Series

Consider again the harmonic oscillator differential equation

$$y''(x) + \lambda y(x) = y''(x) + k^2 y(x) = 0 \qquad (5.75)$$

defined this time on the interval $-L \leq x \leq L$ of length $2L$ with the periodic (5.14) boundary conditions:

$$y(L) = y(-L) \qquad y'(L) = y'(-L). \qquad (5.76)$$

The general solution to equation (5.75) is

$$y(x) = A_k \cos(kx) + B_k \sin(kx).$$

Imposing the periodic boundary conditions (5.76) yields

$$
\begin{aligned}
A_k \cos(kL) + B_k \sin(kL) &= A_k \cos(kL) - B_k \sin(kL), \\
-A_k k \sin(kL) + B_k k \cos(kL) &= A_k k \sin(kL) - B_k k \cos(kL),
\end{aligned}
$$

which can only be satisfied for non-zero A_k and B_k if

$$\sin(kL) = 0 \quad \longrightarrow \quad k = \frac{n\pi}{L}, \quad n = 0, 1, 2, 3, \ldots.$$

Accordingly,

$$\{1, \cos(\frac{n\pi x}{L}), \sin(\frac{n\pi x}{L})\}, \quad n = 1, 2, 3, \ldots \qquad (5.77)$$

is the set of eigenfunctions satisfying the Sturm-Liouville problem (5.75)-(5.76).

Now, we investigate their orthogonality properties described by the formulas:

$$\int_{-L}^{L} 1 \cdot 1 \, dx = 2L, \tag{5.78a}$$

$$\int_{-L}^{L} 1 \cdot \cos(\frac{n\pi x}{L}) \, dx = 0, \quad \int_{-L}^{L} 1 \cdot \sin(\frac{n\pi x}{L}) \, dx = 0 \tag{5.78b}$$

$$\int_{-L}^{L} \cos(\frac{n\pi x}{L}) \cdot \cos(\frac{m\pi x}{L}) \, dx = \frac{1}{2} \int_{-L}^{L} \cos(\frac{(n+m)\pi x}{L}) + \cos(\frac{(n-m)\pi x}{L}) \, dx$$

$$= \begin{cases} 0 & \text{if } n \neq m, \\ L & \text{if } n = m, \end{cases} \tag{5.78c}$$

$$\int_{-L}^{L} \sin(\frac{n\pi x}{L}) \cdot \sin(\frac{m\pi x}{L}) \, dx = \frac{1}{2} \int_{-L}^{L} - \cos(\frac{(n+m)\pi x}{L}) + \cos(\frac{(n-m)\pi x}{L}) \, dx$$

$$= \begin{cases} 0 & \text{if } n \neq m, \\ L & \text{if } n = m, \end{cases} \tag{5.78d}$$

$$\int_{-L}^{L} \cos(\frac{n\pi x}{L}) \cdot \sin(\frac{m\pi x}{L}) \, dx = \frac{1}{2} \int_{-L}^{L} \sin(\frac{(n+m)\pi x}{L}) + \sin(\frac{(n-m)\pi x}{L}) \, dx$$

$$= 0. \tag{5.78e}$$

The eigenfunction expansion in the eigenfunctions from equation (5.77) takes a form

$$f(x) = \frac{a_0}{2} + \sum_{n=1}^{\infty} \left[a_n \cos(\frac{n\pi x}{L}) + b_n \sin(\frac{n\pi x}{L}) \right], \quad -L \leq x \leq L. \tag{5.79}$$

This expansion is called a Fourier series representation of the function $f(x)$ on the interval $-L \leq x \leq L$ with the Fourier coefficients given by a_n and b_n. Sometimes a_n are called the Fourier cosine coefficients and b_n are called the Fourier sine coefficients. We find from relation (5.29) and the above orthogonality properties that the values of the Fourier coefficients are

$$a_0 = \frac{1}{L} \int_{-L}^{L} f(x) \, dx, \tag{5.80a}$$

$$a_n = \frac{1}{L} \int_{-L}^{L} f(x) \cos(\frac{n\pi x}{L}) \, dx, \quad n = 1, 2, 3, \ldots, \tag{5.80b}$$

$$b_n = \frac{1}{L} \int_{-L}^{L} f(x) \sin(\frac{n\pi x}{L}) \, dx, \quad n = 1, 2, 3, \ldots. \tag{5.80c}$$

5.2.1   Complex Form of Fourier Series

By substituting formulas

$$\cos\alpha = \frac{e^{i\alpha} + e^{-i\alpha}}{2}, \quad \sin\alpha = \frac{e^{i\alpha} - e^{-i\alpha}}{2i}$$

into the Fourier series (5.79) we can rewrite it in a complex form

$$
\begin{aligned}
f(x) &= \frac{a_0}{2} + \frac{1}{2}\sum_{n=1}^{\infty}\left[(a_n - i\,b_n)e^{\frac{n\pi x}{L}} + (a_n + i\,b_n)e^{-\frac{n\pi x}{L}}\right] \\
&= \frac{a_0}{2} + \frac{1}{2}\sum_{n=1}^{\infty}(a_n - i\,b_n)e^{\frac{n\pi x}{L}} + \frac{1}{2}\sum_{n=-1}^{-\infty}(a_{-n} + i\,b_{-n})e^{\frac{n\pi x}{L}},
\end{aligned}
\tag{5.81}
$$

where in the second term we let the running index n go to $-n$ and defined $a_{-n} = a_n$ and $b_{-n} = -b_n$. Furthermore, defining

$$c_n = \frac{a_n - i\,b_n}{2}, \quad c_{-n} = \frac{a_{-n} - i\,b_{-n}}{2}, \quad n > 1, \quad c_0 = \frac{a_0}{2}, \tag{5.82}$$

we can rewrite the complex Fourier series (5.81) in a compact form

$$\boxed{f(x) = \sum_{-\infty}^{\infty} c_n e^{\frac{i n\pi x}{L}}.}\tag{5.83}$$

The coefficients c_n and c_{-n} are recovered from

$$
\begin{aligned}
c_n &= \frac{a_n - i\,b_n}{2} = \frac{1}{2L}\int_{-L}^{L} f(x)\left[\cos(\frac{n\pi x}{L}) - i\sin(\frac{n\pi x}{L})\right]dx \\
&= \frac{1}{2L}\int_{-L}^{L} f(x)e^{-\frac{i n\pi x}{L}}\,dx \\
c_{-n} = c_{-n} &= \frac{a_{-n} - i\,b_{-n}}{2} = \frac{1}{2L}\int_{-L}^{L} f(x)\left[\cos(\frac{n\pi x}{L}) + i\sin(\frac{n\pi x}{L})\right]dx \\
&= \frac{1}{2L}\int_{-L}^{L} f(x)e^{\frac{i n\pi x}{L}}\,dx,
\end{aligned}
$$

which can be summarized as

$$c_n = \frac{1}{2L}\int_{-L}^{L} f(x)e^{-\frac{i n\pi x}{L}}\,dx = \frac{1}{2L}\langle e^{\frac{i n\pi x}{L}} \mid f(x)\rangle \tag{5.84}$$

for $n = 0, \pm1, \pm2, \ldots$.

☞ **Example 5.2.1.** Let r be a real constant and function f defined by:

$$f(x) = e^{rx}, \quad -\pi < x < \pi.$$

The corresponding coefficients of the complex Fourier series are calculated using relation (5.84) as

$$
\begin{aligned}
c_n &= \frac{1}{2\pi} \int_{-\pi}^{\pi} e^{rx} e^{-\mathrm{i}nx} \, \mathrm{d}x = \frac{1}{2\pi} \int_{-\pi}^{\pi} e^{(r-\mathrm{i}n)x} \, \mathrm{d}x \\
&= \frac{1}{2\pi} \frac{1}{r - \mathrm{i}n} \left(e^{(r-\mathrm{i}n)\pi} - e^{-(r-\mathrm{i}n)\pi} \right) = \frac{1}{2\pi} \frac{1}{r - \mathrm{i}n} (-1)^n \left(e^{r\pi} - e^{-r\pi} \right) \\
&= \frac{(-1)^n}{\pi} \sinh(r\pi) \frac{r + \mathrm{i}n}{r^2 + n^2} \, .
\end{aligned}
$$

The complex Fourier series becomes

$$
e^x = \frac{1}{\pi} \sinh(r\pi) \sum_{-\infty}^{\infty} (-1)^n \frac{r + \mathrm{i}n}{r^2 + n^2} e^{\mathrm{i}nx}, \quad -\pi < x < \pi .
$$

5.2.2 Mean-Square Convergence of Fourier Series

We need to address the question of how well is a function f with a finite norm

$$
\|f\|^2 = \frac{1}{2L} \int_{-L}^{L} |f(x)|^2 \, \mathrm{d}x < \infty,
$$

approximated by its Fourier series. Let us define a partial sum of the Fourier series (5.83):

$$
S_N(x) = \sum_{n=-N}^{N} c_n e^{\frac{\mathrm{i}n\pi x}{L}}, \tag{5.85}
$$

with coefficient c_n defined as in (5.84). As a measure of accuracy of a Fourier approximation we consider convergence of the norm of $f(x) - S_N(x)$ (mean-square convergence) to zero given by:

$$
\lim_{N \to \infty} \|f - S_N\|^2 = \lim_{N \to \infty} \int_{-L}^{L} |f(x) - S_N(x)|^2 \, \mathrm{d}x = 0 . \tag{5.86}
$$

In the case when (5.86) holds we say that the Fourier series mean-square converges to the function f. When such convergence occurs it implies the following Parseval's equation:

$$
\frac{1}{2L} \int_{-L}^{L} |f(x)|^2 \, \mathrm{d}x = \sum_{n=-\infty}^{\infty} |c_n|^2 = \frac{1}{4} a_0^2 + \frac{1}{2} \sum_{n=1}^{\infty} \left(|a_n|^2 + |b_n|^2 \right) . \tag{5.87}
$$

The argument goes as follows. First, consider the following inequality

$$
\|f - S_N\|^2 = \langle f - S_N \,|\, f - S_N \rangle = \langle f \,|\, f \rangle - \langle f \,|\, S_N \rangle - \langle S_N \,|\, f \rangle + \langle S_N \,|\, S_N \rangle \geq 0, \tag{5.88}
$$

valid for any N. The last term $\langle S_N \mid S_N \rangle$ can be evaluated using identity

$$\langle e^{\frac{i n \pi x}{L}}, e^{\frac{i m \pi x}{L}} \rangle = \int_{-L}^{L} e^{\frac{i(m-n)\pi x}{L}} \, dx = \begin{cases} 0 & m \neq n \\ 2L & m = n \end{cases}.$$

It follows that

$$\langle S_N \mid S_N \rangle = \sum_{n=-N}^{N} c_n \sum_{m=-N}^{N} c_m^* \langle e^{\frac{i n \pi x}{L}}, e^{\frac{i m \pi x}{L}} \rangle$$

$$= 2L \sum_{n=-N}^{N} |c_n|^2.$$

Next, from the definition (5.84) we obtain:

$$\langle f \mid S_N \rangle = \sum_{n=-N}^{N} c_n \langle f(x), e^{\frac{i m \pi x}{L}} \rangle = 2L \sum_{n=-N}^{N} |c_n|^2 \tag{5.89}$$

$$\langle S_N \mid f \rangle = \sum_{n=-N}^{N} c_n^* \langle e^{\frac{i m \pi x}{L}}, f(x) \rangle = 2L \sum_{n=-N}^{N} |c_n|^2. \tag{5.90}$$

Now, we can rewrite the inequality (5.88) as

$$\|f - S_N\|^2 = \langle f \mid f \rangle - 2L \sum_{n=-N}^{N} |c_n|^2 - 2L \sum_{n=-N}^{N} |c_n|^2 + 2L \sum_{n=-N}^{N} |c_n|^2$$

$$= \langle f \mid f \rangle - 2L \sum_{n=-N}^{N} |c_n|^2 \geq 0$$

and obtain in the $N \to \infty$ limit:

$$\frac{1}{2L} \langle f \mid f \rangle \geq \sum_{n=-\infty}^{\infty} |c_n|^2, \tag{5.91}$$

which bears the name of Bessel's inequality.

For a function f such that $\|f\|^2 < \infty$ it follows from a general Sturm-Liouville setup that the Fourier series is complete, meaning that it converges in the mean to f as shown below:

$$0 = \lim_{N \to \infty} \|f - S_N\|^2 = \lim_{N \to \infty} \left[\langle f \mid f \rangle - 2L \sum_{n=-N}^{N} |c_n|^2 \right], \tag{5.92}$$

which implies Parseval's equation (5.87). Note, however, that the mean-square convergence of Fourier series does not imply the point-wise convergence. We discuss the point-wise convergence in subsection 5.2.4.

☞ **Example 5.2.2.** Applying Parseval's relation (5.87) to Example 5.2.1 with $f(x) = \exp(rx)$ and $L = \pi$ we get

$$\frac{1}{2\pi} \int_{-\pi}^{\pi} |e^{2rx}|^2 \, dx = \frac{1}{2r\pi} \sinh(2r\pi)$$

and

$$\sum_{n=-\infty}^{\infty} |c_n|^2 = \frac{1}{\pi^2} \sinh^2(r\pi) \left(\frac{1}{r^2} + 2 \sum_{n=1}^{\infty} \frac{1}{r^2 + n^2} \right)$$

We conclude from above relations that

$$\frac{1}{r^2} + 2 \sum_{n=1}^{\infty} \frac{1}{r^2 + n^2} = \frac{\pi}{r} \coth(r\pi),$$

which leads to

$$\sum_{n=1}^{\infty} \frac{1}{r^2 + n^2} = \frac{1}{2} \left(\frac{\pi}{r} \coth(r\pi) - \frac{1}{r^2} \right),$$

reproducing the well-known identity.

5.2.3 Even and Odd Functions

A function $f(x)$ is called an *odd* function on $-L \leq x \leq L$ if

$$f(-x) = -f(x), \quad -L \leq x \leq L.$$

A function $f(x)$ is called an *even* function on $-L \leq x \leq L$ if

$$f(-x) = f(x), \quad -L \leq x \leq L.$$

Note, that even function is symmetric under reflection around the y-axis. Odd function is symmetric under combined reflections around the x- and y-axes.

☞ **Example 5.2.3.** $\sin x$, $\tan x$, x and x^3 are odd functions. $\cos x$, x^2 and $|x|$ are even functions.

☞ **Example 5.2.4.** The product of two even functions $f_{even}(x) g_{even}(x)$ is an even function. The product of two odd functions $f_{odd}(x) g_{odd}(x)$ is also an even function. The product of an even function and an odd function $f_{even}(x) g_{odd}(x)$ is an odd function.

☞ **Example 5.2.5.** For an odd function the integral over $-L \leq x \leq L$ interval vanishes due to

$$\int_{-L}^{L} f_{odd}(x) \, dx = \int_{0}^{L} f_{odd}(x) \, dx + \int_{-L}^{0} f_{odd}(x) \, dx$$

$$= \int_{0}^{L} f_{odd}(x) \, dx + \int_{L}^{0} f_{odd}(-x) \, d(-x)$$

$$= \int_{0}^{L} f_{odd}(x) \, dx + \int_{0}^{L} f_{odd}(-x) \, dx = 0 \, .$$

For an even function the integral over the interval $-L \leq x \leq L$ can be expressed by an integral over $0 \leq x \leq L$ interval as follows:

$$\int_{-L}^{L} f_{even}(x) \, dx = \int_{0}^{L} f_{even}(x) \, dx + \int_{-L}^{0} f_{even}(x) \, dx$$

$$= \int_{0}^{L} f_{even}(x) \, dx + \int_{0}^{L} f_{even}(-x) \, dx$$

$$= 2 \int_{0}^{L} f_{even}(x) \, dx \, .$$

If the function $f(x)$ is odd then $f(x) \cos(n\pi x/L)$ is odd too and according to Example (5.2.5) all the Fourier cosine coefficients a_n are zero. In this case, Fourier series (5.79) becomes a sine series:

$$f(x) = \sum_{n=1}^{\infty} b_n \sin(\frac{n\pi x}{L}), \quad -L \leq x \leq L, \tag{5.93a}$$

where

$$b_n = \frac{1}{L} \int_{-L}^{L} f(x) \sin(\frac{n\pi x}{L}) \, dx = \frac{2}{L} \int_{0}^{L} f(x) \sin(\frac{n\pi x}{L}) \, dx \quad n = 1, 2, 3, \dots . \tag{5.93b}$$

☞ **Example 5.2.6.** Consider, $f(x) = x$ given on the interval $-\pi \leq x \leq \pi$ as shown in Figure (5.1). Since, $f(x)$ is odd all cosine coefficients $a_n = (1/\pi) \int_{-\pi}^{\pi} x \cos(nx) \, dx$ vanish. The remaining sine coefficients are evaluated using integration by parts as follows:

$$b_n = \frac{2}{\pi} \int_{0}^{\pi} x \sin(nx) \, dx = \frac{2}{\pi n} \int_{0}^{\pi} x \left(-\frac{d}{dx} \cos(nx) \right) \, dx$$

$$= \frac{2}{\pi n} \int_{0}^{\pi} \cos(nx) \, dx - \frac{2}{\pi n} \pi \cos(n\pi)$$

$$= \frac{2}{n} (-1)^{n+1}$$

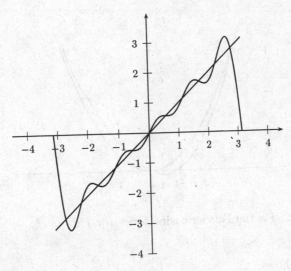

Figure 5.1: Partial Fourier sum for $f(x) = x$, $-\pi \leq x \leq \pi$.

for $n = 1, 2, 3. \ldots$

If the function $f(x)$ is an even function then $f(x) \sin(n\pi x/L)$ is odd and according to Example (5.2.5) all the sine coefficients b_n are zero. Then, Fourier series (5.79) becomes a cosine series:

$$f(x) = \frac{a_0}{2} + \sum_{n=1}^{\infty} a_n \cos(\frac{n\pi x}{L}), \quad -L \leq x \leq L, \tag{5.94a}$$

where

$$a_n = \frac{1}{L} \int_{-L}^{L} f(x) \cos(\frac{n\pi x}{L}) \, \mathrm{d}x = \frac{2}{L} \int_{0}^{L} f(x) \cos(\frac{n\pi x}{L}) \, \mathrm{d}x \quad n = 1, 2, 3, \ldots \tag{5.94b}$$

and

$$a_0 = \frac{1}{L} \int_{-L}^{L} f(x) \, \mathrm{d}x = \frac{2}{L} \int_{0}^{L} f(x) \, \mathrm{d}x. \tag{5.94c}$$

☞ **Example 5.2.7.** Consider, $f(x) = x^2$ given on the interval $-\pi \leq x \leq \pi$ as shown in Figure (5.2). Since, $f(x)$ is even all sine coefficients $b_n = (1/\pi) \int_{-\pi}^{\pi} x^2 \sin(nx) \, \mathrm{d}x$ vanish. The remaining cosine coefficients are readily

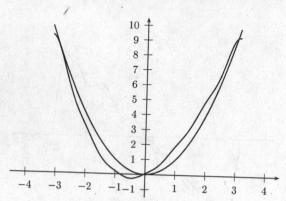

Figure 5.2: Partial Fourier cosine series for $f(x) = x^2$, $-\pi \le x \le \pi$

found as follows:

$$a_n = \frac{2}{\pi} \int_0^\pi x^2 \cos(nx)\,dx$$

$$= \frac{2}{\pi n} \int_0^\pi x^2 \left(\frac{d}{dx}\sin(nx)\right) dx$$

$$= -\frac{2 \cdot 2}{\pi n} \int_0^\pi x \sin(nx)\,dx$$

$$= \frac{4}{n^2}(-1)^n$$

and $a_0 = (2/\pi) \int_0^\pi x^2\,dx = 2\pi^2/3$.

Any function $f(x)$ can be written as $f(x) = f_{even}(x) + f_{odd}(x)$, with $f_{even}(x) = \frac{1}{2}(f(x) + f(-x))$ and $f_{odd}(x) = \frac{1}{2}(f(x) - f(-x))$. Representing, $f_{even}(x)$ by a cosine series (5.94a) and $f_{odd}(x)$ by a sine series (5.93a) we recover the Fourier series (5.79).

5.2.4 Periodic Functions. Point-wise Convergence of Fourier Series

A function $f(x)$, is called periodic of period P if $f(x + P) = f(x)$ for all x. A periodic function of period $P = 2L$ needs only to be specified within the interval $-L \le x \le L$. An arbitrary point y outside this interval can always be expressed as $y = x + kP$ with some integer k, positive or negative, and the value of the periodic function f at y is simply given by $f(y) = f(x + kP) = f(x)$. Hence, we only need to know a periodic function within the interval $-P/2 \le x \le P/2$ where P is its period.

Conversely, let function f be known on the interval $-L \le x < L$. We can expand f beyond this interval by extending it to the periodic function of period

$2L$. This works in the following way: For y that, let say, is inside $L \leq y \leq 3L$, the expanded periodic function f is defined so that $f(y) = f(y - 2L)$, and for y which let say is inside $kL \leq y \leq (k+2)L$ we set $f(y) = f(y - (k+1)L)$. This successively defines a periodic function with period $2L$.

On the other hand a function f on the interval $-L < x \leq L$ has the Fourier series representation on this interval. Since Fourier expansion (5.79) is periodic with the period $2L$ it extends to the entire x-axis by covering it with adjacent copies of the original interval $-L < x \leq L$. This process practically defines a procedure of a periodic expansion. Thus the periodic extension is accomplished by expanding given function in Fourier series on the primary interval $-L < x \leq L$ and through the use of periodicity of Fourier series it extends it to the whole axis.

In what follows we will address the issue of the point-wise convergence of Fourier series. First, we need to introduce a concept of piecewise continuity.

Definition 5.2.1. *A function F is called piecewise continuous on an interval $-L < x < L$ if it is continuous everywhere except at finitely many points $x_1, x_2, ..., x_k$ and the left-hand and right-hand limits of f exist at each of the points $x_1, x_2, ..., x_k$.*

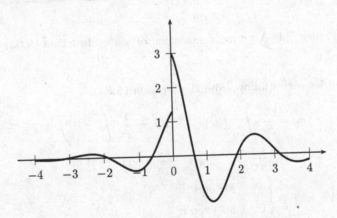

Figure 5.3: A function with the left-right limits.

Now, let function f and its derivative f' (which exists except possibly at finitely many points) be piecewise continuous in an interval $-L < x < L$. See Figure 5.3 for a graph of a function exhibiting such behavior. Furthermore, let f be periodic of period $2L$ outside $-L < x < L$. Then its Fourier series is convergent everywhere and its limit is equal to:

① $f(x)$, if f is continuous at x.

② $\lim_{\epsilon \to +0} \frac{1}{2}\left(f(x_i + \epsilon) + f(x_i - \epsilon)\right)$, which are arithmetic mean values of f, at the points x_i, $i = 1, 2, ..., k$ of discontinuity of f.

The statement implies that $\lim_{N\to\infty} S_N(x) = f(x)$ at a point of continuity of f and

$$\lim_{N\to\infty} S_N(x) = \lim_{\epsilon\to+0} \frac{1}{2}\left(f(x_i + \epsilon) + f(x_i - \epsilon)\right)$$

at the points of discontinuity.

☞ **Example 5.2.8.** Let $f(x)$ be a step function

$$f(x) = \begin{cases} 1 & \text{for } 0 \le x < \pi \\ 0 & \text{for } -\pi \le x < 0 \end{cases} \tag{5.95}$$

See Figure (5.4) for a periodic extension of this function. The coefficients of

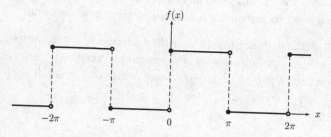

Figure 5.4: A periodic extension of a step function (5.95).

the Fourier series follow from the formula (5.84):

$$c_n = \frac{1}{2\pi}\int_{-\pi}^{\pi} f(x)e^{-inx}\,\mathrm{d}x = \frac{1}{2\pi}\int_0^\pi e^{-inx}\,\mathrm{d}x$$

$$= \frac{i}{2n\pi}\left[e^{-inx}\right]_0^\pi = \frac{i}{2n\pi}\left[(-1)^n - 1\right]$$

$$= \begin{cases} \frac{-i}{n\pi} & \text{for odd } n \\ 0 & \text{for even } n \\ \frac{1}{2} & n = 0 \end{cases}.$$

Thus

$$f(x) = \sum_{n=-\infty}^{\infty} c_n e^{inx} = \frac{1}{2} + \frac{2}{\pi}\sum_{l=0}^{\infty} \frac{\sin(2l+1)x}{(2l+1)}.$$

We note that at the point, $x = 0$, of discontinuity the Fourier series converges to the value

$$\frac{1}{2} = \lim_{\epsilon\to+0}\frac{1}{2}\left(f(0+\epsilon) + f(0-\epsilon)\right).$$

5.2.5 Half-Range Expansion

In many cases given is initially a function f defined on the half-range interval $0 < x < L$ and the aim is to periodically extend it to the whole x-axis. The sine and cosine series are particularly useful when representing functions defined on half of the original interval, i.e., on the interval $0 < x < L$ instead of on the interval $-L < x < L$ since sine and cosine coefficients in (5.93b), (5.94b) (5.94c) only involve integrals over $0 \le x \le L$ interval. There are two basic choices concerning extension of the function beyond the half-range and they involve either sine- or cosine-series.

We can extend function f to the interval $-L < x < 0$ as an even function by reflecting the graph of the function about the y-axis or by setting $f(x) = f(-x)$ for x in $-L < x < 0$.

☞ **Example 5.2.9.** Let function

$$f(x) = x$$

be defined on the interval $0 < x < \pi$. We choose to extend it to $-\pi < x < \pi$

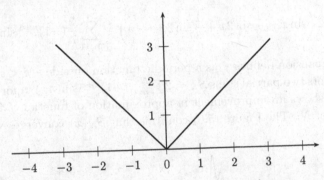

Figure 5.5: $f(x) = |x|$ on $-\pi < x < \pi$.

interval as an even function: $f(x) = |x|$ on $-\pi < x < \pi$ (as seen in Figure (5.5)) and then apply periodic extension of period 2π to extend it to the whole x-axis. This function will therefore be represented by the cosine-series. The coefficients a_n are found to be:

$$a_n = \frac{2}{\pi} \int_0^{\pi} x \cos(nx)\, \mathrm{d}x = \frac{2}{\pi}\frac{(-1)^n - 1}{n^2} = \begin{cases} 0 & n \text{ even} \\ -\frac{4}{\pi n^2} & n \text{ odd} \end{cases} \quad n = 1, 2, 3, \dots$$

and $a_0 = \pi$. Thus, the cosine series representation of the $|x|$ function appears to be given by:

$$|x| = \frac{\pi}{2} - \frac{4}{\pi} \sum_{k=0}^{\infty} \frac{\cos(2k+1)x}{(2k+1)^2}.$$

From the fact that Fourier series will converge at $x = 0$ to $f(0) = 0$ we find the result

$$\sum_{k=0}^{\infty} \frac{1}{(2k+1)^2} = \frac{\pi^2}{8}.$$

Another possibility is to extend function f to the interval $-L < x < 0$ as an odd function by reflecting the graph of the function about the y-axis and then the negative x-axis or alternatively by setting $f(x) = -f(-x)$ for x in $-L < x < 0$.

☞ **Example 5.2.10.** Let $f(x) = x$ be given on the interval $0 < x < \pi$. We now extend $f(x) = x$ to the whole interval $-\pi < x < \pi$. This clearly defines an odd extension, x, as opposed to the even extension described in Example (5.2.9). The Fourier coefficients for that function have already been calculated in Example (5.2.6) and we find the Fourier series to be

$$x = 2\left(\sin x - \frac{1}{2}\sin 2x + \frac{1}{3}\sin 3x + \cdots\right) = \sum_{n=1}^{\infty} \frac{2}{n}(-1)^{n+1}\sin nx.$$

This expansion defines f as a periodic function outside $-\pi < x < \pi$. Figure (5.6) plots two partial sums $S_N = 2\sum_{n=1}^{N}(-1)^{n+1}\sin(nx)/n$ for two values of N and shows an improvement in approximation of function f for increasing values of N. The Figure (5.6) confirms that $S_N(x)$ converges where $f(x)$ is

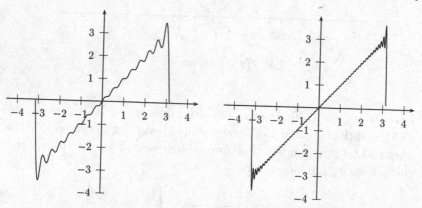

Figure 5.6: Partial sums S_N for $f(x) = x$ with $N = 15$ and $N = 55$.

continuous. Furthermore, $S_N \to 0$ at $x = \pi$; the average value of discontinuity at this point. In addition, we also notice the so-called Gibbs phenomenon, which manifests itself in form of spikes near the discontinuity points. Gibbs phenomenon results in overshooting near $x = \pi$, which persists for large values of N.

☞ **Example 5.2.11.** Let function

$$f(x) = \cos x$$

be defined on the interval $0 < x < \pi$. Let us extend the cosine function defined on the interval $0 < x < \pi$ to an odd function on $-\pi < x < \pi$ such that:

$$f(x) = \begin{cases} \cos x, & 0 < x < \pi \\ -\cos x, & -\pi < x < 0. \end{cases} \tag{5.96}$$

Note, that this extension has a discontinuity at $x = 0$. We then periodically extend this function to the whole x-axis. Accordingly, we will represent it by the sine series (5.93a). The formula (5.93b) for the b_n coefficients yields

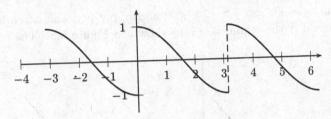

Figure 5.7: A periodic extension of function (5.96).

$$b_n = \frac{2}{\pi} \int_0^\pi \cos(x) \sin(nx) \quad n = 1, 2, 3, \ldots \tag{5.97}$$

To evaluate the integral it is helpful to use the trigonometric identity $2\cos\alpha\sin\beta = \sin(\alpha+\beta) - \sin(\alpha-\beta)$. The result is

$$b_n = \begin{cases} \frac{4n}{\pi(n^2-1)}, & n = \text{even} \\ 0, & n = \text{odd} \end{cases}$$

and the sine series representation of the cosine function (extended as an odd function) appears to be given by:

$$\cos x = \frac{8}{\pi} \sum_{k=0}^\infty \frac{k}{(4k^2-1)} \sin(2kx) = \frac{8}{3\pi} \sin 2x$$

$$+ \frac{16}{15\pi} \sin 4x + \ldots, \quad 0 < x < \pi.$$

Note that at $x = 0$, where the function is discontinuous, the sine series converges to $\frac{1}{2}(f(+0) + f(-0)) = \frac{1}{2} + (-\frac{1}{2}) = 0$ in agreement with the fact that the sine series represents an odd function.

We can therefore create an even/odd function (over $-L < x < L$) out of a function that initially was defined on the half-interval $0 < x < L$. This is accomplished by representing $f(x)$ in terms of cosine/sine series on $0 < x < L$.

There remains a third possibility to perform periodic extension, this time with the period L. For function defined on $0 < x < L$ we first extend it to x on $-L < x < 0$ by imposing condition $f(x) = f(x + L)$ and then periodically extend f to the whole x-axis. The resulting function will have the period L.

☞ **Example 5.2.12.** Given is again the function

$$f(x) = x, \quad \text{for} \quad 0 < x < \pi.$$

Set $f(x) = x + \pi$ on $-\pi < x < 0$ and apply periodic extension of period π to extend it the whole x-axis as shown in Figure 5.8. The corresponding

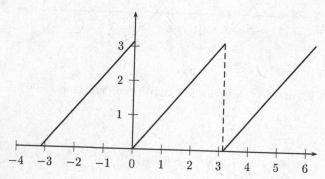

Figure 5.8: A periodic extension of period π of $f(x) = x$.

Fourier series must also have the period equal to π. The Fourier expansion (5.79) defined on the interval $-L < x < L$ needs now to be defined on the shifted interval $0 < x < 2L$ with $L = \pi/2$ and correspondingly the generic terms $\cos(\frac{n\pi x}{L})$ and $\sin(\frac{n\pi x}{L})$ of expansion (5.79) become $\cos(2nx)$ and $\sin(2nx)$ upon substitution $L = \pi/2$. The modified Fourier expansion reads, therefore,

$$f(x) = \frac{a_0}{2} + \sum_{n=1}^{\infty} [a_n \cos(2nx) + b_n \sin(2nx)], \quad 0 < x < 2L = \pi, \tag{5.98}$$

with the Fourier coefficients

$$a_0 = \frac{1}{L}\int_0^{2L} f(x)\,\mathrm{d}x = \frac{2}{\pi}\int_0^{\pi} x\,\mathrm{d}x = \pi\,,$$

$$a_n = \frac{1}{L}\int_0^{2L} f(x)\cos(\frac{n\pi x}{L})\,\mathrm{d}x = \frac{2}{\pi}\int_0^{\pi} x\cos(2nx)\,\mathrm{d}x = 0\,,$$

and

$$b_n = \frac{1}{L}\int_0^{L} f(x)\sin(\frac{n\pi x}{L})\,\mathrm{d}x = \frac{2}{\pi}\int_0^{\pi} x\cos(2nx)\,\mathrm{d}x = -\frac{1}{n}\,.$$

The Fourier series (5.98) on $0 < x < \pi$ becomes

$$x = \frac{\pi}{2} - \sum_{n=1}^{\infty}\frac{1}{n}\sin(2nx)\,.$$

The period is equal to π and the limit of the Fourier series at $x = 0$ is equal to $\pi/2$ that agrees with the value of $\frac{1}{2}(f(+0) + f(-0))$.

In all three cases considered above we were able to obtain periodic extensions beyond the initial interval $0 < x < L$. The first two examples concerned even and odd $2L$-periodic expansions. The last example was an L-periodic expansion.

◢ Problems: Fourier series.

Problem 5.2.1. Find the Fourier series of the function

$$f(x) = \begin{cases} L - x & \text{if} & 0 < x < L \\ 0 & \text{if} & L \leq x < 2L \\ \frac{L}{2} & \text{if} & x = 2L \end{cases}$$

of period $2L$.

Problem 5.2.2. Find the Fourier series of the function

$$f(x) = \begin{cases} \sin x & \text{if} & 0 < x < \pi \\ 0 & \text{if} & \pi \leq x \leq 2\pi \end{cases}$$

of period 2π.

Problem 5.2.3. Find the Fourier series of the function

$$f(x) = \begin{cases} \sin x & \text{if} & 0 < x < \pi \\ x & \text{if} & -\pi < x \leq 0 \end{cases}$$

of period 2π.

Problem 5.2.4. Apply Parseval's relation to function $f(x) = x^2$ on $-\pi < x < \pi$ and its Fourier series derived in Example 5.2.7. Use the result to evaluate

$$\sum_{n=1}^{\infty} \frac{1}{n^4}.$$

Problem 5.2.5. Given is function $f(x) = |x|/x$ defined on the interval $0 < x < L$.
(a) Find extension of $f(x)$ to $-L < x < L$ as an odd function and expand it in the Fourier series.
(b) Use Parseval's identity to find the sum of the series

$$\sum_{n=0}^{\infty} \frac{1}{(2n+1)^2}.$$

Problem 5.2.6. Find the Fourier series of the function

$$f(x) = x \sin x, \quad (-\pi < x < \pi)$$

which is assumed to have the period 2π.

Problem 5.2.7. Find the Fourier series of the function

$$f(x) = \begin{cases} x(\pi - x) & \text{if} \quad 0 < x < \pi/2 \\ -x(\pi + x) & \text{if} \quad -\pi/2 < x \le 0 \end{cases}$$

of period π. Use your result to find the sum of the series $\sum_{n=1}^{\infty} \frac{1}{n^2}$.

Problem 5.2.8. Find the Fourier series of period $2L$ of $f(x) = x^2$
(a) on the interval $-L < x \le L$.
(b) on the interval $0 < x \le 2L$.
(c) on the interval $L < x \le 3L$.

Problem 5.2.9. Find the Fourier series of function

$$f(x) = \begin{cases} 0 & \text{if} \quad 1 < x < L \\ 1 & \text{if} \quad -1 < x < 1 \\ 0 & \text{if} \quad -L < x < -1 \end{cases}$$

of period $2L$ with $L > 1$.

Problem 5.2.10. Find the complex Fourier series of $f(x) = x+1$ in on the interval $-\pi < x < \pi$.

Problem 5.2.11. Find the Fourier series of period $\pi/2$ of $f(x) = \sin x$ on $0 < x < \pi/2$.

5.3   Dirac Delta Function and Fourier Integral Transform

5.3.1   Dirac Delta Function

Here, we consider an orthonormal expansion of function $f(x)$

$$f(x) = \sum_{n=0}^{\infty} a_n \varphi_n(x),$$ (5.99)

with respect to the orthonormal basis of functions φ_n, $n = 0, 1, 2, \ldots$ such that

$$\langle \varphi_n \, | \, \varphi_m \rangle = \delta_{nm}.$$ (5.100)

The coefficients a_n can then be obtained by taking an inner product of orthonormal expansion (5.99) with a function φ_n:

$$\langle \varphi_n \, | \, f \rangle = \sum_{m=0}^{\infty} a_m \langle \varphi_n \, | \, \varphi_m \rangle = \sum_{m=0}^{\infty} a_m \delta_{nm} = a_n,$$

where we used the orthonormality property (5.100). Substituting the coefficients a_n back into expansion (5.99), we obtain

$$f(x) = \sum_{n=0}^{\infty} \int_{-\infty}^{\infty} \varphi_n^*(y) f(y) \varphi_n(x) \, \mathrm{d}y$$

$$= \int_{-\infty}^{\infty} f(y) \left[\sum_{n=0}^{\infty} \varphi_n^*(y) \varphi_n(x) \right] \mathrm{d}y,$$ (5.101)

where in the last equation we had interchanged the sum with the integral. Expression in the square bracket defines the so-called Dirac delta function $\delta(x - y)$:

$$\delta(y - x) = \sum_{n=0}^{\infty} \varphi_n^*(y) \varphi_n(x),$$ (5.102)

with the basic property $f(x) = \int_{-\infty}^{\infty} f(y) \delta(y - x) \, \mathrm{d}y$, which we first encountered in Chapter 1 in the more general three-dimensional integral (1.177).

Equation (5.102) is called the completeness relation and it is a basic property of the orthonormal set of functions which solve the Sturm-Liouville problem. The completeness relation for the orthonormal functions satisfying orthonormality condition (5.100) guarantees the orthonormal expansion (5.99).

Now, we will try to gain some graphical intuition about the Dirac delta function $\delta(x)$. Since, the basic property of the Dirac delta function reads $f(0) = \int_{-\infty}^{\infty} f(y) \delta(y) \, \mathrm{d}y$ it follows that $\delta(x)$ is zero everywhere on the x axis with exception of the point $x = 0$. The area under the graph of the $\delta(x)$ function is equal to one according to condition $\int_{-\infty}^{\infty} \delta(x) \, \mathrm{d}x = 1$, obtained by plugging $f(x) = 1$ into $f(0) = \int_{-\infty}^{\infty} f(y) \delta(y) \, \mathrm{d}y$. Therefore, we can visualize the $\delta(x)$ function as a very narrow "spike" centered around x with a unity area. The finite area requirement requires that the spike must be of an infinite height. For that reason the Dirac delta function is not a proper function. The proper way to make sense of it is to define it as a limit of a sequence of finite functions that become more peaked as their parameter increases. We will represent the $\delta(x)$ function as limit of the functions $\delta_n(x)$ defined as

$$\delta_n(x) = \begin{cases} 2n & \text{for } -\frac{1}{n} < x < \frac{1}{n} \\ 0 & \text{for } |x| \geq \frac{1}{n}. \end{cases} \tag{5.103}$$

The area of the rectangle in Figure (5.9) is normalized to one for any n, as seen from the integral:

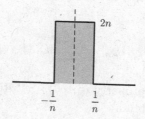

Figure 5.9: $\delta_n(x)$ functions.

$$\int_{-\infty}^{\infty} \delta_n(x)\, \mathrm{d}x = \int_{-\frac{1}{n}}^{\frac{1}{n}} 2n\, \mathrm{d}x = 2n\left(\frac{1}{n} - \left(-\frac{1}{n}\right)\right) = 1.$$

As n increases the width $2/n$ decreases and the height increases to maintain unity integral. So the height of $\delta_n(x)$ becomes infinite in the $n \to \infty$ limit and the function is not well-defined in this limit. However, the integral

$$\int_{-\infty}^{\infty} f(x)\delta_n(x)\, \mathrm{d}x$$

remains finite for the function f, which is continuous around $x = 0$. To verify this property we use the mean value theorem and calculate the integral as follows

$$\int_{-\frac{1}{n}}^{\frac{1}{n}} f(x)2n\, \mathrm{d}x = f(\xi)\, 2n\left(\frac{1}{n} - \left(-\frac{1}{n}\right)\right) = f(\xi),$$

where ξ is some point in the interval $-\frac{1}{n} < x < \frac{1}{n}$. As n increases ξ converges to zero and the limiting procedure

$$\lim_{n \to \infty} \int_{-\infty}^{\infty} f(x)\delta_n(x)\, \mathrm{d}x = \lim_{\xi \to 0} f(\xi) = f(0)$$

involves only finite quantities and is well-defined. Therefore, it makes sense to define the Dirac delta function, $\delta(x)$, through the following limiting procedure,

$\int_{-\infty}^{\infty} f(x)\delta(x) \, \mathrm{d}x = \lim_{n\to\infty} \int_{-\infty}^{\infty} f(x)\delta_n(x) \, \mathrm{d}x$. Accordingly, the Dirac delta function $\delta(x)$ is entirely defined by the integral:

$$\int_{-\infty}^{\infty} f(x)\delta(x) \, \mathrm{d}x = f(0). \tag{5.104}$$

or the corresponding integral

$$\int_{-\infty}^{\infty} f(x)\delta(x - x_0) \, \mathrm{d}x = f(x_0). \tag{5.105}$$

obtained by moving the "spike" from $x = 0$ to $x = x_0$. By taking function f such that $f(x) = 1$ for $a < x < b$ and $f(x) = 0$ elsewhere we obtain

$$\int_a^b \delta(x - x_0) \, \mathrm{d}x = \begin{cases} 1 & \text{for } a < x < b \\ 0 & \text{elsewhere}. \end{cases}$$

This equation characterizes completely the delta function. It shows that the δ-function vanishes when its argument differs from zero and it is infinite in the point where its argument is zero to maintain the unit area under the graph.

Here, we list other properties of $\delta(x)$ following from definition (5.105)

$$\delta(-x) = \delta(x) \tag{5.106a}$$
$$x\delta(x) = 0 \tag{5.106b}$$
$$x\delta'(x) = -\delta(x) \tag{5.106c}$$
$$\delta(ax) = \frac{1}{|a|}\delta(x) \tag{5.106d}$$
$$\delta(x^2 - a^2) = \frac{1}{2|a|}\left(\delta(x - a) + \delta(x + a)\right) \tag{5.106e}$$
$$\delta(f(x)) = \sum_{\substack{i=1, \\ f(x_i)=0}}^{n} \left|\frac{\mathrm{d}f}{\mathrm{d}x}(x_i)\right|^{-1} \delta(x - x_i). \tag{5.106f}$$

Now, consider $\Theta(x)$ being a step function:

$$\Theta(x) = \begin{cases} 1 & x > 0 \\ 0 & x < 0. \end{cases} \tag{5.106g}$$

It holds that the derivative of $\Theta(x)$ equals the delta function:

$$\frac{\mathrm{d}\Theta(x)}{\mathrm{d}x} = \delta(x). \tag{5.106h}$$

☞ **Example 5.3.1.** For illustration we will prove identities (5.106e) and (5.106h). First, we consider the relation (5.106e). In the integral

$$\int_{-\infty}^{\infty} f(x)\delta(x^2 - a^2)\,dx$$

the non-zero contributions originate from only two points, $x = \pm a$, where the argument of the delta function vanishes. Assume for a moment that a is positive. To separate these two contributions we split the integral into two parts:

$$\int_{-\infty}^{\infty} f(x)\delta(x^2 - a^2)\,dx = \int_{-\infty}^{0} f(x)\delta\left((x-a)(x+a)\right)\,dx$$

$$+ \int_{0}^{\infty} f(x)\delta\left((x+a)(x-a)\right)\,dx.$$

The contribution to the first integral over the negative part of the x axis comes from the point $-a$. In the vicinity of $-a$ we have $x \sim -a$ or $x - a \sim -2a$ and, therefore, $\delta\left((x-a)(x+a)\right) = \delta\left((-2a)(x+a)\right) = \delta(x+a)|2a|^{-1}$, where we used property (5.106d). The second integrand differs from zero only around the point $x = a$. In a vicinity of a we have $x + a \sim 2a$ and we can write $\delta\left((x+a)(x-a)\right) = \delta\left((2a)(x-a)\right) = \delta(x-a)|2a|^{-1}$. Plugging these two results back into the integral yields

$$\int_{-\infty}^{\infty} f(x)\delta(x^2 - a^2)\,dx = \int_{-\infty}^{0} f(x)\delta(x+a)|2a|^{-1}\,dx$$

$$+ \int_{0}^{\infty} f(x)\delta(x-a)|2a|^{-1}\,dx$$

$$= |2a|^{-1}\left((f(-a) + f(a)\right).$$

The result is not affected by letting $a \to -a$. The right hand side agrees with the integral $\int_{-\infty}^{\infty} f(x)|2a|^{-1}\left((\delta(x-a) + \delta(x+a)\right)\,dx$ which completes the proof of identity (5.106e).

Next, we turn our attention to relation (5.106h). Multiplying both sides by a test function f, which goes to zero at infinity, and integrating from $-\infty$ to ∞ yields after integration by parts and use of definition (5.106g):

$$\int_{-\infty}^{\infty} \Theta'(x)\,f(x)\,dx = \left[\Theta(x)f(x)\right]_{-\infty}^{\infty} - \int_{-\infty}^{\infty} \Theta(x)\,f'(x)\,dx$$

$$= -\int_{0}^{\infty} f'(x)\,dx = f(0) = \int_{-\infty}^{\infty} \delta(x)f(x)\,dx,$$

which implies relation (5.106h).

5.3.2   Fourier Integral Transform

Now, let us consider a new representation for the Dirac δ-function given by:

$$\delta(x) = \frac{1}{2\pi} \int_{-\infty}^{\infty} d\,k\, e^{-i\,kx} . \qquad (5.107)$$

To establish this relation we will evaluate the integral:

$$
\begin{aligned}
\int_{-\infty}^{\infty} d\,k\, e^{-i\,kx} e^{-\epsilon|k|} &= \int_{-\infty}^{0} d\,k\, e^{(\epsilon - i\,x)k} + \int_{0}^{\infty} d\,k\, e^{-(\epsilon + i\,x)k} \\
&= \frac{1}{\epsilon - i\,x}\left[e^{(\epsilon - i\,x)k} \right]_{-\infty}^{k=0} + \frac{1}{-\epsilon - i\,x}\left[e^{-(\epsilon + i\,x)k} \right]_{0}^{\infty} \\
&= \frac{1}{\epsilon - i\,x} + \frac{1}{\epsilon + i\,x} = \frac{(\epsilon + i\,x) + (\epsilon - i\,x)}{(\epsilon - i\,x)(\epsilon + i\,x)} \\
&= \frac{2\epsilon}{\epsilon^2 + x^2} ,
\end{aligned}
\qquad (5.108)
$$

in which we inserted a regularizing term $\exp(-\epsilon|k|)$ to ensure convergence. The integral on the right hand side of equation (5.107) can be obtained by taking the limit $\epsilon \to 0$ in the end of calculation. In this way we obtain:

$$\frac{1}{2\pi} \int_{-\infty}^{\infty} d\,k\, e^{-i\,kx} = \lim_{\epsilon \to 0} \frac{1}{\pi} \frac{\epsilon}{\epsilon^2 + x^2} . \qquad (5.109)$$

Since,

$$\lim_{\epsilon \to 0} \frac{1}{\pi} \frac{\epsilon}{\epsilon^2 + x^2} = 0, \quad \text{for} \ \ x \neq 0, \qquad (5.110)$$

the function defined by equation (5.109) vanishes for a non-zero x. Moreover, the area under the graph of the function $\epsilon/\pi(\epsilon^2 + x^2)$ is equal to one as seen from:

$$\text{area} = \int_{-\infty}^{\infty} \frac{1}{\pi} \frac{\epsilon}{\epsilon^2 + x^2} \, dx = \left[\frac{1}{\pi} \arctan \frac{x}{\epsilon} \right]_{-\infty}^{\infty} = \frac{1}{\pi}\left(\frac{\pi}{2} - \left(-\frac{\pi}{2}\right) \right) = 1 . \qquad (5.111)$$

These are the two basic properties which define the Dirac delta function and therefore we have proved a new representation of the Dirac delta function:

$$\delta(x) = \lim_{\epsilon \to 0} \frac{1}{\pi} \frac{\epsilon}{\epsilon^2 + x^2} = \frac{1}{2\pi} \int_{-\infty}^{\infty} d\,k\, e^{-i\,kx} . \qquad (5.112)$$

Definition 5.3.1. *Define*

$$\widetilde{f}(k) = \frac{1}{\sqrt{2\pi}} \int_{-\infty}^{\infty} d x f(x)\, e^{-i\,kx} , \qquad (5.113)$$

as a Fourier transform of the function $f(x)$.

The integral

$$\frac{1}{\sqrt{2\pi}} \int_{-\infty}^{\infty} \mathrm{d}\,k\,\tilde{f}(k)e^{\mathrm{i}\,kx} = \int_{-\infty}^{\infty} \frac{\mathrm{d}\,k}{\sqrt{2\pi}} \left[\int_{-\infty}^{\infty} \frac{\mathrm{d}\,x'}{\sqrt{2\pi}} f(x')\,e^{-\mathrm{i}\,kx'} \right] e^{\mathrm{i}\,kx} \qquad (5.114)$$

reproduces function $f(x)$ due to the relation:

$$\frac{1}{2\pi} \int_{-\infty}^{\infty} \mathrm{d}x' \int_{-\infty}^{\infty} \mathrm{d}\,k\, f(x')\,e^{-\mathrm{i}\,k(x'-x)} = \int_{-\infty}^{\infty} \mathrm{d}x' f(x')\delta(x'-x) = f(x)\,, \quad (5.115)$$

where we interchanged the order of k and x integrations and used relation (5.112). Hence, the inverse Fourier transformation takes a form:

$$f(x) = \frac{1}{\sqrt{2\pi}} \int_{-\infty}^{\infty} \mathrm{d}\,k\,\tilde{f}(k)e^{\mathrm{i}\,kx}\,. \qquad (5.116)$$

☞ **Example 5.3.2.** Consider a rectangular pulse of height 1 and width $2L$ described by

$$h_L(x) = \begin{cases} 1 & -L < x < L \\ 0 & \text{elsewhere}. \end{cases} \qquad (5.117)$$

The function $h_L(x)$ has the following Fourier transform

$$\begin{aligned}
\widetilde{h}_L(k) &= \frac{1}{\sqrt{2\pi}} \int_{-\infty}^{\infty} \mathrm{d}x f(x)\,e^{-\mathrm{i}\,kx} = \frac{1}{\sqrt{2\pi}} \int_{-L}^{L} \mathrm{d}x\, e^{-\mathrm{i}\,kx} \\
&= \frac{1}{\sqrt{2\pi}} \frac{1}{-\mathrm{i}\,k} \left[e^{-\mathrm{i}\,kx} \right]_{-L}^{L} = \frac{1}{\sqrt{2\pi}} \frac{1}{-\mathrm{i}\,k} \left(e^{-\mathrm{i}\,kL} - e^{\mathrm{i}\,kL} \right) \\
&= \sqrt{\frac{2}{\pi}} \frac{\sin(kL)}{k}\,.
\end{aligned}$$

Figure 5.10 shows a graph of $\widetilde{h}_l(k)$. Two features of this graph are particularly important. The height of the curve is equal to $\widetilde{h}_l(0) = L\sqrt{2}/\sqrt{\pi}$ as obtained from L'Hospital's rule, $\sin(kL)/k \to L$ for $k \to 0$. Also, we note that the length $2\pi/L$ of the interval $-\pi/L \le k \le \pi/L$ is a measure of the width of the function $\widetilde{h}_l(k)$. As the pulse $h_L(x)$ gets broader and its width, L, increases, the width of its Fourier transform $\widetilde{h}_l(k)$ gets smaller in accordance with the uncertainty relation (5.125) to be discussed below.

As a consequence of definition (5.113) we find that

$$\begin{aligned}
\mathrm{i}\,k\widetilde{f}(k) &= -\frac{1}{\sqrt{2\pi}} \int_{-\infty}^{\infty} \mathrm{d}x\, f(x) \frac{\mathrm{d}}{\mathrm{d}x}e^{-\mathrm{i}\,kx} \\
&= -\frac{1}{\sqrt{2\pi}} \left[f(x)\,e^{-\mathrm{i}\,kx} \right]_{-\infty}^{\infty} + \frac{1}{\sqrt{2\pi}} \int_{-\infty}^{\infty} \mathrm{d}x\, f'(x)\,e^{-\mathrm{i}\,kx} \qquad (5.118) \\
&= \widetilde{f'}(k)
\end{aligned}$$

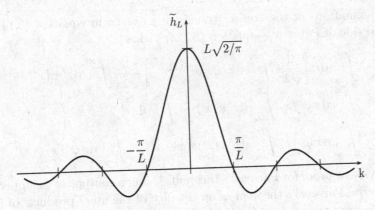

Figure 5.10: Fourier transform of a rectangular pulse.

valid for functions $f(x)$ which vanish as $x \to \pm\infty$. Similarly,

$$(\widetilde{-\mathrm{i}\,xf})(k) = \frac{1}{\sqrt{2\pi}} \int_{-\infty}^{\infty} \mathrm{d}x\, f(x) \frac{\mathrm{d}}{\mathrm{d}k} e^{-\mathrm{i}kx} = \frac{\mathrm{d}}{\mathrm{d}k} \widetilde{f}(k)\,.$$

Some other basic properties of the Fourier transform are:

$$
\begin{aligned}
(\widetilde{e^{\mathrm{i}\,ax}f})(k) &= \widetilde{f}(k-a) & (5.119)\\
(\widetilde{D_a f})(k) &= e^{\mathrm{i}ka}\widetilde{f}(k), \quad (D_a f)(x) \equiv f(x+a) & (5.120)\\
\widetilde{f_a}(k) &= \frac{1}{|a|}\widetilde{f}\!\left(\frac{k}{a}\right), \quad f_a(x) \equiv f(ax) & (5.121)\\
\widetilde{f^*}(k) &= \widetilde{f}^*(-k) & (5.122)
\end{aligned}
$$

for a being a real non-zero constant and $f^*(x)$ a complex conjugate of $f(x)$. These relations readily follow from the definition (5.113). Relation (5.121) shows that as $f(x)$ gets more narrow around $x = 0$, $\widetilde{f}(k)$ gets wider around $k = 0$, and vice versa.

The identity

$$\int_{-\infty}^{\infty} \varphi^*(x)\psi(x)\,\mathrm{d}x = \int_{-\infty}^{\infty} \widetilde{\varphi}^*(k)\widetilde{\psi}(k)\,\mathrm{d}k \qquad (5.123)$$

valid for two functions φ, ψ and their Fourier transforms $\widetilde{\varphi}^*$ and $\widetilde{\psi}$, is called Parseval's theorem.

Inserting definitions of the Fourier transform as given in equation (5.113) on the right hand side of Parseval's equation (5.123) yields

$$
\int_{-\infty}^{\infty} dk \left[\frac{1}{\sqrt{2\pi}} \int_{-\infty}^{\infty} dx\, \varphi(x)\, e^{-i\,kx} \right]^* \left[\frac{1}{\sqrt{2\pi}} \int_{-\infty}^{\infty} dx'\, \psi(x')\, e^{-i\,kx'} \right]
$$

$$
= \int_{-\infty}^{\infty} dx\, \varphi^*(x) \int_{-\infty}^{\infty} dx'\, \psi(x') \frac{1}{2\pi} \int_{-\infty}^{\infty} dk\, e^{i\,k(x-x')}
$$

$$
= \int_{-\infty}^{\infty} dx\, \varphi^*(x) \int_{-\infty}^{\infty} dx'\, \psi(x') \delta(x-x') = \int_{-\infty}^{\infty} dx\, \varphi^*(x)\psi(x), \quad (5.124)
$$

which provides a proof for Parseval's theorem. Using a concept of the inner product we can rewrite Parseval's theorem as an equality of the inner products of functions and their Fourier transforms. Explicitly,

$$
\langle \varphi \,|\, \psi \rangle = \langle \widetilde{\varphi} \,|\, \widetilde{\psi} \rangle \,,
$$

where $\langle \widetilde{\varphi} \,|\, \widetilde{\psi} \rangle = \int_{-\infty}^{\infty} dk\, \widetilde{\varphi}^*(k)\widetilde{\psi}(k)$. In particular, the norms of the function and its Fourier transform are equal as expressed by: $\|\varphi\| = \|\widetilde{\varphi}\|$.

We will use Parseval's theorem to prove the fundamental Heisenberg uncertainty relation:

$$
\boxed{(\Delta_f x)(\Delta_f k) \geq \frac{1}{2}\,,}
\qquad (5.125)
$$

where $(\Delta_f x)$ and $(\Delta_f k)$ are uncertainties (or standard deviations) defined as:

$$
(\Delta_f x)^2 = \frac{\int_{-\infty}^{\infty} dx\, (x - \langle x \rangle)^2 |f(x)|^2}{\int_{-\infty}^{\infty} dx\, |f(x)|^2} = \frac{\int_{-\infty}^{\infty} dx\, (x - \langle x \rangle)^2 |f(x)|^2}{\|f\|^2} \quad (5.126)
$$

$$
(\Delta_f k)^2 = \frac{\int_{-\infty}^{\infty} dk\, (k - \langle k \rangle)^2 |\widetilde{f}(k)|^2}{\int_{-\infty}^{\infty} dk\, |\widetilde{f}(k)|^2} = \frac{\int_{-\infty}^{\infty} dk\, (k - \langle k \rangle)^2 |\widetilde{f}(k)|^2}{\|f\|^2} \,, (5.127)
$$

where

$$
\langle x \rangle = \langle f, xf \rangle = \int_{-\infty}^{\infty} dx\, x |f(x)|^2 \qquad (5.128a)
$$

and

$$
\langle k \rangle = \langle \widetilde{f}, k\widetilde{f} \rangle = \int_{-\infty}^{\infty} dk\, k |\widetilde{f}(k)|^2 \qquad (5.128b)
$$

are the average values or expectation values of x and k, respectively.

In order to ensure convergence of the integral in (5.128a) we choose f that goes to zero at infinity sufficiently fast in order for $x|f(x)|^2 \to 0$, when $x \to \pm\infty$. For such function f it holds that

$$
-\int_{-\infty}^{\infty} dx\, x \left(\frac{df}{dx} f^* + f \frac{df^*}{dx} \right) = \int_{-\infty}^{\infty} dx\, |f(x)|^2 \qquad (5.129)
$$

as follows by integration by parts and the fact that

$$\int_{-\infty}^{\infty} dx \frac{d}{dx} \left(x|f(x)|^2 \right) = \left[x|f(x)|^2 \right]_{-\infty}^{\infty} = 0.$$

For simplicity, we consider the case of a function f such that the expectation values in (5.128a) and (5.128a) vanish and, thus, $\langle x \rangle = 0$ and $\langle k \rangle = 0$. In that case

$$\begin{aligned}
\|f\|^2 \|f\|^2 (\Delta_f x)^2 (\Delta_f k)^2 &= \int_{-\infty}^{\infty} dx\, x^2 |f(x)|^2 \int_{-\infty}^{\infty} dk\, k^2 |\widetilde{f}(k)|^2 \\
&= \int_{-\infty}^{\infty} dx\, x^2 |f(x)|^2 \int_{-\infty}^{\infty} dk\, |-ik\widetilde{f}(k)|^2 \\
&= \int_{-\infty}^{\infty} dx\, x^2 |f(x)|^2 \int_{-\infty}^{\infty} dk\, |\widetilde{f'}(k)|^2 \\
&= \int_{-\infty}^{\infty} dx\, x^2 |f(x)|^2 \int_{-\infty}^{\infty} dx\, |f'(x)|^2 ,
\end{aligned}$$

where use was made of the identity (5.118) and of the Parseval's theorem, $\|\varphi\| = \|\widetilde{\varphi}\|$.

The above equation can be rewritten using the concept of norm as

$$\|f\|^2 \|f\|^2 (\Delta_f x)^2 (\Delta_f k)^2 = \|xf\|^2 \|f'\|^2 .$$

Due to Cauchy-Schwartz inequality (5.20), $|\langle f \mid g \rangle| \leq \|f\|\|g\|$, we obtain

$$\begin{aligned}
(\Delta_f x)^2 (\Delta_f k)^2 &= \frac{1}{(\|f\|^2)^2} \|xf\|^2 \|f'\|^2 \\
&\geq \frac{1}{(\|f\|^2)^2} |\langle xf \mid f' \rangle|^2 = \frac{1}{(\|f\|^2)^2} \left| \int_{-\infty}^{\infty} dx\, xf^*(x)f'(x) \right|^2 ,
\end{aligned}$$

thanks to $(xf^*(x)f'(x))^* = xf^{*\prime}f$, this equals to

$$\begin{aligned}
&= \frac{1}{(\|f\|^2)^2} \left| \int_{-\infty}^{\infty} dx\, \frac{x}{2} \left(f^*(x)f'(x) + f^{*\prime}(x)f(x) \right) \right|^2 \\
&= \frac{1}{(\|f\|^2)^2} \frac{1}{4} \left| \int_{-\infty}^{\infty} dx\, |f(x)|^2 \right|^2 \\
&= \frac{1}{4} ,
\end{aligned}$$

where we used (5.129).

Hence, we have shown that

$$(\Delta_f x)(\Delta_f k) \geq \frac{1}{2}$$

as desired.

☞ **Example 5.3.3.** The function

$$g(x) = \frac{1}{\sqrt{a}(2\pi)^{\frac{1}{4}}} e^{-\frac{x^2}{4a^2}}$$ (5.130)

is called a Gaussian function while it's square

$$P(x) = |g(x)|^2 = g^2(x) = \frac{1}{a\sqrt{2\pi}} e^{-\frac{x^2}{2a^2}}$$ (5.131)

is its Gaussian distribution. It is normalized in such a way that the area

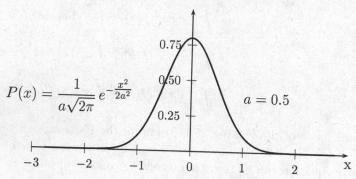

Figure 5.11: A Gaussian distribution with $a = \frac{1}{2}$.

under its graph is unity:

$$\int_{-\infty}^{\infty} dx\, P(x) = 1,$$

as follows from the integral formulas (5.74). The average value of x vanishes

$$\langle x \rangle = \langle g, xg \rangle = \int_{-\infty}^{\infty} dx\, x\, P(x) = \frac{1}{a\sqrt{2\pi}} \int_{-\infty}^{\infty} dx\, x e^{-\frac{x^2}{2a^2}} = 0,$$

since the above integrand is an odd function. Expression $\langle g, x^2 g \rangle$ involves an even integrand and yields a non-zero integral that can be obtained from the formula (5.74) as

$$\langle g, x^2 g \rangle = \int_{-\infty}^{\infty} dx\, x^2\, P(x) = \frac{1}{a\sqrt{2\pi}} \int_{-\infty}^{\infty} dx\, x^2 e^{-\frac{x^2}{2a^2}} = a^2.$$

Therefore,

$$(\Delta_g x)^2 = \frac{\int_{-\infty}^{\infty} dx\, (x - \langle x \rangle)^2 |g(x)|^2}{\|g\|^2} = \int_{-\infty}^{\infty} dx\, x^2\, P(x) = a^2$$

and a^2 is a measure of uncertainty (or spreading) of the Gaussian distribution $P(x)$ as shown on the Figure (5.11) with $a = 1/2$. Thus the factor appearing in the denominator of the exponential of the Gaussian distribution determines its width or spreading.

The Fourier transform corresponding to the Gaussian function is

$$\widetilde{g}(k) = \frac{1}{\sqrt{a}(2\pi)^{\frac{3}{4}}} \int_{-\infty}^{\infty} dx\, e^{-\frac{x^2}{4a^2}} e^{-i\,kx}\,.$$

Completing the square according to:

$$e^{-\frac{x^2}{4a^2}} e^{-ikx} = e^{-\frac{x^2}{4a^2}} e^{-i\,kx} e^{a^2 k^2} e^{-a^2 k^2}$$

$$= e^{-\left(\frac{x}{2a}+i\,ka\right)^2} e^{-a^2 k^2}$$

and replacing x with the integration variable $\xi = x/2a + i\,ka$ leads, using the formula (4.84), to the Fourier transform

$$\widetilde{g}(k) = \frac{2a\exp(-a^2 k^2)}{\sqrt{a}(2\pi)^{\frac{3}{4}}} \int_{-\infty}^{\infty} d\xi\, e^{-\xi^2} = \frac{2a\exp(-a^2 k^2)}{\sqrt{a}(2\pi)^{\frac{3}{4}}} \sqrt{\pi}$$

$$= \sqrt{\frac{2a}{\sqrt{2\pi}}}\, e^{-a^2 k^2}\,,$$

(5.132)

which itself is Gaussian. Let

$$P(k) = \widetilde{g}^2(k) = \frac{2a}{\sqrt{2\pi}} e^{-2a^2 k^2} = \frac{2a}{\sqrt{2\pi}} \exp\left(-k^2 / \left(2\frac{1}{4a^2}\right)\right)$$

be a Gaussian distribution of $\widetilde{g}$. The area under the graph of $P(k)$ remains normalized to unity, $\int_{-\infty}^{\infty} dk\, P(k) = 1$, due to Parseval's theorem (5.123). Also, the average value of k vanishes, $\int_{-\infty}^{\infty} dk\, kP(k) = 0$, since the integrand is an odd function. However, comparing the exponential functions $P(x)$ and $P(k)$ we see that the spreading of the Fourier transform of the Gaussian distribution is an inverse of the spreading of the original Gaussian distribution. A factor a^2 in the denominator of the exponential of $P(x)$ has been replaced by a factor $1/4a^2$ at the corresponding position in the exponential of $P(k)$. This is a basic principle of the uncertainty relation. Indeed, calculation of the uncertainty in k gives

$$(\Delta_g k)^2 = \int_{-\infty}^{\infty} dk\, k^2 P(k) = \frac{2a}{\sqrt{2\pi}} \int_{-\infty}^{\infty} dk\, k^2 e^{-2a^2 k^2} = \frac{1}{(2a)^2}$$

(5.133)

and as a Gaussian function g gets more sharply peaked its Fourier transform $\widetilde{g}$ gets wider and vice versa (see Figs. (5.12) and (5.13)). This is in agreement with a general property (5.121) of the Fourier transform.

For the product of uncertainties we obtain

$$(\Delta_g x)(\Delta_g k) = a\frac{1}{2a} = \frac{1}{2}\,.$$

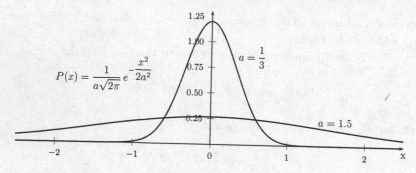

$$P(x) = \frac{1}{a\sqrt{2\pi}} e^{-\frac{x^2}{2a^2}}$$

Figure 5.12: Examples of Gaussian distributions with $a = 1.5, 1/3$.

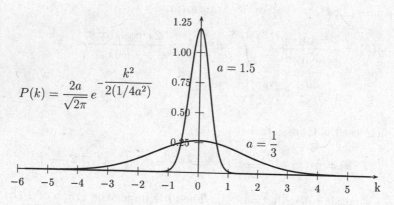

$$P(k) = \frac{2a}{\sqrt{2\pi}} e^{-\frac{k^2}{2(1/4a^2)}}$$

Figure 5.13: Examples of Fourier transforms of Gaussian distributions with $a = 1.5, 1/3$.

Thus, comparing with the Heisenberg inequality relation we see that for the Gaussian distribution the equality has been reached. This is an unique property of a Gaussian function.

◈ **Problems: Dirac Delta function and Fourier integral transform.**

Problem 5.3.1. Show that for the functions $g_n(x)$ defined as

$$g_n(x) = \begin{cases} n & 0 < x < \frac{1}{n} \\ 0 & \text{elsewhere} \end{cases}$$

it holds that $\lim_{n\to\infty} g_n(x) = \delta(x)$.

Problem 5.3.2. Show that

$$\int_{-\infty}^{\infty} \delta'(x) f(x)\, dx = -f'(0)\,.$$

Problem 5.3.3. Show that

$$\delta\left((x-a)(x-b)\right) = \frac{1}{|a-b|}\left(\delta(x-a) + \delta(x-b)\right)$$

for $a \neq b$.

Problem 5.3.4. Calculate the integral:

$$\int_{-\infty}^{\infty} f(x)\delta(-ax + b)\, dx\,,$$

where a and b are real, positive constants.

Problem 5.3.5. Calculate the integral:

$$\int_{-3}^{0} \left(e^{-x^2} + 3x^3\right) \delta(x-1)\, dx\,.$$

Problem 5.3.6. Calculate the integral:

$$\int_{\pi/4}^{3\pi/4} x^2 \delta(\cos x)\, dx\,.$$

Problem 5.3.7. Calculate the integral:

$$\int_{0}^{\infty} \delta\left(\cos(\pi x)\right) \sin(2x)\, dx\,.$$

Problem 5.3.8. Find the Fourier transform for $f(x) = \sin(\pi x)$.

Problem 5.3.9. Derive the integral Fourier transform for the function

$$f(x) = \begin{cases} x & -1 < x < 1 \\ 0 & \text{elsewhere.} \end{cases}$$

Problem 5.3.10. Derive the integral Fourier transform for functions

(a) $e^{-r|x|},\ r > 0$, (b) $e^{-r|x|}\sin(2x),\ r > 0$,

(c) $xe^{-r|x|},\ r > 0$, (d) $e^{-x^2}\cos x$.

Problem 5.3.11. Verify Parseval's relation $\|h_L\| = \|\widetilde{h}_L\|$ for the rectangular pulse defined in equation (5.117).

Problem 5.3.12. Derive the integral Fourier transform for the function

$$f(x) = \begin{cases} e^{-ax} & x \geq 0 \\ 0 & x < 0 \end{cases}$$

for some positive constant a. Verify Parseval's relation $\|f\| = \|\tilde{f}\|$ for this function.

Problem 5.3.13. Derive the integral Fourier transform for the function

$$f(x) = \begin{cases} x & 0 \leq x < 1 \\ -x & -1 \leq x < 0 \\ 0 & \text{elsewhere}. \end{cases}$$

Problem 5.3.14. Show that the Fourier transform of the so-called convolution integral $F(x) = \int_{-\infty}^{\infty} g(x - x')f(x')\,dx'$ of two functions $g(x)$ and $f(x)$ is given by

$$\tilde{F}(k) = \tilde{g}(k)\tilde{f}(k)$$

in terms the Fourier transforms $\tilde{g}(k)$, $\tilde{f}(k)$ of $g(x)$ and $f(x)$.

Problem 5.3.15. Prove properties (5.119)-(5.122) of the Fourier transform.

Chapter 6

Linear Systems of Differential Equations

SYSTEMS of linear differential equations can be conveniently reformulated in a matrix formalism. The resulting matrix equation has the form of a single differential equation. All the unknown functions fit into one column vector and all the constant coefficients end up in one single coefficient matrix. Consequently, the concepts of matrices, their eigenvalues and eigenvectors, and most importantly the application of eigenvalue problems to the diagonalization of matrices all play fundamental roles in finding solutions to linear differential equations with constant coefficients. Knowledge of matrix and linear algebra is also instrumental in formulating the underlying existence and uniqueness theorem. Eigenvalues of the coefficient matrix are invaluable tools in determining the qualitative behavior of solutions as time evolves. Difficult cases arise for coefficient matrices that cannot be diagonalized. A method for handling repeated eigenvalues is presented for such cases.

6.1 Homogeneous Systems of Differential Equations

Here, we consider a system of constant coefficient differential equations involving functions $x_1(t)$, $x_2(t)$, ..., $x_n(t)$ depending on one independent variable t. We usually view t as a time. The system is said to be linear if the differential equations are linear. A linear system of n unknown functions will have a general form:

$$\frac{dx_1}{dt} = a_{11}x_1 + a_{12}x_2 + \cdots + a_{1n}x_n + r_1(t),$$

$$\frac{dx_2}{dt} = a_{21}x_1 + a_{22}x_2 + \cdots + a_{2n}x_n + r_2(t),$$

$$\vdots$$

$$\frac{dx_n}{dt} = a_{n1}x_1 + a_{n2}x_2 + \cdots + a_{nn}x_n + r_n(t). \tag{6.1}$$

It is convenient to employ a linear algebra method and rewrite such system as

$$\frac{d}{dt}\begin{bmatrix} x_1(t) \\ x_2(t) \\ \vdots \\ x_n(t) \end{bmatrix} = \begin{bmatrix} a_{11} & a_{12} & \cdots & a_{1n} \\ a_{21} & a_{22} & \cdots & a_{2n} \\ \vdots & \vdots & \cdots & \vdots \\ a_{n1} & a_{n2} & \cdots & a_{nn} \end{bmatrix} \begin{bmatrix} x_1(t) \\ x_2(t) \\ \vdots \\ x_n(t) \end{bmatrix} + \begin{bmatrix} r_1(t) \\ r_2(t) \\ \vdots \\ r_n(t) \end{bmatrix}. \tag{6.2}$$

Introducing a constant $n \times n$ coefficient matrix $\underline{\underline{A}} = [a_{ij}]_{1 \le i \le n, 1 \le j \le n}$ and two vector-valued functions

$$\underline{x}(t) = \begin{bmatrix} x_1(t) \\ x_2(t) \\ \vdots \\ x_n(t) \end{bmatrix}, \qquad \underline{r}(t) = \begin{bmatrix} r_1(t) \\ r_2(t) \\ \vdots \\ r_n(t) \end{bmatrix}$$

allows us to recast (6.2) in a matrix form:

$$\boxed{\underline{\dot{x}}(t) = \underline{\underline{A}}\,\underline{x}(t) + \underline{r}(t),} \tag{6.3}$$

where $\dot{x} = dx/dt$.

Using the same terminology as in Chapter 3, the system of equations (6.3) is called homogeneous if the source $\underline{r}(t)$ vanishes (meaning that all components $r_i(t)$ are zero), otherwise it is called non-homogeneous.

We begin with the homogeneous system of differential equations

$$\underline{\dot{x}}(t) = \underline{\underline{A}}\,\underline{x}(t), \tag{6.4}$$

with the initial condition

$$\underline{x}(0) = \underline{c},$$

taken here (without loss of generality) at the time $t = 0$. One easily verifies that a formal solution to equation (6.4) is given by

$$\underline{x}(t) = e^{t\underline{\underline{A}}}\,\underline{x}(0). \tag{6.5}$$

In equation (6.5) we have used an exponential of a matrix. It has been introduced in equation (2.159), as a power series expansion:

$$e^{\underline{\underline{A}}} = \underline{\underline{I}} + \underline{\underline{A}} + \frac{1}{2!}\underline{\underline{A}}^2 + \cdots + \frac{1}{n!}\underline{\underline{A}}^n + \cdots = \sum_{k=0}^{\infty} \frac{1}{k!}\underline{\underline{A}}^k. \tag{6.6}$$

Note that the above expression mimics term by term the corresponding Taylor expansion of the exponential function of a number. This matrix expansion starts with the identity matrix $\underline{\underline{I}}$, which formally agrees with $\underline{\underline{A}}^0$.

If matrices $\underline{\underline{A}}$ and $\underline{\underline{B}}$ commute, i.e. they satisfy $\underline{\underline{A}}\,\underline{\underline{B}} = \underline{\underline{B}}\,\underline{\underline{A}}$, then

$$e^{\underline{\underline{A}}+\underline{\underline{B}}} = e^{\underline{\underline{A}}} e^{\underline{\underline{B}}} = e^{\underline{\underline{B}}} e^{\underline{\underline{A}}}. \tag{6.7}$$

The above identity can be derived from the definition (6.6) using the following argument

$$e^{\underline{\underline{A}}+\underline{\underline{B}}} = \sum_{n=0}^{\infty} \frac{1}{n!}\left(\underline{\underline{A}}+\underline{\underline{B}}\right)^n = \sum_{n=0}^{\infty}\sum_{k=0}^{n} \frac{1}{k!(n-k)!}\underline{\underline{A}}^k\underline{\underline{B}}^{n-k} = \sum_{k=0}^{\infty}\sum_{j=0}^{\infty} \frac{1}{k!}\underline{\underline{A}}^k\frac{1}{j!}\underline{\underline{B}}^j.$$

It follows, that for any $n \times n$ matrix $\underline{\underline{A}}$ the corresponding matrix exponential $\exp(\underline{\underline{A}})$ is non-singular and its inverse is given by:

$$\left(e^{\underline{\underline{A}}}\right)^{-1} = e^{-\underline{\underline{A}}}.$$

Going back to matrices of the type $\exp\left(t\underline{\underline{A}}\right)$ introduced in equation (6.5) we find from the identity (6.7) the following multiplication law:

$$e^{(t+s)\underline{\underline{A}}} = e^{t\underline{\underline{A}}} e^{s\underline{\underline{A}}}. \tag{6.8}$$

This multiplication rule might be used to compute a derivative of the power series $\exp\left(t\underline{\underline{A}}\right)$, with respect to t, as follows

$$\frac{d}{dt}e^{t\underline{\underline{A}}} = \lim_{\epsilon \to 0} \frac{1}{\epsilon}\left(e^{(\epsilon+t)\underline{\underline{A}}} - e^{t\underline{\underline{A}}}\right) = \lim_{\epsilon \to 0} \frac{1}{\epsilon}\left(e^{\epsilon\underline{\underline{A}}} - \underline{\underline{I}}\right)e^{t\underline{\underline{A}}}$$

$$= \lim_{\epsilon \to 0} \frac{1}{\epsilon}\left(\epsilon\underline{\underline{A}} + \frac{1}{2!}\epsilon^2\underline{\underline{A}}^2 + \cdots + \frac{1}{n!}\epsilon^n\underline{\underline{A}}^n + \cdots\right)e^{t\underline{\underline{A}}}$$

$$= \underline{\underline{A}}\,e^{t\underline{\underline{A}}}.$$

In particular, taking a derivative of the column vector $\underline{x}(t)$ given in (6.5) results in

$$\frac{d}{dt}\underline{x}(t) = \frac{d}{dt}e^{t\underline{\underline{A}}}\,\underline{x}(0) = \underline{\underline{A}}e^{t\underline{\underline{A}}}\,\underline{x}(0) = \underline{\underline{A}}\,\underline{x}(t),$$

which establishes $\underline{x}(t)$ as a solution to the homogeneous system (6.4).

☞ **Example 6.1.1.** We consider a two-dimensional case with a homogeneous system (6.4) having the explicit form

$$\frac{d}{dt}x_1 = x_1, \quad \frac{d}{dt}x_2 = -x_2$$

or for $\underline{x} = \begin{bmatrix} x_1 & x_2 \end{bmatrix}^T$

$$\frac{d}{dt}\underline{x} = \sigma_3\underline{x}, \quad \text{where} \quad \sigma_3 = \begin{bmatrix} 1 & 0 \\ 0 & -1 \end{bmatrix}.$$

According to equation (6.5), the solution is given by

$$\underline{x}(t) = e^{t\sigma_3}\underline{x}(0).$$

Since,

$$\sigma_3^2 = \underline{I}, \quad \sigma_3^3 = \sigma_3, \quad \ldots, \quad \sigma_3^{2k} = \underline{I}, \quad \sigma_3^{2k+1} = \sigma_3,$$

the power series expansion for $\exp(t\sigma_3)$ can be computed explicitly as

$$
\begin{aligned}
e^{t\sigma_3} &= \underline{I} + t\sigma_3 + \frac{1}{2!}t^2\sigma_3^2 + \cdots + \frac{1}{n!}t^n\sigma_3^n \cdots \\
&= \underline{I}\left((1 + \frac{1}{2!}t^2 + \cdots + \frac{1}{(2k)!}t^{2k} + \cdots \right) \\
&\quad + \sigma_3\left(t + \frac{1}{3!}t^3 + \cdots + \frac{1}{(2k+1)!}t^{2k+1} + \cdots \right) \\
&= \frac{1}{2}\underline{I}\left(e^t + e^{-t}\right) + \frac{1}{2}\sigma_3\left(e^t - e^{-t}\right) \\
&= \begin{bmatrix} e^t & 0 \\ 0 & e^{-t} \end{bmatrix}.
\end{aligned}
$$

It follows that

$$\frac{d}{dt}e^{t\sigma_3} = \frac{d}{dt}\begin{bmatrix} e^t & 0 \\ 0 & e^{-t} \end{bmatrix} = \begin{bmatrix} e^t & 0 \\ 0 & -e^{-t} \end{bmatrix} = \begin{bmatrix} 1 & 0 \\ 0 & -1 \end{bmatrix}\begin{bmatrix} e^t & 0 \\ 0 & e^{-t} \end{bmatrix}$$

or

$$\frac{d}{dt}e^{t\sigma_3} = \sigma_3 e^{t\sigma_3}$$

and subsequently $d\underline{x}(t)/dt = \sigma_3\,\underline{x}(t)$ for

$$\underline{x}(t) = \begin{bmatrix} e^t & 0 \\ 0 & e^{-t} \end{bmatrix} \cdot \begin{bmatrix} x_1(0) \\ x_2(0) \end{bmatrix} = \begin{bmatrix} e^t x_1(0) \\ e^{-t} x_2(0) \end{bmatrix}.$$

Now, we will investigate under what condition the exponential vector-function $\underline{x}(t) = e^{\lambda t} \underline{v}$, with a constant column vector $\underline{v}$, is a (non-zero) solution to equation (6.4). Upon substitution of $\underline{x}(t) = e^{\lambda t} \underline{v}$ into the equation (6.4) we obtain

$$\frac{\mathrm{d}}{\mathrm{d}t} e^{\lambda t} \underline{v} = \underline{\underline{A}} e^{\lambda t} \underline{v} = e^{\lambda t} \underline{\underline{A}} \underline{v} = \lambda e^{\lambda t} \underline{v} \quad \text{if} \quad \underline{\underline{A}} \underline{v} = \lambda \underline{v}.$$

Thus, $\underline{x}(t) = e^{\lambda t} \underline{v}$ is a solution of the homogeneous linear system of differential equations (6.4) if and only if $\underline{v}$ is an eigenvector of the coefficient matrix $\underline{\underline{A}}$ with the eigenvalue λ.

Let $\underline{v}_1, \ldots, \underline{v}_k$ be a set of k linearly independent eigenvectors of the coefficient matrix $\underline{\underline{A}}$. Then the linear combination

$$\underline{x}(t) = \sum_{i=1}^{k} c_i e^{\lambda_i t} \underline{v}_i,$$

with k arbitrary constants c_i is also a solution of equation (6.4) as follows from linearity of the differential equations (6.1).

Next, we will quote without a proof *a uniqueness theorem* concerning the the the homogeneous linear system of differential equations.

If the vector functions $\underline{x}_1(t), \ldots, \underline{x}_n(t)$ are linearly independent solutions of the homogeneous system of n linear differential equations, then every solution $\underline{x}(t)$ can be written in an unique way as a linear combination

$$\underline{x}(t) = \sum_{i=1}^{n} c_i \underline{x}_i(t).$$

In other words, the linearly independent solutions $\underline{x}_1(t), \ldots, \underline{x}_n(t)$ span the solution space and it is said that they form *the fundamental solution set*. The above linear combination is often referred to as a *superposition principle*.

It simplifies discussion to construct out of fundamental solutions a non-singular matrix:

$$\underline{\underline{X}}(t) = \begin{bmatrix} \underline{x}_1(t) & \underline{x}_2(t) & \cdots & \underline{x}_n(t) \end{bmatrix}.$$

Matrix $\underline{\underline{X}}(t)$ is called a fundamental matrix, its i-th column is the solution $\underline{x}_i$.

Thus, a linear combination of $\underline{x}_i$ solutions can be written in a matrix form as

$$\underline{x}(t) = \underline{\underline{X}}(t) \underline{c}, \qquad \underline{c} = \begin{bmatrix} c_1 \\ c_2 \\ \vdots \\ c_n \end{bmatrix}.$$

The coefficients c_i are fixed by the initial condition as seen by taking $t = 0$ in the above equation. This yields the following expression

$$\underline{x}(0) = \underline{\underline{X}}(0)\,\underline{c} \quad \longrightarrow \quad \underline{c} = \underline{\underline{X}}^{-1}(0)\,\underline{x}(0)$$

for the coefficient vector $\underline{c}$.

Plugging the above equation back into the superposition equation we see that knowledge of the fundamental matrix allows us to obtain a closed expression:

$$\underline{x}(t) = \underline{\underline{X}}(t)\,\underline{\underline{X}}^{-1}(0)\,\underline{x}(0)$$

for the general solution of the homogeneous system of equation.

We notice that the above equation yields an alternative expression for the matrix exponential:

$$e^{t\underline{\underline{A}}} = \underline{\underline{X}}(t)\,\underline{\underline{X}}^{-1}(0)\,. \tag{6.9}$$

We would like to construct the fundamental solution set in terms of the linearly independent eigenvectors of the coefficient matrix $\underline{\underline{A}}$. The condition for that is that the $n \times n$ coefficient matrix $\underline{\underline{A}}$ has n linearly independent eigenvectors which is, as we recall from Chapter 2, a maximum number of linearly independent eigenvectors that $n \times n$ matrix can have. This condition is automatically satisfied by Hermitian (or real symmetric) $n \times n$ matrix. We have also seen that the matrices which have n distinct eigenvalues satisfy this condition. Recall, that the $n \times n$ matrix possesses n linearly independent eigenvectors if and only if is diagonalizable.

6.1.1   A Diagonalizable Coefficient Matrix

Now, we assume that $\underline{v}_1, \ldots, \underline{v}_n$ are linearly independent eigenvectors of $\underline{\underline{A}}$. Then, n distinct exponential solutions $e^{\lambda_1 t}\underline{v}_1, \ldots, e^{\lambda_n t}\underline{v}_n$ form the fundamental solution set of equation (6.4). According to the uniqueness theorem, each solution $\underline{x}(t)$ of equation (6.4) can be expressed as a linear combination

$$\underline{x}(t) = \sum_{i=1}^{n} c_i e^{\lambda_i t}\underline{v}_i \tag{6.10}$$

in exactly one way and there are no other solutions than those in (6.10).

We denote in that case the fundamental matrix by $\underline{\underline{V}}(t)$. It is equal to

$$\underline{\underline{V}}(t) = \begin{bmatrix} e^{\lambda_1 t}\underline{v}_1 & e^{\lambda_2 t}\underline{v}_2 & \cdots & e^{\lambda_n t}\underline{v}_n \end{bmatrix}. \tag{6.11}$$

The superposition in (6.10) can now be rewritten in the matrix form as

$$\underline{x}(t) = \underline{\underline{V}}(t)\,\underline{c}, \qquad \underline{c} = \begin{bmatrix} c_1 \\ c_2 \\ \vdots \\ c_n \end{bmatrix}. \tag{6.12}$$

Recall, that an $n \times n$ matrix $\underline{\underline{A}}$ which possesses n linearly independent eigenvectors is diagonalizable by a similarity transformation

$$\underline{\underline{A}} \; \rightarrow \; \underline{\underline{V}}^{-1}\underline{\underline{A}}\underline{\underline{V}} \equiv \underline{\underline{D}}, \tag{6.13}$$

with matrix $\underline{\underline{V}} = \underline{\underline{V}}(0)$ whose columns are eigenvectors of $\underline{\underline{A}}$:

$$\underline{\underline{V}} = \begin{bmatrix} \underline{v}_1 & \underline{v}_2 & \cdots & \underline{v}_n \end{bmatrix}$$

and $\underline{\underline{D}}$ is the diagonal matrix

$$\underline{\underline{D}} = \begin{bmatrix} \lambda_1 & 0 & 0 & 0 \\ 0 & \lambda_2 & 0 & 0 \\ 0 & 0 & \ddots & 0 \\ 0 & 0 & 0 & \lambda_n \end{bmatrix},$$

with eigenvalues of the coefficient matrix $\underline{\underline{A}}$ along the diagonal.

We can rewrite the matrix $\underline{\underline{V}}(t)$ defined in (6.11) as

$$\underline{\underline{V}}(t) = \begin{bmatrix} \underline{v}_1 & \underline{v}_2 & \cdots & \underline{v}_n \end{bmatrix} \begin{bmatrix} e^{\lambda_1 t} & 0 & 0 & 0 \\ 0 & e^{\lambda_2 t} & 0 & 0 \\ 0 & 0 & \ddots & 0 \\ 0 & 0 & 0 & e^{\lambda_n t} \end{bmatrix} \tag{6.14}$$

$$= \underline{\underline{V}} \sum_{i=0}^{\infty} \frac{t^i}{i!} \begin{bmatrix} \lambda_1 & 0 & 0 & 0 \\ 0 & \lambda_2 & 0 & 0 \\ 0 & 0 & \ddots & 0 \\ 0 & 0 & 0 & \lambda_n \end{bmatrix}^i = \underline{\underline{V}}\, e^{t\underline{\underline{D}}}.$$

The similarity transformation which maps $\underline{\underline{A}}$ to $\underline{\underline{V}}^{-1}\underline{\underline{A}}\underline{\underline{V}}$ also transforms the exponential matrix $\exp(t\underline{\underline{A}})$ into $\underline{\underline{V}}^{-1} \exp(t\underline{\underline{A}})\, \underline{\underline{V}}$ as shown by the following calculation

$$e^{t\underline{\underline{D}}} = e^{t\underline{\underline{V}}^{-1}\underline{\underline{A}}\underline{\underline{V}}} = \sum_{n=0}^{\infty} \frac{t^n}{n!} \left(\underline{\underline{V}}^{-1}\underline{\underline{A}}\underline{\underline{V}} \right)^n$$

$$= \sum_{n=0}^{\infty} \frac{t^n}{n!} \underbrace{\underline{\underline{V}}^{-1}\underline{\underline{A}}\underline{\underline{V}}\,\underline{\underline{V}}^{-1}\underline{\underline{A}}\underline{\underline{V}} \cdots \underline{\underline{V}}^{-1}\underline{\underline{A}}\underline{\underline{V}}\,\underline{\underline{V}}^{-1}\underline{\underline{A}}\underline{\underline{V}}}_{n}$$

$$= \sum_{n=0}^{\infty} \frac{t^n}{n!}\underline{\underline{V}}^{-1} \underbrace{\underline{\underline{A}}\,\underline{\underline{A}} \cdots \underline{\underline{A}}\,\underline{\underline{A}}}_{n}\, \underline{\underline{V}} = \underline{\underline{V}}^{-1} \left(\sum_{n=0}^{\infty} \frac{t^n}{n!}\underline{\underline{A}}^n \right) \underline{\underline{V}}$$

$$= \underline{\underline{V}}^{-1}\, e^{t\underline{\underline{A}}}\, \underline{\underline{V}},$$

where in the second line we used the fact that the factor $\underline{\underline{V}}\,\underline{\underline{V}}^{-1}$ between every two $\underline{\underline{A}}$ matrices is equal to the identity matrix.

Thus, the similarity transformation (6.13), which diagonalizes the matrix $\underline{\underline{A}}$ also diagonalizes the exponential matrix $\exp\left(t\underline{\underline{A}}\right)$ and we can rewrite it, using (6.14), as

$$e^{t\underline{\underline{A}}} = \underline{\underline{V}}\,e^{t\underline{\underline{D}}}\,\underline{\underline{V}}^{-1} = \underline{\underline{V}}(t)\,\underline{\underline{V}}^{-1},$$

which agrees with the general expression (6.9). Using that the fundamental matrix $\underline{\underline{V}}(t)$ is related to the exponential matrix $\exp\left(t\underline{\underline{A}}\right)$ through

$$\underline{\underline{V}}(t) = e^{t\underline{\underline{A}}}\,\underline{\underline{V}}, \tag{6.15}$$

we can now rewrite equation (6.12) as

$$\underline{x}(t) = \underline{\underline{V}}(t)\,\underline{c} = e^{t\underline{\underline{A}}}\,\underline{\underline{V}}\,\underline{c}. \tag{6.16}$$

Thus, comparing with expression (6.5) we find that the connection between the coefficient vector $\underline{c}$ and the vector function $\underline{x}(t)$ at $t = 0$ is

$$\underline{\underline{V}}\,\underline{c} = \underline{x}(0)$$

or

$$\underline{c} = \underline{\underline{V}}^{-1}\,\underline{x}(0). \tag{6.17}$$

Let us now summarize the few steps required to find a solution to a homogeneous linear system of differential equations (6.4) for the case of a diagonalizable $n \times n$ coefficient matrix $\underline{\underline{A}}$.

① The first step is to solve the eigenvalue problem of associated $\underline{\underline{A}}$, which yields all the eigenvectors $\underline{v}_i$ and their eigenvalues λ_i.

② The next step is to construct the fundamental matrix $\underline{\underline{V}}(t)$ according to its definition (6.11).

③ The general solution is then given by $\underline{x}(t) = \underline{\underline{V}}(t)\,\underline{c}$, where $\underline{c}$ is the constant column vector determined by initial conditions through relation (6.17).

☞ **Example 6.1.2.** Now, we illustrate the above procedure in case of the system of equations:

$$\dot{x}_1 = 4x_1 + 2x_2, \quad \dot{x}_2 = -x_1 + x_2, \tag{6.18}$$

with initial conditions, at $t = 0$,

$$x_1(0) = -1, \quad x_2(0) = 3. \tag{6.19}$$

These equations enter into the matrix differential equation

$$\underline{\dot{x}} = \begin{bmatrix} \dot{x}_1 \\ \dot{x}_2 \end{bmatrix} = \begin{bmatrix} 4 & 2 \\ -1 & 1 \end{bmatrix} \begin{bmatrix} x_1 \\ x_2 \end{bmatrix} = \underline{\underline{A}}\,\underline{x}\,.$$

First, we solve the eigenvalue problem. The characteristic equation of $\underline{\underline{A}}$ is

$$\begin{vmatrix} 4-\lambda & 2 \\ -1 & 1-\lambda \end{vmatrix} = (\lambda - 2)(\lambda - 3) = 0\,.$$

Thus, the eigenvalues are 2 and 3. Solving $(\underline{\underline{A}} - \lambda\underline{\underline{I}})\,\underline{v} = 0$ for $\lambda_1 = 2$ and $\lambda_2 = 3$ yields two linearly independent eigenvectors:

$$(\underline{\underline{A}} - \lambda_1\underline{\underline{I}})\,\underline{v}_1 = \begin{bmatrix} 2 & 2 \\ -1 & -1 \end{bmatrix} \underline{v}_1 = 0 \quad \rightarrow \quad \underline{v}_1 = \begin{bmatrix} 1 \\ -1 \end{bmatrix}$$

$$(\underline{\underline{A}} - \lambda_2\underline{\underline{I}})\,\underline{v}_2 = \begin{bmatrix} 1 & 2 \\ -1 & -2 \end{bmatrix} \underline{v}_2 = 0 \quad \rightarrow \quad \underline{v}_2 = \begin{bmatrix} 2 \\ -1 \end{bmatrix}\,.$$

We have gathered enough information to construct the matrix $\underline{V}(t)$ as:

$$\underline{V}(t) = \begin{bmatrix} e^{2t}\underline{v}_1 & e^{3t}\underline{v}_2 \end{bmatrix} = \begin{bmatrix} e^{2t} & 2e^{3t} \\ -e^{2t} & -e^{3t} \end{bmatrix}\,.$$

From equation (6.16) we find a general solution:

$$\begin{bmatrix} x_1(t) \\ x_2(t) \end{bmatrix} = \underline{V}(t)\,\underline{c} = \begin{bmatrix} e^{2t}c_1 + 2e^{3t}c_2 \\ -e^{2t}c_1 - e^{3t}c_2 \end{bmatrix}$$

to the system of equations (6.18), given in terms of arbitrary constants c_1, c_2. The initial conditions fix these constants through

$$\underline{c} = \underline{V}^{-1}\underline{x}(0) = \begin{bmatrix} -1 & -2 \\ 1 & 1 \end{bmatrix} \begin{bmatrix} -1 \\ 3 \end{bmatrix} = \begin{bmatrix} -5 \\ 2 \end{bmatrix}\,,$$

where $\underline{V}^{-1}$ is an inverse of

$$\underline{V} = \underline{V}(t=0) = \begin{bmatrix} 1 & 2 \\ -1 & -1 \end{bmatrix}\,.$$

Plugging the above value of $\underline{c}$ into an expression for $\underline{x}(t)$ we find that

$$\begin{bmatrix} x_1(t) \\ x_2(t) \end{bmatrix} = \begin{bmatrix} -5e^{2t} + 4e^{3t} \\ 5e^{2t} - 2e^{3t} \end{bmatrix}$$

solves the homogeneous system of the two differential equations (6.18) with the initial conditions (6.19).

Note, that $\underline{\underline{V}}$ diagonalizes $\underline{\underline{A}}$ according to

$$\underline{\underline{V}}^{-1}\underline{\underline{A}}\,\underline{\underline{V}} = \begin{bmatrix} 2 & 0 \\ 0 & 3 \end{bmatrix}.$$

Also, the exponential matrix $\exp\left(t\underline{\underline{A}}\right)$ is easily found as

$$e^{t\underline{\underline{A}}} = \underline{\underline{V}}(t)\,\underline{\underline{V}}^{-1} = \begin{bmatrix} -e^{2t} + 2\,e^{3t} & -2\,e^{2t} + 2\,e^{3t} \\ e^{2t} - e^{3t} & 2\,e^{2t} - e^{3t} \end{bmatrix}.$$

6.1.2 Generalized Eigenvectors

The above procedure works only if the coefficient matrix is diagonalizable. Next example goes beyond this limitation by illustrating a method of calculating the exponential matrix in the case of an 2×2 coefficient matrix with only one eigenvector.

☞ **Example 6.1.3.** Consider the linear system of two differential equations

$$\frac{\mathrm{d}x_1}{\mathrm{d}t} = x_1 - x_2, \qquad \frac{\mathrm{d}x_2}{\mathrm{d}t} = x_2,$$

with initial conditions $x_1(0) = a_0$ and $x_2(0) = b_0$. Adopting the matrix notation we find a solution

$$\begin{bmatrix} x_1(t) \\ x_2(t) \end{bmatrix} = e^{t\underline{\underline{A}}}\begin{bmatrix} a_0 \\ b_0 \end{bmatrix}, \quad \text{with} \quad \underline{\underline{A}} = \begin{bmatrix} 1 & -1 \\ 0 & 1 \end{bmatrix}. \tag{6.20}$$

The characteristic equation $|\underline{\underline{A}} - \lambda\underline{\underline{I}}| = (\lambda - 1)^2 = 0$ has a double root $\lambda = 1$ for which we only find one independent eigenvector that solves equation

$$(\underline{\underline{A}} - \underline{\underline{I}})\,\underline{v} = \begin{bmatrix} 0 & -1 \\ 0 & 0 \end{bmatrix}\underline{v} = 0 \quad \rightarrow \quad \underline{v} = \begin{bmatrix} 1 \\ 0 \end{bmatrix}.$$

Hence, the 2×2 matrix $\underline{\underline{A}}$ is not diagonalizable and the procedure developed above fails in this case. We will solve the differential matrix equation (6.20) by directly calculating an expression for the exponential matrix $\exp\left(t\underline{\underline{A}}\right)$. Fortunately, it is easy to find a closed expression for an arbitrary power of $\underline{\underline{A}}$ by successively applying matrix multiplications

$$\underline{\underline{A}}^2 = \begin{bmatrix} 1 & -2 \\ 0 & 1 \end{bmatrix}, \quad \underline{\underline{A}}^3 = \begin{bmatrix} 1 & -3 \\ 0 & 1 \end{bmatrix}, \quad \dots, \quad \underline{\underline{A}}^n = \begin{bmatrix} 1 & -n \\ 0 & 1 \end{bmatrix}.$$

Knowing every term of the power expansion of the matrix exponential of $\underline{\underline{A}}$

we can proceed as follows

$$\sum_{n=0}^{\infty} \frac{1}{n!} t^n \underline{\underline{A}}^n = \sum_{n=0}^{\infty} \frac{1}{n!} t^n \begin{bmatrix} 1 & -n \\ 0 & 1 \end{bmatrix} = \begin{bmatrix} \sum_{n=0}^{\infty} t^n/n! & -t \sum_{n=1}^{\infty} t^{n-1}/(n-1)! \\ 0 & \sum_{n=0}^{\infty} t^n/n! \end{bmatrix}$$

$$= \begin{bmatrix} e^t & -te^t \\ 0 & e^t \end{bmatrix}.$$

Therefore, the solution can be written as

$$\underline{x}(t) = \begin{bmatrix} x_1(t) \\ x_2(t) \end{bmatrix} = \begin{bmatrix} e^t & -te^t \\ 0 & e^t \end{bmatrix} \begin{bmatrix} a_0 \\ b_0 \end{bmatrix}$$

or

$$x_1(t) = a_0 e^t - b_0 t e^t, \quad x_2(t) = b_0 e^t.$$

The two independent vector functions

$$\underline{x_1}(t) = \begin{bmatrix} 1 \\ 0 \end{bmatrix} e^t, \quad \underline{x_2}(t) = \left(t \begin{bmatrix} -1 \\ 0 \end{bmatrix} + \begin{bmatrix} 0 \\ 1 \end{bmatrix} \right) e^t$$

form a fundamental set of solutions to equation $\underline{\dot{x}} = \underline{\underline{A}} \, \underline{x}$. In this basis, the general solution $\underline{x}(t)$ reads as

$$\underline{x}(t) = a_0 \underline{x_1}(t) + b_0 \underline{x_2}(t).$$

Note that the above solution employed, in addition to the eigenvector $\underline{v}$ of matrix $\underline{\underline{A}}$, vector $\begin{bmatrix} -1 & 0 \end{bmatrix}^T$, that is related to $\underline{v}$ via

$$\begin{bmatrix} -1 \\ 0 \end{bmatrix} = \underline{\underline{A}} \, \underline{v} = \begin{bmatrix} 1 & -1 \\ 0 & 1 \end{bmatrix} \begin{bmatrix} 1 \\ 0 \end{bmatrix}.$$

Such vectors will be referred to, in what follows, as generalized eigenvectors.

The above example suggests that we need a new notion of generalized eigen-vectors in order to generalize a method for solving the homogeneous system of differential differential equation to the system with the $n \times n$ coefficient matrix $\underline{\underline{A}}$ that has less than n independent eigenvectors. Generalized eigenvectors will provide a convenient basis for calculations of the exponentials of matrices. First, we need few definitions.

Definition 6.1.1. *The algebraic multiplicity of an eigenvalue λ_i of matrix $\underline{\underline{A}}$ is the multiplicity, k_i, of λ_i as a root of the characteristic polynomial* $\mid \underline{\underline{A}} - \lambda \underline{\underline{I}} \mid = (\lambda - \lambda_1)^{k_1} \cdots (\lambda - \lambda_i)^{k_i} \cdots (\lambda - \lambda_\gamma)^{k_\gamma}.$

Here, $\lambda_1, \ldots, \lambda_\gamma$ with $1 \leq \gamma \leq n$ are distinct numbers appearing as (repeated) eigenvalues of $\underline{\underline{A}}$.

In the example (6.1.3), the eigenvalue $\lambda = 1$ was a double root of the characteristic polynomial or in other words had an algebraic multiplicity equal to 2.

For an $n \times n$ matrix the sum of algebraic multiplicities of all eigenvalues equals n,

$$k_1 + k_2 + \cdots + k_\gamma = n \,,$$

since the order of the characteristic polynomial of the $n \times n$ matrix is equal to n.

Definition 6.1.2. *The geometric multiplicity of an eigenvalue λ_i of matrix $\underline{\underline{A}}$ is the number, d_i, of linearly independent eigenvectors associated with λ_i.*

In other words, d_i is a dimension of the vector space containing all vectors $\underline{v}$ which satisfy $\left(\underline{\underline{A}} - \lambda_i \underline{\underline{I}}\right) \underline{v} = 0$. This space is often referred to as a null-space or kernel of the matrix $\underline{\underline{A}} - \lambda_i \underline{\underline{I}}$ and is denoted by $\ker\left(\underline{\underline{A}} - \lambda_i \underline{\underline{I}}\right)$. In physics, we refer to such eigenvalue λ_i as being degenerate or more precisely as being d_i-fold degenerate.

The geometric multiplicity of an eigenvalue is always less or equal to its algebraic multiplicity and as expressed by relation

$$1 \leq d_i \leq k_i, \quad i = 1, \ldots, \gamma \,.$$

That this relation holds can be seen by performing a change of basis to a new basis containing eigenvectors from $\ker\left(\underline{\underline{A}} - \lambda_i \underline{\underline{I}}\right)$. That change of basis simultaneously rotates matrix $\underline{\underline{A}}$ to a matrix which has λ_i on the d_i places along the diagonal, hence, λ_i appears as a root of the characteristic polynomial at least d_i times.

If, $d_i = k_i$, for all $1 \leq i \leq \gamma$ the matrix is diagonalizable by a similarity transformation (6.13) and the basis of all the kernels $\ker\left(\underline{\underline{A}} - \lambda_i \underline{\underline{I}}\right)$, $i = 1, \ldots, \gamma$ is n-dimensional.

If the geometric multiplicity is smaller than algebraic multiplicity, $d_i < k_i$, for any of the eigenvalues λ_i then the matrix will have less than n linearly independent eigenvectors. In such case the matrix is called *defective*. In Example 6.1.3, there was only one eigenvector associated to $\lambda = 1$, which caused the diagonalization method to fail. For defective matrices we have to look beyond the eigenvectors from kernels $\ker\left(\underline{\underline{A}} - \lambda_i \underline{\underline{I}}\right)$, $i = 1, \ldots, \gamma$ to span the whole underlying vector space.

Definition 6.1.3. *If λ_i is an eigenvalue of $\underline{\underline{A}}$ and*

$$\left(\underline{\underline{A}} - \lambda_i \underline{\underline{I}}\right)^p \underline{v} = 0$$

for some integer $1 \leq p \leq k_i$ then $\underline{v}$ is called a generalized eigenvector associated with λ_i.

Denote by $V_p(\lambda_i) = \ker\left(\underline{\underline{A}} - \lambda_i \underline{\underline{I}}\right)^p$ a space of all solutions to

$$\left(\underline{\underline{A}} - \lambda_i \underline{\underline{I}}\right)^p \underline{v} = 0 \,.$$

One can show that there is always an integer p, $1 \leq p \leq k_i$ such that the dimension of $V_p(\lambda_i)$ is equal to the algebraic multiplicity k_i. Thus, it is possible to find among the generalized eigenvectors a k_i-dimensional basis. Since, the sum of all algebraic multiplicities equals n the generalized eigenvectors associated with all eigenvalues will span the whole vector space.

In the following, we fix the index i and show how to construct a chain of linearly independent generalized eigenvectors in a quite common case when $\left(\underline{\underline{A}} - \lambda_i \underline{\underline{I}}\right)^{k_i}$ is different from zero and with only one eigenvector $\underline{v}_i^{(1)}$ such that $\left(\underline{\underline{A}} - \lambda_i \underline{\underline{I}}\right) \underline{v}_i^{(1)} = 0$ (the arguments can be extended on case by case basis beyond these assumptions).

In this case, we explicitly construct a basis in $V_{k_i}(\lambda_i)$ as a chain of generalized eigenvectors $\underline{v}_i^{(1)}, \underline{v}_i^{(2)}, \ldots, \underline{v}_i^{(k_i)}$ with $\underline{v}_i^{(1)}$ being an ordinary eigenvector of $\underline{\underline{A}}$ with the eigenvalue λ_i, $\left(\underline{\underline{A}} - \lambda_i \underline{\underline{I}}\right) \underline{v}_i^{(1)} = 0$, and the remaining vectors related to each other through

$$\left(\underline{\underline{A}} - \lambda_i \underline{\underline{I}}\right) \underline{v}_i^{(2)} = \underline{v}_i^{(1)}$$
$$\left(\underline{\underline{A}} - \lambda_i \underline{\underline{I}}\right) \underline{v}_i^{(3)} = \underline{v}_i^{(2)}$$
$$\vdots \qquad \vdots$$
$$\left(\underline{\underline{A}} - \lambda_i \underline{\underline{I}}\right) \underline{v}_i^{(k_i)} = \underline{v}_i^{(k_i - 1)}.$$

These relations can be summarized as

$$\left(\underline{\underline{A}} - \lambda_i \underline{\underline{I}}\right) \underline{v}_i^{(1)} = 0$$
$$\left(\underline{\underline{A}} - \lambda_i \underline{\underline{I}}\right) \underline{v}_i^{(r)} = \underline{v}_i^{(r-1)}, \quad r = 2, \ldots, k_i. \tag{6.21}$$

It follows that

$$\left(\underline{\underline{A}} - \lambda_i \underline{\underline{I}}\right)^r \underline{v}_i^{(r)} = \left(\underline{\underline{A}} - \lambda_i \underline{\underline{I}}\right)^{r-1} \underline{v}_i^{(r-1)} = \ldots = \left(\underline{\underline{A}} - \lambda_i \underline{\underline{I}}\right) \underline{v}_i^{(1)} = 0$$

for $r = 1, 2, \ldots, k_i$. Thus, all the generalized eigenvectors defined through the chain of relations (6.21) satisfy

$$\left(\underline{\underline{A}} - \lambda_i \underline{\underline{I}}\right)^{k_i} \underline{v}_i^{(r)} = 0, \quad r = 1, 2, \ldots, k_i$$

and, therefore, belong to $V_{k_i}(\lambda_i)$. Also, they are all linearly independent. Suppose, namely that a linear combination $\sum_{r=1}^{k_i} c_r \underline{v}_i^{(r)}$ vanishes. It follows then that all the coefficients c_r must be zero. First, we will show that $c_{k_i} = 0$. Assume, that the matrix $\left(\underline{\underline{A}} - \lambda_i \underline{\underline{I}}\right)^{k_i - 1} \neq 0$ and act with it on $\sum_{r=1}^{k_i} c_r \underline{v}_i^{(r)} = 0$. This eliminates all but one of the generalized eigenvectors:

$$0 = \sum_{r=1}^{k_i} c_r \left(\underline{\underline{A}} - \lambda_i \underline{\underline{I}}\right)^{k_i - 1} \underline{v}_i^{(r)} = c_{k_i} \left(\underline{\underline{A}} - \lambda_i \underline{\underline{I}}\right)^{k_i - 1} \underline{v}_i^{(k_i)} = c_{k_i} \underline{v}_i^{(1)} \quad \rightarrow \quad c_{k_i} = 0.$$

Applying, the matrix $\left(\underline{\underline{A}} - \lambda_i \underline{\underline{I}}\right)^{k_i - 2}$ on $\sum_{r=1}^{k_i - 1} c_r \underline{v}_i^{(r)} = 0$ we obtain $c_{k_i - 1} = 0$ and so on.

Since, the dimension of $V_{k_i}(\lambda_i)$ cannot be greater than the algebraic multiplicity k_i it follows that k_i linearly independent vectors $\underline{v}_i^{(1)}, \underline{v}_i^{(2)}, \ldots, \underline{v}_i^{(k_i)}$ from equation (6.21) form a basis of $V_{k_i}(\lambda_i)$. Thus the dimension of $V_{k_i}(\lambda_i)$ is equal to k_i.

Generally, it holds that the vector space $\mathbb{C}^n$ is a sum of $V_{k_i}(\lambda_i)$ spaces and any vector $\underline{v}$ from the $\mathbb{C}^n$ can be written as a linear combination $\underline{v} = \sum_{i=1}^{\gamma} \sum_{r_i=1}^{k_i} c_i^{(r_i)} \underline{v}_i^{(r_i)}$ of the generalized eigenvectors $\underline{v}_i^{(r_i)}$ from the Definition (6.1.3)

☞ **Example 6.1.4.** Here, we show an example of a defective 3×3 matrix $\underline{\underline{A}}$ with the geometric multiplicity of one of the eigenvalues being smaller than its algebraic multiplicity. Suppose,

$$\underline{\underline{A}} = \begin{bmatrix} 1 & 1 & 1 \\ 0 & 2 & 2 \\ 0 & 0 & 2 \end{bmatrix} \rightarrow \underline{\underline{A}} - \lambda \underline{\underline{I}} = \begin{bmatrix} 1-\lambda & 1 & 1 \\ 0 & 2-\lambda & 2 \\ 0 & 0 & 2-\lambda \end{bmatrix}.$$

The characteristic polynomial $|\underline{\underline{A}} - \lambda \underline{\underline{I}}| = (1-\lambda)(2-\lambda)^2$ has roots 1 and 2 with $\lambda = 2$ having algebraic multiplicity 2. The corresponding eigenvectors satisfy

$$(\underline{\underline{A}} - \underline{\underline{I}})\,\underline{v} = \begin{bmatrix} 0 & 1 & 1 \\ 0 & 1 & 2 \\ 0 & 0 & 1 \end{bmatrix} \begin{bmatrix} v_1 \\ v_2 \\ v_3 \end{bmatrix} = 0 \rightarrow \underline{v}_1 = \begin{bmatrix} 1 \\ 0 \\ 0 \end{bmatrix}$$

$$(\underline{\underline{A}} - 2\underline{\underline{I}})\,\underline{v} = \begin{bmatrix} -1 & 1 & 1 \\ 0 & 0 & 2 \\ 0 & 0 & 0 \end{bmatrix} \begin{bmatrix} v_1 \\ v_2 \\ v_3 \end{bmatrix} = 0 \rightarrow \underline{v}_2^{(1)} = \begin{bmatrix} 1 \\ 1 \\ 0 \end{bmatrix}$$

and, thus, there is only one eigenvector for $\lambda = 2$. However

$$(\underline{\underline{A}} - 2\underline{\underline{I}})^2 \begin{bmatrix} v_1 \\ v_2 \\ v_3 \end{bmatrix} = \begin{bmatrix} 1 & -1 & 1 \\ 0 & 0 & 0 \\ 0 & 0 & 0 \end{bmatrix} \underline{v} = 0 \rightarrow \begin{bmatrix} v_1 \\ v_2 \\ v_3 \end{bmatrix} = a \begin{bmatrix} 1 \\ 1 \\ 0 \end{bmatrix} + b \begin{bmatrix} -1 \\ 0 \\ 1 \end{bmatrix}$$

and the space of the generalized eigenvectors associated to $\lambda = 2$ is two-dimensional (a plane orthogonal to a vector $\begin{bmatrix} 1 & -1 & 1 \end{bmatrix}^T$). A choice

$$\underline{v}_2^{(2)} = \frac{1}{2} \begin{bmatrix} -1 \\ 0 \\ 1 \end{bmatrix}$$

as a second generalized eigenvector ensures the relation $(\underline{\underline{A}} - 2\underline{\underline{I}})\,\underline{v}_2^{(2)} = \underline{v}_2^{(1)}$.

It is easy to see that $\underline{v}_1, \underline{v}_2^{(1)}, \underline{v}_2^{(2)}$ are linearly independent and form a basis of the three-dimensional space $\mathbb{R}^3$.

Recall now equation (6.5) and write $\underline{x}(0) = \sum_{i=1}^{\gamma} \sum_{r_i=1}^{k_i} c_i^{(r_i)} v_i^{(r_i)}$ in a basis of the generalized eigenvectors. Using the identity

$$e^{t\underline{\underline{A}}} = e^{t\lambda_i \underline{\underline{I}}}\, e^{t(\underline{\underline{A}}-\lambda_i\underline{\underline{I}})},$$

we obtain

$$
\begin{aligned}
\underline{x}(t) = e^{t\underline{\underline{A}}}\,\underline{x}(0) &= \sum_{i=1}^{\gamma} \sum_{r_i=1}^{k_i} e^{t\underline{\underline{A}}}\, c_i^{(r_i)} v_i^{(r_i)} \\
&= \sum_{i=1}^{\gamma} e^{t\lambda_i\underline{\underline{I}}} \sum_{r_i=1}^{k_i} e^{t(\underline{\underline{A}}-\lambda_i\underline{\underline{I}})}\, c_i^{(r_i)} v_i^{(r_i)} \\
&= \sum_{i=1}^{\gamma} e^{t\lambda_i\underline{\underline{I}}} \sum_{r_i=1}^{k_i} \sum_{m=0}^{\infty} \frac{t^m}{m!}\left(\underline{\underline{A}} - \lambda_i\underline{\underline{I}}\right)^m c_i^{(r_i)} v_i^{(r_i)} \\
&= \sum_{i=1}^{\gamma} e^{t\lambda_i\underline{\underline{I}}} \sum_{r_i=1}^{k_i} \sum_{m=0}^{k_i-1} \frac{t^m}{m!}\left(\underline{\underline{A}} - \lambda_i\underline{\underline{I}}\right)^m v_i^{(r_i)} c_i^{(r_i)},
\end{aligned}
\tag{6.22}
$$

where we used that $\left(\underline{\underline{A}} - \lambda_i\underline{\underline{I}}\right)^l v_i^{(r_i)} = 0, r_i = 1, \ldots, k_i$ for $l \geq r_i$ to conclude that the power series in equation (6.22) truncates. Due to this truncation the method does not require calculating the infinite power series defining the exponential matrix $\exp\left(t\underline{\underline{A}}\right)$. Equation (6.22) shows how to obtain the solution $\underline{x}(t)$ in terms of a finite number of terms containing the finite powers of the well-known matrices $\left(\underline{\underline{A}} - \lambda_i\underline{\underline{I}}\right)$ applied on the basis elements chosen from $V_{k_i}(\lambda_i)$ vector spaces.

☞ **Example 6.1.5.** Let

$$\underline{\underline{A}} = \begin{bmatrix} 1 & 0 & 0 \\ 0 & 4 & 1 \\ -1 & -1 & 2 \end{bmatrix} \rightarrow \underline{\underline{A}} - \lambda\underline{\underline{I}} = \begin{bmatrix} 1-\lambda & 0 & 0 \\ 0 & 4-\lambda & 1 \\ -1 & -1 & 2-\lambda \end{bmatrix}.$$

The characteristic polynomial is $\mid \underline{\underline{A}} - \lambda\underline{\underline{I}} \mid = (\lambda-1)(\lambda-3)^2$ so the root $\lambda = 3$ has the algebraic multiplicity 2. We find that the eigenvector $\underline{v}_1$ associated to $\lambda = 1$ satisfies

$$\left(\underline{\underline{A}} - 1\underline{\underline{I}}\right)\underline{v}_1 = \begin{bmatrix} 0 & 0 & 0 \\ 0 & 3 & 1 \\ -1 & -1 & 1 \end{bmatrix}\begin{bmatrix} v_1 \\ v_2 \\ v_3 \end{bmatrix} = 0 \rightarrow \underline{v}_1 = \begin{bmatrix} 4 \\ -1 \\ 3 \end{bmatrix}.$$

The eigenvector $\underline{v}_2^{(1)}$ associated with $\lambda = 3$ is found from

$$\left(\underline{\underline{A}} - 3\underline{\underline{I}}\right)\underline{v} = \begin{bmatrix} -2 & 0 & 0 \\ 0 & 1 & 1 \\ -1 & -1 & -1 \end{bmatrix}\begin{bmatrix} v_1 \\ v_2 \\ v_3 \end{bmatrix} = 0 \rightarrow \underline{v}_2^{(1)} = \begin{bmatrix} 0 \\ 1 \\ -1 \end{bmatrix}.$$

The generalized eigenvector $\underline{v}_2^{(2)}$ is obtained as a solution to

$$\left(\underline{A} - 3\underline{I}\right)\begin{bmatrix} v_1 \\ v_2 \\ v_3 \end{bmatrix} = \begin{bmatrix} 0 \\ 1 \\ -1 \end{bmatrix} \quad \rightarrow \quad \underline{v}_2^{(2)} = \frac{1}{2}\begin{bmatrix} 0 \\ 1 \\ 1 \end{bmatrix}.$$

It follows that $\underline{v}_2^{(2)}$ also satisfies

$$\left(\underline{A} - 3\underline{I}\right)^2\begin{bmatrix} v_1 \\ v_2 \\ v_3 \end{bmatrix} = \begin{bmatrix} 4 & 0 & 0 \\ -1 & 0 & 0 \\ 3 & 0 & 0 \end{bmatrix}\underline{v}_2^{(2)} = 0.$$

For

$$\underline{x}(0) = \sum_{i=1}^{2}\sum_{r_i=1}^{k_i} c_i^{(r_i)} v_i^{(r_i)} = c_1 v_1 + c_2^{(1)} v_2^{(1)} + c_2^{(2)} v_2^{(2)}$$

the solution according to (6.22) is

$$\underline{x}(t) = e^{t\underline{I}}\sum_{m=0}^{\infty}\frac{t^m}{m!}\left(\underline{A} - \underline{I}\right)^m c_1 v_1 + e^{3t\underline{I}}\sum_{r=1}^{2}\sum_{m=0}^{\infty}\frac{t^m}{m!}\left(\underline{A} - 3\underline{I}\right)^m c_2^{(r)} v_2^{(r)}$$

$$= e^t c_1\underline{v}_1 + e^{3t}c_2^{(1)}\underline{v}_2^{(1)} + e^{3t}c_2^{(2)}\underline{v}_2^{(2)} + te^{3t}c_2^{(2)}\underline{v}_2^{(1)},$$

where we used that $\underline{v}_2^{(2)}$ satisfies $\left(\underline{A} - 3\underline{I}\right)\underline{v}_2^{(2)} = \underline{v}_2^{(1)}$. Vectors

$$e^t\,\underline{v}_1, \quad e^{3t}\,\underline{v}_2^{(1)}, \quad e^{3t}\left(\underline{v}_2^{(2)} + t\underline{v}_2^{(1)}\right), \qquad (6.23)$$

form a fundamental set of solutions to $\underline{\dot{x}} = \underline{A}\,\underline{x}$. It is clear that the first two vectors are members of a fundamental set of solutions. Applying the matrix $\underline{A}$ on the third vector we obtain

$$e^{3t}\underline{A}\left(\underline{v}_2^{(2)} + t\underline{v}_2^{(1)}\right) = e^{3t}\left(3\underline{v}_2^{(2)} + \underline{v}_2^{(1)} + 3t\underline{v}_2^{(1)}\right)$$

$$= \frac{d}{dt}\left[e^{3t}\left(\underline{v}_2^{(2)} + t\underline{v}_2^{(1)}\right)\right]$$

$$= 3e^{3t}\left(\underline{v}_2^{(2)} + t\underline{v}_2^{(1)}\right) + e^{3t}\,\underline{v}_2^{(1)},$$

which verifies that also the third vector satisfies the equation $\underline{\dot{x}} = \underline{A}\,\underline{x}$. All three vectors are linearly independent and enter the fundamental matrix:

$$\underline{X}(t) = \begin{bmatrix} 4\,e^t & 0 & 0 \\ -e^t & e^{3t} & (\frac{1}{2} + t)\,e^{3t} \\ 3\,e^t & -e^{3t} & (\frac{1}{2} - t)\,e^{3t} \end{bmatrix},$$

as columns. For $t = 0$ the fundamental matrix equals

$$\underline{\underline{X}} = \underline{\underline{X}}(0) = \begin{bmatrix} 4 & 0 & 0 \\ -1 & 1 & \frac{1}{2} \\ 3 & -1 & \frac{1}{2} \end{bmatrix}.$$

From these expressions we derive the matrix exponential

$$e^{t\underline{\underline{A}}} = \underline{\underline{X}}(t)\,\underline{\underline{X}}^{-1} =$$

$$\begin{bmatrix} e^t & 0 & 0 \\ -\dfrac{e^t}{4} + \dfrac{e^{3t}}{2} - \dfrac{(\frac{1}{2}+t)\,e^{3t}}{2} & \dfrac{e^{3t}}{2} + (\frac{1}{2}+t)\,e^{3t} & -\dfrac{e^{3t}}{2} + (\frac{1}{2}+t)\,e^{3t} \\ \dfrac{3\,e^t}{4} - \dfrac{e^{3t}}{2} - \dfrac{(\frac{1}{2}-t)\,e^{3t}}{2} & -\dfrac{e^{3t}}{2} + (\frac{1}{2}-t)\,e^{3t} & \dfrac{e^{3t}}{2} + (\frac{1}{2}-t)\,e^{3t} \end{bmatrix}.$$

Let $v_i^{(1)}, v_i^{(2)}, \ldots, v_i^{(k_i)}$ be a basis of generalized eigenvectors for $V_{k_i}(\lambda_i)$, which satisfy the algorithm (6.21). These vectors give rise to following k_i elements of the fundamental set of solutions to $\dot{\underline{x}} = \underline{\underline{A}}\,\underline{x}$:

$$x_i^{(1)} = e^{\lambda_i t}\, v_i^{(1)}$$

$$x_i^{(2)} = e^{\lambda_i t}\left(\underline{v}_i^{(2)} + t\underline{v}_i^{(1)} \right)$$

$$x_i^{(3)} = e^{\lambda_i t}\left(\underline{v}_i^{(3)} + t\underline{v}_2^{(2)} + \frac{t^2}{2}\underline{v}_2^{(1)} \right)$$

$$\vdots \qquad \vdots$$

$$x_i^{(k_i)} = e^{\lambda_i t} \sum_{j=1}^{k_i} \frac{t^{k_i - j}}{(k_i - j)!}\, \underline{v}_i^{(j)}$$

for the eigenvalue λ_i with algebraic multiplicity equal to k_i and geometric multiplicity equal to one. These expressions generalize construction of solutions presented in equation (6.23).

☞ **Example 6.1.6.** Consider again the 2×2 matrix:

$$\underline{\underline{A}} = \begin{bmatrix} 1 & -1 \\ 0 & 1 \end{bmatrix}$$

from Example (6.1.3). Matrix $\underline{\underline{A}}$ has an eigenvalue $\lambda = 1$ with the algebraic multiplicity equal to 2. We have seen that

$$\underline{v}^{(1)} = \begin{bmatrix} 1 \\ 0 \end{bmatrix}$$

satisfies $\left(\underline{A} - \underline{I}\right) \underline{v}^{(1)} = 0$. Since $\left(\underline{A} - \underline{I}\right)^2 = 0$, all the two-dimensional vectors belong to $\ker \left(\underline{A} - \underline{I}\right)^2$ and the geometric multiplicity is equal to 2. We choose as a second generalized eigenvector

$$\underline{v}^{(2)} = \begin{bmatrix} 0 \\ -1 \end{bmatrix}$$

related to $\underline{v}^{(1)}$ via $\left(\underline{A} - \underline{I}\right) \underline{v}^{(2)} = \underline{v}^{(1)}$, which is a special case of relation (6.21). The initial condition can be written in the basis of generalized eigenfunctions as

$$\underline{x}(0) = \begin{bmatrix} a_0 \\ b_0 \end{bmatrix} = a_0 \underline{v}^{(1)} - b_0 \underline{v}^{(2)}.$$

Applying the exponential matrix $\exp\left(t\underline{A}\right)$ on $\underline{x}(0)$ we get from the truncated power series expansion:

$$\underline{x}(t) = e^{t\underline{A}}\underline{x}(0) = e^t \left(a_0 e^{t\underline{A}-t}\underline{v}^{(1)} - b_0 e^{t\underline{A}-t}\underline{v}^{(2)} \right)$$

$$= e^t \left(a_0 \underline{v}^{(1)} - b_0 \underline{v}^{(2)} - b_0 t \left(\underline{A} - \underline{I}\right) \underline{v}^{(2)} \right)$$

$$= e^t \left(a_0 \underline{v}^{(1)} - b_0 \underline{v}^{(2)} - b_0 t \underline{v}^{(1)} \right)$$

reproducing the result of Example (6.1.3).

6.2   The Non-Homogeneous Linear Differential Equations

Consider now the non-homogeneous linear system of equations (6.3) with $\underline{r} \neq 0$ and the diagonalizable constant coefficient matrix $\underline{A}$. Let $\underline{V}$ be a matrix which diagonalizes $\underline{A}$ by a similarity transformation as in equation (6.13). For $\underline{x}(t)$ being a solution to equation (6.3) the vector function $\underline{y}(t) = \underline{V}^{-1}\underline{x}(t)$ satisfies

$$\frac{d}{dt}\underline{y}(t) = \underline{V}^{-1}\frac{d}{dt}\underline{x}(t) = \underline{V}^{-1}\underline{A}\,\underline{V}\,\underline{y}(t) + \underline{V}^{-1}\underline{r}(t)$$

$$= \underline{D}\,\underline{y}(t) + \bar{\underline{r}}(t),$$

where $\bar{\underline{r}}(t) = \underline{V}^{-1}\underline{r}(t)$.

Due to the fact that $\underline{D}$ is the diagonal matrix we can decompose the above equation into its components

$$\frac{dy_k(t)}{dt} - \lambda_k y_k(t) = \bar{r}_k(t), \quad k = 1, \ldots, n. \tag{6.24}$$

For every k this equation has a form of the first order linear non-homogeneous differential equation involving an unknown function $y_k(t)$. Solution to (6.24) can be obtained from the well-known formula (3.40):

$$y_k(t) = c_k e^{\lambda_k t} + e^{\lambda_k t} \int e^{-\lambda_k t} \bar{r}_k(t) \, \mathrm{d}t, \qquad k = 1, \ldots, n, \qquad (6.25)$$

where c_k is an integration constant. In the matrix form expressions (6.25) become

$$\underline{y} = e^{t\underline{D}} \underline{c} + e^{t\underline{D}} \int e^{-t\underline{D}} \underline{\bar{r}}(t) \, \mathrm{d}t.$$

After multiplication of both sides by $\underline{V}$ the solution is expressed in terms of the original variable $\underline{x}(t)$ as

$$\underline{x}(t) = e^{t\underline{A}} \underline{V} \underline{c} + e^{t\underline{A}} \int e^{-t\underline{A}} \underline{r}(t) \, \mathrm{d}t.$$

The first term represents a solution to the homogeneous system while the second term:

$$\underline{x}_P(t) = e^{t\underline{A}} \int e^{-t\underline{A}} \underline{r}(t) \, \mathrm{d}t \qquad (6.26)$$

is a particular solution to the non-homogeneous system of differential equations.

The same result can be obtained in a more general way, which does not rely on assumption that $\underline{A}$ is diagonalizable. In a spirit of a method leading to expression (3.95) we replace a constant vector $\underline{c}$ in expression $\underline{x}(t) = \underline{V}(t) \underline{c}$ by an unknown vector-function $\underline{u} = \begin{bmatrix} u_1 & \cdots & u_n \end{bmatrix}^T$ to be determined below. Thus, we search for a solution to the non-homogeneous system of the form:

$$\underline{x}_P(t) = \underline{V}(t) \underline{u}(t). \qquad (6.27)$$

Substituting $\underline{x}_P(t)$ into equation (6.3) yields

$$\frac{\mathrm{d}}{\mathrm{d}t} \underline{x}_P(t) = \left(\frac{\mathrm{d}}{\mathrm{d}t} \underline{V}(t) \right) \underline{u} + \underline{V}(t) \frac{\mathrm{d}}{\mathrm{d}t} \underline{u} = \underline{A} \underline{V}(t) \underline{u} + \underline{r}(t).$$

The matrix $\underline{V}(t)$ satisfies $\dot{\underline{V}}(t) = \underline{A} \underline{V}(t)$ and, therefore,

$$\underline{A} \underline{V}(t) \underline{u} + \underline{V}(t) \frac{\mathrm{d}}{\mathrm{d}t} \underline{u} = \underline{A} \underline{V}(t) \underline{u} + \underline{r}(t).$$

Canceling the common term $\underline{A} \underline{V}(t) \underline{u}$ on both sides of the above equation we arrive at the equation

$$\frac{\mathrm{d}}{\mathrm{d}t} \underline{u} = \underline{V}^{-1}(t) \underline{r},$$

which involves a first order derivative of the unknown function $\underline{u}$. After an integration we obtain

$$\underline{u}(t) = \int \underline{V}^{-1}(t) \underline{r}(t) \, \mathrm{d}t.$$

Thus,

$$\underline{x}_P(t) = \underline{\underline{V}}(t) \int \mathrm{d}t \, \underline{\underline{V}}^{-1}(t) \, \underline{r}(t)$$

$$= \exp\left(t\underline{\underline{A}}\right) \int \exp\left(-t\underline{\underline{A}}\right) \underline{r}(t) \, \mathrm{d}t , \qquad (6.28)$$

which agrees with the result (6.26) obtained above using an additional assumption that $\underline{\underline{A}}$ was diagonalizable.

☞ **Example 6.2.1.** For the source vector function

$$\underline{r}(t) = e^{3t} \begin{bmatrix} 1 - 3t \\ 3 \end{bmatrix}$$

and matrix

$$\underline{\underline{A}} = \begin{bmatrix} 1 & -1 \\ 0 & 1 \end{bmatrix}$$

taken from Example (6.1.3) we will solve the non-homogeneous equation $\dot{\underline{x}} = \underline{\underline{A}}\,\underline{x} + \underline{r}$ with the initial condition

$$\underline{x}(0) = \begin{bmatrix} 1 \\ 3 \end{bmatrix} .$$

First, we express $\underline{r}$ in terms of the generalized eigenvectors $v^{(1)}, v^{(2)}$ introduced in Example (6.1.6)

$$\underline{r}(t) = (1 - 3t)e^{3t}v^{(1)} - 3e^{3t}v^{(2)} .$$

Now, we are able to calculate the action of the matrix exponential

$$e^{-t\underline{\underline{A}}} = e^{-t} e^{-t(\underline{\underline{A}} - \underline{\underline{I}})} = e^{-t}\left(\underline{\underline{I}} - t(\underline{\underline{A}} - \underline{\underline{I}})\right)$$

on $\underline{r}$ (note, that the power series expansion of $\exp\left(-t\underline{\underline{A}}\right)$ terminates due to the fact that $\left(\underline{\underline{A}} - \underline{\underline{I}}\right)^2 = 0$). We find

$$e^{-t\underline{\underline{A}}}\underline{r}(t) = (1 - 3t)e^{2t}v^{(1)} - 3e^{2t}v^{(2)} + 3te^{2t}v^{(1)} = e^{2t} \begin{bmatrix} 1 \\ 3 \end{bmatrix} ,$$

where we used that $(\underline{\underline{A}} - \underline{\underline{I}})v^{(2)} = v^{(1)}$. After integration, we obtain

$$\int^t \mathrm{d}t \, e^{-t\underline{\underline{A}}}\underline{r}(t) = \frac{1}{2}e^{2t} \begin{bmatrix} 1 \\ 3 \end{bmatrix} + \underline{c}.$$

For $t = 0$ the initial condition implies that

$$\frac{1}{2} \begin{bmatrix} 1 \\ 3 \end{bmatrix} + \underline{c} = \begin{bmatrix} 1 \\ 3 \end{bmatrix} .$$

This result fixes the integration constant to

$$\underline{c} = \begin{bmatrix} 1/2 \\ 3/2 \end{bmatrix}.$$

Now, we can rewrite

$$\int^t dt\, e^{-t\underline{\underline{A}}} \underline{r}(t) = \frac{1}{2} e^{2t} \begin{bmatrix} 1 \\ 3 \end{bmatrix} + \frac{1}{2} \begin{bmatrix} 1 \\ 3 \end{bmatrix} = \frac{1}{2} \left(v^{(1)} - 3v^{(2)} \right) \left(1 + e^{2t} \right)$$

and, thus,

$$\underline{x}_P(t) = e^{t\underline{\underline{A}}} \int^t dt\, e^{-t\underline{\underline{A}}} \underline{r}(t) = \frac{1}{2} (e^t + e^{3t}) \left((1 - 3t) v^{(1)} - 3v^{(2)} \right)$$

$$= \frac{1}{2} (e^t + e^{3t}) \begin{bmatrix} 1 - 3t \\ 3 \end{bmatrix},$$

where we used that

$$e^{t(\underline{\underline{A}} - \underline{\underline{I}})} \left(v^{(1)} - 3v^{(2)} \right) = \left(\underline{\underline{I}} + t(\underline{\underline{A}} - \underline{\underline{I}}) \right) \left(v^{(1)} - 3v^{(2)} \right)$$

$$= v^{(1)} - 3v^{(2)} - 3tv^{(1)}.$$

6.3 Stability of Linear Systems

For the linear system $\dot{\underline{x}} = \underline{\underline{A}}\,\underline{x}$ the equilibrium point or the critical point is a constant vector such that $\dot{\underline{x}} = 0$. Thus, the equilibrium point is a point where the system is at rest.

For the two-dimensional homogeneous linear systems with non-singular constant coefficient matrix $\underline{\underline{A}}$ the origin $(0,0)$ is the only critical point. Due to uniqueness theorem for the homogeneous linear systems the only solution which passes through the origin $(0,0)$ is the zero solution $\underline{x} = 0$, since it both satisfies the differential equation and initial condition $x_1(t_0) = 0$, $x_2(t_0) = 0$.

If every solution converges to $(0,0)$ as $t \to \infty$ a system is said to be *asymptotically stable* and $(0,0)$ is said to be asymptotically stable equilibrium. If the trajectories are all bound as $t \to \infty$ then a system is called *stable*, otherwise it is called *unstable*.

The stability theory studies properties of critical points to find out how the solution reacts to perturbations from an equilibrium position. A big question is whether it stays inside a bounded region, returns to the critical point or wanders away? Correspondingly, we talk about stable, asymptotically stable or unstable equilibria. From properties of critical points the stability theory is able to predict behavior of the general solutions as $t \to \infty$.

To illustrate the concept of stability and asymptotic stability of linear differential equations we first consider a simple one-dimensional differential equation.

☞ **Example 6.3.1.** Equation

$$\dot{x}(t) = \lambda\, x(t)$$

possesses a well-known exponential solution:

$$x(t) = e^{\lambda t}\, x(0)\,.$$

In particular, if the starting point is taken to be $x(0) = 0$ then $x(t) = 0$ is a solution which is called an equilibrium point. If once $x(t)$ is equal to the equilibrium point, it remains equal to the equilibrium point for ever.

We will investigate what will happen for other than zero starting points. Figure 6.1 shows the three possible outcomes for $\lambda > 0$, $\lambda < 0$ and $\lambda = 0$ for a positive $x(0)$. If $\lambda < 0$ (or $\mathrm{Re}\,\lambda < 0$ if λ is complex) then we say that the

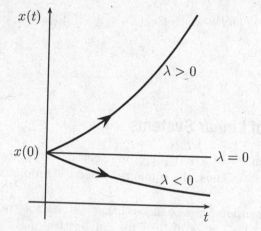

Figure 6.1: If $\lambda > 0$ then $x \to \infty$ and if $\lambda < 0$ then $x \to 0$ for $t \to \infty$ and if $\lambda = 0$ then $x(t) = $ constant.

exponential solution is *decaying* and it follows $x(t) \to 0$ as $t \to \infty$. The zero solution is said in this case to be asymptotically stable.

For $\lambda = 0$ all solutions are constants. In that case, if one starts near zero solution one stays for ever close to the zero solution. In such case, we say that the zero solution is stable (but not asymptotically stable).

Finally, in the case of $\lambda > 0$ (or $\mathrm{Re}\,\lambda > 0$ if λ is complex) the solution $x(t)$ goes to infinity as t goes to infinity. The zero solution is in this case said to be unstable.

Now, we turn our attention back to the question of stability of the linear system of differential equations $\dot{\underline{x}} = \underline{\underline{A}}\,\underline{x}$ for an 2×2 real and non-singular matrix $\underline{\underline{A}}$. Because our discussion will involve eigenvalues and eigenvectors of $\underline{\underline{A}}$ it is easiest to carry it out by means of a similarity transformation:

$$\underline{\underline{V}}^{-1}\,\underline{\underline{A}}\,\underline{\underline{V}} = \underline{\underline{F}}$$

by a 2×2 real matrix $\underline{\underline{V}}$. In general, a similarity transformation is used to diagonalize a matrix. If the matrix $\underline{\underline{A}}$ is defective the simplest possible result of the diagonalization attempt is a matrix $\underline{\underline{F}}$ in the so-called Jordan form (see equation (6.29) below). We will distinguish between three main cases.

①

$$\underline{\underline{F}} = \begin{bmatrix} \lambda_1 & 0 \\ 0 & \lambda_2 \end{bmatrix}$$

In this case the matrix $\underline{\underline{A}}$ is diagonalizable and has two real eigenvalues λ_1, λ_2. Recall that in such case the matrix $\underline{\underline{V}}$ can conveniently be chosen to

$$\underline{\underline{V}} = \begin{bmatrix} \underline{v}_1 & \underline{v}_2 \end{bmatrix},$$

where $\underline{v}_i, i = 1, 2$ are eigenvectors of $\underline{\underline{A}}$: $\underline{\underline{A}}\underline{v}_i = \lambda_i \underline{v}_i$.

②

$$\underline{\underline{F}} = \begin{bmatrix} \lambda & 1 \\ 0 & \lambda \end{bmatrix}, \quad \text{real } \lambda. \tag{6.29}$$

In this case $\underline{\underline{A}}$ is defective (non-diagonalizable) matrix with a single defective eigenvalue λ with algebraic multiplicity 2 and geometric multiplicity 1. If $v^{(1)}$ is an eigenvector of $\underline{\underline{A}}$ with the eigenvalue λ and $v^{(2)}$ is a generalized eigenvector such that $\left(\underline{\underline{A}} - \lambda\underline{\underline{I}}\right) v^{(2)} = v^{(1)}$ then

$$\underline{\underline{V}} = \begin{bmatrix} \underline{v}^{(1)} & \underline{v}^{(2)} \end{bmatrix}. \tag{6.30}$$

We refer the reader to Problem 6.3.3 for the proof.

③

$$\underline{\underline{F}} = \begin{bmatrix} \mu & \nu \\ -\nu & \mu \end{bmatrix}, \quad \text{real } \mu, \nu.$$

For the matrix $\underline{\underline{A}}$ with two conjugate complex eigenvalues $\mu \pm i\nu$, its eigenvectors $\underline{v}_1, \underline{v}_2$ are given in terms of two real vectors $\underline{u}, \underline{w}$, such that

$$\underline{\underline{A}}\,\underline{u} = \mu\underline{u} - \nu\underline{w}, \quad \underline{\underline{A}}\,\underline{w} = \nu\underline{u} + \mu\underline{w},$$

as $\underline{v}_1 = \underline{u} + i\,\underline{w}$ and $\underline{v}_2 = \underline{u} - i\,\underline{w}$. Now, it follows (see Problem 6.3.4) that

$$\underline{\underline{V}} = \begin{bmatrix} \underline{u} & \underline{w} \end{bmatrix}.$$

The vector function $\underline{y}(t) = \underline{\underline{V}}^{-1}\,\underline{x}(t)$ satisfies the linear differential equation

$$\frac{\mathrm{d}}{\mathrm{d}t}\underline{y} = \underline{\underline{V}}^{-1}\frac{\mathrm{d}}{\mathrm{d}t}\underline{x}(t) = \underline{\underline{V}}^{-1}\,\underline{\underline{A}}\,\underline{x}(t)$$

$$= \underline{\underline{V}}^{-1}\,\underline{\underline{A}}\,\underline{\underline{V}}\,\underline{y}(t) \tag{6.31}$$

$$= \underline{\underline{F}}\,\underline{y}(t)\,.$$

In the case ①, the matrix $\underline{\underline{F}}$ is diagonal and equation for the vector function $\underline{y}(t)$ splits into two independent scalar differential equations:

$$\frac{\mathrm{d}y_1(t)}{\mathrm{d}t} = \lambda_1\,y_1(t), \qquad \frac{\mathrm{d}y_2(t)}{\mathrm{d}t} = \lambda_2\,y_2(t)\,,$$

for the components $y_1(t), y_2(t)$. There are two exponential solutions

$$y_1(t) = y_1(0)e^{\lambda_1 t}, \quad y_2(t) = y_2(0)e^{\lambda_2 t}\,. \tag{6.32}$$

It is convenient to depict a solution $y_1(t), y_2(t)$ as a curve in the $y_1\,y_2$-plane. The $y_1(t), y_2(t)$ solution curve is called a *trajectory* or an *orbit* while the $y_1\,y_2$-plane is called *the phase plane*. *The phase plot* or as it is also called *the phase diagram* depicts trajectories as plots of $y_1(t)$ versus $y_2(t)$. It is useful to accompany the phase plot by a phase vector diagram showing tangent vectors to the trajectories in the $y_1\,y_2$-plane. This provides an important information about direction of each trajectory. Although the direct reference to time t is absent in the phase plot, the phase plot still gives a good picture of how the system evolves in time or as one often says flows according to the differential equation.

If,

$$\lambda_1 < 0, \quad \lambda_2 < 0\,,$$

then the solutions of equation (6.32) are parabolic curves in the $y_1\,y_2$-plane which converge toward origin as shown in Figure 6.2. Transforming the system back to the original x_1, x_2 coordinates will have an effect of rotating the parabolic curves from Figure 6.2 into Figure 6.3. The parabolic curves from Figure 6.3 exhibit the same behavior as the parabolic curves from Figure 6.2 from the point of view of convergence to the origin.

In the special case of a degenerate negative eigenvalue $\lambda_1 = \lambda_2 < 0$ (with the 2-dimensional eigenspace) the slope $y_2(t)/y_1(t) = y_2(0)/y_1(0)$ becomes constant and the trajectories are straight lines converging toward origin, see Figure 6.4. If one of the real eigenvalues is positive,

$$\lambda_2 < 0 < \lambda_1\,,$$

the solution curves become the hyperbolic curves shown in Figure 6.5. In this case only solutions starting on the y_2-axis with initial conditions $y_1(0) = 0, y_2(0) \neq 0$ will move toward the origin along the y_2-axis, all other solutions will eventually move

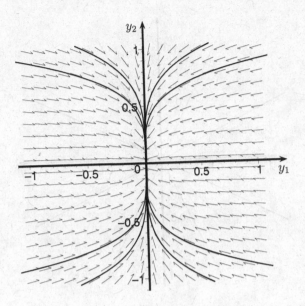

Figure 6.2: Parabolic curves. $\lambda_1 < 0, \lambda_2 < 0$ and $\lambda_1 \neq \lambda_2$.

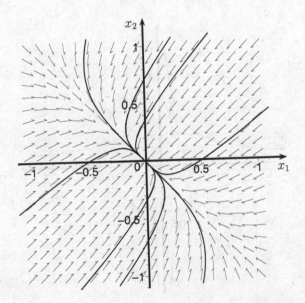

Figure 6.3: Parabolic curves in the $x_1 x_2$- plane.

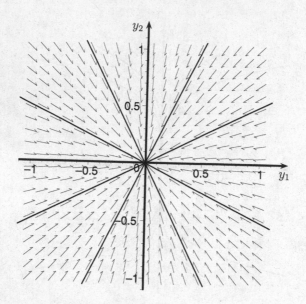

Figure 6.4: Straight lines trajectories for the degenerate negative eigenvalue $\lambda = \lambda_1 = \lambda_2 < 0$.

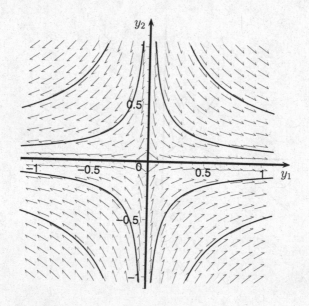

Figure 6.5: Hyperbolic curves for $\lambda_2 < 0 < \lambda_1$.

away from the origin toward $y_1 = \pm\infty$. We say that the origin is a hyperbolic fixed point or a saddle point. We say that the system is stable in one direction (y_2-axis) and unstable in another (y_1-axis).

Let us consider case ② with a defective negative eigenvalue

$$\lambda < 0.$$

This is the case of the defective coefficient matrix with one-dimensional eigenspace associated to λ. Equation (6.31) reads in components

$$\frac{dy_1(t)}{dt} = \lambda y_1(t) + y_2(t), \quad \frac{dy_2(t)}{dt} = \lambda y_2(t).$$

Plugging solution $y_2(t) = \exp(\lambda t)y_2(0)$ into the first equation yields the first-order non-homogeneous equation:

$$\frac{dy_1(t)}{dt} = \lambda y_1(t) + e^{\lambda t}y_2(0),$$

which is solved by

$$y_1(t) = e^{\lambda t}y_1(0) + te^{\lambda t}y_2(0).$$

The resulting trajectory is shown in Figure 6.6 and referred to as an improper stable node. All trajectories are curved lines that move toward the origin. The y_1-axis is the direction of the eigenvector $v^{(1)} = [1 \ 0]^T$ associated to the eigenvalue λ and it is the direction along which solutions approach the origin.

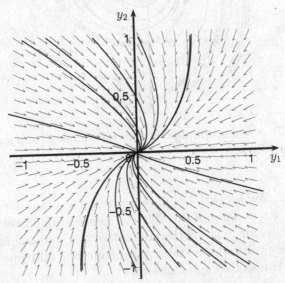

Figure 6.6: Improper stable node for the defective eigenvalue λ.

Finally, in case ③ we deal with a set of differential equations

$$\frac{dy_1(t)}{dt} = \mu y_1(t) + \nu y_2(t), \quad \frac{dy_2(t)}{dt} = -\nu y_1(t) + \mu y_2(t).$$

These equations simplify to

$$\frac{d\rho}{dt} = \mu\rho, \quad \frac{d\phi}{dt} = -\nu, \qquad (6.33)$$

when expressed in terms of the polar coordinates

$$\rho = \sqrt{y_1^2 + y_2^2}, \quad \tan\phi = \frac{y_2}{y_1}.$$

Both equations in (6.33) can easily be integrated to yield $\phi(t) = -\nu t + \phi_0$ and

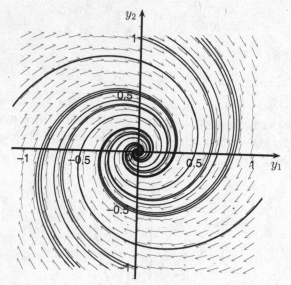

Figure 6.7: A stable spiral point for two complex conjugate eigenvalues with $\operatorname{Re}\lambda_i > 0$.

$\rho(t) = \rho_0 \exp(\mu t)$. We see that the distance from the origin increases for $\mu > 0$ and decreases for $\mu > 0$. Thus, if two different complex conjugate eigenvalues have non-zero, negative real part $\mu < 0$ then the solution is spiraling toward the origin. In that case the origin is called a stable spiral point and all trajectories eventually converge to the origin as they spiral inward.

For $\mu = 0$ the radius is constant for the given fixed initial conditions. Thus, for the purely imaginary complex conjugate eigenvalues all the trajectories are closed orbits around the origin, which is then called a center. The above results link the

Linear Systems of Differential Equations

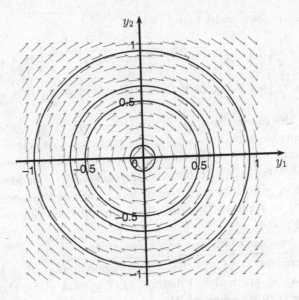

Figure 6.8: A center for two complex conjugate eigenvalues with $\operatorname{Re}\lambda_i = 0$.

stability of the linear system with an 2×2 constant coefficient matrix $\underline{\underline{A}}$ to the value of its eigenvalues. The situation can be summarized as follows.

An equilibrium solution $(0,0)$ for the system

$$\frac{\mathrm{d}\underline{x}(t)}{\mathrm{d}t} = \underline{\underline{A}}\,\underline{x}(t)$$

is asymptotically stable if every eigenvalue of $\underline{A}$ has a negative real part and is unstable if $\underline{A}$ has at least one eigenvalue with a positive real part.

In case of the 2×2 matrices we can determine the stability of the linear system directly in terms of the determinant and trace of the coefficient matrix $\underline{A}$ without invoking its diagonalization. Recall, that to find the eigenvalues of the 2×2 coefficient matrix

$$\underline{\underline{A}} = \begin{bmatrix} a & b \\ c & d \end{bmatrix},$$

we need to solve

$$\begin{vmatrix} a - \lambda & b \\ c & d - \lambda \end{vmatrix} = \lambda^2 - (a+d)\lambda + ad - bc = \lambda^2 - \operatorname{Tr}\underline{\underline{A}}\,\lambda + |\underline{\underline{A}}| = 0\,.$$

The eigenvalues are then easily found to be

$$\lambda_1 = \frac{1}{2}\left\{a + d + \sqrt{(a+d)^2 - 4(ad - bc)}\right\} = \frac{1}{2}\left\{\text{Tr}\,\underline{\underline{A}} + \sqrt{\Delta}\right\} \quad (6.34)$$

$$\lambda_2 = \frac{1}{2}\left\{a + d - \sqrt{(a+d)^2 - 4(ad - bc)}\right\} = \frac{1}{2}\left\{\text{Tr}\,\underline{\underline{A}} - \sqrt{\Delta}\right\}, \quad (6.35)$$

where the Δ is the discriminant $\Delta = (\text{Tr}\,\underline{\underline{A}})^2 - 4|\underline{\underline{A}}|$. The sign of Δ determines whether an eigenvalue is real or complex and plays a major role in classification of the two-dimensional, linear, homogeneous, constant coefficient systems with respect to their stability. We distinguish between three main cases depending on whether the discriminant Δ is positive, negative or zero. Each of the cases outlined below will be illustrated by a phase diagram, which plots the solution trajectories in the $x_1(t)\,x_2(t)$-phase plane.

I | $\Delta > 0$, the case of *distinct real eigenvalues*.

If the discriminant Δ of the characteristic polynomial of $\underline{\underline{A}}$ is positive then the eigenvalues λ_1, λ_2 are real and distinct. Let $\underline{v}_1$ and $\underline{v}_2$ be the corresponding eigenvectors. The corresponding solution is a superposition

$$\underline{x}(t) = c_1 e^{\lambda_1 t}\underline{v}_1 + c_2 e^{\lambda_2 t}\underline{v}_2. \quad (6.36)$$

The asymptotic stability takes place only when the exponential terms are decaying. The possible cases are as follows.

Ia $\text{Tr}\,\underline{\underline{A}} = a + d > 0$, $|\underline{\underline{A}}| = ad - bc > 0$.

Then $\lambda_1 > \lambda_2 > 0$. The eigenvalues of $\underline{\underline{A}}$ are of equal positive sign. For a very large time t the exponential $e^{\lambda_1 t}$ is much greater then $e^{\lambda_2 t}$. The trajectory will therefore eventually end up going to infinity along the $\underline{v}_1$-direction. The critical point is an unstable node.

Ib $\text{Tr}\,\underline{\underline{A}} = a + d < 0$, $|\underline{\underline{A}}| = ad - bc > 0$.

Then $0 > \lambda_1 > \lambda_2$. Both eigenvalues are negative and in this case the solution $\underline{x}(t)$ tends to $(0,0)$ as $t \to \infty$. The critical point is *a stable node*. The exponential $e^{\lambda_2 t}$ decreases much faster than the exponential $e^{\lambda_1 t}$. As time grows the term with $e^{\lambda_2 t}$ becomes negligible as compared to the first term (we assume $c_1 \neq 0, c_2 \neq 0$) in eq. (6.36). Thus, $\underline{x}(t)$ is well approximated by $c_1 e^{\lambda_1}\underline{v}_1$ when t is large. The solution will approach the origin $(0,0)$ along the $\underline{v}_1$-direction. Conversely, far from $(0,0)$, when $t \to -\infty$, the solution will go off to ∞ along the $\underline{v}_2$-direction.

Ic $|\underline{\underline{A}}| = ad - bc < 0$.

Then $\lambda_1 > 0 > \lambda_2$. The eigenvalues have different signs. The critical point is *a saddle point*. There is one line (parallel to the $\underline{v}_2$-direction) along which the trajectory moves toward the origin and another line (parallel to the $\underline{v}_1$-direction) along which it moves away from the origin. For $c_1 \neq 0$ and $c_2 \neq 0$ in equation (6.36) the trajectories move away from the origin at both $t \to +\infty$ and $t \to -\infty$

☞ **Example 6.3.2.** Let us see an example of an unstable node. Consider the system

$$\dot{\underline{x}} = \underline{\underline{A}}\,\underline{x} = \begin{bmatrix} 3 & -\frac{1}{2} \\ -2 & 3 \end{bmatrix} \underline{x}.$$

Here, $\Delta = 4 > 0$, the trace is positive and, thus, the eigenvalues are positive. They are given by $\lambda_1 = 4, \lambda_2 = 2$. After finding corresponding eigenvectors we are able to write a general solution as

$$\underline{x} = c_1 e^{4t} \begin{bmatrix} 1 \\ -2 \end{bmatrix} + c_2 e^t \begin{bmatrix} 1 \\ 2 \end{bmatrix}.$$

Since both exponentials are positive this solution goes away from the origin to infinity as time becomes large $t \to \infty$. For a very large time we can neglect the second term and $\underline{x}$ becomes approximately parallel to the eigenvector of the largest eigenvalue of $\underline{\underline{A}}$:

$$\underline{x} \approx c_1 e^{4t} \begin{bmatrix} 1 \\ -2 \end{bmatrix}.$$

Next, we consider an example of a stable node. The system is taken this time to be

$$\dot{\underline{x}} = \underline{\underline{A}}\,\underline{x} = \begin{bmatrix} -1 & \frac{1}{4} \\ 1 & -1 \end{bmatrix} \underline{x}.$$

Here, $\operatorname{Tr}\underline{\underline{A}} = -2$, and, $\Delta = 1$. The eigenvalues are $\lambda_1 = -1/2$ and $\lambda_2 = -3/2$ and the general solution is

$$\underline{x} = c_1 e^{-t/2} \begin{bmatrix} 1 \\ 2 \end{bmatrix} + c_2 e^{-3t/2} \begin{bmatrix} 1 \\ -2 \end{bmatrix}.$$

The corresponding phase plot is shown in Figure (6.9). Since both eigenvalues are negative the directions of arrows are inward pointing toward the origin. Far from the origin the trajectories become parallel to the direction of the eigenvector $\begin{bmatrix} 1 & -2 \end{bmatrix}^T$ associated with the more negative eigenvalue λ_2.

Next, consider the system

$$\dot{\underline{x}} = \underline{\underline{A}}\,\underline{x} = \begin{bmatrix} -1 & 4 \\ 1 & -1 \end{bmatrix} \underline{x}.$$

The eigenvalues $\lambda_1 = 1, \lambda_2 = -3$ have opposite signs (in agreement with the determinant, $|\underline{\underline{A}}| = -3$, being negative). The general solution is

$$\underline{x} = c_1 e^t \underline{v}_1 + c_2 e^{-3t} \underline{v}_2 = c_1 e^t \begin{bmatrix} 2 \\ 1 \end{bmatrix} + c_2 e^{-3t} \begin{bmatrix} -2 \\ 1 \end{bmatrix}$$

and goes to infinity at both large positive and large negative times (in general case $c_1, c_2 \neq 0$). As shown in Figure (6.10) the solutions approach $\underline{v}_1$-

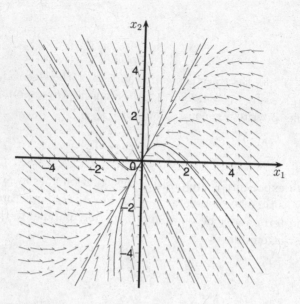

Figure 6.9: A stable node. The system with two negative eigenvalues.

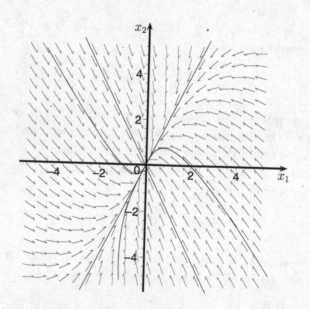

Figure 6.10: A saddle point. Two eigenvalues of opposite sign. One attractive direction and one repelling.

direction (associated with the positive eigenvalue) as $t \to \infty$. The trajectories start as $t \to -\infty$ along the $\underline{v}_2$-direction (associated with the negative eigenvalue). In both limits the solutions are unbounded and the saddle point is unstable. The exception is the case of $c_1 = 0$ for which the solution converges toward the origin.

II | $\Delta < 0$. Complex distinct eigenvalues. |

The eigenvalues of $\underline{A}$ are complex conjugates

$$\lambda_1 = \mu + i\nu, \ \lambda_1 = \mu - i\nu, \quad \nu = \frac{1}{2}\sqrt{-\Delta}, \quad \mu = \frac{1}{2}(a+d) = \frac{1}{2}\operatorname{Tr}\underline{A}.$$

The corresponding eigenvectors form a complex and mutually conjugate pair:

$$\underline{v}_1 = \underline{u} + i\underline{w}, \quad \underline{v}_2 = \underline{u} - i\underline{w},$$

where two real and linearly independent vectors $\underline{u}, \underline{w}$ are scalar multiples of

$$\begin{bmatrix} 1 \\ (d-a)/(2b) \end{bmatrix}, \quad \begin{bmatrix} 1 \\ \sqrt{-\Delta}/(2b) \end{bmatrix}.$$

One of the solutions of the linear system is then

$$\underline{x}_1(t) = e^{\lambda_1 t}\underline{v}_1 = e^{(a+d)t/2}e^{i\nu t}(\underline{u} + i\underline{w}) = e^{(a+d)t/2}(\underline{u}\cos(\nu t) - \underline{w}\sin(\nu t))$$
$$+ i e^{(a+d)t/2}(\underline{u}\sin(\nu t) + \underline{w}\cos(\nu t)) = \underline{y}_1(t) + i\underline{y}_2(t),$$

while the remaining solution is $\underline{x}_2(t) = \underline{y}_1(t) - i\underline{y}_2(t)$. Thus two real, linearly independent solutions of $\underline{\dot{x}} = \underline{A}\,\underline{x}$ can be written as

$$\underline{y}_1(t) = e^{(a+d)t/2}\left[\underline{u}\sin(\nu t) + \underline{w}\cos(\nu t)\right], \tag{6.37}$$
$$\underline{y}_2(t) = e^{(a+d)t/2}\left[\underline{u}\sin(\nu t) + \underline{w}\cos(\nu t)\right].$$

These solutions oscillate in sign. The trajectories are spirals if $\operatorname{Tr}\underline{A} = (a+d)/2 \neq 0$ and ellipses if $\operatorname{Tr}\underline{A} = (a+d)/2 = 0$.

There are three sub-cases depending on the sign of the trace of $\underline{A}$.

IIa $2\operatorname{Tr}\underline{A} = a+d > 0$. The real part of the eigenvalues is positive. Solutions move away from zero in an outward spiral. The point is an unstable spiral.

IIb $2\operatorname{Tr}\underline{A} = a+d < 0$. The real part of the eigenvalues is negative. Solutions approach zero in an inward spiral and the point is a stable spiral.

IIc $2\operatorname{Tr}\underline{A} = a+d = 0$ and the eigenvalues of $\underline{A}$ are purely imaginary $\lambda_\pm = \pm i\nu$. Trajectories are closed ellipses and the solution is a *center*.

☞ **Example 6.3.3.** The system

$$\underline{\dot{x}} = \begin{bmatrix} -1 & -1 \\ 1 & -1 \end{bmatrix}\underline{x}$$

provides an example of a spiral. Here, $\Delta = -4 < 0$, the trace is negative and two complex eigenvalues are given by $\lambda_1 = -1 + i, \lambda_2 = -1 - i$. The corresponding eigenvectors are

$$\underline{v}_1 = \begin{bmatrix} 1 \\ -i \end{bmatrix}, \quad \underline{v}_2 = \begin{bmatrix} 1 \\ i \end{bmatrix}.$$

The solution is then

$$\underline{x} = c_1 e^{-t} e^{i t} \begin{bmatrix} 1 \\ -i \end{bmatrix} + c_2 e^{-t} e^{-i t} \begin{bmatrix} 1 \\ i \end{bmatrix}.$$

Let initial conditions be $x_1(0) = 1, x_2(0) = 0$. Then the coefficients c_1, c_2 have to satisfy

$$\begin{bmatrix} 1 \\ 0 \end{bmatrix} = \underline{x}(0) = c_1 \begin{bmatrix} 1 \\ -i \end{bmatrix} + c_2 \begin{bmatrix} 1 \\ i \end{bmatrix},$$

which yields $c_1 = c_2 = 1/2$. Hence, for that particular initial condition the solution is

$$\underline{x} = e^{-t} \frac{1}{2} \left(e^{i t} \begin{bmatrix} 1 \\ -i \end{bmatrix} + e^{-i t} \begin{bmatrix} 1 \\ i \end{bmatrix} \right) = e^{-t} \begin{bmatrix} \cos t \\ \sin t \end{bmatrix}.$$

Since the real part of the eigenvalues is negative, the spiral is inward. The phase plot is shown in Figure (6.11). To show an example of a center we

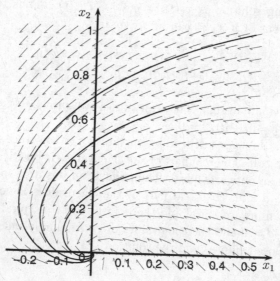

Figure 6.11: A spiral point with the inward spiral trajectories.

consider a system

$$\dot{\underline{x}} = \begin{bmatrix} -2 & 2 \\ -4 & 2 \end{bmatrix} \underline{x}.$$

The coefficient matrix in this equation is traceless and both eigenvalues, $\lambda = \pm 2i$, are imaginary. The corresponding eigenvectors are

$$\underline{v}_1 = \begin{bmatrix} 1 \\ 1+i \end{bmatrix}, \quad \underline{v}_2 = \begin{bmatrix} 1 \\ 1-i \end{bmatrix}$$

and the general solution can be expressed in terms of the periodic functions as

$$\underline{x} = c_1 e^{2i t} \begin{bmatrix} 1 \\ 1+i \end{bmatrix} + c_2 e^{-2i t} \begin{bmatrix} 1 \\ 1-i \end{bmatrix} = (c_1 + c_2) \begin{bmatrix} \cos 2t \\ \cos 2t - \sin 2t \end{bmatrix}$$

$$+ i\,(c_1 - c_2) \begin{bmatrix} \sin 2t \\ \cos 2t + \sin 2t \end{bmatrix}.$$

As seen from the phase diagram in Figure (6.12) the trajectories are ellipses

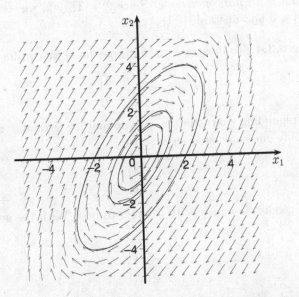

Figure 6.12: A center with two imaginary eigenvalues.

and the origin is stable, but not asymptotically stable.

III $\boxed{\Delta = 0. \text{ Double real eigenvalue}}$

Here the discriminant Δ is zero and $(a+d)^2 = 4(ad-bc)$. The matrix $\underline{\underline{A}}$ has a double real eigenvalue

$$\lambda_1 = \lambda_2 = \frac{1}{2}\,\mathrm{tr}\,\underline{\underline{A}}.$$

IIIa $b = c = 0$, $a = d \neq 0$ or $\underline{\underline{A}} = \lambda\underline{\underline{I}}$.

The system has a general solution

$$x(t) = c_1 e^{\lambda t} \begin{bmatrix} 1 \\ 0 \end{bmatrix} + c_2 e^{\lambda t} \begin{bmatrix} 0 \\ 1 \end{bmatrix}.$$

The critical point is a *proper node* or a *star*. It is stable if $a < 0$ ($\lambda < 0$) and unstable if $a > 0$ ($\lambda > 0$).

IIIb $\underline{\underline{A}} \neq \lambda\underline{\underline{I}}$.

Let $\underline{v}_1$ be an eigenvector of $\underline{\underline{A}}$ with eigenvalue λ. In addition, there exists a generalized eigenvector $\underline{v}_2$ such that $(\underline{\underline{A}} - \lambda\underline{\underline{I}})\,\underline{v}_2 \neq 0$ and $\underline{v}_1, \underline{v}_2$ are linearly independent. The solution takes a form

$$x(t) = c_1 e^{\lambda t}\underline{v}_1 + c_2 e^{\lambda t}\left[\underline{v}_2 + t\underline{v}_1\right].$$

The critical point is an *improper node*. Since $\lambda = \mathrm{Tr}\,\underline{\underline{A}}/2$, we see that the system is stable if $\mathrm{Tr}\,\underline{\underline{A}} < 0$ and unstable if $\mathrm{Tr}\,\underline{\underline{A}} > 0$.

☞ **Example 6.3.4.** To illustrate an improper node we consider the system

$$\dot{x} = \underline{\underline{A}}\,x = \begin{bmatrix} 1 & -4 \\ 1 & -3 \end{bmatrix} x.$$

The corresponding characteristic polynomial has a double root $\lambda = -1$ but there is only one associated eigenvector

$$v_1 = \begin{bmatrix} 2 \\ 1 \end{bmatrix}.$$

Thus the geometric multiplicity is equal to one. With the generalized eigenvector

$$v_2 = \begin{bmatrix} 1 \\ 0 \end{bmatrix}$$

such that $(\underline{\underline{A}} + \underline{\underline{I}})\,v_2 = v_1$ we can describe the general solution as

$$x = c_1 e^{-t}\underline{v}_1 + c_2\left[e^{-t}t\underline{v}_1 + e^{-t}\underline{v}_2\right].$$

For large time and $c_2 \neq 0$ it is the term $e^{-t}t v_1$ which dominates the above expression. Accordingly, the trajectories will approach the origin along the $\underline{v}_1$-direction as shown in Figure (6.13).

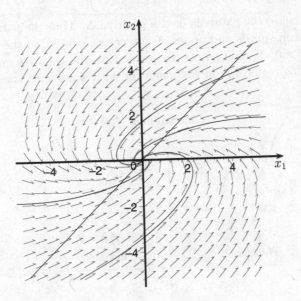

Figure 6.13: Improper node with one eigenvector and one generalized eigenvector.

Below, in the Table 6.1 we summarize all the different possibilities reviewed above. Another way to summarize the position of all stability regions with respect

Eigenvalues	Discriminant	Type of Critical Point
$\lambda_2 > \lambda_1 > 0$	$\Delta > 0$	unstable node
$\lambda_1 < \lambda_2 < 0$	$\Delta > 0$	stable node
$\lambda_1 < 0 < \lambda_2$	$\Delta > 0$	unstable saddle point
$\lambda = \mu \pm i\nu,\ \mu > 0$	$\Delta < 0$	unstable spiral point
$\lambda = \mu \pm i\nu,\ \mu < 0$	$\Delta < 0$	stable spiral point
$\lambda = \pm i\nu$	$\Delta < 0$	linear center
$\lambda_1 = \lambda_2 > 0$	$\Delta = 0$	unstable star
$\lambda_1 = \lambda_2 < 0$	$\Delta = 0$	stable star

Table 6.1: Critical points and the corresponding eigenvalues.

to possible values of $\operatorname{Tr} \underline{\underline{A}}$ and $|\underline{\underline{A}}|$ is shown in Figure 6.14. Here the vertical axis denotes the value of $|\underline{\underline{A}}|$ and the horizontal axis refers to the value of $\operatorname{Tr} \underline{\underline{A}}$. The parabola is given by a curve $(\operatorname{Tr} \underline{\underline{A}})^2 - 4|\underline{\underline{A}}| = 0$ on which discriminant Δ is zero and two eigenvalues coincide. The points above the parabola have a negative Δ

and the points below the parabola have a positive Δ. Thus the parabola defines the boundary between spirals and nodes. Centers are located on the positive vertical axis where $|\underline{A}| > 0$. Saddle points are located on the negative vertical axis. The asymptotic stable region lies in the quadrant with a negative trace and a positive determinant.

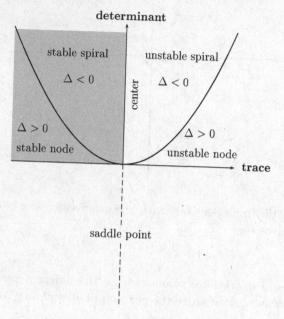

Figure 6.14: Regions of stability.

☞ **Example 6.3.5.** The Lorentz force $\vec{F}$ acting on a charge q moving with velocity $\vec{v}$ in a magnetic field $\vec{B}$ is $\vec{F} = q(\vec{v} \times \vec{B})$. For an electron with the negative charge $q = -e$ the Lorentz force gives rise to the equations of motion:

$$\frac{d\vec{v}}{dt} = \frac{e}{m}\vec{B} \times \vec{v},$$

which in matrix notation (see Chapter 2) become

$$\frac{d\underline{v}}{dt} = \frac{e}{m}\underline{\underline{B}}\,\underline{v} = \frac{e}{m}\begin{bmatrix} 0 & -B_3 & B_2 \\ B_3 & 0 & -B_1 \\ -B_2 & B_1 & 0 \end{bmatrix}\begin{bmatrix} v_1 \\ v_2 \\ v_3 \end{bmatrix}.$$

We can always rotate the coordinate system so that the magnetic field is oriented along the z-axis and $\vec{B} = B\hat{k}$. Therefore, without losing generality we can rewrite the equations of motion as

$$\frac{d}{dt}\begin{bmatrix} v_1 \\ v_2 \\ v_3 \end{bmatrix} = \frac{eB}{m}\begin{bmatrix} 0 & -1 & 0 \\ 1 & 0 & 0 \\ 0 & 0 & 0 \end{bmatrix}\begin{bmatrix} v_1 \\ v_2 \\ v_3 \end{bmatrix}. \tag{6.38}$$

Obviously, finding solutions to the equations of motion requires finding the eigenvalues of the matrix:

$$\underline{\underline{A}} = -i\,\underline{\underline{J_3}} = \begin{bmatrix} 0 & -1 & 0 \\ 1 & 0 & 0 \\ 0 & 0 & 0 \end{bmatrix}$$

given in terms of the generator $\underline{\underline{J_3}}$ of the infinitesimal rotations defined in formula (2.185). The characteristic polynomial of $\underline{\underline{A}}$ is $\lambda(\lambda^2 + 1) = 0$. One root is $\lambda = 0$ and the two remaining roots, $\lambda = \pm i$, are purely imaginary indicating a center. For $\lambda = 0$, the eigenvalue problem reads

$$\begin{bmatrix} 0 & -1 & 0 \\ 1 & 0 & 0 \\ 0 & 0 & 0 \end{bmatrix}\begin{bmatrix} v_1 \\ v_2 \\ v_3 \end{bmatrix} = 0 \implies \underline{v} = \begin{bmatrix} 0 \\ 0 \\ 1 \end{bmatrix}.$$

For $\lambda = \pm i$ we find

$$\begin{bmatrix} \mp i & -1 & 0 \\ 1 & \pm i & 0 \\ 0 & 0 & \mp i \end{bmatrix}\begin{bmatrix} v_1 \\ v_2 \\ v_3 \end{bmatrix} = 0 \implies \underline{v}_\pm = \begin{bmatrix} \pm i \\ 1 \\ 0 \end{bmatrix}$$

from which we derive two corresponding solutions

$$\underline{v}_+(t) = e^{it}\begin{bmatrix} i \\ 1 \\ 0 \end{bmatrix} = \begin{bmatrix} -\sin t \\ \cos t \\ 0 \end{bmatrix} + i\begin{bmatrix} \cos t \\ \sin t \\ 0 \end{bmatrix}$$

and

$$\underline{v}_-(t) = \underline{v}_+^*(t) = e^{-it}\begin{bmatrix} -i \\ 1 \\ 0 \end{bmatrix} = \begin{bmatrix} -\sin t \\ \cos t \\ 0 \end{bmatrix} - i\begin{bmatrix} \cos t \\ \sin t \\ 0 \end{bmatrix}.$$

The real and imaginary parts

$$\underline{v}_1(t) = \begin{bmatrix} -\sin t \\ \cos t \\ 0 \end{bmatrix}, \quad \underline{v}_2(t) = \begin{bmatrix} \cos t \\ \sin t \\ 0 \end{bmatrix}.$$

of the velocity $\underline{v}_+(t)$ are expressed in terms of the periodic functions and represent an equivalent set of solutions. The complete formula for the velocity of the particle described by equation (6.38) is obtained by multiplying t by the factor $\omega = eB/m$ from equation (6.38) and by taking a linear combination of all the above solutions:

$$\underline{v}(t) = c_1 \begin{bmatrix} 0 \\ 0 \\ 1 \end{bmatrix} + c_2 \begin{bmatrix} -\sin(\omega t) \\ \cos(\omega t) \\ 0 \end{bmatrix} + c_3 \begin{bmatrix} \cos(\omega t) \\ \sin(\omega t) \\ 0 \end{bmatrix} . \tag{6.39}$$

Integration of $\underline{v}(t)$ yields the trajectory $\underline{x}(t)$, which can be conveniently written as

$$\underline{x}(t) = \begin{bmatrix} K\cos(\omega t - \alpha) \\ K\sin(\omega t - \alpha) \\ c_1 t \end{bmatrix} .$$

To simplify the above expression possible integration constants were ignored. Equating $d\underline{x}/dt$ with $\underline{v}$ leads to relations

$$K\omega\cos\alpha = c_2, \quad K\omega\sin\alpha = c_3$$

between the amplitude K, phase α and coefficients c_2, c_3. Thus, the tra-

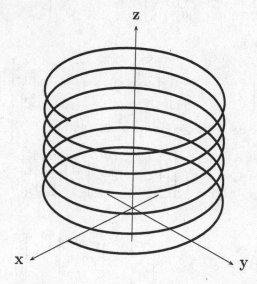

Figure 6.15: Trajectory of a charged particle in a magnetic field.

jectory of an electron in a magnetic field is a superposition of an uniform linear motion along the direction of the magnetic field (here the z-axis) and an uniform circular motion in the plane perpendicular to the direction of the magnetic field. The result of this superposition is a circular helix shown in Figure 6.15. See Problem 6.3.12 for an alternative derivation of this result using an angle and axis parametrization (2.175) of a skew-symmetric matrix discussed in Chapter 2.

☜ Linear Systems of Differential Equations: Problems

Problem 6.3.1. Calculate the exponential matrix $\exp\left(t\underline{\underline{A}}\right)$ for

(a) $\underline{\underline{A}} = \begin{bmatrix} \lambda_1 & 0 \\ 0 & \lambda_2 \end{bmatrix}$

(b) $\underline{\underline{A}} = \begin{bmatrix} \lambda & 1 \\ 0 & \lambda \end{bmatrix}$

(c) $\underline{\underline{A}} = \begin{bmatrix} a & b \\ -b & a \end{bmatrix}$

Problem 6.3.2. Show that the determinant $|\underline{\underline{V}}(t)|$ of the fundamental matrix $\underline{\underline{V}}(t)$ satisfies

$$\frac{d}{dt} |\underline{\underline{V}}(t)| = |\underline{\underline{V}}(t)| \operatorname{Tr}\underline{\underline{A}}$$

and, therefore, is equal to $|\underline{\underline{V}}(t)| = \exp\left(t\operatorname{Tr}\underline{\underline{A}}\right)|\underline{\underline{V}}(0)|$.

Problem 6.3.3. Let $\underline{\underline{A}}$ be an 2×2 matrix with only one eigenvalue λ (of algebraic multiplicity 2 but geometric multiplicity 1) and with one eigenvector $\underline{v}^{(1)}$. Let $\underline{v}^{(2)}$ be a generalized eigenvalue of $\underline{\underline{A}}$ such that $(\underline{\underline{A}} - \lambda\underline{\underline{A}})\,\underline{v}^{(2)} = \underline{v}^{(1)}$. For matrix $\underline{\underline{V}}$ constructed out of $\underline{v}^{(1)}, \underline{v}^{(2)}$ as

$$\underline{\underline{V}} = \begin{bmatrix} \underline{v}^{(1)} & \underline{v}^{(2)} \end{bmatrix}$$

show that

$$\underline{\underline{A}}\,\underline{\underline{V}} = \lambda\underline{\underline{V}} + \begin{bmatrix} 0 & \underline{v}^{(1)} \end{bmatrix}$$

and use it to prove that the similarity transformation $\underline{\underline{A}} \to \underline{\underline{V}}^{-1}\underline{\underline{A}}\,\underline{\underline{V}}$ does not fully diagonalize matrix $\underline{\underline{A}}$ but instead brings it to the so-called Jordan matrix form

$$\underline{\underline{V}}^{-1}\underline{\underline{A}}\,\underline{\underline{V}} = \begin{bmatrix} \lambda & 1 \\ 0 & \lambda \end{bmatrix}.$$

Construct matrix $\underline{\underline{V}}$ and verify the above construction for

$$\underline{\underline{A}} = \begin{bmatrix} -3 & 1 \\ -1 & -1 \end{bmatrix}.$$

Problem 6.3.4. Let $\underline{\underline{A}}$ be the real 2×2 matrix with two conjugate complex eigenvalues λ_1, λ_2 and associated eigenvectors $\underline{v}_1, \underline{v}_2$. Show, that you can write $\underline{v}_1 = \underline{u} + i\underline{w}, \underline{v}_2 = \underline{u} - i\underline{w}$ in terms of real vectors $\underline{u}, \underline{w}$ such that

$$\underline{\underline{A}}\,\underline{u} = \mu\underline{u} - \nu\underline{w}, \qquad \underline{\underline{A}}\,\underline{v} = \nu\underline{u} + \mu\underline{w},$$

where the real numbers μ, ν are real and imaginary parts of λ_1.
Also, show that

$$\underline{\underline{V}} = \begin{bmatrix} \underline{u} & \underline{w} \end{bmatrix}$$

transforms $\underline{\underline{A}}$ by a similarity transformation according to

$$\underline{\underline{V}}^{-1} \underline{\underline{A}} \underline{\underline{V}} = \begin{bmatrix} \mu & \nu \\ -\nu & \mu \end{bmatrix}.$$

Problem 6.3.5. For an $n \times n$ matrix $\underline{\underline{A}}$ with only one eigenvalue λ_1, the characteristic polynomial is given by $p_A(\lambda) = (\lambda - \lambda_1)^n$. Due to the so-called Cayley-Hamilton theorem it holds that $\left(\underline{\underline{A}} - \lambda_1 \underline{\underline{I}} \right)^n = 0$.

(a) Show that the exponential matrix is in that case given by

$$e^{t\underline{\underline{A}}} = e^{t\lambda_1} \left(\underline{\underline{I}} + t \left(\underline{\underline{A}} - \lambda_1 \underline{\underline{I}} \right) + \frac{t^2}{2} \left(\underline{\underline{A}} - \lambda_1 \underline{\underline{I}} \right)^2 + \cdots \right.$$
$$\left. + \frac{t^{n-1}}{(n-1)!} \left(\underline{\underline{A}} - \lambda_1 \underline{\underline{I}} \right)^{n-1} \right).$$

(b) Use the above formula to solve the homogeneous system of differential equations (6.4) for

$$\underline{\underline{A}} = \begin{bmatrix} 1 & 0 & 0 \\ 0 & 3 & 2 \\ 2 & -2 & -1 \end{bmatrix},$$

with the initial condition $\underline{x}(0) = \begin{bmatrix} 1 & -1 & 2 \end{bmatrix}^T$. Find a fundamental set of solutions.

Problem 6.3.6. Find $\exp\left(t\underline{\underline{N}} \right) \underline{v}$ for the matrix

$$\underline{\underline{N}} = \begin{bmatrix} 1 & 1 & -1 \\ -4 & 5 & -2 \\ 0 & 0 & 3 \end{bmatrix}$$

and vector $\underline{v} = \begin{bmatrix} 1 & 3 & 2 \end{bmatrix}^T$ by expressing $\underline{v}$ in terms of the generalized eigenvectors of $\underline{\underline{N}}$.

Problem 6.3.7. Find $\exp\left(t\underline{\underline{M}} \right)$ for the matrix

$$\underline{\underline{M}} = \begin{bmatrix} 2 & 0 & 0 \\ 1 & -1 & -1 \\ 3 & 0 & -1 \end{bmatrix}$$

using a basis of its generalized eigenvectors.

Problem 6.3.8. For each of the two-dimensional systems listed below
   1) find eigenvalues and eigenvectors
   2) Is the critical point stable, asymptotically stable or unstable?
   3) sketch a phase plot.

(a) $\dot{\underline{x}} = \begin{bmatrix} 1 & 2 \\ 2 & 1 \end{bmatrix} \underline{x}$
$\qquad\qquad\qquad$
(b) $\dot{\underline{x}} = \begin{bmatrix} 1 & -5 \\ 1 & -3 \end{bmatrix} \underline{x}$

(c) $\dot{\underline{x}} = \begin{bmatrix} 1 & -1 \\ 2 & 4 \end{bmatrix} \underline{x}$

Problem 6.3.9. Given the matrix exponential

$$e^{t\underline{\underline{C}}} = \begin{bmatrix} e^{2t} + t\,e^{2t} & t\,e^{2t} \\ -t\,e^{2t} & e^{2t} - t\,e^{2t} \end{bmatrix}.$$

Find the matrix $\underline{\underline{C}}$ and find the fundamental matrix $\underline{\underline{V}}(t)$ corresponding to the homogeneous linear system $\dot{\underline{x}} = \underline{\underline{C}}\,\underline{x}$

Problem 6.3.10. Given the matrix exponential

$$e^{t\underline{\underline{T}}} = \begin{bmatrix} \dfrac{1}{2}\,e^{-t} + \dfrac{1}{2}\,e^{3t} & \dfrac{1}{2}\,e^{3t} - \dfrac{1}{2}\,e^{-t} \\[2mm] \dfrac{1}{2}\,e^{3t} - \dfrac{1}{2}\,e^{-t} & \dfrac{1}{2}\,e^{-t} + \dfrac{1}{2}\,e^{3t} \end{bmatrix}$$

find matrix $\underline{\underline{T}}$. Find the particular solution to the non-homogeneous linear system

$$\dot{\underline{x}} = \underline{\underline{T}}\,\underline{x} + \begin{bmatrix} e^{2t} \\ 1 \end{bmatrix}.$$

Problem 6.3.11. Consider the system of two differential equations

$$\dot{x}_1 = -4x_2, \quad \dot{x}_1 = x_1.$$

Find a general solution and sketch the corresponding phase plot. Is the system stable, asymptotically stable or unstable?

Problem 6.3.12. In Example 6.3.5, we have derived the equation of motion

$$\frac{d\underline{v}}{dt} = \frac{e}{m}\,\underline{\underline{B}}\,\underline{v} \tag{6.40}$$

for an electron in a magnetic field $\vec{B} = (B_1, B_2, B_3)$ represented by a skew-symmetric matrix

$$\underline{\underline{B}} = \begin{bmatrix} 0 & -B_3 & B_2 \\ B_3 & 0 & -B_1 \\ -B_2 & B_1 & 0 \end{bmatrix} = -\mathrm{i}\left(\vec{B} \cdot \underline{\underline{\vec{J}}}\right).$$

(a) Use an angle and axis parametrization of an exponential of a skew-symmetric matrix $\underline{\underline{\Omega}} = \underline{\underline{B}}$ introduced in (2.175) to rewrite a general solution

$$\underline{v}(t) = \underline{v}_0 \exp\left(\frac{e}{m}\underline{\underline{B}}\,t\right)$$

to the differential equation (6.40) as

$$\vec{v}(t) = \vec{v}_0 + \frac{e}{m\omega}\vec{B}\times\vec{v}_0\,\sin(\omega t) + \frac{e^2}{m^2\omega^2}\vec{B}\times\left(\vec{B}\times\vec{v}_0\right)(1-\cos(\omega t)),$$

where

$$\omega = \frac{e}{m}\sqrt{\vec{B}^2}$$

is the so-called cyclotron frequency.

(b) Show that for $\vec{B} = B\hat{k}$ and the initial velocity $\vec{v}_0 = v_2^{(0)}\hat{j} + v_3^{(0)}\hat{k}$ the particle's trajectory is given by

$$x = \frac{m}{eB}v_2^{(0)}\left(1-\cos(\omega t)\right)+x_0, \quad y = \frac{m}{eB}v_2^{(0)}\sin(\omega t)+y_0, \quad z = v_3^{(0)}t+z_0$$

where $\vec{r}_0 = x_0\hat{i}+y_0\hat{j}+z_0\hat{k}$ is the particle's position at $t = 0$. Compare with results of Example 6.3.5, which showed that the charged particle will follow a circular helix curve along the magnetic field lines.

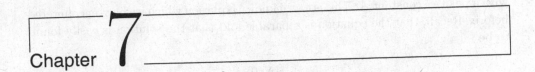

Chapter 7

Nonlinear Differential Equations

FOR a linear differential equation a linear combination of two solutions is also a solution. This linearity property allows us to split a linear problem into smaller parts and solve each part separately or inversely superimpose the smaller solutions into a more complete solution. Elegant mathematical techniques like Fourier analysis have been invented to explore the linearity. There is no superposition principle for equations which are nonlinear. Adding two solutions of the nonlinear differential equation does not produce another solution. An attempt to solve the nonlinear differential equation has to come to terms with its full complexity. Also, a linear system responds to small variations in its parameters in a way which typically is proportional to the initial changes. In contrast, in nonlinear systems a tiny change in the input can result in a big change in the output.

Not surprisingly in view of the above the nonlinear equations are rarely solvable with the exception of a class of integrable nonlinear equations having soliton solutions. Fortunately, the advent of computers has brought about progress in the study of nonlinear differential equations and has revealed a variety of novel behaviors including chaos.

7.1 One-Dimensional Nonlinear Differential Equations

To introduce the concept of equilibrium points and their (asymptotic) stability in the setting of nonlinear differential equations we start by considering a standard one-dimensional example of a fish hatchery. Suppose $N(t)$ is the number of fish at time t. We assume that fish multiply at a fixed rate per fish, i.e., the rate of increase of the number of fish divided by all fish is constant. Similarly, we assume that the death rate is also at a fixed rate per fish. Thus, with no outside disturbance and an infinite supply of food and water we have that the rate with which $N(t)$ is changing is governed by the differential equation:

$$\frac{\mathrm{d}N}{\mathrm{d}t} = AN(t), \tag{7.1}$$

with a positive constant A. This is a well known equation which we have encountered before. Recall, that the equation is separable and that the solution is easily found to be

$$N(t) = N(0)e^{At}.$$

Recognizing that in real life there are limits on available food and space we have to correct the above oversimplified picture. We shall assume that due to competition for food and space in a fish hatchery the death rate per fish is proportional to the number of fish present. Subsequently, we add the total death rate $-BN(t)^2$ to equation (7.1), resulting in

$$\frac{\mathrm{d}N}{\mathrm{d}t} = AN(t) - BN(t)^2. \tag{7.2}$$

In addition, let us assume that the annual commercial catch removes a constant number of fish per year. We set that number to C. This modifies the above equation to

$$\frac{\mathrm{d}N}{\mathrm{d}t} = AN(t) - BN(t)^2 - C. \tag{7.3}$$

We can always divide both sides by B and rescale the time $t \to Bt$. It is therefore sufficient to work with $B = 1$. Without losing generality we will consider an example:

$$\frac{\mathrm{d}N}{\mathrm{d}t} = 7N(t) - N(t)^2 - 10 = -\left(N(t) - 2\right)\left(N(t) - 5\right), \tag{7.4}$$

where $N(t)$ measures the number of fish in thousands per a year. Thus, the number 10 in equation (7.4) stands for $10,000$ fish caught per year. Now, we are ready to ask the most crucial question: For what initial number of fish $N_0 = N(t = 0)$ will the fishery be able to run a sustainable harvest activity? The answer will reveal the significance of stable critical points.

The equation is separable and an explicit solution can be found in terms of the initial data, N_0. We separate variables to get

$$\frac{\mathrm{d}N}{(N(t) - 2)(N(t) - 5)} = -\,\mathrm{d}t \longrightarrow \int \frac{\mathrm{d}N}{(N(t) - 2)(N(t) - 5)} = -t + c_0.$$

The integrand can be rewritten using partial fractions:

$$\frac{1}{(N(t) - 2)(N(t) - 5)} = \frac{1}{3}\left(\frac{1}{N(t) - 5} - \frac{1}{N(t) - 2}\right).$$

Thus, by substitution

$$\int \frac{dN}{3} \left(\frac{1}{N(t) - 5} - \frac{1}{N(t) - 2} \right) = -t + c_0$$

or

$$\ln(N(t) - 5) - \ln(N(t) - 2) = -3t + 3c_0 .$$

By applying the exponential function to both sides we obtain

$$\frac{N(t) - 5}{N(t) - 2} = e^{-3t + 3c_0} = Ke^{-3t} ,$$

with the constant, $K = \exp(3c_0)$. For time $t = 0$ we find an expression for the constant K:

$$K = \frac{N_0 - 5}{N_0 - 2} ,$$

where $N_0 = N(0)$. This leaves us with the equation

$$\frac{N - 5}{N - 2} = \frac{N_0 - 5}{N_0 - 2} e^{-3t} ,$$

which we are now ready to solve for $N = N(t)$. After multiplying both sides of the above equation by $(N_0 - 2)(N - 2)$ and grouping all the terms with N on the left hand side of the equation we obtain

$$N \left(N_0 - 2 - (N_0 - 5)e^{-3t} \right) = 5(N_0 - 2) - 2(N_0 - 5)e^{-3t}$$

leading to the following expression for N:

$$N(t) = \frac{5(N_0 - 2) - 2(N_0 - 5)e^{-3t}}{N_0 - 2 - (N_0 - 5)e^{-3t}} . \tag{7.5}$$

If N_0 is in the interval $2 < N_0 < 5$ then $N(t) \to 5$ as $t \to \infty$. However, if $N_0 < 2$ then $N(t)$ becomes zero for a finite t such that

$$e^{-3t} = \frac{5}{2} \frac{2 - N_0}{5 - N_0}, \quad \text{or} \quad t = \frac{1}{3} \ln \frac{2(5 - N_0)}{5(2 - N_0)} .$$

Thus, if the fish hatchery started with less than $2,000$ fish it would reach null population in finite time. For a starting population of greater than $2,000$ the fish population would eventually reach $5,000$ and the hatchery, in principle, would be able to co-exist with the annual harvest. See, Figure 7.1 for the behavior of $N(t)$ for various initial conditions.

Only in rare cases are we interested in finding how long it takes for the solution to reach equilibrium. Such information can be derived from the analytic expressions as the ones obtained above. More frequently, we would like to study the qualitative behavior of solutions, which follows from the structure of differential equations and the position of critical points. This information is available from a plot showing

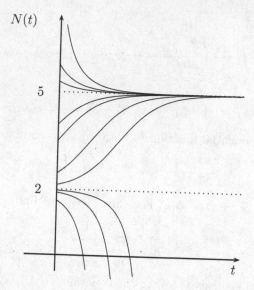

Figure 7.1: Solution curves $N(t)$ from eq. (7.5) for various values of N_0.

the derivative dN/dt versus N. Such a plot is an example of a phase portrait. Therefore, rather than studying the plots of N and dN/dt versus time t, we turn our attention to trajectories in the dN/dt-N phase plane as depicted in Figure 7.2. Despite the absence of time t on the phase plane the form of trajectories around the equilibria points indicates how the system will develop as the time progresses. This graphical method is especially useful in more complicated cases since it does not require obtaining detailed solutions in the analytic form. A task that becomes increasingly difficult as the differential equations get more complicated.

A phase portrait for the fish harvest is seen in Figure 7.3. Let us analyze the behavior of the derivative dN/dt versus N as shown on the phase portrait 7.3. One sees that $dN(t)/dt > 0$ for $2 < N(t) < 5$, while $dN(t)/dt < 0$ for $N(t) < 2$ and $N(t) > 5$. This implies that $N(t)$ is increasing for $N(t)$ inside the interval $2 \leq N \leq 5$ while it is decreasing outside that interval. This means that asymptotically all solutions with $N(t) > 2$ at any time t will converge to the final position 5. Also, if the solution $N(t)$ assumes a value which is in the range $0 < N(t) < 2$, then $N(t)$ will eventually reach 0 as time t increases.

In terms of the fish example the value $5,000$ is a maximum fish population which the environment will support. An increase beyond $5,000$ fish results in a death rate increase which causes a decline in the population back to $5,000$.

Figure 7.3 reproduces these results graphically. It shows how $N(t)$ evolves in time (flows) by drawing arrows pointing to the right in the interval between 2 and 5 where $dN(t)/dt > 0$. Similarly, in the interval where $dN(t)/dt < 0$, the arrows

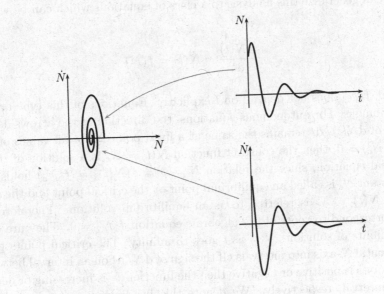

Figure 7.2: Transition to the phase portrait.

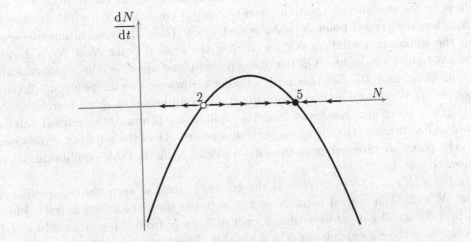

Figure 7.3: Phase portrait for the fish harvest.

point to the left. The directions of arrows indicate the flow of $N(t)$.

We can generalize this analysis to a class of equations which can be written as:

$$\frac{dN(t)}{dt} = \dot{N}(t) = f(N) \tag{7.6}$$

such that $f(N)$ does not depend on t explicitly. Equations of this type are said to be autonomous. For autonomous equations, the direction of the arrows defined by the sign of $dN(t)/dt$ remains the same at a fixed point N for all values of t.

If $f(N_c) = 0$, then the constant function $N(t) = N_c$ is a solution of the above differential equation, since the relation $\dot{N} = 0 = f(N(t)) = f(N_c)$ holds forever. The constant N_c is called an equilibrium point or the critical point and the constant function $N(t) = N_c$ is referred to as an equilibrium solution. Therefore critical points correspond to roots of the algebraic equation $f(N) = 0$. They are the only possible limits of solutions $N(t)$ as t goes to infinity. The critical points partition the horizontal N-axis into intervals. If the sign of dN/dt on an interval between two critical points is positive or negative then the function N is increasing or decreasing on that interval, respectively. We denote this by drawing right directed or left directed arrows along the horizontal line in each interval. This allows us to make qualitative sketches of solution curves in the $dN(t)/dt$-N-plane by hand, without solving differential equations.

Imagine a solution with the initial value N_0 which differs slightly from one of the roots N_c of $f(N) = 0$. An important question is whether such a solution will stay close to the equilibrium solution or will move away. If the solution remains close to the equilibrium solution given the initial condition N_0, which is close to N_c, then the critical point $N = N_c$ is said to be stable. If the solution converges to the equilibrium solution as $t \to \infty$ then the critical point $N = N_c$ is said to be asymptotically stable. On the phase portrait one identifies the asymptotically stable point from the fact that both adjacent arrows next to the point are directed toward it. We denote an asymptotically stable equilibrium by a solid circle.

If the solution moves away from the equilibrium solution, the critical point is said to be unstable. On the phase portrait, one or both of the adjacent arrows next to the point are directed away from it. We denote the unstable equilibrium by an open circle.

For $f(N) = -(N-2)(N-5)$ the critical points, as seen above, are $N = 2$ and $N = 5$. The critical point $N = 2$ is unstable since the solution $N(t)$, which is initially smaller or greater than 2, will drift away from 2 for large values of t. Conversely, the critical point at $N = 5$ is stable. This is indicated by arrows pointing toward this point from the left and the right. In fact, $N = 5$ is asymptotically stable since $N(t) \to 5$ for $t \to \infty$ if and only if $N_0 > 2$. In Figure 7.3, we followed the convention of representing unstable fixed points (like $N = 2$) by open dot and stable fixed points (like $N = 5$) by solid black dot. Thus, on the phase portraits the arrows point toward the solid black dots from the left and right directions and point away from the open dots from the left and right directions.

☞ **Example 7.1.1.** Consider differential equation:

$$\frac{dx}{dt} = -x(x-1)(x-2)(x-3)(x-4).$$

A phase portrait is shown in Figure 7.4. It shows arrows between the four

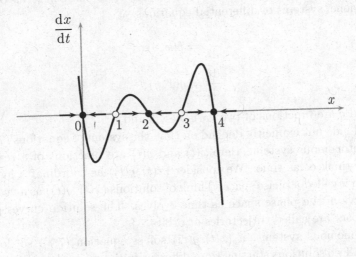

Figure 7.4: Phase portrait for $\dot{x} = -x(x-1)(x-2)(x-3)(x-4)$.

critical points $x_c = 0, 1, 2, 3, 4$ for which the right hand side of the differential equation is zero. The directions of arrows show that the critical points $x_c = 0, 2, 4$ are asymptotically stable (black solid dots), while the remaining critical points $x_c = 1, 3$ are unstable (open dots).

⚐ One-Dimensional Nonlinear Differential Equations: Problems

Problem 7.1.1. Draw the phase portrait and determine the stability of each critical point of the equation

$$\frac{dx}{dt} = x(1-x).$$

Confirm your results by solving explicitly for $x(t)$.

Problem 7.1.2. Find critical points of the following equations and determine whether they are stable or unstable by drawing the phase portrait.

(a) $\dot{x} = -(x-2)^2$ (b) $\dot{x} = -x^2 + 1$

(c) $\dot{x} = x^2(3-x)$ (d) $\dot{x} = \sin x$

7.2   Two-Dimensional Nonlinear Differential Equations

In this section, we will study problems which can be formulated in terms of nonlinear two-dimensional systems of differential equations:

$$\frac{dx}{dt} = f(x,y),$$

$$\frac{dy}{dt} = g(x,y), \tag{7.7}$$

where f and g are functions of two variables, x and y.

If f and g do not explicitly depend on time the system of equations (7.7) is said to be an autonomous system. Here $x(t)$ and $y(t)$ are functions of a real variable t, which we think of as time. We consider $(x(t), y(t))$ as coordinates in the (x, y) space, which is called a phase space. Think of solutions $(x(t), y(t))$ as a vector which traces a curve in the phase space as time evolves. The solution curves plotted in the phase space are called trajectories or orbits.

For autonomous systems, if $(x(t), y(t))$ solves equation (7.7), so does $(x(t+c), y(t+c))$. Two solutions starting from different initial points located on the same trajectory will trace identical paths differing only by a shift in time t. Thus, the trajectories depend on the position in the phase space and not on the time they are being traced. In other words, the point of a trajectory in phase space completely specifies the system's behavior.

A plot of all the qualitatively different trajectories is called a phase portrait. The trajectory evolves in such a way that the vector field $(\dot{x}(t), \dot{y}(t)) = (f(x), g(x))$ remains tangent to the trajectory at each point (x, y). For the autonomous system (7.7) the vector field $(\dot{x}(t), \dot{y}(t))$ depends on position (x, y) only and not on time. Thus, the trajectory will always have the same direction when it passes through a given (non-critical) point (x, y). For that reason the trajectories never intersect.

The slope of a trajectory is given by the ratio dy/dx. To calculate it we use the chain rule

$$\frac{dy}{dx} = \frac{\dot{y}}{\dot{x}} = \frac{g}{f}. \tag{7.8}$$

Definition 7.2.1. *A point (x_c, y_c) is a critical point of the system of differential equations $\dot{x} = f(x,y)$, $\dot{y} = g(x,y)$ if*

$$f(x_c, y_c) = 0, \quad g(x_c, y_c) = 0.$$

The critical points are also called equilibria because if a trajectory begins at a critical point it will remain there forever due to the uniqueness of the system of differential equations. Thus, the critical points are important because if the system ever reaches one of equilibrium points it will stay there.

Definition 7.2.2. *Solutions through critical points are called equilibrium solutions.*

Equilibria determine the appearance of the phase portrait.

☞ **Example 7.2.1.** Consider the system

$$\frac{dx}{dt} = y, \qquad \frac{dy}{dt} = -x, \tag{7.9}$$

It is easy to solve this system by eliminating the variable, y. Then we obtain a one-variable linear oscillator equation $\ddot{x} + x = 0$. The solution is

$$x(t) = a\cos t + b\sin t.$$

Substituting this into the equation $\dot{x} = y$ gives $y(t) = -a\sin t + b\cos t$, where a, b are constants fixed by a specific point on the trajectory taken at a specific time. In principle, we are able to find the trajectory traced by the solution $(x(t), y(t))$ as time evolves.

A more direct way to find the shape of trajectories is as follows. Consider the slope from relation (7.8):

$$\frac{dy}{dx} = \frac{g}{f} = -\frac{x}{y}$$

or

$$y\,dy + x\,dx = 0.$$

This is a separable equation which can be easily integrated to yield $x^2 + y^2 = C$ with C being an integration constant. What this means is that the function

$$P(t) = r(t)^2 = x(t)^2 + y(t)^2$$

remains constant along the trajectory. We can double-check it by differentiating P with respect to t to get

$$\frac{dP}{dt} = 2\frac{dx(t)}{dt}x(t) + 2\frac{dy(t)}{dt}y(t).$$

Plugging this in equation (7.9) and eliminate derivatives we get

$$\frac{dP}{dt} = 2\left(y(t)x(t) - x(t)y(t)\right) = 0,$$

which confirms that the function P is constant along the solution curve.

We next rewrite equation (7.9) in the form of a vector equation

$$\frac{\partial}{\partial t} \begin{bmatrix} x \\ y \end{bmatrix} = \begin{bmatrix} y \\ -x \end{bmatrix}$$

and associate the vector field

$$\vec{V}(x,y) = \begin{bmatrix} y \\ -x \end{bmatrix}$$

to every point (x, y) on the plane. In order to show the direction of motion along the allowed trajectories it is useful to draw arrows tangent to the trajectories. Figure 7.5 shows orbits of the form of the closed circles with the

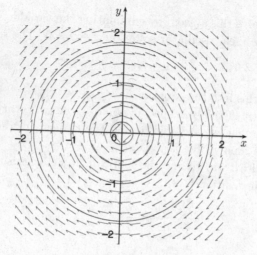

Figure 7.5: Orbits and vector fields of the linear oscillator model (7.9).

arrows depicting the tangential vector fields. The direction of the vector field is clockwise which is consistent with the above formula for $\vec{V}(x, y)$. Take, for instance, the region with $x > 0, y > 0$. Then $V_x = \dot{x}$ is positive and the arrows will point from left to right. Simultaneously, $V_y = \dot{y}$ is negative and so the arrows will point downward. This confirms the directions of the vector fields associated to all orbits in Figure 7.5.

Initial conditions select just one particular circle as the trajectory of the problem. For instance with the initial conditions

$$x(0) = 1 \quad \text{and} \quad y(0) = 0,$$

the curve will start at the point $(0, 1)$, which forces the function P to take value $P(0) = 1$ at $t = 0$. Hence, the orbit must be the unit circle. Obviously, any circle would be a solution curve of the differential equations.

The study of nonlinear systems in the plane starts with a study of their equilibria and the trajectories near the equilibrium positions. One can learn a great deal about the behavior of linear systems in a neighborhood of an equilibrium point from a study of their linearized versions around that equilibrium point.

In order to investigate the stability of a fixed point, one has to linearize the equation of motion around a fixed point. Suppose (x, y) is in a vicinity of the critical point (x_c, y_c) and let

$$\delta x = x - x_c, \quad \delta y = y - y_c.$$

By definition, $(\delta x, \delta y)$ is a small displacement of (x, y) from the critical point (x_c, y_c). Next, we Taylor expand the function $f(x, y)$ from equation (7.7) around (x_c, y_c):

$$f(x, y) = f(x_c, y_c) + (x - x_c)\frac{\partial f}{\partial x}(x_c, y_c) + (y - y_c)\frac{\partial f}{\partial y}(x_c, y_c) + \cdots$$

$$\approx \delta x \frac{\partial f}{\partial x}(x_c, y_c) + \delta y \frac{\partial f}{\partial y}(x_c, y_c),$$

where we took into account that $f(x_c, y_c) = 0$ and neglected the higher order terms in the Taylor expansion. Similarly,

$$g(x, y) \approx \delta x \frac{\partial g}{\partial x}(x_c, y_c) + \delta y \frac{\partial g}{\partial y}(x_c, y_c).$$

Inserting the last two relations back into equation (7.7) and taking into account that $\dot{\delta x} = \dot{x}$ and $\dot{\delta y} = \dot{y}$ we get, in matrix notation

$$\begin{bmatrix} \dot{\delta x} \\ \dot{\delta y} \end{bmatrix} = \begin{bmatrix} \frac{\partial f}{\partial x} & \frac{\partial f}{\partial y} \\ \frac{\partial g}{\partial x} & \frac{\partial g}{\partial y} \end{bmatrix}\Bigg|_{(x_c, y_c)} \begin{bmatrix} \delta x \\ \delta y \end{bmatrix}. \tag{7.10}$$

Accordingly, we arrived at a linearized system with a constant coefficient matrix

$$\underline{J}(x_c, y_c) = \begin{bmatrix} \frac{\partial f}{\partial x}(x_c, y_c) & \frac{\partial f}{\partial y}(x_c, y_c) \\ \frac{\partial g}{\partial x}(x_c, y_c) & \frac{\partial g}{\partial y}(x_c, y_c) \end{bmatrix}.$$

The matrix $\underline{J}(x_c, y_c)$ is called the Jacobian or linearization matrix. Its matrix elements are partial derivatives $\frac{\partial f}{\partial x}, \frac{\partial f}{\partial y}, \frac{\partial g}{\partial x}, \frac{\partial g}{\partial y}$ taken at (x_c, y_c). This matrix governs the behavior of the linear system around the critical point (x_c, y_c). The powerful Hartman-Grobman theorem (which we will not prove) states that the dynamics of any nonlinear system is similar to the dynamics of a linear system locally, in the neighborhood of the critical point, which has negative or positive real parts. What this means is that if $\underline{J}(x_c, y_c)$ is the Jacobian matrix of the linearized system at the critical point then

- if both eigenvalues of $\underline{\underline{J}}(x_c, y_c)$ have a negative real part, then (x_c, y_c) is a stable equilibrium of the nonlinear system.

- if $\underline{\underline{J}}(x_c, y_c)$ has at least one eigenvalue with a positive real part, then (x_c, y_c) is not a stable equilibrium of the nonlinear system.

Thus, all the eigenvalues of the Jacobian matrix must have negative real parts in order to ensure asymptotic stability.

☞ **Example 7.2.2.** Consider the nonlinear system of equations

$$\dot{x} = -x + x^3 \,,$$
$$\dot{y} = -y \,. \tag{7.11}$$

In order for $\dot{x}, \dot{y}$ to be zero it must hold that $y = 0$ and $x(1 - x^2) = 0$. Thus, there are three critical points $(0,0), (\pm 1, 0)$ and they all lie on the x-axis. To investigate their stability we examine the Jacobian of the system

$$\underline{\underline{J}}(x, y) = \begin{bmatrix} -1 + 3x^2 & 0 \\ 0 & -1 \end{bmatrix} \,.$$

For the equilibrium at the origin we find

$$\underline{\underline{J}}(0,0) = \begin{bmatrix} -1 & 0 \\ 0 & -1 \end{bmatrix} \,,$$

which has a degenerate eigenvalue -1. The origin is stable and the trajectories of the linearized system are straight lines approaching $(0,0)$ along the x and y-axes.

For the remaining two critical points, $(\pm 1, 0)$, the Jacobian becomes

$$\underline{\underline{J}}(\pm 1, 0) = \begin{bmatrix} 2 & 0 \\ 0 & -1 \end{bmatrix} \,.$$

With one positive $\lambda = 2$ and one negative eigenvalue $\lambda = -1$, the two critical points are saddles.

The original system is easy to solve since both directions are separated. Figure 7.6 shows the trajectories of the original nonlinear system and confirms the results obtained by the linearization method.

Up to now we have dealt with the so-called *hyperbolic* critical points.

Definition 7.2.3. *The critical point (x_c, y_c) is said to be hyperbolic if its Jacobian matrix $\underline{\underline{J}}(x_c, y_c)$ has no eigenvalues with zero real part ($\operatorname{Re} \lambda_i \neq 0$ for all eigenvalues of $\underline{\underline{J}}(x_c, y_c)$).*

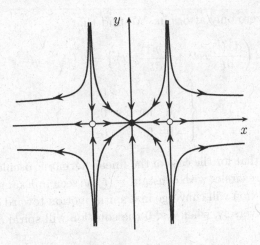

Figure 7.6: Phase portrait of the system (7.11).

The case of a non-hyperbolic equilibrium point has to be dealt with separately. In such a case the behavior of the nonlinear systems may be dependent on the higher-order terms in the Taylor series expansion that we neglected when linearizing the system. Examples of non-hyperbolic equilibria are centers with purely imaginary eigenvalues. Thus, for centers, the linearization approach fails to predict stability. The next example will illustrate this point.

☞ **Example 7.2.3.** Consider the system of autonomous equations

$$\dot{x} = y,$$
$$\dot{y} = -x - a(x^2 + y^2)y. \tag{7.12}$$

Note that for $a = 0$ the above equations reduce to the linear harmonic oscillator in equation (7.9).

The critical point for (7.12) is obviously at origin. Following the linearization approach we expand around the origin to obtain the Jacobian matrix

$$\underline{\underline{J}}(0,0) = \begin{bmatrix} 0 & 1 \\ -1 & 0 \end{bmatrix}.$$

The eigenvalues $\lambda_1 = i$, $\lambda_2 = -i$ are purely imaginary. Thus, $(0,0)$ is a non-hyperbolic equilibrium and, therefore, the linearization test is inconclusive.

As in Example 7.2.1 we consider the positive function

$$P(t) = r(t)^2 = x(t)^2 + y(t)^2,$$

which measures the distance of the point on a trajectory to the origin and therefore is zero only at origin. We find that

$$\frac{dP}{dt} = \frac{dr^2}{dt} = 2\left(\frac{dx(t)}{dt}x(t) + \frac{dy(t)}{dt}y(t)\right) = 2\left(xy + y(-x - a(x^2 + y^2)y)\right)$$

$$= -2a(x^2 + y^2)y^2 = \begin{cases} < 0 & \text{for } a > 0. \\ > 0 & \text{for } a < 0. \\ = 0 & \text{for } a = 0. \end{cases}$$

We conclude that for the case of the linear harmonic oscillator with $a = 0$ the trajectories are circles with constant $r^2(t)$ as seen in Example 7.2.1. Next, for $a > 0$ the solution will converge in a spiral motion toward equilibrium, which is stable. Conversely, when $a < 0$ the solution will spiral away to infinity.

We have seen in Example 7.2.3 that the linearization approach did not predict the stability for the center $(0,0)$ and its dependence on a parameter a. The three different cases we have obtained were examined by means of one auxiliary positive function P. The behavior of P determined the stability of the non-hyperbolic equilibrium point. This gives rise to the idea of the Liapunov function, which can be applied independently of the type of stability of the underlying equilibria. The method based on the Liapunov function provides a means for examining the stability of a non-hyperbolic equilibria in two and higher dimensions. Consider, again, the system (7.7) and suppose that (x_0, y_0) is its non-hyperbolic critical point. Let $P(x, y)$ be a function of two-variables, which satisfies two conditions

$$P(x_0, y_0) = 0$$

and

$$P(x, y) > 0, \text{ if } (x, y) \neq (x_0, y_0) .$$

Then

① If on all trajectories of the phase space

$$\frac{dP(x(t), y(t))}{dt} < 0, \quad \longrightarrow \quad (x_0, y_0) \text{ is asymptotically stable.}$$

② If on all trajectories of the phase space

$$\frac{dP(x(t), y(t))}{dt} \leq 0, \quad \longrightarrow \quad (x_0, y_0) \text{ is stable.}$$

③ If on all trajectories of the phase space

$$\frac{dP(x(t), y(t))}{dt} > 0, \quad \longrightarrow \quad (x_0, y_0) \text{ is unstable.}$$

There is no general prescription on how to find the Liapunov function P. Its key property is revealed by the following calculation of the directional derivative of P:

$$\frac{dP(x(t), y(t))}{dt} = \frac{\partial P(x, y)}{\partial x} \dot{x} + \frac{\partial P(x, y)}{\partial y} \dot{y} = \frac{\partial P(x, y)}{\partial x} f(x, y) + \frac{\partial P(x, y)}{\partial y} g(x, y)$$

$$= \vec{\nabla} P(x, y) \cdot (f(x, y), g(x, y)) .$$

If $dP(x(t), y(t))/dt$ vanishes for all solutions of the system then it implies that all contour curves of the Liapunov function P are trajectories of the phase space. If this derivative is negative for all solutions of the system, this means that the trajectories are not contour curves of P. As a point moving along a trajectory intersects a contour curve, it moves from the exterior of the contour curve into the interior. Thus, (x_0, y_0) is a stable equilibrium.

If $dP(x(t), y(t))/dt$ is positive for all solutions, the trajectories are not contour curves of P and as a point moving along a trajectory intersects a contour curve, it moves from the interior of the contour curve out to the exterior.

☞ **Example 7.2.4.** Consider

$$\dot{x} = -2y - x^3 , \tag{7.13}$$
$$\dot{y} = x - y^3 .$$

The Jacobian matrix at the critical point $(0, 0)$ equals:

$$\underline{\underline{J}}(0, 0) = \begin{bmatrix} 0 & -2 \\ 1 & 0 \end{bmatrix} .$$

The eigenvalues obey $\lambda_1 + \lambda_2 = 0$ and $\lambda_1 \lambda_2 = 2$. The solution is purely imaginary: $\lambda_{1/2} = \pm i \sqrt{2}$. The Liapunov function $P = x^2 + 2y^2$ satisfies:

$$\dot{P} = \frac{\partial P}{\partial x} \dot{x} + \frac{\partial P}{\partial y} \dot{y} = 2x(-2y - x^3) + 4y(x - y^3) = -2x^4 - 4y^4 < 0$$

for $(x, y) \neq (0, 0)$. The solution is spiraling toward the stable critical point at the origin as shown in Figure 7.7.

Quite frequently the conservation of energy in the underlying physical system suggests a particular choice of a Liapunov function. In such cases, the conserved total energy E is a convenient choice for the Liapunov function $P = E$ and results in the Liapunov function which obeys $\dot{P} = 0$. In this case, the non-hyperbolic equilibrium is stable but not asymptotically stable and is the center for the closed

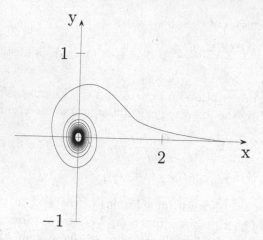

Figure 7.7: A solution curve of the model (7.13).

trajectories which lie on the contour curves of the P function. We encountered such a case in the linear oscillator of Example 7.2.1

7.2.1 Nonlinear Pendulum

Consider a pendulum consisting of a light rod of length l attached at one end to a mass m and at the other to a frictionless pivot, so that the pendulum can swing freely in a vertical plane as shown in Figure 7.8. We use polar coordinates φ, ρ in a

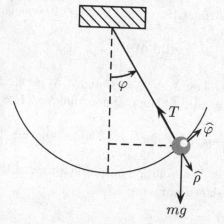

Figure 7.8: Pendulum.

vertical plane to describe the pendulum's motion. The tangential velocity of a ball at a distance $\rho = l$ from the pivot is $\vec{v}_\varphi = l\dot{\varphi}\hat{\varphi}$, and the tangential acceleration is

$\vec{a}_l = \ddot{\varphi}\hat{\varphi}$. The component of the gravitational force $\vec{F} = -mg\hat{k}$ perpendicular to rod is $F_\varphi = -mg\sin\varphi$. Thus, the equation of motion is

$$ml\ddot{\varphi} = -gm\sin\varphi \tag{7.14}$$

or:

$$\boxed{\begin{aligned} &\ddot{\varphi} = -\frac{g}{l}\sin\varphi = -\omega^2\sin\varphi, \\ &\omega = \sqrt{\frac{g}{l}}. \end{aligned}} \tag{7.15}$$

This equation is a nonlinear ordinary differential equation. It is nonlinear because of the term $\sin\varphi$. If the oscillations are so small that we can use an approximation $\sin\varphi = \varphi$, the pendulum satisfies the linear equation

$$\ddot{\varphi} = -\omega^2\varphi. \tag{7.16}$$

This system is called a linear pendulum and its solution is:

$$\varphi = c_1\cos(\omega t) + c_2\sin(\omega t). \tag{7.17}$$

We will present an alternative approach to the linear pendulum problem (7.16). It is based on the observation that a non-autonomous system in the form of the second order differential equation (7.16) can be recast as an autonomous system (7.7) for the price of adding one extra dimension. Namely, define the variables

$$y = \dot{\varphi}, \quad x = \varphi. \tag{7.18}$$

We can rewrite the second order differential equation (7.16) as a system of two 1-st order differential equations, in terms of coordinates x and y as

$$\dot{x} = y, \quad \dot{y} = -\omega^2 x \tag{7.19}$$

or

$$\underline{\dot{x}} = \begin{bmatrix} 0 & 1 \\ -\omega^2 & 0 \end{bmatrix}\underline{x}, \quad \underline{x} = \begin{bmatrix} x \\ y \end{bmatrix}. \tag{7.20}$$

This model coincides with Example 7.2.1 for $\omega^2 = 1$. The phase plane for the pendulum will now lie on the xy-plane. The coefficient matrix in this equation is traceless and the eigenvalues $\lambda = \pm i\omega$ are imaginary. The corresponding eigenvectors are

$$\underline{v}_1 = \begin{bmatrix} 1 \\ i\omega \end{bmatrix}, \quad \underline{v}_2 = \begin{bmatrix} 1 \\ -i\omega \end{bmatrix}.$$

As in Example 6.3.3, which described a center with two imaginary eigenvalues, we find here that the general solution is given by:

$$\underline{x} = c_1 e^{i\omega t} \begin{bmatrix} 1 \\ i\omega \end{bmatrix} + c_2 e^{-i\omega t} \begin{bmatrix} 1 \\ -i\,w \end{bmatrix} = (c_1 + c_2) \begin{bmatrix} \cos \omega t \\ \omega \sin \omega t \end{bmatrix}$$
$$+ i\,(c_1 - c_2) \begin{bmatrix} \sin \omega t \\ \omega \cos \omega t \end{bmatrix}.$$

The solutions are periodic functions, which means that trajectories repeat periodically and must, therefore, close. This situation is known from Chapter 6 as a center. In that particular case, the allowed phase curves in the phase plane are ellipses and the critical point $x = 0, y = 0$ is stable, but not asymptotically stable.

It is useful to introduce another method to obtain the form of allowed trajectories, which generalizes to the case of the non-linear pendulum. Using the chain rule, we are able to eliminate the time variable from this system:

$$\frac{dy}{dx} = \frac{\dot{y}}{\dot{x}} = -\omega^2 \frac{x}{y}.$$

It is obvious that the solutions of $y\,dy + \omega^2 x\,dx = 0$ are closed curves which satisfy

$$y^2 + \omega^2 x^2 = C \tag{7.21}$$

for any constant C. For each integration constant C corresponds one particular ellipse. The trajectories are traced out in periodic motion. The direction of flows for increasing time t can be obtained from equation (7.19). For $x, y > 0$ we have $\dot{x} = y > 0$ and $\dot{y} = -\omega^2 x < 0$ and, thus, the ellipses are traced in the clockwise direction as shown in Figure 7.9.

For the nonlinear pendulum from equation (7.15) we repeat the above steps applied to the linear pendulum finding the autonomous system of equations:

$$\dot{x} = y, \quad \dot{y} = -\omega^2 \sin x \tag{7.22}$$

given in terms of x and y coordinates as defined in (7.18).

The chain rule gives:

$$\frac{dy}{dx} = \frac{\dot{y}}{\dot{x}} = -\omega^2 \frac{\sin x}{y}.$$

This a separable equation which can be integrated to yield:

$$\frac{y^2}{2} - \omega^2 \cos x = C. \tag{7.23}$$

This equation defines the allowed phase curves for any integration constant C.

Critical points (x_c, y_c) must be such that

$$\sin x_c = 0, \quad y_c = 0. \tag{7.24}$$

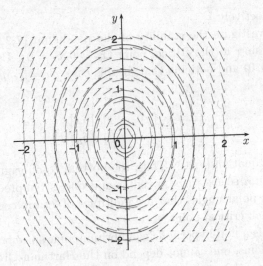

Figure 7.9: Elliptic orbits and vector fields of the linear oscillator model $\dot{x} = y, \dot{y} = -\omega^2 x$ with $\omega^2 = 1.8$.

Thus, $x_c = n\pi$ (for an integer n). Clearly, critical points x_c that are even multiples of π are equivalent to the zero angle $x_c = \varphi = 0$. This is the expected down-hanging equilibrium position of the pendulum. The critical points x_c that are odd multiples of π are equivalent to the upside-down equilibrium position of the pendulum. Hence, the pendulum only has two distinctive stationary solutions, namely $x_c = \varphi = 0$ and $x_c = \varphi = 180°$.

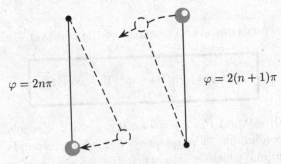

$$\varphi = 2n\pi \qquad\qquad \varphi = 2(n+1)\pi$$

Figure 7.10: Pendulum's equilibria positions.

One expects these equilibria to be different from the physical point of view. For $x_c = \varphi = 0$, the mass is hanging down and the position represents a stable minimum. For $x_c = \varphi = 180°$ the pendulum is inverted and the slightest disturbance causes the pendulum to move away from this point as shown in Figure 7.10. Mathematical examination of the stability of these critical points will confirm that they are stable and unstable, respectively.

Let us first investigate the stability of the pendulum (7.22) with $f(x,y) = y$ and $g(x,y) = -\omega^2 \sin x$ at the critical point $(x_c = 0, y_c = 0)$. Using the expansion, $\sin x \approx x$, about $(0,0)$ and noticing that, in this case, $\delta x = x, \delta y = y$ we arrive at a special case of (7.10):

$$\begin{bmatrix} \dot{\delta x} \\ \dot{\delta y} \end{bmatrix} = \begin{bmatrix} 0 & 1 \\ -\omega^2 & 0 \end{bmatrix} \begin{bmatrix} \delta x \\ \delta y \end{bmatrix},$$

which agrees with equation (7.20). Thus, the bottom equilibrium position $(0,0)$ of a nonlinear pendulum seems to be a center and one is tempted to conclude that all the equilibria at polar angles that are multiples of 2π are centers with periodic solutions around each critical point.

However, $(0,0)$ is a non-hyperbolic equilibrium point since both eigenvalues are purely imaginary. Thus, one cannot depend on the Hartman-Grobman theorem to predict the behavior of the nonlinear pendulum around this equilibrium and we have to find a different method to confirm our physical intuition about these equilibria. The remedy is to use the law of conservation of energy for the non-linear pendulum. Equations (7.22) of the model imply that

$$\omega^2 \dot{x} \sin x = \omega^2 y \sin x,$$
$$y\dot{y} = -\omega^2 y \sin x.$$

Hence,

$$0 = y\dot{y}(t) + \omega^2 \dot{x} \sin x(t) = \frac{d}{dt}\left[\frac{1}{2}y^2(t) + \omega^2\left(C - \cos x(t)\right)\right],$$

where C is an arbitrary constant. The choice of the constant $C = 1$ in

$$\boxed{P(x,y) = \frac{1}{2}y^2(t) + \omega^2\left(1 - \cos x(t)\right)} \tag{7.25}$$

ensures that $P(0,0) = 0$ and $P(x,y) > 0$ for $0 < x < 2\pi$. Especially, $P(\pi,0) = 2\omega^2$ for the inverted pendulum. The quantity $P(x,y)$ corresponds to the total energy for the pendulum which is zero when the system is at its minimum position. In agreement with the law of conservation of energy in physics it holds that $dP(x(t),y(t))/dt = 0$. Thus, the energy is constant along any trajectory in phase space and trajectories are contour curves of P, as follows explicitly from comparing trajectories given by equation (7.23) with the expression for P. Therefore, according to Liapunov method, the origin $(0,0)$ is stable (but not asymptotically stable). Furthermore, the trajectories lie on ellipses (the contour curves of P) with their centers at the origin. Thus, the behavior near the lowest position of the linearized pendulum is oscillatory as expected.

The trajectories given by equation (7.23) are drawn explicitly in Figure 7.11 showing the phase space of the nonlinear pendulum. The Figure confirms the above conclusion about equilibria at $x = 0, 2n\pi$ being centers.

For another equilibrium at $(x, y) = (\pi, 0)$, we choose $\delta x = x - \pi$ and $\delta y = y$ and make the following expansion:

$$\sin x = \sin(\delta x + \pi) = -\sin \delta x \approx -\delta x.$$

The linearized system (7.10) becomes

$$\begin{bmatrix} \dot{\delta x} \\ \dot{\delta y} \end{bmatrix} = \begin{bmatrix} 0 & 1 \\ \omega^2 & 0 \end{bmatrix} \begin{bmatrix} \delta x \\ \delta y \end{bmatrix}.$$

The difference from the case of the equilibrium at the bottom is the sign in front of ω^2, which is positive. Accordingly, the determinant of the Jacobian matrix is equal to $-\omega^2 < 0$. The Jacobian matrix is traceless and, therefore, the matrix has two purely real eigenvalues $\lambda_1 = \omega, \lambda_2 = -\omega$ of opposite sign. The corresponding eigenvectors are

$$\lambda_1 = \omega \; : \; \begin{bmatrix} -\omega & 1 \\ \omega^2 & -\omega \end{bmatrix} \underline{v}_1 = 0 \longrightarrow \underline{v}_1 = \begin{bmatrix} 1 \\ \omega \end{bmatrix},$$

$$\lambda_2 = -\omega \; : \; \begin{bmatrix} \omega & 1 \\ \omega^2 & \omega \end{bmatrix} \underline{v}_2 = 0 \longrightarrow \underline{v}_2 = \begin{bmatrix} 1 \\ -\omega \end{bmatrix}.$$

Thus, the critical point $(\pi, 0)$ together with all equivalent equilibria $((2n + 1)\pi, 0)$ are saddles. The trajectory moves toward the saddle point along a line parallel to the $\underline{v}_2$-vector and moves away from the saddle point along a direction parallel to $\underline{v}_1$. The fact these two eigenvalues were real ensures that the equilibrium points are

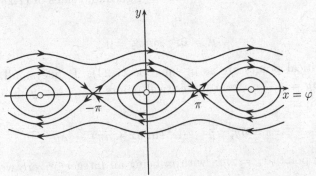

Figure 7.11: Pendulum's phase space.

hyperbolic, and our conclusion about the equilibria being saddles is relevant for the corresponding nonlinear pendulum system.

Note the three types of trajectories in Figure 7.11. Closed curves around the origin and other centers located at $2n\pi$ correspond to the pendulum that oscillates periodically between maximum angles φ_{max} and $-\varphi_{max}$ with $\varphi_{max} < \pi$. These trajectories correspond to $P < 2\omega^2$, the energy less than the energy of the pendulum at the upright position at $\varphi = \pi$. When $P > 2\omega^2$, the pendulum possesses energy greater than the critical energy corresponding to the pendulum pointing vertically upward at $\varphi = \pi$. It rotates around the pivot indefinitely in a circulating motion. Finally, there is a borderline case, $P = 2\omega^2$. In this case, the trajectory passes through the unstable equilibrium points. Here, the pendulum starts upright with $\varphi = \pi$ having just enough energy to swing back up to the upright $\varphi = \pi$ position. The curve corresponding to the borderline case along which $P = 2\omega^2$ is known as the separatrix, as it divides the phase space into two types of motion.

Next, we consider a pendulum with frictional damping $-c\dot{\varphi}$ force acting on it. Since the pendulum is now able to lose energy (through friction) we expect that it will become more stable.

The undamped pendulum equation (7.14) is modified by an extra term, $-c\dot{\varphi}$,

$$ml\ddot{\varphi} = -gm\sin\varphi - c\dot{\varphi} \tag{7.26}$$

or equivalently,

$$\ddot{\varphi} + \omega^2\sin\varphi + \gamma\dot{\varphi} = 0, \quad \gamma = \frac{c}{ml}. \tag{7.27}$$

In terms of variables introduced in equation (7.18) we can recast the damping pendulum equation of motion as a two-dimensional non-linear system

$$\dot{x} = y, \quad \dot{y} = -\omega^2\sin x - \gamma y. \tag{7.28}$$

Equilibrium points are found by setting the right hand sides of (7.28) to zero. This yields

$$y_c = 0, \quad \sin x_c = 0,$$

so that the critical points are as in equation (7.24). Define small displacements $\delta x = x - x_c, \delta y = y - y_c$ from the critical points $(x_c = n\pi, y_c = 0)$ and rewrite equations (7.28) as

$$\dot{x} = \delta y, \quad \dot{y} = -\omega^2\sin(\delta x + n\pi) - \gamma\delta y. \tag{7.29}$$

For the critical points $x_c = 2m\pi$ with m being an integer or zero we will be using the expansion

$$\sin(\delta x + 2m\pi) = \sin\delta x \approx \delta x.$$

Thus, in this case, the linearized damped pendulum is described by the system

$$\dot{\delta x} = \delta y, \quad \dot{\delta y} = -\omega^2\delta x - \gamma\delta y$$

or

$$\begin{bmatrix} \dot{\delta x} \\ \dot{\delta y} \end{bmatrix} = \begin{bmatrix} 0 & 1 \\ -\omega^2 & -\gamma \end{bmatrix} \begin{bmatrix} \delta x \\ \delta y \end{bmatrix}.$$

The eigenvalues λ_1, λ_2 of the Jacobian

$$\begin{bmatrix} 0 & 1 \\ -\omega^2 & -\gamma \end{bmatrix}$$

must satisfy relations

$$\lambda_1 + \lambda_2 = -\gamma, \quad \lambda_1 \lambda_2 = \omega^2,$$

where the right hand sides are equal to the trace and the determinant of the Jacobian. It is easy to check that quantities

$$\lambda_1 = -\frac{\gamma}{2} + i\sqrt{\omega^2 - \frac{\gamma^2}{4}}, \quad \lambda_2 = -\frac{\gamma}{2} - i\sqrt{\omega^2 - \frac{\gamma^2}{4}}$$

solve the above equations. Stability depends on the sign of the discriminant $D = \omega^2 - \gamma^2/4$, which determines the character of damping. Like in subsection 3.2.8, where we considered damped oscillations described by the second-order differential equation, we distinguish between three situations.

- $D < 0$, thus λ_1, λ_2 are real, distinct and both negative. Accordingly, the linearized system has a stable (improper) node at $(x_c = 2m\pi, y_c = 0)$. This conclusion applies to the non-linear system as well. The pendulum is referred to as over-damped.

- $D = 0$, therefore $\lambda_1 = \lambda_2 < 0$, thus the linearized system has a stable node at $(x_c = 2m\pi, y_c = 0)$. The pendulum is critically damped.

- $D > 0$, the eigenvalues λ_1, λ_2 are complex with real negative parts, hence the linearized system has a stable spiral at $(x_c = 2m\pi, y_c = 0)$. The pendulum is under-damped.

Since the real part of both eigenvalues is negative in all three cases all the equilibria are asymptotically stable.

Now, we turn to a discussion on critical points $(x_c = (2m + 1)\pi, y_c = 0)$ with m being an integer. These equilibria are odd multiples of π and correspond to the upright position of the pendulum. For $\delta x = x - (2m + 1)\pi$, we expand the sine function in the vicinity of the critical point as follows

$$\sin x = \sin(\delta x + (2m + 1)\pi) = \sin(\delta x + \pi) = -\sin \delta x \approx -\delta x.$$

This gives rise to the linear system

$$\begin{bmatrix} \dot{\delta x} \\ \dot{\delta y} \end{bmatrix} = \begin{bmatrix} 0 & 1 \\ \omega^2 & -\gamma \end{bmatrix} \begin{bmatrix} \delta x \\ \delta y \end{bmatrix}.$$

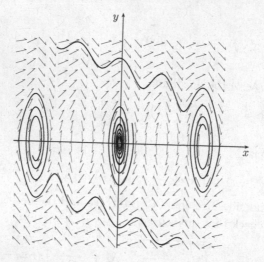

Figure 7.12: Phase portrait of the nonlinear pendulum in the under-damped case with three stable spirals at $\pm 2\pi, 0$.

Note. this time the plus sign is in front of ω^2. Accordingly, the eigenvalues λ_1, λ_2 must now satisfy

$$\lambda_1 + \lambda_2 = -\gamma, \quad \lambda_1 \lambda_2 = -\omega^2.$$

The solution

$$\lambda_{1/2} = -\frac{\gamma}{2} \pm \sqrt{\omega^2 + \frac{\gamma^2}{4}}$$

consists of two real eigenvalues of opposite signs. This is the signature of a saddle point. Thus, the critical points $(\pm\pi, 0), (\pm 3\pi, 0), \ldots$ of odd multiples of π are again saddle points independently of the strength of damping.

7.2.2   Two-Dimensional Population Models

Recall our one dimensional model of a fish hatchery. In two dimensions we are able make a step toward building more realistic models by including interactions between two populations. Consider a simple ecological model of two competing species.

We imagine an environment in which a certain population denoted by x lives on some type of food (like plankton in case of fish). It is assumed that food is present in unlimited quantities. Also present is a population of predators, denoted by y, for which the only means of survival is eating the first species. The model assumes that, if left alone, the growth of x population would increase while that of y population would decrease. This is described by equations

$$\dot{x} = ax, \quad \dot{y} = -by, \quad a >, b > 0.$$

The solutions are $x(t) = x_0 \exp(at)$ and $y(t) = y_0 \exp(-bt)$. Thus, the prey population grows without limit by making use of unlimited resources and the predator population starves in the absence of available prey.

In the presence of interaction the prey decrease in number, while the predators increase. The effect of predation on both populations is proportional to the number of encounters of predators with prey which is proportional to the product xy. This results in a Lotka-Volterra predator-prey model, governed by:

$$\dot{x} = ax - cxy, \qquad\qquad (7.30)$$
$$\dot{y} = -by + cxy, \quad c > 0.$$

Here c is positive to reflect the following two situations. As predator population y increases, the net growth rate of the prey population x decreases. Vice versa, if a prey population x grows, the prey species have more food and this tends to promote an increase of the predator population y. The first step in the analysis of the system (7.30) consists of finding the critical points (x_c, y_c) for which the right hand sides of equations (7.30) vanish:

$$(a - cy_c)x_c = 0 \quad \text{and} \quad (-b + cx_c)y_c = 0.$$

We find two solutions

$$(x_c = 0, y_c = 0),$$
$$(x_c = \frac{b}{c}, y_c = \frac{a}{c}).$$

Our objective is to determine the stability of these critical points. To accomplish this task we apply the linearization approach:

$$\frac{d}{dt}\begin{bmatrix} x \\ y \end{bmatrix} = \begin{bmatrix} f(x,y) \\ g(x,y) \end{bmatrix} \quad\longrightarrow\quad \frac{d}{dt}\begin{bmatrix} \delta x \\ \delta y \end{bmatrix} = \begin{bmatrix} \frac{\partial f}{\partial x} & \frac{\partial f}{\partial y} \\ \frac{\partial g}{\partial x} & \frac{\partial g}{\partial y} \end{bmatrix}\Bigg|_{(x_c, y_c)} \begin{bmatrix} \delta x \\ \delta y \end{bmatrix}.$$

First, we find the partial derivatives appearing in the Jacobian matrix:

$$f(x,y) = ax - cxy \quad\longrightarrow\quad \frac{\partial f}{\partial x} = a - cy, \quad \frac{\partial f}{\partial y} = -cx,$$

$$g(x,y) = -by + cxy \quad\longrightarrow\quad \frac{\partial g}{\partial x} = cy, \quad \frac{\partial g}{\partial y} = -b + cx.$$

Let us start with $(x_c = 0, y_c = 0)$ for which the effective linearized problem becomes

$$\frac{d}{dt}\begin{bmatrix} \delta x \\ \delta y \end{bmatrix} = \begin{bmatrix} \frac{\partial f}{\partial x} & \frac{\partial f}{\partial y} \\ \frac{\partial g}{\partial x} & \frac{\partial g}{\partial y} \end{bmatrix}\Bigg|_{(0,0)} \begin{bmatrix} \delta x \\ \delta y \end{bmatrix} = \begin{bmatrix} a & 0 \\ 0 & -b \end{bmatrix}\begin{bmatrix} \delta x \\ \delta y \end{bmatrix}.$$

The eigenvalues $a, -b$ of the Jacobian matrix are real and of opposite sign. The equilibrium is a saddle. The eigenvector corresponding to $\lambda = a$ is

$$\begin{bmatrix} 0 & 0 \\ 0 & -b-a \end{bmatrix} \underline{v} = 0 \quad \longrightarrow \quad \underline{v} = \begin{bmatrix} 1 \\ 0 \end{bmatrix}.$$

Similarly, we find the eigenvector corresponding to $\lambda = -b$ through

$$\begin{bmatrix} a-b & 0 \\ 0 & 0 \end{bmatrix} \underline{v} = 0 \quad \longrightarrow \quad \underline{v} = \begin{bmatrix} 0 \\ 1 \end{bmatrix}.$$

Now, we consider the critical point at $(x_c = \frac{b}{c}, y_c = \frac{a}{c})$.

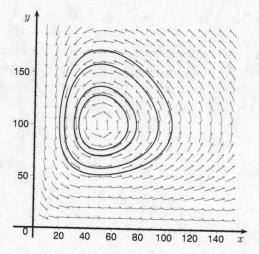

Figure 7.13: Phase portrait of the predator-prey model (7.30) with $a = -0.1, b = -0.2, c = -0.002$. Note the stable center at $(100, 50)$ and saddle at the origin.

The Jacobian matrix is

$$\begin{bmatrix} \frac{\partial f}{\partial x} & \frac{\partial f}{\partial y} \\ \frac{\partial g}{\partial x} & \frac{\partial g}{\partial y} \end{bmatrix} \Bigg|_{(\frac{b}{c}, \frac{a}{c})} = \begin{bmatrix} 0 & -b \\ a & 0 \end{bmatrix}.$$

The eigenvalues satisfy the following relations:

$$\lambda_1 + \lambda_2 = 0, \quad \lambda_1 \lambda_2 = ab.$$

The solution given by two imaginary eigenvalues $i\sqrt{ab}, -i\sqrt{ab}$ yields a center at the equilibrium point $(b/c, a/c)$. The corresponding eigenvectors are obtained from the eigenvalue equations

$$\lambda = i\sqrt{ab} : \begin{bmatrix} -i\sqrt{ab} & -b \\ a & -i\sqrt{ab} \end{bmatrix} \underline{v} = 0; \quad \longrightarrow \quad \underline{v} = \begin{bmatrix} \sqrt{ab} \\ -ia \end{bmatrix},$$

$$\lambda = -i\sqrt{ab} : \begin{bmatrix} i\sqrt{ab} & -b \\ a & i\sqrt{ab} \end{bmatrix} \underline{v} = 0; \longrightarrow \underline{v} = \begin{bmatrix} \sqrt{ab} \\ ia \end{bmatrix}.$$

The complete solution is

$$\begin{bmatrix} x(t) \\ y(t) \end{bmatrix} = \begin{bmatrix} b/c \\ a/c \end{bmatrix} + c_1 \begin{bmatrix} \sqrt{ab} \\ -ia \end{bmatrix} e^{i\sqrt{ab}} + c_2 \begin{bmatrix} \sqrt{ab} \\ ia \end{bmatrix} e^{-i\sqrt{ab}}$$

$$= \begin{bmatrix} b/c \\ a/c \end{bmatrix} + k_1 \begin{bmatrix} \sqrt{ab}\cos\sqrt{ab}\,t \\ a\sin\sqrt{ab}\,t \end{bmatrix} + k_2 \begin{bmatrix} \sqrt{ab}\sin\sqrt{ab}\,t \\ -a\cos\sqrt{ab}\,t \end{bmatrix},$$

where we have replaced constants $(c_1 + c_2), i(c_1 - c_2)$ by k_1, k_2. Introducing additional constants C and α such that

$$C\cos\alpha = k_1 c\sqrt{\frac{a}{b}}, \quad C\sin\alpha = -k_2 c\sqrt{\frac{a}{b}},$$

helps to simplify expressions for the population of prey $x(t)$ and predator $y(t)$ to:

$$x(t) = \frac{b}{c}\left(1 + C\cos\left(\sqrt{ab}\,t + \alpha\right)\right),$$

$$y(t) = \frac{a}{c}\left(1 + C\sqrt{\frac{b}{a}}\sin\left(\sqrt{ab}\,t + \alpha\right)\right).$$

Thus, the predator population is lagging behind the prey by 90°. This describes the

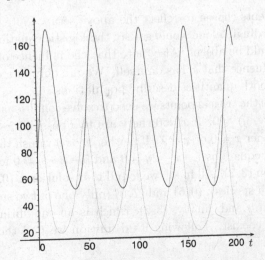

Figure 7.14: Plots of solutions $x(t)$ (black) and $y(t)$ (gray) of the predator-prey model (7.30) with $a = -0.1, b = -0.2, c = -0.002$. Note, that the solutions are 90° out of phase.

out of phase oscillations of prey/predator populations: If prey increase, predators will also increase until prey decrease. However, the predator population will still be growing for some time after the prey population has begun to decline and this is responsible for the out of phase fluctuation. Eventually, the predators will starve and the prey population will start to increase and so on.

☞ **Example 7.2.5.** In this example, we will study a Lotka-Volterra model for two competing species. It describes two species competing for a shared limited resource (think about rabbits x and sheep y competing for grass). Each population interferes with the other's access to resources and the fate of one population will affect the growth of the other one.

We assume that the interaction consists of rabbits-sheep encounters with the rate of encounters being proportional to the product of both populations, xy. These encounters reduce the growth rate for each of the two species but the effects those encounters have on the two species is not the same and the rate constants differ between these two species. The rabbits are more likely to loose confrontations accompanying competition for grass. On the other hand rabbits have a higher individual growth rate. The competition model is then governed by a set of equations:

$$\dot{x} = x\left(7 - x - 2y\right),$$
$$\dot{y} = y\left(5 - x - y\right), \tag{7.31}$$

with coefficients chosen to reflect the above setup. Note, that in absence of y, the x coordinate alone would satisfy the logistic equation. The same holds for y if x would be absent. Also, note that the influence of y on x is stronger than the influence that x has on itself. We are only interested in $x \geq 0$ and $y \geq 0$ since both quantities describe populations.

To locate the critical points we need to solve conditions $x\left(7 - x - 2y\right) = 0$ and $y\left(5 - x - y\right) = 0$. If $x = 0$ there are two possibilities $y = 0$ and $y = 5$. For $y = 0$ either $x = 0$ or $x = 7$. If neither x or y vanish the problem reduces to two linear equations $7 - x - 2y = 0$ and $5 - x - y = 0$ for which we have a single solution $(3, 2)$. All in all, we found four solutions: $(0, 0)$ (the extinction point for both species), $(0, 5)$ and $(7, 0)$ (only one of two species exists at any of these points) and, finally, $(3, 2)$, which is an equilibrium point at which both species coexist. Following its definition we find the Jacobian matrix $\underline{\underline{J}}(x, y)$ to be:

$$\underline{\underline{J}}(x, y) = \begin{bmatrix} 7 - 2x - 2y & -2x \\ -y & 5 - x - 2y \end{bmatrix}.$$

We are now able to determine the stability of all the critical points of the linearized Lotka competing model by examining properties of Jacobian matrix for each of the critical points:

$$\underline{\underline{J}}(0,0) = \begin{bmatrix} 7 & 0 \\ 0 & 5 \end{bmatrix} \longrightarrow \left(\lambda_1 = 7, \underline{v} = \begin{bmatrix} 1 \\ 0 \end{bmatrix} \right), \left(\lambda_2 = 5, \underline{v} = \begin{bmatrix} 0 \\ 1 \end{bmatrix} \right).$$

In this case, both eigenvalues are positive. Thus, the extinction point $(0,0)$ is an unstable (repelling) node.

The Jacobian matrix for the solution $(0,5)$ is given by

$$\underline{\underline{J}}(0,5) = \begin{bmatrix} -3 & 0 \\ -5 & -5 \end{bmatrix} \longrightarrow \left(\lambda_1 = -3, \underline{v} = \begin{bmatrix} 2 \\ -5 \end{bmatrix} \right), \left(\lambda_2 = -5, \underline{v} = \begin{bmatrix} 0 \\ 1 \end{bmatrix} \right).$$

Here, both eigenvalues are negative. Thus $(0,5)$ is stable (attracting) node.

The Jacobian matrix for the solution $(7,0)$ is given by

$$\underline{\underline{J}}(7,0) = \begin{bmatrix} -7 & -14 \\ 0 & -2 \end{bmatrix} \longrightarrow \left(\lambda_1 = -2, \underline{v} = \begin{bmatrix} -14 \\ 5 \end{bmatrix} \right), \left(\lambda_2 = -7, \underline{v} = \begin{bmatrix} 1 \\ 0 \end{bmatrix} \right).$$

Again, both eigenvalues are negative. Thus, $(7,0)$ is stable (attracting) node.

Finally, for the critical point of co-existence it holds that

$$\underline{\underline{J}}(3,2) = \begin{bmatrix} -3 & -6 \\ -2 & -2 \end{bmatrix} \longrightarrow \left(\lambda_1 = 1, \underline{v} = \begin{bmatrix} 3 \\ -2 \end{bmatrix} \right), \left(\lambda_2 = -6, \underline{v} = \begin{bmatrix} 2 \\ 1 \end{bmatrix} \right).$$

This time, the eigenvalues have opposite signs. Thus, $(3,2)$ is a saddle. Trajectories in the direction of the eigenvector corresponding to $\lambda_2 = -5$, will initially approach the fixed point, but will move away from the fixed point as time increases and their behavior becomes dominated by the positive (unstable) $\lambda_1 = 1$ eigenvalue.

Since all these equilibria are hyperbolic the Hartman-Grobman implies that in the vicinity of these points the phase portrait of the nonlinear competition model looks like the phase portrait of its linearized counterparts. Combining these points reveals a picture like that in Figure 7.15. As shown in Figure 7.15 trajectories start at the origin or at infinity ($t = -\infty$) and approach the stable equilibria $(0,5), (7,0)$. In Figure 7.15 we notice that the line connecting the origin $(0,0)$ and saddle at $(3,2)$ divides trajectories into two categories: those which converge toward $(0,5)$ (point of extinction for x) and those which converge toward $(7,0)$ (point of extinction for y). This example illustrates a principle of competitive exclusion. According to this principle, two populations that compete for the same limited resource typically do not coexist.

⇔ Two-Dimensional Nonlinear Differential Equations: Problems

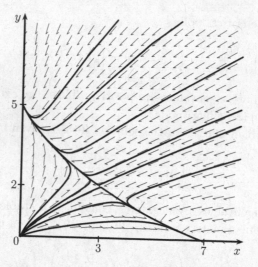

Figure 7.15: Phase portrait of the competition model (7.31).

Problem 7.2.1. (a) For the autonomous system

$$\dot{x} = -x + 2y, \quad \dot{y} = -x - y^3,$$

find the critical point and determine its stability by linearization.
(b) Find the value of the positive constant k in

$$P(x,y) = x^2 + ky^2,$$

such that $\dot{P} \leq 0$ everywhere. Does this confirm your result from (a)?

Problem 7.2.2. For the system of nonlinear differential equations

$$\dot{x} = -x + y^2, \quad \dot{y} = x - 2y + y^2,$$

find the critical points and their stability. Sketch the trajectories near each critical point and try to combine these sketches to complete the phase portrait.

Problem 7.2.3. For the system of nonlinear differential equations

$$\dot{x} = x^2 - y^2, \quad \dot{y} = x^2 + y^2 - 1,$$

find the critical points and their stability. Sketch the trajectories near each critical point.

Problem 7.2.4. Which of the critical points $(0,0)$, $(3,0)$, $(0,1)$ and $(1,2)$ of the system

$$\dot{x} = -3x + x^2 + xy, \quad \dot{y} = y - y^2 + xy,$$

is stable ?

Problem 7.2.5. Consider a nonlinear oscillator governed by equation:

$$\ddot{x} + ax^2\dot{x} + x = 0\,.$$

This example is an extension of the linear harmonic oscillator from Example 7.2.1. Transform this equation into the two-dimensional autonomous system by adding one dimension $y = \dot{x}$. Find the critical point and examine its stability. Can you use Hartman-Grobman theorem?

Problem 7.2.6. For the autonomous system

$$\dot{x} = y^2 - x, \quad \dot{y} = y\,,$$

find the critical point and determine its stability by linearization. Solve the equations explicitly and find a set of points such that they converge to equilibrium for $t \to \infty$.

Problem 7.2.7. Consider the so-called van der Pol equation:

$$\dot{x} = y, \quad \dot{y} = -x + \mu(1 - x^2)y\,,$$

where μ is a constant. Find the critical point and determine its stability by linearization for various values of μ.

Problem 7.2.8. Consider the so-called Duffing equation

$$\ddot{x} + x + ax^3 = 0\,,$$

where a is a constant.
(a) Transform this equation into a two-dimensional system of autonomous differential equations and find the critical points for $a > 0$ and $a < 0$. Apply the linearization approach whenever possible.
(b) Integrate equation (7.8) for this model and use the result to carry out your analysis based on the Liapunov function. Sketch the phase portraits for $a > 0$ and $a < 0$.

Problem 7.2.9. The interaction of two species of animals is modeled by

$$\dot{x} = x(4 - x - y), \quad \dot{y} = y(x - 2) \quad x > 0, \ y > 0\,.$$

Find the critical points and use linearization to determine their stabilities. Sketch the phase portrait.

Problem 7.2.10. Consider the system:

$$\dot{x} = x(y - 1), \quad \dot{y} = y(3 - 2x - y) \quad x > 0, \ y > 0\,.$$

Find the critical points and use linearization to determine their stabilities. Sketch the phase portrait.

Problem 7.2.11. Given is the system of equations

$$\dot{x} = \alpha x - y - 2(x^2 + y^2)x,$$
$$\dot{y} = x - \alpha y - 2(x^2 + y^2)y,$$

with positive constant α such that $\alpha \neq 1$. Clearly, $(0,0)$ is an equilibrium.
(a) For what values of α can you use linearization approach to determine the stability of $(0,0)$? Is $(0,0)$ stable for these values of α?
(b) Use $P(x,y) = x^2 - 2\alpha xy + y^2$ as a Liapunov function to determine the stability of $(0,0)$ for the remaining values of α.
(c) Determine the stability of the hyperbolic equilibrium at the origin of the system of equations:

$$\dot{x} = -y - 2(x^2 + y^2)x,$$
$$\dot{y} = x - 2(x^2 + y^2)y.$$

7.3 Three-Dimensional Models; The Lorenz equations

The Lorenz model is a three-dimensional autonomous system defined by a set of equations:

$$\boxed{\begin{aligned} \dot{x} &= \sigma\,(-x + y), \\ \dot{y} &= rx - y - xz, \\ \dot{z} &= -bz + xy. \end{aligned}} \tag{7.32}$$

The constants σ, r, b are all real and positive. The system serves as a model of weather forecasts. Hence, it provides a simple example of a chaotic system.
To find the critical points we set

$$\dot{x} = 0, \quad \dot{y} = 0, \quad \dot{z} = 0.$$

Accordingly, the right-hand sides of the differential equation (7.32) must all equal zero:

$$0 = \sigma\,(-x + y),$$
$$0 = rx - y - xz,$$
$$0 = -bz + xy.$$

These equations determine the equilibrium points. The first equation implies that $x = y$. The last two equation then become

$$0 = x\,(r - 1 - z),$$
$$bz = x^2.$$

or

$$xr - x - \frac{1}{b}x^3 = 0, \quad \longrightarrow \quad x = 0, x = \pm\sqrt{b(r-1)}.$$

Thus, one critical point is $(x_c, y_c, z_c) = (0, 0, 0)$ and the others are

$$(x_c, y_c, z_c) = \left(\sqrt{b(r-1)}, \sqrt{b(r-1)}, r-1\right),$$

$$(x_c, y_c, z_c) = \left(-\sqrt{b(r-1)}, -\sqrt{b(r-1)}, r-1\right), \tag{7.33}$$

for $r > 1$ (we search for real solutions only). To examine the stability of the system, we linearize the Lorenz equation around the critical point (x_c, y_c, z_c). In matrix notation:

$$\begin{bmatrix} \dot{\delta x} \\ \dot{\delta y} \\ \dot{\delta z} \end{bmatrix} = \begin{bmatrix} -\sigma & \sigma & 0 \\ r-z & -1 & -x \\ y & x & -b \end{bmatrix} \begin{bmatrix} \delta x \\ \delta y \\ \delta z \end{bmatrix}, \tag{7.34}$$

where $\delta x = x - x_c, \delta y = y - y_c, \delta z = z - z_c$. For the equilibrium point at the origin

$$\begin{bmatrix} \dot{x} \\ \dot{y} \\ \dot{z} \end{bmatrix} = \begin{bmatrix} -\sigma & \sigma & 0 \\ r & -1 & 0 \\ 0 & 0 & -b \end{bmatrix} \begin{bmatrix} x \\ y \\ z \end{bmatrix} = \underline{J}(0,0,0) \begin{bmatrix} x \\ y \\ z \end{bmatrix}, \tag{7.35}$$

which can also be obtained by neglecting quadratic terms in (7.32) as shown here

$$\dot{x} = \sigma(-x + y),$$
$$\dot{y} = rx - y,$$
$$\dot{z} = -bz.$$

The characteristic equation $|\underline{J}(0,0) - \lambda\underline{I}| = 0$ is equivalent to

$$(-b - \lambda)[(-\sigma - \lambda)(-1 - \lambda) - r\sigma] = 0,$$

with solutions

$$\lambda = -b,$$

$$\lambda = -\frac{1}{2}(1 + \sigma) + \frac{1}{2}\sqrt{(1 + \sigma)^2 + 4\sigma(r-1)},$$

$$\lambda = -\frac{1}{2}(1 + \sigma) - \frac{1}{2}\sqrt{(1 + \sigma)^2 + 4\sigma(r-1)}.$$

As long as $r < 1$ the last two eigenvalues are either real and negative or complex conjugates of each other. The outcome depends on the relation between the positive constants r and σ. In any case, the equilibrium $(0, 0, 0)$ is asymptotically stable for $r < 1$. For $r > 1$ the last two eigenvalues are real but acquire opposite signs signaling a saddle with one unstable direction associated with the eigenvector of positive eigenvalue. Thus, the origin becomes unstable.

When $r > 1$ the system possesses two other critical points as given in equation (7.33). To examine their stability we need to substitute their values from equation (7.33) into the Jacobian:

$$\underline{\underline{J}}(x, y, z) = \begin{bmatrix} -\sigma & \sigma & 0 \\ r - z & -1 & -x \\ y & x & -b \end{bmatrix}. \tag{7.36}$$

This results in

$$\underline{\underline{J}}(x_c, y_c, z_c) = \begin{bmatrix} -\sigma & \sigma & 0 \\ 1 & -1 & \mp\sqrt{b(r-1)} \\ \pm\sqrt{b(r-1)} & \pm\sqrt{b(r-1)} & -b \end{bmatrix},$$

where the sign ambiguity in front of some matrix elements reflects different signs in the definition (7.33) of the critical points. It turns out that there is only one characteristic polynomial corresponding to two signs in $\underline{\underline{J}}(x, y, z)$ and it is given by:

$$\lambda^3 - (-b - 1 - \sigma)\lambda^2 - (-b(r-1) - b - b\sigma)\lambda + 2b(r-1)\sigma = 0.$$

We will try to solve this equation for the purely imaginary eigenvalue $\lambda = i\nu$. Substituting ν into the characteristic polynomial yields

$$-i\nu^3 + (-b - 1 - \sigma)\nu^2 + i(br + b\sigma)\nu + 2b(r-1)\sigma = 0.$$

Splitting this equation into the real and imaginary parts results in two equations

$$0 = \nu^3 - (br + b\sigma)\nu$$
$$0 = -(b + 1 + \sigma)\nu^2 + 2b(r-1)\sigma.$$

The first equation has two solutions $\nu = 0$ and $\nu = \sqrt{b(r + \sigma)}$. Only the latter is consistent with the second equation. Inserting it into the second equation yields

$$(b + 1 + \sigma)(r + \sigma) = 2(r - 1)\sigma,$$

which has solution

$$r^* = \frac{\sigma(\sigma + b + 3)}{\sigma - b - 1}.$$

The condition $r^* > 1$ forces $\sigma > b + 1$ so that the denominator is positive.

Thus, we conclude that for $r = r^*$ there exists an imaginary eigenvalue. What about results for other values of r greater than one? For $1 < r < r^*$ the origin is unstable while the equilibria (7.33) are still stable. For $r > r^*$ the complex eigenvalues associated with the equilibria (7.33) develop positive real parts and become unstable spiral nodes. Due to the instability of all equilibria for $r > r^*$ the Lorenz system exhibits some truly amazing dynamics. Following Lorenz we will assign $\sigma = 10$, $b = 8/3$ and $r = 28$, then

$$r^* = 470/19 \approx 24.7368.$$

For some values of $r > r^*$ (including $r = 28$) the trajectories approach a butterfly shaped attractor and wander around two unstable critical points in a random way. This is shown in Figure 7.16 for the initial condition $x(0) = .1, y(0) = .1, z(0) = .1$. Although trajectories are confined to a bounded region instead of diverging to infinity, their behavior is unpredictable. The general shape of trajectories in the

Figure 7.16: Butterfly of the Lorenz model for $r = 28$.

butterfly remain unchanged under slight perturbations of the initial condition but individual trajectories which start very close to each other typically wind up being very distant apart at later times, quite often tracing different wings of the butterfly. This leads to the concept of sensitive dependence upon initial conditions that is characteristic of chaotic motion. This behavior is illustrated in Figure 7.17. It shows a plot of $x(t)$ versus t for two solutions; one with initial conditions $(.1, .1, .1)$ (black line) and another with initial conditions $(.11, .1, .1)$ (gray line). Initially, the difference between these two solutions is very small. In fact, until $t = 20$, it is impossible to distinguish these two curves. Eventually, the solutions behave very differently and there is no evidence that they were once infinitesimally close. Although the system is deterministic in the sense that the solution trajectory to the Lorenz system is determined by physical laws, it is very difficult to make a prediction about solution's future behavior based on the Lorenz model because very small changes in the initial conditions cause very big changes in the output. This is the essence of chaos theory – a tiny change at one time can lead to huge changes at later times. To repeat one of chaos theory's famous metaphors: A butterfly flapping its wings in Brazil can cause a hurricane off the coast of Florida weeks later.

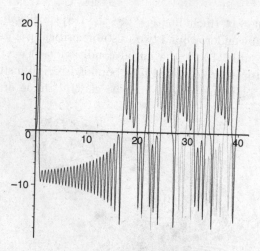

Figure 7.17: Behavior of solution $x(t)$ to the Lorenz equation for $r = 28$ with initial conditions $(.1, .1, .1)$ and $(.11, .1, .1)$.

☞ Three-Dimensional Models; The Lorenz equations: Problems

Problem 7.3.1. Consider the three-dimensional model:

$$\dot{x} = -2y + 2yz - x^3,$$
$$\dot{y} = 2x - xz - y^3,$$
$$\dot{z} = -xy - z^3.$$

(a) Show that the origin is a non-hyperbolic critical point. (b) Use the Liapunov function $P(x, y, z) = x^2 + y^2 + z^2$ to determine whether the origin is stable.

Problem 7.3.2. Consider the Rösler model:

$$\dot{x} = -y - z,$$
$$\dot{y} = x + ay,$$
$$\dot{z} = b + z(x - c),$$

where a, b, c are constants.

(a) Suppose that $c^2 > -4ab$. How many critical points does this model have? Find their coordinates.

(b) Write down the Jacobian matrix in terms of the constants a, b, c.

(c) Determine the stability of the critical points for $a = b = 1/5$ and $c = 5.7$.

(d) Now set $z = 0$ in the Rösler model. Discuss the stability of the reduced model as a function of a.

Problem 7.3.3. Use the Liapunov function,

$$P(x, y, z) = \frac{1}{2}\left(x^2 + \sigma y^2 + \sigma z^2\right),$$

to show that every trajectory solution of the Lorenz model approaches the origin as $t \to \infty$, provided that $r < 1$.

7.4   KdV Equation and Soliton Waves

Despite the complexity of nonlinear differential systems some of the models can be solved under special circumstances. In order for the model to be solvable the dimension of its phase space must match the number of independent constants of motion multiplied by two. Let us illustrate this condition in the case of the nonlinear pendulum.

☞ **Example 7.4.1.** Recall the expression for the conserved energy of a pendulum

$$P(x, y) = E = \frac{1}{2}y^2(t) + \omega^2\left(1 - \cos x(t)\right).$$

We can use the fact that the energy has a constant value to solve for y:

$$y = \sqrt{2\left(E - \omega^2 + \omega^2 \cos x\right)}.$$

Note that we only consider the solution with the positive sign in front of the square root. In general, we need to keep both signs $(+/-)$ in front of the square root to allow us to describe motion back and forth around equilibrium.

Recall the equation $\dot{x} = y$ among two equations of motion (7.22) of the nonlinear pendulum. It can be inverted to give a differential equation

$$\frac{dt}{dx} = \frac{1}{\sqrt{2\left(E - \omega^2 + \omega^2 \cos x\right)}},$$

which can be integrated to yield:

$$t = \int_0^x \frac{dx'}{\sqrt{2\left(E - \omega^2 + \omega^2 \cos x'\right)}} = \sqrt{\frac{1}{2\omega^2}} \int_0^x \frac{dx'}{\sqrt{E/\omega^2 - 1 + \cos x'}},$$

where we have chosen the initial condition, $x(t = 0) = 0$.

A simple trigonometric identity, $\cos x' = 1 - 2\sin^2(x'/2)$, allows us to rewrite the above expression as

$$t = \frac{1}{\omega} \int_0^z \frac{dz'}{\sqrt{E/2\omega^2 - \sin^2 z'}},$$

using the new integration variable, $z = x/2$.

Let us choose:

$$\frac{E}{\omega^2} = (1 - \cos x_{max}) = 2\sin^2(x_{max}/2) < 2\,,$$

where $x_{max} = \varphi_{max} < \pi$ is the maximum angular displacement from the stable equilibrium position. This energy corresponds to oscillations around the stable equilibrium between $-\varphi_{max}$ and φ_{max} without swinging over the inverted position at $\varphi = \pi$.

Then

$$t = \sqrt{\frac{2}{E}} \int_0^z \frac{dz'}{\sqrt{1 - \frac{2\omega^2}{E}\sin^2 z'}} = \sqrt{\frac{2}{E}} \int_0^z \frac{dz'}{\sqrt{1 - \sin^2 z'/\sin^2 z_{max}}}\,,$$

where $z_{max} = x_{max}/2$. As long as the relation $z < z_{max}$ holds the argument of the square root remains positive and time t increases with x. The integral is equal to an Elliptic function of the first kind, defined by:

$$F(\alpha, c) = \int_0^\alpha \frac{d\alpha'}{\sqrt{1 - c^2 \sin^2 \alpha'}}\,.$$

Thus, we found a method to express time t in terms of the polar angle φ via variable $z = x/2 = \varphi/2$:

$$t = \sqrt{\frac{2}{E}} F\left(z, \frac{1}{\sin(x_{max}/2)}\right)\,.$$

We can invert the above relation using Jacobi elliptic function defined by

$$w = F(\alpha, c) \longrightarrow \sin(\alpha) = \mathrm{sn}(w, k)\,.$$

The Jacobi elliptic functions have similar properties to the ordinary trigonometric function and their values can be found in mathematical tables. For the variable $x = \varphi$ of the pendulum we find:

$$\sin(x/2) = \mathrm{sn}\left(t\sqrt{\frac{E}{2}}, \frac{1}{\sin(x_{max}/2)}\right)\,,$$

which gives the angle of the pendulum with respect to the vertical axis as a function of time.

_____

As seen in the above example the nonlinear pendulum is solvable and the constant value of energy determines its solution. For a given system the number of the phase space coordinates divided by two defines degrees of freedom. Therefore, the pendulum has a single degree of freedom (corresponding to one pair consisting of position x and velocity y) and it matches one constant of motion, E. Hence, the condition for solvability is satisfied in this case. Solitons are traveling waves that solve a class of nonlinear partial differential equations. These nonlinear partial differential equations are mathematical models for the time evolution of systems with infinitely many degrees of freedom in the sense that the set of all functions which satisfy the nonlinear partial differential equations constitute an infinite dimensional phase space. In the case of solvable (or integrable) systems their number of dimensions are matched by an infinite number of conservation laws (generalizing the energy conservation law for a pendulum). The time evolution of solvable systems is severely restricted by the existence of an infinite number of conservation laws. These conserved quantities are responsible for time-independence of parameters which characterize the solitons. Consequently, solitons are stable and maintain their shape and speed.

Along with chaos, solitons are recognized as basic nonlinear phenomena. The theory of soliton waves employs an enormous range of theories and methods in various fields of physics and mathematics and finds numerous applications in many areas of natural sciences from physics to biology and engineering. The concept of solitons proved to have a great impact on nuclear and particle physics, nonlinear molecular and solid state physics, and nonlinear optics, where the goal is to use nonlinear solitary waves as the information-carrying "bits" in optical fibers.

A remarkable discovery of solitons can be traced back to August 1832 when naval architect John Scott Russell observed the bow wave of a barge that had suddenly stopped in a Union Canal near Edinburgh. The bow wave detached from the boat assuming the form of large solitary wave pulse and moved forward with a velocity of about eight miles per hour. Scott Russel who followed the wave on horseback noticed that it maintained its shape and speed as it rolled along the channel. Russell became fascinated with his discovery, and made further experiments with such waves in a wave tank he constructed. Eventually, he was able to derive the relation for a soliton's speed as a function of height.

This phenomenon remained largely a mystery until Korteweg and de Vries derived in 1885 a non-linear wave equation (the KdV equation) to describe long-wave propagation on shallow water. Norman Zabusky and Martin Kruskal started to investigate the system numerically in the early sixties. From a detailed numerical study they found that stable pulse-like waves could exist in a system described by the KdV equation. A remarkable property of these solitary waves was their ability to interact and leave these interactions without any change of shape or speed. It was this particle-like nature that led Zabusky and Kruskal to propose for these waves the name of solitons.

The solitary waves appear only in the wave equations which are nonlinear (i.e. equations that contain terms of quadratic or higher powers of the unknown function) contrary to the ordinary linear equations we have studied in chapters 3 and 4. One of the most famous soliton equations is the KdV equation:

$$
\frac{\partial u}{\partial t} = \underbrace{6\,u\,\frac{\partial u}{\partial x}}_{nonlinear\ term} + \underbrace{\frac{\partial^3 u}{\partial x^3}}_{dispersive\ term}\,. \tag{7.37}
$$

In the context of water waves, $u = u(x,t)$ is the displacement of the water surface from its equilibrium position at location x and time t.

The first and second term on the right hand side of the equation are the nonlinear and the dispersive term, respectively.

Generally, all waves are characterized by an effect called dispersion. We will investigate this effect by neglecting the nonlinear term in the KdV equation. This leaves us with a linear equation:

$$
\frac{\partial u}{\partial t} = \frac{\partial^3 u}{\partial x^3}\,, \tag{7.38}
$$

with the dispersive term on the right hand side of equation. Because this equation is linear we take a superposition of simple harmonic waves described by a Fourier integral transform

$$
u(x,t) = u(kx + \omega t) = \int u(k) e^{i\,(kx+\omega t)}\,\mathrm{d}k\,,
$$

with wave number k that is related the wavelength λ (the minimum distance between any two points on a wave that behave identically) through

$$
k = \frac{2\pi}{\lambda}.
$$

Plugging $u(x,t)$ into the linear equation (7.38) we get a relation $\omega = -k^3$ involving angular frequency ω and wave number k. This is a dispersion relation. It describes how plane wave components with different wave numbers travel with respect to each other. It is useful to rewrite the argument $kx + \omega t$ as $k(x + v_p t)$ where we have introduced the phase velocity

$$
v_p = \frac{\omega}{k} = -k^2
$$

that is equal to the speed of the point of constant phase. The above phase velocity depends on a wave number in such a way that the wave components with shorter wavelengths (large wave number k) will move faster than the wave components with longer wavelengths (small wave number k) causing a widening of a wave packet . This phenomenon causes a wave composed of a superposition of elementary components with different wavelengths to spread. Such an event is called dispersion as it has its origin in the dispersion relation describing the dependence of frequency on wave number.

Now, we focus on the role of the nonlinear term. We discard the dispersive term in the KdV equation which leaves:

$$\frac{\partial u}{\partial t} = 6u\frac{\partial u}{\partial x}.\tag{7.39}$$

We seek a traveling wave solution to equation (7.39) of the form

$$u(x,t) = f(x + \mathsf{v}t).\tag{7.40}$$

Substituting expression (7.40) into (7.39) we find

$$\mathsf{v}f' = 6ff',$$

where

$$f' = \frac{\mathrm{d}f(s)}{\mathrm{d}s}$$

is a first order derivative with respect to f's argument. For $f' \neq 0$ we find that the speed v is equal to the amplitude f of the wave pulse. Thus, each part of the wave moves at its own speed. The part of the wave with larger amplitude moves faster then the part of the wave with smaller amplitude. As a result, the wave "breaks"; that is, portions of the wave undergoing greater displacements move faster than, and therefore overtake, those undergoing smaller displacements. This splitting is a result of the nonlinearity and, like dispersion, leads to a change in shape as the wave propagates.

A remarkable property of the KdV equation is that the opposite effects caused by dispersive linear widening and nonlinear wave "breaking" balance each other (see Figure 7.18) allowing for the existence of localized traveling wave solutions that propagate without changing form. Let us consider functions of the form shown in expression (7.40) to verify that they form soliton solutions of the KdV equation, which maintain their localized shape as they travel to the left at the speed v.

Plugging (7.40) into the KdV equation (7.37) gives the differential equation:

$$\mathsf{v}f' - 6ff' - f''' = 0.$$

Integrating it once with respect to x produces an integration constant, K_1:

$$\mathsf{v}f - 3f^2 - f'' = K_1.$$

Multiply both sides of the above equation by f' to get

$$K_1 f' = f'\left(\mathsf{v}f - 3f^2 - f''\right)$$

$$= \mathsf{v}ff' - 3f^2 f' - f''f' = \left(\frac{\mathsf{v}}{2}f^2 - f^3 - \frac{1}{2}(f')^2\right)'.$$

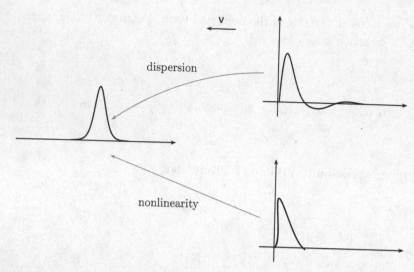

Figure 7.18: For solutions to the KdV equation, dispersion and nonlinear breaking balance each other. Thus, the soliton wave travels without changing its shape.

Integrating once more we arrive at

$$K_1 f + K_2 = \frac{\mathsf{v}}{2} f^2 - f^3 - \frac{1}{2}(f')^2 .$$

A simple solution is obtained for both integration constants K_1, K_2 set to zero. The solution agrees with conditions for a localized wave with f and its derivatives vanishing at $\pm\infty$. In this case, the problem boils down to solving a first order differential equation

$$f' = \frac{\mathrm{d}f}{\mathrm{d}s} = \pm\sqrt{\mathsf{v}f^2 - 2f^3} .$$

This equation is separable and can be cast in the following form

$$\frac{\mathrm{d}f}{f(\mathsf{v} - 2f)^{1/2}} = \pm\,\mathrm{d}s .$$

Based on the identity

$$\frac{\mathrm{d}}{\mathrm{d}x}\operatorname{arctanh}\left(g(x)\right) = \frac{1}{1 - g^2(x)}\frac{\mathrm{d}g(x)}{\mathrm{d}x} ,$$

the above equation can be integrated to yield

$$-\frac{2}{\sqrt{\mathsf{v}}}\operatorname{arctanh}\left(\sqrt{1 - \frac{2}{\mathsf{v}}f}\right) = \pm\,(x + \mathsf{v}t + \delta) ,$$

where δ is an integration constant that can be thought of as an extra constant phase.

Using the hyperbolic identity, $1 - \tanh^2(y) = 1/\cosh^2(y)$, we can express solution f as

$$f(x,t) = \frac{v}{2}\frac{1}{\cosh^2\left(\frac{\sqrt{v}}{2}(x+vt+\delta)\right)} = \frac{v}{2}\text{sech}^2\left(\frac{\sqrt{v}}{2}(x+vt+\delta)\right). \qquad (7.41)$$

The amplitude in front of the hyperbolic function describes its height. We have

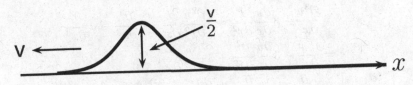

Figure 7.19: A shape of the KdV soliton traveling to the left.

discovered one of the main properties of a soliton: its height is proportional to the speed (see Figure 7.19). Thus tall solitons (with larger amplitudes) move faster than smaller amplitude solitons. Moreover, a soliton's width is inversely proportional to $\sqrt{v}$ and, thus, a soliton's shape also changes with speed: Faster solitons are more narrow than slower solitons.

It is convenient to study properties of soliton systems in terms of a Lax formalism based on the differential operator,

$$L = \frac{\partial^2}{\partial x^2} + u. \qquad (7.42)$$

In addition, we introduce an operator

$$B = 4\frac{\partial^3}{\partial x^3} + 3\left(u\frac{\partial}{\partial x} + \frac{\partial}{\partial x}u\right).$$

These two operators, L and B, are known as a Lax pair with a Lax operator, L. Inserting them into the so-called Lax equation yields

$$\frac{\partial L}{\partial t} = [B, L], \qquad (7.43)$$

where the commutator $[B, L]$ of two operators B and L is equal to $BL - LB$.

Computing that commutator in equation (7.43) produces the following relation:

$$\frac{\partial L}{\partial t} = \frac{\partial u}{\partial t} = \left[4\frac{\partial^3}{\partial x^3} + 3\left(u\frac{\partial}{\partial x} + \frac{\partial}{\partial x} u \right), \frac{\partial^2}{\partial x^2} + u \right]$$

$$= 4\left[\frac{\partial^3}{\partial x^3}, u \right] + 3\left[u\frac{\partial}{\partial x} + \frac{\partial}{\partial x} u, \frac{\partial^2}{\partial x^2} + u \right].$$

(7.44)

Note that $\partial L/\partial t = \partial u/\partial t$ on the left hand side of the above equation. A calculation of the right hand side involves commutation relations:

$$\left[\frac{\partial^3}{\partial x^3}, u \right] = \frac{\partial^3}{\partial x^3} u - u\frac{\partial^3}{\partial x^3}$$

$$= \frac{\partial^3 u}{\partial x^3} + 3\frac{\partial u}{\partial x}\frac{\partial^2}{\partial x^2} + 3\frac{\partial^2 u}{\partial x^2}\frac{\partial}{\partial x} + u\frac{\partial^3}{\partial x^3} - u\frac{\partial^3}{\partial x^3}$$

$$= \frac{\partial^3 u}{\partial x^3} + 3\frac{\partial u}{\partial x}\frac{\partial^2}{\partial x^2} + 3\frac{\partial^2 u}{\partial x^2}\frac{\partial}{\partial x},$$

$$\left[u\frac{\partial}{\partial x}, u \right] = u\frac{\partial}{\partial x} u - u^2\frac{\partial}{\partial x} = u\frac{\partial u}{\partial x} + u^2\frac{\partial}{\partial x} - u^2\frac{\partial}{\partial x} = u\frac{\partial u}{\partial x},$$

(7.45)

$$\left[u\frac{\partial}{\partial x}, \frac{\partial^2}{\partial x^2} \right] = -\frac{\partial^2 u}{\partial x^2}\frac{\partial}{\partial x} - 2\frac{\partial u}{\partial x}\frac{\partial^2}{\partial x^2},$$

$$\left[\frac{\partial}{\partial x} u, \frac{\partial^2}{\partial x^2} \right] = -\frac{\partial^3 u}{\partial x^3} - 2\frac{\partial u}{\partial x}\frac{\partial^2}{\partial x^2} - 3\frac{\partial^2 u}{\partial x^2}\frac{\partial}{\partial x},$$

$$\left[\frac{\partial}{\partial x} u, u \right] = u\frac{\partial u}{\partial x}.$$

Plugging the right hand sides of the above identities into the Lax equation (7.44) reproduces the KdV equation (7.37).

The Lax equation can also be obtained from a pair of eigenvalue equations

$$L\psi(x,t) = \lambda\,\psi(x,t),$$

(7.46)

$$\frac{\partial\psi(x,t)}{\partial t} = B\,\psi(x,t),$$

(7.47)

where $\psi(x,t)$ is a function of x and t and λ is a constant. Let us namely apply a time derivation to both sides of equation (7.46). We get:

$$\frac{\partial}{\partial t} L\psi(x,t) = \frac{\partial L}{\partial t}\psi(x,t) + L\frac{\partial\psi(x,t)}{\partial t} = \frac{\partial L}{\partial t}\psi(x,t) + L\,B\,\psi(x,t),$$

where we used equation (7.47).

On the other hand, equations (7.46) and (7.47) reveal

$$\frac{\partial}{\partial t} L\psi(x,t) = \lambda\frac{\partial\psi(x,t)}{\partial t} = \lambda B\psi(x,t) = B\lambda\psi(x,t) = B\,L\psi(x,t).$$

Subtracting these two equations from each other we get

$$\left(\frac{\partial L}{\partial t} + LB - BL\right)\psi(x,t) = \left(\frac{\partial L}{\partial t} + [L,B]\right)\psi(x,t) = 0, \tag{7.48}$$

valid for all functions $\psi(x,t)$ which satisfy the eigenvalue equations (7.46)-(7.47). The last relation holds if the Lax equation (and therefore also the KdV equation) is satisfied. Accordingly, the Lax equation is being referred to as a compatibility equation of eigenvalue equations (7.46)-(7.47).

Explicitly, the eigenvalue equations (7.46)-(7.47) reads as:

$$\frac{\partial^2 \psi}{\partial x^2} + u\psi = \lambda\psi, \tag{7.49}$$

$$\frac{\partial \psi}{\partial t} = 4\frac{\partial^3 \psi}{\partial x^3} + 3\left(u\frac{\partial \psi}{\partial x} + \frac{\partial u\psi}{\partial x}\right). \tag{7.50}$$

7.4.1 Darboux Transformations and One-Soliton Solution

We have just learned that the function u which satisfies both equations (7.49)-(7.50) will also satisfy their compatibility equation (7.48) and, therefore, the KdV equation itself. Our goal is to design a method to construct solutions of the KdV equation by finding solutions to equations (7.49)-(7.50). The idea is to start with a simple expression of $u(x,t)$ for which it is easy to solve equations (7.49)-(7.50) and then to transform this trivial solution to a non-trivial one. The most simple choice happens to be $u = 0$ for which equations (7.49)-(7.50) simplify to

$$\frac{\partial^2 \psi}{\partial x^2} = \lambda\psi, \tag{7.51}$$

$$\frac{\partial \psi}{\partial t} = 4\frac{\partial^3 \psi}{\partial x^3}. \tag{7.52}$$

It is easy to verify that the hyperbolic functions

$$\psi_0^1(x,t) = \cosh\left(k_1(x - x_1) + 4k_1^3 t\right), \tag{7.53}$$

$$\psi_0^2(x,t) = \sinh\left(k_2(x - x_2) + 4k_2^3 t\right), \tag{7.54}$$

solve equations (7.51) and (7.52) for arbitrary constants x_1, x_2, as long as $k_1^2 = k_2^2 = \lambda$.

Now, we describe a method of Darboux transformation which generates a non-trivial solution to the Lax equation (7.46) out of trivial solutions (7.53) and (7.54) to the eigenvalue equations (7.51) and (7.52).

The Lax operator (7.42) can be rewritten in factorized form as

$$L = \frac{\partial^2}{\partial x^2} + u(x,t) = \left(\frac{\partial}{\partial x} - v(x,t)\right)\left(\frac{\partial}{\partial x} + v(x,t)\right) \tag{7.55}$$

$$= \frac{\partial^2}{\partial x^2} - v^2(x,t) + \frac{\partial v(x,t)}{\partial x}$$

for function v such that

$$u(x,t) = -v^2(x,t) + \frac{\partial v(x,t)}{\partial x} \,. \tag{7.56}$$

Consider function $\psi_0(x,t)$ which satisfies equation

$$\left(\frac{\partial}{\partial x} + v(x,t) \right) \psi_0(x,t) = 0 \,.$$

This is a separable first-order differential equation which is solved by:

$$\psi_0(x,t) = C(t) e^{- \int^x v(x,t)\,dx} \,.$$

Define two first-order differential operators

$$T = \psi_0 \frac{\partial}{\partial x} \psi_0^{-1}, \qquad T^\dagger = \psi_0^{-1} \frac{\partial}{\partial x} \psi_0 \,. \tag{7.57}$$

A short calculation shows that

$$\begin{aligned}
T &= \frac{\partial}{\partial x} - \frac{\partial \psi_0}{\partial x} \psi_0^{-1} = \frac{\partial}{\partial x} + v \\
T^\dagger &= \frac{\partial}{\partial x} + \frac{\partial \psi_0}{\partial x} \psi_0^{-1} = \frac{\partial}{\partial x} - v
\end{aligned}$$

and, thus, the Lax operator can be written as a product of T and $T^\dagger$:

$$L = \frac{\partial^2}{\partial x^2} + u = T^\dagger T \,.$$

Now, we consider the Lax eigenvalue equation (7.46) for some function $\psi_1(x,t)$. It can be rewritten as:

$$T^\dagger T \, \psi_1(x,t) = \lambda \, \psi_1(x,t) \,.$$

Multiply both sides of this equation by T to obtain

$$T\,T^\dagger T \, \psi_1(x,t) = \lambda \, T \, \psi_1(x,t) \,.$$

The function $\psi_2(x,t) = T(\psi_1(x,t))$, obtained as result of acting with the operator T on $\psi_1(x,t)$, will satisfy the eigenvalue equation

$$L' \psi_2(x,t) = \lambda \, \psi_2(x,t) \,,$$

where the new Lax operator L' equals

$$L' = T\,T^\dagger = \left(\frac{\partial}{\partial x} + v \right) \left(\frac{\partial}{\partial x} - v \right) = \frac{\partial^2}{\partial x^2} - v^2 - \frac{\partial v}{\partial x} = \frac{\partial^2}{\partial x^2} + u' \,,$$

and is obtained from L by reversing order of factors, T and $T^\dagger$. The transformation generated by multiplying $\psi_1(x,t)$ with operator T is called the Darboux transformation. We just found that its action is

$$\psi_1(x,t) \xrightarrow{T} \psi_2(x,t) = T(\psi_1(x,t)) = \frac{\partial \psi_1}{\partial x} - \frac{\partial \ln \psi_0}{\partial x}\psi_1,$$

$$u \xrightarrow{T} u' = -v^2 - \frac{\partial v}{\partial x} = u - 2\frac{\partial v}{\partial x} = u + 2\frac{\partial^2 \ln \psi_0}{\partial x^2}, \tag{7.58}$$

where we used relation (7.56) and that $\partial \ln \psi_0/\partial x = -v$.

Let us start with the trivial case, $u = 0$, as in equations (7.51) and (7.52). The solution, $\psi_0(x,t)$, may be chosen to be one of the hyperbolic functions $\psi_0^1(x,t)$ or $\psi_0^2(x,t)$ from equations (7.53), (7.54). Suppose that

$$\psi_0(x,t) = \psi_0^1(x,t) = \cosh\left(k_1(x - x_1) + 4k_1^3 t\right).$$

Then, according to (7.58), the Darboux transformation generates a new non-trivial Lax operator, $\partial^2/\partial x^2 + u'(x,t)$, with

$$u'(x,t) = 2\frac{\partial^2 \ln \psi_0}{\partial x^2} = 2k_1^2 \operatorname{sech}^2\left(k_1(x - x_1) + 4k_1^3 t\right).$$

The function $u'(x,t)$ reproduces the traveling soliton solution (7.41) for the speed v such that $v = 4k_1^2$ or $k_1 = \sqrt{v}/2$.

☞**Example 7.4.2.** Let us apply the Darboux transformation method to the case

$$u(x) = -x^2.$$

The eigenvalue equation (7.46) becomes

$$\frac{d^2\psi}{dx^2} - x^2\psi = \lambda\psi(x,t). \tag{7.59}$$

For a special value $\lambda = -1$, of eigenvalue λ, the above eigenvalue equation can be factorized as

$$\left(\frac{d^2}{dx^2} - x^2 + 1\right)\psi = \left(\frac{d}{dx} - x\right)\left(\frac{d}{dx} + x\right)\psi = T^\dagger T\psi,$$

with

$$T^\dagger = \frac{d}{dx} - x, \quad T = \frac{d}{dx} + x.$$

It is easy to verify that

$$\psi_0(x) = e^{-x^2/2}$$

satisfies $T\psi_0 = 0$ and, therefore, is a solution to the eigenvalue equation (7.59) with $\lambda = -1$.

Applying the Darboux transformation $T^\dagger$ once we obtain

$$\psi_1 = T^\dagger(\psi_0(x)) = \frac{d\psi_0}{dx} + \frac{d\ln\psi_0}{dx}\psi_0(x)$$

$$= \left(\frac{d}{dx} - x\right)\psi_0(x) = -2x\,e^{-x^2/2},$$

$$u_1 = u - 2\frac{\partial^2 \ln\psi_0}{\partial x^2} = -x^2 + 2.$$

Thus, $\psi_1(x)$ is a solution of

$$\frac{d^2\psi_1}{dx^2} + (-x^2 + 2)\psi_1 = -\psi_1(x) \tag{7.60}$$

or

$$\frac{d^2\psi_1}{dx^2} - x^2\psi_1 = \lambda_1\psi_1(x)$$

with eigenvalue $\lambda_1 = -3$.

Applying, successively, the above Darboux transformation $T^\dagger$ produces a sequence of solutions $\psi_n(x)$ with decreasing eigenvalues λ_n:

$$\psi_2 = T^\dagger(\psi_1(x)) = \left(\frac{d}{dx} - x\right)\psi_1(x) = (4x^2 - 2)\,e^{-x^2/2}, \quad \lambda_2 = -5$$

$$\psi_3 = T^\dagger(\psi_2(x)) = \left(\frac{d}{dx} - x\right)\psi_2(x) = (8x^3 - 12x)\,e^{-x^2/2}, \quad \lambda_2 = -7$$

$$\vdots = \vdots$$

$$\psi_n = T^\dagger(\psi_{n-1}(x)) = \left(\frac{d}{dx} - x\right)\psi_{n-1}(x) = \left(\frac{d}{dx} - x\right)^n e^{-x^2/2},$$

with $\lambda_n = -(2n + 1)$. Comparing this chain of relations with the definition of Hermite polynomials

$$H_n(x) = (-1)^n e^{x^2}\frac{d^n}{dx^n}(e^{-x^2})$$

$$= (-1)^n e^{x^2/2}\left(\frac{d}{dx} - x\right)^n (e^{-x^2/2}), \quad n = 1, 2, \ldots \tag{7.61}$$

we observe that ψ_n, generated by the Darboux method, is equal to

$$\psi_n(x) = (-1)^n H_n(x)e^{-x^2/2},$$

which appears in standard quantum mechanics literature as the celebrated harmonic oscillator wave function satisfying the eigenvalue equation

$$\frac{d^2\psi_n}{dx^2} - x^2\psi_n = \lambda_n\psi_n(x), \quad \lambda_n = -(2n + 1).$$

7.4.2 Crum Transformations and the Multi-Soliton Solutions

We have seen that the Darboux transformation maps the Lax operator $L = \partial^2/\partial x^2 + u = T^\dagger T$ into $L' = TT^\dagger$. Behind the Darboux transformation is a similarity transformation of the Lax operator generated by a differential operator. We present a study of such similarity transformations and derive a general method of constructing soliton solutions of the KdV equation in terms of the Wronskian determinant. The Wronskian that we will be working with in this subsection is a determinant of an $n \times n$ matrix and generalizes the Wronskian determinants of the 2×2 and 3×3 matrices we first encountered in Chapter 3 in equations (3.91) and (3.92), while studying solutions to the linear differential equations of the second order.

Let L be the Lax operator as in (7.42). We let it undergo a similarity transformation:

$$L \xrightarrow{U} L_1 = U L U^{-1}, \tag{7.62}$$

generated by the differential operator

$$U = \frac{\partial}{\partial x} - \frac{\partial \ln \Phi}{\partial x} \tag{7.63}$$

defined in terms some yet unspecified function Φ. In terms of

$$f_0 = -\frac{\partial \ln \Phi}{\partial x}$$

the differential operator U and its conjugate $U^\dagger$ become

$$U = \frac{\partial}{\partial x} + f_0, \quad U^\dagger = \frac{\partial}{\partial x} - f_0 \,.$$

The operator U has an inverse U^{-1}:

$$U^{-1} = \left(\frac{\partial}{\partial x} + f_0 \right)^{-1},$$

which appears in the similarity transformation (7.62). According to its definition U^{-1} satisfies the relation $U U^{-1} = 1$.

It turns out that if Φ satisfies the condition that:

$$\frac{\partial}{\partial x} \left(\Phi^{-1} \frac{\partial^2 \Phi}{\partial x^2} \right) + \frac{\partial u}{\partial x} = 0, \tag{7.64}$$

then the operator L_1 obtained as a result of the similarity transformation can be written in the standard Lax form of the differential operator:

$$L_1 = \frac{\partial^2}{\partial x^2} + 2\frac{\partial^2 \ln \Phi}{\partial x^2} + u \,. \tag{7.65}$$

To verify this relation we notice that the differential equation (7.64) can be integrated into:

$$u = -\frac{1}{\Phi}\frac{\partial^2 \Phi}{\partial x^2} - \lambda_0 \,,$$

where λ_0 is an integration constant. In terms of f_0 this relation can be cast into

$$u = -f_0^2 + \frac{\partial f_0}{\partial x} + \lambda_0 \,. \tag{7.66}$$

In the above notation the original Lax operator L can be factorized as follows:

$$L = \frac{\partial^2}{\partial x^2} - f_0^2 + \frac{\partial f_0}{\partial x} + \lambda_0 = \left(\frac{\partial}{\partial x} - f_0\right)\left(\frac{\partial}{\partial x} + f_0\right) + \lambda_0 \,, \tag{7.67}$$

which, in terms of U and $U^\dagger$, is equal to

$$L = U^\dagger U + \lambda_0 \,. \tag{7.68}$$

Now, we can easily see that the similarity transformation (7.62), when applied to L, yields

$$L_1 = ULU^{-1} = UU^\dagger + \lambda_0 = \frac{\partial^2}{\partial x^2} - f_0^2 - \frac{\partial f_0}{\partial x} + \lambda_0 \tag{7.69}$$

and amounts to reversing the order of operators $U, U^\dagger$ in the expression for L, exactly as for the case of the Darboux transformation. Next, we define

$$u_1 = u + 2\frac{\partial^2 \ln \Phi}{\partial x^2} = -f_0^2 - \frac{\partial f_0}{\partial x} + \lambda_0 \tag{7.70}$$

so that the transformed Lax operator can be given a conventional form:

$$L_1 = \frac{\partial^2}{\partial x^2} + u_1 \,.$$

Thus, the condition (7.64) ensures a standard KdV form for the transformed Lax operator L_1.

Introduce a new function $f_1 = -\partial(\ln \Phi_1)/\partial x$ and a new variable λ_1, such that we can rewrite u_1 as

$$u_1 = -f_1^2 + \frac{\partial f_1}{\partial x} + \lambda_1 \,.$$

Alternatively, we write L_1 as $L_1 = U_1^\dagger U_1 + \lambda_1$ with $U_1 = \partial/\partial x + f_1$. Clearly, we have $L_1[\Phi_1] = \lambda_1 \Phi_1$, since $U_1[\Phi_1] = (\partial/\partial x + f_1)\Phi_1 = 0$. The notation we use in this section is such that the differential operator $\mathcal{D}$ acting on function f is denoted by $\mathcal{D}[f]$.

The next step is to reverse the order of $U_1^\dagger$ and U_1 in L_1 by the following Darboux transformation:

$$L_1 \longrightarrow L_2 = U_1 L_1 U_1^{-1} = U_1 U_1^\dagger + \lambda_1 = \frac{\partial^2}{\partial x^2} - f_1^2 - \frac{\partial f_1}{\partial x} + \lambda_1 \,.$$

We then recast the new Lax operator L_2 into the standard form

$$L_2 = U_2^\dagger U_2 + \lambda_2 = \frac{\partial^2}{\partial x^2} - f_2^2 + \frac{\partial f_2}{\partial x} + \lambda_2 \,,$$

with $U_2 = \partial/\partial x + f_2$ and $U_2^\dagger = \partial/\partial x - f_2$, with function f_2 and variable λ_2. The latter two satisfy the relation:

$$-f_1^2 - \frac{\partial f_1}{\partial x} + \lambda_1 = -f_2^2 + \frac{\partial f_2}{\partial x} + \lambda_2 \,.$$

We find that

$$L_2 = \frac{\partial^2}{\partial x^2} + u_2 \,,$$

with u_2 given by

$$u_2 = u_1 + 2\frac{\partial^2 \ln \Phi_1}{\partial x^2} = u_1 - 2\frac{\partial f_1}{\partial x}$$

$$= -f_1^2 - \frac{\partial f_1}{\partial x} + \lambda_1 = u + 2\frac{\partial^2 (\ln \Phi \Phi_1)}{\partial x^2} \,.$$

For Φ_2 such that $f_2 = -\partial \ln \Phi_2/\partial x$ it follows that $U_2[\Phi_2] = 0$ and thus Φ_2 satisfies the eigenvalue problem $L_2[\Phi_2] = \lambda_2 \Phi_2$.

Successive use of the Darboux transformations leads to the Lax operators

$$L_n = \frac{\partial^2}{\partial x^2} + u_n = U_{n-1} L_{n-1} U_{n-1}^{-1} = U_{n-1} \cdots U_1 U \, L \, U^{-1} U_1^{-1} \cdots U_{n-1}^{-1} \quad (7.71)$$

with coefficients

$$u_n = -f_{n-1}^2 - \frac{\partial f_{n-1}}{\partial x} + \lambda_{n-1} = -f_n^2 + \frac{\partial f_n}{\partial x} + \lambda_n = u + 2\frac{\partial^2 \ln(\Phi \Phi_1 \cdots \Phi_{n-1})}{\partial x^2} \,. \quad (7.72)$$

These Lax operators enter a sequence of the eigenvalue problems:

$$L_i[\Phi_i] = \lambda_i \Phi_i \quad \text{or} \quad \left(\frac{\partial^2}{\partial x^2} + u_i - \lambda_i \right) \Phi_i = 0 \qquad i = 0, 1, \dots, \quad (7.73)$$

where, for notational convenience, we have set

$$\Phi_0 = \Phi, \quad u_0 = u, \quad L_0 = L \,.$$

Let $\Psi_1, \ldots, \Psi_n$ be solutions of the related eigenvalue problem involving the function u:

$$L_0[\Psi_i] = L[\Psi_i] = \lambda_i \Psi_i \quad \text{or} \quad \left(\frac{\partial^2}{\partial x^2} + u - \lambda_i\right)\Psi_i = 0 \qquad \Psi_0 \equiv \Phi, \quad i = 0, 1, \ldots$$

$$(7.74)$$

We will now establish a connection between solutions $\Psi_1, \ldots, \Psi_n$ and $\Phi_1, \ldots, \Phi_n$. We start with $n = 1$. It is easy to show that $\Phi_1 = U[\Psi_1]$ since such a function will satisfy the eigenvalue problem $L_1[\Phi_1] = \lambda_1 \Phi_1$ from

$$L_1 U[\Psi_1] = U L U^{-1} U[\Psi_1] = U L[\Psi_1] = U \lambda_1[\Psi_1] = \lambda_1 U[\Psi_1],$$

due to $L[\Psi_1] = \lambda_1 \Psi_1$. More explicitly, the relation between Φ_1 and Ψ_1 is given by:

$$\Phi_1 = U[\Psi_1] = \left(\frac{\partial}{\partial x} + f_0\right)\Psi_1 = \left(\frac{\partial}{\partial x} - \frac{1}{\Phi_0}\frac{\partial \Phi_0}{\partial x}\right)\Psi_1 = \frac{\partial \Psi_1}{\partial x} - \frac{\Psi_1}{\Psi_0}\frac{\partial \Phi_0}{\partial x}$$

$$= \frac{W[\Psi_0, \Psi_1]}{\Psi_0},$$

with the Wronskian:

$$W[f_0, f_1] = f_0\frac{\partial f_1}{\partial x} - f_1\frac{\partial f_0}{\partial x}.$$

More generally, we can express a relation between these two classes of solutions as

$$\Phi_k = U_{k-1}\cdots U_1 U[\Psi_k], \quad k = 0, 1, \ldots, n.$$

This follows from the fact that such Φ_k is a solution of the eigenvalue problem:

$$L_k[\Phi_k] = U_{k-1}\cdots U_1 U L U^{-1}U_1^{-1}\cdots U_{k-1}^{-1}[\Phi_k]$$
$$= U_{k-1}\cdots U_1 U L U^{-1}U_1^{-1}\cdots U_{k-1}^{-1} U_{k-1}\cdots U_1 U[\Psi_k]$$
$$= U_{k-1}\cdots U_1 U L[\Psi_k] = \lambda_k U_{k-1}\cdots U_1[\Psi_k] = \lambda_k \Phi_k.$$

Consider the case $k = n$ for which

$$\Phi_n = U_{n-1}\cdots U_1 U[\Psi_n] = \left(\frac{\partial}{\partial x} + f_{n-1}\right)\cdots\left(\frac{\partial}{\partial x} + f_1\right)\left(\frac{\partial}{\partial x} + f_0\right)[\Psi_n]. \quad (7.75)$$

The key observation is that

$$U_{n-1}\cdots U_1 U = \left(\frac{\partial}{\partial x} + f_{n-1}\right)\cdots\left(\frac{\partial}{\partial x} + f_1\right)\left(\frac{\partial}{\partial x} + f_0\right)$$

is the differential operator of order n which can be cast in the standard form :

$$\mathcal{D}_n = U_{n-1}\cdots U_1 U = \frac{\partial^n}{\partial x^n} + a_{n-1}\frac{\partial^{n-1}}{\partial x^{n-1}} + \ldots + a_1\frac{\partial}{\partial x} + a_0, \quad (7.76)$$

with coefficients $a_i, i = 0, \ldots, n-1$ that depend on functions $f_i, i = 0, \ldots, n-1$. We note that the coefficient in front of the leading term $\partial^n/\partial x^n$ is equal to 1.

The fundamental observation is that the n functions, $\Psi_0, \Psi_1, \ldots \Psi_{n-1}$, are all zero solutions of the differential operator $\mathcal{D}_n$:

$$\mathcal{D}_n[\Psi_i] = 0, \quad i = 0, 1, \ldots, n - 1.$$

For $i = 0$ this equality follows simply from equation (7.63) since $U[\Psi_0] = U[\Phi_0] = U[\Phi] = 0$. For an arbitrary i such that $0 \le i \le n - 1$ the calculation is shown here:

$$\mathcal{D}_n[\Psi_i] = U_{n-1} \cdots U_1 U[\Psi_i] = U_{n-1} \cdots U_i \left(U_{i-1} \cdots U_1 U[\Psi_i] \right) = U_{n-1} \cdots U_i[\Phi_i] = 0$$

since $U_i[\Phi_i] = 0$ due to the definition, $U_i = \partial/\partial x - \partial \ln \Phi_i/\partial x$.

The important point is that a differential operator of order n like $\mathcal{D}_n$ in (7.76) is fully characterized by its n zero solutions. To fully explain how to express $\mathcal{D}_n$ in terms of its zero solutions we need to generalize the concept of Wronskian of two and three functions first encountered in Chapter 3 to a Wronskian $W_n[\psi_1, \ldots, \psi_n]$ of n functions $\psi_1, \ldots, \psi_n$. The object $W_n[\psi_1, \ldots, \psi_n]$ is defined as a determinant of an $n \times n$ matrix with the matrix coefficients:

$$a_{ij} = \frac{\partial^{i-1}\psi_j}{\partial x^{i-1}}, \quad 1 \le i, j \le n.$$

Explicitly,

$$W_n[\psi_1, \ldots, \psi_n] = \begin{vmatrix} \psi_1 & \cdots & \psi_j & \cdots & \psi_n \\ \frac{\partial \psi_1}{\partial x} & \cdots & \frac{\partial \psi_j}{\partial x} & \cdots & \frac{\partial \psi_n}{\partial x} \\ \vdots & \cdots & \vdots & \vdots & \vdots \\ \frac{\partial^{n-1}\psi_1}{\partial x^{n-1}} & \cdots & \frac{\partial^{n-1}\psi_j}{\partial x^{n-1}} & \cdots & \frac{\partial^{n-1}\psi_n}{\partial x^{n-1}} \end{vmatrix}. \tag{7.77}$$

The generalized Wronskian allows us to describe an action of the differential operator $\mathcal{D}_n$ on an arbitrary function f as

$$\mathcal{D}_n[f] = \frac{W_{n+1}[\Psi_0, \Psi_1 \ldots, \Psi_{n-1}, f]}{W_n[\Psi_0, \Psi_1 \ldots, \Psi_{n-1}]}. \tag{7.78}$$

This formula generalizes expressions (3.91) and (3.92) derived in Chapter 3 for the second-order linear differential operator $L[y]$ in terms of Wronskians for 2×2 and 3×3 matrices.

To convince ourselves that the above expression is correct let us substitute one of the functions Ψ_i with $i = 0, 1, \ldots, n - 1$ for f. The nominator in (7.78) vanishes,

$$W_{n+1}[\Psi_0, \ldots, \Psi_i, \ldots, \Psi_{n-1}, \Psi_i] = 0, \quad i = 0, 1, \ldots, n - 1,$$

due to the fact that the determinant of a matrix with two identical columns is identically zero.

We still need to explain the presence of the denominator $W_n[\Psi_0, \Psi_1 \ldots, \Psi_{n-1}]$ in the formula (7.78). It is there to ensure that $\mathcal{D}_n[f]$, defined by the right hand side of (7.78), will be of the form $\partial^n f/\partial x^n + a_{n-1}\partial^{n-1}f/\partial x^{n-1} + \ldots + a_1 \partial f/\partial x + a_0$ as prescribed by relation (7.76). Indeed, the minor expansion along the last column of the Wronskian $W_{n+1}[\Psi_0, \Psi_1 \ldots, \Psi_{n-1}, f]$ yields $(\partial^n f / \partial x^n) W_n[\Psi_0, \Psi_1 \ldots, \Psi_{n-1}] +$ $\ldots$ where $\ldots$ refers to terms with lower derivatives of function f. Thus, dividing by $W_n[\Psi_0, \Psi_1 \ldots, \Psi_{n-1}]$ in (7.78) yields the correct leading term $\partial^n f / \partial x^n$ for $\mathcal{D}_n[f]$.

According to (7.75) it holds that $\Phi_n = \mathcal{D}_n[\Psi_n]$. Plugging the solution, Ψ_n, in place of f in expression (7.78) yields the important result:

The function

$$\Phi_n \equiv \frac{W_{n+1}[\Psi_0, \Psi_1, \ldots, \Psi_n]}{W_n[\Psi_0, \Psi_1, \ldots, \Psi_{n-1}]} \tag{7.79}$$

satisfies the differential equation:

$$\left(\frac{\partial^2}{\partial x^2} + u_n - \lambda_n \right) \Phi_n = 0 \tag{7.80}$$

with

$$u_n = u + 2 \frac{\partial^2 \ln W_n[\Psi_0, \Psi_1, \ldots, \Psi_{n-1}]}{\partial x^2}. \tag{7.81}$$

To understand expression (7.81) recall equation (7.72), which gives u_n in terms of the product $\Phi \Phi_1 \cdots \Phi_{n-1}$. Using result (7.79) for varying values of index n gives

$$\Phi_0 \Phi_1 \cdots \Phi_{n-1} = \Psi_0 \frac{W[\Psi_0, \Psi_1]}{\Psi_0} \frac{W_3[\Psi_0, \Psi_1, \Psi_2]}{W[\Psi_0, \Psi_1]} \cdots \frac{W_n[\Psi_0, \Psi_1, \ldots, \Psi_{n-1}]}{W_{n-1}[\Psi_0, \Psi_1, \ldots, \Psi_{n-2}]}$$
$$= W_n[\Psi_0, \Psi_1, \ldots, \Psi_{n-1}],$$

which turns expression (7.72) into (7.81).

Consider functions $\psi_0^1(x, t)$ and $\psi_0^2(x, t)$ from equations (7.53)-(7.54). This time, consider $k_1 \neq k_2$. Then, $\psi_0^1(x, t)$ and $\psi_0^2(x, t)$ satisfy eigenvalue equations (7.74) with $u = 0$ and $\lambda_1 = k_1^2, \lambda_2 = k_2^2$. Equation (7.81) defines a new KdV eigenvalue problem with the new function

$$u_2(x, t) = 2 \frac{\partial^2 \ln W_2[\psi_0^1, \psi_0^2]}{\partial x^2} = 2 \frac{\partial^2}{\partial x^2} \ln \begin{vmatrix} \psi_0^1 & \psi_0^2 \\ \frac{\partial \psi_0^1}{\partial x} & \frac{\partial \psi_0^2}{\partial x} \end{vmatrix}$$
$$= 2 \frac{\partial^2}{\partial x^2} \ln \begin{vmatrix} \cosh(\gamma_1) & \sinh(\gamma_2) \\ k_1 \sinh(\gamma_1) & k_2 \cosh(\gamma_2) \end{vmatrix}, \tag{7.82}$$

where

$$\gamma_1 = k_1 x - k_1 x_1 + 4 k_1^3 t, \quad \gamma_2 = k_2 x - k_2 x_2 + 4 k_2^3 t.$$

$u_2(x, t)$ is a solution of the KdV equation (7.37). Expression (7.82) is the two soliton solution explicitly given by:

$$u_2(x, t) = 2(k_2^2 - k_1^2) \frac{k_2^2 \cosh^2(\gamma_1) + k_1^2 \cosh^2(\gamma_2) - k_1^2}{[k_2 \cosh(\gamma_1) \cosh(\gamma_2) - k_1 \sinh(\gamma_1) \sinh(\gamma_2)]^2}. \tag{7.83}$$

The two soliton solution (7.83) represents interaction of a soliton of speed $4k_1^2$ with a soliton of speed $4k_2^2$.

Using identities

$$\cosh(\gamma_1)\cosh(\gamma_2) = \frac{1}{2}\left(\cosh(\gamma_1+\gamma_2) + \cosh(\gamma_1-\gamma_2)\right)$$

$$\sinh(\gamma_1)\sinh(\gamma_2) = \frac{1}{2}\left(\cosh(\gamma_1+\gamma_2) - \cosh(\gamma_1-\gamma_2)\right)$$

for the hyperbolic functions we can rewrite the denominator of (7.83) as

$$\left[k_2\cosh(\gamma_1)\cosh(\gamma_2) - k_1\sinh(\gamma_2)\sinh(\gamma_1)\right]^2 =$$
$$\frac{1}{4}\left[(k_2-k_1)\cosh(\gamma_1+\gamma_2) + (k_1+k_2)\cosh(\gamma_1-\gamma_2)\right]^2 .$$

Likewise, the nominator of (7.83) can be rewritten as

$$k_2^2\cosh^2(\gamma_1) + k_1^2\cosh^2(\gamma_2) - k_1^2 = \frac{1}{2}\left(k_2^2\cosh(2\gamma_1) + k_1^2\cosh(2\gamma_2) + k_2^2 - k_1^2\right) .$$

Accordingly, we can rewrite the two-soliton solution in the new form:

$$u_2(x,t) = 4(k_2^2-k_1^2)\frac{k_2^2\cosh(2\gamma_1) + k_1^2\cosh(2\gamma_2) + k_2^2 - k_1^2}{\left[(k_2-k_1)\cosh(\gamma_1+\gamma_2) + (k_1+k_2)\cosh(\gamma_1-\gamma_2)\right]^2} .$$

$$(7.84)$$

☞ **Example 7.4.3.** Let us consider the two-soliton solution with $k_1 = 1$ and $k_2 = 2$. For simplicity, we choose the initial phase shifts x_1 and x_2 to be zero. In this case, $\gamma_1 = x + 4t$ and $\gamma_2 = 2x + 32t$. Thus, expression (7.84) becomes

$$u_2(x,t) = 4 \cdot 3 \frac{4\cosh(2x+8t) + \cosh(4x+64t) + 3}{\left[\cosh(3x+36t) + 3\cosh(x+28t)\right]^2} . \qquad (7.85)$$

In terms of γ_1 equation (7.85) becomes

$$u_2(x,t) = 4 \cdot 3 \frac{4\cosh(2\gamma_1) + \cosh(4\gamma_1+48t) + 3}{\left[\cosh(3\gamma_1+24t) + 3\cosh(\gamma_1+24t)\right]^2} . \qquad (7.86)$$

The same equation in terms of γ_2 can be rewritten as:

$$u_2(x,t) = 4 \cdot 3 \frac{4\cosh(\gamma_2-24t) + \cosh(2\gamma_2) + 3}{\left[\cosh(3\gamma_2/2 - 12t) + 3\cosh(\gamma_2/2 + 12t)\right]^2} . \qquad (7.87)$$

We will now investigate asymptotic behavior of this solution for times $t \to \pm\infty$. Let us apply the limit $t \to \infty$ to equation (7.86) while keeping γ_1 fixed and finite. For very large positive time it holds that $\cosh(a\gamma_1 + bt) \approx$

$\exp(a\gamma_1 + bt)/2$ for $b > 0$ and $\cosh(a\gamma_1 + ct) \approx \exp(-a\gamma_1 - ct)/2$ for $c < 0$. Therefore, in this limit, expression (7.86) becomes

$$u_2(x,t) \approx 12 \frac{\exp(4\gamma_1 + 48t)/2}{[3\exp(\gamma_1 + 24t)/2 + \exp(3\gamma_1 + 24t)/2]^2}$$

$$= \frac{12\exp(4\gamma_1)/2}{[3\exp(\gamma_1)/2 + \exp(3\gamma_1)/2]^2}$$

$$= \frac{2}{[\sqrt{3}\exp(-\gamma_1)/2 + \exp(\gamma_1)/(2\sqrt{3})]^2} = 2\frac{1}{\cosh^2(\gamma_1 - \ln(3)/2)}.$$

Similarly, in the limit $t \to -\infty$ it holds that $\cosh(a\gamma_1 + bt) \approx \exp(-a\gamma_1 - bt)/2$ for $b > 0$ and $\cosh(a\gamma_1 + ct) \approx \exp(a\gamma_1 + ct)/2$ for $c < 0$. Accordingly, equation (7.86) gives in this limit

$$u_2(x,t) \approx 12 \frac{\exp(-4\gamma_1 - 48t)/2}{[3\exp(-\gamma_1 - 24t)/2 + \exp(-3\gamma_1 - 24t)/2]^2}$$

$$= \frac{12\exp(-4\gamma_1)/2}{[3\exp(-\gamma_1)/2 + \exp(-3\gamma_1)/2]^2}$$

$$= \frac{2}{[\sqrt{3}\exp(\gamma_1)/2 + \exp(-\gamma_1)/(2\sqrt{3})]^2} = 2\frac{1}{\cosh^2(\gamma_1 + \ln(3)/2)}.$$

Now, we apply the same approximation to equation (7.87) keeping γ_2 fixed while we let $t \to \pm\infty$. In the limit $t \to \infty$ we obtain

$$u_2(x,t) \approx 12 \frac{2\exp(-\gamma_2 + 24t)}{[3\exp(\gamma_2/2 + 12t)/2 + \exp(-3\gamma_2/2 + 12t)/2]^2}$$

$$= \frac{24\exp(-\gamma_2)}{[3\exp(\gamma_2/2)/2 + \exp(-3\gamma_2/2)/2]^2}$$

$$= \frac{8}{[\sqrt{3}\exp(\gamma_2)/2 + \exp(-\gamma_2)/(2\sqrt{3})]^2} = 8\frac{1}{\cosh^2(\gamma_2 + \ln(3)/2)}.$$

while in the limit $t \to -\infty$ the result is

$$u_2(x,t) = \frac{24\exp(\gamma_2)}{[3\exp(-\gamma_2/2)/2 + \exp(3\gamma_2/2)/2]^2}$$

$$= \frac{8}{[\sqrt{3}\exp(-\gamma_2)/2 + \exp(\gamma_2)/(2\sqrt{3})]^2} = 8\frac{1}{\cosh^2(\gamma_2 - \ln(3)/2)}.$$

Adding the above results to each other we see that, well before collision ($t \to -\infty$) and well after collision ($t \to \infty$), the two-soliton solution is described

Nonlinear Differential Equations

by

$$u_2(x,t) \approx 8 \frac{1}{\cosh^2\left(2(x \pm \ln(3)/4 + 16t)\right)}$$
$$+ 2 \frac{1}{\cosh^2(x \pm (-\ln(3)/2) + 4t)}, \tag{7.88}$$

as $t \to \pm\infty$. Thus, at the limits $t \to \pm\infty$ the two-soliton solution describes two traveling solitons as given by formula (7.41): one moving with speed 16 and amplitude 8 and the other moving with speed 4 and amplitude 2. Comparing the two-soliton solution before ($t \to -\infty$) collision and after ($t \to \infty$) collision we see that each of the two solitons preserves their shapes and speeds.

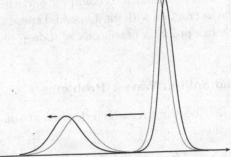

Figure 7.20: Two soliton structure well before the collision. Single non-interacting one-solitons without the phase shifts are shown in gray.

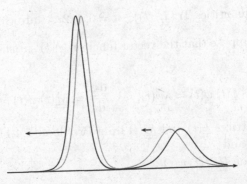

Figure 7.21: Two solitons well after collision. They have the same form as in figure 7.20.

The only change due to interaction appears to be phase shifts. In equation

(7.88), $\ln(3)/4$ is a positive phase shift of the larger soliton and $-\ln(3)/2$ is a negative phase shift of the smaller soliton due to their collision. Thus, a long time before collision, the larger soliton is displaced backward and the smaller soliton is displaced forward (for solitons moving to the left). A long time after collision, the larger soliton (which had overtaken the smaller soliton) is displaced forward and the smaller soliton is displaced backward.

Figures 7.20 and 7.21 show the effect of the phase shifts. It is shown that due to the phase shifts solitons are not in the positions they would have been in without interaction (shown in gray).

We have seen that two solitons emerge from the collision with their shapes unchanged and identities intact, similarly to colliding particles. Collisions in which each object survives the interaction with its shape and speed intact are called elastic collisions. This particle-like property of solitons is of deep physical and mathematical significance.

♖ KdV Equation and Soliton Waves: Problems

Problem 7.4.1. Let $\underline{L}(t)$ be a matrix which dependents on the time variable via a similarity transformation

$$\underline{L}(t) = \underline{U}(t)\, \underline{L}(0)\, \underline{U}^{-1}(t)$$

with a non-singular matrix $\underline{U}(t)$.
(a) Find matrix $\underline{M}(t)$ for which the matrix $\underline{L}(t)$ satisfies the Lax equation

$$\frac{\mathrm{d}\underline{L}}{\mathrm{d}t} = \left[\, \underline{M}(t),\, \underline{L}(t)\, \right].$$

(b) Show that the quantities $\mathrm{Tr}\left(\underline{L}^n(t)\right)$, $n = 0, 1, 2, \ldots$ are time independent.

Problem 7.4.2. Suppose that the vector function $\psi(t)$ satisfies a set of eigenvalue equations

$$\underline{L}(t)\, \psi(t) = \lambda \psi(t), \qquad \frac{\mathrm{d}\psi}{\mathrm{d}t} = \underline{M}(t)\, \psi(t),$$

for a Lax pair of matrices $\underline{L}(t)$ and $\underline{M}(t)$ from Problem (7.4.1). Show that
(a) the eigenvalue λ and
(b) the vector function $\underline{U}^{-1}(t)\, \psi(t)$
are time independent.

Problem 7.4.3. Suppose that $w = -\partial \ln \psi / \partial x$ with ψ which satisfies the eigenvalue equations (7.51) and (7.52). Show that w satisfies

$$\frac{\partial w}{\partial x} = w^2 - \lambda, \qquad \frac{\partial w}{\partial t} = 4\lambda \frac{\partial w}{\partial x}$$

and use these properties to show that

$$u = 2\frac{\partial^2 \ln \psi}{\partial x^2} = -w^2 - \frac{\partial w}{\partial x} + \lambda$$

obeys the KdV equation.

Problem 7.4.4. (a) Show that if function v, related to function u via equation (7.56), satisfies the so-called modified KdV equation

$$\frac{\partial v}{\partial t} = -6v^2\frac{\partial v}{\partial x} + \frac{\partial^3 v}{\partial x^3},$$

then u obeys the KdV equation (7.37).
(b) Verify that the function $v' = -v$ will satisfy the modified KdV equation and use this fact to prove that u' also satisfies the KdV equation (7.37) if

$$u' = -(v')^2 + \frac{\partial v'}{\partial x} = -v^2 - \frac{\partial v}{\partial x} = u - 2\frac{\partial v}{\partial x}, \quad \text{with} \quad u = -v^2 + \frac{\partial v}{\partial x}.$$

Problem 7.4.5. Let Φ be a solution of the first order differential equation

$$\frac{\partial \Phi}{\partial x} + v\Phi = 0,$$

where v is related to solution u of the KdV equation (7.37) via formula (7.56). Show that Φ satisfies the second order differential equation

$$\frac{\partial^2 \Phi}{\partial x^2} + u\Phi = 0,$$

or, alternatively, that $L\Phi = 0$, where L is the Lax operator (7.42).

Problem 7.4.6. Verify explicitly all commutation relations in equations (7.45).

Problem 7.4.7. Show that the Darboux generators $T, T^\dagger$ from Example (7.4.2) satisfy

$$T^\dagger T = TT^\dagger + 2$$

and use it to conclude that the function $\psi_0(x)$ such that $T\psi_0 = 0$, satisfies

$$TT^\dagger \psi_0 = -2\psi_0.$$

Show that $\psi_1 = T^\dagger(\psi_0)$ satisfies

$$TT^\dagger \psi_1 = -4\psi_1.$$

Problem 7.4.8. Prove the second equality in equation (7.61), which defines the Hermite polynomials.

Problem 7.4.9. Prove identities

$$U \frac{\partial^2}{\partial x^2} U^{-1} = \frac{\partial^2}{\partial x^2} + 2 \frac{\partial^2 \ln \Phi}{\partial x^2} + \dot{\Phi} \frac{\partial}{\partial x} \left(\Phi^{-1} \frac{\partial^2 \Phi}{\partial x^2} \right) \left(\frac{\partial}{\partial x} \right)^{-1} \Phi^{-1}$$

and

$$U \, u U^{-1} = u + \Phi \frac{\partial u}{\partial x} \left(\frac{\partial}{\partial x} \right)^{-1} \Phi^{-1},$$

used in deriving condition (7.64).

Problem 7.4.10. Show that two linear eigenvalue matrix equations

$$\left(\underline{\underline{I}} \frac{\partial}{\partial x} + \begin{bmatrix} \partial \phi / \partial x & 0 \\ 0 & -\partial \phi / \partial x \end{bmatrix} \right) \underline{\psi} = - \begin{bmatrix} 0 & 1 \\ \lambda & 0 \end{bmatrix} \underline{\psi}, \quad \frac{\partial \underline{\psi}}{\partial t} = \begin{bmatrix} 0 & \lambda^{-1} e^{2\phi} \\ e^{-2\phi} & 0 \end{bmatrix} \underline{\psi},$$

lead to the so-called sinh-Gordon equation

$$\frac{\partial^2 \phi}{\partial x \partial t} = 2 \sinh(2\phi)$$

as a compatibility equation.

Problem 7.4.11. (a) Make a change of variables

$$\phi \to u = -2i\,\phi, \quad x \to \xi = 2x \quad t \to \eta = 2t$$

in the sinh-Gordon equation from Problem 7.4.10 to obtain the sine-Gordon equation

$$\frac{\partial^2 u}{\partial \xi \partial \eta} = \sin(u).$$

(b) Verify that

$$u(\xi, \eta) = 4 \arctan \left(e^{\xi + \eta} \right)$$

solves the sine-Gordon equation.

(c) Find values of $u(\xi, \eta)$ for $x \to \infty$ and $x \to -\infty$

Problem 7.4.12. Let $u(x, t) = u(s)$, with $s = x + vt$ for the constant v be the traveling wave solution to Burgers' equation:

$$\frac{\partial u}{\partial t} + u \frac{\partial u}{\partial x} - \frac{\partial^2 u}{\partial x^2} = 0.$$

(a) Plug $u(x, t) = u(s)$ into Burgers' equation and integrate once to obtain

$$\frac{du}{ds} = A + vu + \frac{1}{2} u^2,$$

where A is an integration constant.

(b) Complete the square on the right hand side of the equation in (a) and use an integration formula

$$\int \frac{dx}{x^2 + a^2} = \frac{1}{a}\arctan\left(\frac{x}{a}\right)$$

to obtain a solution of the form:

$$u(x,t) = -v + \sqrt{2A - v^2}\,\tan\left[\frac{\sqrt{2A - v^2}}{2}(x + vt + B)\right],$$

where B is another integration constant.

Problem 7.4.13. Suppose function $u(x,t)$ satisfies Burgers' equation from Problem (7.4.12). Let $f(x,t)$ be a function related to $u(x,t)$ via the so-called Hopf-Cole relation:

$$u = -2\frac{\partial f}{\partial x} = -2\frac{\partial}{\partial x}\ln f\,.$$

(a) Show that $f(x,t)$ satisfies the equation

$$\frac{\partial f}{\partial t} = \frac{\partial^2 f}{\partial x^2} + C(t)f\,,$$

where $C(t)$ is an integration constant.

(b) Verify that function $F(x,t) = \exp\left(-\int^t C(t)\,dt\right)f(x,t)$ satisfies the linear equation

$$\frac{\partial F}{\partial t} = \frac{\partial^2 F}{\partial x^2}$$

(c) Show that the function

$$F(x,t) = 1 + e^{rx + r^2 t + \delta}$$

solves the equation given in (b) for arbitrary constants r, δ and show that this gives rise to solution

$$u(x,t) = -2r\frac{e^{rx + r^2 t + \delta}}{1 + e^{rx + r^2 t + \delta}}$$

of Burgers' equation.

Problem 7.4.14. Let function $g(x,t)$ be such that

$$\frac{\partial g}{\partial x} = u\,,$$

where $u(x,t)$ satisfies the KdV equation (7.37).

(a) Show that g satisfies equation (ignore the integration constant):

$$\frac{\partial g}{\partial t} = 3\left(\frac{\partial g}{\partial x}\right)^2 + \frac{\partial^3 g}{\partial x^3}\,.$$

(b) Define a so-called tau function τ through the Hopf-Cole relation of the form:

$$g = 2\frac{1}{\tau}\frac{\partial \tau}{\partial x} = 2\frac{\partial}{\partial x}\ln \tau$$

and show that τ satisfies the *bilinear* equation:

$$\tau\left(\frac{\partial^2 \tau}{\partial x \partial t} - \frac{\partial^4 \tau}{\partial x^4}\right) - \frac{\partial \tau}{\partial x}\left(\frac{\partial \tau}{\partial t} - \frac{\partial^3 \tau}{\partial x^3}\right) + 3\left(\frac{\partial \tau}{\partial x}\frac{\partial^3 \tau}{\partial x^3} - \left(\frac{\partial^2 \tau}{\partial x^2}\right)^2\right) = 0. \qquad (7.89)$$

(c) Show that the function

$$\tau(x, t) = 1 + e^{kx + k^3 t + \delta}$$

solves the bilinear equation (7.89) for arbitrary constants k, δ and show that this solution reproduces the one-soliton solution (7.41) for

$$u = \frac{\partial g}{\partial x} = 2\frac{\partial^2}{\partial x^2}\ln \tau$$

with $k^2 = v$.

(d) Show that

$$\tau = 1 + h_1 + h_2 + c_{12}h_1 h_2,$$

with

$$h_i = e^{k_i x + k_i^3 t + \delta_i}, \quad i = 1, 2, \quad c_{12} = \left(\frac{k_1 - k_2}{k_1 + k_2}\right)^2$$

is also a solution of the bilinear equation (7.89). It is associated with the two soliton solution.

(e) Show that the τ function given in (d) can be rewritten as a determinant of the 2×2 matrix $\underline{\underline{A}}$ with matrix elements

$$a_{nm} = \delta_{nm} + 2\frac{k_m}{k_n + k_m}h_m, \quad n, m = 1, 2.$$

Problem 7.4.15. Show that the wave function ψ which satisfies eigenvalue equations (7.51) and (7.52) will also satisfy the bilinear equation (7.89) and, therefore,

$$u = 2\frac{\partial^2 \ln \psi}{\partial x^2}$$

will satisfy the KdV equation.

References to Chapter 7

1. David Ho, "Supplement no 2, Introduction to stability", lecture notes available at http://www.math.gatech.edu/~bourbaki/math2403/pdf/

2. Nathan Kutz, "Class Notes on Introduction to Differential Equations and Applications", lecture notes available at http://www.amath.washington.edu/courses/351-winter-2002/

3. David K. Campbell, "Nonlinear Science, from Paradigms to Practicalities", Los Alamos Science, Number 15, Special Issue 1987, pg 218

4. Peter S. Lomdahl, "What Is a Soliton?", Los Alamos Science, Number 10, Spring 1984, pg 24

5. M. J. Ablowitz and P. A. Clarkson, "Solitons, nonlinear evolution equations and inverse scattering", London Mathematical Society lecture note series 149, Cambridge University Press, 1991.

Part II

MAPLE

How to read Part II

- You should read this part next to a computer running Maple. All the routines presented here were checked on Maple 9, 9.5 and 10. With few exceptions, they should also run on Maple 7 and 8.

- Maple has three interfaces: a command line interface, a classical interface and a Java interface. You should use Maple's classical interface. It has full functionality, it starts fast, and it does not waste system resources.

- If you are not familiar with Maple, then read first the Appendix B. It contains a short introduction to Maple. The Maple connoisseur can omit this appendix entirely.

- To take full advantage of this book, follow actively the flow of each chapter. Work explicitly each example. The chapters in part two are mirroring the chapters of part one. However, there is a significant difference: in part two, the emphasis is put on the Maple implementation of the mathematical techniques described in part one.

- Finally, remember that the human brain is the ultimate mathematical machine. Any mathematical software should be used intelligently, and only as a tool. Maple, does not replace the mathematical knowledge. It only helps in performing tedious computations, visualizing solutions, and checking hypotheses.

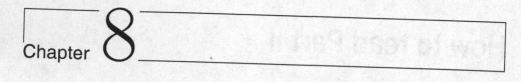

Chapter 8

Vectors and Vector Calculus

M APLE has a couple of dedicated packages for performing calculations that involve vectors and matrices. For Maple 8 and higher, it is recommended to use the packages LinearAlgebra and VectorCalculus. There exists another package linalg which is obsolete. It is kept mainly for backward compatibility.

The LinearAlgebra package contains a large set of routines for performing abstract algebra calculations such as matrix manipulations (adding, multiplication, finding eigenvalues and eigenvectors, matrix inversion, etc.), routines for vector manipulations (scalar product, cross product, orthonormalization of vector sets), and many more. For a detailed list of all routines type ?LinearAlgebra at the command prompt.

The VectorCalculus package operates in general with three-dimensional vectors. The package contains routines for calculating curl, divergence, gradient, vector potentials, to name just a few. The package LinearAlgebra is more suited for abstract algebraic manipulations, while VectorCalculus is mostly used for three-dimensional vector and vector field manipulations.

8.1 Vectors

We illustrate the main concepts regarding vectors and matrices using the case of three-dimensional real vectors. Therefore we begin by using the VectorCalculus package. This package will be also useful for computations involving vector differential operators in three dimensions. Later, for abstract vector manipulations, we will illustrate the use of LinearAlgebra package.

436

8.1.1 Vector Definition and Basic Operations with Vectors

```
> restart;
> with(VectorCalculus):
```

We can define a vector using either the `Vector` command

```
> v := Vector( [a,b,c] );
```

$$v := a\,\mathbf{e_x} + b\,\mathbf{e_y} + c\,\mathbf{e_z}$$

or we can simply write

```
> v1:=<a,b,c>;
```

$$v1 := a\,\mathbf{e_x} + b\,\mathbf{e_y} + c\,\mathbf{e_z}$$

```
> attributes(v);
```

$$coords = cartesian$$

By default, the `VectorCalculus` package assumes the cartesian system of coordinates. This assumption can be modified with `SetCoordinates` command, as we will show below. Let us change to spherical coordinates

```
> SetCoordinates( spherical );
```

$$spherical$$

```
> w := <r,r^2,sin(theta*phi)>;
```

$$w := r\,\mathbf{e_r} + r^2\,\mathbf{e_\phi} + \sin(\theta\,\phi)\,\mathbf{e_\theta}$$

```
> attributes(w);
```

$$coords = spherical$$

Now, let us consider the two-dimensional polar coordinates,

```
> SetCoordinates( polar );
```

$$polar$$

```
> u:=<r,1/r>;
```

$$u := r\,\mathbf{e_r} + \frac{1}{r}\,\mathbf{e_\theta}$$

Next, we change back to the cartesian coordinates, and exemplify the basic operations with vectors in `VectorCalculus` package

```
> SetCoordinates( cartesian );
```

$$cartesian$$

Let us now consider two arbitrary vectors

```
> v1:=<x1,y1,z1>;   v2:=<x2,y2,z2>;
```

$$v1 := x1\,\mathbf{e_x} + y1\,\mathbf{e_y} + z1\,\mathbf{e_z}$$
$$v2 := x2\,\mathbf{e_x} + y2\,\mathbf{e_y} + z2\,\mathbf{e_z}$$

Then the usual operations with vectors are defined as following:

- *Multiplication of a vector with a scalar*

```
> k * v1;
```

$$k\,x1\,\mathbf{e_x} + k\,y1\,\mathbf{e_y} + k\,z1\,\mathbf{e_z}$$

- *Vector addition*
  > v1 + v2;

$$(x1 + x2)\,\mathbf{e_x} + (y1 + y2)\,\mathbf{e_y} + (z1 + z2)\,\mathbf{e_z}$$

- *Scalar (dot) product*
  > v1 . v2;

$$x1\,x2 + y1\,y2 + z1\,z2$$

- *Cross product*
  > v1 &x v2;

$$(y1\,z2 - z1\,y2)\,\mathbf{e_x} + (z1\,x2 - x1\,z2)\,\mathbf{e_y} + (x1\,y2 - y1\,x2)\,\mathbf{e_z}$$

Let us check Gibbs formula:

$$\vec{v}_1 \times (\vec{v}_2 \times \vec{v}_3) = (\vec{v}_1 \cdot \vec{v}_3)\,\vec{v}_2 - (\vec{v}_1 \cdot \vec{v}_2)\,\vec{v}_3 . \tag{8.1}$$

We have:

  > v3:=<x3,y3,z3>;

$$v3 := x3\,\mathbf{e_x} + y3\,\mathbf{e_y} + z3\,\mathbf{e_z}$$

  > term1:=simplify(v1 &x (v2 &x v3));

$$term1 := (y1\,x2\,y3 - y1\,y2\,x3 - z1\,z2\,x3 + z1\,x2\,z3)\,\mathbf{e_x} +$$
$$(z1\,y2\,z3 - z1\,z2\,y3 - x1\,x2\,y3 + x1\,y2\,x3)\,\mathbf{e_y} +$$
$$(x1\,z2\,x3 - x1\,x2\,z3 - y1\,y2\,z3 + y1\,z2\,y3)\,\mathbf{e_z}$$

  > term2:=simplify((v1.v3) * v2 - (v1.v2) * v3);

$$term2 := (y1\,x2\,y3 - y1\,y2\,x3 - z1\,z2\,x3 + z1\,x2\,z3)\,\mathbf{e_x} +$$
$$(z1\,y2\,z3 - z1\,z2\,y3 - x1\,x2\,y3 + x1\,y2\,x3)\,\mathbf{e_y} +$$
$$(x1\,z2\,x3 - x1\,x2\,z3 - y1\,y2\,z3 + y1\,z2\,y3)\,\mathbf{e_z}$$

  > term1 - term2;

$$0\,\mathbf{e_x}$$

As another example, let us check the Jacobi identity

$$\vec{v}_1 \times (\vec{v}_2 \times \vec{v}_3) + \vec{v}_3 \times (\vec{v}_1 \times \vec{v}_2) + \vec{v}_2 \times (\vec{v}_3 \times \vec{v}_1) = 0 . \tag{8.2}$$

We have

  > simplify(v1 &x (v2 &x v3)
  >   + v3 &x (v1 &x v2) + v2 &x (v3 &x v1));

$$0\,\mathbf{e_x}$$

Note that the zero vector is represented by Maple as $0\,\mathbf{e_x}$.

In the following example let us check the Lagrange identity

$$|\vec{v}_1 \times \vec{v}_2|^2 + (\vec{v}_1 \cdot \vec{v}_2)^2 = |\vec{v}_1|^2\,|\vec{v}_2|^2 . \tag{8.3}$$

We start by defining a procedure to calculate the squared norm of a real vector

```
> norm2:=proc(a) a . a end proc;
```

$$norm2 := \mathbf{proc}(a)\ VectorCalculus : -`.`(a,\ a)\ \mathbf{end\ proc}$$

We have

```
> norm2(v1 &x v2) + (v1 . v2)^2 -norm2(v1)*norm2(v2);
```

$$(y1\ z2 - z1\ y2)^2 + (z1\ x2 - x1\ z2)^2 + (x1\ y2 - y1\ x2)^2$$
$$+ (x1\ x2 + y1\ y2 + z1\ z2)^2 - (x1^2 + y1^2 + z1^2)(x2^2 + y2^2 + z2^2)$$

which we can further simplify to obtain the expected result

```
> simplify(%);
```

$$0$$

An alternate solution makes use of the `LinearAlgebra` package, and the built-in commands `VectorNorm`, `CrossProduct` and respectively `DotProduct`. We have

```
> with(LinearAlgebra):
> VectorNorm(CrossProduct(v1, v2), 2, conjugate=false)^2
>    + DotProduct(v1,v2,conjugate=false)^2
>    - VectorNorm(v1, 2, conjugate=false)^2
>    * VectorNorm(v2, 2,conjugate=false)^2;
```

$$y1^2\ z2^2 - 2\ y1\ z2\ z1\ y2 + z1^2\ y2^2 + z1^2\ x2^2 - 2\ z1\ x2\ x1\ z2 + x1^2\ z2^2 + x1^2\ y2^2$$
$$- 2\ x1\ y2\ y1\ x2 + y1^2\ x2^2 + (x1\ x2 + y1\ y2 + z1\ z2)^2$$
$$- (x1^2 + y1^2 + z1^2)(x2^2 + y2^2 + z2^2)$$

```
> simplify(%);
```

$$0$$

If $\vec{s_1}$ and $\vec{s_2}$ are two arbitrary vectors, then their cross product $\vec{s_3} = \vec{s_1} \times \vec{s_2}$ is another vector perpendicular to the plane spanned by $\vec{s_1}$ and $\vec{s_2}$. In the following example we visualize this property.

```
> s1:=<1,1,0>;
```

$$s1 := \mathbf{e_x} + \mathbf{e_y}$$

```
> s2:=<-cos(Pi/4),sin(Pi/4),0>;
```

$$s2 := -\frac{\sqrt{2}}{2}\,\mathbf{e_x} + \frac{\sqrt{2}}{2}\,\mathbf{e_y}$$

```
> s3:= s1 &x s2;
```

$$s3 := \sqrt{2}\,\mathbf{e_z}$$

```
> with(plots):
```

Warning, the name changecoords has been redefined

```
>  ss1:=arrow(s1,
>     shape=cylindrical_arrow,color=gray,
>     width=[1/10,relative=false]):
>  ss2:=arrow(s2,
>     shape=cylindrical_arrow,color=gray,
>     width=[1/10,relative=false]):
>  ss3:=arrow(s3,
>     shape=cylindrical_arrow,color=black,
>     width=[1/10,relative=false]):
>  display(ss1,ss2,ss3,
>     scaling=constrained, axes=normal,
>     orientation=[-12,59],labels=[x,y,z]);
```

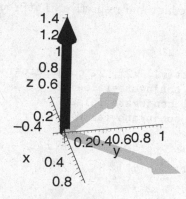

8.1.2   The Action of Matrices on Vectors

A matrix can be written simply as

```
>  A:=<<-1,0,2>|<0,1,0>|<-1,0,1>>;
```

$$A := \begin{bmatrix} -1 & 0 & -1 \\ 0 & 1 & 0 \\ 2 & 0 & 1 \end{bmatrix}$$

where inside the brackets < > we list the columns of the matrix. More about matrix definition will be discussed in chapter 9.

Let us consider a vector $\vec{u}$ given by

```
>  u:=<1,-1,1>;
```

$$u := \mathbf{e_x} - \mathbf{e_y} + \mathbf{e_z}$$

Then the action of matrix A on the vector $\vec{u}$ yields another vector denoted by $\vec{u}_1$. Let us find $\vec{u}_1 = A\,\vec{u}$ and graphically represent the two vectors $\vec{u}$ and $\vec{u}_1$. The action of a matrix on a vector is represented in Maple by the dot product

```
>  u1:= A.u;
```

$$u1 := (-2)\,\mathbf{e_x} - \mathbf{e_y} + 3\,\mathbf{e_z}$$

```
>  with(plots):
>  p1:=arrow(u,
>      shape=cylindrical_arrow,color=gray,
>      width=[1/10,relative=false]):
>  p2:=arrow(u1,shape=cylindrical_arrow,color=black,
>      width=[1/10,relative=false]):
>  display(p1,p2, scaling=constrained,
>      axes=boxed, orientation=[-112,47],labels=[x,y,z]);
```

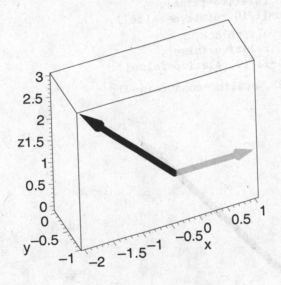

In the above plot we have represented $\vec{u}$ by a gray arrow, and the transformed vector $\vec{u}_1$ by a black arrow. In this way, we clearly illustrate the action of matrices on vectors: they tranform one vector into another vector.

Rotations are another example of operators that are represented by matrices. Rotations don't change the magnitude of vectors but only their orientation. Let us illustrate this property in the case of two-dimensional vectors. We consider a unit vector $\vec{w}$

```
>  w:=<cos(Pi/12),sin(Pi/12)>;
```

$$w := \cos(\frac{\pi}{12})\,\mathbf{e_x} + \sin(\frac{\pi}{12})\,\mathbf{e_y}$$

Let us write the rotation matrix R corresponding to a counter-clock rotation with $\pi/6$

```
>  R:=<<cos(Pi/6), sin(Pi/6)>|<-sin(Pi/6),cos(Pi/6)>>;
```

$$R := \begin{bmatrix} \dfrac{\sqrt{3}}{2} & \dfrac{-1}{2} \\ \dfrac{1}{2} & \dfrac{\sqrt{3}}{2} \end{bmatrix}$$

Then, acting with R on $\vec{w}$, we get a new vector $\vec{w}_1 = R\vec{w}$, given by

```
>   w1:= R. w;
```

$$w1 := (\frac{1}{2}\sqrt{3}\cos(\frac{\pi}{12}) - \frac{1}{2}\sin(\frac{\pi}{12}))\,\mathbf{e_x} + (\frac{1}{2}\cos(\frac{\pi}{12}) + \frac{1}{2}\sqrt{3}\sin(\frac{\pi}{12}))\,\mathbf{e_y}$$

Now, we represent $\vec{w}$ by a gray arrow, and the rotated vector $\vec{w}_1$ by a black arrow as shown below

```
>  r1:=arrow(w, color=gray,
>      width=[1/45,relative=false],
>      head_length=[1/10,relative=false]):
>  r2:=arrow(w1, color=black,
>      width=[1/45,relative=false],
>      head_length=[1/10,relative=false]):
>  display(r1, r2 , scaling=constrained);
```

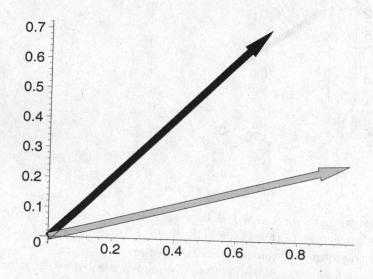

We can readily check that rotation preserves the length of the vector

```
>  simplify(w.w);
```

1

```
>  simplify(w1.w1);
```

$$1$$

8.1.3 Abstract Vector Manipulation

For abstract vector manipulations, in 2, 3, 4 or higher dimensions we should use the package LinearAlgebra. Much of the syntax used in the package LinearAlgebra is the same as in VectorCalculus package, thus we can start immediately using it.

For example, let us calculate the angle between two vectors $\vec{u}_1$ and $\vec{u}_2$ in four-dimensional space, where the vectors are

```
>  with(LinearAlgebra):
```

```
>  u1:=<1,2,3,4>:  u2:=<4,3,2,1>:
```

The angle between u_1 and u_2 is given by the VectorAngle command

```
>  VectorAngle(u1,u2);
```

$$\arccos(\frac{2}{3})$$

We can immediately check this result using formula

$$\cos\alpha = \frac{u_1 \cdot u_2}{\|u_1\|\|u_2\|} \,. \tag{8.4}$$

We have

```
>  norm1:=Norm(u1,2);
```

$$norm1 := \sqrt{30}$$

```
>  norm2:=Norm(u2,2);
```

$$norm2 := \sqrt{30}$$

```
>  scalar_prod:=u1.u2;
```

$$scalar\_prod := 20$$

```
>  alpha:=arccos(scalar_prod/(norm1*norm2));
```

$$\alpha := \arccos(\frac{2}{3})$$

Let us use the LinearAlgebra package to check whether the vectors

$$\vec{A} = \frac{1}{\sqrt{2}}(-1,0,1), \quad \vec{B} = \frac{1}{5}(3,-4,5), \quad \vec{C} = \frac{1}{3}(2,2,-1),$$

are linearly independent.

```
>  with(LinearAlgebra):
```

```
>  A:=<-1,0,1>/sqrt(2):  B:=<3,-4,5>/5:  C:=<2,2,-1>/3:
```

```
>  Basis([A,B,C]);
```

$$\left[\left[\begin{array}{c} -\dfrac{\sqrt{2}}{2} \\[4pt] 0 \\[4pt] \dfrac{\sqrt{2}}{2} \end{array}\right], \left[\begin{array}{c} \dfrac{3}{5} \\[4pt] \dfrac{-4}{5} \\[4pt] 1 \end{array}\right], \left[\begin{array}{c} \dfrac{2}{3} \\[4pt] \dfrac{2}{3} \\[4pt] \dfrac{-1}{3} \end{array}\right]\right]\right]$$

The routine Basis returns a basis for a vector space spanned by $\vec{A}, \vec{B}$ and $\vec{C}$. In this case the base is the set of vectors $\{\vec{A}, \vec{B}, \vec{C}\}$ itself, proving their linear independence. Another way to verify the linear independence of $\vec{A}, \vec{B}$, and $\vec{C}$ is to show that mixed product $\vec{A} \cdot (\vec{B} \times \vec{C}) \neq 0$.

```
> A.CrossProduct(B,C);
```

$$\frac{2\sqrt{2}}{3}$$

We have used here the CrossProduct command instead of short-hand notation &x

```
> A.(B &x C);
```

$$\frac{2\sqrt{2}}{3}$$

Note that the parentheses around the cross product are optional

```
> A. B &x C;
```

$$\frac{2\sqrt{2}}{3}$$

The geometric meaning of the product $\vec{A} \cdot (\vec{B} \times \vec{C})$ is the volume of the parallelepiped generated by these three vectors. A non-zero volume, means that the three vectors are not in the same plane, hence they are linearly independent. Let us visualize these three vectors.

```
> with(plots):

> v1:=arrow(A, shape=cylindrical_arrow):

> v2:=arrow(B, shape=cylindrical_arrow):

> v3:=arrow(C, shape=cylindrical_arrow):

> display( v1,v2,v3, scaling=constrained,
>     orientation=[22,56], axes=boxed, labels=[x,y,z] );
```

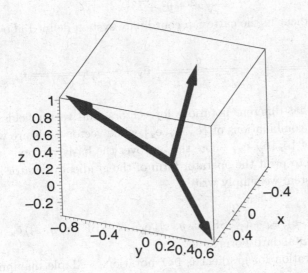

8.2   Vector Differential Operators

For vector calculus calculations, we should use the package `VectorCalculus`. This package outputs vectors and vector fields as linear combinations of basis vectors.

8.2.1   Gradient

To calculate the gradient of a scalar function `f`, we can use any of the equivalent commands `Gradient`, `Del` or `Nabla`.

```
> restart:
> with(VectorCalculus):
```

Warning, the assigned names <,> and <|> now have a global binding

Warning, these protected names have been redefined and unprotected: *, +, ., D, Vector, diff, int, limit, series

To perform calculations within `VectorCalculus` package one must declare explicitly a coordinate system. There are 37 coordinate systems predefined in Maple. See the help page `?SetCoordinates` for more details.

```
> SetCoordinates( cartesian[x,y,z]);
```
$$cartesian_{x,y,z}$$

Let us now consider the following scalar function

```
> f:=sqrt(x^2+y^2+z^2);
```

$$f := \sqrt{x^2 + y^2 + z^2}$$

and calculate its gradient in the cartesian coordinate system defined above.

```
> Gradient(f);
```

$$\frac{x}{\sqrt{x^2 + y^2 + z^2}}\, \bar{\mathbf{e}}_\mathbf{x} + \frac{y}{\sqrt{x^2 + y^2 + z^2}}\, \bar{\mathbf{e}}_\mathbf{y} + \frac{z}{\sqrt{x^2 + y^2 + z^2}}\, \bar{\mathbf{e}}_\mathbf{z}$$

Remark

Observe that Maple has different notations for vectors, and vector fields. Vectors are denoted as linear combinations of $\{\mathbf{e_x}, \mathbf{e_y}, \mathbf{e_z}\}$ while vector fields are written as linear combinations of $\{\bar{\mathbf{e}}_\mathbf{x}, \bar{\mathbf{e}}_\mathbf{y}, \bar{\mathbf{e}}_\mathbf{z}\}$ –note the bar over the basis vectors.

If we want Maple to print the operator form of the gradient operator in the chosen coordinate system we simply write

```
> Gradient();
```

$$\left(\tfrac{\partial}{\partial x}\, \mathrm{SF}\, (x, y, z)\right) \bar{\mathbf{e}}_\mathbf{x} + \left(\tfrac{\partial}{\partial y}\, \mathrm{SF}\, (x, y, z)\right) \bar{\mathbf{e}}_\mathbf{y} + \left(\tfrac{\partial}{\partial z}\, \mathrm{SF}\, (x, y, z)\right) \bar{\mathbf{e}}_\mathbf{z}$$

where SF stands for "Standard Form".

An equivalent notation for gradient is Del notation. Maple manipulates Del operator exactly as $\vec{\nabla}$ operator in vector calculus. We have

```
> Del(f);
```

$$\frac{x}{\sqrt{x^2 + y^2 + z^2}}\, \bar{\mathbf{e}}_\mathbf{x} + \frac{y}{\sqrt{x^2 + y^2 + z^2}}\, \bar{\mathbf{e}}_\mathbf{y} + \frac{z}{\sqrt{x^2 + y^2 + z^2}}\, \bar{\mathbf{e}}_\mathbf{z}$$

Similar results are obtained using Nabla operator. Try it!

8.2.2 Directional Derivative, and the Geometrical Interpretation of the Gradient

The geometrical interpretation of the gradient is related to the directional derivative of a scalar function along a direction $\vec{u}$. Maple has built-in routines to calculate directional derivatives.

Let us consider a scalar field $f = 3xyz + z^2$, and calculate its directional derivative along the vector $\vec{u} = (\mathbf{e_x} + \mathbf{e_y} + \mathbf{e_z})/\sqrt{3}$. We have

```
> f:=3*x*y*z+z^2;
```

$$f := 3\,x\,y\,z + z^2$$

```
> u:=<1,1,1>/sqrt(3);
```

$$u := \frac{\sqrt{3}}{3}\, \mathbf{e_x} + \frac{\sqrt{3}}{3}\, \mathbf{e_y} + \frac{\sqrt{3}}{3}\, \mathbf{e_z}$$

```
> DirectionalDiff(f, u, [x,y,z]);
```

$$y\,z\,\sqrt{3} + x\,z\,\sqrt{3} + \frac{(3\,x\,y + 2\,z)\,\sqrt{3}}{3}$$

At this point it is instructive to check that indeed the directional derivative of f along $\vec{u}$ is given by

$$D_u(f) = \vec{\nabla} f \cdot \hat{u}\,, \tag{8.5}$$

where $\hat{u} = \frac{\vec{u}}{|\vec{u}|}$. In our case $\vec{u}$ is already normalized, thus we have

```
> Del(f) . u;
```

Error, (in DotProd) cannot take the dot product of a Vector and a
VectorField

Maple issues an error message, signaling that we must be careful on how we manipulate the objects. We cannot mix vectors and vector fields. In eq. (8.5), $\hat{u}$ has the meaning of a vector field. We can immediately fix this problem by defining a vector field out of $\vec{u}$ using the VectorField command.

```
> uF:=VectorField(u);
```

$$uF := \frac{\sqrt{3}}{3}\,\bar{e}_x + \frac{\sqrt{3}}{3}\,\bar{e}_y + \frac{\sqrt{3}}{3}\,\bar{e}_z$$

```
> Del(f) . uF;
```

$$y\,z\,\sqrt{3} + x\,z\,\sqrt{3} + \frac{(3\,x\,y + 2\,z)\,\sqrt{3}}{3}$$

As expected, we get the same result as before.

Remark

The gradient field is perpendicular to constant level surfaces of the scalar field. The following example is a visualization of this property.

```
> with(plots):
```

```
> g:=x*y*z;
```

$$g := x\,y\,z$$

```
> Del(g);
```

$$y\,z\,\bar{e}_x + x\,z\,\bar{e}_y + x\,y\,\bar{e}_z$$

The surfaces of constant level $g(x, y, z) = 1$ and $g(x, y, z) = 3$ are given by

```
> Surface1:=1/(x*y); Surface3:=3/(x*y);
```

$$Surface1 := \frac{1}{x\,y}$$

$$Surface3 := \frac{3}{x\,y}$$

```
> p1:=gradplot3d(g,x=0..1,y=0..1,z=0..10,
>     thickness=2,color=black):
```

```
> p2:=plot3d({Surface1, Surface3},x=0..1,y=0..1,view=0..10):
```

```
> display(p1,p2,orientation=[21,129],axes=normal);
```

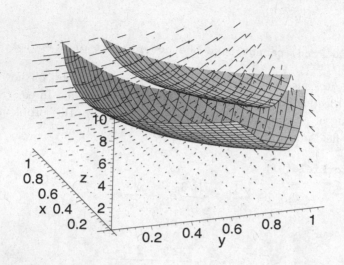

Observe that the gradient field lines are perpendicular to the surfaces of constant level.

We can use the above property to find the unit vector normal to a give surface. For example, find the normal vector normal to

$$z = x^2 + xy - 2$$

at the point $A = (1, -1, -2)$. To this end, we first define the scalar field $h = -z + x^2 + xy - 2$. Then the surface above, is given by the level surface $h = 0$.

```
>   restart;
>   with(VectorCalculus):
>   SetCoordinates( cartesian[x,y,z]);
```
$$cartesian_{x,y,z}$$
```
>   h:=-z + x^2 + x*y - 2;
```
$$h := -z + x^2 + xy - 2$$
```
>   A := <1,-1,-2>;
```
$$A := e_x - e_y - 2\,e_z$$

The normal to the surface in point A, is just **Del(h)** evaluated in vector A, since the gradient of h is normal to h's level surfaces. The gradient of h is
```
>   Gh:=Del (h);
```
$$Gh := (2\,x + y)\,\bar{e}_x + x\,\bar{e}_y - \bar{e}_z$$

We evaluate the gradient at A with **evalVF** function
```
>   Gh0:=evalVF( Gh, A);
```

$$Gh0 := \mathbf{e_x} + \mathbf{e_y} - \mathbf{e_z}$$

The vector we have obtained so far, is perpendicular to the surface at the point $A = (1, -1, -2)$. To normalize it, we use the routine `Normalize` from the `LinearAlgebra`. To save memory, we don't load the entire `LinearAlgebra` package just to access one routine; instead we can use only the desired function using the following syntax

```
> Gh0_unit := LinearAlgebra :- Normalize(Gh0, 2);
```

$$Gh0\_unit := \begin{bmatrix} \dfrac{\sqrt{3}}{3} \\[2mm] \dfrac{\sqrt{3}}{3} \\[2mm] -\dfrac{\sqrt{3}}{3} \end{bmatrix}$$

Let us plot the result. We apply the same "memory conservation" philosophy as above. We use only two functions `arrow` and `display` from `plots` package, in order to avoid loading the entire package. The syntax is

```
> one:=plots[arrow](Gh0_unit, color=blue, thickness=3):
> S:=plot3d(h+z, x =0..2, y=-2..1,
>     view=-3..2, style=wireframe, axes=boxed):
> plots[display](S, one,
>     scaling=unconstrained, orientation=[-27,67]);
```

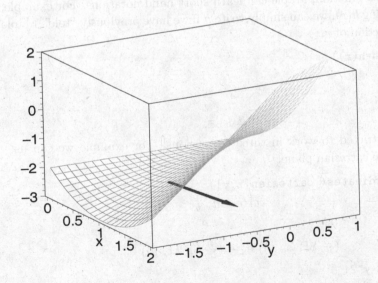

Click and drag on the graph to rotate the three-dimensional configuration and observe it from different angles.

8.2.3   Changing the Coordinate System

The cartesian coordinate system is just one of the many coordinate systems already built in the `VectorCalculus` package. Let us explore the `Del` operator in spherical coordinates.

```
> SetCoordinates( spherical[r,theta,phi]);
```

$$spherical_{r,\,\theta,\,\phi}$$

```
> Del();
```

$$\left(\tfrac{\partial}{\partial r} \mathrm{SF}\ (r,\ \theta,\ \phi)\right) \bar{\mathbf{e}}_{\mathbf{r}} + \frac{\tfrac{\partial}{\partial \theta} \mathrm{SF}\ (r,\ \theta,\ \phi)}{r}\ \bar{\mathbf{e}}_{\theta} + \frac{\tfrac{\partial}{\partial \phi} \mathrm{SF}\ (r,\ \theta,\ \phi)}{r \sin(\theta)}\ \bar{\mathbf{e}}_{\phi}$$

```
> g:=r^2;
```

$$g := r^2$$

```
> Del(g);
```

$$2\,r\,\bar{\mathbf{e}}_{\mathbf{r}}$$

```
> g1:=r*cos(phi) - r*sin(theta);
```

$$g1 := r\cos(\phi) - r\sin(\theta)$$

```
> simplify(Del(g1));
```

$$(\cos(\phi) - \sin(\theta))\ \bar{\mathbf{e}}_{\mathbf{r}} - \cos(\theta)\ \bar{\mathbf{e}}_{\theta} - \frac{\sin(\phi)}{\sin(\theta)}\ \bar{\mathbf{e}}_{\phi}$$

Using the `alias` function, Maple can learn short hand notations. For example, instead of writing $h(r)$, we can simply write h if we have previously "told" Maple that h is a function of r.

```
> alias(h=h(r));
```

$$h$$

```
> Del(h);
```

$$\left(\tfrac{\partial}{\partial r} h\right) \bar{\mathbf{e}}_{\mathbf{r}}$$

We are not restricted to work in three dimensions. For example we can use the two-dimensional cartesian plane.

```
> SetCoordinates( cartesian[x,y]);
```

$$cartesian_{x,\,y}$$

```
> Del();
```

$$\left(\tfrac{\partial}{\partial x} \mathrm{SF}\ (x,\ y)\right) \bar{\mathbf{e}}_{\mathbf{x}} + \left(\tfrac{\partial}{\partial y} \mathrm{SF}\ (x,\ y)\right) \bar{\mathbf{e}}_{\mathbf{y}}$$

```
> f2:=x^2+y^2;
```

$$f2 := x^2 + y^2$$

```
> Del(f2);
```

$$2\,x\,\bar{\mathbf{e}}_{\mathbf{x}} + 2\,y\,\bar{\mathbf{e}}_{\mathbf{y}}$$

8.2.4 Manipulating Vector Fields

To define vector fields we use the `VectorField` command.

```
> F:=VectorField( <x^2, y, -z>);
```
$$F := x^2\,\bar{e}_x + y\,\bar{e}_y - z\,\bar{e}_z$$

```
> attributes(F);
```
$$vectorfield;\ coords = cartesian_{x,\,y,\,z}$$

With the help of the `attributes` command we can inspect all the properties of a previously defined object. In this case, we learn that F is a vector field written in the cartesian coordinate system. Let us use `evalVF` to evaluate the vector field F for a particular vector $< 2, a, 3 >$.

```
> evalVF(F, <2,a,3>);
```
$$4\,e_x + a\,e_y - 3\,e_z$$

The last aspect we discuss here is how to convert a vector or a vector field to other coordinate system. To change the coordinate system we use the `MapToBasis` command.

```
> MapToBasis(F, cylindrical[r, theta, z]);
```
$$(r^2\cos(\theta)^3 + r\sin(\theta)^2)\,\bar{e}_r + (-r^2\cos(\theta)^2\sin(\theta) + r\sin(\theta)\cos(\theta))\,\bar{e}_\theta - z\,\bar{e}_z$$

8.2.5 Divergence

Let us define a vector field by

```
> F1:=VectorField( <sin(x^y),z^2,x*y*z>);
```
$$F1 := \sin(x^y)\,\bar{e}_x + z^2\,\bar{e}_y + x\,y\,z\,\bar{e}_z$$

Then the divergence of F_1 is defined simply as the dot product of `Del` (or `Nabla`) operator with the vector field `F1`

```
> Del .  F1;
```
$$\frac{\cos(x^y)\,x^y\,y}{x} + x\,y$$

```
> Nabla .  F1;
```
$$\frac{\cos(x^y)\,x^y\,y}{x} + x\,y$$

Let us change to spherical coordinates

```
> SetCoordinates( spherical[r, theta, phi]);
```
$$spherical_{r,\,\theta,\,\phi}$$

and consider the following vector field

```
> G:=VectorField(<1,0,1>);
```
$$G := \bar{e}_r + \bar{e}_\phi$$

Then, its divergence is

```
> Del .  G;
```

$$\frac{2}{r}$$

As an exercise, let us convert G to cartesian system, and calculate again its divergence. We should get the same result.

```
> Gxyz:=MapToBasis(G, cartesian[x,y,z]);
```

$$Gxyz := (\frac{x}{\sqrt{x^2+y^2+z^2}} - \frac{y}{\sqrt{x^2+y^2}})\,\bar{e}_x$$
$$+(\frac{y}{\sqrt{x^2+y^2+z^2}} + \frac{x}{\sqrt{x^2+y^2}})\,\bar{e}_y + \frac{z}{\sqrt{x^2+y^2+z^2}}\,\bar{e}_z$$

```
> SetCoordinates( cartesian[x,y,z]);
```

$$cartesian_{x,y,z}$$

```
> simplify(Del .   Gxyz);
```

$$\frac{2}{\sqrt{x^2+y^2+z^2}}$$

As expected, we get the same result.

8.2.6   Curl

To calculate the curl of a vector field V, we can use one of the three equivalent commands: `Curl(V)`, `Del &x V` or `CrossProduct(Del, V)`. Let us consider for example the following vector field

```
> V:=VectorField( <y^3, -x, z> );
```

$$V := y^3\,\bar{e}_x - x\,\bar{e}_y + z\,\bar{e}_z$$

Then we have

```
> Curl (V);
```

$$(-1 - 3\,y^2)\,\bar{e}_z$$

```
> Del &x V;
```

$$(-1 - 3\,y^2)\,\bar{e}_z$$

```
> CrossProduct(Del, V);
```

$$(-1 - 3\,y^2)\,\bar{e}_z$$

The field V is nonconservative because its curl is nonzero. The field V is also called a *rotational* field. We can illustrate this "rotational" structure by visualizing the field.

```
> plots[fieldplot3d](V, x=-1..1, y=-1..1,
>     z=-1..1, scaling=constrained, shading=zhue);
```

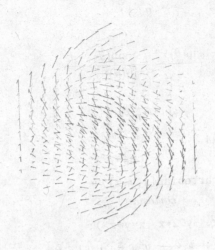

8.2.7 Successive Applications of ∇ Operator

The command `Del` is flexible, as it allows to be manipulated as a formal vector. Therefore we can have new operators by taking the dot product `Del . Del`, or the cross product `Del &x Del`, or other combinations. Thus, the `VectorCalculus` package is extremely useful in writing relative complicated expressions in the `Del` operator using a natural notation.

The following examples illustrate these features.

```
> restart;
> with(VectorCalculus):
```

Warning, the assigned names <,> and <|> now have a global binding

Warning, these protected names have been redefined and unprotected: *, +, ., D, Vector, diff, int, limit, series

```
> SetCoordinates( cartesian[x,y,z]);
```
$$cartesian_{x,y,z}$$

```
> f:=x*y*z^2;
```
$$f := x\,y\,z^2$$

```
> Del . Del (f);
```
$$2\,x\,y$$

```
> SetCoordinates( spherical[r,theta,phi]);
```
$$spherical_{r,\theta,\phi}$$

```
>  g:=r^n;
```

$$g := r^n$$

```
>  Del . Del (g);
```

$$\frac{\sin(\theta)\, r^n\, n + \sin(\theta)\, r^n\, n^2}{r^2 \sin(\theta)}$$

```
>  simplify(%);
```

$$r^{(n-2)}\, n\, (1+n)$$

```
>  Del &x (Del (g) );
```

$$0\, \bar{e}_r$$

```
>  SetCoordinates( cartesian[x,y,z]);
```

$$cartesian_{x,\,y,\,z}$$

```
>  V:=VectorField(<x, -y^2*x, x*y*z>);
```

$$V := x\, \bar{e}_x - y^2\, x\, \bar{e}_y + x\, y\, z\, \bar{e}_z$$

```
>  Del &x V;
```

$$x\, z\, \bar{e}_x - y\, z, \bar{e}_y - y^2\, \bar{e}_z$$

```
>  Del . (Del &x V);
```

$$0$$

Let us check the following identity

$$\vec{\nabla} \times \left(\vec{\nabla} \times \vec{U} \right) = \vec{\nabla} \left(\vec{\nabla} \cdot \vec{U} \right) - \left(\vec{\nabla} \cdot \vec{\nabla} \right) \vec{U} \,, \tag{8.6}$$

for an arbitrary vector field $U = (f_1(x,y,z), f_2(x,y,z), f_3(x,y,z))$. We have

```
>  U:=VectorField(<f1(x,y,z), f2(x,y,z),f3(x,y,z)>);
```

$$U := \mathrm{f1}(x,\, y,\, z)\, \bar{e}_x + \mathrm{f2}(x,\, y,\, z)\, \bar{e}_y + \mathrm{f3}(x,\, y,\, z)\, \bar{e}_z$$

```
>  Term1:= Del &x (Del &x U):
>  Term2:= Del(Del . U) - (Del . Del)(U):
>  simplify(Term1 - Term2);
```

$$0\, \bar{e}_x$$

8.3 Curvilinear Coordinates

In a curvilinear coordinate system, the position of a point P is described by the intersection of three surfaces $q_1 = $ constant, $q_2 = $ constant and $q_3 = $ constant. Therefore we can identify the position of P as $P(q_1, q_2, q_3)$. The curvilinear coordinates are given by transformations

$$x = x(q_1, q_2, q_3) \,, \quad y = y(q_1, q_2, q_3) \,, \quad z = z(q_1, q_2, q_3) \,. \tag{8.7}$$

Let us consider the case of toroidal coordinates to exemplify the main Maple concepts related to curvilinear coordinates. The toroidal coordinates (v, u, ζ, a) are defined in function of a parameter a via the transformations:

$$x = \frac{a \cos \zeta \sinh v}{-\cos u + \cosh v} \ , \ y = \frac{a \sin \zeta \sinh v}{-\cos u + \cosh v} \ , \ z = \frac{a \sin u}{-\cos u + \cosh v} \ . \tag{8.8}$$

In most cases we take $a = 1$. The coordinate v runs from 0 to ∞ while the variables u and ζ run from 0 to 2π. The coordinate u is also known as the *poloidal* angle.

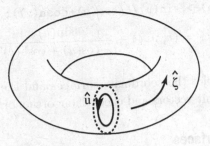

Figure 8.1: Toroidal coordinates

To visualize the toroidal coordinates we need the tools included in the `plots` package. Therefore, we start the worksheet by calling this package.

```
>  restart;
>  with(plots):
```

Warning, the name changecoords has been redefined

```
>  coordplot3d(toroidal,orientation=[163,125]);
```

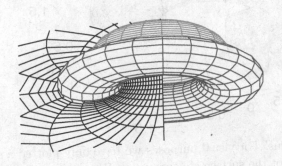

A Short Course in Mathematical Methods with Maple

Let us assume that $a = 1$. Then we define the toroidal coordinates as functions of v, u, ζ via

```
> x :=(v,u,zeta)-> cos(zeta)*sinh(v)/(-cos(u)+cosh(v));
```

$$x := (v, u, \zeta) \to \frac{\cos(\zeta)\sinh(v)}{-\cos(u) + \cosh(v)}$$

```
> y :=(v,u,zeta)-> sin(zeta)*sinh(v)/(-cos(u)+cosh(v)) ;
```

$$y := (v, u, \zeta) \to \frac{\sin(\zeta)\sinh(v)}{-\cos(u) + \cosh(v)}$$

```
> z :=(v,u,zeta)-> sin(u)/(-cos(u)+cosh(v));
```

$$z := (v, u, \zeta) \to \frac{\sin(u)}{-\cos(u) + \cosh(v)}$$

Next we are going to plot the coordinate surfaces and the coordinate lines. Then we will calculate the unit vectors and check their orthonormality.

8.3.1 Coordinate Surfaces

Surfaces of constant v are tori with major radii $R_v = \frac{a}{\tanh v}$ and minor radii $r_v = \frac{a}{\sinh v}$. To plot the surface of constant v we write

```
> plot3d([x(1.5,u,zeta),y(1.5,u,zeta),z(1.5,u,zeta)],
>     u=0..2*Pi,zeta=0..3*Pi/2,scaling=constrained,
>     orientation=[-72,59],axes=frame);
```

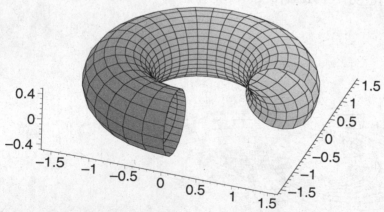

As expected we got a torus. For visual purposes we have only plotted a section of the torus. Next, let us plot the surface of constant u. We get

```
> plot3d([x(v,Pi/3,zeta),y(v,Pi/3,zeta),z(v,Pi/3,zeta)],
>     v=0..2,zeta=0..3*Pi/2,scaling=constrained,
>     orientation=[-72,59],axes=frame);
```

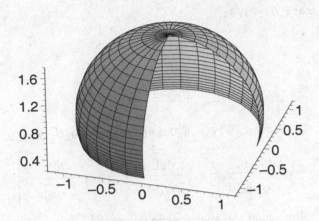

Finally, the surface of constant ζ is a plane

```
>  plot3d([x(v,u,Pi/4),y(v,u,Pi/4),z(v,u,Pi/4)],
>     v=0..2,u=0..2*Pi, scaling=constrained,
>     orientation=[-108,70],axes=boxed);
```

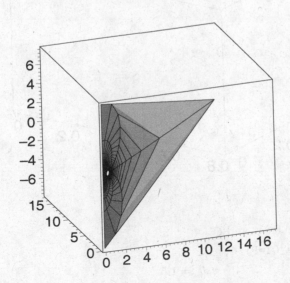

8.3.2 Coordinate Lines

Let us calculate (and visualize) the v-lines, u-lines and ζ-lines. The v-lines are the curves $r[v, u, \zeta]$ with u and ζ held fixed. Consider for example

```
> u_0:=Pi/3; zeta_0:=Pi/4;
```

$$u\_0 := \frac{1}{3}\pi$$

$$zeta\_0 := \frac{1}{4}\pi$$

Then,

```
> v_line:=[x(v,u_0,zeta_0),y(v,u_0,zeta_0),z(v,u_0,zeta_0)];
```

$$v\_line := \left[\frac{1}{2}\frac{\sqrt{2}\sinh(v)}{-\frac{1}{2}+\cosh(v)}, \ \frac{1}{2}\frac{\sqrt{2}\sinh(v)}{-\frac{1}{2}+\cosh(v)}, \ \frac{1}{2}\frac{\sqrt{3}}{-\frac{1}{2}+\cosh(v)} \right]$$

Next, let us plot the v-line using the spacecurve command

```
> spacecurve(v_line, v=0..2,
>    orientation=[145,60], axes=boxed,thickness=2);
```

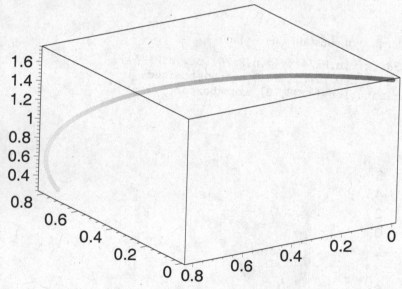

To obtain the u-line, we fix v, and ζ.

```
> v_0:=1.5; zeta_0:=Pi/4;
```

$$v\_0 := 1.5$$

$$zeta\_0 := \frac{1}{4}\pi$$

Vectors and Vector Calculus

> ```
u_line:=[x(v_0,u,zeta_0),y(v_0,u,zeta_0),z(v_0,u,zeta_0)];
```

$$u\_line := [1.064639728\,\frac{\sqrt{2}}{-\cos(u)+2.352409615},\ 1.064639728$$

$$\frac{\sqrt{2}}{-\cos(u)+2.352409615},\ \frac{\sin(u)}{-\cos(u)+2.352409615}]$$

Now, we can plot the *u*-line

> ```
spacecurve(u_line,u=0..2*Pi,axes=boxed,
    orientation=[145,60],thickness=2);
```

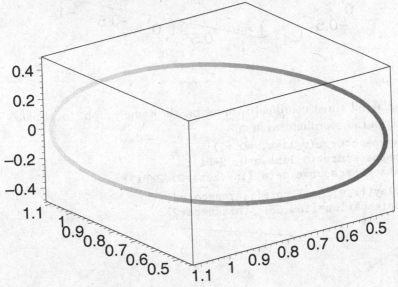

Finally, we obtain the ζ-line keeping v and u constants.

> ```
zeta_line:=[x(v_0,u_0,zeta),y(v_0,u_0,zeta),z(v_0,u_0,zeta)];
```
$$zeta\_line := [1.149464696\cos(\zeta),\ 1.149464696\sin(\zeta),\ 0.2699187026\,\sqrt{3}]$$

> ```
spacecurve(zeta_line,zeta=0..2*Pi,axes=boxed,
    orientation=[145,60],thickness=2);
```

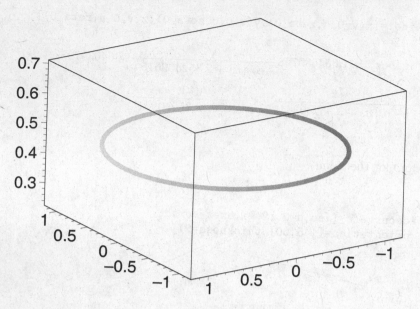

We can display all three coordinate curves on the same graph, to visualize the orthogonality of the coordinate system.

```
>  l_v:=spacecurve(v_line,v=0..2):
>  l_u:=spacecurve(u_line,u=0..2*Pi):
>  l_zeta:=spacecurve(zeta_line,zeta=0..2*Pi):
>  display({l_v,l_u,l_zeta}, axes=boxed,
>      orientation=[159,68], thickness=2);
```

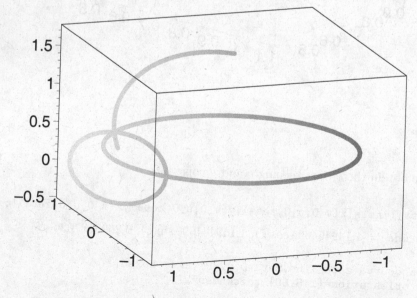

8.3.3 Unit Vectors

We conclude this part by calculating the unit vectors associated with the toroidal coordinates. In addition to the `LinearAlgebra` package we also need the package `RealDomain` in order to discard the unnecessary solutions in the complex domain that Maple may consider.

```
> with(LinearAlgebra):with(RealDomain):
```

```
Warning, these protected names have been redefined and
unprotected: Im, Re, ^, arccos, arccosh, arccot, arccoth, arccsc,
arccsch, arcsec, arcsech, arcsin, arcsinh, arctan, arctanh, cos,
cosh, cot, coth, csc, csch, eval, exp, expand, limit, ln, log,
sec, sech, signum, simplify, sin, sinh, solve, sqrt, surd, tan,
tanh
```

Denote by r the position vector of an arbitrary point.

```
> r:=<x(v,u,zeta),y(v,u,zeta),z(v,u,zeta)>;
```

$$r := \begin{bmatrix} \dfrac{\cos(\zeta)\sinh(v)}{-\cos(u)+\cosh(v)} \\[2mm] \dfrac{\sin(\zeta)\sinh(v)}{-\cos(u)+\cosh(v)} \\[2mm] \dfrac{\sin(u)}{-\cos(u)+\cosh(v)} \end{bmatrix}$$

Then the normal vector to v-surface is $r_v = \frac{\partial}{\partial v} r$

```
> r_v:=map(diff,r,v);
```

$$r\_v := \begin{bmatrix} \dfrac{\cos(\zeta)\cosh(v)}{-\cos(u)+\cosh(v)} - \dfrac{\cos(\zeta)\sinh(v)^2}{(-\cos(u)+\cosh(v))^2} \\[3mm] \dfrac{\sin(\zeta)\cosh(v)}{-\cos(u)+\cosh(v)} - \dfrac{\sin(\zeta)\sinh(v)^2}{(-\cos(u)+\cosh(v))^2} \\[3mm] -\dfrac{\sin(u)\sinh(v)}{(-\cos(u)+\cosh(v))^2} \end{bmatrix}$$

We have used the `map` command to apply the `diff` operator on each component of the vector. The scale coefficient h_v is the norm of r_v.

```
> h_v:=simplify(VectorNorm(r_v,2)) assuming cosh(v) > cos(u);
```

$$h\_v := \frac{1}{-\cos(u)+\cosh(v)}$$

Therefore, the unit vector e_v is the ratio of r_v to its norm h_v.

```
> e_v:=map(simplify, ScalarMultiply(r_v,(1/h_v)));
```

$$e\_v := \begin{bmatrix} \dfrac{\cos(\zeta)\,(\cosh(v)\cos(u) - 1)}{\cos(u) - \cosh(v)} \\[2ex] -\dfrac{\sin(\zeta)\,(\cosh(v)\cos(u) - 1)}{-\cos(u) + \cosh(v)} \\[2ex] \dfrac{\sin(u)\sinh(v)}{\cos(u) - \cosh(v)} \end{bmatrix}$$

Similarly we get:

```
> r_u:=map(diff,r,u);
```

$$r\_u := \begin{bmatrix} -\dfrac{\cos(\zeta)\sinh(v)\sin(u)}{(-\cos(u) + \cosh(v))^2} \\[2ex] -\dfrac{\sin(\zeta)\sinh(v)\sin(u)}{(-\cos(u) + \cosh(v))^2} \\[2ex] \dfrac{\cos(u)}{-\cos(u) + \cosh(v)} - \dfrac{\sin(u)^2}{(-\cos(u) + \cosh(v))^2} \end{bmatrix}$$

```
> h_u:=simplify(VectorNorm(r_u,2)) assuming cosh(v) > cos(u);
```

$$h\_u := \frac{1}{-\cos(u) + \cosh(v)}$$

```
> e_u:=map(simplify, ScalarMultiply(r_u,(1/h_u)));
```

$$e\_u := \begin{bmatrix} \dfrac{\cos(\zeta)\sinh(v)\sin(u)}{\cos(u) - \cosh(v)} \\[2ex] \dfrac{\sin(\zeta)\sinh(v)\sin(u)}{\cos(u) - \cosh(v)} \\[2ex] -\dfrac{\cosh(v)\cos(u) - 1}{\cos(u) - \cosh(v)} \end{bmatrix}$$

and respectively

```
> r_zeta:=map(diff,r,zeta);
```

$$r\_zeta := \begin{bmatrix} -\dfrac{\sin(\zeta)\sinh(v)}{-\cos(u) + \cosh(v)} \\[2ex] \dfrac{\cos(\zeta)\sinh(v)}{-\cos(u) + \cosh(v)} \\[2ex] 0 \end{bmatrix}$$

```
> h_zeta:= simplify(VectorNorm(r_zeta,2))
>    assuming cosh(v) > cos(u);
```

$$h\_zeta := \frac{\sqrt{\cosh(v)^2 - 1}}{-\cos(u) + \cosh(v)}$$

```
> e_zeta:=map(simplify, ScalarMultiply(r_zeta,(1/h_zeta)));
```

$$e\_zeta := \begin{bmatrix} -\dfrac{\sin(\zeta)\sinh(v)}{\sqrt{\cosh(v)^2 - 1}} \\[3ex] \dfrac{\cos(\zeta)\sinh(v)}{\sqrt{\cosh(v)^2 - 1}} \\[2ex] 0 \end{bmatrix}$$

Now, we can easily check now the orthonormality of the unit vectors:

```
> simplify( DotProduct(e_v,e_v) );
```
$$1$$

```
> simplify( DotProduct(e_u,e_u) );
```
$$1$$

```
> simplify( DotProduct(e_zeta,e_zeta) ) assuming cosh(v) > 1;
```
$$1$$

```
> simplify( DotProduct(e_v,e_u) );
```
$$0$$

```
> simplify( DotProduct(e_v,e_zeta) );
```
$$0$$

```
> simplify( DotProduct(e_u,e_zeta) );
```
$$0$$

We observe indeed that the vectors we have calculated are orthogonal and of length one.

8.4 Scalar and Vector Potentials

8.4.1 Scalar Potential

The scalar potential of a vector field $\vec{F}$ is a scalar field ϕ such that

$$\vec{\nabla}\phi = \vec{F} . \tag{8.9}$$

Maple has the built-in routine called `ScalarPotential` that calculates the scalar potential of a vector field.

Let us work the following example

```
> restart;
> with(VectorCalculus):
> SetCoordinates( cartesian[x,y,z]);
```
$$cartesian_{x,y,z}$$

```
> F:=VectorField(<x,y,z>);
```
$$F := x\,\bar{e}_x + y\,\bar{e}_y + z\,\bar{e}_z$$

```
> f:=ScalarPotential(F);
```

$$f := \frac{x^2}{2} + \frac{y^2}{2} + \frac{z^2}{2}$$

We can immediately check that f is indeed the scalar potential for the vector field $\vec{F}$.

```
>  Del(f);
```

$$x \, \bar{e}_x + y \, \bar{e}_y + z \, \bar{e}_z$$

As an exercise, let us change the coordinate system to the cylindrical system, and repeat the above procedure for a different vector field.

```
>  SetCoordinates( cylindrical[r,theta,z]);
```

$$cylindrical_{r,\,\theta,\,z}$$

```
>  F1:=VectorField(<r, 0,1>);
```

$$F1 := r \, \bar{e}_r + \bar{e}_z$$

```
>  f1:=ScalarPotential(F1);
```

$$f1 := \frac{1}{2} \, r^2 \cos(\theta)^2 + \frac{1}{2} \, r^2 \sin(\theta)^2 + z$$

```
>  Del(f1);
```

$$(r \cos(\theta)^2 + r \sin(\theta)^2) \, \bar{e}_r + e \, \bar{e}_z$$

```
>  simplify(%);
```

$$r \, \bar{e}_r + \bar{e}_z$$

In the case that there is no scalar field for a given vector field, Maple returns the *NULL* result (i.e. it doesn't print anything)

```
>  F2:=VectorField(<sin(theta), r, 1>);
```

$$F2 := \sin(\theta) \, \bar{e}_r + r \, \bar{e}_\theta + \bar{e}_z$$

```
>  f2:=ScalarPotential(F2);
```

$$f2 :=$$

8.4.2 Vector Potential

Given a vector field $\vec{F}$, we define its vector potential $\vec{U}$ as another vector field such that

$$\vec{F} = \vec{\nabla} \times \vec{U} . \tag{8.10}$$

Note that the vector potential is not unique. Indeed, because the curl of a gradient is zero, $\vec{\nabla} \times \left(\vec{\nabla} \phi \right) = 0$, the vector field $\vec{U}$ is defined up to the gradient of an arbitrary scalar ϕ.

The Maple routine VectorPotential calculates the vector potential of a given vector field. Its use is exemplified below.

```
>  restart;
```

```
>  with(VectorCalculus):
```

```
>  SetCoordinates( cartesian[x,y,z]);
```
$$cartesian_{x,y,z}$$

```
>  F:=VectorField(<-y,0,x>);
```
$$F := -y\,\bar{e}_x + x\,\bar{e}_z$$

```
>  U:=VectorPotential(F);
```
$$U := -x\,y\,\bar{e}_x + y\,z\,\bar{e}_y$$

Let us check that the curl of $\vec{U}$ gives $\vec{F}$ as expected

```
>  Del &x U;
```
$$-y\,\bar{e}_x + x\,\bar{e}_z$$

Finally, we add an arbitrary gradient field to the vector potential $\vec{U}$, $\vec{U}_a = \vec{U} + \vec{\nabla}\phi$ and verify that indeed $\vec{\nabla} \times \vec{U}_a$ gives the same vector field $\vec{F}$ as above. Note the use of **alias** command.

```
>  alias(phi=phi(x,y,z));
```
$$\phi$$

```
>  Ua:=U + Del(phi);
```
$$Ua := (-x\,y + (\tfrac{\partial}{\partial x}\,\phi))\,\bar{e}_x + (y\,z + (\tfrac{\partial}{\partial y}\,\phi))\,\bar{e}_y + (\tfrac{\partial}{\partial z}\,\phi)\,\bar{e}_z$$

```
>  Del &x Ua;
```
$$-y\,\bar{e}_x + x\,\bar{e}_z$$

If the vector potential does not exist, then Maple returns the *NULL* result (i.e. it prints nothing).

```
>  F1:=VectorField(<x*y, z^2, x-y>);
```
$$F1 := x\,y\,\bar{e}_x + z^2\,\bar{e}_y + (x - y)\,\bar{e}_z$$

```
>  U1:=VectorPotential(F1);
```
$$U1 :=$$

which confirms that there is no vector potential for the vector field $\vec{F}_1$.

8.5 Integrals

Using the **VectorCalculus** package we can calculate different types of integrals such as line integrals, surface integrals, as well as more complicated integral operations such as the flux of a vector field through a surface. These tools will prove useful in later calculations involving the Divergence Theorem, Gauss and Stokes Theorems, etc.

8.5.1   Line Integrals

We consider in the beginning simple two-dimensional line integrals. After we gain insight into these integration methods, we can easily generalize to the three-dimensional case.

For our calculations we need `VectorCalculus`, and `plots` packages, and we explicitly set up the two-dimensional cartesian coordinate system. We have

```
> restart; with(VectorCalculus): with(plots):
```

```
> SetCoordinates( cartesian[x,y]);
```

$$cartesian_{x,y}$$

Let us consider two integration paths. One path, which we denote by C_1, is the straight line from $(0,0)$ to $(1,1)$. The second path, C_2, is formed by two line segments: $(0,0)$ to $(1,0)$ and $(1,0)$ to $(1,1)$.

```
> C1:=Line(<0,0>, <1,1>);
```

$$C1 := \text{Line}(0\,ex,\ ex + ey)$$

```
> C2:=LineSegments(<0,0>,<1,0>,<1,1>);
```

$$C2 := \text{LineSegments}(0\,ex,\ ex,\ ex + ey)$$

We use `polygonplot` command to visualize the two integration paths

```
> path:=[[0,0],[1,0],[1,1]];
```

$$path := [[0, 0],\ [1, 0],\ [1, 1]]$$

```
> polygonplot(path, thickness=3);
```

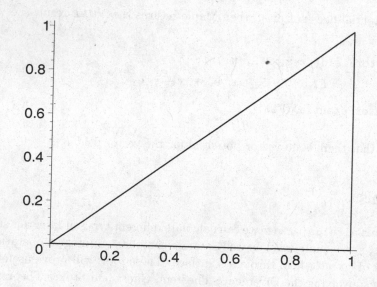

Vectors and Vector Calculus

Let us consider the vector field $\vec{F} = 2y\,\bar{e}_x + x\,\bar{e}_y$, and integrate it along the two paths C_1 and respectively C_2. We have

```
> F:=VectorField(<2*y,x>);
```
$$F := 2y\,\bar{e}_x + x\,\bar{e}_y$$

```
> LineInt(F,C1);
```
$$\frac{3}{2}$$

```
> LineInt(F,C2);
```
$$1$$

We observe that the two integrals are different. The fact that the integral depends on the path, is an indication that the vector field $\vec{F}$ cannot be expressed as the gradient of a scalar field. Indeed

```
> U:=ScalarPotential(F);
```
$$U :=$$

Maple returns the *NULL* answer, which signals the absence of a scalar potential for the field.

Let us consider another field, that has a scalar potential. We expect that its integral along the two different paths C_1 and C_2 will yield the same result.

```
> F1:=VectorField(<3*x^2*z, 2*y>);
```
$$F1 := 3x^2\,z\,\bar{e}_x + 2y\,\bar{e}_y$$

First, we check that the field has a scalar potential

```
> U1:=ScalarPotential(F1);
```
$$U1 := x^3\,z + y^2$$

and then, we calculate the line integrals

```
> LineInt(F1, C1);
```
$$z + 1$$

```
> LineInt(F1, C2);
```
$$z + 1$$

As expected the line integral of $\vec{F}_1$ is path independent.

Remarks

1) Line integrals can be calculated on some predefined curves. For example

```
> LineInt( VectorField(<y,-x>), Ellipse( x^2/25+y^2/4-1 ).);
```
$$-20\,\pi$$

2) If a formal expression is desired then use the 'inert' option

```
>  LineInt( VectorField( <y,-x> ),
>  Ellipse( x^2/25+y^2/4-1 ), 'inert' );
```

$$\int_0^{2\pi} -10\sin(t)^2 - 10\cos(t)^2 \, dt$$

For other options in defining the line integration see the help file with the `?LineInt` command.

8.5.2 Surface Integrals

There are two powerful routines in `VectorCalculus` package that allow surface integration: `Flux` and `SurfInt`.

Calculating the Flux of a Vector Field

The `Flux` routine calculate the flux of a vector field through a given surface.

```
>  restart; with(VectorCalculus):  with(plots):
```

```
>  SetCoordinates( cartesian[x,y,z]);
```
$$cartesian_{x,y,z}$$

For exemplification purposes we choose the vector field

$$\vec{F} = 3x\,\bar{e}_x + xyz\,\bar{e}_y\,,+(y-z)\,\bar{e}_z$$

and calculate its flux through the cylindric surface centered around z-axis having the radius 2 and height 2.
The field is given by

```
>  F:=VectorField(<3*x, x*y*z, y-z>);
```
$$F := 3x\,\bar{e}_x + xyz\,\bar{e}_y + (y-z)\,\bar{e}_z$$

while the surface is parameterized by

```
>  cyl:=Surface(<2*cos(u), 2*sin(u),v>, u=0..2*Pi, v=-1..1);
```
$$cyl := \text{Surface}(2\cos(u)\,ex + 2\sin(u)\,ey + v\,ez,\ u = 0..2\pi,\ v = -1..1)$$

Let us plot the vector field

```
>  fieldplot3d(F,x=-2..2, y=-2..2,z=-1..1,axes=boxed);
```

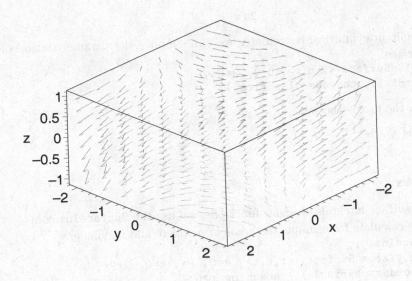

and the surface

```
> plot3d([2*cos(u), 2*sin(u),v],u=0..2*Pi,v=-1..1,
>     shading=zgrayscale, thickness=1, axes=boxed);
```

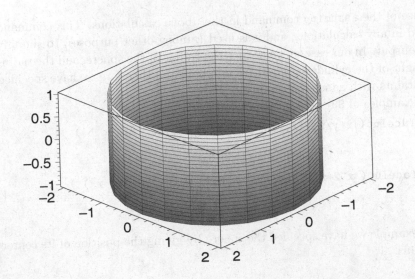

To calculate the flux we simply write

```
> Flux(F,cyl);
```

$$24\pi$$

Using the built in cylindrical coordinates we can simplify the parametrization of the same surface.

```
>   cyl_1:=Surface(<2,u,v>, u=0..2*Pi,
>       v=-1..1, coords=cylindrical):
```

As expected, the flux is the same

```
> Flux(F,cyl_1);
```

$$24\pi$$

Surface Integrals

To calculate surface integrals of scalar functions, we use the `SurfaceInt` command. Let us calculate for example the areas of a sphere and a cylinder

```
>  SurfaceInt( 1,
>      [x,y,z] = Surface( <r,s,t>, s=0..Pi, t=0..2*Pi,
>      coords=spherical ) ) assuming r>0;
```

$$4\pi r^2$$

```
> SurfaceInt( 1,
>      [x,y,z]= Surface( <r, u, v>, u=0..2*Pi, v=0..h,
>      coords=cylindrical) ) assuming h > 0 assuming r>0;
```

$$2rh\pi$$

Remark

Note the use of the `assuming` command in the above calculations. This command can be used in any calculations, and it is used (among other purposes) to simplify the Maple output. In our case we knew that the radius of the sphere, and the radius and the height of the cylinder are positive numbers and therefore we have specified these constraints with `assuming` command.

Other examples of `SurfaceInt` command are

```
> SurfaceInt( x*y/z, [x,y,z] = Box( 1..2, 2..3, 4..5 ) );
```

$$-30\ln(2) + 15\ln(5) + \frac{27}{16}$$

```
> SurfaceInt( x^2*z^2, [x,y,z] = Sphere( <0,0,0>, r ) );
```

$$\frac{4r^6\pi}{15}$$

In the last example we have specified the sphere by giving the position of its center and the radius.

8.5.3   Divergence Theorem

The divergence theorem states that the flux of a vector field $\vec{F}$ through a closed surface S is equal to the volume integral of the field's divergence $\vec{\nabla} \cdot \vec{F}$ calculated over the volume V, enclosed by the surface S

$$\oint_S \vec{F} \cdot d\vec{A} = \int_V \vec{\nabla} \cdot \vec{F} \, dV . \tag{8.11}$$

Let us exemplify the divergence theorem considering a concrete vector field

$$\vec{F} = -5\,x^2\,\bar{\mathbf{e}}_x + y\,\bar{\mathbf{e}}_y + z\,x\,\bar{\mathbf{e}}_z \,,$$

and a sphere of radius 3 centered in the origin. We have

```
> SetCoordinates( cartesian[x,y,z]);
```
$$cartesian_{x,\,y,\,z}$$

```
> F:=VectorField(<-5*x^2, y, z*x>);
```
$$F := -5\,x^2\,\bar{\mathbf{e}}_x + y\,\bar{\mathbf{e}}_y + z\,x\,\bar{\mathbf{e}}_z$$

```
> S:=Sphere(<0,0,0>,3);
```
$$S := \text{Sphere}(0\,ex,\,3)$$

```
> Flux(F,S);
```
$$36\,\pi$$

Remark
Note the use of the `Flux` command, and the `Sphere` option in defining the integration surface. See `?Flux` help file for more options.

Now, let us calculate the divergence of $\vec{F}$

```
> divF:=Del .   F;
```
$$divF := 1 - 9\,x$$

and then calculate its volume integral. For this, we rewrite the $\vec{\nabla} \cdot \vec{F}$ in spherical coordinates

```
> divF_s:=subs({x=R*sin(u)*cos(v),
>    y=R*sin(u)*sin(v),
>    z=R*cos(u)},
>    divF);
```
$$divF\_s := 1 - 9\,R\sin(u)\cos(v)$$

In spherical coordinates, the integration becomes

```
> intV:=int(
>    int( int(divF_s*R^2*sin(u), R=0..3),
>    u=0..Pi), v=0..2*Pi);
```
$$intV := 36\,\pi$$

We obtain the same result as above, in accord with the divergence theorem.

8.5.4   Stokes Theorem

According to Stokes theorem, the circulation of vector field $\vec{F}$ along a closed curve C, is equal to the flux of the curl of the field $\vec{\nabla} \times \vec{F}$, through any open surface S bounded by the closed curve C

$$\oint_C \vec{F} \cdot \vec{dl} = \int_S \vec{\nabla} \times \vec{F} \cdot \vec{dA} . \tag{8.12}$$

Let us consider a concrete example

```
>  restart:  with(VectorCalculus):  with(plots):
>  SetCoordinates( cartesian[x,y,z]);
```
$$cartesian_{x,y,z}$$
```
>  F:=VectorField( <y^3, -x^3,0>);
```
$$F := y^3\,\bar{e}_x - x^3\,\bar{e}_y$$

We want to check Stokes theorem for $\vec{F}$, along the unit circle, in xy plane with the center at origin. Let us visualize the vector field and the contour of integration.

```
>  display3d(fieldplot3d
>     (F, x=-1..1,y=-1..1, z=-0.5..0.5, grid=[5,5,5],
>      scaling=constrained,thickness=2),
>      spacecurve([cos(t),sin(t),0], t=0..2*Pi,
>      axes=boxed, color=black, thickness=3,
>      orientation=[65,43]) );
```

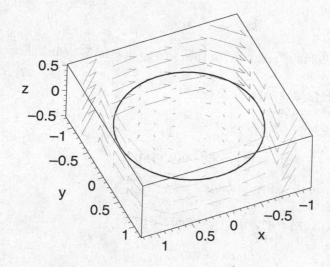

We calculate the circulation of the field with `LineInt` command, where we specify the circle by its center, radius and normal vector. We obtain

```
> Stokes_left:=LineInt( F, Circle3D(<0,0,0>, 1, <0,0,1>));
```

$$Stokes\_left := -\frac{3\pi}{2}$$

```
> curlF:=Del &x F;
```

$$curlF := (-3x^2 - 3y^2)\,\bar{e}_z$$

As the surface for the flux integration we choose the disc in xy plane with unit radius and centered at the origin. We get (using cylindrical coordinates to parameterize the disk)

```
> Stokes_right:=Flux(curlF, Surface(<r,u,0>, r=0..1,u=0..2*Pi,
> coords=cylindrical));
```

$$Stokes\_right := -\frac{3\pi}{2}$$

Observe that the two results are the same, as expected from the Stokes theorem.

8.6 Problems

Problem 8.6.1. Apply the Gram-Schmidt orthonormalization procedure to orthonormalize the following three vectors

$$w_1 = \begin{bmatrix} 2 \\ 1 \\ 0 \\ -1 \end{bmatrix}, \quad w_2 = \begin{bmatrix} 1 \\ 0 \\ 2 \\ -1 \end{bmatrix}, \quad w_3 = \begin{bmatrix} 0 \\ -1 \\ 1 \\ 0 \end{bmatrix}.$$

Verify that the resultant vectors are orthonormal.

Solution

```
> restart;
> ?LinearAlgebra[GramSchmidt]
> with(LinearAlgebra):
> w1 := <2,1,0,-1>;
```

$$w1 := \begin{bmatrix} 2 \\ 1 \\ 0 \\ -1 \end{bmatrix}$$

```
> w2 := <1,0,2,-1>;
```

$$w2 := \begin{bmatrix} 1 \\ 0 \\ 2 \\ -1 \end{bmatrix}$$

```
>  w3 := <0,-1,1,0>;
```

$$w3 := \begin{bmatrix} 0 \\ -1 \\ 1 \\ 0 \end{bmatrix}$$

```
>  new := GramSchmidt([w1,w2,w3], normalized);
```

$$new := \left[\begin{bmatrix} \dfrac{\sqrt{6}}{3} \\ \dfrac{\sqrt{6}}{6} \\ 0 \\ -\dfrac{\sqrt{6}}{6} \end{bmatrix} , \begin{bmatrix} 0 \\ -\dfrac{\sqrt{2}}{6} \\ \dfrac{2\sqrt{2}}{3} \\ -\dfrac{\sqrt{2}}{6} \end{bmatrix} , \begin{bmatrix} \dfrac{1}{2} \\ \dfrac{-5}{6} \\ \dfrac{-1}{6} \\ \dfrac{1}{6} \end{bmatrix} \right]$$

```
>  new[1] . new[1];
```

$$1$$

```
>  new[1] . new[2];
```

$$0$$

```
>  new[1] . new[3];
```

$$0$$

The other orthonormality relations can be computed similarly.

Problem 8.6.2. Find the values of α for which the three vectors

$$\vec{v}_1 = \begin{bmatrix} 1 \\ 1+\alpha \\ 1 \end{bmatrix}, \quad \vec{v}_2 = \begin{bmatrix} \alpha \\ 0 \\ 1 \end{bmatrix}, \quad \vec{v}_3 = \begin{bmatrix} -1 \\ -2 \\ 0 \end{bmatrix},$$

are linearly dependent.

Problem 8.6.3. Find the projection $\text{Proj}_A \vec{B}$ of vector $\vec{B} = -\mathbf{e_x} + 2\,\mathbf{e_y} + \mathbf{e_z}$ along the vector $\vec{A} = \mathbf{e_x} - \mathbf{e_y}$.

Problem 8.6.4. For four arbitrary three-dimensional vectors $\vec{A}, \vec{B}, \vec{C}$ and $\vec{D}$ show by direct computation that

$$\left(\vec{A} \times \vec{B}\right) \cdot \left(\vec{C} \times \vec{D}\right) = \left(\vec{A} \cdot \vec{C}\right)\left(\vec{B} \cdot \vec{D}\right) - \left(\vec{A} \cdot \vec{D}\right)\left(\vec{B} \cdot \to C\right).$$

Problem 8.6.5. Show that

$$\vec{\nabla} \cdot \left(g\vec{\nabla}f \times f\vec{\nabla}g\right) = 0$$

Problem 8.6.6. Find potential ϕ of the vector field:

$$(x^2 + y^2 + z^2)^{-3/2} (x\,\bar{e}_x + y\,\bar{e}_y + z\,\bar{e}_z).$$

Problem 8.6.7.
(a) Find the curl and div for the vector field $\vec{F} = x^2\,\bar{e}_x + e^y\,\bar{e}_y + sin(z)\,\bar{e}_z$
(b) Is there a scalar field whose gradient is $\vec{F}$?

Problem 8.6.8. The velocity vector in the uniform circular motion is given by $\vec{v} = \vec{\omega} \times \vec{r}$ where $\vec{\omega}$ is the constant rotation vector. Show that
(a) $\vec{\nabla} \cdot \vec{v} = 0$
(b) $\vec{\nabla} \times \vec{v} = 2\,\vec{\omega}$
(c) $\vec{\nabla} (\vec{\omega} \cdot \vec{r}) = \vec{\omega}$

Problem 8.6.9. Consider the scalar field $f = z + x^3 - 4xyz$. Compute and plot the gradient field $\vec{\nabla} f$.

Problem 8.6.10.
(a) Plot the following vector fields $\vec{F} = xyz\,\bar{e}_x - z^3\,\bar{e}_z$ and $\vec{\nabla} \times \vec{F}$.
(b) Find the scalar potential of $\vec{\nabla} \times \left(\vec{\nabla} \times \vec{F} \right)$.

Problem 8.6.11. Calculate the divergence and the curl of $\vec{F} = u\,\bar{e}_u + v\,\bar{e}_v$ in toroidal (u, v, ζ) coordinates. (*Hint:* Use the SetCoordinates command)

Problem 8.6.12. Use the AddCoordinates command to define your own cylindric coordinate system as

$$x = r\sin(\theta) , \ y = r\cos(\theta) , \ z = z .$$

In the new system, calculate find the action of $\vec{\nabla} \cdot$ and $\vec{\nabla} \times$ operators on the vector field $\vec{F} = r^2\,\bar{e}_r - r\sin(\theta)\,\bar{e}_z$.

Problem 8.6.13. Check the divergence theorem for the vector field $\vec{F} = r^7 \sin(\theta)\,\bar{e}_r$ using as volume the sphere of radius ρ centered in the origin.

Problem 8.6.14. Check Stokes' theorem for the vector field $\vec{F} = xy\,\bar{e}_x + 3yz\,\bar{e}_y - zx\,\bar{e}_z$ using the contour defined by the triangle $(0, 0, 0), (2, 0, 0), (0, 0, 3)$.

Problem 8.6.15. Check that for an arbitrary vector field $\vec{F}$, we have $\oint_S \vec{\nabla} \times \vec{F} \cdot \overrightarrow{dA} = 0$ for any closed surface S.

Chapter 9

Matrices and Rotations

IN Maple one can define matrices by using the Matrix palette from the user graphic interface, or we can directly write the matrix elements with the help of the angle-bracket (< >) notation. Other ways of defining matrices include the **Matrix** command, user defined procedures to specify directly the matrix elements, etc. Maple manipulates matrices as objects, and many matrix operations are intuitively defined.

9.1   Matrix Definition

Using the package **LinearAlgebra**, we can define a matrix either by specifying each column, or by specifying each row. For example, let us consider the 2×3 matrix

$$A = \begin{bmatrix} 1 & 2 & 3 \\ 4 & 5 & 6 \end{bmatrix}. \tag{9.1}$$

After loading the **LinearAlgebra** package, we can either write the matrix by listing its columns

```
>   with(LinearAlgebra):
>   A:=<<1,4>|<2,5>|<3,6>>;
```

$$A := \begin{bmatrix} 1 & 2 & 3 \\ 4 & 5 & 6 \end{bmatrix}$$

or we can write the same matrix by listing its rows

```
>   A1:=<<1|2|3>,<4|5|6>>;
```

$$A1 := \begin{bmatrix} 1 & 2 & 3 \\ 4 & 5 & 6 \end{bmatrix}$$

Remark

The matrix definition syntax comes basically from the way the vector are defined within the `LinearAlgebra` package: a column vector is written as `<a,b,c>` while a row vector is written as `<a|b|c>`. Then, in our example, in the first case the matrix A is defined as a *row vector* whose elements are the *column vectors* `<1,4>`, `<2,5>` and respectively `<3,6>`, while in the second case, the same matrix is seen as a *column vector* whose elements are the *row vectors* `<1|2|3>` and respectively `<4|5|6>`.

9.2   Operations with Matrices

The `LinearAlgebra` package performs all the usual matrix operations. Many commands are implemented redundantly, by specifying either a short hand notation, or a long name. For example the matrix multiplication of two matrices A and B can be written either intuitively as `A.B`, or in the long form `MatrixMatrixMultiply(A,B)`. In this chapter we will use mainly the short hand notation.

9.2.1   Matrix Multiplication

A $k \times n$ matrix multiplying a n-dimensional vector yields a k-dimensional vector. In other words we can see a $k \times n$ matrix as an operator that maps a n-dimensional vector into a k-dimensional vector. We arrive at the same conclusion using the matrix multiplication rules. Multiplying a $k \times n$ matrix with a $n \times 1$ matrix we get a $k \times 1$ matrix. For exemplification, let us consider

```
>   v:=<a,b,c>;
```

$$v := \begin{bmatrix} a \\ b \\ c \end{bmatrix}$$

To multiply the matrix A from equation (9.1) above with the vector v we write simply

```
>   A.v;
```

$$\begin{bmatrix} a + 2b + 3c \\ 4a + 5b + 6c \end{bmatrix}$$

More generally, for two matrices

```
>   M1:=<<1,0,0,2>|<0,1,1,5>|<1,2,3,4>>;
```

$$M1 := \begin{bmatrix} 1 & 0 & 1 \\ 0 & 1 & 2 \\ 0 & 1 & 3 \\ 2 & 5 & 4 \end{bmatrix}$$

```
> M2:=<<0,0,2>|<1,-2,1>|<0,0,-1>|<1,0,0>>;
```

$$M2 := \begin{bmatrix} 0 & 1 & 0 & 1 \\ 0 & -2 & 0 & 0 \\ 2 & 1 & -1 & 0 \end{bmatrix}$$

we have

```
> M1.M2;
```

$$\begin{bmatrix} 2 & 2 & -1 & 1 \\ 4 & 0 & -2 & 0 \\ 6 & 1 & -3 & 0 \\ 8 & -4 & -4 & 2 \end{bmatrix}$$

and respectively

```
> M2.M1;
```

$$\begin{bmatrix} 2 & 6 & 6 \\ 0 & -2 & -4 \\ 2 & 0 & 1 \end{bmatrix}$$

9.2.2 Complex Conjugation

The complex conjugation of a matrix returns another matrix whose entries are the complex conjugates of the initial matrix. Let us consider a 2×2 matrix with complex entries

$$C = \begin{bmatrix} 1 + 2i & 1 - i \\ 3 & 5 \end{bmatrix}. \tag{9.2}$$

In Maple we write:

```
> C:=<<1+2*I, 3>|<1-I, 5>>;
```

$$C := \begin{bmatrix} 1 + 2I & 1 - I \\ 3 & 5 \end{bmatrix}$$

If we apply the `conjugate` command on the entire matrix we get

```
> conjugate(C);
```

```
Error, invalid input: simpl/conjugate expects its 1st argument, a, to
be of type algebraic, but received Matrix(2, 2, [[...],[...]],
datatype = anything, storage = rectangular, order = Fortran_order,
shape = [])
```

The error message suggests that we should apply the `conjugate` command on each matrix entry rather than on the whole matrix. One way of doing this is to use the `map` command applies (or maps) a procedure to each operand of an expression. We have

```
> map(conjugate,C);
```

$$\begin{bmatrix} 1-2I & 1+I \\ 3 & 5 \end{bmatrix}$$

9.2.3 Matrix Transposition

To calculate the transpose of C from equation (9.2) we use the `Transpose` command

> `Ct:=Transpose(C);`

$$Ct := \begin{bmatrix} 1+2I & 3 \\ 1-I & 5 \end{bmatrix}$$

9.2.4 Hermitian Conjugation

The Hermitian conjugate of C (or the adjoint matrix) is obtained be taking the complex conjugate of its transposed form. We can accomplish this in Maple in two ways. Either we take the complex conjugate of Ct matrix above

> `Ch:=map(conjugate,Ct);`

$$Ch := \begin{bmatrix} 1-2I & 3 \\ 1+I & 5 \end{bmatrix}$$

or we use the built-in command `HermitianTranspose`

> `HermitianTranspose(C);`

$$\begin{bmatrix} 1-2I & 3 \\ 1+I & 5 \end{bmatrix}$$

As we have expected the two results are the same.

9.2.5 The Inverse of a Matrix

To calculate the inverse matrix C^{-1} we simply write:

> `C1:=MatrixInverse(C);`

$$C1 := \begin{bmatrix} \dfrac{10}{173} - \dfrac{65}{173}I & \dfrac{11}{173} + \dfrac{15}{173}I \\ \dfrac{-6}{173} + \dfrac{39}{173}I & \dfrac{28}{173} - \dfrac{9}{173}I \end{bmatrix}$$

Now we can immediately check that $C\,C^{-1} = C^{-1}C = 1$, where 1 is the 2×2 identity matrix.

> `C1.C;`

$$\begin{bmatrix} 1 & 0 \\ 0 & 1 \end{bmatrix}$$

> `C.C1;`

$$\begin{bmatrix} 1 & 0 \\ 0 & 1 \end{bmatrix}$$

A shorthand notation for the inverse matrix is `1/C`.

> `C_inv:=1/C;`

$$C\_inv := \begin{bmatrix} \dfrac{10}{173} - \dfrac{65}{173}I & \dfrac{11}{173} + \dfrac{15}{173}I \\ \dfrac{-6}{173} + \dfrac{39}{173}I & \dfrac{28}{173} - \dfrac{9}{173}I \end{bmatrix}$$

We can easily check that `C1` and `C_inv` are the same matrix

> `C_inv - C1;`

$$\begin{bmatrix} 0 & 0 \\ 0 & 0 \end{bmatrix}$$

Note that in the `LinearAlgebra` package, many operations with matrices are written in a natural form, close to the way we write them on paper. For example, adding (or subtracting) two matrices is written simply as $A + B$ (or $A - B$). Multiplying two matrices is written as $A.B$ and so on.

9.2.6   The Determinant of a Matrix

The determinant of C from equation (9.2) is calculated with

> `Determinant(C);`

$$2 + 13I$$

Let us check some properties of determinants.

Property 1

The determinant of a matrix doesn't change if we add to a column another column multiplied with a constant.

Therefore we expect that the matrix

$$M = \begin{bmatrix} a & e & h \\ b & f & i \\ c & g & j \end{bmatrix}, \qquad (9.3)$$

and the matrix

$$M_1 = \begin{bmatrix} a + ke & e & h \\ b + kf & f & i \\ c + kg & g & j \end{bmatrix}, \qquad (9.4)$$

will have the same determinant.

Let us check this property in Maple. First consider the matrix M and calculate its determinant.

```
>   M:=<<a,b,c>|<e,f,g>|<h,i,j>>;
```

$$M := \begin{bmatrix} a & e & h \\ b & f & i \\ c & g & j \end{bmatrix}$$

```
>   dM:=Determinant(M);
```

$$dM := afj - aig + bgh - bej + cei - cfh$$

Next, write the matrix M_1, calculate its determinant, and compare it with the determinant of M.

```
>   M1:=<<a+k*e,b+k*f,c+k*g>|<e,f,g>|<h,i,j>>;
```

$$M1 := \begin{bmatrix} a+ke & e & h \\ b+kf & f & i \\ c+kg & g & j \end{bmatrix}$$

```
>   dM1:=Determinant(M);
```

$$dM1 := afj - aig + bgh - bej + cei - cfh$$

```
>   dM - dM1;
```

$$0$$

As expected the two determinants coincide. The same rule holds for rows.

Property 2

If we exchange the position of two rows, or two columns in a matrix, the determinant of the resulting matrix differs only by a sign from the determinant of the initial matrix.

```
>   M:=<<a,b,c>|<e,f,g>|<h,i,j>>;
```

$$M := \begin{bmatrix} a & e & h \\ b & f & i \\ c & g & j \end{bmatrix}$$

```
>   M1:=<<a,b,c>|<h,i,j>|<e,f,g>>;
```

$$M1 := \begin{bmatrix} a & h & e \\ b & i & f \\ c & j & g \end{bmatrix}$$

```
>   simplify( Determinant(M)/Determinant(M1));
```

$$-1$$

Property 3

For any $n \times n$ matrix M and any scalar k we have $|kM| = k^n |M|$, where $|M|$ is the determinant of M.

```
>   M2 := k*M;
```

$$M2 := k \begin{bmatrix} a & e & h \\ b & f & i \\ c & g & j \end{bmatrix}$$

```
> simplify(Determinant(M2)/Determinant(M));
```

$$k^3$$

9.3 Eigenvectors and Eigenvalues

Given a matrix G, we say that a nonzero vector $\vec{v}$ is an eigenvector of G, if there exist a $\lambda \neq 0$ such that $G\,\vec{v} = \lambda\,\vec{v}$.

Let us consider a 3×3 matrix

$$G = \begin{bmatrix} 1 & 1 & 0 \\ 1 & 2 & 1 \\ 0 & 3 & 1 \end{bmatrix} \tag{9.5}$$

To find its eigenvalues and eigenvectors in Maple we will use the **Eigenvectors** command

```
> G:=<<1,1,0>|<1,2,1>|<0,3,1>>;
```

$$G := \begin{bmatrix} 1 & 1 & 0 \\ 1 & 2 & 3 \\ 0 & 1 & 1 \end{bmatrix}$$

```
> Eigenvectors(G);
```

$$\begin{bmatrix} \frac{3}{2} + \frac{\sqrt{17}}{2} \\ \frac{3}{2} - \frac{\sqrt{17}}{2} \\ 1 \end{bmatrix}, \begin{bmatrix} 1 & 1 & -3 \\ \frac{1}{2} + \frac{\sqrt{17}}{2} & \frac{1}{2} - \frac{\sqrt{17}}{2} & 0 \\ 1 & 1 & 1 \end{bmatrix}$$

The **Eigenvectors(G)** command solves the eigenvalue problem $G\,\vec{v} = \lambda\,\vec{v}$ by returning an expression sequence whose first member is a vector containing the eigenvalues of G, and whose second member is a matrix whose columns are the eigenvectors of G. For example, in our case the third eigenvalue is $\lambda_3 = 1$ and the corresponding eigenvector is $\vec{v}_3 = (-3, 0, 1)$ which is the third column of the returned matrix.

Let us check by direct computation that indeed v_3 is an eigenvector of G with the eigenvalue 1. In first step we extract the eigenvector matrix which is the second member of the sequence returned by the **Eigenvectors(G)** command. Note the use of the **ditto** operator (the percent % sign) to simplify the command line

```
> G_vects:=%[2];
```

$$G\_vects := \begin{bmatrix} 1 & 1 & -3 \\ \dfrac{1}{2} + \dfrac{\sqrt{17}}{2} & \dfrac{1}{2} - \dfrac{\sqrt{17}}{2} & 0 \\ 1 & 1 & 1 \end{bmatrix}$$

By this command we have assigned the name `G_vects` to the matrix containing the eigenvectors of `G` matrix. In the next step, we extract the third column of `G_vects`, which is exactly the eigenvector we were looking for.

```
>   v3:=G_vects[1..3,3];
```

$$v3 := \begin{bmatrix} -3 \\ 0 \\ 1 \end{bmatrix}$$

Now we can check directly that indeed v_3 is an eigenvector of G with the corresponding eigenvalue $\lambda_3 = 1$.

```
>   G.v3;
```

$$\begin{bmatrix} -3 \\ 0 \\ 1 \end{bmatrix}$$

If we need the floating point approximation for the eigenvector problem we should use the `evalf` command

```
>   evalf(Eigenvectors(G));
```

$$\begin{bmatrix} 3.561552813 \\ -0.561552813 \\ 1. \end{bmatrix}, \begin{bmatrix} 1. & 1. & -3. \\ 2.561552813 & -1.561552813 & 0. \\ 1. & 1. & 1. \end{bmatrix}$$

9.3.1 Diagonalization of Matrices

As we have seen in the first part of the book, a square $n \times n$ matrix A, which possesses n linearly independent vectors, can be diagonalized via the transformation

$$A \mapsto V^{-1}AV \equiv D. \tag{9.6}$$

where the matrix V is constructed in such a way that its columns are the eigenvectors of A.

The following worksheet diagonalizes a 4×4 matrix, using the recipe described above.

```
>   restart;
>   with(LinearAlgebra):
>   A:=<<-32, 6, 10,13>| <-111,22,31,42>|
>       <19,-4,-1,-7>| <-44,8,12,19>>;
```

$$A := \begin{bmatrix} -32 & -111 & 19 & -44 \\ 6 & 22 & -4 & 8 \\ 10 & 31 & -1 & 12 \\ 13 & 42 & -7 & 19 \end{bmatrix}$$

```
> EA:=Eigenvectors(A);
```

$$EA := \begin{bmatrix} 2 \\ 3 \\ -1 \\ 4 \end{bmatrix}, \begin{bmatrix} -4 & \dfrac{-11}{2} & \dfrac{-11}{2} & -5 \\ 1 & 1 & 1 & 1 \\ 1 & \dfrac{3}{2} & \dfrac{3}{2} & 1 \\ 1 & \dfrac{5}{2} & 2 & 2 \end{bmatrix}$$

```
> V:=EA[2];
```

$$V := \begin{bmatrix} -4 & \dfrac{-11}{2} & \dfrac{-11}{2} & -5 \\ 1 & 1 & 1 & 1 \\ 1 & \dfrac{3}{2} & \dfrac{3}{2} & 1 \\ 1 & \dfrac{5}{2} & 2 & 2 \end{bmatrix}$$

```
> Diag1:=V^(-1) .A. V;
```

$$Diag1 := \begin{bmatrix} 2 & 0 & 0 & 0 \\ 0 & 3 & 0 & 0 \\ 0 & 0 & -1 & 0 \\ 0 & 0 & 0 & 4 \end{bmatrix}$$

Another (and more direct) way of diagonalizing a matrix without calculating its eigenvectors explicitly is to calculate the Jordan form of the matrix using the JordanForm command

```
> Q := JordanForm(A, output='Q');
```

$$Q := \begin{bmatrix} 11 & -4 & -11 & 5 \\ -2 & 1 & 2 & -1 \\ -3 & 1 & 3 & -1 \\ -4 & 1 & 5 & -2 \end{bmatrix}$$

Then, we have

```
> Diag2:=Q^(-1) .A. Q;
```

$$\begin{bmatrix} -1 & 0 & 0 & 0 \\ 0 & 2 & 0 & 0 \\ 0 & 0 & 3 & 0 \\ 0 & 0 & 0 & 4 \end{bmatrix}$$

We arrive at the same type of diagonal matrix as before, with the difference that the order of the eigenvalues on the diagonal may have changed.

9.4 Special Matrices

9.4.1 Orthogonal Matrices

By definition, an orthogonal matrix B is a matrix whose transpose is equal with its inverse

$$B^T = B^{-1} . \tag{9.7}$$

For example the matrix

$$\frac{1}{3}\begin{bmatrix} 2 & -2 & 1 \\ 1 & 2 & 2 \\ 2 & 1 & -2 \end{bmatrix} \tag{9.8}$$

is orthogonal. We can easily check the orthogonality of B using Maple. First we define the matrix B as

```
> B:=(1/3).<<2,1,2>|<-2,2,1>|<1,2,-2>>;
```

$$B := \begin{bmatrix} \frac{2}{3} & \frac{-2}{3} & \frac{1}{3} \\ \frac{1}{3} & \frac{2}{3} & \frac{2}{3} \\ \frac{2}{3} & \frac{1}{3} & \frac{-2}{3} \end{bmatrix}$$

then, we calculate its transpose

```
> Bt:=Transpose(B);
```

$$Bt := \begin{bmatrix} \frac{2}{3} & \frac{1}{3} & \frac{2}{3} \\ \frac{-2}{3} & \frac{2}{3} & \frac{1}{3} \\ \frac{1}{3} & \frac{2}{3} & \frac{-2}{3} \end{bmatrix}$$

We can immediately check that $B \cdot B^T = 1$

```
> B.Bt;
```

$$\begin{bmatrix} 1 & 0 & 0 \\ 0 & 1 & 0 \\ 0 & 0 & 1 \end{bmatrix}$$

Let us check by direct computation that indeed the transpose and the inverse are equal. The inverse of B is

```
>  B1:=1/B;
```

$$B1 := \begin{bmatrix} \dfrac{2}{3} & \dfrac{1}{3} & \dfrac{2}{3} \\[2mm] \dfrac{-2}{3} & \dfrac{2}{3} & \dfrac{1}{3} \\[2mm] \dfrac{1}{3} & \dfrac{2}{3} & \dfrac{-2}{3} \end{bmatrix}$$

and therefore we have as expected

```
>  Bt -B1;
```

$$\begin{bmatrix} 0 & 0 & 0 \\ 0 & 0 & 0 \\ 0 & 0 & 0 \end{bmatrix}$$

Maple has a built in command `IsOrthogonal` to check whether or not a matrix is orthogonal.

```
>  IsOrthogonal(B);
```

$$true$$

Finally, let us check that the column vectors that form an orthogonal matrix are orthonormal. We use the command `Column(M,n)` that extracts the column n from matrix M.

```
>  v1:=Column(B,1);
```

$$v1 := \begin{bmatrix} \dfrac{2}{3} \\[2mm] \dfrac{1}{3} \\[2mm] \dfrac{2}{3} \end{bmatrix}$$

```
>  v2:=Column(B,2);
```

$$v2 := \begin{bmatrix} \dfrac{-2}{3} \\[2mm] \dfrac{2}{3} \\[2mm] \dfrac{1}{3} \end{bmatrix}$$

```
>  v3:=Column(B,3);
```

$$v3 := \begin{bmatrix} \dfrac{1}{3} \\[2mm] \dfrac{2}{3} \\[2mm] \dfrac{-2}{3} \end{bmatrix}$$

Then the scalar products of $\vec{v}_1$, $\vec{v}_2$ and $\vec{v}_3$ are respectively

```
> v1.v2;
```
$$0$$

```
> v1.v3;
```
$$0$$

```
> v2.v3;
```
$$0$$

```
> v1.v1;
```
$$1$$

```
> v2.v2;
```
$$1$$

```
> v3.v3;
```
$$1$$

as expected for an orthogonal matrix.
The eigenvalues of B are the roots of its characteristic polynomial.

```
> CharacteristicPolynomial(B,lambda);
```
$$\lambda^3 - \frac{2}{3}\lambda^2 - \frac{2}{3}\lambda + 1$$

```
> solve(%,lambda);
```
$$-1, \frac{5}{6} + \frac{1}{6}I\sqrt{11}, \frac{5}{6} - \frac{1}{6}I\sqrt{11}$$

Equivalently, the `Eigenvalues(B)` command produces the same result

```
> Eigenvalues(B);
```
$$\begin{bmatrix} -1 \\ \dfrac{5}{6} + \dfrac{1}{6}I\sqrt{11} \\ \dfrac{5}{6} - \dfrac{1}{6}I\sqrt{11} \end{bmatrix}$$

9.4.2 Unitary Matrices

A nonsingular matrix U whose Hermitian conjugate equals its inverse is called unitary

$$U^\dagger = U^{-1} . \tag{9.9}$$

If U is a real matrix, then its Hermitian conjugate is just its transpose, hence a real unitary matrix is an orthogonal one. Let us analyze the following matrix

```
> U:=<<0,sqrt(2)/2,0,I*sqrt(2)/2>|
>     <sqrt(2)/2,0,I*sqrt(2)/2,0>|<0,I*sqrt(2)/2,0,sqrt(2)/2>|
>     <I*sqrt(2)/2,0,sqrt(2)/2,0>>;
```

$$U := \begin{bmatrix} 0 & \dfrac{\sqrt{2}}{2} & 0 & \dfrac{1}{2} I \sqrt{2} \\[2mm] \dfrac{\sqrt{2}}{2} & 0 & \dfrac{1}{2} I \sqrt{2} & 0 \\[2mm] 0 & \dfrac{1}{2} I \sqrt{2} & 0 & \dfrac{\sqrt{2}}{2} \\[2mm] \dfrac{1}{2} I \sqrt{2} & 0 & \dfrac{\sqrt{2}}{2} & 0 \end{bmatrix}$$

First using the `IsUnitary` command, we check immediately if the matrix is unitary or not

```
> IsUnitary(U);
```

$$true$$

Next, we check explicitly that $U^\dagger = U^{-1}$

```
> Uh:=HermitianTranspose(U);
```

$$Uh := \begin{bmatrix} 0 & \dfrac{\sqrt{2}}{2} & 0 & \dfrac{-1}{2} I \sqrt{2} \\[2mm] \dfrac{\sqrt{2}}{2} & 0 & \dfrac{-1}{2} I \sqrt{2} & 0 \\[2mm] 0 & \dfrac{-1}{2} I \sqrt{2} & 0 & \dfrac{\sqrt{2}}{2} \\[2mm] \dfrac{-1}{2} I \sqrt{2} & 0 & \dfrac{\sqrt{2}}{2} & 0 \end{bmatrix}$$

```
> U.Uh;
```

$$\begin{bmatrix} 1 & 0 & 0 & 0 \\ 0 & 1 & 0 & 0 \\ 0 & 0 & 1 & 0 \\ 0 & 0 & 0 & 1 \end{bmatrix}$$

Finally, we check that the determinant of U is ± 1 and that its eigenvalues are the roots of the unity

```
> Determinant(U);
```

$$1$$

```
> CharacteristicPolynomial(U,lambda);
```

$$\lambda^4 + 1$$

```
> solve(%,lambda);
```

$$\dfrac{\sqrt{2}}{2} + \dfrac{1}{2} I \sqrt{2}, \; -\dfrac{\sqrt{2}}{2} + \dfrac{1}{2} I \sqrt{2}, \; -\dfrac{\sqrt{2}}{2} - \dfrac{1}{2} I \sqrt{2}, \; \dfrac{\sqrt{2}}{2} - \dfrac{1}{2} I \sqrt{2}$$

Using the command `Eigenvalues` we obtain the same result.

> Eigenvalues(U);

$$\begin{bmatrix} \dfrac{\sqrt{2}}{2} + \dfrac{1}{2}I\sqrt{2} \\[2ex] -\dfrac{\sqrt{2}}{2} + \dfrac{1}{2}I\sqrt{2} \\[2ex] \dfrac{\sqrt{2}}{2} - \dfrac{1}{2}I\sqrt{2} \\[2ex] -\dfrac{\sqrt{2}}{2} - \dfrac{1}{2}I\sqrt{2} \end{bmatrix}$$

9.4.3 Hermitian Matrices

A nonsingular matrix H is called Hermitian if it is equal with its own Hermitian conjugate

$$H = (H^T)^* . \tag{9.10}$$

It is known that the eigenvalues of a Hermitian matrix are all real, and that the eigenvectors corresponding to different eigenvalues are orthogonal. Let us check these properties, using a concrete example. We have

> H:=<<0,I,I>|<-I,1,0>|<-I,0,1>>;

$$H := \begin{bmatrix} 0 & -I & -I \\ I & 1 & 0 \\ I & 0 & 1 \end{bmatrix}$$

> Hh:=HermitianTranspose(H);

$$Hh := \begin{bmatrix} 0 & -I & -I \\ I & 1 & 0 \\ I & 0 & 1 \end{bmatrix}$$

Therefore, H is a Hermitian matrix. Next, we check that it is a nonsingular matrix, and that its eigenvalues are all real

> Determinant(H);

$$-2$$

> P:=CharacteristicPolynomial(H,lambda);

$$P := \lambda^3 - 2\lambda^2 - \lambda + 2$$

To see that all the roots of the characteristic polynomial of a hermitian matrix are real, it is very intuitive to represent graphically its characteristic polynomial. We have

> plot(P,lambda=-2..3,y=-5..5,color=black,thickness=3);

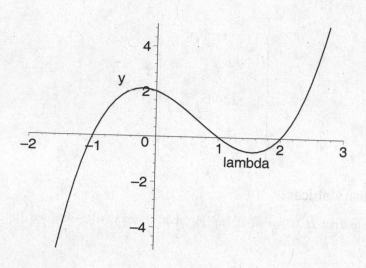

We see that there are three real solutions in the vicinity of $\lambda = -1, \lambda = 1$ and $\lambda = 3$. Indeed, solving the equation $P = 0$ we get

```
>  solve(P=0,lambda);
```

$$-1, 1, 2$$

Equivalently, we can use the **Eigenvectors** command

```
>  EV:=Eigenvectors(H);
```

$$EV := \begin{bmatrix} 2 \\ -1 \\ 1 \end{bmatrix}, \begin{bmatrix} 1 & 1 & 0 \\ I & \dfrac{-1}{2}I & -1 \\ I & \dfrac{-1}{2}I & 1 \end{bmatrix}$$

Note that the **Eigenvectors** command returns a list with two members. The first member is a column vector containing the eigenvalues of the matrix, while the second term is a matrix that has on the first column the first eigenvector, on the second column the second eigenvector, etc.

To isolate the eigenvectors we first extract from the previous result the eigenvector matrix

```
>  Vects:=EV[2];
```

$$Vects := \begin{bmatrix} 1 & 1 & 0 \\ I & \dfrac{-1}{2}I & -1 \\ I & \dfrac{-1}{2}I & 1 \end{bmatrix}$$

Now we can isolate each eigenvector of H using the `Column` command

> `v1:=Column(Vects,1);`

$$v1 := \begin{bmatrix} 1 \\ I \\ I \end{bmatrix}$$

> `v2:=Column(Vects,2);`

$$v2 := \begin{bmatrix} 1 \\ \dfrac{-1}{2}I \\ \dfrac{-1}{2}I \end{bmatrix}$$

> `v3:=Column(Vects,3);`

$$v3 := \begin{bmatrix} 0 \\ -1 \\ 1 \end{bmatrix}$$

Finally, let us calculate the dot products $\vec{v}_i \cdot \vec{v}_j$ for $i \neq j$

> `v1.v2;`

$$0$$

> `v1.v3;`

$$0$$

> `v2.v3;`

$$0$$

As expected, the eigenvectors $\vec{v}_1$, $\vec{v}_2$ and $\vec{v}_3$ having different (real) eigenvalues, are orthogonal to each other.

9.5 Defining Matrices in Maple: General Properties

The `Matrix` command within `LinearAlgebra` package allows for a great flexibility in defining a matrix. The general syntax is

> `Matrix(r, c, init, ro, sym, sc, sh, st, o, dt, f, a);`

where the optional parameters are

r – row dimension of Matrix; a non-negative integer, or integer range.

c – column dimension of Matrix; a non-negative integer, or integer range.

init – initial values for the Matrix; Maple procedure, table, array, list, Array, Matrix, Vector, set of equations, or expression of type algebraic

ro – equation of the form readonly=true or false; specifies whether Matrix entries can be changed

sym – equation of the form symbol=name; specifies the symbolic name to be used for the Matrix entries

sc – equation of the form scan=name or scan=list specifying the structure and/or data order for interpreting initial values; interpreter for initial values in parameter init

sh – equation of the form shape=name or shape=list specifying one or more built-in or user-defined indexing functions; storage allocation for Matrix entries

st – equation of the form storage=name, where name is a permitted storage mode; storage requirements for Matrix entries

o – equation of the form order=name, where name is either C_order or Fortran_order; specifies if Matrix entries are stored by rows or columns

dt – equation of the form datatype=name, where name is any Maple type, complex[8], or integer[n] for n=1,2,4, or 8; type of data stored in Matrix

f – equation of the form fill=value, where value is of the type specified by the dt parameter; specifies Matrix entries at locations not otherwise set

a – equation of the form attributes=list, where list specifies permitted attributes; specifies additional mathematical properties of the Matrix

give some examples of using the `Matrix` command.

```
>   restart;
>   with(LinearAlgebra):
>   # Matrix(r, c, init, ro, sym, sc, sh, st, o, dt, f, a)
```

The command `Matrix(r)` defines a square $r \times r$ matrix with all entries given by the `fill` parameter (default=0).

```
>   Matrix(3);
```

$$\begin{bmatrix} 0 & 0 & 0 \\ 0 & 0 & 0 \\ 0 & 0 & 0 \end{bmatrix}$$

Similarly, the command `Matrix(r,c)` defines a matrix with `r` rows and `c` columns whose entries are given by `fill` parameter (default=0).

```
>   Matrix(2,5);
```

$$\begin{bmatrix} 0 & 0 & 0 & 0 & 0 \\ 0 & 0 & 0 & 0 & 0 \end{bmatrix}$$

We can define the `fill` parameter to be simply a number, for example

```
> Matrix(2,5,fill=7);
```

$$\begin{bmatrix} 7 & 7 & 7 & 7 & 7 \\ 7 & 7 & 7 & 7 & 7 \end{bmatrix}$$

If the first two options r and c are given, then the fill option is considered automatically, i.e. there is not need to specify it separately

```
> Matrix(2,5,7);
```

$$\begin{bmatrix} 7 & 7 & 7 & 7 & 7 \\ 7 & 7 & 7 & 7 & 7 \end{bmatrix}$$

Even if it looks easier not to use the full option with the fill parameter, for clarity, it is a good idea to use the fill option in full.

The Matrix(r,c,init) command constructs a matrix with r rows and c columns whose entries are determined by the parameter init (and parameter fill if not all the entries are described by init). The parameter init can be any of the following: *procedure, expression of type algebraic, table, set of equations, array or Array, Matrix, list or list of lists*, and *vector*. For example, the init can be a function

```
> f:= (i,j) -> x^(i+j-4);
```

$$f := (i,\, j) \rightarrow x^{(i+j-4)}$$

```
> Matrix(2,5,f);
```

$$\begin{bmatrix} \dfrac{1}{x^2} & \dfrac{1}{x} & 1 & x & x^2 \\ \dfrac{1}{x} & 1 & x & x^2 & x^3 \end{bmatrix}$$

or a procedure

```
> g:= (i,j) -> if i<3 and j < 4 then int(x^(i+j-4),x) else 0 fi;
```

$$g := \mathbf{proc}(i,\, j)$$
$$\textbf{option } operator,\ arrow;$$
$$\textbf{if } i < 3 \textbf{ and } j < 4 \textbf{ then } \mathrm{int}(x^{(i+j-4)},\, x) \textbf{ else } 0 \textbf{ end if}$$
$$\textbf{end proc}$$

```
> Matrix(3,5,g);
```

$$\begin{bmatrix} -\dfrac{1}{x} & \ln(x) & x & 0 & 0 \\ \ln(x) & x & \dfrac{x^2}{2} & 0 & 0 \\ 0 & 0 & 0 & 0 & 0 \end{bmatrix}$$

A matrix can be defined as a diagonal matrix, whose entries (on the diagonal) are defined by a vector. For example

```
> vec:=Vector([1,2,3,a,b]);
```

$$vec := \begin{bmatrix} 1 \\ 2 \\ 3 \\ a \\ b \end{bmatrix}$$

```
> Matrix(vec,shape=diagonal);
```

$$\begin{bmatrix} 1 & 0 & 0 & 0 & 0 \\ 0 & 2 & 0 & 0 & 0 \\ 0 & 0 & 3 & 0 & 0 \\ 0 & 0 & 0 & a & 0 \\ 0 & 0 & 0 & 0 & b \end{bmatrix}$$

If we want to modify only one element of a matrix, it is easy to do so by just redefining it

```
> A:=Matrix(2,4,fill=a);
```

$$A := \begin{bmatrix} a & a & a & a \\ a & a & a & a \end{bmatrix}$$

```
> A[1,3]:=b;
```

$$A_{1,3} := b$$

```
> A;
```

$$\begin{bmatrix} a & a & b & a \\ a & a & a & a \end{bmatrix}$$

In some situations, we do not want to alter the entries in a matrix. Then, we can define the matrix to be read only.

```
> A:=Matrix(2,4,fill=a,readonly=true);
```

$$A := \begin{bmatrix} a & a & a & a \\ a & a & a & a \end{bmatrix}$$

If we want to change an entry

```
> A[1,3]:=b;
```

```
Error, cannot assign to a read-only Matrix
```

we get an error message, reminding us that the matrix was "locked" as read only. This property is useful when we define matrices that we don't want to change throughout the calculations. The read only property, as well as other matrix properties can be added later, using the MatrixOptions command. For example, let us define the three-dimensional delta Kronecker symbol as

```
> delta:=Matrix(Vector([1,1,1]),shape=diagonal);
```

$$\delta := \begin{bmatrix} 1 & 0 & 0 \\ 0 & 1 & 0 \\ 0 & 0 & 1 \end{bmatrix}$$

If we call `MatrixOptions` with the name of the matrix as the only argument, we get a list of the properties of that matrix

```
> MatrixOptions(delta);
```

shape = [*diagonal*], *datatype* = *anything*, *storage* = *diagonal*, *order* = *Fortran_order*

Now, we use `MatrixOptions` command to add the *read-only* property to the matrix entries

```
> MatrixOptions(delta, readonly=true);
> delta[1,1];
```

$$1$$

```
> delta[1,1] :=0;
```

```
Error, cannot assign to a read-only Matrix
```

We can see now, that the matrix δ got a new property, namely the *read-only*

```
> MatrixOptions(delta);
```

shape = [*diagonal*], *datatype* = *anything*, *storage* = *diagonal*, *order* = *Fortran_order*, *readonly*

9.5.1 Matrix Function

The `MatrixFunction` command defines the function $f(A)$ for a square matrix A.

```
> restart; with(LinearAlgebra):
> A:=<<1,0,0>|<0,0,a>|<a,1,0>>;
```

$$A := \begin{bmatrix} 1 & 0 & a \\ 0 & 0 & 1 \\ 0 & a & 0 \end{bmatrix}$$

```
> MatrixFunction(A, 1+x-x^2, x);
```

$$1 + \begin{bmatrix} 1 & 0 & a \\ 0 & 0 & 1 \\ 0 & a & 0 \end{bmatrix} - \begin{bmatrix} 1 & 0 & a \\ 0 & 0 & 1 \\ 0 & a & 0 \end{bmatrix}^2$$

The first argument is the matrix itself, the second one is the function, and the third argument specifies the variable. Note that by default the `MatrixFunction` command does not simplify the result. But we can always use the `simplify` command.

```
> simplify(%);
```

$$\begin{bmatrix} 1 & -a^2 & 0 \\ 0 & 1-a & 1 \\ 0 & a & 1-a \end{bmatrix}$$

In the next example we consider a function with two variables. Now we see the relevance of the third argument in the syntax of the `MatrixFunction` command. We have

> `MatrixFunction(A, 1+t*x^2, t);`

$$1 + \begin{bmatrix} 1 & 0 & a \\ 0 & 0 & 1 \\ 0 & a & 0 \end{bmatrix} x^2$$

> `simplify(%);`

$$\begin{bmatrix} x^2+1 & 0 & x^2 a \\ 0 & 1 & x^2 \\ 0 & x^2 a & 1 \end{bmatrix}$$

which gives a different result than

> `MatrixFunction(A, 1+t*x^2, x);`

$$1 + t \begin{bmatrix} 1 & 0 & a \\ 0 & 0 & 1 \\ 0 & a & 0 \end{bmatrix}^2$$

> `simplify(%);`

$$\begin{bmatrix} t+1 & t\,a^2 & t\,a \\ 0 & t\,a+1 & 0 \\ 0 & 0 & t\,a+1 \end{bmatrix}$$

Finally, let us see some less trivial examples

> `B:=<<a,0>|<a,1>>;`

$$B := \begin{bmatrix} a & a \\ 0 & 1 \end{bmatrix}$$

> `MatrixFunction(B, sin(u), u);`

$$\begin{bmatrix} \sin(a) & \dfrac{a\,(\sin(a)-\sin(1))}{a-1} \\ 0 & \sin(1) \end{bmatrix}$$

> `MatrixFunction(B, exp(y), y);`

$$\begin{bmatrix} e^a & \dfrac{a\,(e^a-e)}{a-1} \\ 0 & e \end{bmatrix}$$

In addition, we can apply matrix options such as **readonly=true** etc., similar to the options of the **Matrix** command. For more information, see **?MatrixFunction** help file.

9.6 Matrix Exponential and the Group of Rotations

A set G is called a *group* if there exists an operation $\circ : G \times G \mapsto G$ such that: (1) the product $g_1 \circ g_2$ is in G for any g_1, g_2 in G (closure); (2)$g_1 \circ (g_2 \circ g_3) = (g_1 \circ g_2) \circ g_3$ for any g_1, g_2, g_3 in G (associativity); (3) there exist an unique element e in G such that $g \circ e = e \circ g = g$ for any g in G (identity element); and (4) for any g there exist an inverse g^{-1} in G such that $g \circ g^{-1} = g^{-1} \circ g = e$ (inverse element).

For example, the set of all invertible $n \times n$ matrices form a group. The closure is automatically satisfied as the product of two $n \times n$ matrices is a $n \times n$ matrix, the associativity can be checked immediately from the matrix product rules, the identity element is the $n \times n$ identity matrix, and the inverse is satisfied by definition. This group is called the general linear group GL_n. All other examples of matrix groups are subgroups of GL_n.

We know that when acting with a matrix A on a vector $\vec{v}$, we get another vector $\vec{u}$. We write $\vec{u} = A\vec{v}$. If $|\vec{v}| = |\vec{u}|$ (their lengths are the same) we say that A is a rotation matrix. The reader can check immediately that the rotation matrices form a group.

We can characterize the rotation of the coordinate system by the angle ϕ around the axis $\vec{\omega} = (\omega_1, \omega_2, \omega_3)$ via the rotation matrix $R(\omega, \phi)$:

$$\vec{r'} = R(\omega, \phi)\vec{r} \ . \tag{9.11}$$

Here $\vec{r'}$ represents the position vector in the new (rotated) coordinate system, $\vec{r}$ is the old vector, and the rotation matrix is given by

$$R(\omega, \phi) = e^{-\phi\Omega} \ , \quad \text{where } \Omega = \begin{bmatrix} 0 & -\omega_3 & \omega_2 \\ \omega_3 & 0 & -\omega_1 \\ -\omega_2 & \omega_1 & 0 \end{bmatrix} \ . \tag{9.12}$$

Remember that the exponential of the matrix A, is a short hand form for the infinite series $e^A = 1 + A + 1/2! \, A^2 + \cdots$

Let us use Maple to explore these ideas.

```
>  restart;

>  with(LinearAlgebra):

>  Omega:=
      <<0, omega[3], -omega[2]>|
      <-omega[3], 0, omega[1]>|
      <omega[2], -omega[1], 0>>;
```

$$\Omega := \begin{bmatrix} 0 & -\omega_3 & \omega_2 \\ \omega_3 & 0 & -\omega_1 \\ -\omega_2 & \omega_1 & 0 \end{bmatrix}$$

The general form of the rotation matrix for an arbitrary Ω is quite complicated, therefore, we first calculate the rotation matrix around the z direction, where $\vec{\omega} = (0, 0, 1)$. Denote

```
> Omega_z:=subs({omega[1]=0, omega[2]=0, omega[3]=1},Omega);
```

$$Omega\_z := \begin{bmatrix} 0 & -1 & 0 \\ 1 & 0 & 0 \\ 0 & 0 & 0 \end{bmatrix}$$

To calculate the rotation matrix we use `MatrixExponential` command which calculates the exponential of a given matrix. We obtain

```
> R_z:=MatrixExponential(-phi*Omega_z);
```

$$R\_z := \begin{bmatrix} \cos(\phi) & \sin(\phi) & 0 \\ -\sin(\phi) & \cos(\phi) & 0 \\ 0 & 0 & 1 \end{bmatrix}$$

Note that the `MatrixExponential` command understands the product of a scalar with a matrix ($-\phi\Omega$ in our case).

It is instructive to visualize this rotation for a concrete example. If we consider the rotation with $60°$ around the z axis, we have

```
> R_z1:=subs(phi=Pi/3, R_z);
```

$$R\_z1 := \begin{bmatrix} \dfrac{1}{2} & \dfrac{\sqrt{3}}{2} & 0 \\[2mm] -\dfrac{\sqrt{3}}{2} & \dfrac{1}{2} & 0 \\[2mm] 0 & 0 & 1 \end{bmatrix}$$

We apply this rotation to the following vector

```
> old_vec:=<1,1,1>;
```

$$old\_vec := \begin{bmatrix} 1 \\ 1 \\ 1 \end{bmatrix}$$

After rotation, we have

```
> new_vec := R_z1 . old_vec;
```

$$new\_vec := \begin{bmatrix} \cos(\frac{\pi}{3}) + \sin(\frac{\pi}{3}) \\[2mm] -\sin(\frac{\pi}{3}) + \cos(\frac{\pi}{3}) \\[2mm] 1 \end{bmatrix}$$

Finally, let us visualize this rotation

```
> with(plots):
```

```
>   ov:=arrow(old_vec, shape=cylindrical_arrow, color=green,
>       width=[1/10, relative=false]):

>   nv:=arrow(new_vec, shape=cylindrical_arrow, color=red,
>       width=[1/10, relative=false]):

>   display(ov,nv, axes=normal, orientation=[20,76]);
```

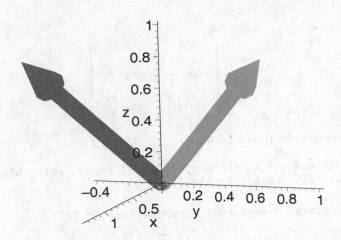

Let us analyze one more example, by considering the rotation around x axis. In this case $\vec{\omega} = (1,0,0)$, and consequently the Ω matrix becomes

```
>   Omega_x:=subs({omega[1]=1, omega[2]=0, omega[3]=0},Omega);
```

$$Omega\_x := \begin{bmatrix} 0 & 0 & 0 \\ 0 & 0 & -1 \\ 0 & 1 & 0 \end{bmatrix}$$

The corresponding rotation matrix is

```
>   R_x:=MatrixExponential(-phi*Omega_x);
```

$$R\_x := \begin{bmatrix} 1 & 0 & 0 \\ 0 & \cos(\phi) & \sin(\phi) \\ 0 & -\sin(\phi) & \cos(\phi) \end{bmatrix}$$

As in the previous example, we choose the rotation angle $\phi = 60^{o}$. We obtain

```
>   R_x1:=subs(phi=Pi/3, R_x);
```

$$R\_x1 := \begin{bmatrix} 1 & 0 & 0 \\ 0 & \dfrac{1}{2} & \dfrac{\sqrt{3}}{2} \\ 0 & -\dfrac{\sqrt{3}}{2} & \dfrac{1}{2} \end{bmatrix}$$

The transformed vector reads

```
>  new_vec_x:= R_x1 .  old_vec;
```

$$new\_vec\_x := \begin{bmatrix} 1 \\ \cos(\dfrac{\pi}{3}) + \sin(\dfrac{\pi}{3}) \\ -\sin(\dfrac{\pi}{3}) + \cos(\dfrac{\pi}{3}) \end{bmatrix}$$

Finally, let us visualize the rotation with 60° around the x axis

```
>  nvx:=arrow(new_vec_x, shape=cylindrical_arrow, color=blue,
>     width=[1/10, relative=false]):

>  display(ov,nvx, axes=normal, orientation=[20,76]);
```

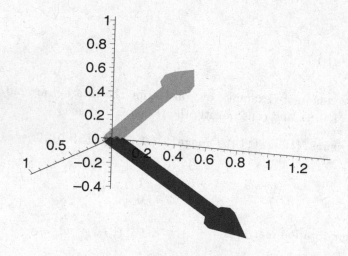

Similar results are obtained by considering the rotation around y-axis. The reader is encouraged to work completely this example.

9.7 Problems

Problem 9.7.1. As an example, let us consider the matrix

$$A := \begin{bmatrix} 0 & -1 & -2 & -3 \\ 1 & 0 & -1 & -2 \\ 2 & 1 & 0 & -1 \end{bmatrix}$$

(a) Examine the matrix entries, and use the `Matrix` command in conjunction with a carefully designed function of matrix indices $f(i,j)$ to generate this matrix.
(b) Define $B = A A^T$ and calculate its eigenvalues, determinant, and its rank.
(c) Find the eigenvectors of B and represent them graphically.

Solution.

By inspecting the matrix A we observe that its entries $A_{ii} = 0$. This suggest that a possible function that generates the entries is $f(i,j) = i - j$. A quick test, shows that indeed this is the case. Then we have

```
> restart: with(LinearAlgebra):
> f:=(i,j)-> i -j;
```

$$f := (i, j) \rightarrow i - j$$

```
> A:=Matrix(3,4,f);
```

$$A := \begin{bmatrix} 0 & -1 & -2 & -3 \\ 1 & 0 & -1 & -2 \\ 2 & 1 & 0 & -1 \end{bmatrix}$$

Note, how efficient is the command `Matrix`. Next, let's calculate B

```
> B:=A . Transpose (A);
```

$$B := \begin{bmatrix} 14 & 8 & 2 \\ 8 & 6 & 4 \\ 2 & 4 & 6 \end{bmatrix}$$

```
> Eigenvalues(B);
```

$$\begin{bmatrix} 0 \\ 20 \\ 6 \end{bmatrix}$$

One zero eigenvalue signals the possible degeneracy of the matrix. Indeed, the determinant of B is

```
> Determinant(B);
```

$$0$$

and its rank is

```
> Rank(B);
```

Next, let us find the eigenvectors of B. We get

```
> EV:=Eigenvectors(B);
```

$$EV := \begin{bmatrix} 6 \\ 0 \\ 20 \end{bmatrix}, \begin{bmatrix} \dfrac{-1}{2} & 1 & 3 \\ \dfrac{1}{4} & -2 & 2 \\ 1 & 1 & 1 \end{bmatrix}$$

To extract the eigenvectors of B we have to extract the columns of the matrix which is the second object on the list returned by the `Eigenvectors` command. Let us accomplish this task by making use of the power of programming language built in Maple. The one-line command reads

```
>  for i to 3 do V||i := Column(EV[2],i) od;
```

$$V1 := \begin{bmatrix} \dfrac{-1}{2} \\ \dfrac{1}{4} \\ 1 \end{bmatrix}$$

$$V2 := \begin{bmatrix} 1 \\ -2 \\ 1 \end{bmatrix}$$

$$V3 := \begin{bmatrix} 3 \\ 2 \\ 1 \end{bmatrix}$$

The `for` command provides the ability to execute a statement sequence repeatedly, either for a counted number of times (using the `for...to` clauses) or until a condition is satisfied (using the `while` clause). Both forms of clauses can be present simultaneously. For more details, call the Maple help file using `?for` command. In the line above, we have closed the `do` execution with `od` which is an handy archaism coming from old Maple versions. The reader is encouraged to use (for consistency) the modern `end do` command instead.

The next command `V||i` is an example of a *concatenation* operator. The result of `V||1` is simply `V1`. Similarly, `V||2` yields `V2` and so on. The reader should learn about the concatenation operator from the examples provided in the corresponding Maple help file. See `?||` for more details.

Let us check if the three vectors are linearly independent. For this it is enough to calculate the product $\vec{V}_1 \cdot \vec{V}_2 \times \vec{V}_3$

```
>  V1 .  V2 &x V3;
```

$$\frac{-21}{2}$$

Equivalently, we can check if the **Basis** commands returns a valid result. We obtain

> `Basis([V1, V2, V3]);`

$$\left[\left[\begin{array}{c} \dfrac{-1}{2} \\ \dfrac{1}{4} \\ 1 \end{array}\right], \left[\begin{array}{c} 3 \\ 2 \\ 1 \end{array}\right], \left[\begin{array}{c} 1 \\ -2 \\ 1 \end{array}\right]\right]$$

which confirms the linear independence of $\vec{V}_1, \vec{V}_2$ and $\vec{V}_3$.

In the last step, we will represent graphically the three vectors.

> `with(plots):`

Warning, the name changecoords has been redefined

> `for j to 3 do ar||j := arrow(V||j, shape=cylindrical_arrow,`
> `    width=[1/10,relative=false]) od:`

> `display(ar1, ar2, ar3, axes=normal, scaling=unconstrained);`

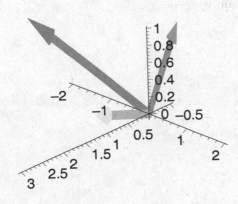

Problem 9.7.2. Given is the 3×3 matrix

$$B = \frac{1}{2}\left[\begin{array}{ccc} 1 & \sqrt{2} & 1 \\ -\sqrt{2} & 0 & \sqrt{2} \\ 1 & -\sqrt{2} & 1 \end{array}\right]$$

(a) Show that B is a rotation matrix.

(b) Find eigenvector of B corresponding to the eigenvalue $\lambda = 1$.

(c) Find the axis of rotation $\hat{n}$ and the angle of rotation Φ about $\hat{n}$ corresponding to B.

Problem 9.7.3. Find the value of the parameter m such that $det(A) = 1$ where the matrix A is given by

$$A = \begin{bmatrix} 1 & m+2 & 0 & m+3 \\ 1-m & 1 & m & 0 \\ m & -1 & m & -2 \\ 0 & 0 & 1 & 0 \end{bmatrix}$$

Problem 9.7.4. Given the matrix

$$A = \begin{bmatrix} 1 & 4 & 2 & 2 \\ 2 & 3 & 1 & 3 \\ 3 & 2 & 3 & 1 \\ 4 & 1 & 4 & 4 \end{bmatrix}$$

(a) Write a Maple program that extracts the columns of the matrix and labels them v1, v2, v3 and respectively v4.
(b) Check if the four vectors above are linearly independent.
(c) Construct another matrix $B = [\vec{v}_1, \vec{v}_4, \vec{v}_3, \vec{v}_2]$ and find the ratio of $det(A)/det(B)$. Explain your result.

Chapter 10

Differential Equations

E SSENTIALLY, Maple solves most differential equations using only one single command: dsolve. If the differential equation cannot be solved exactly, the user can call upon a rich repertoire of numerical methods to solve the equation.

10.1 First Order Differential Equations

A differential equation relates a function with its derivatives up to the n-th order. The simplest differential equation is the first order equation $y' = F(x, y)$.

10.1.1 Separable Equations

A first order differential equation is separable if $y' = -M(x,y)/N(x,y)$. For example, the

```
>   deq:=diff(y(x),x) = x/(1+x^2)/(1+y(x));
```

$$deq := \frac{d}{dx} y(x) = \frac{x}{(1+x^2)(1+y(x))}$$

is separable.

In general Maple takes over the task of solving the differential equations, with little or no input from the user. The command used is dsolve.

```
>   dsolve(deq,y(x));
```

$$y(x) = -1 + \sqrt{1 + \ln(1+x^2) + 2\_C1}, \ y(x) = -1 - \sqrt{1 + \ln(1+x^2) + 2\_C1}$$

Note that Maple found two solutions, and also introduced the arbitrary integration constant _C1. We can further specify the solution if we impose initial conditions.

```
>   ic:= y(0)=1;
```

$$ic := y(0) = 1$$

Then,

```
> dsolve({deq, ic}, y(x));
```

$$y(x) = -1 + \sqrt{4 + \ln(1 + x^2)}$$

By specifying the initial condition Maple determines the integration constant, and in this case it also isolates one solution. Observe the syntax for the `dsolve` command: the initial condition is introduced inside the curly brackets together with the differential equation.

Maple has a powerful tool to characterize differential equations, called `odeadvisor`, which resides inside the `DEtools` package.

```
> with(DEtools):
> odeadvisor(deq);
```

$$[\_separable]$$

We learn that the above equation is a separable equation. Given an ordinary differential equation, the `odeadvisor` command's main goal is to classify it according to standard text books, and to display a help page including related information for solving it (when the word 'help' is given as an extra argument).

```
> odeadvisor(deq,help);
```

$$[\_separable]$$

In this case, Maple produces a help file with examples and book references on separable differential equations.

An interesting equation is the the skydiver equation: when a skydiver jumps out of an airplane, it will feel the combined effects of gravity and wind resistance. If $v(t)$ represents the vertical velocity, a good model for this situation is

```
> deq1:= diff(v(t),t) = g - k*v(t);
```

$$deq1 := \frac{d}{dt} v(t) = g - k v(t)$$

The equation is separable, and a look at `odeadvisor`

```
> odeadvisor(deq1, help);
```

$$[\_quadrature]$$

tells us that this equation is just an integral in disguised format, that can be solved mainly by integrating both sides.

```
> dsolve(deq1, v(t));
```

$$v(t) = \frac{g}{k} + e^{(-k t)} \_C1$$

Differential Equations

We can fix the integration constant imposing the initial condition

```
>   ic1:= v(0) = v0;
```
$$ic1 := v(0) = v0$$

Then, the solution reads

```
>   dsolve({deq1, ic1}, v(t));
```
$$v(t) = \frac{g}{k} - \frac{e^{(-kt)}(g - k\,v0)}{k}$$

Observe that when t is very large, the exponential term becomes insignificantly small. Therefore, the speed of the skydiver after enough time becomes a constant, also known as *terminal velocity*. We can graphically represent the variation of speed for a skydiver. Observe that the terminal velocity is independent of the initial conditions. We can visualize this by plotting the velocity for two different initial conditions

```
>   vel:=rhs(%);
```
$$vel := \frac{g}{k} - \frac{e^{(-kt)}(g - k\,v0)}{k}$$

```
>   p1:=plot( subs({g=9.8, v0=0, k=0.55}, vel),
>         t=0..12, thickness=3, color=blue, numpoints=100):
>   p2:=plot(subs({g=9.8, v0=10.0, k=0.55}, vel),
>         t=0..12, thickness=3, color=red, numpoints=100):

>   plots[display](p1,p2);
```

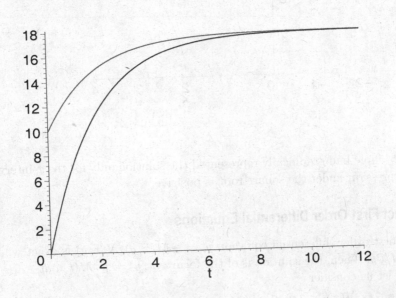

Let us solve another differential equation.

```
>   deq2:=diff(y(x),x) = (7*x^2+x-12)/(2*y(x) - 5);
>        ic2:= y(2) = 3;
```

$$deq2 := \frac{d}{dx}\,y(x) = \frac{7\,x^2 + x - 12}{2\,y(x) - 5}$$

$$ic2 := y(2) = 3$$

```
>   odeadvisor(deq2);
```

$$[\_separable]$$

```
>   sol2:=dsolve({deq2,ic2}, y(x)) ;
```

$$sol2 := y(x) = \frac{5}{2} + \frac{\sqrt{129 + 84\,x^3 + 18\,x^2 - 432\,x}}{6}$$

```
>   plot(rhs(sol2), x=-3..4, thickness=2, color=black,
>        numpoints=1000);
```

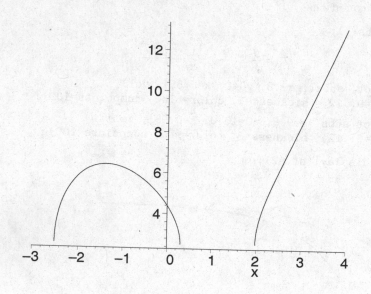

Observe that Maple had graphically represented the solution only for the x-interval where the expression under the square root is positive.

10.1.2   Exact First Order Differential Equations

A separable first order differential equation $y' = -M(x,y)/N(x,y)$ is exact when $\partial N/\partial x = \partial M/\partial y$. Then, the solution is of the form $\phi(x,y) = \int^x M(t,y)dt + c_1(y)$. For example, let us consider

```
>   deq3:= 2*x*y(x) - 3*x^2 + (y(x) + x^2 +1)*diff(y(x), x)=0;
```

$$deq3 := 2\,x\,\mathrm{y}(x) - 3\,x^2 + (\mathrm{y}(x) + x^2 + 1)\left(\tfrac{d}{dx}\,\mathrm{y}(x)\right) = 0$$

```
> odeadvisor(deq3);
```

$$[\_exact, \_rational, [\_1st\_order, \_with\_symmetry\_[F(x), G(x)]], \ldots]$$

For this equation the command `odeadvisor` produces a wealth of information. We will neglect for the moment the information on the symmetry of the equation, and observe that the equation is *exact*. Before asking Maple for an automatic solution, let us check that the exactness condition $\partial N/\partial x = \partial M/\partial y$ is satisfied. First, we identify M and N.

```
> exact:=solve(deq3, diff(y(x),x));
```

$$exact := \frac{x\,(-2\,\mathrm{y}(x) + 3\,x)}{\mathrm{y}(x) + x^2 + 1}$$

```
> exact1:=subs(y(x)=y1, exact);
```

$$exact1 := \frac{x\,(-2\,y1 + 3\,x)}{y1 + x^2 + 1}$$

We want to treat $y(x)$ and x as independent variables, therefore we made the substitution $y(x) = y_1$. This will allow Maple to correctly calculate the partial derivatives with respect to y and x.

```
> M:=-numer(exact1);
```

$$M := -x\,(-2\,y1 + 3\,x)$$

```
> diff(M, y1);
```

$$2\,x$$

```
> N:=denom(exact1);
```

$$N := y1 + x^2 + 1$$

Similarly, the derivative of N with respect to x yields

```
> diff(N, x);
```

$$2\,x$$

We have verified that the differential equation is exact. Note the use of the two commands `numer` that extracts the numerator of the fraction, and `denom`, which extracts the denominator. Finally, let us solve the equation

```
> dsolve(deq3, y(x));
```

$$y(x) = -x^2 - 1 + \sqrt{x^4 + 2\,x^2 + 1 + 2\,x^3 - 2\,\_C1},$$
$$y(x) = -x^2 - 1 - \sqrt{x^4 + 2\,x^2 + 1 + 2\,x^3 - 2\,\_C1}$$

We see that Maple effortlessly solves the equation, and presents us the two possible solutions. We identify a particular solution by imposing the initial condition

```
> ic3:= y(2)= 1;
```

$$ic3 := \mathrm{y}(2) = 1$$

```
> sol3:=dsolve({deq3, ic3}, y(x));
```

$$sol3 := \mathrm{y}(x) = -x^2 - 1 + \sqrt{x^4 + 2\,x^2 - 4 + 2\,x^3}$$

10.1.3   Linear First Order Differential Equations

A first order linear differential equation can be written as $y'(x) + p(x)y(x) = q(x)$ where $p(x)$ and $q(x)$ are arbitrary functions of x. The function $q(x)$ is also called the "source" term. When $q = 0$, the equation is called *homogeneous*, and when $q \neq 0$, the equation is *non-homogeneous*. Let us solve this equation.

```
> restart;
> deq:= diff(y(x),x) + p(x)*y(x) = q(x);
```

$$deq := (\tfrac{d}{dx}\,\mathrm{y}(x)) + \mathrm{p}(x)\,\mathrm{y}(x) = \mathrm{q}(x)$$

```
> dsolve(deq, y(x));
```

$$\mathrm{y}(x) = \left(\int \mathrm{q}(x)\, e^{(\int \mathrm{p}(x)\,dx)}\, dx + \_C1 \right) e^{(\int -\mathrm{p}(x)\,dx)}$$

This is the general solution. Given the initial conditions, and the particular functions $p(x)$ and $q(x)$ we can find the concrete solution. For example:

```
> deq1:= diff(y(x),x) + (1 - 1/x)*y(x) = x;
```

$$deq1 := (\tfrac{d}{dx}\,\mathrm{y}(x)) + (1 - \frac{1}{x})\,\mathrm{y}(x) = x$$

```
> dsolve(deq1, y(x));
```

$$\mathrm{y}(x) = x + x\,e^{(-x)}\,\_C1$$

We use this simple example as a toy model to explore further uses of the `odeadvisor` command from the `DEtools` package.

```
> DEtools[odeadvisor](deq1);
```

$$[\_linear]$$

Using the information outputted by the `odeadvisor` we can indicate to `dsolve` to use a specific sequence of methods to tackle the ODE. In this example, the method for `linear` ODEs is indicated; in addition the optional argument `useInt` makes `dsolve` not evaluate the integral appearing in the answer.

```
> dsolve(deq1, [linear], useInt);
```

$$\mathrm{y}(x) = \left(\int x\,e^{(\int \frac{x-1}{x}\,dx)}\, dx + \_C1 \right) e^{(\int -1 + \frac{1}{x}\,dx)}$$

This is important when we want to see the structure of the solution before performing all the integrals. One obtain the final solution using the `value` command

```
> value(%);
```
$$y(x) = (e^x + \_C1)\, e^{(\ln(x)-x)}$$

which simplifies to the same result

```
> simplify(%);
```
$$y(x) = (e^x + \_C1)\, x\, e^{(-x)}$$

For example, let us solve the following equation

$$y' - y/x = 3x^2 - 4x + 2$$

```
> ode1:=diff(y(x),x) -y(x)/x = 3*x^2 -4*x +2;
```
$$ode1 := (\tfrac{d}{dx}\, y(x)) - \frac{y(x)}{x} = 3x^2 - 4x + 2$$

```
> dsolve(ode1, y(x));
```
$$y(x) = (\frac{3x^2}{2} - 4x + 2\ln(x) + \_C1)\, x$$

Let us consider now the following problem:

A particle is moving vertically in a medium whose resistance is characterized by $\frac{1}{2}kv^2 e^{-\frac{z}{a}}$ per unit of mass, where v is the velocity of the particle at height z above the ground ($z = 0$), and a, k are constants. If the initial velocity is $v_0 = 20m/s$, find the maximum height reached by the particle when $a = 0.3m$ and $k = 0.5m^{-1}$.

Solution. When the particle reaches the maximum height, its speed is $v = 0$. Therefore, we have to find a solution for v as a function of z, and then solve for z when $v(z) = 0$.

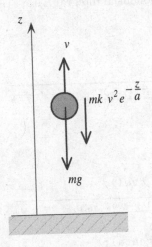

Gravity and the friction forces act in the negative direction of the z- axis, while the velocity is oriented in the positive direction of z. Therefore, the equation of motion reads

$$m\frac{dv}{dt} = -mg - \frac{1}{2}mkv^2\, e^{-\frac{z}{a}} \quad \text{where} \quad v = \frac{dz}{dt}\,.$$

We can transform this equation into a simple differential equation if we observe that

$$\frac{dv}{dt} = \frac{dv}{dz}\frac{dz}{dt} = v\frac{dv}{dz} = \frac{1}{2}\frac{d(v^2)}{dz}\,.$$

Therefore, we obtain

$$\frac{1}{2}\frac{d(v^2)}{dz} = -g - \frac{1}{2}kv^2\, e^{-\frac{z}{a}},$$

or, by making the substitution $v^2 = u$, and rearranging the terms

$$\frac{du}{dz} + ke^{-\frac{z}{a}}u = -2g \, ,$$

which can be immediately solved. Observe that the initial condition $v(0) = v_0$ becomes $u(0) = v_0^2$ in this case. We will use Maple to find the solution to this problem. We have

```
> ode2:= diff(u(z),z) + k*exp(-z/a)*u(z) = -2*g;
```

$$ode2 := \left(\frac{d}{dz}\,u(z)\right) + k\,e^{(-\frac{z}{a})}\,u(z) = -2\,g$$

We will use the `assume` command to force the condition $k > 0$ and $a > 0$. This is a good practice when solving physical problems because it discourages Maple in finding the most general solution. In our case we don't need Maple to assume that k or a are complex parameters.

```
> assume(a>0); assume(k>0);

> ic2:=u(0) = v0^2;
```

$$ic2 := u(0) = v0^2$$

```
> dsolve({ode2, ic2}, u(z));
```

$$u(z) = \left(-2\,g\,a^{\sim}\,\mathrm{Ei}(1, k^{\sim}\,a^{\sim}\,e^{(-\frac{z}{a^{\sim}})}) + \frac{2\,e^{(a^{\sim}\,k^{\sim})}\,g\,a^{\sim}\,\mathrm{Ei}(1, a^{\sim}\,k^{\sim}) + v0^2}{e^{(a^{\sim}\,k^{\sim})}}\right)\,e^{(k^{\sim}\,a^{\sim}\,e^{(-\frac{z}{a^{\sim}})})}$$

The $\sim$ symbol next to the variables k and respectively a, is the way Maple announces that these variables have been assigned some properties via the `assume` command. The solution is not trivial. By `Ei` Maple denoted the exponential integral $Ei(a, z) = \int_1^\infty e^{-tz}t^{-a}dt$. For more details, see `?Ei`.

Finally, one obtains the formula for velocity

```
> vel:=sqrt(rhs(%));
```

$$vel := \sqrt{\left(-2\,g\,a^{\sim}\,\mathrm{Ei}(1, k^{\sim}\,a^{\sim}\,e^{(-\frac{z}{a^{\sim}})}) + \frac{2\,e^{(a^{\sim}\,k^{\sim})}\,g\,a^{\sim}\,\mathrm{Ei}(1, a^{\sim}\,k^{\sim}) + v0^2}{e^{(a^{\sim}\,k^{\sim})}}\right)\,e^{(k^{\sim}\,a^{\sim}\,e^{(-\frac{z}{a^{\sim}})})}}$$

For our problem we have

```
> vel1:=subs({v0=20, a=0.3, k = 0.5, g=9.8}, vel);
```

$$vel1 := \sqrt{\left(-5.88\,\mathrm{Ei}(1, 0.15\,e^{(-3.33\ldots z)}) + \frac{5.88\,e^{0.15}\,\mathrm{Ei}(1, 0.15) + 400}{e^{0.15}}\right)\,e^{(0.15\,e^{(-3.33\ldots z)})}}$$

```
> plot(vel1, z=0..20, thickness=3, color=magenta);
```

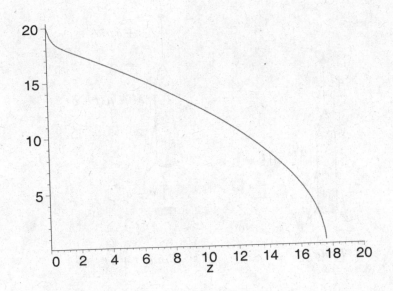

From the graph, we see that the solution for our problem ($v = 0$) lays in between $z = 16$ and $z = 18\ m$. To find the numerical solution of this equation we use `fsolve`

```
>  h_max:=fsolve(vel1, z=18);
```
$$h\_max := 17.60883611$$

Therefore, the particle will reach a maximum altitude $h_{max} := 17.60883611\ m$

Another interesting example is the following problem.

A rocket fired vertically upwards in the air, experiences a frictional force F_f proportional with its speed and its acceleration $F_f = rv + k\,dv/dt$, where r and k are positive constants. The initial mass of the rocket including its fuel is M. Considering that the fuel is ejected at a constant rate of mass μ per second with the constant backward speed u relative to the rocket, find the velocity of the rocket v as a function of time t. Plot v versus t when $r = 0.15\ kg/m$, $k = 1.0\ kg\,s/m$, $\mu = 0.1\ kg/s$, $M = 5\ kg$, $u = 550\ m/s$, $g = 9.8\ m/s^2$.

Solution. Suppose that the rocket is launched at $t = 0$. The remaining mass of the rocket plus the fuel at time t is $M - \mu t$. It has an upward velocity v and a momentum $p(t) = (M - \mu t)\,v$. After a small time interval δt, the amount of $\mu\,\delta t$ of fuel was ejected backwards with speed u relative to the rocket and $v - u$ with respect with the ground. As result, the speed of the rocket increases to $v + \delta v$. It has now a new mass $M - \mu\,(t + \delta t)$ and momentum $[M - \mu\,(t + \delta t)](v + \delta v)$. Therefore, the net momentum of the system at time $t + \delta t$ is $p(t + \delta t) = [M - \mu\,(t + \delta t)](v + \delta v) + \mu\,\delta t (v - u)$. In first order approximation, the net change in momentum is $p(t + \delta t) - p(t) = [M - \mu\,(t + \delta t)](v + \delta v) + \mu\,\delta t\,(v - u) - (M - \mu t)v = (M - \mu t)\delta v - \mu u\,\delta t$. But

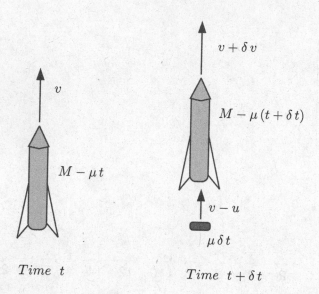

the variation of momentum in time equals the net force when we make the time variation infinitesimally small. We have $\lim_{\delta t \to 0}[p(t + \delta t) - p(t)]/\delta t = F_{net}$. In our case the net force is the sum of the weight of the rocket and its remaining fuel $-(M - \mu t)g$, and the frictional force $-r\,v - k\,dv/dt$. We obtain

$$(M - \mu t)\frac{dv}{dt} - \mu u = -(M - \mu t)\,g - r\,v - k\,\frac{dv}{dt}\;.$$

Initially the rocket was at rest, thus the initial condition reads $v(0) = 0$. The Maple worksheet that solves this equation reads

```
>   restart;
>   ode:=(M-mu*t)*diff(v(t),t) - mu*u=-(M-mu*t)*g
>       -(k*diff(v(t),t) + r*v(t));
```

$$ode := (M - \mu t)\left(\tfrac{d}{dt}\,\mathrm{v}(t)\right) - \mu u = -(M - \mu t)\,g - k\left(\tfrac{d}{dt}\,\mathrm{v}(t)\right) - r\,\mathrm{v}(t)$$

```
>   ic:=v(0)=0;
```

$$ic := \mathrm{v}(0) = 0$$

```
>   vel:=dsolve({ode,ic},v(t));
```

$$vel := \mathrm{v}(t) = \frac{(-\mu t + M + k)^{\left(\frac{r}{\mu}\right)}(\mu^2 u + g\mu k - r\mu u + r M g)}{-(M + k)^{\left(\frac{r}{\mu}\right)} r\mu + (M + k)^{\left(\frac{r}{\mu}\right)} r^2}$$
$$-\;\frac{r M g - \mu t g r + \mu^2 u - r\mu u + g\mu k}{r\,(-\mu + r)}$$

```
>   vv:=subs({k=1, r=0.15, M=5, mu=0.1, u=550, g=9.8}, rhs(vel)):
>   plot(vv, t=1..50, thickness=3, color=blue);
```

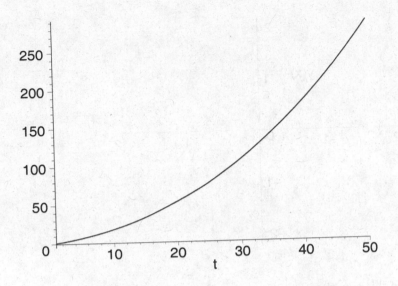

As an exercise, decrease the speed u (speed of the ejected gas) from $550\ m/s$ to $250\ m/s$ and plot again v versus time t. Explain the new graph.

10.1.4   Vector Fields Associated to First Order Differential Equations

In general, for a first order differential equation

$$\frac{dy(x)}{dx} = F(x,y) \tag{10.1}$$

the function $F(x,y)$ can be thought of as giving the gradient of the solution curve at the point (x,y). Thus, geometrically speaking, the solutions of the differential equation are curves representing the field lines of the corresponding vector field. In Maple we use the command DEplot to graphically represent this vector field. Let us represent the vector field associated a simple first order differential equation.

```
>  deq:=diff(y(x),x)=x*exp(x);
```

$$deq := \tfrac{d}{dx}\,\mathrm{y}(x) = x\,e^x$$

```
>  with(DEtools):  with(plots):
>  vf:=DEplot(deq, y(x), x=-2..2, y=-2..2, arrows=medium,
>       color=magenta):

>  display(vf);
```

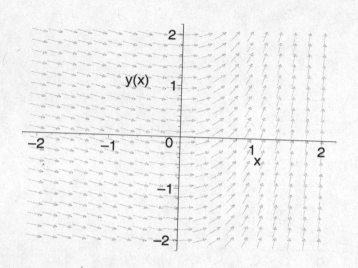

The solution of this equation depends on a integration constant, which can be fixed by imposing initial conditions. Let us look at few possible initial conditions, and their corresponding solution.

```
>   ic1:=y(0)=-2; ic2:= y(0)=-1; ic3:= y(0)=0; ic4:=y(0)=1;
```
$$ic1 := y(0) = -2$$
$$ic2 := y(0) = -1$$
$$ic3 := y(0) = 0$$
$$ic4 := y(0) = 1$$

We can make·use of the `concatenation` operator || together with `seq` command to obtain in one line all three solutions. Observe that we made use of the `rhs` command to extract the right hand side of the solution.

```
>   for i to 4 do sol||i:=rhs(dsolve({deq,ic||i},y(x))) end do;
```
$$sol1 := x\,e^x - e^x - 1$$
$$sol2 := x\,e^x - e^x$$
$$sol3 := x\,e^x - e^x + 1$$
$$sol4 := x\,e^x - e^x + 2$$

Let us use the same technique to quickly calculate the three corresponding plot, and then to display the solutions together with the vector field

```
>   for i to 4 do p||i:=plot(sol||i, x=-2..2, thickness=3) end do:
>   display(vf, p1, p2, p3 ,p4);
```

Differential Equations

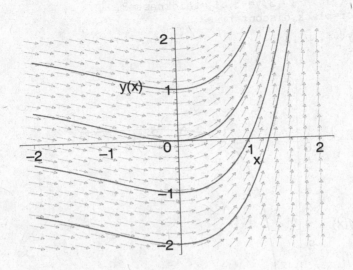

One can see that the solutions of the equation are nothing but field lines to the vector field that characterizes the first order differential equation.

Consider the equation

$$\frac{dy(x)}{dx} + \left(1 + \frac{1}{x}\right) y(x) = \frac{1}{x}, \tag{10.2}$$

with the initial condition $y(1) = 0$. The Maple worksheet that completely solves this equation is:

```
> restart;
> deq:=diff(y(x),x)+(1+1/x)*y(x)=1/x;
```

$$deq := \left(\frac{d}{dx} y(x)\right) + \left(1 + \frac{1}{x}\right) y(x) = \frac{1}{x}$$

```
> ic:=y(1)=0;
```

$$ic := y(1) = 0$$

```
> sol:=dsolve({deq,ic},y(x));
```

$$sol := y(x) = \frac{1}{x} - \frac{e^{(-x)}}{x\, e^{(-1)}}$$

```
> combine(sol,exp);
```

$$y(x) = \frac{1}{x} - \frac{e^{(-x+1)}}{x}$$

```
> y1:=rhs(%);
```

$$y1 := \frac{1}{x} - \frac{e^{(-x+1)}}{x}$$

```
> with(DEtools): with(plots):
> vf:=DEplot(deq,y(x), x=-1..3, y=-3..1):
```

```
>  p:=plot(y1,x=-1..3,y=-3..1, thickness=3,
>     color=black,discont=true):
>  display(vf, p);
```

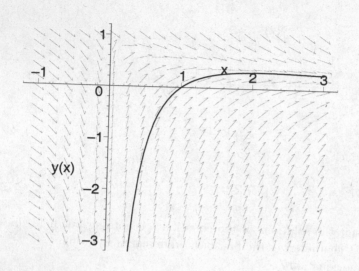

Note the use of the `combine` command in the worksheet. This command applies transformations which combine terms in sums, products, and powers into a single term. Here we used it to combine the ratio of the two exponentials into one single exponential. The `combine` command can be mapped to lists, matrices or sets. For more information on `combine` type `?combine` at the command prompt. Also, note the option `discont=true` used to avoid the singularity in the origin when plotting the solution.

10.2   Compact Form of Displaying Differential Equations

We can display differential equations in compact form, i.e. y' instead of $\frac{dy(x)}{dx}$ using the command `declare` within PDEtools package.

```
>  restart;
>  with(PDEtools):
>  declare(y(x), prime=x);
```

$y(x)$ *will now be displayed as* y

derivatives with respect to x *of functions of one variable will now be displayed with* '

During the calculations, we can check the declarations already made by

```
>  declare();
```

Declared :

$y(x)$ *to be displayed as* y

derivatives with respect to x *of functions of one variable are being displayed with* '

For example, let us consider a differential equation

```
>  ode:=diff(y(x),x) *y(x)^2 + f(x)*y(x) = r(x);
```

$$ode := y' \, y^2 + f(x) \, y = r(x)$$

If we want to see the full expression, we can use the command OFF

```
>  OFF;
>  ode;
```

$$(\tfrac{d}{dx} \, y(x)) \, y(x)^2 + f(x) \, y(x) = r(x)$$

One switch back to the compact form using ON command

```
>  ON;
>  ode;
```

$$y' \, y^2 + f(x) \, y = r(x)$$

If we want to see the equation in standard form, but to keep the switch ON, we use show command

```
>  show;
```

$$(\tfrac{d}{dx} \, y(x)) \, y(x)^2 + f(x) \, y(x) = r(x)$$

```
>  ode;
```

$$y' \, y^2 + f(x) \, y = r(x)$$

Suppose that we want to display in a compact form the functions $f(x)$ and $r(x)$, then we can declare them as well

```
>  declare(f(x), r(x));
```

$f(x)$ *will now be displayed as* f

$r(x)$ *will now be displayed as* r

Now, the same equation reads

```
>  ode;
```

$$y' \, y^2 + f \, y = r$$

Let us test the new notations by defining another differential equation by taking the derivative of the equation with respect to x

```
>  ode1:=diff(ode,x);
```

$$ode1 := y'' \, y^2 + 2 \, y'^2 \, y + f' \, y + f \, y' = r'$$

Note that Maple computes the differentiation, and writes the equation in a compact form. The standard form is much longer

```
>  show;
```

$(\frac{d^2}{dx^2}\,y(x))\,y(x)^2 + 2\,(\frac{d}{dx}\,y(x))^2\,y(x) + (\frac{d}{dx}\,f(x))\,y(x) + f(x)\,(\frac{d}{dx}\,y(x)) = \frac{d}{dx}\,r(x)$

We can turn off the display of y' instead of $dy(x)/dx$ using the undeclare command

```
>  undeclare(prime);
```

> *There is no more prime differentiation variable; all derivatives*
> *will be displayed as indexed functions*

```
>  ode1;
```

$$y_{x,x}\,y^2 + 2\,y_x{}^2\,y + f_x\,y + f\,y_x = r_x$$

```
>  ode;
```

$$y_x\,y^2 + f\,y = r$$

Further, we return to the standard display using undeclare(all) command

```
>  undeclare(all);
```

> $y(x)$ *will now be displayed* ∗ *as is*∗
> $f(x)$ *will now be displayed* ∗ *as is*∗
> $r(x)$ *will now be displayed* ∗ *as is*∗

```
>  ode;
```

$$y_x\,y(x)^2 + f(x)\,y(x) = r(x)$$

```
>  OFF;
>  ode;
```

$$(\frac{d}{dx}\,y(x))\,y(x)^2 + f(x)\,y(x) = r(x)$$

The index notation of differentiation is useful in displaying partial differential equations

```
>  pde:=a(x)*diff(f(x,y),y)+b(y)*diff(f(x,y),x)=f(x,y)^2;
```

$$pde := a(x)\,(\tfrac{\partial}{\partial y}\,f(x,\,y)) + b(y)\,(\tfrac{\partial}{\partial x}\,f(x,\,y)) = f(x,\,y)^2$$

```
>  ON;
>  pde;
```

$$a(x)\,f_y + b(y)\,f_x = f(x,\,y)^2$$

The above equation can be displayed in a more compact form, by using the declare command

```
>  declare(f(x,y));
```

> $f(x,\,y)$ *will now be displayed as* f

```
>  pde;
```

$$a(x)\,f_y + b(y)\,f_x = f^2$$

As an added commodity, Maple can declare simultaneously all the functions in a differential (or partial differential) equation

> `declare(pde);`

$$a(x) \; \textit{will now be displayed as} \; a$$
$$f(x, y) \; \textit{will now be displayed as} \; f$$
$$b(y) \; \textit{will now be displayed as} \; b$$

> `pde;`

$$a \, f_y + b \, f_x = f^2$$

> `declare(ode1);`

$$f(x) \; \textit{will now be displayed as} \; f$$
$$r(x) \; \textit{will now be displayed as} \; r$$
$$y(x) \; \textit{will now be displayed as} \; y$$

> `ode1;`

$$y_{x,x} \, y^2 + 2 \, y_x{}^2 \, y + f_x \, y + f \, y_x = r_x$$

10.3 Second Order Differential Equations

A second order differential equation has the general form

$$\frac{d^2 y}{dx^2} = F\left(x, \, y, \, \frac{dy}{dx}\right) . \tag{10.3}$$

As before, the `dsolve` command will be instrumental in solving differential equations that are particular cases of equation (10.3).

10.3.1 Second Order Homogeneous Linear Differential Equations with Constant Coefficients

For the second order differential equation with constant coefficients

$$a \frac{d^2 y(x)}{dx^2} + b \frac{dy(x)}{dx} + c \, y(x) = 0 , \tag{10.4}$$

the general solution reads

$$y(x) = C_1 \exp\left(\frac{-b + \sqrt{b^2 - 4\,ac}}{2\,a} x\right) + C_2 \exp\left(\frac{b + \sqrt{b^2 - 4\,ac}}{2\,a} x\right) ,$$

where the integration constants C_1 and C_2 are fixed by specifying the initial conditions $y(0)$, $y'(0)$.

For example, let us solve the following differential equation

> `restart:`

```
> deq:=a*diff(y(x),x,x)+b*diff(y(x),x)+c*y(x);
```

$$deq := a\left(\tfrac{d^2}{dx^2}\,y(x)\right) + b\left(\tfrac{d}{dx}\,y(x)\right) + c\,y(x)$$

```
> sol:=dsolve(deq,y(x));
```

$$sol := y(x) = \_C1\,e^{\left(-\tfrac{(b-\sqrt{b^2-4\,c\,a})\,x}{2\,a}\right)} + \_C2\,e^{\left(-\tfrac{(b+\sqrt{b^2-4\,c\,a})\,x}{2\,a}\right)}$$

Note that Maple introduces two integration constants denoted as `_C1` and `_C2`. These constants are automatically evaluated when we choose the initial conditions.

Let us particularize the above equation by choosing its coefficients to be $a = b = 1$ and $c = 2$. We also fix the initial conditions $y(0) = 0$, $y'(0) = 1$. Then, we have successively

```
> deq1:=subs({a=1, b=1, c=2},deq);
```

$$deq1 := \left(\tfrac{d^2}{dx^2}\,y(x)\right) + \left(\tfrac{d}{dx}\,y(x)\right) + 2\,y(x)$$

```
> dsolve({deq1, y(0)=0, D(y)(0)=1},{y(x)});
```

$$y(x) = \tfrac{2}{7}\,\sqrt{7}\,e^{(-\tfrac{x}{2})}\,\sin\!\left(\tfrac{\sqrt{7}\,x}{2}\right)$$

Next, we plot the solution we just have obtained.

```
> y1:=rhs(%);
```

$$y1 := \tfrac{2}{7}\,\sqrt{7}\,e^{(-\tfrac{x}{2})}\,\sin\!\left(\tfrac{\sqrt{7}\,x}{2}\right)$$

```
> plot(y1, x=-2.5..7, thickness=2);
```

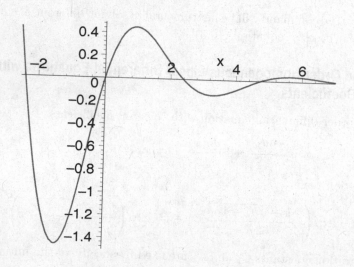

If we are just interested in plotting the solution without solving explicitly the equation we can use the **DEplot** command within the **DEtools** package. We have

```
>  with(DEtools):
>  DEplot(deq1,y(x),x=-2.5..7,
>       [[y(0)=0,D(y)(0)=1]], stepsize=.05,
>       linecolor=black);
```

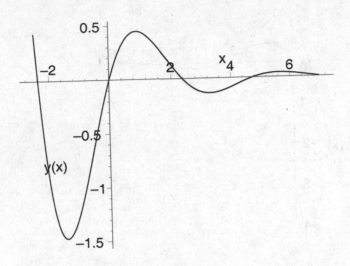

Remarks

1. The command DEplot is very convenient in the general case of equation (10.3) when a closed analytical solution may not be possible. Then this command can still be used to visualize the solution of the equation.

2. In our example observe the use of **stepsize** option. The value of the step size in the numeric algorithm is chosen experimentally in order to produce a "smooth" graph.

3. The command DEplot has also the advantage of producing directly multiple graphs corresponding to multiple sets of initial conditions.

Let us consider the following two different sets of initial conditions

```
>  ic1:=[y(0)=0,D(y)(0)=1];
```
$$ic1 := [y(0) = 0, \, D(y)(0) = 1]$$

```
>  ic2:=[y(0)=1,D(y)(0)=1];
```
$$ic2 := [y(0) = 1, \, D(y)(0) = 1]$$

We have

```
> DEplot(deq1,y(x),x=-2.5..7,
>    [ic1,ic2],stepsize=.05,
>    linecolor=[red,blue]);
```

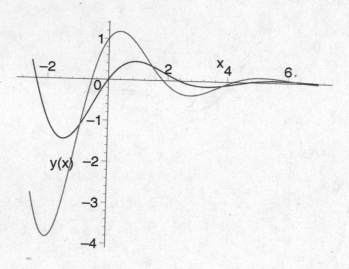

Solve $y'' + 2y' - 3y = 0$ with $y(0) = 0$ and $y'(0) = 2$.
Solution.

```
> restart;
```

```
> de1:=diff(y(x),x$2) + 2*diff(y(x),x)-3*y(x) = 0;
```

$$de1 := (\tfrac{d^2}{dx^2} y(x)) + 2(\tfrac{d}{dx} y(x)) - 3y(x) = 0$$

```
> ic1:=y(0) = 0, D(y)(0)=2;
```

$$ic1 := y(0) = 0, D(y)(0) = 2$$

```
> dsolve({de1,ic1},{y(x)});
```

$$y(x) = -\frac{1}{2} e^{(-3x)} + \frac{1}{2} e^x$$

```
> y1:=rhs(%);
```

$$y1 := -\frac{1}{2} e^{(-3x)} + \frac{1}{2} e^x$$

```
> plot(y1, x=-1.5..4,y=-30..30);
```

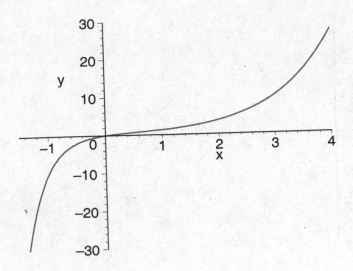

Solve $y'' + 2y' + 3y = 0$ with $y(0) = 2\sqrt{2}$ and $y'(0) = \sqrt{2}$.

Solution.

```
> restart;
```

```
> de2:=diff(y(x),x$2) + 2*diff(y(x),x)+3*y(x) = 0;
```
$$de2 := (\tfrac{d^2}{dx^2}\,\mathrm{y}(x)) + 2\,(\tfrac{d}{dx}\,\mathrm{y}(x)) + 3\,\mathrm{y}(x) = 0$$

```
> ic2:=y(0) = 2*sqrt(2), D(y)(0)=sqrt(2);
```
$$ic2 := \mathrm{y}(0) = 2\sqrt{2},\ \mathrm{D}(y)(0) = \sqrt{2}$$

```
> dsolve({de2,ic2},{y(x)});
```
$$\mathrm{y}(x) = 3\,e^{(-x)}\sin(\sqrt{2}\,x) + 2\sqrt{2}\,e^{(-x)}\cos(\sqrt{2}\,x)$$

```
> y2:=rhs(%);
```
$$y2 := 3\,e^{(-x)}\sin(\sqrt{2}\,x) + 2\sqrt{2}\,e^{(-x)}\cos(\sqrt{2}\,x)$$

```
> plot(y2, x=-1..6,y=-3..3,thickness=2,numpoints=1000);
```

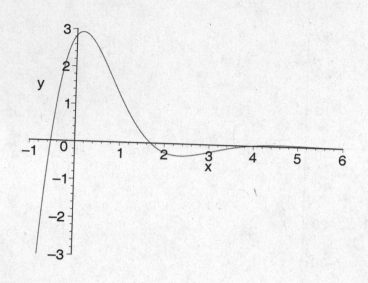

Solve $y'' + 2y' + y = 0$ with $y(0) = 2$ and $y'(0) = 2$.
Solution.

```
> restart;
```

```
> de3:=diff(y(x),x$2) + 2*diff(y(x),x)+ y(x) = 0;
```
$$de3 := \left(\frac{d^2}{dx^2}\, y(x)\right) + 2\left(\frac{d}{dx}\, y(x)\right) + y(x) = 0$$

```
> ic3:=y(0) = 2, D(y)(0)=2;
```
$$ic3 := y(0) = 2, \; D(y)(0) = 2$$

```
> dsolve({de3,ic3},{y(x)});
```
$$y(x) = 2\, e^{(-x)} + 4\, e^{(-x)}\, x$$

```
> y3:=rhs(%);
```
$$y3 := 2\, e^{(-x)} + 4\, e^{(-x)}\, x$$

```
> plot(y3, x=-1..6,y=-4..3,thickness=2,numpoints=1000);
```

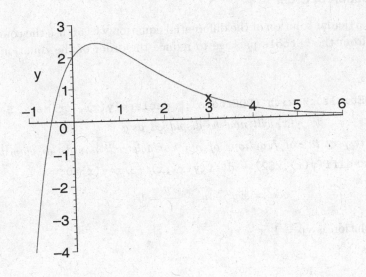

Comments

1. We have illustrated in these examples the three fundamental types of second order homogeneous linear differential equations with constant coefficients. The characteristic polynomial of each equation had different types of roots. The characteristic polynomial of the first equation had real roots, the second polynomial had complex conjugate roots, while the characteristic polynomial of the third equation had two identical roots.

2. We can use again Maple to quickly check the statements made above. We have successively:

```
>  p1:=r^2 +2*r -3;
```
$$p1 := r^2 + 2r - 3$$

```
>  solve(p1,r);
```
$$1, -3$$

```
>  p2:=r^2+2*r+3;
```
$$p2 := r^2 + 2r + 3$$

```
>  solve(p2,r);
```
$$-1 + \sqrt{2}\,I, \; -1 - \sqrt{2}\,I$$

```
>  p3:=r^2+2*r+1;
```
$$p3 := r^2 + 2r + 1$$

```
>  solve(p3,r);
```
$$-1, -1$$

10.3.2   Reduction of Order

If we know a particular solution of the differential equation we can use the command `reduceOrder` from the `DEtools` package to reduce the order of the equation or to solve it.

```
>  restart;
```

```
>  with(DEtools):  with (PDEtools):  declare(y(x), prime=x):
```
$$y(x) \text{ will now be displayed as } y$$
derivatives with respect to x of functions of one variable will now be displayed with '

```
>  ode:= x*diff(y(x),x$2) - diff(y(x),x) -3/x*y(x)=0;
```

$$ode := x\,y'' - y' - \frac{3\,y}{x} = 0$$

A particular solution is $yp = 1/x$

```
>  yp:=1/x;
```

$$yp := \frac{1}{x}$$

Indeed we can readily check

```
>  eval(subs(y(x)=yp, ode));
```
$$0 = 0$$

An equivalent way of testing a solution is given by the `odetest` command

```
>  odetest(y(x)=yp, ode);
```
$$0$$

Now we reduce the order of the equation

```
>  reduceOrder(ode, y(x), yp);
```

$$\frac{1}{x} \int e^{\left(\int \frac{1}{x}\, dx\right)} x^2 \, dx$$

```
>  y1:=value(%);
```

$$y1 := \frac{x^3}{4}$$

Maple has returned directly a solution. Let us check it out

```
>  odetest(y(x)=y1, ode);
```
$$0$$

The general solution will be a linear combination of these two solutions

```
>  sol:=C1*yp+C2*y1;
```

$$sol := \frac{C1}{x} + \frac{C2\,x^3}{4}$$

```
>  odetest(y(x)=sol, ode);
```

0

For this example one see immediately that indeed this is the solution by using directly the `dsolve` command

```
> dsolve(ode, y(x));
```

$$y = \frac{\_C1}{x} + \_C2\, x^3$$

For higher order equations, the reducing procedure yields a reduced equation if the particular solution is give. Let us consider for example

```
> ode1 := diff(y(x),x$3) - 2*diff(y(x),x$2) -diff(y(x),x) +
> 2*y(x)=0;
```

$$ode1 := y''' - 2\,y'' - y' + 2\,y = 0$$

A particular solution is

```
> yp1:=exp(-x);
```

$$yp1 := e^{(-x)}$$

```
> odetest(y(x)=yp1, ode1);
```

0

Reducing the order, one gets the reduced equation

```
> ode_red:=reduceOrder(ode1, y(x), yp1);
```

$$ode\_red := y'' - 5\,y' + 6\,y$$

As a side comment note that Maple finds immediately the general solution for this equation.

```
> dsolve(ode1, y(x));
```

$$y = \_C1\, e^x + \_C2\, e^{(2\,x)} + \_C3\, e^{(-x)}$$

10.3.3 Non-Homogeneous Second Order Linear Differential Equations

Non-homogeneous equations

$$y''(x) + p(x)y'(x) + q(x)y(x) = r(x)\,, \tag{10.5}$$

have the general solution of the form

$$y(x) = y_p(x) + y_h(x)\,, \tag{10.6}$$

where $y_p(x)$ is a particular solution of the equation, and $y_h(x)$ is a solution of the homogeneous equation $y''(x) + p(x)y'(x) + q(x)y(x) = 0$.

In general, Maple solves immediately non-homogeneous differential equations, using `dsolve` with no supplementary options.

```
> restart;
```

```
> with(DEtools): with(PDEtools): declare(y(x), prime=x):
```
$$y(x) \text{ will now be displayed as } y$$
derivatives with respect to x of functions of one variable will now be displayed with '
```
> ode:=diff(y(x),x$2)+4*diff(y(x),x)-5*y(x)=sin(x);
```
$$ode := y'' + 4\,y' - 5\,y = \sin(x)$$
```
> dsolve(ode, y(x));
```

$$y = e^x \, \_C2 + e^{(-5\,x)} \, \_C1 - \frac{1}{13}\cos(x) - \frac{3}{26}\sin(x)$$

10.3.4 General Method for Particular Solutions

If $y_1(x)$ and $y_2(x)$ form a basis of solutions for the homogeneous equation $y''(x) + p(x)y'(x) + q(x)y(x) = 0$, and if we denote by $W(x)$ their Wronskian,

$$W(x) = \begin{vmatrix} y_1(x) & y_2(x) \\ y_1'(x) & y_2'(x) \end{vmatrix} = y_1(x)y_2'(x) - y_1'(x)y_2(x) \,,$$

then a particular solution of the non-homogeneous differential equation $y''(x) + p(x)y'(x) + q(x)y(x) = r(x)$ is given by

$$y_p(x) = -y_1(x)\int^x \frac{y_2(t)\,r(t)}{W(t)}dt + y_2(x)\int^x \frac{y_1(t)\,r(t)}{W(t)}dt \,. \tag{10.7}$$

Let us exemplify the above mechanism for the following second order non-homogeneous differential equation

$$y'' - 2y' + y = e^{-x}\sin(x) \,. \tag{10.8}$$

First, we solve it directly using the **dsolve** command. We have
```
> restart; with(DEtools):with(PDEtools):
```
```
> declare(y(x), prime=x):
```
$$y(x) \text{ will now be displayed as } y$$
derivatives with respect to x of functions of one variable will now be displayed with '
```
> ode:=diff(y(x),x$2)-2*diff(y(x),x)+y(x)=exp(-x)*sin(x);
```
$$ode := y'' - 2\,y' + y = e^{(-x)}\sin(x)$$
```
> odeadvisor(ode);
```
$$[[\_2nd\_order, \_linear, \_nonhomogeneous]]$$
```
> dsolve(ode, y(x));
```

$$y = e^x \, \_C2 + e^x \, x \, \_C1 + \frac{1}{25}\left(4\cos(x) + 3\sin(x)\right)e^{(-x)}$$

Next, let us use the procedure defined in formula (10.7). We search for a basis of solutions of the homogeneous equation $y'' - 2y' + y = 0$, using the `constcoeffsols` routine.

```
>   ode_h:=diff(y(x),x$2)-2*diff(y(x),x)+y(x)=0;
```

$$ode\_h := y'' - 2\,y' + y = 0$$

```
>   S:=constcoeffsols(ode_h , y(x) );
```

$$S := [e^x, \ e^x\,x]$$

We learn that the basis of the space of solutions is composed of two functions e^x and $e^x x$ respectively. We can apply now directly the formula (10.7) to calculate the particular solution of the equation. We have

```
>   y1:=S[1]; y2:=S[2];r:=exp(-x)*sin(x);
```

$$y1 := e^x$$

$$y2 := e^x\,x$$

$$r := e^{(-x)}\sin(x)$$

```
>   with(VectorCalculus):with(LinearAlgebra):
```

Warning, the names &x, CrossProduct and DotProduct have been rebound

Warning, the names &x, CrossProduct and DotProduct have been rebound

```
>   wr:=Determinant(Wronskian([y1,y2],x));
```

$$wr := (e^x)^2$$

```
>   yp:=simplify(-y1*int(y2*r/wr,x)+y2*int(y1*r/wr,x));
```

$$yp := \frac{1}{25}\left(4\cos(x) + 3\sin(x)\right)e^{(-x)}$$

Therefore, the general solution reads

```
>   y:=C1*y1+C2*y2+yp;
```

$$y := C1\,e^x + C2\,e^x\,x + \frac{1}{25}\left(4\cos(x) + 3\sin(x)\right)e^{(-x)}$$

which essentially is the same as the solution found directly by `dsolve`.

Maple has a built in routine called `varparam` that applies automatically the above mechanism . Once we obtained the basis of the homogeneous equation with `constcoeffsols` command, we can use the `varparam` command to find the solution. The first argument of `varparam` command is a vector containing the basis of the homogeneous equation, the second parameter is the source term, and the third parameter is the variable. We have

```
>   varparam(S,exp(-x)*sin(x),x);
```

$$\_C_1\,e^x + \_C_2\,e^x\,x + \frac{4}{25}\,e^{(-x)}\cos(x) + \frac{3}{25}\,e^{(-x)}\sin(x)$$

which again, yields the same solution as the `dsolve` command.

Higher order differential equation can be solved in a similar way. The last example in this section deals with a non-homogeneous third order differential equation. We will solve it directly with `dsolve` and explicitly via the `varparam` routine.

```
>   ode1:= diff(y(x),x$3) + 9* diff(y(x),x$2)
>        +27*diff(y(x),x)+27*y(x)
>        =exp(-3*x)*arcsin(x);
```

$$ode1 := y''' + 9\,y'' + 27\,y' + 27\,y = e^{(-3x)}\arcsin(x)$$

```
>   odeadvisor(ode1);
```

$$[[\_3rd\_order, \_linear, \_nonhomogeneous]]$$

```
>   dsolve(ode1, y(x));
```

$$y = \frac{1}{6}\,e^{(-3x)}\,x^3\arcsin(x) + \frac{11}{36}\,e^{(-3x)}\,x^2\sqrt{1-x^2}+$$
$$\frac{1}{9}\,e^{(-3x)}\sqrt{1-x^2} + \frac{1}{4}\,e^{(-3x)}\,x\arcsin(x)$$
$$+\, \_C1\,e^{(-3x)} + \_C2\,e^{(-3x)}\,x + \_C3\,e^{(-3x)}\,x^2$$

Let us illustrate again the use of `varparam`. First we use `constcoeffsols` to find a fundamental set of solutions to the corresponding homogeneous equation, and then `varparm` to implement the method of variation of parameters to obtain the general solution to the equation.

```
>   ode1_h:=diff(y(x),x$3) + 9* diff(y(x),x$2)
>        +27*diff(y(x),x)+27*y(x) =0;
```

$$ode1\_h := y''' + 9\,y'' + 27\,y' + 27\,y = 0$$

```
>   S1:=constcoeffsols(ode1_h , y(x) );
```

$$S1 := [e^{(-3x)}, e^{(-3x)}\,x, e^{(-3x)}\,x^2]$$

```
>   varparam(S1, exp(-3*x)*arcsin(x), x);
```

$$\frac{1}{6}\,e^{(-3x)}\,x^3\arcsin(x) + \frac{11}{36}\,e^{(-3x)}\,x^2\sqrt{1-x^2}+$$
$$\frac{1}{9}\,e^{(-3x)}\sqrt{1-x^2} + \frac{1}{4}\,e^{(-3x)}\,x\arcsin(x)$$
$$+\, \_C_1\,e^{(-3x)} + \_C_2\,e^{(-3x)}\,x + \_C_3\,e^{(-3x)}\,x^2$$

We conclude this section by observing the power of the `dsolve` routine. Although we have presented explicitly different additional ways of solving ordinary differential equations, in most cases one suffices to just use `dsolve`.

10.4   Numerical Solutions of Differential Equations

Numerical solutions of differential equations can be obtained using `dsolve` together with the `type=numeric` option.

```
>   restart;
>   de:=diff(y(x),x) = tan(x - exp(y(x))); ic:=y(0)=-0.5;
```

$$de := \tfrac{d}{dx}\, y(x) = \tan(x - e^{y(x)})$$
$$ic := y(0) = -0.5$$

```
>  nsol:=dsolve({de,ic},y(x),type=numeric);
```

$$nsol := \mathbf{proc}(x\_rkf45) \ldots \mathbf{end\ proc}$$

We can use the above result to find the numerical approximation for the solution of the differential equation for given values of x.

```
>  nsol(1);
```

$$[x = 1., y(x) = -0.547701497462655840]$$

Observe that the above command returns a list with two objects $[x, y(x)]$. To isolate only the $y(x)$, we select the second term of the list

```
>  nsol(1)[2];
```

$$y(x) = -0.547701497462655840$$

We can graph the solution using the 'plot' command

```
>  plot( 'rhs( nsol(x)[2] )', x=0..4, thickness=3, color=black);
```

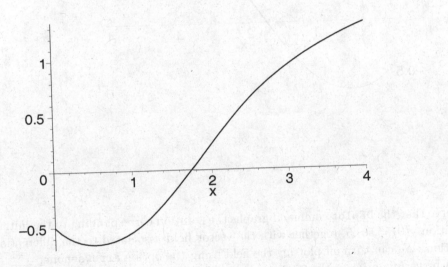

Observe that in the **plot** command we have enclosed **rhs(nsol(x)[x])** in single quotes (' ') to delay the evaluation of **rhs(nsol(x)[2]** .

We can also plot the solution of our differential equation using the the **odeplot** command within the **plots** package.

Using

```
> with(plots): odeplot( nsol, [x, y(x)], x=0..4):
```

one obtains the same plot as before.

Another command that can be used in graphing the solutions of differential equations is DEplot from DEtools package. As we have already seen in the previous section, in this case we obtain only the graph, not the explicit numerical solutions as before.

```
> with(DEtools): DEplot( de, y(x), x=0..4,
>     {[0,-0.5]}, color = gray,
>       linecolor=black, stepsize=0.01);
```

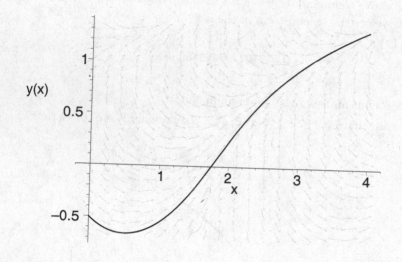

Observe that the DEplot command graphs the solution corresponding to the initial condition $y(0) = -0.5$, together with the vector field associated to the differential equation. We can turn off plotting the field using the option arrows=none.

A powerful option of the DEplot command consists in allowing us to plot solutions of the differential equations under changing initial conditions.

```
> ics:=seq([0,i/2],i=-2..3);
```

$$ics := [0, -1], [0, \frac{-1}{2}], [0, 0], [0, \frac{1}{2}], [0, 1], [0, \frac{3}{2}]$$

```
> DEplot( de, y(x), x=0..4, {ics}, color = gray,
>     linecolor=black, stepsize=0.01);
```

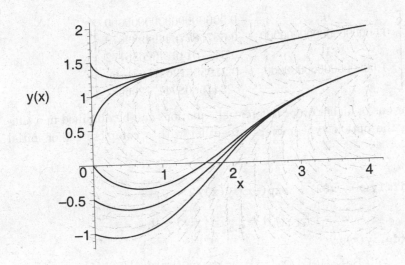

How can we get an array of numerical values for a given differential equation? If we want to tabulate the numerical $y(x)$ for a series of points $x_1, x_2, \ldots x_n$, we use the `output=array` option. We define a list with the desired values of x

```
> x_range:=[seq(i/2, i=0..4)];
```

$$x\_range := [0, \frac{1}{2}, 1, \frac{3}{2}, 2]$$

Then we obtain
```
> nsol1:=dsolve({de,ic},y(x),
>     type=numeric,
>     output=array(x_range));
```

$$nsol1 := \begin{bmatrix} \begin{bmatrix} [x, \text{y}(x)] \\ \begin{bmatrix} 0. & -0.500000000000000000 \\ 0.500000000000000000 & -0.657587784494989670 \\ 1. & -0.547701497462656172 \\ 1.50000000000000000 & -0.219556126275631896 \\ 2. & 0.234156089447120658 \end{bmatrix} \end{bmatrix} \end{bmatrix}$$

If we want to isolate only the array with the numerical values, we should select the element [2,1] from the above array.

```
> nvals:=nsol1[2,1];
```

$$nvals := nsol1_{2,1}$$

Note that Maple doesn't print automatically the values of the array. One can easily see them by using the 'print' command.

```
> print(nvals);
```

$$\begin{bmatrix} 0. & -0.500000000000000000 \\ 0.500000000000000000 & -0.657587784494989670 \\ 1. & -0.547701497462656172 \\ 1.50000000000000000 & -0.219556126275631896 \\ 2. & 0.234156089447120658 \end{bmatrix}$$

Numerical solutions of higher order differential equations can be obtained in a similar way, using the option `type=numeric` together with a complete set of initial conditions.

```
> restart:
> de:= diff(y(x), x$2) + exp(-x)*y(x)/2 = 0;
```

$$de := (\tfrac{d^2}{dx^2}\, y(x)) + \frac{1}{2}\, e^{(-x)}\, y(x) = 0$$

```
> dsolve(de, y(x));
```

$$y(x) = \_C1\,\mathrm{BesselJ}(0,\, \sqrt{2}\, e^{(-\frac{x}{2})}) + \_C2\,\mathrm{BesselY}(0,\, \sqrt{2}\, e^{(-\frac{x}{2})})$$

The analytic solution is given in terms of the Bessel functions $J_\nu(x)$ and respectively $Y_\nu(x)$. In chapter 11, we will discuss in detail these special functions.

```
> ic:= y(0) = 1, D(y)(0)=1;
```

$$ic := y(0) = 1,\, \mathrm{D}(y)(0) = 1$$

```
> dsolve({de, ic}, y(x));
```

$$y(x) = -\frac{(\mathrm{BesselY}(0,\sqrt{2})\,\sqrt{2} - \mathrm{BesselY}(1,\sqrt{2}))\,\mathrm{BesselJ}(0,\sqrt{2}\,e^{(-\frac{x}{2})})}{-\mathrm{BesselJ}(1,\sqrt{2})\,\mathrm{BesselY}(0,\sqrt{2}) + \mathrm{BesselY}(1,\sqrt{2})\,\mathrm{BesselJ}(0,\sqrt{2})}$$
$$+ \frac{(-\mathrm{BesselJ}(1,\sqrt{2}) + \sqrt{2}\,\mathrm{BesselJ}(0,\sqrt{2}))\,\mathrm{BesselY}(0,\sqrt{2}\,e^{(-\frac{x}{2})})}{-\mathrm{BesselJ}(1,\sqrt{2})\,\mathrm{BesselY}(0,\sqrt{2}) + \mathrm{BesselY}(1,\sqrt{2})\,\mathrm{BesselJ}(0,\sqrt{2})}$$

Numerically we have

```
> nsol:= dsolve( {de, ic}, y(x), type = numeric);
```

$$nsol := \mathbf{proc}(x\_rkf45) \ldots \mathbf{end\ proc}$$

Entering

```
> nsol(4);
```

$$[x = 4.,\, y(x) = 2.71369053117378911,\, \tfrac{d}{dx}\, y(x) = 0.200712064730237866]$$

returns an ordered triple that corresponds to $x = 4$, $y'(x)$ in $x = 4$, and $y''(x)$ in $x = 4$.

```
> nsol(4)[2];
```

$$y(x) = 2.71369053117378911$$

```
> with(plots):
> odeplot( nsol, [x,y(x)], 0..4);
```

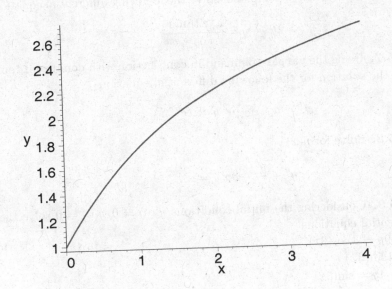

10.5 Problems

Problem 10.5.1. Solve the following differential equation subject to the initial condition $y(0) = -1$

$$y' = \frac{y + x}{y - x} \cos 2x \ .$$

Problem 10.5.2. Solve the following differential equation

$$x^2 \, y' + (2 + x) \, y = 0 \ .$$

Problem 10.5.3. Solve the differential equation

$$y' + y \cot x = \frac{1}{\sin x} \ .$$

Problem 10.5.4. Solve the differential equation

$$\frac{x}{y} + \frac{1}{y'} = 0 \ .$$

Problem 10.5.5. The radioactive decay of two different nuclides is described by a system of equations

$$\frac{dN_1}{dt} = -\lambda_1 \, N_1 \ , \qquad \frac{dN_2}{dt} = \lambda_1 \, N_1 - \lambda_2 \, N_2$$

Find N_2 as a function of time for initial conditions $N_1(0) = N_0$ and $N_2(0) = 0$.

Problem 10.5.6. Using `dsolve` command, solve the following differential equation

$$y'' + 2y' + y = \frac{e^{-x}\tan(x)}{\cos^{15} x} .$$

Problem 10.5.7. Using the `varpar` command in conjunction with `constcoeffsol` command find the solution for the following differential equation

$$y'' + ay' + \frac{1}{2}a^2 y = xe^{bx} .$$

Problem 10.5.8. Solve for $y(x)$

$$y'' + \frac{y'}{x} - 3xy = x .$$

Problem 10.5.9. Considering the initial conditions $y'(0) = 0, y(0) = 0$ solve the following differential equations
(a) $y'' - 6y' + 9y = 4\cos(3x)$
(b) $y'' + 9y = 4\sin(3x)$
(c) $y'' - 13y' + 4y = \sin(5x)$

Chapter 11

Power Series Solutions of Differential Equations

$\mathbf{N}$OT all differential equations admit solutions expressible in terms of known functions, and there are cases when we want more information than we can obtain from the numerical solution of such equations. In such cases, we use the power series method. The method leads to a solution to the differential equation given as a series

$$y(x) = \sum_{n=0}^{\infty} a_n (x - x_0)^n = a_0 + a_1(x - x_0) + a_2(x - x_0)^2 + \cdots . \qquad (11.1)$$

We solve for the unknown coefficients a_n by plugging (11.1) back into the differential equation and identifying the terms with the same power of $x - x_0$.

11.1 Power Series Expansion

Maple uses the command `series(f, x=x0, n)` to expand a given function f in series around the point $x = x_0$ till order n.

$$y(x) = a_0 + a_1(x - x_0) + a_2(x - x_0)^2 + \cdots + O(n) \qquad (11.2)$$

11.1.1 Order of Expansion

In the example below, we expand $y(x) = \cos x / (1 + x^3)$ around $x = 0$ till order $n = 10$

```
>  y:=cos(x)/(1+x^3);
```

$$y := \frac{\cos(x)}{1 + x^3}$$

```
> series(y, x=0, 10);
```

$$1 - \frac{1}{2}x^2 - x^3 + \frac{1}{24}x^4 + \frac{1}{2}x^5 + \frac{719}{720}x^6 - \frac{1}{24}x^7 - \frac{20159}{40320}x^8 - \frac{719}{720}x^9 + O(x^{10})$$

Note the syntax of the **series** command. The first argument is the function to be expanded in series, the second argument is the variable and the fixed point $x = x_0$. When expanding around $x_0 = 0$ we can use the abbreviated command **series(y, x, 12)** instead of **series(y, x=0, 12)**. The third argument is optional, and it represents the *order of the expansion*. If we want to expand many functions till a given order, we can fix the expansion order using the environmental variable **Order**. Thus, we can omit the optional argument for the expansion order.

```
> Order:=8;
```

$$Order := 8$$

```
> y1:=x*(1-exp(x)); y2:=1/(2-sin(x));
```

$$y1 := (1 - e^x)\,x$$

$$y2 := \frac{1}{2 - \sin(x)}$$

```
> series(y1,x);
```

$$-x^2 - \frac{1}{2}x^3 - \frac{1}{6}x^4 - \frac{1}{24}x^5 - \frac{1}{120}x^6 - \frac{1}{720}x^7 + O(x^8)$$

```
> series(y2,x);
```

$$\frac{1}{2} + \frac{1}{4}x + \frac{1}{8}x^2 + \frac{1}{48}x^3 - \frac{1}{96}x^4 - \frac{13}{960}x^5 - \frac{43}{5760}x^6 - \frac{193}{80640}x^7 + O(x^8)$$

11.1.2   Leading Term of a Series

Sometimes, we have a complicated function, and we only need to know the leading term. In this case we use the "leadterm" option in the **series** command. For example

```
> g:=int(x*exp(x^5),x);
```

$$g := -\frac{1}{5}(-1)^{(3/5)}\left(\frac{x^2\,(-1)^{(2/5)}\,\pi\,\csc(\frac{2\pi}{5})}{\Gamma(\frac{3}{5})\,(-x^5)^{(2/5)}} - \frac{x^2\,(-1)^{(2/5)}\,\Gamma(\frac{2}{5}, -x^5)}{(-x^5)^{(2/5)}}\right)$$

```
> series('leadterm'(%), x=0);
```

$$\frac{1}{2}x^2$$

11.1.3 Radius of Convergence of a Series

The radius of convergence R of the series of eq. (11.1) is given by the range of x values for which the expansion is valid. Mathematically this means that for all x such that $|x - x_0| < R$, the series (11.1) converges. Two general formulas for R are

$$R = \frac{1}{\lim_{n \to \infty} \left| \frac{a_{n+1}}{a_n} \right|} \quad \text{or} \quad R = \frac{1}{\lim_{n \to \infty} \sqrt[n]{|a_n|}} . \tag{11.3}$$

As an example, let us consider the following series

```
>  S:=Sum((-1)^n/(n+3)*(x-3)^n, n=1..infinity);
```

$$S := \sum_{n=1}^{\infty} \frac{(-1)^n (x - 3)^n}{n + 3}$$

Note that we have used the command Sum instead of sum. Usually, capitalized commands are not evaluated in Maple. Therefore, Maple returns just a "pretty print" of the input. Later on, when we evaluate the above series, we can use the value command in conjunction with Sum.

Let us identify the coefficients a_n in the series S defined above. By inspection we get

```
>  a:=n->(-1)^n/(n+3);
```

$$a := n \to \frac{(-1)^n}{n + 3}$$

We will use the first formula from (11.3) to calculate the radius of convergence. First we have to calculate the ratio $|a_{n+1}/a_n|$. We get

```
>  ratio:=abs(a(n+1)/a(n));
```

$$ratio := \left| \frac{(-1)^{(n+1)} (n + 3)}{(n + 4) (-1)^n} \right|$$

Observe that Maple is unaware that n is an integer, and consequently it doesn't simplify too much the result. We use the command assuming to further restrict the variable n.

```
>  ratio1:=abs(a(n+1)/a(n)) assuming n::integer;
```

$$ratio1 := \left| \frac{n + 3}{n + 4} \right|$$

The result reads better, but still, Maple kept the absolute value symbol, signaling that it considers that n can take negative integer values as well. (At this point, the reader can readily check that the same result could be obtained by using simplify(ratio) command.) We can go further by nesting another assuming command, imposing n to be positive. In this way we obtain the desired result.

```
>  ratio2:=abs(a(n+1)/a(n)) assuming n::integer assuming n>0;
```

$$ratio2 := \frac{n+3}{n+4}$$

Then the radius of convergence is

```
> Radius:=1/Limit(ratio2, n=infinity);
```

$$Radius := \frac{1}{\displaystyle\lim_{n\to\infty}\frac{n+3}{n+4}}$$

For illustration purposes, we have used the inert form `Limit` of the command `limit` in the formula of the radius of convergence. (The active form `limit` would have directly produced the result.) Finally, we get the numerical value of R by applying the `value` command

```
> R:=value(Radius);
```

$$R := 1$$

Let us evaluate the above series. Maple is able to provide a close formula in many cases. We have

```
> f:=value(S);
```

$$f := -\frac{1}{3}\frac{x^3 - \dfrac{21\,x^2}{2} + 39\,x - 3\ln(-2+x) - \dfrac{99}{2}}{(x-3)^3}$$

We can indeed check that by expanding the above function in series we get our initial series.

```
> series(f,x=3,8);
```

$$-\frac{1}{4}(x-3) + \frac{1}{5}(x-3)^2 - \frac{1}{6}(x-3)^3 + \frac{1}{7}(x-3)^4 + O((x-3)^5)$$

```
> S_partial:=Sum((-1)^n/(n+3)*(x-3)^n, n=1..M);
```

$$S\_partial := \sum_{n=1}^{M}\frac{(-1)^n(x-3)^n}{n+3}$$

```
> value(subs(M=5,S_partial));
```

$$-\frac{x}{4} + \frac{3}{4} + \frac{(x-3)^2}{5} - \frac{(x-3)^3}{6} + \frac{(x-3)^4}{7} - \frac{(x-3)^5}{8}$$

11.1.4   Taylor Series

The Taylor series expansion of a function $f(x)$ in $x = x_0$ is given by

$$
\begin{aligned}
f(x) &= f(x_0) + \frac{f'(x_0)}{1!}(x-x_0) + \cdots + \frac{f^{(n)}(x_0)}{n!}(x-x_0)^n + \cdots \\
&= \sum_{n=0}^{\infty}\frac{f^{(n)}(x_0)}{n!}(x-x_0)^n,
\end{aligned}
\tag{11.4}
$$

for all x such that $|x - x_0| < R$ where R is the radius of convergence of the power series. The Taylor expansion is a particular case of the series expansion.

Taylor expansions can be obtained via the `taylor` command.

```
> restart;
```

```
> taylor(exp(x),x=a,4);
```

$$e^a + e^a\,(x-a) + \frac{1}{2}\,e^a\,(x-a)^2 + \frac{1}{6}\,e^a\,(x-a)^3 + \mathrm{O}((x-a)^4)$$

Similar to `series` command, in the `taylor` command, the first parameter is the function to be expanded, the second parameter is the point around which the function is expanded, and the third parameter (optional) represents the truncation order of the series.

11.2 Power Series Method: a Step-by-Step Example

Let us employ the power series expansion to solve differential equations. We will consider a simple example.

```
> restart;
```

```
> deq:=diff(y(x),x,x)-y(x)=0;
```

$$deq := (\tfrac{d^2}{dx^2}\,\mathrm{y}(x)) - \mathrm{y}(x) = 0$$

To gain intuition, we first solve this equation using the `dsolve` command.

```
> y1:=rhs(dsolve(deq,y(x)));
```

$$y1 := \_C1\,e^{(-x)} + \_C2\,e^x$$

For simplicity, we replace $\_C1$ and $\_C2$ with A and respectively B.

```
> y1a:=subs({_C1=A, _C2=B},y1);
```

$$y1a := A\,e^{(-x)} + B\,e^x$$

For later reference, we expand in series the solution. (The truncation order was previously setup to 6.)

```
> y1s:=series(y1a,x);
```

$$y1s := (A+B) + (-A+B)\,x + (\frac{A}{2} + \frac{B}{2})\,x^2 + (\frac{B}{6} - \frac{A}{6})\,x^3$$

$$+(\frac{A}{24} + \frac{B}{24})\,x^4 + (-\frac{A}{120} + \frac{B}{120})\,x^5 + \mathrm{O}(x^6)$$

Now, let us solve the same equation, using the `type=series` option in the `dsolve` command. We have

```
> y2:=dsolve(deq,y(x),type=series);
```

$$y2 := y(x) =$$

$$y(0) + D(y)(0)\, x + \frac{1}{2}\, y(0)\, x^2 + \frac{1}{6}\, D(y)(0)\, x^3$$

$$+ \frac{1}{24}\, y(0)\, x^4 + \frac{1}{120}\, D(y)(0)\, x^5 + O(x^6)$$

If we specify the initial conditions as following

```
>   ic:=y(0)=M, D(y)(0)=N;
```

$$ic := y(0) = M,\ D(y)(0) = N$$

we have

```
>   y2:=rhs(dsolve({deq,ic},y(x),type=series));
```

$$y2 := M + N\,x + \frac{M}{2}\,x^2 + \frac{N}{6}\,x^3 + \frac{M}{24}\,x^4 + \frac{N}{120}\,x^5 + O(x^6)$$

which is exactly the previous solution provided that M=A+B and N=A-B.

We can convert a series to a polynomial, using the `convert/polynom` command

```
>   convert(y2, polynom);
```

$$M + N\,x + \frac{1}{2}\,M\,x^2 + \frac{1}{6}\,N\,x^3 + \frac{1}{24}\,M\,x^4 + \frac{1}{120}\,N\,x^5$$

This conversion is useful when we want to freely manipulate only the first terms of the series solution.

11.3   Power Series Method, Expansion Around the Regular Point

Consider the homogeneous differential equation of second order

$$\frac{d^2\, y(x)}{dx^2} + p(x)\frac{d\, y(x)}{dx} + q(x) = 0 \ . \tag{11.5}$$

If the functions $p(x)$ and $q(x)$ are analytic at x_0, then there exists a general power solution $y(x) = C_1\, y_1(x) + C_2\, y_2(x)$ with linearly independent solutions y_1 and y_2 being analytic at x_0. The point x_0 is called a regular point.

For example, let us consider the equation

$$\left(x^2 - 9\right)\, y''(x) + \left(x^2 + 3\right)\, y'(x) + (x+2)\, y(x) = 0 \ , \tag{11.6}$$

subject to the initial conditions $y(0) = 0, y'(0) = 1$. The coefficients p and q are respectively $p(x) = \frac{x^2+3}{x^2-9}$ and $q(x) = \frac{x+2}{x^2-9}$. The points $x = \pm 3$ are singular. The coefficient functions $p(x)$ and $q(x)$ are analytic around $x = 0$, $|x| < 3$.

```
>   restart;

>   deq:=(x^2-9)*diff(y(x),x$2)+(x^2+3)*diff(y(x),x)+(x+2)*y(x)=0;
```

$$deq := (x^2 - 9)\left(\tfrac{d^2}{dx^2}\, y(x)\right) + (x^2 + 3)\left(\tfrac{d}{dx}\, y(x)\right) + (x + 2)\, y(x) = 0$$

```
> ic:=y(0)=0, D(y)(0)=1;
```
$$ic := y(0) = 0,\; D(y)(0) = 1$$

```
> dsolve({deq,ic},y(x),type=series,order=8);
```
$$y(x) = x + \frac{1}{6}\, x^2 + \frac{1}{18}\, x^3 + \frac{19}{648}\, x^4 + \frac{7}{972}\, x^5 + \frac{2}{729}\, x^6 + \frac{689}{734832}\, x^7 + O(x^8)$$

Note the `order=8` option in `dsolve` command. It allows us to choose the truncation order in the series solution.

If we want to plot the solution, we have to convert the truncated series solution into a polynomial, and isolate the right-hand-side. We have successively

```
> convert(%,polynom);
```
$$y(x) = x + \frac{1}{6}\, x^2 + \frac{1}{18}\, x^3 + \frac{19}{648}\, x^4 + \frac{7}{972}\, x^5 + \frac{2}{729}\, x^6 + \frac{689}{734832}\, x^7$$

```
> y:=rhs(%);
```
$$y := x + \frac{1}{6}\, x^2 + \frac{1}{18}\, x^3 + \frac{19}{648}\, x^4 + \frac{7}{972}\, x^5 + \frac{2}{729}\, x^6 + \frac{689}{734832}\, x^7$$

```
> plot(y,x=-2.9..2.9,thickness=2,color=black);
```

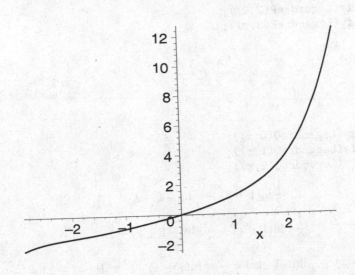

Note that the `dsolve` command without any additional options returns a blank line, which signals that Maple cannot find a closed form solution.

```
> dsolve({deq,ic},y(x));
```

11.3.1   Legendre's Differential Equation; Legendre's Polynomials

We write Legendre's differential equation as following

```
>   restart;
>   deq:=(1-x^2)*diff(y(x),x,x) -
>   2*x*diff(y(x),x)
>   + n*(n+1)*y(x) = 0 assuming n::real ;
```

$$deq := (1 - x^2)\left(\tfrac{d^2}{dx^2}\, y(x)\right) - 2\,x\left(\tfrac{d}{dx}\, y(x)\right) + n\,(n+1)\,y(x) = 0$$

```
>   dsolve(deq,y(x));
```

$$y(x) = \_C1 \, \text{LegendreP}(n,\, x) + \_C2 \, \text{LegendreQ}(n,\, x)$$

which is as expected a combination of Legendre's functions. Note the use of the assuming command. (Here we assume that n is real). We use the syntax assuming n::real, which is a module allocation procedure. (See ?module, for more details.)

```
>   simplify(LegendreP(0,x));
```

$$1$$

```
>   simplify(LegendreP(1,x));
>   simplify(LegendreP(2,x));
>   simplify(LegendreP(3,x));
```

$$x$$

$$-\frac{1}{2} + \frac{3\,x^2}{2}$$

$$\frac{5}{2}\,x^3 - \frac{3}{2}\,x$$

```
>   simplify(LegendreQ(0,x));
>   simplify(LegendreQ(1,x));
>   simplify(LegendreQ(2,x));
```

$$\frac{1}{2}\ln(1+x) - \frac{1}{2}\ln(-1+x)$$

$$\frac{1}{2}\,x\ln(1+x) - \frac{1}{2}\,x\ln(-1+x) - 1$$

$$-\frac{1}{4}\ln(1+x) + \frac{1}{4}\ln(-1+x) + \frac{3}{4}\,x^2\ln(1+x) - \frac{3}{4}\,x^2\ln(-1+x) - \frac{3\,x}{2}$$

Let us plot these functions. For brevity we use the **alias** command.

```
>   alias(Lp= LegendreP);
```

$$Lp$$

```
>   plot({Lp(0,x),Lp(1,x),Lp(2,x),Lp(3,x)},x=-1..1,color=black,
>        thickness=2);
```

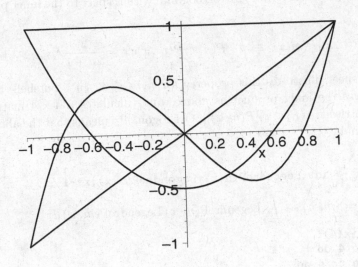

```
>   alias(Lq= LegendreQ);
```

$$Lp,\ Lq$$

```
>   plot({Lq(0,x),Lq(1,x),Lq(2,x)},x=1..2.5,y=0..1,color=black,
>   thickness=2);
```

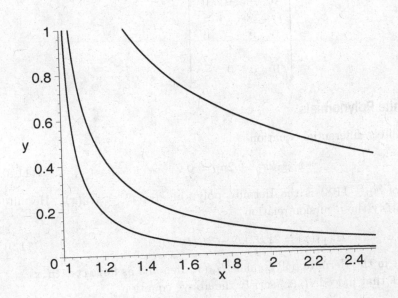

The Legendre Polynomials $P(n, x)$ are orthogonal with respect to the inner product

$$\langle P(n, x) | P(m, x) \rangle \equiv \int_{-1}^{1} P(n, x) P(m, x) dx = \frac{2}{2n+1} \delta_{mn} \qquad (11.7)$$

The next worksheet illustrates this property for $n, m \leq 4$. In a nutshell, the worksheet defines the inner product as above, and calculates a 4×4 matrix whose (n, m) entries are $\langle P(n, x) | P(m, x) \rangle$. Orthogonality means that the above matrix is diagonal.

```
>  restart;
>  M:=(m,n)-> int(LegendreP(n,x)*LegendreP(m,x),x=-1..1);
```

$$M := (m, n) \rightarrow \int_{-1}^{1} \text{LegendreP}(n, x) \, \text{LegendreP}(m, x) \, dx$$

```
>  A:=Matrix(4):
>  for m to 4 do
>      for n to 4 do
>          A[m,n]:=M(m,n)
>      end do
>  end do;
>  A;
```

$$\begin{bmatrix} \dfrac{2}{3} & 0 & 0 & 0 \\[2mm] 0 & \dfrac{2}{5} & 0 & 0 \\[2mm] 0 & 0 & \dfrac{2}{7} & 0 \\[2mm] 0 & 0 & 0 & \dfrac{2}{9} \end{bmatrix}$$

11.3.2 Hermite Polynomials

Consider Hermite's differential equation

$$y'' - 2xy' + 2ny = 0 . \qquad (11.8)$$

One solution of eq. (11.8) is the Hermite polynomial $y(x) = H_n(x)$. Hermite polynomials satisfy the recursion relation

$$H_{n+1}(x) = 2xH_n(x) - 2nH_{n-1}(x) . \qquad (11.9)$$

Maple has built in the Hermite polynomials $H_n(x)$ denoted as HermiteH(n,x).
   Let us check that indeed $H_n(x)$ satisfy the above equation.

```
>  restart;
>  de:=diff(y(x),x$2) - 2*x*diff(y(x),x)+2*n*y(x);
```

$$de := (\tfrac{d^2}{dx^2}\, y(x)) - 2\, x\, (\tfrac{d}{dx}\, y(x)) + 2\, n\, y(x)$$

```
> step1:=subs(y(x)=HermiteH(n,x), de);
```

$$step1 := (\tfrac{\partial^2}{\partial x^2}\, \text{HermiteH}(n,\, x)) - 2\, x\, (\tfrac{\partial}{\partial x}\, \text{HermiteH}(n,\, x))$$
$$+ 2\, n\, \text{HermiteH}(n,\, x)$$

```
> step2:=simplify(step1);
```

$$step2 := -2\, n\, (-2\, \text{HermiteH}(n - 2,\, x)\, n + 2\, \text{HermiteH}(n - 2,\, x)$$
$$+ 2\, x\, \text{HermiteH}(n - 1,\, x) - \text{HermiteH}(n,\, x))$$

We can now use the recursion formula (11.9) to further simplify the equation

```
> step3:= subs(HermiteH(n,x) =
> 2*x*HermiteH(n-1,x)
>         -2*(n-1)*HermiteH(n-2,x), step2);
```

$$step3 := -2\, n\, (-2\, \text{HermiteH}(n - 2,\, x)\, n + 2\, \text{HermiteH}(n - 2,\, x)$$
$$+ 2\, (n - 1)\, \text{HermiteH}(n - 2,\, x))$$

Observe that Maple makes the substitution, but it doesn't simplify the above expression by default. We get the final result, using either `simplify` or `expand`.

```
> expand(%);
```

$$0$$

Let us plot the first four Hermite polynomials. To simplify the notation, we first call the `orthopoly` package. In this way we can write `H(n,x)` instead of `HermiteH(n,x)`.

```
> with(orthopoly);
```

$$[G,\, H,\, L,\, P,\, T,\, U]$$

Then the first four polynomials are

```
> H_n:= H(0,x), H(1,x), H(2,x), H(3,x);
```

$$H\_n := 1,\, 2\, x,\, -2 + 4\, x^2,\, 8\, x^3 - 12\, x$$

```
> plot({H_n},x=-1.5..1.5, color=black,thickness=2);
```

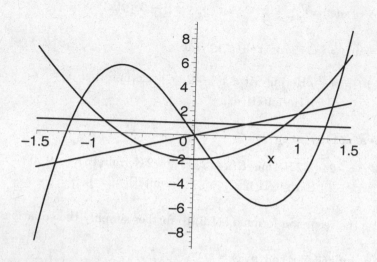

The Hermite polynomials are orthogonal to each other, satisfying

$$\int_{-\infty}^{\infty} H_i(x)H_j(x)\, e^{-x^2}\, dx = 2^j\, j!\, \sqrt{\pi}\, \delta_{ij}\ .$$

This property can be readily illustrated by calculating a matrix A whose elements are $A_{ij} = \langle H_i(x)\,|\, H_j(x)\rangle$. We have

```
>   M:= (i,j)->      int(
>   HermiteH(i,x)*HermiteH(j,x)*exp(-x^2),
>       x=-infinity..infinity);
```

$$M := (i,\, j) \rightarrow \int_{-\infty}^{\infty} \mathrm{HermiteH}(i,\, x)\, \mathrm{HermiteH}(j,\, x)\, e^{(-x^2)}\, dx$$

```
>   A:=Matrix(4):
>   for i to 4 do        for j to 4 do
>        A[i,j]:=M(i,j)         end do
>   end do;
>   A;
```

$$\begin{bmatrix} 2\sqrt{\pi} & 0 & 0 & 0 \\ 0 & 8\sqrt{\pi} & 0 & 0 \\ 0 & 0 & 48\sqrt{\pi} & 0 \\ 0 & 0 & 0 & 384\sqrt{\pi} \end{bmatrix}$$

Hermite polynomials can be generated directly using Rodrigues differential formula

$$H_n(x) = (-1)^n e^{x^2} \frac{d^n}{dx^n}\left(e^{-x^2}\right)\ . \tag{11.10}$$

As an exercise, let us implement this formula in Maple.

```
> my_H:=(n,x)->expand( (-1)^n*exp(x^2)*diff(exp(-x^2),x$n));
```

$$my\_H := (n, x) \rightarrow expand((-1)^n e^{(x^2)} (\tfrac{d^n}{dx^n} (e^{(-x^2)})))$$

```
> my_H(0,x); my_H(1,x); my_H(2,x); my_H(3,x);
```

Error, (in my_H) wrong number (or type) of parameters in function diff

$$2x$$
$$-2 + 4x^2$$
$$-12x + 8x^3$$

Observe that for $n = 0$ the naive implementation fails due to the fact that Maple cannot compute the zero-th derivative of a function. We can easily bypass this difficulty by using the conditional **if** statement.

```
> my_H:=(n,x)-> if n = 0 then 1         else
> expand( (-1)^n*exp(x^2)*diff(exp(-x^2),x$n))
> end if ;
```

$$my\_H := \mathbf{proc}(n, x)$$
$$\mathbf{option}\ operator,\ arrow;$$
$$\mathbf{if}\,n = 0\,\mathbf{then}\,1\,\mathbf{else}$$
$$expand((-1)^n * exp(x^2) * diff(exp(-x^2), x\,\$\,n))\,\mathbf{end\ if}$$
$$\mathbf{end\ proc}$$

```
> my_H(0,x); my_H(1,x); my_H(2,x); my_H(3,x);
```

$$1$$
$$2x$$
$$-2 + 4x^2$$
$$-12x + 8x^3$$

11.3.3 Laguerre Polynomials

Laguerre polynomials satisfy the differential equation

$$xy'' + (1 - x)y' + ny = 0 . \tag{11.11}$$

One solution of eq. (11.11) is given in terms of Laguerre polynomials $L_n(x)$. One of their recursion formulas is

$$nL_n(x) = (2n - 1 - x)L_{n-1}(x) - (n - 1)L_{n-2}(x) . \tag{11.12}$$

The Maple built-in function for Laguerre polynomials is **LaguerreL**. The following Maple worksheet shows that indeed Laguerre polynomials satisfy equation (11.11) and illustrates some other properties of these polynomials.

```
> restart;
> de:=x*diff(y(x),x,x)+(1-x)*diff(y(x),x)+n*y(x)=0;
```

$$de := x\left(\frac{d^2}{dx^2}\,y(x)\right) + (1-x)\left(\frac{d}{dx}\,y(x)\right) + n\,y(x) = 0$$

```
> de1:=subs(y(x)=LaguerreL(n,x),de);
```

$$de1 := x\left(\frac{\partial^2}{\partial x^2}\,\text{LaguerreL}(n,\,x)\right)$$
$$+(1-x)\left(\frac{\partial}{\partial x}\,\text{LaguerreL}(n,\,x)\right) + n\,\text{LaguerreL}(n,\,x) = 0$$

```
> step1:=expand(de1);
```

$$step1 := \frac{n^2\,\text{LaguerreL}(n,\,x)}{x} - \frac{2\,n^2\,\text{LaguerreL}(n-1,\,x)}{x}$$
$$+ \frac{n\,\text{LaguerreL}(n-1,\,x)}{x} + \frac{n^2\,\text{LaguerreL}(n-2,\,x)}{x}$$
$$- \frac{\text{LaguerreL}(n-2,\,x)\,n}{x} + n\,\text{LaguerreL}(n-1,\,x) = 0$$

We use the recursion formula

$$n\,L_n(x) = (2n-1-x)\,L_{n-1}(x) - (n-1)\,L_{n-2}(x)\,,$$

to further simplify the result

```
> Ln:=(2*n-1-x)*LaguerreL(n-1,x)/n-(n-1)*LaguerreL(n-2,x)/n;
```
$$Ln := \frac{(2\,n-1-x)\,\text{LaguerreL}(n-1,\,x)}{n} - \frac{(n-1)\,\text{LaguerreL}(n-2,\,x)}{n}$$

```
> expand(subs(LaguerreL(n,x)=Ln, step1));
```
$$0 = 0$$

Let us list and plot the first four Laguerre polynomials. To simplify notation, we use the `orthopoly` package.

```
> with(orthopoly);
```
$$[G,\,H,\,L,\,P,\,T,\,U]$$

```
> L_n:=L(0,x), L(1,x), L(2,x), L(3,x);
```
$$L\_n := 1,\, 1-x,\, 1-2\,x + \frac{1}{2}\,x^2,\, 1-3\,x + \frac{3}{2}\,x^2 - \frac{1}{6}\,x^3$$

```
> plot({L_n},x=-1..7,y=-4..8, color=black, thickness=2);
```

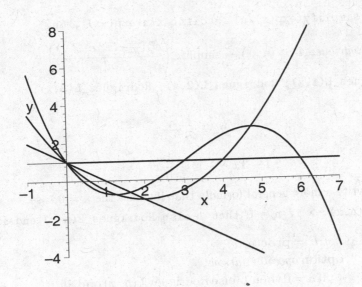

The Laguerre polynomials are orthogonal, satisfying

$$\int_0^\infty e^{-x} L_n(x) L_m(x) dx = \delta_{m,n} \; .$$

```
> M:= (m,n) ->
> int(LaguerreL(m,x)*LaguerreL(n,x)*exp(-x),
> x=0..infinity);
```

$$M := (m, n) \to \int_0^\infty \text{LaguerreL}(m, x) \, \text{LaguerreL}(n, x) \, e^{(-x)} \, dx$$

```
> A:=Matrix(4):
> for i to 4 do        for j to 4 do
>                 A[i,j]:=M(i,j)        end do;
> end do;
> A;
```

$$\begin{bmatrix} 1 & 0 & 0 & 0 \\ 0 & 1 & 0 & 0 \\ 0 & 0 & 1 & 0 \\ 0 & 0 & 0 & 1 \end{bmatrix}$$

The last property that we investigate in this section, is the Rodrigues differential formula, which generates the Laguerre polynomials according to

$$L_n(x) = \frac{e^x}{n!} \frac{d^n}{dx^n} \left(x^n e^{-x} \right) \; .$$

We have

```
> Rodrigues_L:=(n,x) ->
>        simplify(exp(x)/n!  * diff( x^n*exp(-x) ,x$n));
```

$$Rodriguez\_L := (n,\, x) \rightarrow simplify\left(\frac{e^x \left(\frac{d^n}{dx^n}\left(x^n\, e^{(-x)}\right)\right)}{n!}\right)$$

```
> Rodrigues_L(1,x); Rodrigues_L(2,x); Rodrigues_L(3,x);
```

$$1 - x$$

$$1 - 2x + \frac{1}{2}x^2$$

$$1 - 3x + \frac{3}{2}x^2 - \frac{1}{6}x^3$$

If we want to write a more general formula that includes the $n = 0$ case, we write

```
> my_L:=(n,x) -> if n = 0 then 1 else Rodrigues_L(n,x) end if;
```

$$my\_L := \mathbf{proc}(n,\, x)$$
$$\mathbf{option}\ operator,\ arrow;$$
$$\mathbf{if}\, n = 0\, \mathbf{then}\, 1\ \mathbf{else}\, \text{Rodrigues\_}L(n,\, x)\, \mathbf{end\ if}$$
$$\mathbf{end\ proc}$$

```
> my_L(0,x); my_L(1,x); my_L(2,x); my_L(3,x);
```

$$1$$

$$1 - x$$

$$1 - 2x + \frac{1}{2}x^2$$

$$1 - 3x + \frac{3}{2}x^2 - \frac{1}{6}x^3$$

11.3.4   Chebyshev Polynomials

Chebyshev polynomials are orthogonal polynomials that satisfy the differential equation

$$(1 - x^2)\, y'' - x\, y' + n^2 y = 0 \,. \tag{11.13}$$

In this subsection we will address only the Chebyshev polynomials of first kind, denoted by T_n.

```
> restart;
> de:=(1-x^2)*diff(y(x),x,x) -x*diff(y(x),x)+n^2*y(x)=0;
```

$$de := (1 - x^2)\left(\frac{d^2}{dx^2}\, y(x)\right) - x\left(\frac{d}{dx}\, y(x)\right) + n^2\, y(x) = 0$$

```
> de1:=subs(y(x)=ChebyshevT(n,x),de);
```

$$de1 := (1 - x^2)\left(\frac{\partial^2}{\partial x^2}\, \text{ChebyshevT}(n,\, x)\right) - x\left(\frac{\partial}{\partial x}\, \text{ChebyshevT}(n,\, x)\right)$$
$$+ n^2\, \text{ChebyshevT}(n,\, x) = 0$$

```
> step1:=simplify(de1);
```

$step1 := n(\mathrm{ChebyshevT}(n,\,x) + 2\,n\,\mathrm{ChebyshevT}(n-1,\,x)\,x$
$- 2\,x\,\mathrm{ChebyshevT}(n-1,\,x) - n\,\mathrm{ChebyshevT}(n-2,\,x) + \mathrm{ChebyshevT}(n-2,\,x)$
$- n\,\mathrm{ChebyshevT}(n,\,x))/(-1+x^2) = 0$

To further simplify, let us use the recursion formula $T_n(x) = 2\,x\,T_{n-1}(x) - T_{n-2}(x)$. We have

```
>   Tn:=2*x*ChebyshevT(n-1,x)-ChebyshevT(n-2,x);
```
$$Tn := 2\,x\,\mathrm{ChebyshevT}(n-1,\,x) - \mathrm{ChebyshevT}(n-2,\,x)$$

```
>   expand(subs(ChebyshevT(n,x)=Tn, step1));
```
$$0 = 0$$

Next, let us list the first four Chebyshev polynomials and plot them

```
>   with(orthopoly);
```
$$[G,\,H,\,L,\,P,\,T,\,U]$$

```
>   T_n:=T(0,x), T(1,x), T(2,x), T(3,x);
```
$$T\_n := 1,\,x,\,-1+2\,x^2,\,4\,x^3 - 3\,x$$

```
>   plot({T_n},x=-1.2..1.2,y=-2..2, color=black, thickness=2);
```

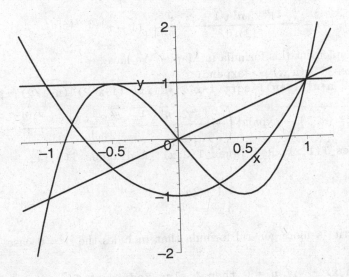

Chebyshev polynomials of the first kind, satisfy the orthogonality condition

$$\int_{-1}^{1} \frac{T_m(x)T_n(x)}{\sqrt{1-x^2}}\,dx = \begin{cases} \dfrac{\pi}{2}\,\delta_{mn}, & \text{for } m \neq 0,\, n \neq 0 \\[2mm] \pi, & \text{for } m = n = 0 \end{cases} \qquad (11.14)$$

```
>    M:= (m,n) -> int(ChebyshevT(m,x)*ChebyshevT(n,x)/sqrt(1-x^2),
>         x=-1..1);
```

$$M := (m, n) \to \int_{-1}^{1} \frac{\text{ChebyshevT}(m, x)\, \text{ChebyshevT}(n, x)}{\sqrt{1 - x^2}}\, dx$$

```
>    A:=Matrix(5):
>    for i to 4 do
>       for j to 4 do
>          A[i,j]:=M(i-1,j-1)
>       end do;
>    end do;

>    A;
```

$$\begin{bmatrix} \pi & 0 & 0 & 0 & 0 \\ 0 & \dfrac{\pi}{2} & 0 & 0 & 0 \\ 0 & 0 & \dfrac{\pi}{2} & 0 & 0 \\ 0 & 0 & 0 & \dfrac{\pi}{2} & 0 \\ 0 & 0 & 0 & 0 & 0 \end{bmatrix}$$

The Rodrigues formula that generates Chebyshev polynomials can be written as follows

$$T_n(x) = \frac{(-1)^n\, 2^n\, n!\, \sqrt{1 - x^2}}{(2\,n)!} \frac{d^n}{dx^n}\left[\left(1 - x^2\right)^{n-1/2}\right].$$

We can easily implement this formula in Maple. We have

```
>    Rodrigues_T:=(n,x) -> expand(
>    (-1)^n*2^n*n!/(2*n)!*sqrt(1-x^2)*diff( (1-x^2)^(n-1/2),x$n));
```

$$Rodrigues\_T := (n, x) \to \text{expand}\left(\frac{(-1)^n\, 2^n\, n!\, \sqrt{1 - x^2}\, (\frac{d^n}{dx^n}\, (1 - x^2)^{(n-1/2)})}{(2\,n)!}\right)$$

```
>    Rodrigues_T(1,x); Rodrigues_T(2,x); Rodrigues_T(3,x);
```

$$x$$
$$-1 + 2\,x^2$$
$$4\,x^3 - 3\,x$$

If we want to write a more general formula that includes the $n = 0$ case, we write

```
>    my_T:=(n,x) -> if n = 0 then 1 else Rodrigues_T(n,x) end if;
```

$$my\_T := \mathbf{proc}(n, x)$$
$$\mathbf{option}\ operator,\ arrow;$$
$$\mathbf{if}\,n = 0\,\mathbf{then}\,1\ \mathbf{else}\,Rodrigues\_T(n, x)\,\mathbf{end\ if}$$
$$\mathbf{end\ proc}$$

```
>    my_T(0,x); my_T(1,x); my_T(2,x); my_T(3,x);
```

1

$$x$$
$$-1 + 2\,x^2$$
$$4\,x^3 - 3\,x$$

11.4 The Maple Package `powseries`

The `powseries` package contains routines that allow for creation and manipulation of formal power series

```
> restart;
```

```
> with(powseries);
```

> [*compose, evalpow, inverse, multconst, multiply, negative, powadd,*
> *powcos, powcreate, powdiff, powexp, powint, powlog, powpoly, powsin,*
> *powsolve, powsqrt, quotient, reversion, subtract, template, tpsform*]

To create a formal series we use the `powcreate` command. In the next example we will build a series such that the coefficient of x^n is $\sqrt{n}$. Some individual coefficients can be explicitly specified.

```
> powcreate(s(k)=sqrt(k!), s(3)=Pi, s(5)=1);
```

```
> s(0), s(1), s(2), s(3), s(4), s(5), s(6);
```

$$1, 1, \sqrt{2}, \pi, \sqrt{24}, 1, \sqrt{720}$$

We can visualize the series using the `tpsform` command which prints the truncated part of a power series. In our case we want to print the terms of the series until order 8.

```
> tpsform(s, x, 8);
```

$$1 + x + \sqrt{2}\,x^2 + \pi\,x^3 + \sqrt{24}\,x^4 + x^5 + \sqrt{720}\,x^6 + \sqrt{5040}\,x^7 + O(x^8)$$

Another way to create a series is to specify the recursion relation between its coefficients together with the initial conditions

```
> powcreate(r(n)=2*r(n-1) - r(n-2), r(0)=2, r(1)=1);
```

The first 7 coefficients are

```
> seq( r(j), j=0..6);
```

$$2, 1, 0, -1, -2, -3, -4$$

and the corresponding truncated power series reads

```
> tpsform(r, z, 7);
```

$$2 + z - z^3 - 2\,z^4 - 3\,z^5 - 4\,z^6 + O(z^7)$$

Observe that we are free to choose the variable of the series expansion.

In the next example we create a series that has only even powers of x. It is a good practice to unassign the name of the variable used for the series. This can be done using either **unassign** command or t='t' command.

```
>   t:='t';
```

$$t := t$$

```
>   powcreate(t(n)=x^n);
>   tpsform(t, x, 10);
```

$$1 + x^2 + x^4 + x^6 + x^8 + O(x^{10})$$

Let us discuss other functions from the **powseries** package: **powdiff** for formal derivative of a series, **powadd** for adding two series, **multiply** for multiplication of two series, and **multconst** for multiplication of a series with a constant.

```
>   restart; with(powseries):
```

Given a general series s

```
>   powcreate(s(n)=a[n]);
>   y0:=tpsform(s, x, 5);
```

$$y0 := a_0 + a_1 x + a_2 x^2 + a_3 x^3 + a_4 x^4 + O(x^5)$$

we can take the derivative of the series via the **powdiff** command

```
>   s1:=powdiff(s);
```

$$s1 := \textbf{proc}(powparm) \dots \textbf{end proc}$$

```
>   y1:=tpsform(s1, x, 5);
```

$$y1 := a_1 + 2 a_2 x + 3 a_3 x^2 + 4 a_4 x^3 + 5 a_5 x^4 + O(x^5)$$

We can add two series with **powadd**

```
>   a:= powadd(s,s1);
```

$$a := \textbf{proc}(powparm) \dots \textbf{end proc}$$

```
>   tpsform(a,x);
```

$$(a_0 + a_1) + (a_1 + 2 a_2) x + (a_2 + 3 a_3) x^2 + (a_3 + 4 a_4) x^3$$
$$+ (a_4 + 5 a_5) x^4 + (a_5 + 6 a_6) x^5 + O(x^6)$$

or we can multiply two series

```
>   m:=multiply(s,s1);
```

$$m := \textbf{proc}(powparm) \dots \textbf{end proc}$$

```
>   tpsform(m,x);
```

$$a_0 a_1 + (2 a_0 a_2 + a_1{}^2) x + (3 a_0 a_3 + 3 a_1 a_2) x^2 + (4 a_0 a_4 + 4 a_1 a_3 + 2 a_2{}^2) x^3 +$$
$$(5 a_0 a_5 + 5 a_1 a_4 + 5 a_2 a_3) x^4 + (6 a_0 a_6 + 6 a_1 a_5 + 6 a_2 a_4 + 3 a_3{}^2) x^5 + O(x^6)$$

Note that if we want to multiply the series s with x, one first have to define formally x as a series, and then use the `multiply` command to multiply the two series.

```
> powcreate(p(n)=0,p(1)=1);
> sx:=tpsform(p,x);
```

$$sx := x$$

The above command defines x as a formal series in x with all the coefficients equal to zero, except $a_1 = 1$. Now, we can go ahead and multiply the two series

```
> b:=multiply(s,p);
```

$$b := \mathbf{proc}(powparm) \ldots \mathbf{end\ proc}$$

```
> tpsform(b,x,5);
```

$$a_0\,x + a_1\,x^2 + a_2\,x^3 + a_3\,x^4 + \mathrm{O}(x^5)$$

We can multiply a series with a constant using `multconst` command

```
> multconst(s, 17);
```

$$\mathbf{proc}(powparm) \ldots \mathbf{end\ proc}$$

```
> tpsform(%,x);
```

$$17\,a_0 + 17\,a_1\,x + 17\,a_2\,x^2 + 17\,a_3\,x^3 + 17\,a_4\,x^4 + 17\,a_5\,x^5 + \mathrm{O}(x^6)$$

11.5 Expansion Around the Singular Point: Frobenius Method

A differential equation of the form

$$y''(x) + \frac{p_0(x)}{x}y'(x) + \frac{q_0(x)}{x^2}y(x) = 0 , \qquad (11.15)$$

where $p_0(x)$ and $q_0(x)$ are analytic at origin, has at least one solution of the form

$$y(x) = x^r \sum_{m=0}^{\infty} a_m x^m , \quad r \in \mathbb{R}, a_0 \neq 0 . \quad (\textit{Frobenius theorem}) \qquad (11.16)$$

As an application of the Frobenius theorem, let us solve the Euler equation

```
> restart;
> de:=x^2*diff(y(x),x,x)+p0*x*diff(y(x),x)+q0*y(x)=0;
```

$$de := x^2 \left(\tfrac{d^2}{dx^2}\,y(x) \right) + p0\,x\left(\tfrac{d}{dx}\,y(x) \right) + q0\,y(x) = 0$$

```
> dsolve(de,y(x));
```

$$y(x) = \_C1\,x^{\left(-\frac{p0}{2}+1/2+\frac{\sqrt{p0^2-2\,p0+1-4\,q0}}{2}\right)} + \_C2\,x^{\left(-\frac{p0}{2}+1/2-\frac{\sqrt{p0^2-2\,p0+1-4\,q0}}{2}\right)}$$

One can immediately see that Maple did not find all the solutions. For example, for the particular case when $q_0 = (p_0 - 1)^2/4$, we have

```
> de1:=x^2*diff(y(x),x,x)+p0*x*diff(y(x),x)+(p0-1)^2/4*y(x)=0;
```

$$de1 := x^2 \left(\frac{d^2}{dx^2}\, \mathrm{y}(x)\right) + p0\, x \left(\frac{d}{dx}\, \mathrm{y}(x)\right) + \frac{1}{4}\,(p0-1)^2\, \mathrm{y}(x) = 0$$

This Euler equation has a different type of solution, than the one previously found by Maple

```
>  dsolve(de1,y(x));
```

$$\mathrm{y}(x) = \_C1\, x^{(-\frac{p0}{2}+1/2)} + \_C2\, x^{(-\frac{p0}{2}+1/2)} \ln(x)$$

To find the most general solution we will employ the Frobenius theorem. To this end, we call the `powseries` package which allows for creation and manipulation of formal power series.

```
>  with(powseries):
```

Next, we create a general series s, whose coefficients are unspecified constants $a_0, a_1, a_2, \ldots$

```
>  powcreate(s(n)=a[n]);
```

Now, we create the "truncated power series" out of the series s, by using the `tpsform` command. We choose x as expansion variable, and for brevity we truncate the series at the 5-th.

```
>  y0:=tpsform(s,x,5);
```

$$y0 := a_0 + a_1\, x + a_2\, x^2 + a_3\, x^3 + a_4\, x^4 + \mathrm{O}(x^5)$$

We now convert the above series to a polynomial using the `convert` command, and use the Frobenius theorem to define the solution as

```
>  ys:=x^r*convert(y0,polynom);
```

$$ys := x^r \left(a_0 + a_1\, x + a_2\, x^2 + a_3\, x^3 + a_4\, x^4\right)$$

Then the equation becomes

```
>  eq:=factor(x^2*diff(ys,x$2)+x*p0*diff(ys,x)+q0*ys);
```

$$\begin{aligned}
eq := x^r (& q0\, a_4\, x^4 + r^2\, a_1\, x + 5\, r\, a_3\, x^3 + 3\, r\, a_2\, x^2 + 7\, r\, a_4\, x^4 \\
& + 3\, p0\, a_3\, x^3 + r\, a_1\, x + q0\, a_1\, x + r^2\, a_2\, x^2 + r^2\, a_3\, x^3 + r^2\, a_4\, x^4 \\
& + p0\, r\, a_0 + p0\, a_1\, x2\, p0\, a_2\, x^2 + 4\, p0\, a_4\, x^4 + q0\, a_2\, x^2 + q0\, a_3\, x^3 + 2\, a_2\, x^2 \\
& + 6\, a_3\, x^3 + 12\, a_4\, x^4 p0\, r\, a_1\, x + p0\, r\, a_2\, x^2 + p0\, r\, a_3\, x^3 + p0\, r\, a_4\, x^4 \\
& + r^2\, a_0 - r\, a_0 + q0\, a_0)
\end{aligned}$$

To isolate the coefficient corresponding to the lowest order, we factor out the x^r term. One option to use the `op` command to extract the second term

```
>  term1:=op(2,eq);
```

$$term1 := q0\,a_4\,x^4 + r^2\,a_1\,x + 5\,r\,a_3\,x^3 + 3\,r\,a_2\,x^2 + 7\,r\,a_4\,x^4 + 3\,p0\,a_3\,x^3$$
$$+\,r\,a_1\,x + q0\,a_1\,x + r^2\,a_2\,x^2 + r^2\,a_3\,x^3 + r^2\,a_4\,x^4 + p0\,r\,a_0 + p0\,a_1\,x$$
$$+\,2\,p0\,a_2\,x^2 + 4\,p0\,a_4\,x^4 + q0\,a_2\,x^2 + q0\,a_3\,x^3 + 2\,a_2\,x^2 + 6\,a_3\,x^3$$
$$+\,12\,a_4\,x^4 + p0\,r\,a_1\,x + p0\,r\,a_2\,x^2 + p0\,r\,a_3\,x^3 + p0\,r\,a_4\,x^4$$
$$+\,r^2\,a_0 - r\,a_0 + q0\,a_0$$

Then, we use `coeff` command in conjunction with `factor` to extract the lowest order coefficient and to factorize it

```
>  c0:=factor(coeff(term1,x,0));
```
$$c0 := a_0\,(r\,p0 + r^2 - r + q0)$$

Because a_0 cannot be zero, the only remaining option is to demand that r is a solution of the equation $r p_0 + r^2 - r + q_0 = 0$. That is

```
>  term2:=op(2,c0);
```
$$term2 := r\,p0 + r^2 - r + q0$$

```
>  sols:=solve(term2,r);
```
$$sols := -\frac{p0}{2} + \frac{1}{2} + \frac{\sqrt{p0^2 - 2\,p0 + 1 - 4\,q0}}{2},$$
$$-\frac{p0}{2} + \frac{1}{2} - \frac{\sqrt{p0^2 - 2\,p0 + 1 - 4\,q0}}{2}$$

```
>  r1:=sols[1];r2:=sols[2];
```
$$r1 := -\frac{p0}{2} + \frac{1}{2} + \frac{\sqrt{p0^2 - 2\,p0 + 1 - 4\,q0}}{2}$$
$$r2 := -\frac{p0}{2} + \frac{1}{2} - \frac{\sqrt{p0^2 - 2\,p0 + 1 - 4\,q0}}{2}$$

The next term we consider is the coefficient of x^{r+1}. We have

```
>  c1:=factor(coeff(op(2,eq),x,1));
```
$$c1 := a_1\,(r^2 + r + q0 + p0 + r\,p0)$$

With the value of the constant r fixed above, we see that `c1` cannot be zero, unless $a_1 = 0$.

```
>  simplify(subs(r=r1,c1));
```
$$a_1\,(1 + \sqrt{p0^2 - 2\,p0 + 1 - 4\,q0})$$

```
>  simplify(subs(r=r2,c1));
```
$$-a_1\,(-1 + \sqrt{p0^2 - 2\,p0 + 1 - 4\,q0})$$

Similarly, we conclude that all the coefficients $a_n = 0$ for $n > 0$. Depending on the roots r_1, r_2 we have three different cases. For real roots, $r_1 \neq r_2$, we obtain exactly the solution that Maple found in the first case. For $r_1 = r_2 = r$, we have the second solution (corresponding to the equation de1). Finally, for complex roots $r_{1,2} = \alpha \pm i\beta$, where alpha and beta are real constants we have

```
> assume (alpha>0); assume(beta > 0);
> de2:=subs({p0=1-2*alpha, q0=(-2*alpha)^2/4+beta^2},de);
```

$$de2 := x^2 \left(\tfrac{d^2}{dx^2} y(x)\right) + (1 - 2\,\alpha^\sim)\,x\left(\tfrac{d}{dx} y(x)\right) + (\alpha^{\sim 2} + \beta^{\sim 2})\,y(x) = 0$$

```
> dsolve(de2,y(x));
```

$$y(x) = \_C1\, x^{\alpha^\sim} \sin(\beta^\sim \ln(x)) + \_C2\, x^{\alpha^\sim} \cos(\beta^\sim \ln(x))$$

This is the third explicit type of solution for the Euler equation.

In general, we have to remember that Maple cannot provide the most general answer. Maple is just a tool (albeit a powerful one) which we should manipulate intelligently in order to solve our problem.

11.5.1   Bessel's Differential Equation

Bessel's differential equation is encountered in many cases where the system studied exhibits cylindrical symmetry. The general form of Bessel's equation is

$$x^2\, y'' + x\, y' + \left(x^2 - \nu^2\right) y = 0 ,\tag{11.17}$$

and its solution is given by

$$y(x) = C_1\, J_\nu(x) + C_2\, Y_\nu(x) ,\tag{11.18}$$

where $J_\nu(x)$ and $Y_\nu(x)$ are Bessel functions of first and second order respectively. Let us use Maple to study some of the properties of Bessel functions.

```
> restart;
> de:=x^2*diff(y(x),x$2)+x*diff(y(x),x)+(x^2-nu^2)*y(x)=0;
```

$$de := x^2 \left(\tfrac{d^2}{dx^2} y(x)\right) + x\left(\tfrac{d}{dx} y(x)\right) + (x^2 - \nu^2)\,y(x) = 0$$

```
> dsolve(de, y(x));
```

$$y(x) = \_C1\, \text{BesselJ}(\nu,\, x) + \_C2\, \text{BesselY}(\nu,\, x)$$

To gain intuition on Bessel functions, let us expand them in series around the origin.

```
> series(BesselJ(0, x), x,8);
```

$$1 - \frac{1}{4} x^2 + \frac{1}{64} x^4 - \frac{1}{2304} x^6 + \mathrm{O}(x^8)$$

```
> series(BesselJ(1,x),x,8);
```

$$\frac{1}{2} x - \frac{1}{16} x^3 + \frac{1}{384} x^5 - \frac{1}{18432} x^7 + \mathrm{O}(x^8)$$

```
> series(BesselJ(2,x),x,8);
```

$$\frac{1}{8} x^2 - \frac{1}{96} x^4 + \frac{1}{3072} x^6 + \mathrm{O}(x^8)$$

Power Series Solutions of Differential Equations

We see that for integer values of ν the functions alternate between "odd" or "even" behavior, and that $J_0(x)$ is the only one that is not zero at origin. If we plot these functions we get

```
> plot( {BesselJ(0,x), BesselJ(1,x),
> BesselJ(2,x)},
>    x=0..17, color=[magenta, blue, black], thickness=2);
```

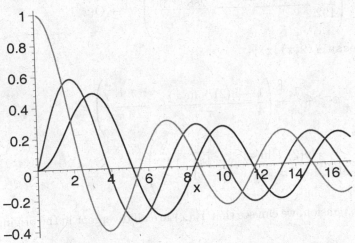

For half integer values of ν we have an interesting connection with more familiar trigonometric functions

```
> BesselJ(1/2,x);
```

$$\frac{\sqrt{2}\sin(x)}{\sqrt{\pi}\sqrt{x}}$$

```
> BesselJ(-1/2,x);
```

$$\frac{\sqrt{2}\cos(x)}{\sqrt{\pi}\sqrt{x}}$$

Let us follow a similar approach for the Bessel functions of second order, $Y_n(x)$.

```
> series(BesselY(0,x), x);
```

$$\left(\frac{2\,(-\ln(2)+\ln(x))}{\pi}+\frac{2\,\gamma}{\pi}\right)+\left(-\frac{1}{2}\,\frac{-\ln(2)+\ln(x)}{\pi}-\frac{-\frac{1}{2}+\frac{\gamma}{2}}{\pi}\right)x^2$$

$$+\left(\frac{1}{32}\,\frac{-\ln(2)+\ln(x)}{\pi}-\frac{\frac{3}{64}-\frac{\gamma}{32}}{\pi}\right)x^4+O(x^6)$$

```
> series(BesselY(1,x),x);
```

$$-\frac{2}{\pi}x^{-1} + \left(\frac{-\ln(2)+\ln(x)}{\pi} - \frac{-2\gamma+1}{2\pi}\right)x$$

$$+\left(-\frac{1}{8}\frac{-\ln(2)+\ln(x)}{\pi} - \frac{-\frac{5}{16}+\frac{\gamma}{4}}{2\pi}\right)x^3 +$$

$$\left(\frac{1}{192}\frac{-\ln(2)+\ln(x)}{\pi} - \frac{\frac{5}{288}-\frac{\gamma}{96}}{2\pi}\right)x^5 + O(x^7)$$

```
>   series(BesselY(2,x),x);
```

$$-\frac{4}{\pi}x^{-2} - \frac{1}{\pi} + \left(\frac{1}{4}\frac{-\ln(2)+\ln(x)}{\pi} - \frac{-\gamma+\frac{3}{4}}{4\pi}\right)x^2 +$$

$$\left(-\frac{1}{48}\frac{-\ln(2)+\ln(x)}{\pi} - \frac{-\frac{17}{144}+\frac{\gamma}{12}}{4\pi}\right)x^4 + O(x^6)$$

From their series expansion, we can see that $Y_\nu(x)$ are all divergent in the origin. Indeed

```
>   plot({BesselY(0,x), BesselY(1,x),
>   BesselY(2,x)}, x=0..12,
>        y=-4..1,  color=[magenta, blue, black],thickness=2);
```

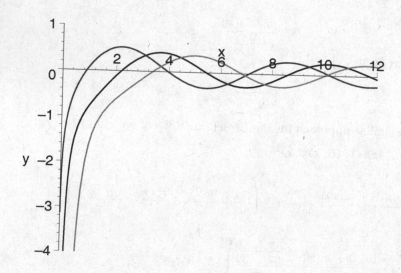

The Bessel function of second order $Y_\nu(x)$ is defined in terms of Bessel function of first order $J_\nu(x)$ via

$$Y_\nu(x) = \frac{2}{\pi} \left[J_\nu(x) \cos(\nu\pi) - J_{-\nu}(x) \right] . \qquad (11.19)$$

We can check numerically this equation

```
> Bess_Y:=(nu, x) ->
> (BesselJ(nu,x)*cos(nu*Pi) - BesselJ(-nu,x))/sin(nu*Pi);
```

$$Bess\_Y := (\nu, x) \rightarrow \frac{\mathrm{BesselJ}(\nu,\, x) \cos(\nu\,\pi) - \mathrm{BesselJ}(-\nu,\, x)}{\sin(\nu\,\pi)}$$

```
> zero:=expand(Bess_Y(nu,x) - BesselY(nu,x));
```

$$zero := \frac{\mathrm{BesselJ}(\nu,\, x) \cos(\nu\,\pi)}{\sin(\nu\,\pi)} - \frac{\mathrm{BesselJ}(-\nu,\, x)}{\sin(\nu\,\pi)} - \mathrm{BesselY}(\nu,\, x)$$

```
> evalf(subs({nu=2.4, x=10.3}, zero));
```

$$-0.30\,10^{-9}$$

Let us explore Maple's ability to expand $J_\nu(x)$ in series.

```
> series(BesselJ(nu,x),x);
```

```
Error, (in series/function) unable to compute series
```

We see that Maple cannot expand in series this Bessel function. The problem might be that Maple cannot determine the truncation order for an arbitrary ν. One way of bypassing this difficulty is to use the powseries package in order to explicitly construct the series expansion of $J_\nu(x)$. We know that

$$J_\nu(x) = \sum_{k=0}^{\infty} \frac{(-1)^k}{k!\,\Gamma(\nu+k+1)} \left(\frac{x}{2} \right)^{2k+\nu} , \qquad (11.20)$$

where Γ is the well-known Gamma function (or generalized factorial) defined as $\Gamma(\nu) = \int_0^\infty t^{\nu-1} e^{-t} dt$. The basic properties of the Gamma function include $\Gamma(\nu + 1) = \nu\,\Gamma(\nu)$ and $\Gamma(1) = 1$. We see that for ν a positive integer, Gamma function coincides with the usual factorial.

Let us implement eq. (11.20) in Maple.

```
> restart;
> with(powseries):
> powcreate(s(k)=(-1)^k*2^(-2*k-nu)/(k!*Gamma(nu+k+1))*x^k);
> x^nu*tpsform(s, x,8);
```

$$x^\nu \left(\frac{2^{(-\nu)}}{\Gamma(\nu+1)} - \frac{2^{(-2-\nu)}}{\Gamma(\nu+2)}\, x^2 + \frac{1}{2}\frac{2^{(-4-\nu)}}{\Gamma(\nu+3)}\, x^4 - \frac{1}{6}\frac{2^{(-6-\nu)}}{\Gamma(\nu+4)}\, x^6 + O(x^8) \right)$$

```
> simplify(%,power);
```

$$x^\nu \left(\frac{2^{(-\nu)}}{\Gamma(\nu+1)} - \frac{2^{(-2-\nu)}}{\Gamma(\nu+2)}\, x^2 + \frac{2^{(-5-\nu)}}{\Gamma(\nu+3)}\, x^4 - \frac{1}{6}\frac{2^{(-6-\nu)}}{\Gamma(\nu+4)}\, x^6 + O(x^8) \right)$$

Note that the truncation order applies to the series in parenthesis, not to the whole expression. This is natural, because we don't know the value of ν.

11.6 Problems

Problem 11.6.1. Consider the relativistic formula for kinetic energy

$$K = \frac{mc^2}{\sqrt{1 - \frac{v^2}{c^2}}} - mc^2$$

(a) Using the series expansion for K show that

$$K = \frac{m}{2}v^2 + \frac{3m}{8c^2}v^4 + \frac{5m}{16c^4}v^6 + O(v^8)$$

(b) What happens with the series expansion in the limit of small speeds $v \ll c$?

Problem 11.6.2. (a) Solve the following equation via two methods: exact (with dsolve) and using the series expansion

$$\left(1 - x^2\right)y'' - 6y' - 6y = 0 \ .$$

(b) Use the convert command to convert the series to a polynomial and graph both the exact and the approximate (series) solution on the same graph.
(c) What is the radius of convergence for the series solution?

Problem 11.6.3. The study of the hydrostatic equilibrium of white dwarf stars, leads to the Lane-Emden equation

$$\frac{1}{x^2}\frac{d}{dx}\left(x^2\frac{df}{dx}\right) = -f^n \ .$$

where f is the normalized density of the matter inside the star. This equation has explicit analytic solutions only for $n = 0, 1, 5$.
(a) Use dsolve to find the exact analytic solutions for $n = 0$ and $n = 1$.
(b) Solve the equation for the physically interesting cases $n = 3/2$ and respectively $n = 3$, satisfying the initial conditions $f(0) = 1, f'(0) = 0$. (*Hint:* You may have to use numerical techniques.)

Problem 11.6.4. Show that
(a) $J_{n+2}J_{-n} - J_{-(n+2)}J_n = \frac{4(n+1)}{\pi x^2}\sin(n+1)\pi$.
(b) $J_{n+2}Y_n - Y_{n+2}J_n = \frac{2}{x^2}(n+1)$.

Chapter 12

Special Functions and Generalized Fourier Series

SPECIAL functions play an important role in solving many problems in mathematical physics. As we have seen in the previous chapter, Maple has built-in routines to directly manipulate most of the special functions, such as Hermite, Laguerre, Legendre, Jacobi, and respectively Chebyshev polynomials.

12.1 The Sturm-Liouville Theory

12.1.1 Introduction

The Sturm-Liouville equation

$$\frac{d}{dx}\left(p(x)\frac{dy(x)}{dx}\right) + q(x)y(x) + \lambda w(x)y(x) = 0 ; \qquad (12.1)$$

can be written as the eigenvalue problem

$$\mathcal{L}y = -\lambda wy , \qquad (12.2)$$

where $\mathcal{L}$ is the differential operator

$$\mathcal{L} = p(x)\frac{d^2}{dx^2} + \frac{dp(x)}{dx}\frac{d}{dx} + q(x) , \qquad (12.3)$$

The functions $p(x), q(x), w(x)$ are real functions defined on $[a, b]$. The function $w(x)$, which is also called the *weight function*, is assumed to be positive.

```
>  restart:
>  with(PDEtools): declare(prime=x, p(x),q(x),w(x),y(x));
```

derivatives with respect to x of functions of one variable will now be displayed with '

$$p(x) \text{ will now be displayed as } p$$
$$q(x) \text{ will now be displayed as } q$$
$$w(x) \text{ will now be displayed as } w$$
$$y(x) \text{ will now be displayed as } y$$

The Sturm-Liouville equation is

```
> ode:=Diff(p(x)*Diff(y(x),x),x)+q(x)*y(x)+lambda*w(x)*y(x)=0;
```
$$ode := \left(\frac{\partial}{\partial x}\left(p\left(\frac{\partial}{\partial x}\,y\right)\right)\right) + q\,y + \lambda\,w\,y = 0$$

```
> ode_SL:=value(ode);
```
$$ode\_SL := p'\,y' + p\,y'' + q\,y + \lambda\,w\,y = 0$$

Sturm-Liouville equation can be cast as the eigenvalue problem $\mathcal{L}y = -\lambda w y$, where $\mathcal{L}$ is the differential operator

```
> L:=proc(f, p, q)
> p*diff(f,x$2) + diff(p,x)*diff(f,x)+q*f
> end proc;
```
$$L := \mathbf{proc}$$
$$(f,\,p,\,q)\,p * \operatorname{diff}(f,\,x\,\$\,2) + \operatorname{diff}(p,\,x) * \operatorname{diff}(f,\,x) + q * f$$
$$\mathbf{end\ proc}$$

```
> L(y(x), p(x), q(x)) = - lambda*w(x)*y(x);
```
$$p\,y'' + p'\,y' + q\,y = -\lambda\,w\,y$$

For example Legendre's equation, can be cast in a Sturm-Liouville form, with $p = 1 - x^2$, $q = 0$, $w = 1$ and $\lambda = n(n+1)$.

```
> n:='n':
```

```
> L(y(x),(1-x^2),0) = -n*(n+1)*y(x);
```
$$(1 - x^2)\,y'' - 2\,x\,y' = -n\,(n+1)\,y$$

The Sturm-Liouville operator is self-adjoint, i.e.

$$\int_a^b f^* \mathcal{L}(g)dx = \int_a^b g\mathcal{L}(f^*)dx \ , \tag{12.4}$$

which allows for the definition of the inner product of two functions

$$\langle f|g\rangle = \int_a^b f^*(x)g(x)w(x)dx \ , \tag{12.5}$$

where $w(x)$ is the weight function. The eigenvalues of $\mathcal{L}$ are real, and the corresponding eigenfunctions are orthogonal with respect to the inner product (12.5)

$$\mathcal{L}(y_n) = -\lambda_n w y_n \quad , \quad \lambda_n \in \mathbf{R} \; ,$$
$$\langle y_n | y_m \rangle = 0 \quad , \quad \text{if } m \neq n \; .$$

It can be shown that the eigenfunctions y_n of the self-adjoint operator $\mathcal{L}$, from a complete set of functions, which allows for the series expansion of a function $f(x)$

$$f(x) = \sum_{n=0}^{\infty} a_n \, y_n(x) \; , \quad a_n = \frac{\langle y_n | f \rangle}{\langle y_n | y_n \rangle} \tag{12.6}$$

provided that $f(x)$ is at least piecewise continuous.

Because Legendre's equation can be written as a Sturm-Liouville equation, it follows that we can use Legendre polynomials expand functions in series. Thus, we obtain the Fourier-Legendre expansion. Similarly, the Fourier-Bessel follows expansion from the fact that Bessel's equation fits into a Sturm-Liouville framework.

12.1.2 Inner Product and Gram-Schmidt Orthogonalization of Functions

The following Maple procedure implements the inner product with respect to the weight function $w(x)$ of two arbitrary functions $f(x)$ and $g(x)$, defined on the $[a, b]$ interval.

```
>    ip:=proc(a,b,f,g,w)
>        int(conjugate(f)*g*w,x=a..b);
>    end proc;
```

$$ip := \mathbf{proc}(a, b, f, g, w)$$
$$\text{int}(\text{conjugate}(f) * g * w, x = a..b)$$
$$\mathbf{end\ proc}$$

For example let us calculate the inner product of $\cos^2 x$ with $\cos x$ over the $[-\pi/2, \pi/2]$ interval with weight function $w(x) = 1$. We have

```
>    ip(-Pi/2,Pi/2,cos(x)^2,cos(x),1);
```

$$\frac{4}{3}$$

Because $\langle f | f \rangle \geq 0$, we can define the norm of a function as

$$\|f\| = \sqrt{\langle f | f \rangle} \; .$$

The Maple implementation is straightforward, using the ip procedure defined previously

```
>    nf:=proc(a,b,f,w) sqrt(ip(a,b,f,f,w)); end proc;
```

$$nf := \mathbf{proc}(a, b, f, w) \, \text{sqrt}(\text{ip}(a, b, f, f, w)) \, \mathbf{end\ proc}$$

```
>    nf(0,2*Pi,sin(x),1);
```

$$\sqrt{\pi}$$

As an application, let us check the Cauchy-Schwartz inequality

$$|\langle f|g \rangle| \leq ||f|| \, ||g||$$

for a concrete example

```
>   a:=0:  b:=Pi:  f:=x^2*sin(x):  g:=sin(x):
>   term1:=abs(ip(a,b,f,g,1));
```

$$term1 := -\frac{1}{4}\pi + \frac{1}{6}\pi^3$$

```
>   term2:=nf(a,b,f,1)*nf(a,b,g,1);
```

$$term2 := \frac{\sqrt{75\,\pi + 10\,\pi^5 - 50\,\pi^3}\,\sqrt{2}\,\sqrt{\pi}}{20}$$

```
>   is(term1 < term2);
```

$$true$$

```
>   evalf(term1); evalf(term2);
```

$$4.382314620$$
$$5.236247570$$

Note that we use the command `is` to test the assumption that `term1 < term2`. The reader is encouraged to investigate the alternative command `coulditbe`.

Using the inner product defined above, we can build a set of orthogonal functions, and use them to series expand any function as linear combination of the base functions. Let us exemplify the procedure of orthonormalization for the basis of functions $1, x, x^2, x^3, \ldots$.

```
>   for i from 0 to 4 do g[i]:= x^i end do;
```

$$g_0 := 1$$
$$g_1 := x$$
$$g_2 := x^2$$
$$g_3 := x^3$$
$$g_4 := x^4$$

We show that these functions are not orthogonal with respect to the inner product $\langle f|g \rangle = \int_{-1}^{1} f(x)^* g(x)dx$, and then we use the Gram-Schmidt procedure to build an orthogonal set of functions. The elements of the basis constructed in this way turn out to be proportional to Legendre's polynomials.

We can visually check orthogonality by checking whether the matrix A, whose entries are $A_{ij} = \langle g_i|g_j \rangle$ is diagonal. We obtain

```
>   A:=Matrix(5):
```

```
>   for k to 5 do
>       for j to 5 do
>           A[k,j]:=ip(-1,1,g[k-1],g[j-1],1):
>       end do;
>   end do;

>   A;
```

$$\begin{bmatrix} 2 & 0 & \dfrac{2}{3} & 0 & \dfrac{2}{5} \\[2mm] 0 & \dfrac{2}{3} & 0 & \dfrac{2}{5} & 0 \\[2mm] \dfrac{2}{3} & 0 & \dfrac{2}{5} & 0 & \dfrac{2}{7} \\[2mm] 0 & \dfrac{2}{5} & 0 & \dfrac{2}{7} & 0 \\[2mm] \dfrac{2}{5} & 0 & \dfrac{2}{7} & 0 & \dfrac{2}{9} \end{bmatrix}$$

We immediately see that the functions $g_i = x^i$, $i = 0, 1, 2, \ldots$ are not orthogonal. Let us apply the Gram-Schmidt orthogonalization procedure. We build a series of orthogonal functions f_j out of g_j via the recursion formulas

$$f_0 = g_0, \quad f_n = g_n - \sum_{j=0}^{n-1} \frac{\langle f_j | g_n \rangle}{\langle f_j | f_j \rangle} f_j$$

Then, we orthonormalize f_j simply by dividing each f_j by its norm $\|f_j\|$. We obtain

```
>   f[0]:=g[0]; fn[0]:=simplify(f[0]/nf(-1,1,f[0],1));
```

$$f_0 := 1$$

$$fn_0 := \frac{\sqrt{2}}{2}$$

```
>   f[1]:=g[1]
>   -ip(-1,1,f[0],g[1],1)/ip(-1,1,f[0],f[0],1)*f[0];
>   fn[1]:=simplify(f[1]/nf(-1,1,f[1],1));
```

$$f_1 := x$$

$$fn_1 := \frac{x\sqrt{6}}{2}$$

```
>   f[2]:=g[2]
>   -ip(-1,1,f[0],g[2],1)/ip(-1,1,f[0],f[0],1)*f[0]
>   -ip(-1,1,f[1],g[2],1)/ip(-1,1,f[1],f[1],1)*f[1];
>   fn[2]:=simplify(f[2]/nf(-1,1,f[2],1));
```

$$f_2 := x^2 - \frac{1}{3}$$

$$fn_2 := \frac{(3x^2 - 1)\sqrt{10}}{4}$$

```
>   f[3]:=g[3]
>   -ip(-1,1,f[0],g[3],1)/ip(-1,1,f[0],f[0],1)*f[0]
>   -ip(-1,1,f[1],g[3],1)/ip(-1,1,f[1],f[1],1)*f[1]
>   -ip(-1,1,f[2],g[3],1)/ip(-1,1,f[2],f[2],1)*f[2];
>   fn[3]:=simplify(f[3]/nf(-1,1,f[3],1));
```

$$f_3 := x^3 - \frac{3}{5}x$$

$$fn_3 := \frac{x(5x^2 - 3)\sqrt{14}}{4}$$

```
>   f[4]:=g[4] -ip(-1,1,f[0],g[4],1)/ip(-1,1,f[0],f[0],1)*f[0]
>   -ip(-1,1,f[1],g[4],1)/ip(-1,1,f[1],f[1],1)*f[1]
>   -ip(-1,1,f[2],g[4],1)/ip(-1,1,f[2],f[2],1)*f[2]
>   -ip(-1,1,f[3],g[4],1)/ip(-1,1,f[3],f[3],1)*f[3];
>   fn[4]:=simplify(f[4]/nf(-1,1,f[4],1));
```

$$f_4 := x^4 + \frac{3}{35} - \frac{6}{7}x^2$$

$$fn_4 := \frac{3(35x^4 + 3 - 30x^2)\sqrt{2}}{16}$$

Now, let's check that the new functions fn[j] are orthonormal to each other

```
>   B:=Matrix(5):
>   for k to 5 do
>       for j to 5 do
>           B[k,j]:=ip(-1,1,fn[k-1],fn[j-1],1):
>       end do;
>   end do;

>   B;
```

$$\begin{bmatrix} 1 & 0 & 0 & 0 & 0 \\ 0 & 1 & 0 & 0 & 0 \\ 0 & 0 & 1 & 0 & 0 \\ 0 & 0 & 0 & 1 & 0 \\ 0 & 0 & 0 & 0 & 1 \end{bmatrix}$$

An interesting fact is that the polynomials we have obtained are proportional to Legendre polynomials.

```
>   with(orthopoly):

>   map(simplify ,[seq(fn[i]/P(i,x), i=0..4)]);
```

$$[\frac{\sqrt{2}}{2}, \frac{\sqrt{6}}{2}, \frac{\sqrt{10}}{2}, \frac{\sqrt{14}}{2}, \frac{3\sqrt{2}}{2}]$$

The reader may have noticed that the orthonormalization implementation was quite laborious. We can spare the extra work by writing a short routine to automatize the Gram-Schmidt orthonormalization procedure. But there are few subtleties, in the way we implement such a procedure. First, we make sure that the indices used in summation don't have preassigned values. We set

```
>   j:='j':  n:='n':  M:=4:

>   f[0]:=g[0]:
>   for n from 1 to M do
>   f[n]:=g[n]
>       - sum(ip(-1,1,f[j],g[n],1)/ip(-1,1,f[j],f[j],1)*f[j],
>       j=0..n-1):
>   end do:

>   seq([f[n]],n=0..4);
```

$$[1], [x], [x^2 - \frac{2}{3}], [x^3], [x^4 - \frac{4}{5}]$$

Even though the formula appears to be correct, the evaluation procedure gives a wrong result. The reason is the particular way in which Maple evaluates the sum. The sum command is designed to return a formula for an indefinite (or definite) sum or product, rather than an explicit sum or product.

To first evaluate each term, and then perform the addition, one must use the add command

```
>   M:=4:j:='j':  n:='n':

>   f[0]:=g[0]:
>   for n from 1 to M do
>   f[n]:=g[n]
>       -add(ip(-1,1,f[j],g[n],1)/ip(-1,1,f[j],f[j],1)*f[j],
>       j=0..n-1):
>   end do:

>   seq([f[n]],n=0..4);
```

$$[1], [x], [x^2 - \frac{1}{3}], [x^3 - \frac{3}{5}x], [x^4 + \frac{3}{35} - \frac{6}{7}x^2]$$

This is the expected result. To normalize these functions, we simply divide each term by its norm

```
>   seq([simplify(f[k]/nf(-1,1,f[k],1))], k=0..M);
```

$$[\frac{\sqrt{2}}{2}], [\frac{x\sqrt{6}}{2}], [\frac{(3x^2-1)\sqrt{10}}{4}], [\frac{x(5x^2-3)\sqrt{14}}{4}], [\frac{3(35x^4+3-30x^2)\sqrt{2}}{16}]$$

12.1.3 Fourier-Legendre Series

Legendre equation can be cast in Sturm-Liouville form with $p = 1-x^2$, $q = 0$, $w = 1$ and $\lambda = n(n+1)$. Consequently, the Legendre polynomials form a complete and orthogonal set of functions satisfying

$$\langle P_n | P_m \rangle = \int_{-1}^{1} P_n P_m \, d\dot{x} = \frac{2}{2n+1} \delta_{nm} \, ,$$

that can be used to expand functions in series.

Let us check the orthogonality

```
>  restart;
>  with(orthopoly);
```

$$[G, H, L, P, T, U]$$

```
>  b:=(m,n)->int(P(n,x)*P(m,x),x=-1..1);
```

$$b := (m,\, n) \to \int_{-1}^{1} P(n,\, x)\, P(m,\, x)\, dx$$

```
>  M:=5:
>  A:=Matrix(M):
>  for i from 0 to M-1 do
>      for j from 0 to M-1 do
>          A[i+1,j+1]:=b(i,j):
>      end do
>  end do;
>  A;
```

$$\begin{bmatrix} 2 & 0 & 0 & 0 & 0 \\ 0 & \dfrac{2}{3} & 0 & 0 & 0 \\ 0 & 0 & \dfrac{2}{5} & 0 & 0 \\ 0 & 0 & 0 & \dfrac{2}{7} & 0 \\ 0 & 0 & 0 & 0 & \dfrac{2}{9} \end{bmatrix}$$

Because Legendre polynomials $P_n(x)$ form a complete set of functions with respect to the weight function $w(x) = 1$ over the interval $[-1, 1]$ we can expand a function as follows

$$f(x) = \sum_{n=0}^{\infty} a_n P_n(x) \,,$$

where the coefficients a_n are given by

$$a_n = \frac{\langle Pn | f \rangle}{\langle Pn | Pn \rangle} = \frac{2n+1}{2} \int_{-1}^{1} P_n(x) f(x) dx \,, \quad n = 0, 1, 2, \dots \,.$$

We define the following procedure to calculate the expansion coefficients

```
>  a:=proc(n,f)
>     (2*n+1)/2 * int( P(n,x)*f, x=-1..1)
>  end proc;
```

$$a := \mathbf{proc}(n,\, f)$$
$$1/2 * (2*n + 1) * \text{int}(P(n,\, x) * f,\, x = -1..1)$$
$$\mathbf{end\ proc}$$

For an arbitrary function we get

```
> a(n,f(x));
```

$$\frac{1}{2}(2n+1)\int_{-1}^{1}P(n,\,x)\,f(x)\,dx$$

Now, let us expand $f(x) = x^3 - 2xx^2 + 3$.

```
> f:=x^3-2*x^2+3;
```

$$f := x^3 - 2x^2 + 3$$

The first five coefficients are

```
> for j from 0 to 5 do
>     a||j:=a(j,f);
> end do;
```

$$a0 := \frac{7}{3}$$

$$a1 := \frac{3}{5}$$

$$a2 := \frac{-4}{3}$$

$$a3 := \frac{2}{5}$$

$$a4 := 0$$

$$a5 := 0$$

Note the use of the concatenation operator ||, to simplify notation. We can check the decomposition by explicitly adding the terms of the series

```
> f1:=add(a||k*P(k,x),k=0..5);
```

$$f1 := x^3 - 2x^2 + 3$$

As a second example, let us expand the piecewise continuous function

$$g(x) = \begin{cases} 1, & x \in [0,1] \\ 0, & x \in [-1,0) \end{cases}$$

in the Fourier-Legendre series. The Maple implementation reads

```
> g:=piecewise(x<=0 and x<=1, 1, x>=-1 and x<0,0);
```

$$g := \begin{cases} 1 & x \le 0 \text{ and } x \le 1 \\ 0 & -1 \le x \text{ and } x < 0 \end{cases}$$

The first five coefficients of the series expansion are

```
> seq(a(j,g),j=0..5);
```

$$\frac{1}{2}, \frac{-3}{4}, 0, \frac{7}{16}, 0, \frac{-11}{32}$$

which can be compared with the known result

$$b_n = \frac{2n+1}{2n}P_{n+1}(0).$$

We have

```
> 1/2, seq((2*n+1)/(2*n)*P(n+1,0),n=1..5);
```

$$\frac{1}{2}, \frac{-3}{4}, 0, \frac{7}{16}, 0, \frac{-11}{32}$$

Let us consider a less trivial function $h(x) = x\,e^x \sin x$, $x \in [-1, 1]$.

```
> h:=x*sin(x)*exp(x);
```

$$h := x \sin(x)\, e^x$$

We analyze successive approximations. First, let us consider only the first two terms in the expansion

```
> m:=2:
```

```
> for j from 0 to m
>     do c[j]:= a(j,h):
> end do:
```

```
> j:='j':
```

```
> h1:=sum(c[j]*P(j,x), j=0..m);
```

$$h1 := -\frac{1}{2}\,e^{(-1)}\cos(1) - \frac{1}{4}\sin(1)\,e^{(-1)} + \frac{1}{4}\sin(1)\,e + 3\,e^{(-1)}\cos(1)\,x$$
$$+ (-\frac{95}{8}\,e^{(-1)}\cos(1) + 10\sin(1)\,e^{(-1)} - \frac{15}{8}\,e\cos(1) + \frac{5}{4}\sin(1)\,e)\,(-\frac{1}{2} + \frac{3\,x^2}{2})$$

```
> with(plots):
```

Warning, the name changecoords has been redefined

```
> p:=plot(h, x=-1..1, color=blue ,thickness=2):
```

```
> p1:=plot(h1, x=-1..1, color=red, thickness=3, linestyle=4):
```

```
> display(p,p1);
```

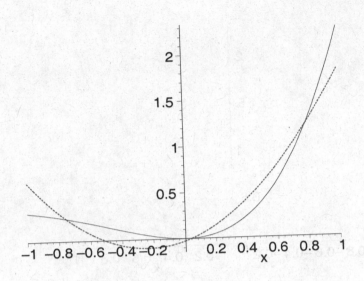

Observe that even if there are only two terms in the series expansion, the approximation is not too far off. The dotted line (the approximate expansion) follows quite closely the general pattern of the exact function (the plain line). We expect to see some improvement when we include the first three terms in the expansion. We have

```
>  m:=3;
```

$$m := 3$$

```
>  for j from 0 to m do c[j]:= a(j,h):  end do:

>  j:='j':

>  h2:=sum(c[j]*P(j,x), j=0..m);
```

$$h2 := -\frac{1}{2} e^{(-1)} \cos(1) - \frac{1}{4} \sin(1) e^{(-1)} + \frac{1}{4} \sin(1) e + 3 e^{(-1)} \cos(1) x$$

$$+ \left(-\frac{95}{8} e^{(-1)} \cos(1) + 10 \sin(1) e^{(-1)} - \frac{15}{8} e \cos(1) + \frac{5}{4} \sin(1) e\right) \left(-\frac{1}{2} + \frac{3 x^2}{2}\right)$$

$$+ \left(\frac{91}{8} e^{(-1)} \cos(1) - \frac{805}{8} \sin(1) e^{(-1)} + \frac{105}{8} e \cos(1) + \frac{35}{8} \sin(1) e\right) \left(\frac{5}{2} x^3 - \frac{3}{2} x\right)$$

```
>  p2:=plot(h2, x=-1..1, color=red, thickness=3, linestyle=4):

>  display(p,p2);
```

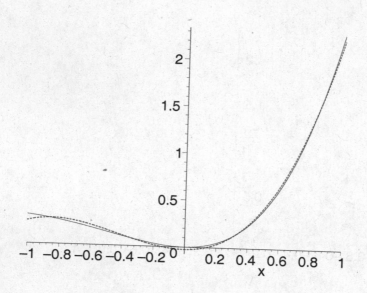

The dotted line represents the sum of the first three terms in the series expansion. It is very close to the exact function. The reader is encouraged to see how well the sum of the first five terms approximates the exact function.

12.1.4 Fourier-Bessel Series

Bessel's differential equation for integers n

$$x^2 J_n''(x) + x J_n'(x) + (x^2 - n^2) J_n(x) = 0 \ . \tag{12.7}$$

can be cast in the Sturm Liouville form

$$\frac{d}{dx}\left(x \frac{d\varphi_r(x)}{dx}\right) - \frac{n^2}{x}\varphi_r(x) + \frac{\beta_{nr}^2}{R^2} x \varphi_r(x) = 0 \ , \quad \text{for } 0 \le x \le R, \tag{12.8}$$

where $\varphi_r(x) = J_n(\beta_{nr} x/R)$, and β_{nr} is the r-th zero of the Bessel function J_n. This is a Sturm-Liouville equation with $p(x) = x$, $q(x) = -n^2/x$, $\lambda = \beta_{rn}^2/R^2$ and respectively $w(x) = x$. Therefore, any well-behaved function on the interval $[0, R]$ can be expanded in terms of $\varphi_r(x)$

$$f(x) = \sum_{r=1}^{\infty} a_r \, \varphi_r(x) = \sum_{r=1}^{\infty} a_r \, J_n(\beta_{nr} x/R) \tag{12.9}$$

where the coefficients a_r are calculated using the orthogonality of $\varphi_r(x)$ with respect to the inner product $\langle \varphi_r | \varphi_p \rangle = \int_0^R \varphi_r^* \varphi_p \, x \, dx$. We have

$$a_r = \frac{\langle \varphi_r | f \rangle}{\langle \varphi_r | \varphi_r \rangle} = \frac{2}{R^2 J_{n+1}^2(\beta_{nr})} \int_0^R J_n(\beta_{nr} x/R) \, f(x) \, x \, dx \ . \tag{12.10}$$

The expansion of eq.(12.9) is known as the Fourier-Bessel expansion.

Let us implement the Fourier-Bessel expansion in Maple

```
> restart;
```

We use BesselJZeros function to find the zeros of Bessel functions. For example, the first 11 zeros of J_0 are

```
> evalf(BesselJZeros(0,1..11));
```

2.404825558, 5.520078110, 8.653727913, 11.79153444, 14.93091771, 18.07106397, 21.21163663, 24.35247153, 27.49347913, 30.63460647, 33.77582021

Next, let's define the following procedure that numerically evaluates the r-th zero of Bessel function J_n

```
> beta:=proc(n,r)
>    evalf(BesselJZeros(n,r));
> end proc;
```

$$\beta := \mathbf{proc}(n, r) \, \text{evalf(BesselJZeros}(n, r)) \, \mathbf{end\ proc}$$

Then, the first two zeros of J_0 and respectively J_1 are

```
> beta(0,1), beta(0,2); beta(1,1), beta(1,2);
```

$$2.404825558, 5.520078110$$
$$3.831705970, 7.015586670$$

The coefficient a_r from eq.(12.10) of the Fourier-Bessel expansion can be simply calculated if n, $f(x)$, and R are given, using the following procedure

```
> a:=proc(r)
>    2*int(BesselJ(n, beta(n,r)*x/R) * f*x,x=0..R)/
>       (R^2*BesselJ(n+1,beta(n,r))^2);
> end proc;
```

$$a := \mathbf{proc}(r)$$
$$2/(R^2 * \text{BesselJ}(n + 1, \beta(n, r))^2) * \text{int(BesselJ}(n, \beta(n, r) * x/R) * f * x, x = 0..R)$$
$$\mathbf{end\ proc}$$

For example, let us illustrate the series expansion of $f(x) = 16 - x^2$ in terms of J_0 (i.e. $n = 0$) for $x \in [0, 4]$.

We have

```
> n:=0;f:=16-x^2; R:=4;
```

$$n := 0$$
$$f := 16 - x^2$$
$$R := 4$$

The first coefficients of the series expansion are

```
> seq(a(k),k=1..11);
```

17.72835619, −2.236440092, 0.7276235288, −0.3358544312, 0.1861798926, −0.1155388098, 0.07740595246, −0.05481086268, 0.04047247842, −0.03088234670, 0.02419531874

Observe that the third coefficient is only equal to 4% of the first coefficient, and the rest are smaller. The rapid convergence of the series is expected given that the function being expanded satisfies the same zero boundary condition as the eigenfunctions φ.

Let us plot the first three partial sums of the expansion

```
> f1:=sum(a(j)*BesselJ(n, beta(n,j)*x/R), j=1..1):
```

```
> plot([f,f1],x=0..R, color=[blue,red],thickness=2);
```

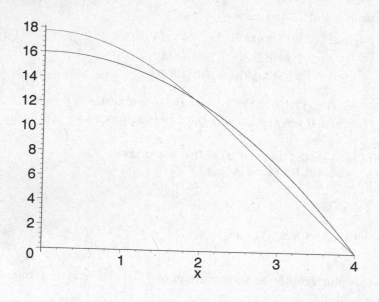

```
> f2:=sum(a(j)*BesselJ(n, beta(n,j)*x/R), j=1..2):
```

```
> plot([f,f2],x=0..R, color=[blue,red],thickness=2);
```

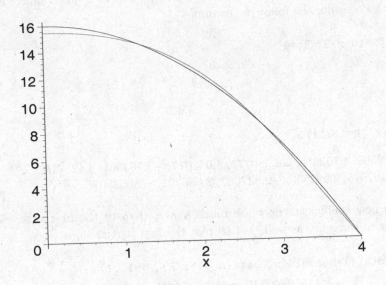

```
>   f3:=sum(a(j)*BesselJ(n, beta(n,j)*x/R), j=1..3):
>   plot([f,f3],x=0..R, color=[blue,red],thickness=2);
```

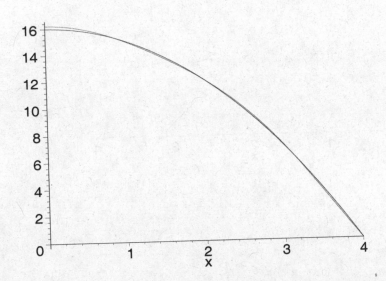

Note that the the last partial sum despite the fact it contains only the first three terms, it represents a very good approximation for our function.

If the function doesn't vanish at the end points, we expect a much slower convergence. Let us consider the following example

```
>   n:=0;f:=10-x^2; R:=4;
```

$$n := 0$$
$$f := 10 - x^2$$
$$R := 4 \ .$$

```
>   seq(a(k),k=1..11);
```

$$8.116508, 4.152355, -4.380771, 4.042017, -3.704961, 3.421718,$$
$$-3.187675, 2.992550, -2.827602, 2.686221, -2.563507$$

Note that the new coefficients decrease much slower than in the first case. We expect a slower convergence as well. Let us plot the first partial sums

```
>   f1:=sum(a(j)*BesselJ(n, beta(n,j)*x/R), j=1..1);
```

$$f1 := 8.116508002 \, \text{BesselJ}(0., 0.6012063895 \, x)$$

```
>   plot([f,f1],x=0..R, color=[blue,red],thickness=2);
```

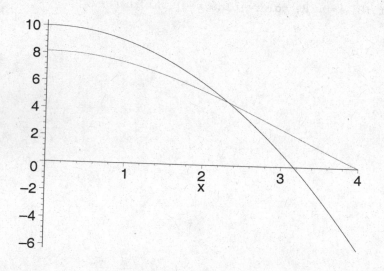

```
>   f2:=sum(a(j)*BesselJ(n, beta(n,j)*x/R), j=1..2):
```

```
>   plot([f,f2],x=0..R, color=[blue,red],thickness=2);
```

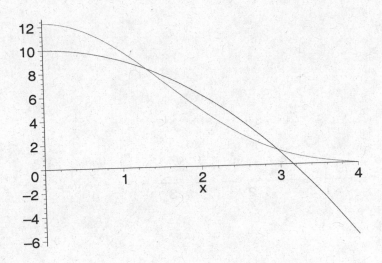

```
> f3:=sum(a(j)*BesselJ(n, beta(n,j)*x/R), j=1..3):
> plot([f,f3],x=0..R, color=[blue,red],thickness=2);
```

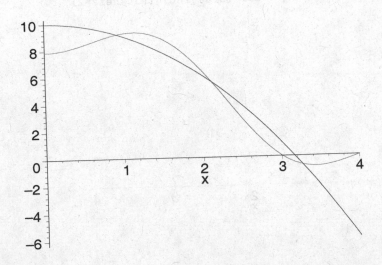

The convergence is clearly slower than in the first example. Let us calculate directly f_{10} (the partial sum of first 10 terms), and respectively f_{20}.

```
> f10:=sum(a(j)*BesselJ(n, beta(n,j)*x/R), j=1..10):
```

```
>  plot([f,f10],x=0..R, color=[blue,red],thickness=2);
```

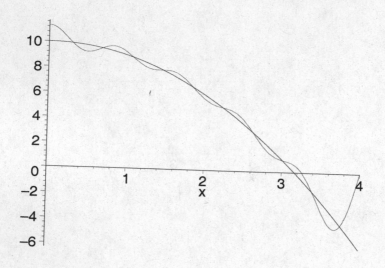

```
>  f20:=sum(a(j)*BesselJ(n, beta(n,j)*x/R), j=1..20):
>  plot([f,f20],x=0..R, color=[blue,red],thickness=2);
```

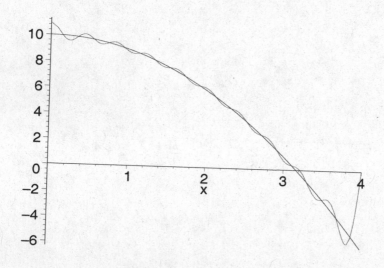

The series is clearly converging, but an interesting phenomenon appears: the partial sum oscillates violently around the end points. This phenomenon, known as the

Gibbs phenomenon, is not restricted to expansions with respect to J_0. For example the partial sum for expansion for J_1, exhibits similar oscillations

```
>   n:=1;f:=10-x^2; R:=4;
```

$$n := 1$$
$$f := 10 - x^2$$
$$R := 4$$

```
>   f20:=sum(a(j)*BesselJ(n, beta(n,j)*x/R), j=1..20):
>   plot([f,f20],x=0..R, color=[blue,red],thickness=2);
```

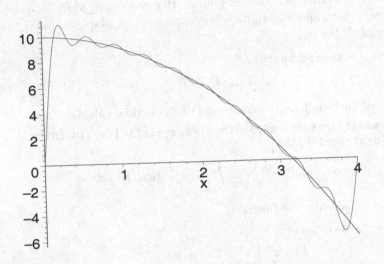

Gibbs phenomenon will also be present for the Fourier expansion, as shown below.

12.1.5 Fourier-Hermite Series

Hermite's differential equation

$$H_n''(x) - 2xH_n'(x) + 2nH_n(x) = 0 ,\qquad(12.11)$$

can be written as a Sturm-Liouville equation

$$\frac{d}{dx}\left(e^{-x^2}\frac{dH_n(x)}{dx}\right) + 2ne^{-x^2}H_n(x) = 0\qquad(12.12)$$

where $p(x) = e^{-x^2}$, $q(x) = 0$, $\lambda = 2n$, and $w(x) = e^{-x^2}$. The eigenfunctions of the Sturm-Liouville operator (in this case the Hermite polynomials) are orthogonal with respect to the inner product $\langle H_n|H_m\rangle = \int_{-\infty}^{\infty} H_n H_m e^{-x^2} dx$. Therefore, a "nice" function $f(x)$ can be expanded as

$$f(x) = \sum_{n=0}^{\infty} a_n H_n(x) \, , \tag{12.13}$$

where

$$a_n = \frac{1}{2^n\, n!\sqrt{\pi}} \int_{-\infty}^{\infty} e^{-x^2} H_n(x) f(x) dx \, . \tag{12.14}$$

The expansion of eq.(12.13) is known as Fourier-Hermite expansion.

To implement the Fourier-Hermite series expansion in Maple we need the orthopoly package. We have

```
> restart; with(orthopoly);
```
$$[G, H, L, P, T, U]$$

The coefficient of Fourier-Hermite expansion (12.14) is then calculated as

```
> a:=n->evalf(int(exp(-x^2)*H(n,x)*f,x=-infinity..infinity)/
>     (2^n*n!*sqrt(Pi)));
```

$$a := n \rightarrow \mathbf{evalf}(\frac{1}{2^n\, n!\,\sqrt{\pi}} \int_{-\infty}^{\infty} e^{(-x^2)}\, \mathrm{H}(n,\,x)\, f\, dx)$$

For example, let us consider the function

```
> f:=x^4-x^3-7*x^2+x+6;
```

$$f := x^4 - x^3 - 7\,x^2 + x + 6$$

```
> seq(a(k),k=0..10);
```

$$3.250, -0.250, -1., -0.1250, 0.0625, 0., 0., 0., 0., 0., 0.$$

We see that we have a finite set of coefficients. We have expected this result because $f(x)$ is a polynomial, and consequently, it can be expressed as a finite linear combination of Hermite polynomials.

Define now a procedure to calculate partial sums up to a given order n.

```
> partsum:=proc(n)
>     add(a(k)*H(k,x),k=0..n)
> end proc;
```

$$partsum := \mathbf{proc}(n)\, \mathrm{add}(a(k) * \mathrm{H}(k,\,x),\, k = 0..n)\, \mathbf{end\ proc}$$

Let us visualize the first partial sums

```
> f0:=partsum(0);
> plot([f,f0],x=-2.5..3.5, y=-15..20, color=[blue, red],
>     thickness=2);
```

$$f0 := 3.250000000$$

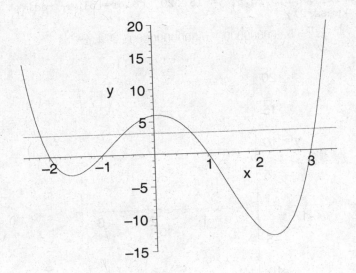

```
>  f1:=partsum(1);
>  plot([f,f1],x=-2.5..3.5, y=-15..20, color=[blue, red],
>      thickness=2);
```
$$f1 := 3.250000000 - 0.5000000000\,x$$

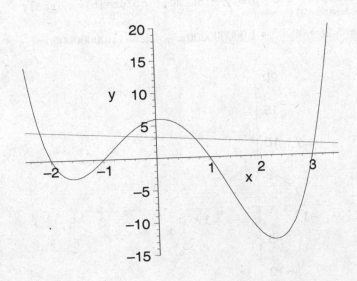

```
>  f2:=partsum(2);
```

```
>  plot([f,f2],x=-2.5..3.5, y=-15..20, color=[blue, red],
>      thickness=2);
```

$$f2 := 5.250000000 - 0.5000000000\,x - 4.\,x^2$$

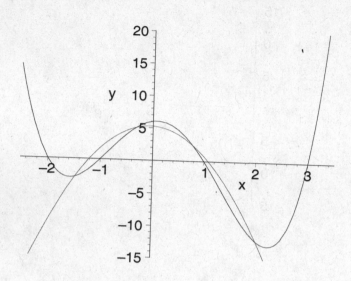

```
>  f3:=partsum(3);
>  plot([f,f3],x=-2.5..3.5,y=-15..20,  color=[blue, red],
>      thickness=2);
```

$$f3 := 5.250000000 + 1.000000000\,x - 4.\,x^2 - 1.000000000\,x^3$$

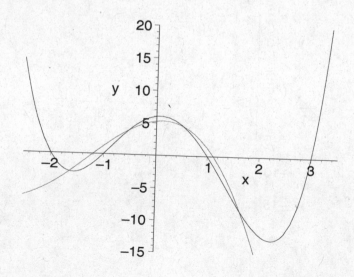

```
>   f4:=partsum(4);
>   plot([f,f4],x=-2.5..3.5, y=-15..20, color=[blue, red],
>       thickness=2);
```

$$f4 := 6.000000000 + 1.000000000\,x - 7.000000000\,x^2 - 1.000000000\,x^3$$
$$+\, 1.000000000\,x^4$$

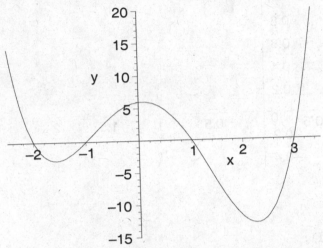

Observe that the fourth partial sum is identical with $f(x)$, and, consequently, its graph coincides with the graph of $f(x)$.

The Fourier-Hermite expansion doesn't converge so rapidly for all functions. In particular, the Fourier-Hermite expansion gives mediocre results for functions with compact support. For example let us consider the rectangular function

$$f(x) = \begin{cases} 1 & \text{for } x \in [0,1]\,, \\ 0 & \text{otherwise}\,. \end{cases} \tag{12.15}$$

The Maple implementation reads

```
>   f:=Heaviside(x)-Heaviside(x-1);
```
$$f := \text{Heaviside}(x) - \text{Heaviside}(x-1)$$

To improve the efficiency of the algorithm, we first calculate all the coefficients a_n

```
>   an:=seq(a(k),k=0..100):
```

We define a new partial sum

```
>   psum:=n->sum(an[k+1]*H(k,x),k=0..n);
```

$$psum := n \rightarrow \sum_{k=0}^{n} an_{k+1}\,\text{H}(k,\,x)$$

and graphically illustrate some partial sums

```
>  f0:=psum(0):
>  plot([f,f0],x=-1..2, y=-0.5..1.5, color=[blue, red],
thickness=2);
```

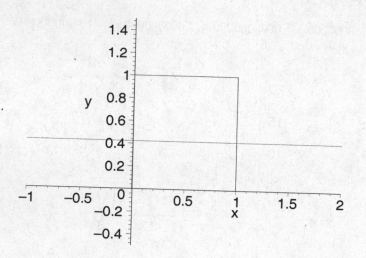

```
>  f1:=psum(1):
>  plot([f,f1],x=-1..2, y=-0.5..1.5, color=[blue, red],
thickness=2);
```

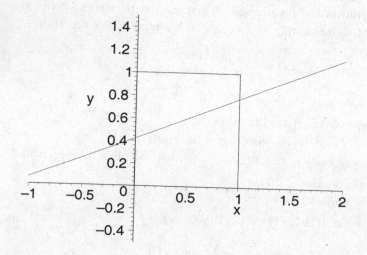

```
>  f3:=psum(3):
>  plot([f,f3],x=-1..2, y=-0.5..1.5, color=[blue, red],
thickness=2);
```

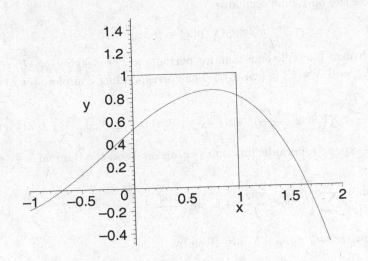

```
>  f100:=psum(100):
>  plot([f,f100],x=-1..2, y=-0.5..1.5, color=[blue, red],
>  thickness=2);
```

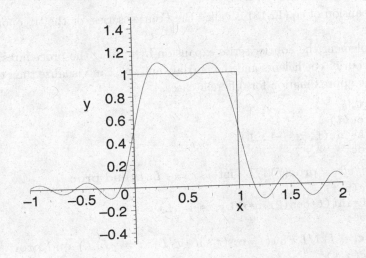

Observe the poor convergence of the series even after one hundred iterations.

12.2   Periodic Sturm-Liouville Problem and the Fourier Series

12.2.1   Trigonometric Form of the Fourier Series

The equation of the harmonic oscillator

$$y''(x) + k^2 y(x) = 0 \; , \tag{12.16}$$

becomes as a Sturm-Liouville equation, for periodic $y(x)$ satisfying $y(-L) = y(L)$, $y'(-L) = y'(L)$, and $\lambda \equiv k^2 = (\pi n/L)^2$. The corresponding complete set of eigenfunctions

$$\{1, \; \cos\left(\frac{\pi n x}{L}\right), \; \sin\left(\frac{\pi n x}{L}\right)\} \; , \quad n = 1, 2, 3, \ldots \tag{12.17}$$

can be used to expand a periodic function $f(x)$ in series on the interval $x \in [-L, L]$

$$f(x) = \frac{a_0}{2} + \sum_{n=1}^{\infty} \left[a_n \cos\left(\frac{\pi n x}{L}\right) + b_n \sin\left(\frac{\pi n x}{L}\right) \right] \; , \quad x \in [-L, L] \; , \tag{12.18}$$

where the coefficients a_0, a_n and b_n are given by

$$a_0 = \frac{1}{L} \int_{-L}^{L} f(x) dx \; ,$$

$$a_n = \frac{1}{L} \int_{-L}^{L} f(x) \cos\left(\frac{\pi n x}{L}\right) dx \; , \quad n = 1, 2, 3, \ldots \; ,$$

$$b_n = \frac{1}{L} \int_{-L}^{L} f(x) \sin\left(\frac{\pi n x}{L}\right) dx \; , \quad n = 1, 2, 3, \ldots \; . \tag{12.19}$$

The series expansion of eq.(12.18) is called the Fourier series, or the trigonometric Fourier series.

Let us implement the Fourier series expansion by writing the procedures to calculate the Fourier coefficients and the partial sum. We can visualize then the accuracy of the approximation for different cases.

```
>  restart;
>  a0:=proc(f)
>      1/L*int(f, x= -L..L)
>  end proc;
```

$$a0 := \mathbf{proc}(f)\, 1/L * \mathrm{int}(f, x = -L..L)\, \mathbf{end\ proc}$$

```
>  a:=proc(n,f)
>      1/L*int(f*cos(Pi*n*x/L),x=-L..L)
>  end proc;
```

$$a := \mathbf{proc}(n, f)\, 1/L * \mathrm{int}(f * \cos(\pi * n * x/L), x = -L..L)\, \mathbf{end\ proc}$$

```
>  b:=proc(n,f)
>      1/L*int(f*sin(Pi*n*x/L),x=-L..L)
>  end proc;
```

$$b := \mathbf{proc}(n, f) \, 1/L * \mathrm{int}(f * \sin(\pi * n * x/L), \, x = -L..L) \ \mathbf{end \ proc}$$

```
>    psum:=proc(k)
>        if k = 0 then a0(f)/2 else
>            a0(f)/2+sum(a(n,f)*cos(Pi*n*x/L)+
>            b(n,f)*sin(Pi*n*x/L),n=1..k)
>        end if
>    end proc;
```

$psum := \mathbf{proc}(k)$

$\quad \mathbf{if}\, k = 0 \, \mathbf{then}\, 1/2 * a0(f)$

$\quad \mathbf{else}\, 1/2 * a0(f) + \mathrm{sum}(a(n,\, f) * \cos(\pi * n * x/L) + b(n,\, f) * \sin(\pi * n * x/L),\, n = 1..k)$

$\quad \mathbf{end \ if}$

$\mathbf{end \ proc}$

Let us study few concrete examples. First, we look at a periodic function

$$f(x) = \begin{cases} -1 & \text{for } x \in [-1,0) \,, \\ 1 & \text{for } x \in [0,1] \,. \end{cases}$$

The Maple implementation reads

```
>    f:=signum(sin(Pi*x));
```

$$f := \mathrm{signum}(\sin(\pi x))$$

```
>    plot(f,x=-2..2,thickness=2, color=blue,numpoints=1000);
```

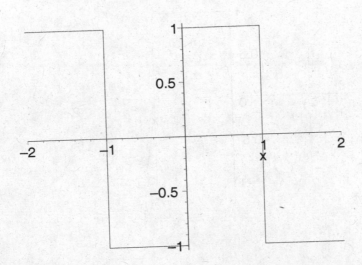

We have $L = 1$. Then, the first partial sums are

```
>    L:=1;
```

$$L := 1$$

```
>  f0:=psum(0);
```

$$f0 := 0$$

```
>  f1:=psum(1);
```

$$f1 := \frac{4\sin(\pi x)}{\pi}$$

```
>  f2:=psum(2);
```

$$f2 := \frac{4\sin(\pi x)}{\pi}$$

```
>  f3:=psum(3);
```

$$f3 := \frac{4\sin(\pi x)}{\pi} + \frac{4}{3}\frac{\sin(3\pi x)}{\pi}$$

```
>  f4:=psum(4);
```

$$f4 := \frac{4\sin(\pi x)}{\pi} + \frac{4}{3}\frac{\sin(3\pi x)}{\pi}$$

```
>  plot([f,f0,f1,f3],x=-2..2,
>     color=[blue,red,magenta,black],
>     thickness=[2,1,1,1], numpoints=1000);
```

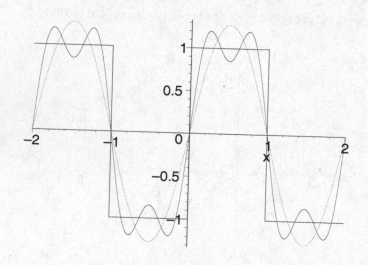

Observe that `f2 = f1` and respectively `f4=f3`, etc. This is a consequence of the symmetry of the function. In our case $f(x)$ is an odd function, which implies $a_n = 0$ for $n = 0, 1, 2, \ldots$. We can easily check this property

```
>  a0(f), seq(a(j,f),j=1..10);
```

$$0, 0, 0, 0, 0, 0, 0, 0, 0, 0, 0$$

In the next example, we consider a periodic function with $L = 1$. On the interval $[-1, 1]$ the function is given by $f(x) = 1 + x$. As before, we will calculate the partial sums, and illustrate graphically the convergence of the corresponding Fourier series

```
>  f:=1+x;
```

$$f := 1 + x$$

```
>  f0:=psum(0);
```

$$f0 := 1$$

```
>  f1:=psum(1);
```

$$f1 := 1 + \frac{2\sin(\pi x)}{\pi}$$

```
>  f2:=psum(2);
```

$$f2 := 1 + \frac{2\sin(\pi x)}{\pi} - \frac{\sin(2\pi x)}{\pi}$$

```
>  f3:=psum(3);
```

$$f3 := 1 + \frac{2\sin(\pi x)}{\pi} - \frac{\sin(2\pi x)}{\pi} + \frac{2}{3}\frac{\sin(3\pi x)}{\pi}$$

```
>  f4:=psum(4);
```

$$f4 := 1 + \frac{2\sin(\pi x)}{\pi} - \frac{\sin(2\pi x)}{\pi} + \frac{2}{3}\frac{\sin(3\pi x)}{\pi} - \frac{1}{2}\frac{\sin(4\pi x)}{\pi}$$

```
>  plot([f,f0,f1,f3,f4], x=-1..1,
>      color=[blue,red,magenta,black,brown],
>      thickness=[3,1,1,1,1], numpoints=1000);
```

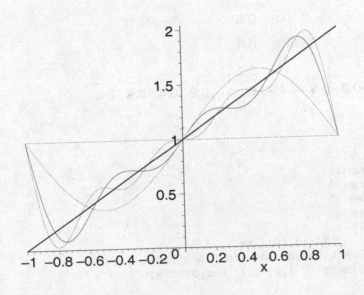

In the last example we will also illustrate the Gibbs phenomenon, which is a series of violent oscillations of the partial sum, which appears near the end points of the interval. We consider a periodic function with $L = 1$ which is defined on the main interval $x \in [-1, 1]$ by

$$f(x) = 1 - x^2 \sin x .$$

We have

```
> f:=1-sin(x)*x^2;
```

$$f := 1 - \sin(x)\, x^2$$

```
> plot(f,x=-1..1,color=blue,thickness=2);
```

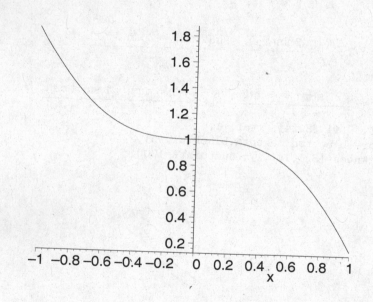

```
> f0:=psum(0):
> f1:=psum(1):
> f2:=psum(2):
> f3:=psum(3):
> f4:=psum(4):

> plot([f,f0,f1,f3,f4], x=-1..1,
>     color=[blue,red,magenta,black,brown],
>     thickness=[2,1,1,1,1], numpoints=1000);
```

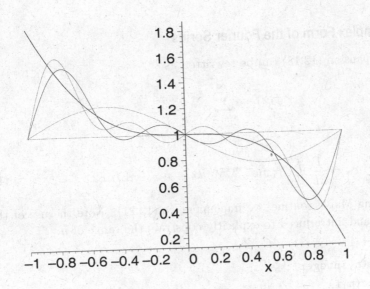

As we increase the number of terms in the partial sum, we see that the overall convergence improves, with the exception of the end points, where the Gibbs oscillations occur. The partial sum of the first twenty five terms is graphed below against the exact function

```
>   f25:=psum(25):
>   plot([f,f25], x=-1..1, color=[blue,red],
>      thickness=[1,2], numpoints=1000);
```

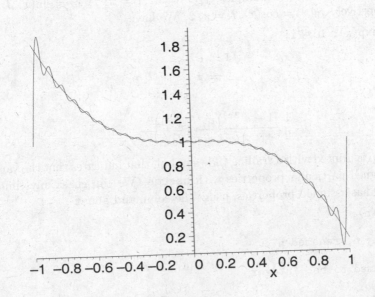

12.2.2 Complex Form of the Fourier Series

The series expansion (12.18) can be rewritten as

$$f(x) = \sum_{n=-\infty}^{\infty} c_n \, e^{in\pi x/L} , \qquad (12.20)$$

where

$$c_n = \frac{1}{2L} \int_{L}^{L} f(x)e^{-in\pi x/L} dx , \quad n = 0, \pm1, \pm2, \dots . \qquad (12.21)$$

In the following Maple routine, we implement eq.(12.21). Note the use of the assume command, introduced to explicitly constrain the range of n.

```
>  restart;

>  assume(n,integer):

>  c:=proc(n,f)
>  1/(2*L)*int(f*exp(-I*n*Pi*x/L),x=-L..L)
>  end proc;
```

$$c := \mathbf{proc}(n, f)$$
$$1/2 * (1/L * \text{int}(f * \exp(-I * n * \pi * x/L), x = -L..L))$$
$$\mathbf{end\ proc}$$

We built the procedure in such a way that the user only needs to provide the function f and the range L.

We will study three simple examples: $f_1(x) = e^x, L = \pi$; $f_2(x) = \sin(x), L = \pi/2$ and respectively $f_3(x) = \cos(x), L = \pi/2$. We have

```
>  f1:=exp(x); L:=Pi;
```

$$f1 := e^x$$
$$L := \pi$$

```
>  c(n,f1);
```

$$\frac{1}{2} \frac{\left(1 - e^{(2\pi)}\right) e^{(-\pi)} (-1)^{n\tilde{}}}{\pi \left(-1 + n\tilde{} \, I\right)}$$

Observe that n is printed with a trailing tilde symbol, that indicates that the variable has been assigned particular properties with assume. We can check any time if a given variable has assumed properties, using the command about

```
>  about(n);
```

```
Originally n, renamed n~:

  is assumed to be: integer
```

```
>  f2:=sin(x); L:=Pi/2;
```
$$f2 := \sin(x)$$
$$L := \frac{\pi}{2}$$

```
>  c(n,f2);
```
$$\frac{4\,I\,n\tilde{\ }\,(-1)^{n\tilde{\ }}}{\pi\,(4\,n^{\tilde{\ }2}-1)}$$

```
>  f3:=cos(x);L:=Pi/2;
```
$$f3 := \cos(x)$$
$$L := \frac{\pi}{2}$$

```
>  c(n,f3);
```
$$\frac{2\,(-1)^{(n\tilde{\ }+1)}}{\pi\,(4\,n^{\tilde{\ }2}-1)}$$

12.3   Dirac Delta Function and the Fourier Transform

12.3.1   Dirac Delta Function

Given a complete and orthonormal set of functions $\varphi_n(x)$ we can expand a function $f(x)$ as

$$f(x) = \sum_{m=0}^{\infty} a_m\,\varphi_m(x) , \qquad (12.22)$$

where the coefficients a_m are given by $a_m = \langle \varphi_m | f \rangle = \int_{-\infty}^{\infty} \varphi_m^*(y) f(y) dy.$
Substituting the expression for a_m back into (12.22) one gets

$$f(x) = \int_{-\infty}^{\infty} f(y) \left[\sum_{m=0}^{\infty} \varphi_m^*(y)\varphi_m(x) \right] dy , \qquad (12.23)$$

which leads to the definition of the Dirac function

$$\delta(y-x) = \sum_{m=0}^{\infty} \varphi_m^*(y)\varphi_m(x) , \qquad (12.24)$$

with the basic property

$$f(x) = \int_{-\infty}^{\infty} f(y)\,\delta(y-x)dy . \qquad (12.25)$$

Let us check few properties of Dirac function:

$$\int_{-\infty}^{\infty} \delta(x) = 1 \tag{12.26}$$

$$\delta(-x) = \delta(x) \tag{12.27}$$

$$x\delta'(x) = -\delta(x) \tag{12.28}$$

$$\delta(ax) = \frac{1}{|a|}\delta(x) \tag{12.29}$$

$$\int_{-\infty}^{x} \delta(x) = \Theta(x) \tag{12.30}$$

$$\frac{d}{dx}\Theta(x) = \delta(x) \tag{12.31}$$

where $\Theta(x)$ is the step function (also known as the Heaviside function)

$$\Theta(x) = \begin{cases} 1 & x > 0 \\ 0 & x < 0 \end{cases} \tag{12.32}$$

```
> restart;
> int(Dirac(x), x=-infinity..infinity);
```
$$1$$
```
> int(Dirac(x-a)*f(x), x=-infinity..infinity);
```
$$f(a)$$

Maple denotes by `Dirac(1,x)` the derivative $\delta'(x)$, and in general by `Dirac(n,x)` the n-derivative of $\delta(x)$.

```
> int(Dirac(1,x)*f(x),x=-infinity..infinity);
```
$$-D(f)(0)$$
```
> int(Dirac(x)*x*f(x),x=-infinity..infinity);
```
$$0$$
```
> int(x*Dirac(1,x)*f(x),x=-infinity..infinity);
```
$$-f(0)$$
```
> int(Dirac(a*x)*f(x),x=-infinity..infinity);
```
$$\frac{f(0)}{|a|}$$

The step function (or Heaviside function) is simply defined in Maple as `Heaviside(x)`. Let us plot it

```
> plot(Heaviside(x), x=-2..2, y=-0.5..1.5 ;
```

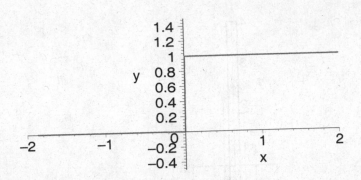

Let us check the relationship between Heaviside and Dirac delta functions

```
>   int(Dirac(x),x);
```
$$\text{Heaviside}(x)$$

```
>   diff(Heaviside(x),x);
```
$$\text{Dirac}(x)$$

Dirac function can be represented as the limit $a \to 0$ of the rectangular function $\delta_a(x)$ with area equal to 1

$$\delta_n(x) = \begin{cases} \frac{1}{a} & \text{for } -\frac{a}{2} < x < \frac{a}{2} \\ 0 & \text{otherwise} \end{cases} \qquad (12.33)$$

We use Maple to visualize $\delta_a(x)$ for few values of a. First, we define it using Heaviside function

```
>   one:=a->1/a*(Heaviside(x+a/2)-Heaviside(x-a/2));
```
$$one := a \to \frac{\text{Heaviside}(x + \frac{1}{2}a) - \text{Heaviside}(x - \frac{1}{2}a)}{a}$$

Let us check that its integral is 1. We have to make explicit the assumption that $a > 0$, for Maple to give us the desired result

```
>   int(one(a),x=-infinity..infinity) assuming a::positive;
```
$$1$$

Graphing $\delta_a(x)$ for $a = \{2, 1, 0.5, 0.25\}$ we get

```
>   plot([one(2), one(1), one(.5), one(0.25)],
>       x=-2..2, y=-0.5..5,
>       scaling=constrained, thickness=2,
>       color =[black,red,magenta,blue]);
```

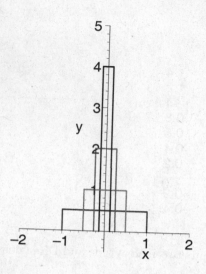

12.3.2   Fourier Integral Transform

The Fourier transform of a function $f(x)$ is defined as

$$\widetilde{f}(k) = \frac{1}{\sqrt{2\pi}} \int_{-\infty}^{\infty} f(x)\, e^{-i\,kx}\, dx \;, \tag{12.34}$$

while the inverse Fourier transform is

$$f(x) = \frac{1}{\sqrt{2\pi}} \int_{-\infty}^{\infty} \widetilde{f}(k)\, e^{i\,kx}\, dk \;. \tag{12.35}$$

Maple uses the package `inttrans` to implement the Fourier transform and its inverse, along with other integral transforms.

```
>  with(inttrans);
```

[*addtable, fourier, fouriercos, fouriersin, hankel, hilbert, invfourier, invhilbert,
invlaplace, invmellin, laplace, mellin, savetable*]

For example, we calculate and visualize the Fourier transform of a rectangular pulse of height 1 and width 2.   We use the `piecewise` command to define this function.   Remember that the syntax for `piecewise` reads `piecewise(cond_1,f_1,...,cond_n,f_n,otherwise)`, where `f_1` is the value of the function when first condition `cond_1` is true, etc. We have

```
>  h:= piecewise(x>-1 and x< 1,1, 0);
```

$$h := \begin{cases} 1 & -1 < x \text{ and } x < 1 \\ 0 & otherwise \end{cases}$$

The Fourier transform is defined `fourier(expr, old_var, new_var)`. In our case, the Fourier transform of the rectangular pulse is

```
>   fh:=fourier(h, x, k);
```

$$fh := \frac{2\sin(k)}{k}$$

```
>   plot(fh,k=-13..13, thickness=3);
```

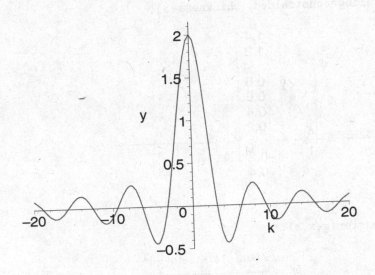

To get back the original function, we apply the inverse Fourier transform

```
>   invfourier(fh, k, x);
```

$$\text{Heaviside}(x+1) - \text{Heaviside}(x-1)$$

The result is exactly our initial function, but expressed in terms of the Heaviside function. Using the `convert` command we can transform the piecewise function into an expression containing Heaviside functions. We get

```
>   convert(h, Heaviside);
```

$$\text{Heaviside}(x+1) - \text{Heaviside}(x-1)$$

Let us analyze one more example.

```
> g:=piecewise(x>=0 and x<1, x, x>=-1 and x<0, -x, 0);
```

$$g := \begin{cases} x & 0 \leq x \text{ and } x < 1 \\ -x & -1 \leq x \text{ and } x < 0 \\ 0 & \textit{otherwise} \end{cases}$$

```
> plot(g, x =-2..2, y=-0.5..1.5,
>    scaling=constrained, thickness=3);
```

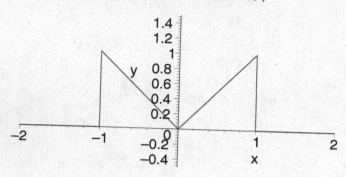

```
> fg:=fourier(g,x,k);
```

$$fg := \frac{2\left(\sin(k)\,k - 2\sin(\frac{k}{2})^2\right)}{k^2}$$

```
> plot(fg, k=-30..30, y=-1..1.25, thickness=3);
```

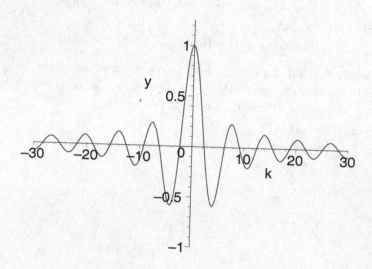

12.4 Problems

Problem 12.4.1. Show that Legendre polynomials satisfy Laplace's formula

$$P_n(x) = \frac{1}{\pi} \int_0^\pi \left(x + \sqrt{x^2 - 1} \, \cos\phi\right)^n d\phi \, .$$

Problem 12.4.2. For all $x > 1$ show that Legendre polynomials satisfy

$$P_0(x) < P_1(x) < \cdots < P_n(x) < \cdots$$

Problem 12.4.3. Prove that Hermite polynomials satisfy

$$(a) \quad \int_{-\infty}^\infty e^{-x^2} H_n^2(x) \, dx = 2n \int_{-\infty}^\infty e^{-x^2} H_{n-1}^2(x) \, dx$$

$$(b) \quad e^{-x^2} H_n(x) = -\frac{d}{dx} \left(e^{-x^2} H_{n-1}(x)\right)$$

Problem 12.4.4. Write a Maple routine that calculates the first 10 terms of the Fourier-Hermite expansion of the following function

$$f(x) = \begin{cases} 1 & \text{for } x > 0 \\ -1 & \text{for } x < 0 \, . \end{cases}$$

Problem 12.4.5. Find the Fourier-Hermite expansion of the following functions: $e^{kx}, \sin kx$ and $\cos kx$.

Problem 12.4.6. Write a Maple routine that calculates the first 25 terms of the Fourier-Bessel expansion of the following function

$$f(x) = \begin{cases} 1 & \text{for } x > 0 \\ 0 & \text{otherwise} \, . \end{cases}$$

Plot the sum of the first 25 terms. Is there a Gibbs phenomenon? Why?

Chapter 13

Linear Systems of Differential Equations

MANY problems in mathematical physics are described by systems of differential equations. If the case of linear systems of differential equations, the problem can be rewritten in matrix formalism. Consequently, tools from linear algebra such as eigenvectors and their use in matrix digitalization, play an important role in finding the solutions of linear systems of differential equations.

From the point of view of Maple, solving systems of differential equations in many cases, is similar to solving ordinary differential equations. We have `dsolve({sys, ics},{vars})`, where `sys` represents the system of differential equations, `ics` are the initial conditions, and `vars` are the unknown functions. For example, to solve the system

$$\frac{dx_1(t)}{dt} = x_2(t) - e^t \, ,$$

$$\frac{dx_2(t)}{dt} = x_1(t) + e^{-t} \, ,$$

$$\frac{dx_3(t)}{dt} = x_1(t) + x_2(t) - x_3(t) + t, \quad (x_1(0) = 0, \ x_2(0) = 1, \ x_3(0) = 0) \, ,$$

(13.1)

we simply write

```
>   sys:=
>        diff(x1(t),t)=x2(t) - exp(t),
>        diff(x2(t),t)=x1(t) + exp(-t),
>        diff(x3(t),t)=x1(t)+ x2(t)-x3(t)+t;
```

$$sys := \tfrac{d}{dt}\,\text{x1}(t) = \text{x2}(t) - e^t, \ \tfrac{d}{dt}\,\text{x2}(t) = \text{x1}(t) + e^{(-t)},$$
$$\tfrac{d}{dt}\,\text{x3}(t) = \text{x1}(t) + \text{x2}(t) - \text{x3}(t) + t$$

```
>   ics:=x1(0)=0, x2(0)=1, x3(0)=0;
```
$$ics := x1(0) = 0, \; x2(0) = 1, \; x3(0) = 0$$

```
>   vars:=x1(t),x2(t),x3(t);
```
$$vars := x1(t), \; x2(t), \; x3(t)$$

```
>   dsolve({sys,ics},{vars});
```
$$\{x1(t) = -\frac{1}{2}e^{(-t)} + \frac{1}{2}e^{t} - \frac{1}{2}e^{(-t)}t - \frac{1}{2}te^{t}, \; x2(t) = e^{t} + \frac{1}{2}e^{(-t)}t - \frac{1}{2}te^{t},$$
$$x3(t) = e^{t} - \frac{1}{2}e^{(-t)}t - \frac{1}{2}te^{t} + t - 1\}$$

The above strategy works for almost all cases. In the next sections, we will look at different types of systems of differential equations, and implement in Maple alternative ways of solving them. In all these instances, one arrives at the same results as in the case of `dsolve` command.

13.1   Homogeneous Systems of Differential Equations

A linear system of differential equations

$$\frac{dx_1(t)}{dt} = a_{11}x_1(t) + a_{12}x_2 + \cdots + a_{1n}x_n(t) + r_1(t)$$

$$\frac{dx_2(t)}{dt} = a_{21}x_1(t) + a_{22}x_2 + \cdots + a_{2n}x_n(t) + r_2(t)$$

$$\vdots \qquad\qquad\qquad\qquad\qquad\qquad (13.2)$$

$$\frac{dx_n(t)}{dt} = a_{n1}x_1(t) + a_{n2}x_2(t) + \cdots + a_{nn}x_n + r_n(t)$$

can be rewritten in matrix form as

$$\frac{dx(t)}{dx} = A\,x(t) + r(t) , \qquad (13.3)$$

where

$$x(t) = \begin{bmatrix} x_1(t) \\ x_2(t) \\ \vdots \\ x_n(t) \end{bmatrix}, \quad A = \begin{bmatrix} a_{11} & a_{12} & \cdots & a_{1n} \\ a_{21} & a_{22} & \cdots & a_{2n} \\ \vdots & \vdots & \cdots & \vdots \\ a_{n1} & a_{n2} & \cdots & a_{nn} \end{bmatrix}, \quad r(t) = \begin{bmatrix} r_1(t) \\ r_2(t) \\ \vdots \\ r_n(t) \end{bmatrix}.$$
$$(13.4)$$

If the source term $r(t) = 0$, the system is called *homogeneous*. In analogy with the ordinary equations case, the solution of a homogeneous system of equations is

$$x(t) = e^{tA}x(0) , \qquad (13.5)$$

where e^{A} is the exponential of the matrix A.

As an example, let us solve the following system of differential equations

$$\frac{dx_1(t)}{dt} = x_2(t)$$

$$\frac{dx_2(t)}{dt} = -x_1(t)$$

$$\frac{dx_3(t)}{dt} = x_1(t) - x_3(t)$$

subject to the initial conditions $x_1(0) = a, x_2(0) = b, x_3(0) = c$.
First, using `dsolve` command we obtain

```
>   de:=diff(x1(t),t)=x2(t),
>       diff(x2(t),t)=-x1(t),
>       diff(x3(t),t)=x1(t)-x3(t);
```

$$de := \tfrac{d}{dt}\mathrm{x1}(t) = \mathrm{x2}(t), \ \tfrac{d}{dt}\mathrm{x2}(t) = -\mathrm{x1}(t), \ \tfrac{d}{dt}\mathrm{x3}(t) = \mathrm{x1}(t) - \mathrm{x3}(t)$$

```
>   ic:=x1(0)=a, x2(0)=b, x3(0)=c;
```

$$ic := \mathrm{x1}(0) = a, \ \mathrm{x2}(0) = b, \ \mathrm{x3}(0) = c$$

```
>   var:=x1(t), x2(t), x3(t);
```

$$var := \mathrm{x1}(t), \ \mathrm{x2}(t), \ \mathrm{x3}(t)$$

```
>   dsolve({de,ic},{var});
```

$$\{\mathrm{x1}(t) = b\sin(t) + a\cos(t),$$

$$\mathrm{x3}(t) = -\frac{1}{2}b\cos(t) + \frac{1}{2}b\sin(t) + \frac{1}{2}a\cos(t) + \frac{1}{2}a\sin(t) + e^{(-t)}\,(\frac{b}{2} - \frac{a}{2} + c),$$

$$\mathrm{x2}(t) = b\cos(t) - a\sin(t)\}$$

Alternatively, we can use the formula of eq.(13.5) to obtain the same result. We write the initial condition vector as

```
>   x0:=<a,b,c>;
```

$$x0 := \begin{bmatrix} a \\ b \\ c \end{bmatrix}$$

To obtain the matrix associated to the system of differential equations (13.6), we use `GenerateMatrix` command from the `LinearAlgebra` package

```
>   with(LinearAlgebra):
>   step1:=GenerateMatrix([de],[var]);
```

$$step1 := \begin{bmatrix} 0 & -1 & 0 \\ 1 & 0 & 0 \\ -1 & 0 & 1 \end{bmatrix}, \begin{bmatrix} -(\tfrac{d}{dt}\mathrm{x1}(t)) \\ -(\tfrac{d}{dt}\mathrm{x2}(t)) \\ -(\tfrac{d}{dt}\mathrm{x3}(t)) \end{bmatrix}$$

The matrix of the system is the first member of the outputted list. Therefore, we write

```
>   A:=step1[1];
```

$$A := \begin{bmatrix} 0 & -1 & 0 \\ 1 & 0 & 0 \\ -1 & 0 & 1 \end{bmatrix}$$

Then, the matrix exponential of A is

```
>   eA:= MatrixExponential(A,t);
```

$$eA := \begin{bmatrix} \cos(t) & -\sin(t) & 0 \\ \sin(t) & \cos(t) & 0 \\ \frac{1}{2}\cos(t) - \frac{1}{2}e^t - \frac{1}{2}\sin(t) & -\frac{1}{2}\sin(t) - \frac{1}{2}\cos(t) + \frac{1}{2}e^t & e^t \end{bmatrix}$$

which generates the desired solution

```
>   eA . x0;
```

$$\begin{bmatrix} b\sin(t) + a\cos(t) \\ b\cos(t) - a\sin(t) \\ (-\frac{1}{2}e^{(-t)} + \frac{1}{2}\cos(t) + \frac{1}{2}\sin(t))\,a + (\frac{1}{2}e^{(-t)} - \frac{1}{2}\cos(t) + \frac{1}{2}\sin(t))\,b + e^{(-t)}\,c \end{bmatrix}$$

As expected, we obtain the same result.

An alternate way of solving homogeneous systems of differential equations is to express the matrix exponential of the system in terms of fundamental matrix. Let A be the matrix of the system and v_i an eigenvector: $Av_i = \lambda_i v_i$. If we denote by $x_i(t) = e^{\lambda_i t} v_i$, and by $X(t)$ the matrix that has as columns the vectors $x_i(t)$, i.e. $X(t) = [x_1(t) \quad x_2(t) \quad \cdots \quad x_n(t)]$ then it can be shown that

$$e^{tA} = X(t)X^{-1}(0) . \tag{13.6}$$

Consequently, the solution of the linear system of differential equations, can be written as

$$x(t) = X(t)X^{-1}(0)\,x(0) . \tag{13.7}$$

As an example, using all three methods, the command `dsolve`, the formula of eq.(13.5) and respectively the formula of eq.(13.7) we will solve the following system of equations

$$\frac{dx_1(t)}{dt} = 4\,x_1(t) - 2\,x_2(t) - 2\,x_3(t)$$

$$\frac{dx_2(t)}{dt} = 4\,x_1(t) - 5\,x_2(t) - 4\,x_3(t)$$

$$\frac{dx_3(t)}{dt} = 4\,x_1(t) - 5\,x_2(t) - 4\,x_3(t)$$

subject to the initial conditions $x(0) = a, x_2(0) = b, x_3(0) = c$.
Using `dsolve` command we have

```
>   restart;
```

```
> des:=
>     diff(x1(t),t)= 4*x1(t)-2*x2(t)-2*x3(t),
>     diff(x2(t),t)= 4*x1(t)-5*x2(t)-4*x3(t),
>     diff(x3(t),t)=4*x1(t)-5*x2(t)-4*x3(t):
> vars:=x1(t), x2(t), x3(t);
```

$$vars := x1(t),\ x2(t),\ x3(t)$$

```
> ics:=x1(0)=a,x2(0)=b, x3(0)=c;
```

$$ics := x1(0) = a,\ x2(0) = b,\ x3(0) = c$$

```
> dsolve([des,ics],[vars]);
```

$$\{x1(t) = (2c + 2b - a)\,e^{(2t)} + 2\,e^{(3t)}\,(-c - b + a),$$
$$x3(t) = (-c - 2b + a)\,e^{(-t)} + (2c + 2b - a)\,e^{(2t)},$$
$$x2(t) = -(-c - 2b + a)\,e^{(-t)} + e^{(3t)}\,(-c - b + a)\}$$

Using the formula of eq. (13.5) we get

```
> with(LinearAlgebra):
> x0:=<a,b,c>;
```

$$x0 := \begin{bmatrix} a \\ b \\ c \end{bmatrix}$$

```
> A:=GenerateMatrix([des],[vars])[1];
```

$$A := \begin{bmatrix} -4 & 2 & 2 \\ -4 & 5 & 4 \\ 3 & -6 & -5 \end{bmatrix}$$

```
> eA:=MatrixExponential(A,t);
```

$$eA := \begin{bmatrix} 2\,e^{(-3t)} - e^{(-2t)} & 2\,e^{(-2t)} - 2\,e^{(-3t)} & 2\,e^{(-2t)} - 2\,e^{(-3t)} \\ e^{(-3t)} - e^{t} & -e^{(-3t)} + 2\,e^{t} & e^{t} - e^{(-3t)} \\ e^{t} - e^{(-2t)} & -2\,e^{t} + 2\,e^{(-2t)} & -e^{t} + 2\,e^{(-2t)} \end{bmatrix}$$

```
> map( collect, eA. x0, exp);
```

$$\begin{bmatrix} (2a - 2b - 2c)\,e^{(-3t)} + (2c + 2b - a)\,e^{(-2t)} \\ (-c - b + a)\,e^{(-3t)} + (c + 2b - a)\,e^{t} \\ (2c + 2b - a)\,e^{(-2t)} + (-c - 2b + a)\,e^{t} \end{bmatrix}$$

Finally, using the formula of eq.(13.7) we have

```
> V:=Eigenvectors(A);
```

$$V := \begin{bmatrix} -2 \\ -3 \\ 1 \end{bmatrix},\ \begin{bmatrix} 1 & 2 & 0 \\ 0 & 1 & -1 \\ 1 & 0 & 1 \end{bmatrix}$$

```
> v1:=Column(V[2],1):  v2:=Column(V[2],2):  v3:=Column(V[2],3):
> lambda1:=V[1][1]:  lambda2:=V[1][2]:  lambda3:=V[1][3]:
```

```
>   x1:=exp(lambda1*t)*v1:
>   x2:=exp(lambda2*t)*v2:
>   x3:=exp(lambda3*t)*v3:

>   X:=<x1|x2|x3>;
```

$$X := \begin{bmatrix} e^{(-2t)} & 2\,e^{(-3t)} & 0 \\ 0 & e^{(-3t)} & -e^t \\ e^{(-2t)} & 0 & e^t \end{bmatrix}$$

```
>   X0:=subs(t=0,X);;
```

$$X0 := \begin{bmatrix} 1 & 2 & 0 \\ 0 & 1 & -1 \\ 1 & 0 & 1 \end{bmatrix}$$

```
>   X01:=1/X0;
```

$$X01 := \begin{bmatrix} -1 & 2 & 2 \\ 1 & -1 & -1 \\ 1 & -2 & -1 \end{bmatrix}$$

```
>   X.X01.x0;
```

$$\begin{bmatrix} (2\,e^{(-3t)} - e^{(-2t)})\,a + (2\,e^{(-2t)} - 2\,e^{(-3t)})\,b + (2\,e^{(-2t)} - 2\,e^{(-3t)})\,c \\ (e^{(-3t)} - e^t)\,a + (-e^{(-3t)} + 2\,e^t)\,b + (e^t - e^{(-3t)})\,c \\ (e^t - e^{(-2t)})\,a + (-2\,e^t + 2\,e^{(-2t)})\,b + (-e^t + 2\,e^{(-2t)})\,c \end{bmatrix}$$

```
>   x_sol:= map (collect, X.X01.x0, exp );
```

$$x\_sol := \begin{bmatrix} (2\,a - 2\,b - 2\,c)\,e^{(-3t)} + (2\,c + 2\,b - a)\,e^{(-2t)} \\ (-c - b + a)\,e^{(-3t)} + (c + 2\,b - a)\,e^t \\ (2\,c + 2\,b - a)\,e^{(-2t)} + (-c - 2\,b + a)\,e^t \end{bmatrix}$$

As expected we have obtained the same result in all three cases.

13.2 Non-Homogeneous Systems of Differential Equations

A non-homogeneous system of differential equations is written in a matrix form as

$$\frac{dx(t)}{dx} = A\,x(t) + r(t) , \tag{13.8}$$

where the source vector $r(t)$ is different from zero. Then, it can be shown that the solution of (13.8) is given by

$$x(t) = e^{tA} \left(c + \int e^{-tA}\,r(t)\,dt \right) , \tag{13.9}$$

where the constant vector c is fixed by the initial conditions.

As an example, let us solve the non-homogeneous system of differential equations

$$\frac{dx_1(t)}{dt} = 2x_2(t) + e^t(1-t)$$

$$\frac{dx_2(t)}{dt} = -x_1(t) + 3x_2(t) + e^t, \quad (x_1(0) = 1, \ x_2(0) = 0).$$

We have

```
>  restart;
>  des:=diff(x1(t),t)=2*x2(t)+r1(t),
>  diff(x2(t),t)=-x1(t)+3*x2(t)+r2(t);
```

$$des := \tfrac{d}{dt}\,x1(t) = 2\,x2(t) + r1(t),\ \tfrac{d}{dt}\,x2(t) = -x1(t) + 3\,x2(t) + r2(t)$$

```
>  r1(t):=exp(t)*(1-t); r2(t):=exp(t);
```

$$r1(t) := e^t\,(1-t)$$

$$r2(t) := e^t$$

```
>  ics:=x1(0)=1,x2(0)=0;
```

$$ics := x1(0) = 1,\ x2(0) = 0$$

```
>  vars:=x1(t),x2(t);
```

$$vars := x1(t),\ x2(t)$$

```
>  dsolve({des,ics},{vars});
```

$$\{x2(t) = -e^t - e^t t + e^{(2t)} - \frac{1}{2}\,e^t t^2,\ x1(t) = (-t - t^2)\,e^t + e^{(2t)}\}$$

```
>  map(collect, %, exp);
```

$$\{x1(t) = (-t - t^2)\,e^t + e^{(2t)},\ x2(t) = e^{(2t)} + (-1 - t - \frac{1}{2}\,t^2)\,e^t\}$$

Now, let's obtain the same result with the help of formula of eq. (13.9)

```
>  with(LinearAlgebra):
>  gm:=GenerateMatrix([des],[vars]);
```

$$gm := \begin{bmatrix} 0 & -2 \\ 1 & -3 \end{bmatrix}, \begin{bmatrix} -(\tfrac{d}{dt}\,x1(t)) + e^t\,(1-t) \\ -(\tfrac{d}{dt}\,x2(t)) + e^t \end{bmatrix}$$

```
>  A:=-gm[1];
```

$$A := \begin{bmatrix} 0 & 2 \\ -1 & 3 \end{bmatrix}$$

```
>  r:=<r1(t),r2(t)>;
```

$$r := \begin{bmatrix} e^t\,(1-t) \\ e^t \end{bmatrix}$$

```
>  eA:=MatrixExponential(A,t);
```

$$eA := \begin{bmatrix} 2\,e^t - e^{(2t)} & 2\,e^{(2t)} - 2\,e^t \\ -e^{(2t)} + e^t & -e^t + 2\,e^{(2t)} \end{bmatrix}$$

```
>  A1:=-t*A;
```

$$A1 := \begin{bmatrix} 0 & -2t \\ t & -3t \end{bmatrix}$$

```
> eA1:=MatrixExponential(A1);
```

$$eA1 := \begin{bmatrix} -e^{(-2t)} + 2e^{(-t)} & -2e^{(-t)} + 2e^{(-2t)} \\ e^{(-t)} - e^{(-2t)} & 2e^{(-2t)} - e^{(-t)} \end{bmatrix}$$

```
> step1:=simplify(eA1.r);
```

$$step1 := \begin{bmatrix} e^{(-t)} + e^{(-t)}t - 2t \\ -t + e^{(-t)} + e^{(-t)}t \end{bmatrix}$$

```
> c:=<c1,c2>;
```

$$c := \begin{bmatrix} c1 \\ c2 \end{bmatrix}$$

```
> step2:=map(collect, simplify(eA.(map(int, step1, t)+c)), exp);
```

$$step2 := \begin{bmatrix} (-c1 + 2\,c2)\,e^{(2t)} + (-t^2 + 2\,c1 - 2 - t - 2\,c2)\,e^t \\ (-c1 + 2\,c2)\,e^{(2t)} + (-2 - t - \frac{1}{2}t^2 + c1 - c2)\,e^t \end{bmatrix}$$

Note the use of the `map` command. First, we have used it to apply the integral `int` to each component of the vector `step1`, and secondly, we have used it to apply the `collect` command on the components of the resulting vector in order to factor out the exponential terms.

```
> zero:=subs(t=0,step2)-<1,0>;
```

$$zero := \begin{bmatrix} c1 - 3 \\ c2 - 2 \end{bmatrix}$$

```
> c1:=3;c2:=2;
```

$$c1 := 3$$
$$c2 := 2$$

```
> solution:=map(collect, step2, exp);
```

$$solution := \begin{bmatrix} (-t - t^2)\,e^t + e^{(2t)} \\ e^{(2t)} + (-1 - t - \frac{1}{2}t^2)\,e^t \end{bmatrix}$$

13.3 Stability of Linear Systems

In this section we explore the stability of linear systems of two differential equations given by

$$\dot{x}(t) = A\,x(t) \tag{13.10}$$

where $x(t) = \begin{bmatrix} x_1(t) \\ x_2(t) \end{bmatrix}$, and $A = \begin{bmatrix} a & b \\ c & d \end{bmatrix}$ is a real and nonsingular matrix. The evolution of the system can be depicted via the phase flows which are two dimensional curves of $(x_1(t), x_2(t))$, where x_1 and x_2 are solutions of eq. (13.10). The critical (or equilibrium) point is the point where $\dot{x} = 0$. The patterns of the phase flows, suggests a characterization of the linear systems, upon the nature of the critical point. This depends on the eigenvalues $\lambda_{1,2}$ of the system matrix A. For a 2×2 matrix, the eigenvalues can be written as $\lambda_{1,2} = 1/2 \, (\text{Tr}\, A \pm \sqrt{\Delta})$, where $\text{Tr}\, A = a + d$ is the trace of the matrix A, and $\Delta = (TrA)^2 - 4 \det(A)$ is the discriminant. For convenience, let us summarize in the table below the different types of fixed points. (You should remember that you have already seen a similar table, as Table 6.2 in the first part of the book.)

	Δ	λ	Fixed Point
1	$\Delta > 0$	$\lambda_1 > \lambda_2 > 0$	unstable node
2	$\Delta > 0$	$\lambda_1 < \lambda_2 < 0$	stable node
3	$\Delta > 0$	$\lambda_1 < 0 < \lambda_2$	unstable saddle (hyperbolic) point
4	$\Delta < 0$	$\lambda_{1,2} = \mu \pm i\nu, \ (\mu, \nu > 0)$	unstable spiral point
5	$\Delta < 0$	$\lambda_{1,2} = -\mu \pm i\nu, \ (\mu, \nu > 0)$	stable spiral point
6	$\Delta < 0$	$\lambda_{1,2} = \pm i\nu$	elliptic fixed point
7	$\Delta = 0$	$\lambda_1 = \lambda_2 > 0$	unstable star
8	$\Delta = 0$	$\lambda_1 = \lambda_2 < 0$	stable star

We will consider each case separately, and use Maple to analyze the stability of the linear system in each case.

13.3.1 Unstable Node

Consider the system

$$\frac{dx_1(t)}{dt} = 8\, x_1(t) + 3\, x_2(t) ,$$
$$\frac{dx_2(t)}{dt} = -6\, x_1(t) - x_2(t) .$$

The unstable node corresponds to $\Delta > 0$ and $\lambda_1 > \lambda_2 > 0$. The matrix of the system is

$$A = \begin{bmatrix} 8 & 3 \\ -6 & -1 \end{bmatrix} .$$

The eigenvalues are $\lambda_1 = 2, \lambda_2 = 5$ and the discriminant is $\Delta = 9 > 0$. Therefore, the conditions for an unstable node in the origin are met. The following Maple worksheet analyzes the system

```
>   restart;
>   with(LinearAlgebra):with(plots):with(DEtools):
```

```
>   des1:=diff(x1(t),t)=8*x1(t)+3*x2(t),
>       diff(x2(t),t)=-6*x1(t)-x2(t);
```

$$des1 := \frac{d}{dt}\,x1(t) = 8\,x1(t) + 3\,x2(t),\ \frac{d}{dt}\,x2(t) = -6\,x1(t) - x2(t)$$

```
>   vars:=x1(t),x2(t);
```

$$vars := x1(t),\ x2(t)$$

Let's use `GenerateMatrix` command to extract the matrix of the system.

```
>   gm1:=GenerateMatrix({des1},{vars});
```

$$gm1 := \begin{bmatrix} -8 & -3 \\ 6 & 1 \end{bmatrix},\ \begin{bmatrix} -(\frac{d}{dt}\,x1(t)) \\ -(\frac{d}{dt}\,x2(t)) \end{bmatrix}$$

Then, the matrix of the system is

```
>   A1:=-gm1[1];
```

$$A1 := \begin{bmatrix} 8 & 3 \\ -6 & -1 \end{bmatrix}$$

The minus sign above was chosen in order to connect our notation with Maple's result. The eigenvalues are

```
>   Eigenvalues(A1);
```

$$\begin{bmatrix} 5 \\ 2 \end{bmatrix}$$

To visualize the phase flows of the system, we use `phaseportrait` command, together with the set of initial conditions

```
>   ics1:=seq([x1(0)=i,x2(0)=i],i=-3..3);
```

$$ics1 := [x1(0) = -3, x2(0) = -3],\ [x1(0) = -2, x2(0) = -2],$$
$$[x1(0) = -1, x2(0) = -1], [x1(0) = 0, x2(0) = 0],$$
$$[x1(0) = 1, x2(0) = 1], [x1(0) = 2, x2(0) = 2], [x1(0) = 3, x2(0) = 3]$$

```
>   phaseportrait([des1],[vars],t=-3..3,[ics1],
>       x1=-5..5, x2=-5..5, stepsize=.05, color=black,
>       linecolor=blue);
```

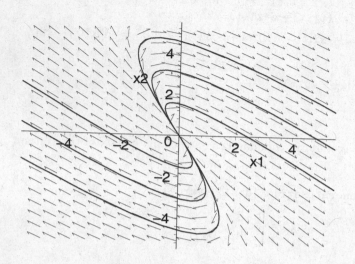

For completeness, let's check that we obtain the curves by solving the system of differential equations and using the parametric plot to graphically represent the evolution of x_1 and x_2. We will represent graphically only the curve corresponding to the initial conditions $x_1(0) = -3, x_2(0) = -3$.

```
> sol1:=dsolve({des1,x1(0)=-3,x2(0)=-3},{vars});
```
$$sol1 := \{x2(t) = 9\,e^{(5t)} - 12\,e^{(2t)},\ x1(t) = -9\,e^{(5t)} + 6\,e^{(2t)}\}$$

```
> one:=rhs(sol1[1]);
```
$$one := 9\,e^{(5t)} - 12\,e^{(2t)}$$

```
> two:=rhs(sol1[2]);
```
$$two := -9\,e^{(5t)} + 6\,e^{(2t)}$$

```
> plot([two,one,t=-3..3],-5..5,-5..5,numpoints=500,thickness=2,
> color=black);
```

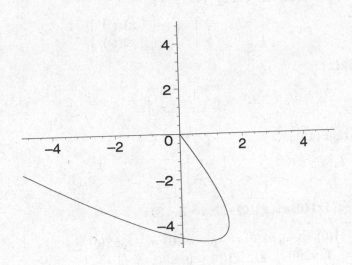

Indeed, we obtained the lower curve in the family of curves produced earlier by the phaseportrait command.

13.3.2 Stable Node

A stable node corresponds to $\Delta > 0$ and $\lambda_1 < \lambda_2 < 0$ respectively. As an example, consider the system

$$\frac{dx_1(t)}{dt} = x_1(t) + 2\,x_2(t)\,,$$
$$\frac{dx_2(t)}{dt} = -4\,x_1(t) - 5\,x_2\;.(t)$$

We have

```
>  restart;
>  with(LinearAlgebra):with(plots):with(DEtools):
```

Warning, the name changecoords has been redefined

Warning, the previous binding of the name RationalCanonicalForm has been removed and it now has an assigned value

```
>  des2:=diff(x1(t),t)=x1(t)+2*x2(t),
>      diff(x2(t),t)=-4*x1(t)-5*x2(t);
```

$$des2 := \tfrac{d}{dt}\,x1(t) = x1(t) + 2\,x2(t),\ \tfrac{d}{dt}\,x2(t) = -4\,x1(t) - 5\,x2(t)$$

```
>  vars:=x1(t),x2(t);
```

$$vars := x1(t),\ x2(t)$$

```
> gm2:=GenerateMatrix({des2},{vars});
```

$$gm2 := \left[\begin{array}{cc} -1 & -2 \\ 4 & 5 \end{array} \right], \left[\begin{array}{c} -(\frac{d}{dt}\,\text{x1}(t)) \\ -(\frac{d}{dt}\,\text{x2}(t)) \end{array} \right]$$

```
> A2:=-gm2[1];
```

$$A2 := \left[\begin{array}{cc} 1 & 2 \\ -4 & -5 \end{array} \right]$$

```
> Eigenvalues(A2);
```

$$\left[\begin{array}{c} -1 \\ -3 \end{array} \right]$$

```
> ics2:=seq([x1(0)=i,x2(0)=i],i=-3..3);
```

$ics2 := [\text{x1}(0) = -3, \text{x2}(0) = -3], [\text{x1}(0) = -2, \text{x2}(0) = -2],$
$[\text{x1}(0) = -1, \text{x2}(0) = -1], [\text{x1}(0) = 0, \text{x2}(0) = 0],$
$[\text{x1}(0) = 1, \text{x2}(0) = 1], [\text{x1}(0) = 2, \text{x2}(0) = 2], [\text{x1}(0) = 3, \text{x2}(0) = 3]$

```
> phaseportrait([des2],[vars],t=-3..3,[ics2],
>    x1=-5..5, x2=-5..5, stepsize=.05, color=black,
>    linecolor=blue);
```

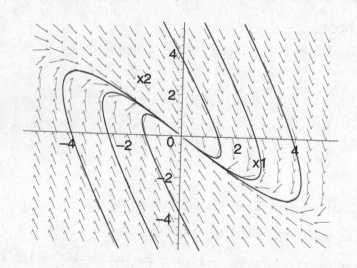

Note how the field lines converge toward the stable node.

13.3.3 Unstable Saddle (Hyperbolic) Point

The conditions for a saddle point are $\Delta > 0$ and $\lambda_1 < 0 < \lambda_2$, respectively. Consider the following example of linear system that exhibits a saddle point

$$\frac{dx_1(t)}{dt} = -8\,x_1(t) - 5\,x_2(t)\,,$$

$$\frac{dx_2(t)}{dt} = 10\,x_1(t) + 7\,x_2(t)\,.$$

The Maple worksheet reads

```
> restart;
```

```
> with(LinearAlgebra):with(plots):with(DEtools):
```

```
Warning, the name changecoords has been redefined
```

```
Warning, the previous binding of the name RationalCanonicalForm has
been removed and it now has an assigned value
```

```
> des3:=diff(x1(t),t)=-8*x1(t)-5*x2(t),
>      diff(x2(t),t)=10*x1(t)+7*x2(t);
```

$$des3 := \tfrac{d}{dt}\,x1(t) = -8\,x1(t) - 5\,x2(t),\ \tfrac{d}{dt}\,x2(t) = 10\,x1(t) + 7\,x2(t)$$

```
> vars:=x1(t),x2(t);
```

$$vars := x1(t),\, x2(t)$$

```
> gm3:=GenerateMatrix({des3},{vars});
```

$$gm3 := \left[\begin{array}{cc} 8 & 5 \\ -10 & -7 \end{array}\right],\ \left[\begin{array}{c} -(\tfrac{d}{dt}\,x1(t)) \\ -(\tfrac{d}{dt}\,x2(t)) \end{array}\right]$$

```
> A3:=-gm3[1];
```

$$A3 := \left[\begin{array}{cc} -8 & -5 \\ 10 & 7 \end{array}\right]$$

```
> Eigenvalues(A3);
```

$$\left[\begin{array}{c} 2 \\ -3 \end{array}\right]$$

The eigenvalues of the system matrix satisfy indeed the required condition $\lambda_1 < 0 < \lambda_2$. Let us plot the phase flows.

```
> ics3:=[x1(0)=-2,x2(0)=3],[x1(0)=2,x2(0)=-3],
>      [x1(0)=3,x2(0)=-1],[x1(0)=-3,x2(0)=1];
```

$$ics3 := [x1(0) = -2,\, x2(0) = 3],\, [x1(0) = 2,\, x2(0) = -3],$$
$$[x1(0) = 3,\, x2(0) = -1],\, [x1(0) = -3,\, x2(0) = 1]$$

```
> phaseportrait([des3],[vars],t=-3..3,[ics3],
>      x1=-5..5, x2=-7..7, stepsize=.05, color=black,
>      linecolor=blue);
```

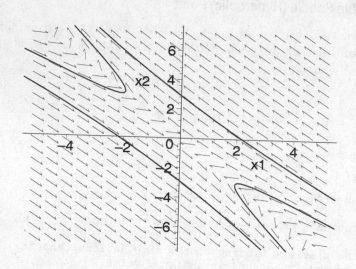

13.3.4   Unstable Spiral Point

The conditions for an unstable spiral point are $\Delta < 0$ and $\lambda_{1,2} = \mu \pm i\nu$, $(\mu, \nu > 0)$. Let us consider the system

$$
\begin{aligned}
\frac{dx_1(t)}{dt} &= x_1(t) + 3\,x_2(t)\ , \\
\frac{dx_2(t)}{dt} &= -2\,x_1(t) + x_2(t)\ .
\end{aligned}
$$

Then, its matrix

$$
A = \begin{bmatrix} 1 & 3 \\ -2 & 1 \end{bmatrix}
$$

has the discriminant $\Delta = -24 < 0$ and the corresponding eigenvalues $\lambda_1 = 1 + i\sqrt{6},\ \lambda_2 = 1 - i\sqrt{6}$. We are in the case of an unstable spiral point. We have

```
>  restart;
>  with(LinearAlgebra):with(plots):with(DEtools):
>  des4:=diff(x1(t),t)=x1(t)+3*x2(t),
>  diff(x2(t),t)=-2*x1(t)+x2(t);
```

$$des4 := \tfrac{d}{dt}\,x1(t) = x1(t) + 3\,x2(t),\ \tfrac{d}{dt}\,x2(t) = -2\,x1(t) + x2(t)$$

```
>  vars:=x1(t),x2(t);
```

$$vars := x1(t),\ x2(t)$$

```
>  gm4:=GenerateMatrix({des4},{vars});
```

$$gm4 := \begin{bmatrix} -1 & -3 \\ 2 & -1 \end{bmatrix}, \begin{bmatrix} -(\frac{d}{dt}\, x1(t)) \\ -(\frac{d}{dt}\, x2(t)) \end{bmatrix}.$$

```
> A4:=-gm4[1];
```

$$A4 := \begin{bmatrix} 1 & 3 \\ -2 & 1 \end{bmatrix}$$

```
> Eigenvalues(A4);
```

$$\begin{bmatrix} 1 + \sqrt{6}\,I \\ 1 - \sqrt{6}\,I \end{bmatrix}$$

```
> ics4:=seq([x1(0)=i,x2(0)=i],i=-3..3);
```

$$\begin{aligned} ics4 := &[x1(0) = -3,\ x2(0) = -3],\ [x1(0) = -2,\ x2(0) = -2], \\ &[x1(0) = -1,\ x2(0) = -1],\ [x1(0) = 0,\ x2(0) = 0], \\ &[x1(0) = 1,\ x2(0) = 1],\ [x1(0) = 2,\ x2(0) = 2], \\ &[x1(0) = 3,\ x2(0) = 3] \end{aligned}$$

```
> phaseportrait([des4],[vars],t=-3..3,
>    [ics4], x1=-5..5, x2=-5..5, stepsize=.05, color=black,
>    linecolor=blue);
```

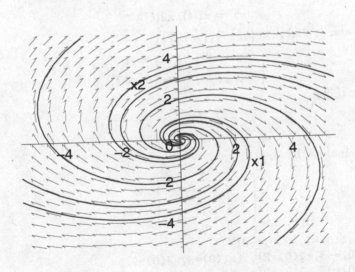

Observe how the field lines spiral out of the critical point, illustrating the unstable point.

13.3.5   Stable Spiral Point

In the case of the stable spiral point we have $\Delta < 0$ and $\lambda_{1,2} = -\mu \pm i\nu$, $(\mu, \nu > 0)$. Consider the system

$$\frac{dx_1(t)}{dt} = -2\,x_1(t) - x_2(t)\,,$$
$$\frac{dx_2(t)}{dt} = 3\,x_1(t) + x_2(t)\,.$$

In this case $\Delta = -3 < 0$ and the eigenvalues of the system matrix are $\lambda_1 = (-1 - i\sqrt{3})/2, \lambda_2 = (-1 + i\sqrt{3})/2$. We have

```
> restart;
```

```
> with(LinearAlgebra):with(plots):with(DEtools):
```

Warning, the name changecoords has been redefined

Warning, the previous binding of the name RationalCanonicalForm has been removed and it now has an assigned value

```
> des5:=diff(x1(t),t)=-2*x1(t)-x2(t),
>       diff(x2(t),t)=3*x1(t)+x2(t);
```

$$des5 := \tfrac{d}{dt}\,x1(t) = -2\,x1(t) - x2(t),\ \tfrac{d}{dt}\,x2(t) = 3\,x1(t) + x2(t)$$

```
> vars:=x1(t),x2(t);
```

$$vars := x1(t),\ x2(t)$$

```
> gm5:=GenerateMatrix({des5},{vars});
```

$$gm5 := \left[\begin{array}{cc} 2 & 1 \\ -3 & -1 \end{array}\right],\ \left[\begin{array}{c} -(\tfrac{d}{dt}\,x1(t)) \\ -(\tfrac{d}{dt}\,x2(t)) \end{array}\right]$$

```
> A5:=-gm5[1];
```

$$A5 := \left[\begin{array}{cc} -2 & -1 \\ 3 & 1 \end{array}\right]$$

```
> Eigenvalues(A5);
```

$$\left[\begin{array}{c} -\dfrac{1}{2} + \dfrac{1}{2}\,I\sqrt{3} \\[2mm] -\dfrac{1}{2} - \dfrac{1}{2}\,I\sqrt{3} \end{array}\right]$$

```
> ics5:=[x1(0)=3,x2(0)=-2],
>       [x1(0)=-3,x2(0)=2], [x1(0)=3,x2(0)=0],
>       [x1(0)=-3,x2(0)=0];
```

$$ics5 := [x1(0) = 3, x2(0) = -2], [x1(0) = -3, x2(0) = 2],$$
$$[x1(0) = 3, x2(0) = 0], [x1(0) = -3, x2(0) = 0]$$

```
> phaseportrait([des5],[vars],t=-3..7,
>       [ics5],x1=-5..5, x2=-5..5,stepsize=.02, color=black,
>       linecolor=blue);
```

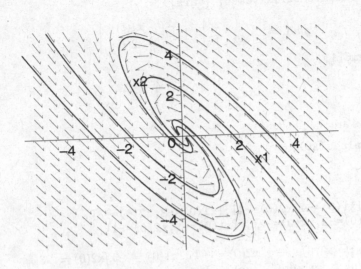

Note that the field lines in the phase flow plot, spiral in, towards the stable critical point.

13.3.6 Elliptic Fixed Point

The elliptic fixed point is characterized by $\Delta < 0$ and $\lambda_{1,2} = \pm i\nu$. The following linear system of differential equations, has an elliptic fixed point.

$$\frac{dx_1(t)}{dt} = x_1(t) - 2\,x_2(t) \,,$$

$$\frac{dx_2(t)}{dt} = 2\,x_1(t) - x_2(t) \,.$$

We have

```
> restart;
> with(LinearAlgebra):with(plots):with(DEtools):
```

Warning, the name changecoords has been redefined

Warning, the previous binding of the name RationalCanonicalForm has been removed and it now has an assigned value

```
> des6:=diff(x1(t),t)=x1(t)-2*x2(t),
> diff(x2(t),t)=2*x1(t)-x2(t);
```

$$des6 := \frac{d}{dt}\,\text{x1}(t) = \text{x1}(t) - 2\,\text{x2}(t),\ \frac{d}{dt}\,\text{x2}(t) = 2\,\text{x1}(t) - \text{x2}(t)$$

```
> vars:=x1(t),x2(t);
```

$$vars := \text{x1}(t),\,\text{x2}(t)$$

```
>  gm6:=GenerateMatrix({des6},{vars});
```

$$gm6 := \begin{bmatrix} -1 & 2 \\ -2 & 1 \end{bmatrix}, \begin{bmatrix} -(\frac{d}{dt}\,x1(t)) \\ -(\frac{d}{dt}\,x2(t)) \end{bmatrix}$$

```
>  A6:=-gm6[1];
```

$$A6 := \begin{bmatrix} 1 & -2 \\ 2 & -1 \end{bmatrix}$$

The eigenvalues are

```
>  Eigenvalues(A6);
```

$$\begin{bmatrix} \sqrt{3}\,I \\ -I\,\sqrt{3} \end{bmatrix}$$

```
>  ics6:=[x1(0)=-4,x2(0)=-4],[x1(0)=-3,x2(0)=-3],
>     [x1(0)=-2,x2(0)=-2], [x1(0)=-1,x2(0)=-1];
```

$$ics6 := [\text{x1}(0) = -4,\, \text{x2}(0) = -4],\, [\text{x1}(0) = -3,\, \text{x2}(0) = -3],$$
$$[\text{x1}(0) = -2,\, \text{x2}(0) = -2], [\text{x1}(0) = -1,\, \text{x2}(0) = -1]$$

The phase flow is given by

```
>  phaseportrait([des6],[vars],t=-3..3,
>     [ics6],x1=-5..5, x2=-5..5, stepsize=.05, color=black,
>     linecolor=blue);
```

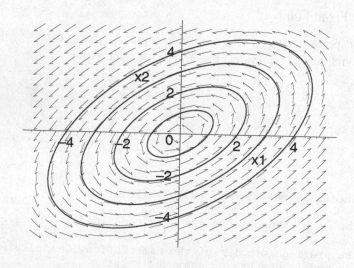

13.3.7 Unstable Star

The unstable star fixed point is characterized by $\Delta = 0$ and $\lambda_1 = \lambda_2 > 0$.

Consider the system

$$\frac{dx_1(t)}{dt} = 4\,x_1(t) - x_2(t)\,,$$
$$\frac{dx_2(t)}{dt} = x_1(t) + 6\,x_2(t)\,.$$

The Maple worksheet investigating this linear system is presented below.

```
> restart;
> with(LinearAlgebra):with(plots):with(DEtools):
```

Warning, the name changecoords has been redefined

Warning, the previous binding of the name RationalCanonicalForm has been removed and it now has an assigned value

```
> des7:=diff(x1(t),t)=4*x1(t)-x2(t),
> diff(x2(t),t)=x1(t)+6*x2(t);
```

$$des7 := \tfrac{d}{dt}\,\mathrm{x1}(t) = 4\,\mathrm{x1}(t) - \mathrm{x2}(t),\ \tfrac{d}{dt}\,\mathrm{x2}(t) = \mathrm{x1}(t) + 6\,\mathrm{x2}(t)$$

```
> vars:=x1(t),x2(t);
```

$$vars := \mathrm{x1}(t),\ \mathrm{x2}(t)$$

```
> gm7:=GenerateMatrix({des7},{vars});
```

$$gm7 := \left[\begin{array}{cc} -4 & 1 \\ -1 & -6 \end{array} \right],\ \left[\begin{array}{c} -(\tfrac{d}{dt}\,\mathrm{x1}(t)) \\ -(\tfrac{d}{dt}\,\mathrm{x2}(t)) \end{array} \right]$$

```
> A7:=-gm7[1];
```

$$A7 := \left[\begin{array}{cc} 4 & -1 \\ 1 & 6 \end{array} \right]$$

```
> Eigenvalues(A7);
```

$$\left[\begin{array}{c} 5 \\ 5 \end{array} \right]$$

Check that the discriminant is zero

```
> Trace(A7)^2-4*Determinant(A7);
```

$$0$$

```
> ics7:=[x1(0)=-4,x2(0)=-4],[x1(0)=-2,x2(0)=-2],
>     [x1(0)=2,x2(0)=-2],[x1(0)=-2,x2(0)=2],
>     [x1(0)=-2,x2(0)=0],[x1(0)=-7,x2(0)=0],
>     [x1(0)=4,x2(0)=4],[x1(0)=2,x2(0)=2],
>     [x1(0)=7,x2(0)=0],[x1(0)=2,x2(0)=0];
```

$$ics7 := [\mathrm{x1}(0) = -4,\, \mathrm{x2}(0) = -4],\ [\mathrm{x1}(0) = -2,\, \mathrm{x2}(0) = -2],\ [\mathrm{x1}(0) = 2,\, \mathrm{x2}(0) = -2],$$
$$[\mathrm{x1}(0) = -2,\, \mathrm{x2}(0) = 2],\ [\mathrm{x1}(0) = -2,\, \mathrm{x2}(0) = 0],\ [\mathrm{x1}(0) = -7,\, \mathrm{x2}(0) = 0],$$
$$[\mathrm{x1}(0) = 4,\, \mathrm{x2}(0) = 4],\ [\mathrm{x1}(0) = 2,\, \mathrm{x2}(0) = 2],\ [\mathrm{x1}(0) = 7,\, \mathrm{x2}(0) = 0],\ [\mathrm{x1}(0) = 2,\, \mathrm{x2}(0) = 0]$$

```
> phaseportrait([des7],[vars],t=-3..3,[ics7],x1=-5..5,
> x2=-5..5,stepsize=.05,color=black,linecolor=blue);
```

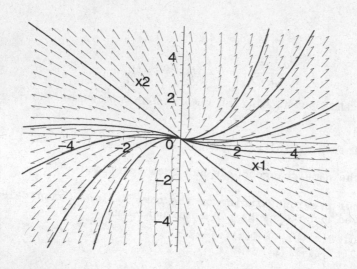

The field lines are all diverging out of the unstable star.

13.3.8 Stable Star

A linear system has a stable star fixed point when $\Delta = 0$ and $\lambda_1 = \lambda_2 < 0$. We use Maple to analyze the following example

$$\frac{dx_1(t)}{dt} = -3\,x_1(t) - x_2(t)\,,$$

$$\frac{dx_2(t)}{dt} = x_1(t) - x_2(t)\,.$$

We have

```
>  restart;
```

```
>  with(LinearAlgebra):with(plots):with(DEtools):
```

Warning, the name changecoords has been redefined

Warning, the previous binding of the name RationalCanonicalForm has been removed and it now has an assigned value

```
>  des8:=diff(x1(t),t)=-3*x1(t)-x2(t),
>  diff(x2(t),t)=x1(t)-x2(t);
```

$$des8 := \tfrac{d}{dt}\,x1(t) = -3\,x1(t) - x2(t),\ \tfrac{d}{dt}\,x2(t) = x1(t) - x2(t)$$

```
>  vars:=x1(t),x2(t);
```

$$vars := x1(t),\ x2(t)$$

```
>  gm8:=GenerateMatrix({des8},{vars});
```

$$gm8 := \left[\begin{array}{cc} 3 & 1 \\ -1 & 1 \end{array}\right],\ \left[\begin{array}{c} -(\tfrac{d}{dt}\,x1(t)) \\ -(\tfrac{d}{dt}\,x2(t)) \end{array}\right]$$

```
>  A8:=-gm8[1];
```

$$A8 := \begin{bmatrix} -3 & -1 \\ 1 & -1 \end{bmatrix}$$

```
>  Eigenvalues(A8);
```

$$\begin{bmatrix} -2 \\ -2 \end{bmatrix}$$

The eigenvalues are identical and negative. The discriminant is zero

```
>  Trace(A8)^2-4*Determinant(A8);
```
$$0$$

For a representative field plot choose the following initial conditions

```
>  ics8:=[x1(0)=-4,x2(0)=-4],[x1(0)=-2,x2(0)=-2],
>      [x1(0)=2,x2(0)=-2],[x1(0)=-2,x2(0)=2],
>      [x1(0)=-2,x2(0)=0],[x1(0)=-7,x2(0)=0],
>      [x1(0)=4,x2(0)=4],[x1(0)=2,x2(0)=2],
>      [x1(0)=7,x2(0)=0],[x1(0)=2,x2(0)=0];
```

$ics8 := [x1(0) = -4, x2(0) = -4], [x1(0) = -2, x2(0) = -2], [x1(0) = 2, x2(0) = -2],$
$[x1(0) = -2, x2(0) = 2], [x1(0) = -2, x2(0) = 0], [x1(0) = -7, x2(0) = 0],$
$[x1(0) = 4, x2(0) = 4], [x1(0) = 2, x2(0) = 2], [x1(0) = 7, x2(0) = 0],$
$[x1(0) = 2, x2(0) = 0]$

Then we have

```
>  phaseportrait([des8],[vars],t=-3..3,
>      [ics8],x1=-5..5, x2=-5..5, stepsize=.05, color=black,
>      linecolor=blue);
```

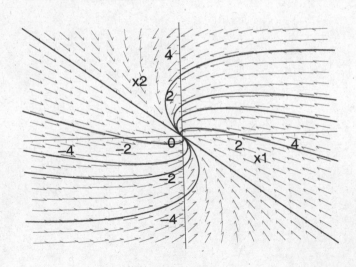

Observe how the field lines are all converging towards the critical point.

13.4   Problems

For each system of differential equations find the eigenvalues and the discriminant of the associated matrix and draw the associated phase flows showing representative solution curves. Decide in each case whether the critical point is stable or not.

Problem 13.4.1. $\frac{dx_1}{dt} = -x_1 + 6x_2$, $\frac{dx_2}{dt} = -2x_2$

Problem 13.4.2. $\frac{dx_1}{dt} = 3, x_1 + 2x_2$, $\frac{dx_2}{dt} = 4x_2$

Problem 13.4.3. $\frac{dx_1}{dt} = x_1 + 2x_2$, $\frac{dx_2}{dt} = 2x_1 + x_2/2$

Problem 13.4.4. $\frac{dx_1}{dt} = 25x_1 - 16x_2$, $\frac{dx_2}{dt} = 32x_1 - 20x_2$

Chapter 14

Nonlinear Differential Equations

MOST nonlinear differential equations cannot be solved exactly. However, we can extract important information about the evolution of their solutions, from the phase portrait analysis. Surprisingly, there exists a class of special nonlinear equations that can be solved exactly. These equations describe the integrable systems, which are systems that admit an infinite number of conserved quantities. Integrable system admit solutions in terms of solitons, or solitary waves.

14.1 One-Dimensional Nonlinear Differential Equations

We consider the example (discussed in the first part of the book), that illustrates the evolution of fish population in a lake, subject to constrains such as limited food supply and commercial catching. The equation describing this system is

$$\frac{dN}{dt} = AN - BN^2 - C,\qquad(14.1)$$

where B is the death rate, and C is the catch rate.

Let us use Maple to analyze this equation. We have

```
>  restart;
>  de:=diff(N(t),t)=A*N(t)-B*N(t)^2-C;
```
$$de := \frac{d}{dt}\,\mathrm{N}(t) = A\,\mathrm{N}(t) - B\,\mathrm{N}(t)^2 - C$$

```
>  sol:=dsolve({de,N(0)=N0},N(t));
```
$$sol := \mathrm{N}(t) = \frac{1}{2}\,\frac{A - \tan(\dfrac{t\,\sqrt{4\,C\,B - A^2}}{2} + \arctan(\dfrac{-2\,B\,N0 + A}{\sqrt{4\,C\,B - A^2}}))\,\sqrt{4\,C\,B - A^2}}{B}$$

We can further simplify the solution. First we isolate the right had side using the `rhs` command, and then expand it, in order to get rid of the nested function. We obtain

```
> R := expand( rhs(sol));
```

$$R := \frac{A}{2B} - \frac{1}{2} \cdot \frac{(\tan(\frac{t\sqrt{4CB-A^2}}{2}) - \frac{2BN0}{\sqrt{4CB-A^2}} + \frac{A}{\sqrt{4CB-A^2}})\sqrt{4CB-A^2}}{B(1-\tan(\frac{t\sqrt{4CB-A^2}}{2})(-\frac{2BN0}{\sqrt{4CB-A^2}} + \frac{A}{\sqrt{4CB-A^2}}))}$$

As we have already seen in the first part of the book, $N(t)$ represents the number of fish in thousands per year. Let us consider $A = 7, B = 1, C = 10$. Thus, $C = 10$ stands for $10,000$ fish caught per year. The solution becomes

```
> Model:=simplify(subs({A=7,B=1,C=10},R));
```

$$Model := \frac{7\tanh(\frac{3t}{2})N0 - 20\tanh(\frac{3t}{2}) + 3N0}{3 + 2\tanh(\frac{3t}{2})N0 - 7\tanh(\frac{3t}{2})}$$

which can be further simplified by collecting the tanh terms

```
> M:=collect(Model, tanh);
```

$$M := \frac{(7N0 - 20)\tanh(\frac{3t}{2}) + 3N0}{3 + (2N0 - 7)\tanh(\frac{3t}{2})}$$

We can plot different solution curves for various values of the initial condition N_0. First, we use **unapply** command to define a function of N_0

```
> f:=unapply(Model,N0);
```

$$f := N0 \rightarrow \frac{7\tanh(\frac{3t}{2})N0 - 20\tanh(\frac{3t}{2}) + 3N0}{3 + 2\tanh(\frac{3t}{2})N0 - 7\tanh(\frac{3t}{2})}$$

Then, we plot various solutions of the equation corresponding to different values of N_0 and carefully analyze the results. For the beginning let us plot $f(1.7)$. We use the plotting option **discont=true** in order to avoid vertical lines in discontinuity points of the graph.

```
> plot(f(1.7), t=0..3, N=0..7,
>    discont=true, thickness=3);
```

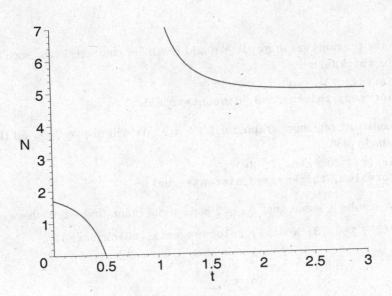

Observe that the plot has two branches. One branch, starts from 1.7 and decays to zero, indicating the extinction of the fish population. Another branch comes from $+\infty$ and goes asymptotically to 5 as t goes from 1 to infinity. This second branch is clearly unphysical, and it should be eliminated. This can be done by constraining the time interval to be smaller than 1.

```
> plot(f(1.7), t=0..1, N=0..7, discont=true,thickness=3);
```

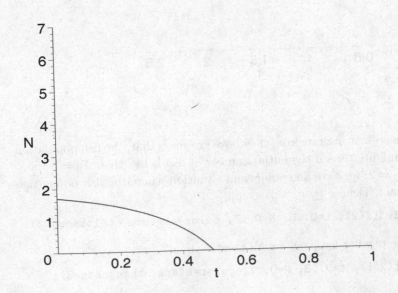

Let us denote the previous graph as g1. We add few more characteristics, such as
`color=red`, and `thickness=3`.

```
>  g1:=plot(f(1.7), t=0..1, N=0..7,
>      color=red, thickness=3, discont=true):
```

Similarly, we construct one more graph, for $N_0 = 1.9$. We will display later all the
graphs into a single plot.

```
>  g2:=plot(f(1.9), t=0..1, N=0..7,
>      color=blue, thickness=3,discont=true):
```

From the above results it seems that $N_0 = 2$ is a special point. Indeed, we have

```
>  plot(f(2), t=0..3, N=0..7, color=magenta, thickness=3);
```

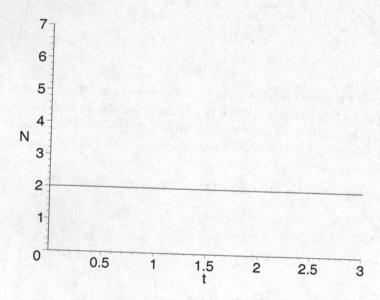

A first conclusion we can draw from the above graphs is that, the fish population
goes to zero in finite time if the initial number of fish is less than $N_0 = 2$ thousands. For $N_0 = 2$, we have an equilibrium situation where the fish population
remains constant. Denote by g3 this graph.

```
>  g3:=plot(f(2), t=0..3, N=0..7, color=magenta, thickness=3):
```

Let us investigate what happens for N_0 greater than 2.

```
>  plot(f(2.1), t=0..3, N=0..7, color=black, thickness=3);
```

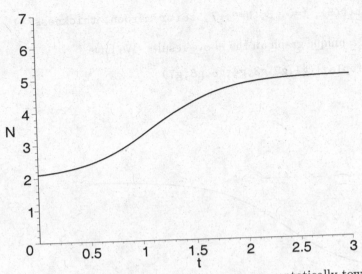

We observe that the fish population increases asymptotically towards $N = 5$ thousands fish. Denote by g4 this graph.

```
>   g4:=plot(f(2.1), t=0..3, N=0..7, color=black, thickness=3):
```

Similarly we have

```
>   g5:=plot(f(3), t=0..3, N=0..7, color=khaki, thickness=3):
>   g6:=plot(f(6), t=0..3, N=0..7, color=coral, thickness=3:
```

For $N_0 = 5$, we also expect to see that N is stationary $(dN/dt = 0)$

```
>   plot(f(5), t=0..3, N=0..7, color=maroon, thickness=3);
```

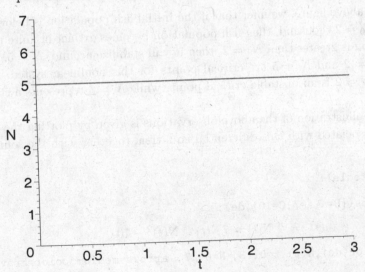

Denote this graph by g7.

```
>  g7:=plot(f(5), t=0..3, N=0..7, color=maroon, thickness=3):
```

Let us collect in a unique graph all the above results. We have

```
>  plots[display](g1,g2,g3,g4,g5,g6,g7);
```

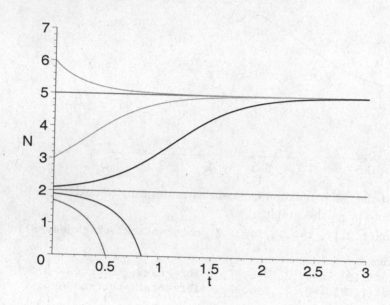

Looking at the above figure, we infer that if the initial fish population is below the critical point $N_0 = 2$ thousand, then fish population becomes extinct in finite time. If fish population is greater than $N_0 = 2$, then it will stabilize around, $N = 5$. The two points, $N = 2$ and $N = 5$ are critical points for this nonlinear system. We conclude that $N = 2$ is an unstable critical point, while $N = 5$ represents a stable critical point.

An intuitive illustration of the above observations is given by plotting the vector field lines associated with this differential equation, together with the solution curves.

```
>  with(DEtools):
```

```
>  de1:=subs({B=1,A=7,C=10},de);
```

$$de1 := \tfrac{d}{dt}\,N(t) = 7\,N(t) - N(t)^2 - 10$$

```
>  vf:=DEplot(de1,N(t),t=0..3, N=0..7, arrows=medium,color=gray):
```

```
>  plots[display](vf,g1,g2,g3,g4,g5,g6,g7);
```

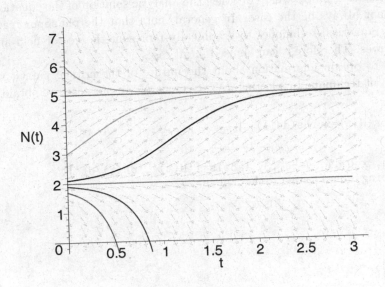

The same result is obtained by using the `phaseportrait` command with the appropriate set of initial conditions. We have

```
>    phaseportrait([de1], N(t), t=0..3,
>        [[N(0)=1.7],[N(0)=1.9],[N(0)=2],[N(0)=2.1],
>        [N(0)=3],[N(0)=5],[N(0)=6]] , N=0..7,
>        color=gray, linecolor=blue, arrows=medium);
```

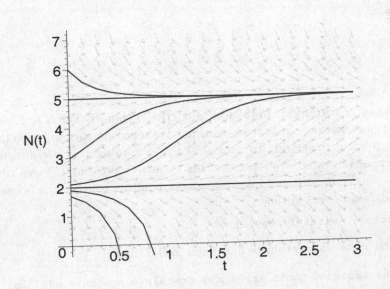

This was a particular example where we knew the analytic solution of the equation. But this may not always be the case. In general, note that the `phaseportrait` command uses numerical techniques to graphically represent the vector field and the solution curves. Therefore it can be used in all cases.

If we want to obtain information about the nature of the critical points, we can use a different technique by plotting dN/dt versus N. In our case, we obtain a parabola

```
>  eq:=subs(N(t)=x,rhs(de1));
```

$$eq := 7\,x - x^2 - 10$$

```
>  plot(eq, x=0..7, y=-2..3, labels=['N(t)','dN(t)/dt'],
>     scaling=constrained, thickness=2);
```

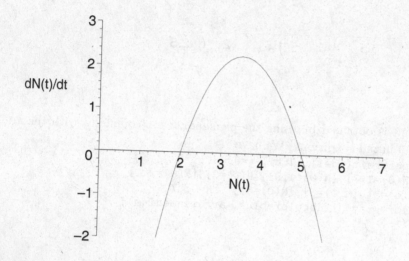

The zeros of the curve correspond to the fixed (or critical) points. For $N < 2$, we have $dN/dt < 0$ and therefore, N decreases in time. We represent this evolution by arrows pointing towards the origin. To the right of $N = 2$ we see that $dN/dt > 0$, therefore N increases in time. We represent this by arrows pointing towards the right. For $N > 5$, $dN/dt < 0$ and we draw arrows pointing towards the origin, symbolizing that N decreases in time.

```
>  with(plots):  with(plottools):
```

Warning, the name changecoords has been redefined

Warning, the assigned names arrow and translate now have a global binding

```
>   p:=plot(eq, x=0..7, y=-2..3, labels=['N(t)','dN(t)/dt'],
>       scaling=constrained, thickness=2):
>   l1:=arrow([0.7, 0], [0.2, 0], .05, .2, .25, color=black):
>   l2:=arrow([1.8, 0], [1.3, 0], .05, .2, .25, color=black):
>   l3:=arrow([2.3, 0], [2.8, 0], .05, .2, .25, color=black):
>   l4:=arrow([3.2, 0], [3.7, 0], .05, .2, .25, color=black):
>   l5:=arrow([4.2, 0], [4.7, 0], .05, .2, .25, color=black):
>   l6:=arrow([5.7, 0], [5.2, 0], .05, .2, .25, color=black):
>   l7:=arrow([6.5, 0], [6, 0], .05, .2, .25, color=black):
>   display(p,l1,l2,l3,l4,l5,l6,l7);
```

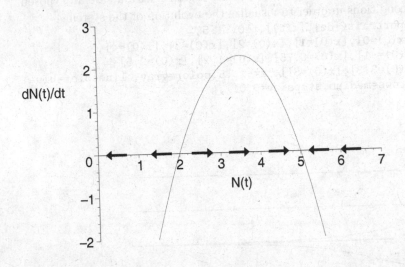

From this representation we immediately see that $N = 2$ is an unstable critical point while $N = 5$ is a stable point.

Let us make use of this technique to analyze another example. We want to find the critical points for the nonlinear system described by the equation

$$\frac{dx}{dt} = -x^5 + 10\,x^4 - 35\,x^3 + 50\,x^2 - 24\,x \,. \tag{14.2}$$

```
>   restart; with(PDEtools):  with(DEtools):
>   declare(x(t));
```
                    $x(t)$ *will now be displayed as* x
```
>   de:= diff(x(t),t)= -x(t)^5 + 10*x(t)^4 - 35*x(t)^3
>       + 50*x(t)^2 - 24*x(t);
```

$$de := x_t = -x^5 + 10\,x^4 - 35\,x^3 + 50\,x^2 - 24\,x$$

The exact solution of this equation is given by

```
>   dsolve({de},x(t));
```

$$\{x = 2 + (4 + \text{RootOf}(-81 + (e^{(24t)}\,\_C1 - 1)\,\_Z^4 + (12\,e^{(24t)}\,\_C1 - 12)\,\_Z^3$$
$$+ (-54 + 48\,e^{(24t)}\,\_C1)\,\_Z^2 + (-108 + 64\,e^{(24t)}\,\_C1)\,\_Z))^{(1/2)}\}, \{x = 2 - (4 +$$
$$\text{RootOf}(-81 + (e^{(24t)}\,\_C1 - 1)\,\_Z^4 + (12\,e^{(24t)}\,\_C1 - 12)\,\_Z^3$$
$$+ (-54 + 48\,e^{(24t)}\,\_C1)\,\_Z^2 + (-108 + 64\,e^{(24t)}\,\_C1)\,\_Z))^{(1/2)}\}$$

Observe that, in this case, the solution is given in terms of the square root of the roots of a fourth order polynomial. This is fairly complicated to manipulate analytically. Therefore we use `phaseportrait` together with a carefully chosen set of initial conditions in order to visualize the evolution of the system.

```
>   phaseportrait([de],[x(t)],t=0..1.5,
>        [[x(0)=0],[x(0)=1],[x(0)=2],[x(0)=3],[x(0)=4],
>        [x(0)=-1],[x(0)=0.75],[x(0)=1.2],[x(0)=2.5],
>        [x(0)=3.3],[x(0)=5]], x=-1..5,color=gray, linecolor=blue,
>        arrows=medium,stepsize=0.01);
```

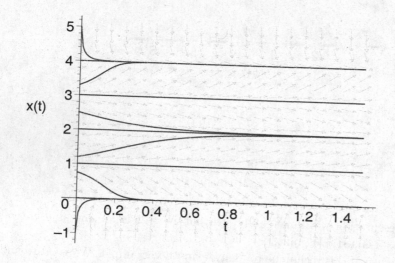

We observe that the critical points are $x = 0$ (stable), $x = 1$ (unstable), $x = 2$ (stable), $x = 3$ (unstable) and, respectively, $x = 4$ (stable).

The same conclusion can be obtained from the phase portrait $(dx/dt, x)$. We can factorize the right hand side of the equation, and then plot dx/dt against x.

```
>   R:=factor(rhs(de));
```

$$R := -x\,(x-1)\,(x-2)\,(x-3)\,(x-4)$$

```
>   plot(R, x=-0.75..4.5, y=-4..4, labels=['x','dx/dt'],
>       numpoints=1000, thickness=2, color=black);
```

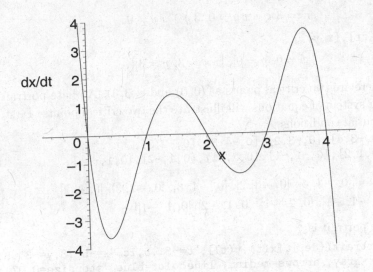

We see that to the left of $x = 0$, dx/dt is positive, therefore x increases, while to the right of $x = 0$, dx/dt is negative, which implies that x decreases. Using the previous representation, this translates to arrows pointing towards $x = 0$, signaling that $x = 0$ is a stable critical point. Similarly we conclude that $x = 1$ is unstable, $x = 2$ is stable, and so on.

14.2 Two-Dimensional Nonlinear Differential Equations

A system of differential equations $dx/dt = f(x,y), dy/dt = g(x,y)$ is called autonomous if f and g do not depend explicitly of t. Critical points are characterized by $dx/dt = 0$ and $dy/dt = 0$. This implies that $f(x,y) = 0$ and $g(x,y) = 0$ simultaneously. Let us consider the autonomous system

$$\frac{dx}{dt} = 3\,x - xy, \qquad \frac{dy}{dt} = 3\,y + xy\,. \tag{14.3}$$

We want to find its critical points and to study the stability of its solutions.

```
>   restart; with(DEtools):
>   des:=diff(x(t),t)=3*x(t)-x(t)*y(t),
>       diff(y(t),t)=3*y(t)+x(t)*y(t);
```

$$des := \frac{d}{dt}\mathrm{x}(t) = 3\mathrm{x}(t) - \mathrm{x}(t)\,\mathrm{y}(t),\ \frac{d}{dt}\mathrm{y}(t) = 3\,\mathrm{y}(t) + \mathrm{x}(t)\,\mathrm{y}(t)$$

Critical points are the solutions of the system of equations

```
>   cr:=3*x-x*y=0, 3*y+x*y=0;
```
$$cr := 3\,x - x\,y = 0,\ 3\,y + x\,y = 0$$
```
>   solve([cr],[x,y]);
```
$$[[x = 0,\ y = 0],\ [x = -3,\ y = 3]]$$

We find that there are two critical points at $(0, 0)$ and $(-3, 3)$. A phase portrait of this nonlinear system of equations will illustrate the two critical points. First, we choose the initial conditions

```
>   ic:=[[0,-3,4],[0,-3,2],[0,-4,3],[0,-2,3],
>       [0,1,2],[0,-1,-1],[0,2,-1],[0,1,-2],[0,1,-1]];
```
$$ic := [[0,\ -3,\ 4],\ [0,\ -3,\ 2],\ [0,\ -4,\ 3],\ [0,\ -2,\ 3],$$
$$[0,\ -1,\ -1],\ [0,\ 2,\ -1],\ [0,\ 1,\ -2],\ [0,\ 1,\ -1]]$$

Then, the phase portrait is

```
>   phaseportrait([des],[x(t),y(t)], t=-3..3,ic, x=-5..3,y=-3..5,
>       color=gray, arrows=medium, linecolor=blue, stepsize=0.07);
```

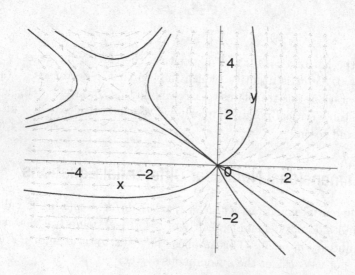

By inspection, we observe that $(0, 0)$ appears to be an unstable star critical point, while $(-3, 3)$ an unstable hyperbolic critical point. As shown in the first part of the book, the characteristics of a critical point can be inferred using the linearization

approach, by calculating the eigenvalues of the Jacobian matrix

$$J = \begin{bmatrix} \dfrac{\partial f}{\partial x} & \dfrac{\partial f}{\partial y} \\ \dfrac{\partial g}{\partial x} & \dfrac{\partial g}{\partial y} \end{bmatrix}$$

evaluated in the critical point. In our example

```
>  f:=3*x-x*y;g:=3*y+x*y;
```

$$f := 3x - xy$$
$$g := 3y + xy$$

```
>  J:=VectorCalculus[Jacobian]([f,g],[x,y]);
```

$$J := \begin{bmatrix} 3 - y & -x \\ y & 3 + x \end{bmatrix}$$

For the critical point $(0,0)$ we have the linearized matrix

```
>  M1:=subs({x=0,y=0},J);
```

$$M1 := \begin{bmatrix} 3 & 0 \\ 0 & 3 \end{bmatrix}$$

```
>  LinearAlgebra[Eigenvalues](M1);
```

$$\begin{bmatrix} 3 \\ 3 \end{bmatrix}$$

The eigenvalues of M_1 are $\lambda_1 = \lambda_2 = 3 > 0$ which implies (as we have seen in the previous chapter) that $(0,0)$ is an unstable star. For the second critical point $(-3,3)$ the Jacobian matrix reads

```
>  M2:=subs({x=-3,y=3},J);
```

$$M2 := \begin{bmatrix} 0 & 3 \\ 3 & 0 \end{bmatrix}$$

```
>  LinearAlgebra[Eigenvalues](M2);
```

$$\begin{bmatrix} 3 \\ -3 \end{bmatrix}$$

Its eigenvalues are $\lambda_1 = -3 < 0 < \lambda_2 = 3$ which corresponds to an unstable hyperbolic (saddle) point. This confirms our initial findings. In particular, our system of equations can be solved exactly, and the same analysis can be done using the exact analytic form of the solutions. The reader is encouraged to use the solutions to reproduce the above results. The solution for arbitrary initial conditions $x(0) = a, y(0) = b$ is

```
>  simplify(dsolve({des,x(0)=a,y(0)=b},[x(t),y(t)]));
```

$$\left\{ \begin{aligned} y(t) &= -\frac{e^{(1/3\,e^{(3\,t)}\,b + 1/3\,e^{(3\,t)}\,a + 3\,t)}\,(b+a)}{-e^{(1/3\,(b+a)\,e^{(3\,t)})} + (-\frac{a^3}{b^3})^{(1/3)}\,e^{(\frac{b}{3}+\frac{a}{3})}}, \\[2ex] x(t) &= \frac{(b+a)\,(-\frac{a^3}{b^3})^{(1/3)}\,(-e^{(1/3\,(e^{(3\,t)}-1)\,(b+a))} + (-\frac{a^3}{b^3})^{(1/3)})\,e^{(3\,t + \frac{2b}{3} + \frac{2a}{3})}}{(-e^{(1/3\,(b+a)\,e^{(3\,t)})} + (-\frac{a^3}{b^3})^{(1/3)}\,e^{(\frac{b}{3}+\frac{a}{3})})^2} \end{aligned} \right\}$$

It is important to note that the critical point analysis is based on the linearized equations, which are obtained as a first order approximation in the Taylor expansion of the nonlinear functions. If the nonlinearity is strong, then second order (or higher order) terms may become important. Consequently, predictions based on the eigenvalues of the Jacobian matrix calculated in the fixed point, will fail to correctly describe the stability of the nonlinear system. For example, consider the system

$$\frac{dx}{dt} = y + x\,(x^2 + y^4), \qquad \frac{dy}{dt} = -x + y\,(x^4 + y^2)\,. \tag{14.4}$$

The critical point analysis based on the linearized matrix of the systems reads

```
>  restart; with(DEtools): with(PDEtools): declare(x(t),y(t)):
```
$$x(t)\ \textit{will now be displayed as } x$$
$$y(t)\ \textit{will now be displayed as } y$$
```
>  des:=diff(x(t),t)=y(t)+x(t)*(x(t)^2+y(t)^4),
>  diff(y(t),t)=-x(t)+y(t)*(x(t)^4+y(t)^2);
```
$$\textit{des} := x_t = y + x\,(x^2 + y^4),\ y_t = -x + y\,(x^4 + y^2)$$
```
>  f:=-y+x*(x^2+y^4);g:=x+y*(x^4+y^2);
```
$$f := -y + x\,(x^2 + y^4)$$
$$g := x + y\,(x^4 + y^2)$$

It is clear that $(0,0)$ is a critical point (f and g vanish simultaneously in the origin). The Jacobian matrix in this point

```
>  M:=subs({x=0,y=0},VectorCalculus[Jacobian]([f,g],[x,y]));
```
$$M := \begin{bmatrix} 0 & -1 \\ 1 & 0 \end{bmatrix}$$

has pure imaginary eigenvalues

```
>  LinearAlgebra[Eigenvalues](M);
```
$$\begin{bmatrix} I \\ -I \end{bmatrix}$$

and consequently, we would expect an elliptic critical point at the origin. However, using the `phaseportrait` command, we get a completely different picture.

```
>   ic:=[[0,1,1],[0,1,-1],[0,-1,-1],[0,-1,1],[0,2,0]];
        ic := [[0, 1, 1], [0, 1, −1], [0, −1, −1], [0, −1, 1], [0, 2, 0]]
>   phaseportrait([des],[x(t),y(t)], t=-5..5,ic, x=-2..2,y=-2..2,
        color=gray,arrows=medium, linecolor=blue, stepsize=0.04);
```

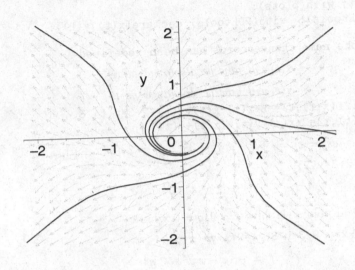

The critical point $(0,0)$ is an unstable spiral point, with the field lines radiating outwards in a clockwise rotation.

The trajectory near a *stable* critical point will remain close to that point. If the trajectory moves closer and closer to the critical point, then we are dealing with an *asymptotically stable* critical point. In the first half of the book we have introduced the Liapunov function which can be used to study the stability of a nonlinear system.

If on all trajectories of the phase space $dP/dt < 0$ then (x_0, y_0) is *asymptotically stable*, if $dP/dt \leq 0$ then (x_0, y_0) is *stable*, and finally, if $dP/dt > 0$ then (x_0, y_0) is *unstable*. In many practical applications, the Liapunov function is the energy of the system, and hence local minima correspond to stable critical points while local maxima correspond to unstable critical points. In general, there is no rule on how to find the Liapunov function. Its key property stems from

$$\frac{dP}{dt} = \frac{\partial P}{\partial x}\dot{x} + \frac{\partial P}{\partial y}\dot{y} = \frac{\partial P}{\partial x}f + \frac{\partial P}{\partial y}g = \vec{\nabla}P \cdot (f,g) . \qquad (14.5)$$

Therefore, if the trajectories of the phase space intersect the contour lines of P going inwards, we have an asymptotically stable point, if they are going outwards, we have an unstable point, while when they coincide with the contour lines of P, we have a stable critical point.

For example, let us study the stability of the critical points of the system

$$\frac{dx}{dt} = -(8\,x + xy^2 + 3\,y^3), \quad \frac{dy}{dt} = 2\,xy\,(x+y)\,, \tag{14.6}$$

given the Liapunov function $P(x,y) = 2\,x^2 + 3\,y^2$. We have

```
>   restart; with(plots):
>   with(DEtools):  with(PDEtools):declare(x(t),y(t)):
```

Warning, the name changecoords has been redefined

$x(t)$ *will now be displayed as* x

$y(t)$ *will now be displayed as* y

```
>   des:=diff(x(t),t)=-8*x(t)-x(t)*y(t)^2-3*y(t)^3,
>   diff(y(t),t)=2*x(t)*y(t)*(x(t)+y(t));
```

$$des := x_t = -8\,x - x\,y^2 - 3\,y^3,\ y_t = 2\,x\,y\,(x+y)$$

```
>   P:=2*x^2+3*y^2;
```

$$P := 2\,x^2 + 3\,y^2$$

To find the critical points we solve $f = 0, g = 0$ where

```
>   f:=-8*x-x*y^2-3*y^3;g:=2*x*y*(x+y);
```

$$f := -8\,x - x\,y^2 - 3\,y^3$$

$$g := 2\,x\,y\,(x+y)$$

```
>   solve([f,g],[x,y]);
```

$$[[x = 0,\ y = 0],\ [x = -2,\ y = 2],\ [x = 2,\ y = -2]]$$

We see that we have three critical points $(0,0), (-2,2)$ and $(2,-2)$. The Jacobian matrix of the system is

```
>   J:=VectorCalculus[Jacobian]([f,g],[x,y]);
```

$$J := \begin{bmatrix} -8 - y^2 & -2\,x\,y - 9\,y^2 \\ 2\,y\,(x+y) + 2\,x\,y & 2\,x\,(x+y) + 2\,x\,y \end{bmatrix}$$

and the corresponding eigenvalues of the Jacobian matrix calculated in each critical point are

```
>   M1:=subs({x=0,y=0},J);
```

$$M1 := \begin{bmatrix} -8 & 0 \\ 0 & 0 \end{bmatrix}$$

```
>   LinearAlgebra[Eigenvalues](M1);
```

$$\begin{bmatrix} 0 \\ -8 \end{bmatrix}$$

```
>   M2:=subs({x=-2,y=2},J);
```

$$M2 := \begin{bmatrix} -12 & -28 \\ -8 & -8 \end{bmatrix}$$

```
> LinearAlgebra[Eigenvalues](M2);
```

$$\begin{bmatrix} -10 + 2\sqrt{57} \\ -10 - 2\sqrt{57} \end{bmatrix}$$

```
> M3:=subs({x=2,y=-2},J);
```

$$M3 := \begin{bmatrix} -12 & -28 \\ -8 & -8 \end{bmatrix}$$

```
> LinearAlgebra[Eigenvalues](M3);
```

$$\begin{bmatrix} -10 + 2\sqrt{57} \\ -10 - 2\sqrt{57} \end{bmatrix}$$

We now use Liapunov function method to study the stability of each critical point. Using formula (14.5) we define

```
> dPdt:=diff(P,x)*f + diff(P,y)*g;
```

$$dPdt := 4x\left(-8x - xy^2 - 3y^3\right) + 12y^2x\left(x + y\right)$$

In order to identify the stability region (where $dP/dt < 0$) let us plot on the same graph dP/dt and the plane $z = 0$. Then, we identify the (x, y) region where dP/dt is below zero. We have

```
> p1:=plot3d(0, x=-5..5,y=-5..5,
>     color=blue, style=patchnogrid, grid=[75,75]):
> p2:=plot3d(dPdt,x=-5..5,y=-5..5, style=patchnogrid,
>     orientation=[55,65], grid=[75,75]):
> plots[display](p1,p2,axes=boxed);
```

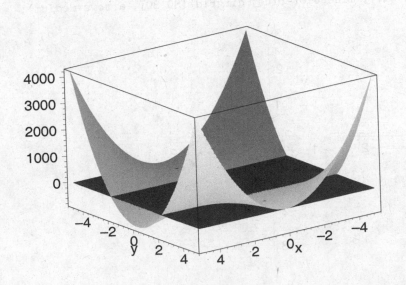

The region of stability seems to be a rectangular band along x axis from $y = -2$ to $y = 2$ together with the line $x = 0$ (which is evident from the analytic form of dP/dt). We can readily visualize this boundary with the help of the `implicitplot` command

```
>   implicitplot(dPdt=0,x=-5..5,y=-5..5, color=blue,
>      thickness=3,numpoints=7000,view=[-3..3,-3..3]);
```

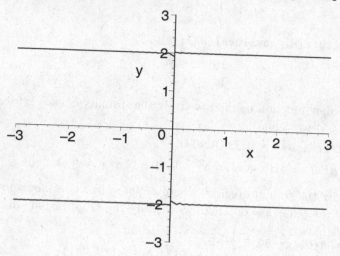

We see that only the origin x=0,y=0 has a region of stability ($dP/dt < 0$) around it. The phase portrait of the system is

```
>   DEplot([des],[x(t),y(t)],t=-10..10,x=-3..3,y=-3..3,
>   color=navy, linecolor=blue, dirgrid=[30,30], arrows=medium);
```

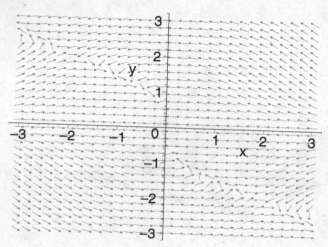

14.3 Three-Dimensional Models; The Lorenz Equations

The Lorenz system of equations

$$\frac{dx}{dt} = \sigma(-x+y),$$

$$\frac{dz}{dt} = rx - y - xz,$$

$$\frac{dt}{dt} = -bz + xy,$$

represents a model of a convecting fluid. It may serve as a simplified model for weather forecast. The parameters σ, r and b are real and positive. Surprisingly, for some values of these parameters, the solutions of the system behave chaotically. This means that a small variation of the initial conditions, may after some time eventually lead to a huge variation of the state of the system. Therefore, even if the system is deterministic, due to inherent errors in measuring the initial state, the state of the system after a long time cannot be predicted at all. Such a system is called chaotic.

```
>   restart;

>   with(DEtools):with(plots):

>   de:=
>       diff(x(t),t)=sigma*(-x(t)+y(t)),
>       diff(y(t),t)=r*x(t) - y(t) - x(t)*z(t),
>       diff(z(t),t)=-b*z(t)+x(t)*y(t);
```

$$de := \frac{d}{dt}x(t) = \sigma(-x(t)+y(t)),$$

$$\frac{d}{dt}y(t) = rx(t) - y(t) - x(t)z(t), \quad \frac{d}{dt}z(t) = -bz(t) + x(t)y(t)$$

At the critical point, the system satisfies simultaneously $dx/dt = 0$, $dy/dt = 0$ and $dz/dt = 0$. This implies

```
>   f:=sigma*(-x+y);
>   g:=r*x-y-x*z;
>   h:=-b*z+x*y;
```

$$f := \sigma(-x+y)$$

$$g := rx - y - xz$$

$$h := -bz + xy$$

```
>   solve({f=0,g=0,h=0},[x,y,z])
>       assuming sigma::positive, r::positive, b::positive;
```

$$[[x = 0, y = 0, z = 0], [x = \mathrm{RootOf}(-br + b + \_Z^2, label = \_L1),$$

$$y = \mathrm{RootOf}(-br + b + \_Z^2, label = \_L1), z = r - 1]]$$

We see that Maple gives the solutions in a general form using the `RootOf` construct. While this answer is elegant, what we need here is a concrete answer. Using the `allvalues` command we obtain

```
>   allvalues(%);
```

$$[[x = 0, y = 0, z = 0], [x = \sqrt{br - b}, y = \sqrt{br - b}, z = r - 1]],$$
$$[[x = 0, y = 0, z = 0], [x = -\sqrt{br - b}, y = -\sqrt{br - b}, z = r - 1]]$$

There are three critical points

```
>   C1:={x=0,y=0,z=0};
>   C2:={x=(b*r-b)^(1/2),y=(b*r-b)^(1/2),z=r-1};
>   C3:={x=-(b*r-b)^(1/2),y=-(b*r-b)^(1/2),z=r-1};
```

$$C1 := \{x = 0, z = 0, y = 0\}$$
$$C2 := \{z = r - 1, x = \sqrt{br - b}, y = \sqrt{br - b}\}$$
$$C3 := \{z = r - 1, x = -\sqrt{br - b}, y = -\sqrt{br - b}\}$$

Note that the last two critical points C_2 and C_3 exist only for $r > 1$. Let us linearize Lorenz's equations near the critical points in order to study their stability. The Jacobian matrix is

```
>   J:=VectorCalculus[Jacobian]([f,g,h],[x,y,z]);
```

$$J := \begin{bmatrix} -\sigma & \sigma & 0 \\ r - z & -1 & -x \\ y & x & -b \end{bmatrix}$$

For the first critical point, we have

```
>   M1:=subs(C1,J);
```

$$M1 := \begin{bmatrix} -\sigma & \sigma & 0 \\ r & -1 & 0 \\ 0 & 0 & -b \end{bmatrix}$$

whose eigenvalues are

```
>   l1:=LinearAlgebra[Eigenvalues](M1);
```

$$l1 := \begin{bmatrix} -b \\ -\dfrac{1}{2} - \dfrac{\sigma}{2} + \dfrac{\sqrt{1 - 2\sigma + \sigma^2 + 4r\sigma}}{2} \\ -\dfrac{1}{2} - \dfrac{\sigma}{2} - \dfrac{\sqrt{1 - 2\sigma + \sigma^2 + 4r\sigma}}{2} \end{bmatrix}$$

Near the other two critical points we have

```
>   M2:=subs(C2,J);
```

$$M2 := \begin{bmatrix} -\sigma & \sigma & 0 \\ 1 & -1 & -\sqrt{br - b} \\ \sqrt{br - b} & \sqrt{br - b} & -b \end{bmatrix}$$

```
>  M3:=subs(C3,J);
```

$$M3 := \begin{bmatrix} -\sigma & \sigma & 0 \\ 1 & -1 & \sqrt{br-b} \\ -\sqrt{br-b} & -\sqrt{br-b} & -b \end{bmatrix}$$

```
>  12:=LinearAlgebra[Eigenvalues](M2):
>  13:=LinearAlgebra[Eigenvalues](M3):
```

The eigenvalues for M_2 and M_3 are too long to display, therefore we terminated the command line by colon (to suppress the output). The chaotic behavior appears for particular values of the parameters. Let us consider $\sigma = 10, b = 8/3$, and $r = 28$. These values particularize the system of equations to

```
>  deL:=subs({sigma=10,b=8/3,r=28},{de});
```

$$deL := \{\tfrac{d}{dt}\, x(t) = -10\, x(t) + 10\, y(t),\ \tfrac{d}{dt}\, z(t) = -\tfrac{8}{3}\, z(t) + x(t)\, y(t),$$

$$\tfrac{d}{dt}\, y(t) = 28\, x(t) - y(t) - x(t)\, z(t)\}$$

To plot the three dimensional evolution of the system, we use DEplot3d command with the initial conditions

```
>  ic:=x(0)=0.1, y(0)=0.1,z(0)=0.1;
```

$$ic := x(0) = 0.1,\ y(0) = 0.1,\ z(0) = 0.1$$

We have
```
>  DEplot3d(deL,{x(t),y(t),z(t)},t=0..25,[[ic]],
>      scene=[x(t),y(t),z(t)], stepsize=0.01, linecolor=blue,
>      orientation=[-73,65]);
```

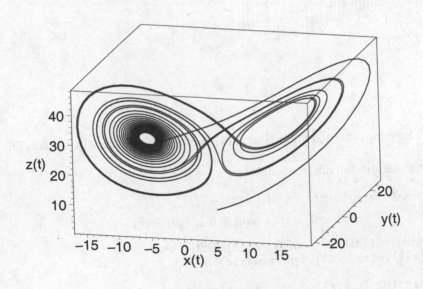

The trajectory described by this system is chaotic, in the sense that it is sensitive to the initial conditions. Small changes in the initial conditions lead to huge changes later in time. Let us illustrate this for only one function, say $x(t)$. The initial conditions are

```
>  ic;
```
$$x(0) = 0.1, \, y(0) = 0.1, \, z(0) = 0.1$$

Then, we have

```
>  p1:=dsolve({deL[1],deL[2],deL[3],ic},
>      {x(t),y(t),z(t)}, type=numeric);
```
$$p1 := \mathbf{proc}(x\_rkf45) \, \dots \, \mathbf{end \, proc}$$

```
>  odeplot(p1,[t,x(t)],0..45,numpoints=1000,color=navy);
```

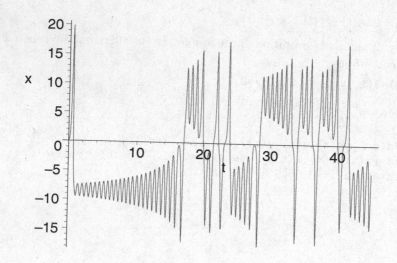

```
>  plot1:=odeplot(p1,[t,x(t)],0..45, numpoints=1000, color=navy):
```

Let us change slightly the initial conditions

```
>  ic2:=x(0) = .11, y(0) = .1, z(0) = .1;
```
$$ic2 := x(0) = 0.11, \, y(0) = 0.1, \, z(0) = 0.1$$

```
>  p2:=dsolve({deL[1],deL[2],deL[3],ic2},
>      {x(t),y(t),z(t)},type=numeric):
```

```
>  odeplot(p2,[t,x(t)],0..45, numpoints=1000,color=red);
```

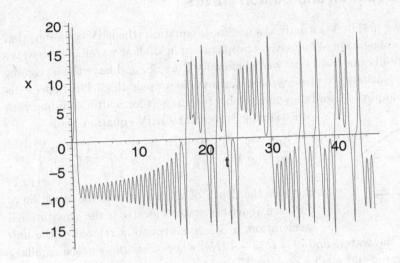

> `plot2:=odeplot(p2,[t,x(t)],0..45,numpoints=1000,color=red):`

Now, if we display the two graphs on the same plot, we obtain

> `display(plot1,plot2);`

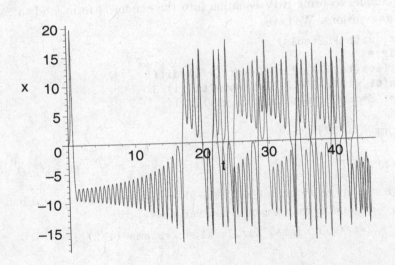

It is easy to see that initially, the two trajectories are quite close, but as time passes, they diverge. After $t = 20$, the two trajectories are in opposite sides of the origin.

14.4   KdV Equation and Soliton Waves

In 1885 Korteweg and de Vries derived a nonlinear equation (the KdV equation) that described the propagation of long waves propagating in shallow water. Those waves were experimentally observed fifty years before by Russell , and have the intriguing property (among others) to preserve their form while propagating. Their speed depends on their amplitude, and they remarkably retain their form after colliding with other similar waves. The KdV equation reads

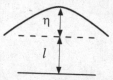

$$\frac{\partial \eta}{\partial t_1} = \frac{3}{2} \sqrt{\frac{g}{l}} \frac{\partial}{\partial x_1} \left(\frac{1}{2} \eta^2 + \frac{2}{3} \alpha \eta + \frac{1}{3} \sigma \frac{\partial^2 \eta}{\partial x_1^2} \right) .$$

(14.7)

Here l is the depth of the water, η is the elevation of the wave above the water's level, g is the gravitational acceleration, α is a small constant related to the uniform motion of the water, and $\sigma = l^3/3 - Tl/\rho g$ where T is the surface capillary tension of the water and ρ is its density. Via transformations

$$t_1 = \frac{2t}{\sqrt{\frac{g}{l\sigma}}}, \ x_1 = x\sqrt{\sigma} \, , \eta = 2u - 2\alpha/3 \, ,$$

the equation (14.7) can be brought into the dimensionless form

$$u_t = 6uu_x + u_{xxx} \, .$$

(14.8)

This is the canonical form of the KdV equation, that we will use from now on.

Let us use Maple, to bring KdV equation into the canonical form, and to study some of its solutions. We have

```
>   restart; with(PDEtools):
>   kdv_orig:=
>      diff(eta(t1,x1),t1)=3/2*sqrt(g/l)*diff(
>      (eta(t1,x1)^2/2 + 2*alpha*eta(t1,x1)/3 +
>      sigma/3*diff(eta(t1,x1),x1,x1)),x1);
```

$kdv\_orig := \frac{\partial}{\partial t1} \eta(t1, x1) =$

$\frac{3}{2} \sqrt{\frac{g}{l}} \left(\eta(t1, x1) \left(\frac{\partial}{\partial x1} \eta(t1, x1) \right) + \frac{2}{3} \alpha \left(\frac{\partial}{\partial x1} \eta(t1, x1) \right) + \frac{1}{3} \sigma \left(\frac{\partial^3}{\partial x1^3} \eta(t1, x1) \right) \right)$

We can use a two step approach to implement the above changes of variables. First, we use **dchange** command to change to the new variables

```
>   tr:={t1 = 2*t/(g/l/sigma)^(1/2), x1 = x*sigma^(1/2)};
```

$$tr := \left\{ t1 = \frac{2t}{\sqrt{\frac{g}{l\sigma}}}, \ x1 = x\sqrt{\sigma} \right\}$$

```
>   k1:=dchange(tr,kdv_orig,[t,x]);
```

$$k1 := \frac{1}{2}\sqrt{\frac{g}{l\,\sigma}}\,(\tfrac{\partial}{\partial t}\,\eta(t,\,x,\,l,\,\sigma,\,g)) = \frac{3}{2}\sqrt{\frac{g}{l}}$$

$$\left(\frac{\eta(t,\,x,\,l,\,\sigma,\,g)\,(\frac{\partial}{\partial x}\,\eta(t,\,x,\,l,\,\sigma,\,g))}{\sqrt{\sigma}} + \frac{2}{3}\,\frac{\alpha\,(\frac{\partial}{\partial x}\,\eta(t,\,x,\,l,\,\sigma,\,g))}{\sqrt{\sigma}} + \frac{1}{3}\,\frac{\frac{\partial^3}{\partial x^3}\,\eta(t,\,x,\,l,\,\sigma,\,g)}{\sqrt{\sigma}}\right)$$

Then, we change the function η to u, using the subs command

```
> kdv1:=simplify(subs(eta(t,x,1,sigma,g)=2*u(t,x)-2*alpha/3,k1))
>     assuming 1::positive, g::positive, sigma::positive;
```

$$kdv1 := \frac{\sqrt{g}\,(\frac{\partial}{\partial t}\,u(t,\,x))}{\sqrt{l}\,\sqrt{\sigma}} = \frac{\sqrt{g}\,(6\,(\frac{\partial}{\partial x}\,u(t,\,x))\,u(t,\,x) + (\frac{\partial^3}{\partial x^3}\,u(t,\,x)))}{\sqrt{l}\,\sqrt{\sigma}}$$

which can be further simplified to

```
> kdv:=sqrt(1)*sqrt(sigma)/sqrt(g)*kdv1;
```

$$kdv := \tfrac{\partial}{\partial t}\,u(t,\,x) = 6\,(\tfrac{\partial}{\partial x}\,u(t,\,x))\,u(t,\,x) + (\tfrac{\partial^3}{\partial x^3}\,u(t,\,x))$$

This is exactly the canonical form (14.8). Maple solves this partial differential equation, and finds one class of solutions

```
> pdsolve(kdv,u(t,x));
```

$$u(t,\,x) = \frac{\_C2 + 8\,\_C3^3}{6\,\_C3} - 2\,\_C3^2\,\tanh(\_C1 + \_C2\,t + \_C3\,x)^2$$

As we have already seen in the first part of the book, this is just one solution out of an infinite number of classes of solutions that satisfy KdV equation. It can be shown (do it!) that by a suitable choice of the integration constants $\_C1$, $\_C2$, and $\_C3$, we can transform the above solution into

```
> u_sol:=v/cosh(sqrt(v)/2*(x+v*t))^2/2;
```

$$u\_sol := \frac{1}{2}\,\frac{v}{\cosh(\frac{\sqrt{v}\,(x + v\,t)}{2})^2}$$

where v is the speed of the wave. Let us check that the above result is a solution of the KdV equation

```
> simplify(eval(subs(u(t,x)=u_sol,kdv)));
```

$$-\frac{1}{2}\,\frac{v^{(5/2)}\,\sinh(\frac{\sqrt{v}\,(x + v\,t)}{2})}{\cosh(\frac{\sqrt{v}\,(x + v\,t)}{2})^3} = -\frac{1}{2}\,\frac{v^{(5/2)}\,\sinh(\frac{\sqrt{v}\,(x + v\,t)}{2})}{\cosh(\frac{\sqrt{v}\,(x + v\,t)}{2})^3}$$

Indeed, we observe that we get the same expression on the both sides of the equation. If we plot the solution we obtained at two different moments in time, we have

```
> p1:=plot(subs({v=2,t=0},u_sol),x=-12..6,u=0..1.2,
>     color=red, thickness=2):
> p2:=plot(subs({v=2,t=3},u_sol),x=-12..6,u=0..1.2,
>     color=blue, thickness=2):
```

```
>  plots[display](p1,p2);
```

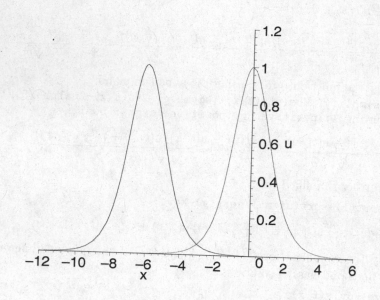

This confirms that the solution is a wave with height $v/2$ moving to the left with speed v. In our example, $v = 2$, hence the wave peaks at $u = 1$ and after 3 seconds the peak arrives at $x = -6$.

14.4.1 Lax Formalism

It is interesting to reformulate the KdV equation in terms of Lax formalism. If we define two differential operators

$$L = \frac{\partial^2}{\partial x^2} + u \,, \qquad (14.9)$$

and

$$B = 4\frac{\partial^3}{\partial x^3} + 3\left(u\frac{\partial}{\partial x} + \frac{\partial}{\partial x}u\right) \,, \qquad (14.10)$$

then from the Lax equation

$$\frac{\partial L}{\partial t} = [B, L] \,, \qquad (14.11)$$

we obtain the KdV equation. In the Lax equation (14.11) the expression $[B, L]$ is the commutator of B and L, defined as $[B, L] = BL - LB$.

Let us check the Lax equation. First, we define the differential operators as members of a differential algebra. For this, we need the Ore_algebra package.

```
> restart;
> with(Ore_algebra): with(PDEtools): declare(u(t,x),v(t,x));
```
$$u(t, x) \; will \; now \; be \; displayed \; as \; u$$
$$v(t, x) \; will \; now \; be \; displayed \; as \; v$$
```
> A:=diff_algebra([Dx,x],[Dt,t], func={u,v});
```
$$A := Ore\_algebra$$

So far we have defined a differential algebra whose non-commutative elements are the differential with respect to x (denoted Dx) and x, and respectively the differential with respect to t (denoted Dt) and t. Also, we have declared that u and v are functions in this algebra, using the option func. Next, we define the differential operators L and B, using the product in this algebra (skew_product)

```
> L:=skew_product(Dx,Dx,A)+u(t,x);
```
$$L := Dx^2 + u$$
```
> B:=4*Dx^3+3*(u(t,x)*Dx + skew_product(Dx,u(t,x),A));
```
$$B := 4 \, Dx^3 + 6 \, u \, Dx + 3 \, u_x$$

Then, to calculate the commutator $[B, L] = BL - LB$, we calculate each product individually, and then subtract them

```
> BL:=skew_product(B,L,A) ;
```
$$BL := 9 \, u_x \, u + 4 \, u_{x,x,x} + (12 \, u_{x,x} + 6 \, u^2) \, Dx + 15 \, u_x \, Dx^2 + 10 \, u \, Dx^3 + 4 \, Dx^5$$
```
> LB:= skew_product(L,B,A);
```
$$LB := 3 \, u_x \, u + 3 \, u_{x,x,x} + (12 \, u_{x,x} + 6 \, u^2) \, Dx + 15 \, u_x \, Dx^2 + 10 \, u \, Dx^3 + 4 \, Dx^5$$

Therefore, the commutator reads

```
> CBL:=BL - LB;
```
$$CBL := 6 \, u_x \, u + u_{x,x,x}$$

Because $\partial L / \partial t$ is just $\partial u / \partial t$

```
> dLdt:=diff(u(t,x),t);
```
$$dLdt := u_t$$

the Lax equation $dL/dt = [B, L]$ yields

```
> dLdt = CBL;
```
$$u_t = 6 \, u_x \, u + u_{x,x,x}$$

which is exactly the KdV equation.

In the first part of the book we have seen that the Lax equation is the compatibility condition for the eigenvalue equations

$$L \psi = \lambda \psi \, , \quad \frac{\partial \psi}{\partial t} = B \psi. \tag{14.12}$$

Another useful command from the `Ore_algebra` package is `applyopr`. The syntax is `applyopr(L,f,A)`. The command computes the result of applying an operator L of the Ore algebra A to a function f. Let us write explicitly the eigenvalue equations using the `applyopr` command. We have

```
>  eq1:=applyopr(L,psi(t,x),A) = lambda* psi(t,x);
```
$$eq1 := u\,\psi(t,\,x) + \psi_{x,\,x} = \lambda\,\psi(t,\,x)$$

and respectively

```
>  eq2:=diff(psi(t,x),t)=applyopr(B,psi(t,x),A);
```
$$eq2 := \psi_t = 3\,u_x\,\psi(t,\,x) + 6\,u\,\psi_x + 4\,\psi_{x,\,x,\,x}$$

The simplest situation corresponds to u=0

```
>  eq3:=subs(u(t,x)=0,eq1);
```
$$eq3 := \psi_{x,\,x} = \lambda\,\psi(t,\,x)$$
```
>  eq4:=subs(u(t,x)=0,eq2);
```
$$eq4 := \psi_t = 4\,\psi_{x,\,x,\,x}$$
```
>  dsolve(eq3, psi(t,x));
```
$$\psi(t,\,x) = \_F1(t)\,e^{(\sqrt{\lambda}\,x)} + \_F2(t)\,e^{(-\sqrt{\lambda}\,x)}$$

These equations will prove to be instrumental in constructing the multi-soliton solutions of the KdV equation.

14.4.2 Multi-soliton Solutions

Solutions of the KdV equation can be systematically obtained from solutions ψ_i of the Lax eigenvalue equation $(u = 0)$,

$$\frac{\partial^2}{\partial x^2}\,\psi_i = \lambda_i\,\psi_i \tag{14.13}$$

via the Wronskian formula

$$u(x,\,t) = 2\,\frac{\partial^2}{\partial x^2}\,\ln(W) \tag{14.14}$$

where

$$W = W(\psi_1,\,\psi_2,\,\ldots,\,\psi_n) \tag{14.15}$$

is the Wronskian determinant composed of solutions $\psi_i(\xi_i)$ of equation (14.13) whereas ξ_i stands for

$$\xi_i = k_i\,(x + 4\,k_i{}^2\,t)\,, \quad \text{for } \lambda_i > 0\,,$$

and respectively

$$\xi_i = k_i\,(x - 4\,k_i{}^2\,t)\,, \quad \text{for } \lambda_i < 0\,.$$

(a) One-soliton solution

We will consider, the case $\lambda > 0$. As we shall see shortly, this case corresponds to soliton solutions that we have already encountered.

```
>   restart; with(LinearAlgebra): with(VectorCalculus):
>   xi:=k*(x+4*k^2*t);
```

$$\xi := k\,(x + 4\,k^2\,t)$$

```
>   psi:=cosh(xi);
```

$$\psi := \cosh(k\,(x + 4\,k^2\,t))$$

```
>   u:=simplify(2*diff(ln(psi),x,x));
```

$$u := \frac{2\,k^2}{\cosh(k\,(x + 4\,k^2\,t))^2}$$

This is exactly the soliton solution we have seen previously. We can check that it satisfies the KdV equation

```
>   KdV:=simplify(diff(u,t)-6*u*diff(u,x)-diff(u,x$3));
```

$$KdV := 0$$

(b) Two-soliton solution

Let us consider two solutions of equation (14.13).

```
>   xi1:=k1*(x+4*k1^2*t);xi2:=k2*(x+4*k2^2*t);
```

$$\xi 1 := k1\,(x + 4\,k1^2\,t)$$

$$\xi 2 := k2\,(x + 4\,k2^2\,t)$$

```
>   psi1:=cosh(xi1);psi2:=sinh(xi2);
```

$$\psi 1 := \cosh(k1\,(x + 4\,k1^2\,t))$$

$$\psi 2 := \sinh(k2\,(x + 4\,k2^2\,t))$$

```
>   w2:=Determinant(Wronskian([psi1,psi2],x));
```

$$w2 := \cosh(k1\,(x + 4\,k1^2\,t))\,\cosh(k2\,(x + 4\,k2^2\,t))\,k2$$
$$- \sinh(k2\,(x + 4\,k2^2\,t))\,\sinh(k1\,(x + 4\,k1^2\,t))\,k1$$

Then, the corresponding KdV solution is

```
>   u2:=simplify(2*diff(ln(w2) ,x,x));
```

$$u2 := -2(k2^2\,\cosh(k1\,(x + 4\,k1^2\,t))^2\,k1^2 - k1^2\,\cosh(k2\,(x$$
$$+4\,k2^2\,t))^2\,k2^2 + k2^2\,k1^2 + \cosh(k2\,(x + 4\,k2^2\,t))^2\,k1^4$$
$$-k1^4 - \cosh(k1\,(x + 4\,k1^2\,t))^2\,k2^4)\Big/\big($$
$$-\cosh(k1\,(x + 4\,k1^2\,t))\,\cosh(k2\,(x + 4\,k2^2\,t))\,k2$$
$$+ \sinh(k2\,(x + 4\,k2^2\,t))\,\sinh(k1\,(x + 4\,k1^2\,t))\,k1\big)^2$$

We can check that indeed it is a solution of the KDV equation

```
> KdV:=simplify(diff(u2,t)-6*u2*diff(u2,x)-diff(u2,x$3));
```

$$KdV := 0$$

Let us plot this solution at three different moments in time

```
> plot(subs({k1=0.5,k2=0.75,t=-10},u2),x=-30..30,
>      thickness=2, color=blue);
```

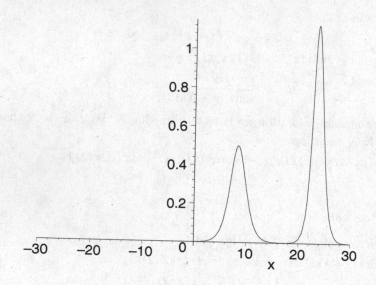

```
> plot(subs({k1=0.5,k2=0.75,t=-2},u2),x=-30..30,
>      thickness=2, color=blue);
```

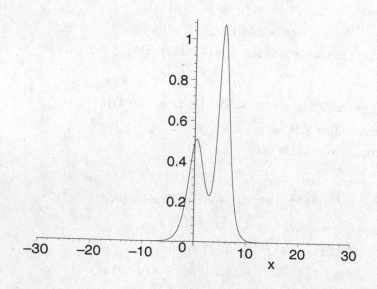

```
>   plot(subs({k1=0.5,k2=0.75,t=10},u2),x=-30..30,
>       thickness=2, color=blue);
```

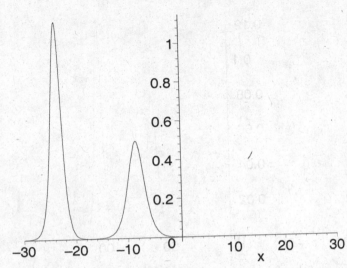

We observe that the soliton with the highest amplitude moves faster, and "collides" with the smaller soliton, creating a rather complicated wave. After collision, both solitons emerge unchanged and continue their propagation to the left.

(c) Three-soliton solution

The process described above can be easily generalized to any number of soliton solutions. Consider for example the case of three-soliton solution

```
>   xi3:=k3*(x+4*k3^2*t);
```

$$\xi 3 := k3\,(x + 4\,k3^2\,t)$$

```
>   psi1; psi2; psi3:=cosh(xi3);
```

$$\cosh(k1\,(x + 4\,k1^2\,t))$$

$$\sinh(k2\,(x + 4\,k2^2\,t))$$

$$\psi 3 := \cosh(k3\,(x + 4\,k3^2\,t))$$

```
>   w3:=Determinant(Wronskian([psi1,psi2,psi3],x)):
>   u3:=2*diff(ln(w3) ,x,x):
```

The solution in this case has a rather long form, therefore we have suppressed the output. Nevertheless, u_3 is a solution of KdV equation

```
>   KdV:=simplify(diff(u3,t)-6*u3*diff(u3,x)-diff(u3,x$3));
```

$$KdV := 0$$

Let us plot the three-soliton solution at two moments in time

```
>   plot(subs({k1=0.15,k2=0.20,k3=0.25,t=-400},u3),
>      x=-130..130, thickness=2, color=blue);
```

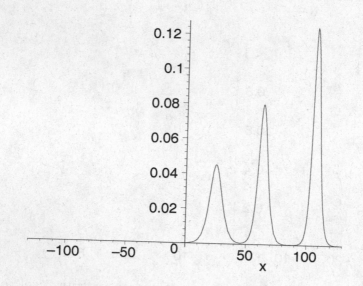

```
>   plot(subs({k1=0.15,k2=0.20,k3=0.25,t=150},u3),
>      x=-130..130, thickness=2, color=blue);
```

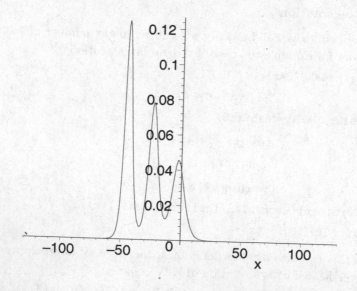

We observe the same behavior we have encountered in the case of the two-soliton solution. The soliton with the highest amplitude moves faster, and with time overtakes the solitons with smaller amplitudes.

We do not want the reader to get the wrong impression that all solutions of KdV equation are pulse-like and propagates towards the left. The reality is quite different. Let us consider for example the case $\lambda < 0$, in the Lax eigenvalue equation (14.13). Then

```
> xi1:=k1*(x-4*k1^2*t);
```
$$\xi1 := k1\,(x - 4\,k1^2\,t)$$

```
> psi1:=sin(xi1);
```
$$\psi1 := \sin(k1\,(x - 4\,k1^2\,t))$$

```
> u1:=simplify(2*diff(ln(psi1),x,x));
```
$$u1 := \frac{2\,k1^2}{-1 + \cos(k1\,(x - 4\,k1^2\,t))^2}$$

```
> KdV:=simplify(diff(u1,t)-6*u1*diff(u1,x)-diff(u1,x$3));
```
$$KdV := 0$$

```
> plot(subs({k1=0.5,t=-5},u1),x=-30..30,y=-10..2,numpoints=500);
```

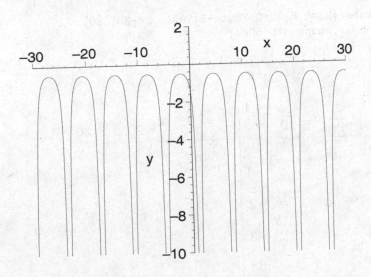

We see that the solution has an infinite number of singularities (the zeros of $\sin \xi_1$) and it moves towards the right. This can be easily visualized using the **animate** command, which is left as an exercise for the reader.

Another type of KdV solution is obtained if we consider two solutions of eq. (14.13)

```
> xi2:=k2*(x-4*k2^2*t); psi2:=cos(xi2);
```
$$\xi2 := k2\,(x - 4\,k2^2\,t)$$
$$\psi2 := \cos(k2\,(x - 4\,k2^2\,t))$$

```
> w2:=Determinant(Wronskian([psi1,psi2],x));
```

$$w2 := -\sin(k1\,(x - 4\,k1^2\,t))\sin(k2\,(x - 4\,k2^2\,t))\,k2$$
$$- \cos(k2\,(x - 4\,k2^2\,t))\cos(k1\,(x - 4\,k1^2\,t))\,k1$$

```
> u2:=simplify(2*diff(ln(w2),x,x));
```

$$u2 := -2\,(k2^4 - \%2^2\,k2^4 - k2^2\,k1^2 - k2^2\,k1^2\,\%1^2 + k2^2\,k1^2\,\%2^2$$
$$+ k1^4\,\%1^2)\,/(k2^2 - k2^2\,\%1^2 - k2^2\,\%2^2 + \%1^2\,k2^2\,\%2^2$$
$$+ 2\,\%2\,\%1\,k2\sin(k2\,(x - 4\,k2^2\,t))\sin(k1\,(x - 4\,k1^2\,t))\,k1$$
$$+ k1^2\,\%2^2\,\%1^2)$$
$$\%1 := \cos(k2\,(x - 4\,k2^2\,t))$$
$$\%2 := \cos(k1\,(x - 4\,k1^2\,t))$$

```
> KdV:=simplify(diff(u2,t)-6*u2*diff(u2,x)-diff(u2,x$3));
```

$$KdV := 0$$

```
> plot(subs({k1=0.5,k2=0.75,t=-5},u2), x=-30..30,
>    y=-5..2,numpoints=500);
```

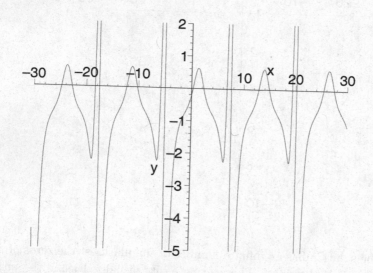

Using the Wronskian methods we can construct a variety of solutions of the KdV equation. Being a nonlinear equation, one cannot guarantee that we will ever find all the solutions.

14.5 Problems

Problem 14.5.1. Find the constants $\_C1$, $\_C2$ and $\_C3$ that transform Maple's solution to the KdV equation

$$u(t, x) = \frac{(8\_C3^3 + \_C2)}{6\_C3} - 2\_C3^2 \tanh(\_C1 + \_C2\,t + \_C3\,x)^2$$

into

$$u\_sol := \frac{v}{2 \cosh\left(\frac{1\,v^{\left(\frac{1}{2}\right)}\,(x+v\,t)}{2}\right)^2}$$

Problem 14.5.2. Using the Wronskian method, construct a seven-soliton solution of the KdV equation. Verify that it is a solution, and plot its time evolution.

A

Complex Variables and Functions

A.1 Complex Numbers

A complex number is an ordered pair (x, y) of real numbers x and y written as

$$z = x + i y, \quad i = \sqrt{-1}.$$

The real numbers x and y are called the real and imaginary parts of the complex number z and denoted as

$$\text{Re } z = x, \quad \text{Im } z = y.$$

In the special case when $y = \text{Im } z = 0$ we write $z = x + i0 = x$ and z is called real. If instead the real part $x = \text{Re } z$ is zero, then $z = 0 + i y$ is called the imaginary number.

Two complex numbers are equal if and only if their real and imaginary parts are equal:

$$x_1 + i y_1 = x_2 + i y_2 \iff x_1 = x_2 \text{ and } y_1 = y_2,$$

for the real numbers x_1, x_2, y_1, y_2. The special case is

$$x + i y = 0 \iff x = 0 \text{ and } y = 0.$$

It is convenient to represent complex numbers graphically using the correspondence

$$x + i y \sim (x, y).$$

This graphical representation is known as the Argand diagram shown in Figure A.1. This diagram represents the complex plane as a two-dimensional plane. The complex number $z = x + i y$ is then associated to the point on a two-dimensional plane with with the real part x being its component on the horizontal (so-called real) axis, and the imaginary part y being its component on the vertical (so-called imaginary) axis. The complex numbers with the positive imaginary part lie in the upper half plane, while those with the negative imaginary part lie in the lower half plane.

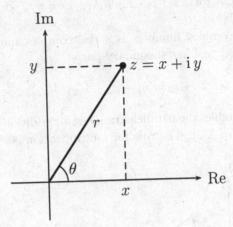

Figure A.1: Argand diagram of complex plane.

Definition A.1.1. *If $z = x + iy$, the absolute value of z is the nonnegative real number*

$$|z| = r = \sqrt{x^2 + y^2}.$$

On the complex plane, the absolute value of z is the distance r from origin to the point z. In polar coordinates:

$$x = r\cos\theta, \quad y = r\sin\theta,$$

$$\theta = \arg(z) = \tan^{-1}\frac{y}{x}, \quad r = |z| = \sqrt{x^2 + y^2},$$

where θ is the angle made by z with the positive x-axis. It holds that $z = r\cos\theta + ir\sin\theta = r(\cos\theta + i\sin\theta)$. This representation of z is called the polar representation. For a real number θ, we define $e^{i\theta}$ by

$$e^{i\theta} = \cos\theta + i\sin\theta$$

and, accordingly, the polar representation becomes

$$z = re^{i\theta}.$$

For $\theta' = \theta + 2n\pi$ where n is an integer it holds that

$$\cos\theta' + i\sin\theta' = \cos\theta + i\sin\theta$$

and, thus, the polar angle θ is unique up to addition of a multiple of 2π. The angle θ is called the argument of z and denoted by $\arg(z)$. The particular argument of z lying in the range $-\pi < \theta \le \pi$ is called the principal argument of z and is denoted by $\mathrm{Arg}(z)$. For example, for $z = i$ one finds $\mathrm{Arg}(z) = \pi/2$ while $\arg(z) = \pi/2 + 2k\pi$ for $k = 0, \pm 1, \pm 2, \ldots$.

The sum of two complex numbers is a real complex numbers obtained by by adding (or subtracting) real and imaginary parts:

$$z_1 + z_2 = x_1 + x_2 + i\,(y_1 + y_2).$$

The addition rule resembles the parallelogram law for addition of two vectors (x_1, y_1) and (x_2, y_2) (see Figure A.2). The rule for multiplication is

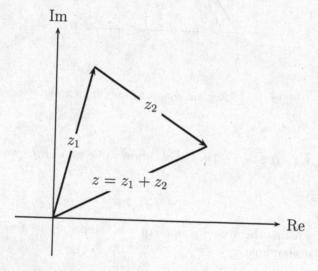

Figure A.2: Addition of two complex numbers.

$$z_1 z_2 = (x_1 + i\,y_1)(x_2 + i\,y_2) = x_1 x_2 - y_1 y_2 + i\,(x_1 y_2 + x_2 y_1)$$

and is illustrated by Figure A.3. In the polar coordinates we have

$$z_1 z_2 = r_1 r_2 \exp i\,(\theta_1 + \theta_2).$$

Using the polar representation the laws of multiplication of the complex numbers can be written as

$$z = r_1 \exp(i\,\theta_1) r_2 \exp(i\,\theta_2) = r_1 r_2 \exp(i\,(\theta_1 + \theta_2)).$$

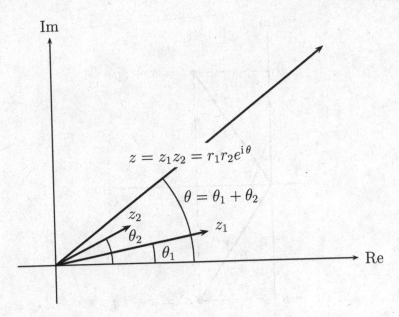

Figure A.3: Multiplication of two complex numbers.

Definition A.1.2. *Suppose $z = x + iy$, the complex conjugate of z is the complex number defined by*

$$z^* = x - iy.$$

The real part and the imaginary part of the the complex conjugate are

$$\text{Re } z^* = x = \frac{1}{2}(z + z^*), \qquad \text{Im } z^* = -y = \frac{-i}{2}(z^* - z).$$

No matter how complicated form z takes a simple rule says that changing the sign of i, wherever it occurs in expression for z, produces z^*. On the complex plane, the complex conjugate of z is obtained by reflecting z in the real axis as shown in Figure A.4. The following properties of the complex conjugate easily follow from the definition:

- $(z_1 \pm z_2)^* = z_1^* \pm z_2^*$,

- $(z_1 z_2)^* = z_1^* z_2^*$,

- $(z_1/z_2)^* = z_1^*/z_2^*$,

- z is real if and only if $z = z^*$,

- $\sqrt{zz^*} = \sqrt{x^2 + y^2}$.

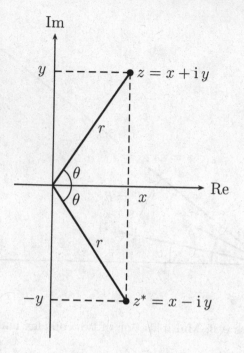

Figure A.4: Obtaining z^* by reflecting z in the real axis.

Note, that division of two complex numbers z_1/z_2, which appeared above, is defined as the product of z_1 by an inverse of z_2. Division is practically performed as follows. Suppose that $z_1 = a + ib$ and $z_2 = c + id$, then the standard expression $z_3 = z_1/z_2 = x + iy$ for their ratio is found by multiplying numerator and denominator by the complex conjugate of the denominator:

$$z_3 = x + iy = \frac{z_1}{z_2} = \frac{a + ib}{c + id}$$

$$= \frac{a + ib}{c + id} \frac{c - id}{c - id}$$

$$= \frac{(ac + bd) + i(-ad + bc)}{c^2 + d^2}$$

$$\rightarrow \quad x = \frac{ac + bd}{c^2 + d^2}, \qquad y = \frac{-ad + bc}{c^2 + d^2}.$$

The complex numbers play an important role in the theory of polynomial equations. Recall, that the quadratic equation

$$z^2 + 2az + b = 0$$

has two roots

$$z_1 = -a + \sqrt{a^2 - b}, \quad z_2 = -a - \sqrt{a^2 - b}.$$

If $a^2 \geq b$ the roots are real if not they are complex.

The most general result is contained in the so-called fundamental theorem of algebra.

The fundamental theorem of algebra states that a n-th order polynomial

$$P(z) \equiv a_n z^n \cdots a_2 z^2 + a_1 z + a_0,$$

where the a's are (real or complex) numbers and $a_n \neq 0$ can be factored as

$$P(z) = a_n(z - z_n)(z - z_{n-1}) \cdots (z - z_1),$$

where n roots

$$z_1, z_2, \cdots, z_n$$

of $P(z) = 0$ may be real or complex.

An important corollary is given by the following statement.

If $P(z)$ is a polynomial with only real coefficients, then its nonreal roots appear in complex conjugate pairs, i.e. if $P(z) = 0$, then also $P(z^*) = 0$.

To prove the above statement consider $P(z) = a_n z^n \cdots a_2 z^2 + a_1 z + a_0 = 0$, such that the coefficients $a_n, \ldots, a_0$ are real. Then $P(z)^* = 0$ and

$$0 = P(z)^* = (a_n z^n)^* + \cdots + (a_2 z^2)^* + (a_1 z)^* + (a_0)^*$$
$$= a_n(z^*)^n + \cdots + a_2(z^*)^2 + a_1 z^* + a_0$$
$$= P(z^*).$$

☞ **Example A.1.1.** The quadratic equation $x^2 + 1 = 0$ has no real roots but only purely imaginary roots $z_1 = i$ and $z_2 = -i$, which are mutually conjugate to each other.

⚲ **Complex numbers: Problems**

Problem A.1.1. Express the following complex numbers in the form $x + iy$, where x and y are real.

(a) $(6 + 8i) + (-4 + 7i)$
(b) $(6 + 8i)(-4 + 7i)$

Problem A.1.2. Express the following complex numbers in the form $x + iy$, where x and y are real.

(a) $(3 - 2i)(1 + i)$

(b) $\frac{(3-2i)}{(1+i)}$

Problem A.1.3. Find $\mathrm{Arg}(z)$ for $z = -1 - i$ and $z = 1 - i$.

Problem A.1.4. Let $z = 1 - 3i$. Find z^*, $|z|$, $\mathrm{Arg}(z)$, and write z in the polar form.

Problem A.1.5. Simplify i^{71}.

Problem A.1.6. (a) Show, that the sum and product of the roots z_i of the polynomial $P(z) = a_n z^n + \cdots + a_2 z^2 + a_1 z + a_0$ are given by

$$\sum_{i=1}^{n} z_i = -\frac{a_{n-1}}{a_n}, \qquad \prod_{i=1}^{n} z_i = \frac{(-1)^n a_0}{a_n}.$$

(b) Use these results to find roots of the quadratic polynomial $P_0(z)$ such that $P(z) = (z - 1)P_0(z) = z^3 - 5z^2 + 17z - 13$.

A.2 Functions of Complex Variables

The elementary functions of real variables can be generalized to functions of complex arguments. One of the most important cases is the exponential function. The exponential of a complex argument can be defined, as for real variables, through the power series expansion

$$\exp(z) = 1 + z + \frac{z^2}{2!} + \cdots \frac{z^n}{n!} + \cdots .$$

With this definition the property

$$\exp(z_1 + z_2) = \exp(z_1)\exp(z_2)$$

can readily be shown to hold. Choosing $z_1 = x$ and $z_2 = iy$ leads to $\exp(x + iy) = \exp(x)\exp(iy)$. Expanding $\exp(iy)$ into the power series yields

$$e^{iy} = \sum_{n=0}^{\infty} i^n \frac{y^n}{n!} = \sum_{k=0}^{\infty} i^{2k} \frac{y^{2k}}{(2k)!} + i \sum_{k=0}^{\infty} i^{2k} \frac{y^{(2k+1)}}{(2k+1)!} = \cos y + i \sin y, \qquad (A.1)$$

due to identity $i^{2k} = (-1)^k$.

The above relation shows that the exponential of a complex argument $z = x + iy$, with x and y real variables can be written as

$$e^z = e^x (\cos y + i \sin y) \tag{A.2}$$

and takes, in particular, values

$$e^0 = e^{2n\pi i} = 1, \quad e^{i\pi/2} = i, \quad e^{i3\pi/2} = -i,, \quad e^{in\pi} = (-1)^n$$

for any integer n.

Conjugating both sides of relation (A.1) gives

$$e^{-iy} = \cos y - i \sin y,$$

which when added to and subtracted from (A.1) reproduces expressions

$$\cos y = \frac{1}{2} \left(e^{iy} + e^{-iy} \right), \quad \sin y = \frac{1}{2i} \left(e^{iy} - e^{-iy} \right).$$

for the trigonometric functions of real variables.

One extends the trigonometric functions to the complex domain by the following definitions:

$$
\begin{aligned}
\sin(z) &= \frac{1}{2i}[\exp(i\,z) - \exp(-i\,z)], \\
\cos(z) &= \frac{1}{2}[\exp(i\,z) + \exp(-i\,z)].
\end{aligned}
$$

The sine and the cosine of a complex variable satisfy the usual identities, as in the real case.

$$
\begin{aligned}
1 &= \sin^2(z) + \cos(z), \\
\sin(z_1 + z_2) &= \sin(z_1)\cos(z_2) + \sin(z_2)\cos(z_1), \\
\sin(-z) &= -\sin(z), \quad \cos(-z) = \cos(z),
\end{aligned}
$$

etc.

Alternatively, the sine and the cosine can be expressed by the power series expansions

$$\sin z = z - \frac{z^3}{3!} + \frac{z^5}{5!} + \cdots = \sum_{n=0}^{\infty} \frac{(-1)^n z^{2n+1}}{(2n+1)!},$$

$$\cos z = 1 - \frac{z^2}{2!} + \frac{z^4}{4!} + \cdots = \sum_{n=0}^{\infty} \frac{(-1)^n z^{2n}}{(2n)!}.$$

The hyperbolic functions sinh, cosh and tanh are closely related to the trigonometric functions. Their definitions are

$$\cosh(x) \equiv \frac{\exp x + \exp(-x)}{2},$$

$$\sinh(x) \equiv \frac{\exp x - \exp(-x)}{2},$$

$$\tanh(x) \equiv \frac{\sinh(x)}{\cosh(x)} = \frac{\exp x - \exp(-x)}{\exp x + \exp(-x)}.$$

The trigonometric and hyperbolic functions are related by through

$$\cosh(i\,z) = \cos(z)\,,$$

$$\sinh(i\,z) = i\,\sin(z)\,.$$

In the case of purely imaginary variable we have

$$\exp(i\,\theta) = 1 + i\,\theta + \frac{(i\,\theta)^2}{2!} + \cdots \frac{(i\,\theta)^n}{n!} + \cdots$$

$$= 1 - \frac{\theta^2}{2!} + \frac{\theta^4}{4!} + \cdots + i\left(\theta - \frac{\theta^3}{3!} + \cdots\right)$$

in agreement with the representation:

$$\exp(i\,\theta) = \cos\theta + i\,\sin\theta\,.$$

Consider powers of z:

$$z^2 = zz = r^2 e^{i\,(\theta+\theta)} = r^2[\cos(2\theta) + i\,\sin(2\theta)]$$

and in general

$$z^n = r^n e^{i\,(n\theta)} = r^n\,[\cos(n\theta) + i\,\sin(2\theta)]\,.$$

For $r = 1$ we derive the important De Moivre's theorem. If n is a positive integer, then

$$\boxed{(\cos\theta + i\,\sin\theta)^n = \cos(n\theta) + i\,\sin(n\theta)\,.}$$

☞ **Example A.2.1.**

$$(\cos\theta + i\,\sin\theta)^3 = \cos(3\theta) + i\,\sin(3\theta)$$

$$= \left(\cos^3\theta - 3\cos\theta\sin^2\theta\right) + i\left(3\cos^2\theta\sin\theta - \sin^3\theta\right)\,.$$

It follows that

$$\implies \begin{cases} \cos(3\theta) = \cos^3\theta - 3\cos\theta\sin^2\theta \\ \sin(3\theta) = 3\cos^2\theta\sin\theta - \sin^3\theta \end{cases}$$

As an application of the De Moivre's theorem we now will describe how to find the roots of the equation $w^n = z$, where z is a given number (complex or real). The solution to that equation requires writing z in polar representation $z = x + i\,y = r\exp(i\,\theta) = r\cos\theta + i\,r\sin\theta$. The advantage of using polar representation is that it is very easy to take the n-th root. We find $w = \sqrt[n]{z} = \sqrt[n]{r}\exp(i\,\theta/n)$. This expression is not unique since the angle θ is determined only up to a multiple of 2π.

That is, in the polar form of z, we could have written $z = r\exp(\mathrm{i}\,(\theta + 2\pi))$ instead of $z = r\exp(\mathrm{i}\,\theta)$. Although, $\exp(\mathrm{i}\,\theta) = \exp(\mathrm{i}\,(\theta + 2\pi))$ it holds that $\exp(\mathrm{i}\,\theta/n) \neq \exp(\mathrm{i}\,(\theta + 2\pi)/n)$ if $n > 1$. To find all n-th roots of z, we must consider the representations $z = r\exp(\mathrm{i}\,(\theta + 2k\pi))$ where k takes the integer values $0, 1, ..., n-1$ (if you continue and reach $k = n$ then the answer will be the same as for $k = 0$). We, therefore, arrive at the following statement.

There are n distinct values of the root $w = z^{1/n}$ given by

$$w = r^{1/n}\left(\cos\frac{\theta + 2k\pi}{n} + \mathrm{i}\,\sin\frac{\theta + 2k\pi}{n}\right), \quad k = 0, 1, ..., n-1. \qquad (A.3)$$

☞ **Example A.2.2.** Things become simpler when we consider the roots of unity given by solutions to $w^n = 1$. All the solutions described by (A.3) with $r = 1$ and $\theta = 0$ lie on the unit circle centered at the origin. The solution with $k = 1$ is nothing but $w = 1$. The remaining solutions are just $n - 1$ other points equally distributed around the unit circle with angle $2\pi/n$ between adjacent solutions. For $n = 3$, the solutions to $w^3 = 1$ are called the cube

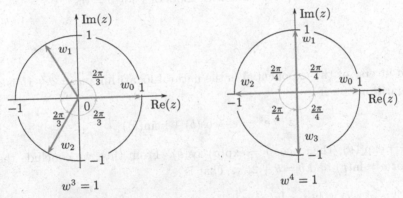

$$w^3 = 1 \qquad\qquad w^4 = 1$$

Figure A.5: The n-th roots of 1 for $n = 3$ and 4.

roots. Obviously, $z = 1 = 1\,(\cos 0 + \mathrm{i}\,\sin 0)$ is an obvious solution for $k = 0$. Altogether there are three solutions to equation $w^3 = 1$ labeled by $k = 0, 1, 2$. They are described by the following formula:

$$w = \cos\frac{2k\pi}{3} + \mathrm{i}\,\sin\frac{2k\pi}{3} = \begin{cases} 1 & k = 0 \\ \cos\frac{2\pi}{3} + \mathrm{i}\,\sin\frac{2\pi}{3} & k = 1 \\ \cos\frac{4\pi}{3} + \mathrm{i}\,\sin\frac{4\pi}{3} & k = 2 \end{cases}$$

and shown in Figure A.5, which also shows four solutions of equation $w^4 = 1$.

They are called the fourth roots of unity and given by

$$w = 1, w = i, w = -1, w = -i .$$

The five roots of the $w^5 = 1$ equation are shown in Figure A.6.

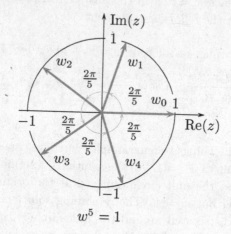

$$w^5 = 1$$

Figure A.6: The n-th roots of 1 for $n = 5$.

The inverse of the exponential is the natural logarithm. $w = a + i b$ is said to be a logarithm of z if

$$z = e^w = e^a \left(\cos(b) + i \sin(b) \right) .$$

Let $z = r \exp(i\theta)$, then $r \cos \theta = \exp(a) \cos(b)$. From this, we conclude that $r = \exp(a)$ or $a = \ln(r)$ and $b = \theta + 2n\pi$. That is

$$w = \ln z = \ln(r \exp(i\theta)) = \ln r + i \operatorname{Arg}(z) + 2i\pi n, \qquad (A.4)$$

for n being an arbitrary integer, Thus, the value of the imaginary part of the logarithm is undetermined up to a multiple of $2i\pi$. This defines $\ln z$ as a multivalued function. Each choice for the integer n in equation (A.4) specifies the so-called branch of the logarithm function with

$$\exp(\ln z) = z$$

being valid no matter, which branch is choosen. The principal value of $\ln z$ corresponds to the choice $n = 0$.

Similar considerations apply to the power function. We define

$$a^z = \exp(z \ln a),$$

where both z and a may be complex. Thus a^z can take on any value

$$a^z = \exp\left(z\left[\ln a + 2\pi n i\right]\right).$$

♠ Functions of Complex Variables: Problems

Problem A.2.1. Using de Moivre's theorem find expressions for $\cos 4\theta$ and $\sin 4\theta$ in terms of of $\cos\theta$ and $\sin\theta$.

Problem A.2.2. Find all solutions to the equation $z^3 = i$.

Problem A.2.3. Find all solutions to the equation $z^3 = -27i$.

Problem A.2.4. Use identity (A.2) to verify the following fundamental properties of the exponential function:

$$e^z \neq 0, \text{ for all } z$$

$$\left|e^z\right| = e^{\mathrm{Re}\, z}, \quad \arg\left(e^z\right) = \mathrm{Im}\, z \text{ up to } 2\pi$$

$$e^z = 1 \iff z = 2ni\,\pi \text{ for any integer } n.$$

Appendix · B

Introducing Maple

THIS appendix introduces basic Maple concepts. The reader familiar with Maple can read directly the last section where we present a concise review of all Maple commands discussed in this appendix.

B.1 First Calculations

When we start a Maple session the first object that appears is the Maple prompt character

```
>
```

Any command in Maple should end with a semicolon. For example by introducing $1 + 1$ at the command prompt, followed by a semicolon, and pressing "ENTER" we obtain

```
>  1 + 1;
```

$$2$$

If the semicolon is forgotten, Maple issues a warning message:

```
>  1 + 1
```

```
Warning, premature end of input
```

We can now type a semicolon at the prompt, and Maple evaluates the expression, to the correct result. To quit a Maple session at any time, use "File/Exit" from the graphic interface.

Maple can be used as a calculator. For example, let us compute $\sqrt[7]{5}$ with 25 digits accuracy. We write

```
>   evalf(5^(1/7),25);
                    1.258498950641826734992787
```

Maple outputs the result immediately. Here we wrote $\sqrt[7]{5}$ as $5^{1/7}$, which in Maple's language, translates as 5^(1/7). The function evalf tells Maple to evaluate the result using the floating-point approximation.

As an exercise, let us use evalf to find the first 500 digits of π. We have

```
>   evalf(Pi,500);
```

3.14159265358979323846264338327950288419716939937510582097 4\
9445923078164062862089986280348253421170679821480865132823 0\
6647093844609550582231725359408128481117450284102701938521 1\
0555964462294895493038196442881097566593344612847564823378 6\
7831652712019091456485669234603486104543266482133936072602 4\
9141273724587006606315588174881520920962829254091715364367 8\
9259036001133053054882046652138414695194151160943305727036 5\
7595919530921861173819326117931051185480744623799627495673 5\
18857527248912279381830119491

Remark

Many commands in Maple have few equivalent forms. In our example, an equivalent form of evalf(Pi,500) is evalf[500](Pi).

B.2 Getting Help

Maple has an extensive help system that can be accessed at any time from the "Help" button. In addition, we can call for help from the command prompt by typing a question mark followed by the command we are looking for. For example, to find help for evalf, we enter

```
>   ?evalf
```

If we need examples on how to use a particular command, then we should enter three question marks followed by the name of the command we are looking for.

```
>   ???evalf
```

Note that the semicolon in this particular case is not necessary.

B.3 Manipulating Symbolic Expressions

One of the most important features of Maple is its ability to manipulate expressions symbolically. Maple can assign numerical or symbolic values to different variables

```
>   a:= 5;
```

$$a := 5$$

```
>   b:= a^2;
```

$$b := 25$$

```
>   f:=(x-y)^2;
```

$$f := (x - y)^2$$

The symbol := is used to *assign* a name to any Maple expression.

The names of the Maple variables can contain any alphanumeric characters and underscores, but they cannot start with a number. In addition, Maple is case sensitive, therefore "a" and "A" are different variables. Examples of good variable names include eqs, B[0], data_in and, respectively, tRaNsPoRt. Examples of *invalid* Maple names are 4th_equation (because it begins with a number), and J&B (because & is not an alphanumeric character).

We can expand the symbolic expression contained in f to obtain the usual binomial expansion

```
>   expand(f);
```

$$x^2 - 2xy + y^2$$

or we can factor it

```
>   factor(%);
```

$$(x - y)^2$$

In the last line, we have introduced a new operator, the so-called "ditto" operator, %. This operator tells Maple that the command (here factor) applies to last evaluated expression (in this case $x^2 - 2xy + y^2$). We should be careful not to abuse the "ditto" operator. It is a good practice to assign variable names to all important steps during a computation.

So far we have introduced three variables: a, b and f. We can quickly check their value by typing

```
>   a; b; f;
```

$$5$$
$$25$$
$$(x - y)^2$$

If we want to clear the value stored in a given variable we can use either the command

```
>   a:='a';
```

$$a := a$$

or we can make use of the unassign operator

```
>   unassign('b','f');
```

Now, if we want to check again the values stored in a, b and f, Maple returns their names which means that there is nothing stored in them.

```
>   a; b; f;
```

$$a$$

$$b$$

$$f$$

Note that if we want to clear more than one variable, unassign command is more convenient. If we want to clear *all* the variables and start with a fresh worksheet, we should use the command restart

```
>   restart;
```

It is a good practice to put the restart command to the start of any new Maple program.

B.4   A Complete Example

Let us discuss a complete example taken from kinematics: the one-dimensional uniform accelerated motion. The example is trivial, therefore we can focus on Maple concepts. The space traveled in the uniform accelerated motion is given by

$$x(t) = \frac{1}{2} at^2 + v_0 t + x_0 \,,$$

where v_0 represents the initial velocity and x_0 the initial position. In the following worksheet we are going to analyze this law of motion and plot the space $x(t)$ for the particular choice of initial conditions $x_0 = 0, v_0 = 0$.

```
>   # Uniform accelerated motion.
>   restart;
>   x:=(t)->1/2*a*t^2 + v_0*t +x_0; # function definition
```

$$x := t \rightarrow \frac{1}{2} a\,t^2 + v\_0\,t + x\_0$$

```
>   x(t1);
```

$$\frac{1}{2} a\,t1^2 + v\_0\,t1 + x\_0$$

```
>   x(3);
```

$$\frac{9}{2} a + 3\,v\_0 + x\_0$$

```
>   x(3.);
```

$$4.500000000\,a + 3.\,v\_0 + x\_0$$

```
>   v:=diff(x(t),t);   #speed
```

$$v := a\,t + v\_0$$

```
>   acc:=diff(x(t),t$2);   #acceleration
```

$$acc := a$$

```
>   int(acc,t)+v_0;
```

$$a\,t + v\_0$$

```
>   int(%,t)+x_0;
```

$$\frac{1}{2}\,a\,t^2 + v\_0\,t + x\_0$$

```
>   x_one:=subs({a=1.5,v_0=0,x_0=0},x(t));
```

$$x\_one := .7500000000\,t^2$$

```
>   plot(x_one,t=0..1);
```

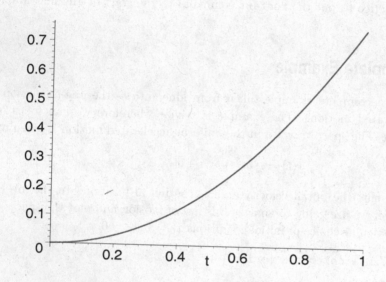

Description. [1]

- The first line starting with # is a comment line. Anything typed after the comment symbol # is ignored by Maple. Comments are useful because they improve the readability of the programs. Comments can be put anywhere within the command line. In this case we announce that the worksheet is an example from kinematics regarding the uniform accelerated motion.

- The next command **restart** initializes Maple, so we can start with a "fresh" environment (erases any previous defined variables). It is a good practice to start any maple worksheet with the **restart** command.

---

[1]Each item refers to one command line

- The third line is an example of function definition in Maple. In this case we define the space x traveled by the object as a function of time t. Note that we enter the mapping operator -> by typing a *minus sign* - followed by a *greater than* symbol > with no space in between them.

- The statement x(t1) asks Maple to evaluate the function x at the particular point in time t=t1.

- If we want the value of x at t=3 then we write simply x(3).

- There is a difference in the way Maple evaluates x(3) and x(3.). In the first case Maple returns the exact value, while in the second case Maple returns the floating-point approximation. As a general rule, whenever in a calculation we enter a number in decimal form, Maple evaluates the result in the floating-point approximation. For example we can check the difference between 15/21 and 15./21. In the first case Maple returns 5/7 while in the second case .7142857143.

- The next line calculates symbolically the speed: $v = dx(t)/dt$. The command v:=diff(x(t),t) is used to calculate the derivative of the function $x(t)$ with respect to the variable t. The same command diff is used to calculate symbolically the mixed derivative, or higher order derivatives. For more examples of diff write ???diff at the command prompt.

- Here we see an example of calculating the acceleration as the second order derivative of $x(t)$ with respect to t, $d^2x(t)/dt^2$ written in Maple as diff(x(t),t$2).

- Integration is the opposite of derivation, and we expect to get back the initial formula of $x(t)$ by computing two successive integrals. Maple can compute indefinite integrals as well as definite integrals. The syntax for the indefinite integral $\int f(x)dx$ is int(f(x),x), while the syntax for the definite integral $\int_a^b f(x)dx$ is int(f(x),x=a..b). Note that Maple doesn't put by default the constants of integration, which in this worksheet are added "by hand". Integrating the acceleration with respect to time we get the speed.

- In the next line we recover the formula of space in uniform accelerated motion by integrating the speed with respect to time and adding the corresponding constant of integration.

- We now want to graphically represent the function $x(t)$, for a particular acceleration $a = 1.5$ and a given choice of initial conditions, let's say $v_0 = 0$ and $x_0 = 0$. For this purpose, we can use the substitution command subs. We define a new variable x_one ,

      x_one := subs({a=1.5, v_0=0, x_0=0}, x(t))

by simultaneously substituting in $x(t) = at^2/2 + v_0 t + x_0$ the desired values for a, v_0 and x_0. When many substitutions are made at once, we should enclose them in curly brackets in the **subs** command.

- Finally, we plot **x_one** over the time interval from 0 to 1. The Maple command is **plot(x_one,t=0..1)**. Note that **0..1** stands for the closed interval $[0, 1]$. The same type of notation for intervals is used in integrals and other Maple commands.

B.5 More on Plots

Maple has the ability of rendering 3-D plots via **plot3d** command. Let us consider the example from the previous section

$$x(t) = at^2/2 + v_0 t + x_0$$

and plot x as a function of time t and acceleration a.

```
>   restart;
>   x:=(t)->1/2*a*t^2 + v_0*t +x_0; # function definition
```

$$x := t \to \frac{1}{2} a t^2 + v\_0 t + x\_0$$

```
>   x_two:=subs({v_0=0,x_0=0},x(t));
```

$$x\_two := \frac{1}{2} a t^2$$

```
>   plot3d(x_two,t=0..1,a=0..1.5,axes=boxed);
```

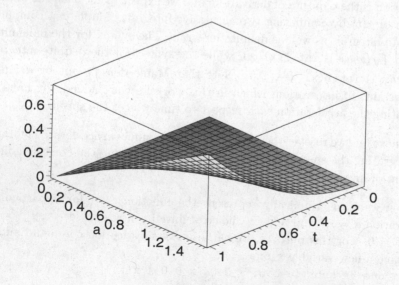

Now, if we click on the graph, we can rotate the three-dimensional graph in any way we like, by keeping the left button of the mouse pressed and moving the mouse. While rotating the graph, observe the change of the angular coordinates of the frame θ and ϕ displayed in the upper left corner of the Maple window.

B.6 More on Expression Manipulation

Maple can be used to solve equations, simplify or cast a given expression to the desired form. For example let us derive Galileo's formula for uniform accelerated motion

$$v^2 = v_0^2 + 2as \ ,$$

where v_0 is the initial speed, v is the final speed, and s is the space traveled with constant acceleration a between the initial and final point.

We have

> `restart;`

We put **restart** as the first command, to "clean-up" the mathematical engine of Maple. Then, we write the formula for speed in the uniform accelerated motion as an equation, say **eq1**. This trick is useful when we want to solve the equation with respect to a given variable.

> `eq1:=v-(v_0 +a*t)=0;`

$$eq1 := v - v\_0 - a\,t = 0$$

In order to eliminate time as a variable, we solve the equation **eq1** with respect to t and store the solution of the equation in the variable **Time**.

> `Time:=solve(eq1,t);`

$$Time := \frac{v - v\_0}{a}$$

Note the syntax of the **solve** command. It is **solve(eqn, var)** where **eqn** is the equation to be solved, and **var** is the variable. In our case we solve **eq1** with respect to t. For more examples on **solve**, write **???solve** at the command prompt. Next we write the formula for space traveled in uniform accelerated motion

> `eq2:=x-(x_0+v_0*t+a*t^2/2)=0;`

$$eq2 := x - x\_0 - v\_0\,t - \frac{1}{2}\,a\,t^2 = 0$$

and eliminate the time by substituting it with the value gotten from **eq1**.

> `step1:=subs(t=Time,eq2);`

$$step1 := x - x\_0 - \frac{v\_0\,(v - v\_0)}{a} - \frac{1}{2}\,\frac{(v - v\_0)^2}{a} = 0$$

We can use now the `simplify` command to simplify the result

```
> step2:=simplify(step1);
```

$$step2 := -\frac{1}{2}\frac{-2\,x\,a + 2\,x\_0\,a - v\_0^2 + v^2}{a} = 0$$

With `expand`, we can further bring the result into a more familiar setting.

```
> step3:=expand(step2);
```

$$step3 := x - x\_0 + \frac{\frac{1}{2}v\_0^2}{a} - \frac{1}{2}\frac{v^2}{a} = 0$$

Many times it is useful to use the succession of commands `simplify` followed by `expand` to "beautify" an expression. At this step we notice the appearance of the group $x - x_0$ which we want to put equal to the space traveled s. One way of doing this is to define some *side-relations* containing the term we are looking for, and then to simplify with respect to these side-relations. Side-relations should be written in a polynomial form. Let us define

```
> siderels:={x-x_0=s};
```

$$siderels := \{x - x\_0 = s\}$$

and simplify

```
> simplify(2*a*step3,siderels,[x]);
```

$$v\_0^2 - v^2 + 2\,a\,s = 0$$

The result is exactly the Galileo's formula we were looking for.

Note that we have to tell Maple what are the variables in the polynomial side-relations (in this case `[x]`). For clarity we have multiplied the expression of `step3` with the common denominator $2a$.

To summarize, this is the worksheet we have just analyzed step by step.

```
> restart;
> eq1:=v-(v_0 +a*t)=0;
```

$$eq1 := v - v\_0 - a\,t = 0$$

```
> Time:=solve(eq1,t);
```

$$Time := \frac{v - v\_0}{a}$$

```
> eq2:=x-(x_0+v_0*t+a*t^2/2)=0;
```

$$eq2 := x - x\_0 - v\_0\,t - \frac{1}{2}\,a\,t^2 = 0$$

```
> step1:=subs(t=Time,eq2);
```

$$step1 := x - x\_0 - \frac{v\_0\,(v - v\_0)}{a} - \frac{1}{2}\frac{(v - v\_0)^2}{a} = 0$$

```
> step2:=simplify(step1);
```

$$step2 := -\frac{1}{2}\frac{-2\,x\,a + 2\,x\_0\,a - v\_0^2 + v^2}{a} = 0$$

```
> step3:=expand(step2);
```

$$step3 := x - x\_0 + \frac{\frac{1}{2}\,v\_0^2}{a} - \frac{1}{2}\frac{v^2}{a} = 0$$

```
> siderels:={x-x_0=s};
```

$$siderels := \{x - x\_0 = s\}$$

```
> simplify(2*a*step3,siderels,[x]);
```

$$v\_0^2 - v^2 + 2\,a\,s = 0$$

B.7 Solving Equations: The "Exact" Way

This section introduces key concepts regarding basic techniques used to solve equations. Tools like `solve`, `fsolve`, `eval` and `map` combined with `simplify`, `expand`, etc., are instrumental in obtaining solutions for a large class of problems.

We use the command `solve` to get solutions for basic algebraic equations. This command takes a set of one or more equations and attempts to solve them exactly for the set of specified variables.

```
> solve({x^2-3*x+2=0},{x});
```

$$\{x = 2\},\ \{x = 1\}$$

```
> solve({a*x+b*y=c,d*x+e*y=f},{x,y});
```

$$\{x = -\frac{-e\,c + f\,b}{-d\,b + a\,e},\ y = \frac{-d\,c + a\,f}{-d\,b + a\,e}\}$$

```
> solve({a*x^2+b*x+c},{x});
```

$$\{x = \frac{1}{2}\frac{-b + \sqrt{b^2 - 4\,a\,c}}{a}\},\ \{x = \frac{1}{2}\frac{-b - \sqrt{b^2 - 4\,a\,c}}{a}\}$$

The last two example emphasize the ability of Maple to solve symbolically equations any time when possible. If instead of an equation, in `solve` is given an expression then Maple automatically assumes that the expression is equal to zero, as shown in the last example above.

Equations and variables can be defined as expressions (or more exactly as lists)

```
> eqs:={x - y +2*z = 2, x-z =5, y+x = 0};
```

$$eqs := \{x - y + 2\,z = 2,\ y + x = 0,\ x - z = 5\}$$

```
> vars:={x,y,z};
```

$$vars := \{x,\,y,\,z\}$$

and then applied the `solve` command.

```
> solve(eqs,vars);
```

$$\{y = -3,\ x = 3,\ z = -2\}$$

This procedure is useful in automatizing the solving process. If we define

```
>  sols:=%;
```

$$sols := \{y = -3,\ x = 3,\ z = -2\}$$

the solutions can be individually extracted via

```
>  sols[1];
```

$$y = -3$$

```
>  sols[2];
```

$$x = 3$$

```
>  sols[3];
```

$$z = -2$$

The solutions of the above equation can also be verified using the two-parameter eval command

```
>  eval(eqs,sols);
```

$$\{5 = 5,\ 2 = 2,\ 0 = 0\}$$

B.8   Solving Equations: Numerical Solutions

Not all equations can be solved exactly. In this case we have to use numerical procedures to find the desired solution. The `fsolve` command finds the root of equations producing approximate numerical answers.

Let us consider the following equation.

$$x = \sin(x)\,e^x \tag{B.1}$$

It is clear that we cannot solve algebraically or in any other closed form this equation. We should therefore consider numerical methods. The next Maple program finds the first two solutions of the above equation.

```
>  restart;

>  eq:=x=sin(x)*exp(x);
```

$$eq := x = \sin(x)\,e^x$$

```
>  plot({x,sin(x)*exp(x)},x=0..3.5,y=-6..8,
>     color=black,thickness=2);
```

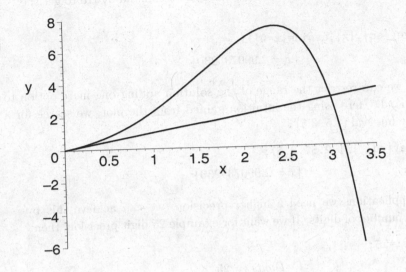

```
>  fsolve({eq},{x});
```
$$\{x = 0.\}$$

```
>  fsolve({eq},{x},avoid={x=0});
```
$$\{x = 2.990735230\}$$

```
>  fsolve({eq},{x},2.5..3.5);
```
$$\{x = 2.990735230\}$$

```
>  Digits:=25;
```
$$Digits := 25$$

```
>  fsolve({eq},{x},2.5..3.5);
```
$$\{x = 2.9907352297318345911116299\}$$

Comment

To see where the solution is, we started the Maple worksheet as usual by resetting Maple's engine with the **restart;** command, then after defining the equation, we plotted on the same graph the left hand side and the right hand side of the equation (B.1). We immediately saw that the first two solutions are one near $x = 0$ and the other one was laying somewhere in between $x = 2.5$ and $x = 3.5$. The **fsolve** command for a general equation searches for a single real root.

```
>  fsolve({eq},{x});
```
$$\{x = 0.\}$$

If we want to avoid the trivial 0 solution, we use the command `avoid` to get the next solution

```
>  fsolve({eq},{x},avoid={x=0});
```

$$\{x = 2.990735230\}$$

Equivalently, we can restrict the range of the solution adding one more option to `fsolve` command. Here using the intuition gained from the plot, we search for a solution in the interval $(2.5, 3.5)$

```
>  fsolve({eq},{x},2.5..3.5);
```

$$\{x = 2.990735230\}$$

If, in some applications we need a higher precision, we can achieve this by modifying the number of digits. If we want for example 25 digit precision, then

```
>  Digits:=25;
```

$$Digits := 25$$

```
>  fsolve({eq},{x},2.5..3.5);
```

$$\{x = 2.990735229731834591116299\}$$

B.9 Maple Procedures

Some computations may require few steps to be completed. If we need to repeat such a computation many times, it is recommended to build a small Maple program or a *procedure*. For exemplification purposes, let us build a simple procedure that adds two numbers.

```
>  plus:=proc(a,b)
>          a+b;
>  end proc;
```

$$plus := \mathbf{proc}(a,\, b)\, a + b\, \mathbf{end\ proc}$$

```
>  plus(1,y);
```

$$1 + y$$

Any procedure begins with a name, in this case `plus`, the keyword `proc` and the arguments of the procedure, in this case `(a,b)`. The procedure ends with the keywords `end proc`. All the commands in a procedure are written between these two mandatory lines. This is a simple example that takes two arguments and returns their sum.

Procedures are able to do more sophisticated computations. In the next example for any function $f(x,y)$ we calculate the quantity $\sqrt{f_x^2 + f_y^2}$, where $f_x = \frac{\partial f}{\partial x}$, and respectively $f_y = \frac{\partial f}{\partial y}$.

```
>   arc:=proc(a)
>         local one,two;
>         one:=diff(a,x);
>         two:=diff(a,y);
>         sqrt(one^2+two^2);
>   end proc;
```

$$arc := \textbf{proc}(a)$$
$$\textbf{local } one, \, two;$$
$$one := \text{diff}(a, \, x); \, two := \text{diff}(a, \, y); \, \text{sqrt}(one^2 + two^2)$$
$$\textbf{end proc}$$

```
>   arc(x+sin(y));
```

$$\sqrt{1 + \cos(y)^2}$$

In this procedure note the declaration of *two local variables* one and two, as intermediate steps in computation. The last command is the output of the procedure.

B.10 Summary

This section contains short descriptions of the main Maple commands introduced in this chapter.

evalf
 Used for numerical evaluation with arbitrary number of digits.
 Example:

```
>   evalf(sqrt(2),15);
```

$$1.41421356237310$$

assign
 Symbolized by := it assigns a value to a given variable.
 Example

```
>   a:=x+y;
```

$$a := x + y$$

expand
 Expands an expression to a sum of terms.
 Example:

```
>   b:=expand((x+y+z)^2);
```

$$b := x^2 + 2\,x\,y + 2\,x\,z + y^2 + 2\,y\,z + z^2$$

factor
 The opposite of expand.
 Example:

```
>   factor(x^2+2*x*y+2*x*z+y^2+2*y*z+z^2);
```

$$(x + y + z)^2$$

ditto operator

Symbolized by %, *ditto operator* is a placeholder for the previous result.
Example:

```
>  expand((a-b)^7);
```

$$a^7 - 7\,a^6\,b + 21\,a^5\,b^2 - 35\,a^4\,b^3 + 35\,a^3\,b^4 - 21\,a^2\,b^5 + 7\,a\,b^6 - b^7$$

```
>  factor(%);
```

$$(a - b)^7$$

unassign

It is used to clear a variable. For a single variable, the short form is `x:= 'x';`
To clear many variables at once, use **unassign**.
Example:

```
>  f:=sin(x):   g:=cos(x):   h:=tan(x):
>  f,g,h;
```

$$\sin(x),\ \cos(x),\ \tan(x)$$

```
>  f:='f';
```

$$f := f$$

```
>  f,g,h;
```

$$f,\ \cos(x),\ \tan(x)$$

```
>  unassign('g','h');
>  f,g,h;
```

$$f,\ g,\ h$$

restart

Use **restart** to clear all the variables in a worksheet and unload all the loaded
packages. Good to use when you start a new computation.

diff

Calculates the derivative (or the partial derivative) of a function.
Example:

```
>  e:=x*sin(y);
```

$$e := x\sin(y)$$

```
>  diff(e,x);
```

$$\sin(y)$$

```
>  diff(e,y);
```

$$x\cos(y)$$

```
>  diff(e,x,y);
```

$$\cos(y)$$

int

Calculates the integral of a function.
Example:

```
>  int(x^2,x);
```

$$\frac{x^3}{3}$$

```
>  int(x^2,x=a..b);
```

$$\frac{b^3}{3} - \frac{a^3}{3}$$

subs

It is used to substitute a subexpression into an expression.

Example:

```
>  expr:=a+b+c;
```

$$expr := a + b + c$$

```
>  subs(a=1-x, expr);
```

$$1 - x + b + c$$

```
>  subs({a=1-x, b=sin(y)}, expr);
```

$$1 - x + \sin(y) + c$$

plot

Plots a 2-D graphs of functions.

Example:

```
>  plot(x-x^7, x=-2..2, y=-1..1, thickness=3);
```

plot3d

Plots in 3-D a surface.

Example:

```
>  plot3d(x*sin(y),x=-1..1, y=0..3*Pi);
```

simplify

Simplifies an expression according to a set of rules.

Example:

```
>  simplify(exp(a-ln(b*exp(c))));
```

$$\frac{e^{(a-c)}}{b}$$

```
>  simplify(sin(x)^3+cos(5*x), trig);
```

$$16\cos(x)^5 - 20\cos(x)^3 + 5\cos(x) + \sin(x) - \sin(x)\cos(x)^2$$

solve

Solves equations or systems of equations.

Example:

```
>  solve(a*x^2+b*x+c=0,x);
```

$$\frac{-b + \sqrt{b^2 - 4ac}}{2a}, \frac{-b - \sqrt{b^2 - 4ac}}{2a}$$

fsolve

> Finds numerical solutions of equations.
> *Example:*
>
> ```
> >  fsolve(1-3*x-x^3+x^5=0, x=0..1);
> ```
> $$0.3232508466$$

Digits

> Specifies the number of digits for numerical calculations.
> *Example:*
>
> ```
> >  Digits:=25;
> ```
> $$Digits := 25$$
> ```
> >  fsolve(1-3*x-x^3+x^5=0,x=0..1);
> ```
> $$0.3232508465859551620795082$$

proc

> Procedures, are used to write Maple programs.
> *Example:* The following procedure that takes two arguments and returns their sum divided by their product
>
> ```
> >  k:=proc(a,b)
> >  (a+b)/(a*b);
> >  end proc;
> ```
> $$k := \mathbf{proc}(a, b)\, (a + b)/(a * b)\, \mathbf{end\ proc}$$
> ```
> >  k(x,y);
> ```
> $$\frac{x + y}{x\, y}$$
> ```
> >  k(1,3);
> ```
> $$\frac{4}{3}$$

B.11 Problems

Problem B.11.1. Evaluate $\sqrt{15}$ with 30 digits.

Problem B.11.2. Factorize: $1 - 2\,x^2 + 2\,y^2 + x^2 - 2\,xy^2 + y^4$

Problem B.11.3. Find the first three positive solutions for the following equation $x \sin x + (1 - x) \cos x = 0$

Problem B.11.4. If the sides of a triangle are a, b and respectively c, write a procedure to calculate its area. Check your answer for the triangle having $a = 3, b = 4, c = 5$.

Problem B.11.5. Calculate the 3rd derivative of x^x with respect to x.

Index for Part I

Index for Part II

Printed in the United States
by Bookmasters